Biofilms in Plant and Soil Health

Biofilms in Plant and Soil Health

Edited by

Iqbal Ahmad
Aligarh Muslim University, Aligarh, India

Fohad Mabood Husain
King Saud University, Saudi Arabia

WILEY Blackwell

This edition first published 2017

Registered Offices
John Wiley & Sons, Inc., 111 River Street, Hoboken, NJ 07030, USA
John Wiley & Sons Ltd, The Atrium, Southern Gate, Chichester, West Sussex, PO19 8SQ, UK

Editorial Office
111 River Street, Hoboken, NJ 07030, USA

For details of our global editorial offices, customer services, and more information about Wiley products visit us at www.wiley.com.

Wiley also publishes its books in a variety of electronic formats and by print-on-demand. Some content that appears in standard print versions of this book may not be available in other formats.

Library of Congress Cataloging-in-Publication Data
Names: Ahmad, Iqbal (Lecturer in agricultural microbiology), editor. |
Husain, Fohad Mabood, editor.
Title: Biofilms in plant and soil health / edited by Dr. Iqbal Ahmad and Dr.
Fohad Mabood Husain.
Description: Hoboken, NJ : John Wiley & Sons, 2017. | Includes
bibliographical references and index. |
Identifiers: LCCN 2017009464 (print) | ISBN 9781119246343 (cloth)
Subjects: LCSH: Soil microbiology. | Biofilms.
Classification: LCC QR111 (ebook) | LCC QR111 .B55 2017 (print) | DDC
579/.1757--dc23
LC record available at https://lccn.loc.gov/2017009464

Cover Design: Wiley
Cover Images: Courtesy of the Editors

Set in 10/12pt WarnockPro by SPi Global, Chennai, India

10 9 8 7 6 5 4 3 2 1

Contents

Preface

Microbes are well known for their diverse metabolic activity and unique survival strategies under various natural ecological niches. The lifestyles of microbial community in diverse environments are now being increasingly explored. Throughout the biological world, microbes thrive predominantly in surface-attached, matrix-enclosed, multicellular communities or biofilms, as opposed to isolated planktonic mode.

Our understanding of microbial interaction with biotic and abiotic surfaces in biofilm growth has been extensively investigated, and its relevance in microbial survival under continuously fluctuating environments is recognized. The physiology, molecular mechanism of biofilm formation, and its regulation have been observed in many bacteria and yeasts. The role of biofilm in medical settings, the food industry, and the environment has also been studied. Recently, the functions of biofilm under more complex habitats associated with soil and plants have been investigated. The nature of these biofilms may be synergistic, associative or pathogenic.

Bacteria causing plant diseases are relatively well understood with respect to the role of biofilm in pathogenesis. However, complex environments like plant root, rhizosphere, and bulk soil have been less explored. Several researchers across the globe are attempting to understand the biofilm in these niches, explore polymicrobial biofilm in these conditions, and exploit these interactions for sustainable agriculture through maintenance of plant and soil health. Developing methods for sustaining crop production and environmental health are of prime importance in feeding global populations on a sustainable basis. Using improved ultrastructure techniques to research molecular biology and biofilm has revolutionized the study on biofilm in complex ecosystem.

Recently, many books on biofilm have been published—mainly on medical, food, and environmental aspects, including bioremediation. However, this is probably the first book that takes a holistic view on biofilms and their significance in plant and soil health.

This book addresses current literature and issues in four sub areas: (i) fundamental significance of biofilm in plant and soil health, and the concept of mono and mixed biofilms by PGPR and fungal biofilms; (ii) biochemical and molecular mechanism in biofilm studies in plant associated bacteria, techniques in studying biofilms and their characterization, gene expression and enhanced antimicrobial resistance in biofilms, and biotic and abiotic factors affecting biofilm *in vitro;* (iii) the ecological significance of soil-associated biofilms and stress management and bioremediation of contaminated soils and degraded ecosystem; and (iv) pathogenic biofilm associated with plants and food and its control measures.

The book is essential for everyone interested in biofilms and their application in agriculture, plant and soil health, and bioremediation, as well as public health concern with environmental pathogenic biofilms. It is recommended to students and researchers of all disciplines of microbiology, biotechnology, and plant and soil sciences and agriculture and environmental biotechnology industry.

With great pleasure, we extend our sincere thanks to all the learned contributors for their timely response, excellent contributions, and consistent support and cooperation.

We would like to sincerely acknowledge the support for scientific evaluation of chapters from learned professors/senior scientists, especially Professor John Pichtel (Ball State University, USA), Dr. Mahipal Singh (Buffalo University USA), Dr. Ashraf A. Khan and Dr. John B. Sutherland (NCTR, USA), Prof. Rumbaugh Kendra (Texas University, USA), and Prof. Elisabeth Grohmann (Germany).

We are grateful to Lt. Gen. (retd.) Zameer Uddin Shah (Vice-Chancellor, AMU), Dr. Bakri bin Matouk Bakri Assas (Rector, UQU, Makkah), Prof. Mohammad Iqbal A. Khan, Dr. Waleed Jameel Altaf, Dr. H. H. Abulreesh, and Dr. Abdullah Safar Althubiani (UQU, Makkah, KSA), Prof. Shamim Ahmad (JNMC, AMU, Aligarh), and Dr. Gurbachan Singh (Director, ASRB, ICAR, New Delhi, India), and Dr. S. Farooq, Director, Himalaya Drug Co., for their encouragement and moral support.

We express our deep sense of gratitude to all our respected teachers, scientific collaborators, biofilm scientific community, and friends for their guidance, support, and healthy criticism. The cooperation received from doctoral research students in the preparation of this book is gratefully acknowledged. The names of a few require special mention: Mohd Musheer Altaf, Mohd Shavez Khan, Menu Maheshwari, and Faizan Abul Qais.

The technical assistance and support rendered from the dynamic Wiley book publishing team is most appreciated and acknowledged.

Many thanks to the members of our families for all the support they have provided. Finally, we acknowledge Almighty God, who provided all the inspirations, insights, positive thoughts, and channels to complete this book project.

We hope that the readers will find the book interesting and informative. We have strived to provide current research trends on this rapidly increasing field, both for instruction and as a motivation for further investigation.

We welcome suggestions and comments by readers for future improvement.

Iqbal Ahmad,
Aligarh, India
Fohad Mabood Husain,
Reyadh, KSA

List of Contributors

Hussein H. Abulreesh
Department of Biology
Faculty of Applied Science
Umm Al-Qura University
Makkah, Saudi Arabia

Iqbal Ahmad
Department of Agricultural Microbiology
Faculty of Agricultural Sciences
Aligarh Muslim University
Aligarh, India

K. Ahmad
Department of Biosciences
Integral University
Lucknow, India

Nasser Abdulatif Al-Shabib
Department of Food Science and Nutrition
College of Food and Agriculture
King Saud University
Riyadh, Saudi Arabia

Mohd Musheer Altaf
Plant Biofilm Research Group
Department of Agricultural Microbiology
Aligarh Muslim University
Aligarh, India

Abdullah Safar Althubiani
Department of Biology
Faculty of Applied Science
Umm Al-Qura University
Makkah, Saudi Arabia

Sravani Ankati
Molecular Plant-Microbe Interaction Lab
Department of Plant Sciences
School of Life Sciences
University of Hyderabad
Hyderabad, India

Firoz Ahmad Ansari
Department of Agricultural Microbiology
Aligarh Muslim University
Aligarh, India

A. Robert Antony
Rhizosphere Biology Laboratory
Department of Microbiology
Bharathidasan University
Tiruchirappalli, India

M. H. Baig
Department of Medical Biotechnology
Yeungnam University
Republic of Korea

Dhamodharan Bakkiyaraj
Department of Biotechnology
Alagappa University
Karaikudi, India

and

Department of Microbiology
Faculty of Science
Prince of Songkla University
Hat Yai, Songkhla, Thailand

Jorge Barriuso
Department of Environmental Biology (Lab 245)
Centro de Investigaciones Biológicas (CIB-CSIC)
Consejo Superior de Investigaciones Científicas
Madrid, Spain

Anirban Basu
Molecular Plant-Microbe Interaction Lab
Department of Plant Sciences
School of Life Sciences
University of Hyderabad
Hyderabad, India

Israel Castillo-Juárez
Colegio de Postgraduados
Campus Montecillo
Posgrado de Botánica
Montecillo, Mexico City, Mexico

Janaina Japiassu de Vasconcelos Cavalcante
Environmental Biotechnology Laboratory
West Zone State University – Uezo
RJ, Brazil

Naybi Muñoz-Cazares
Colegio de Postgraduados
Campus Montecillo
Posgrado de Botánica
Montecillo, Mexico City, Mexico

K.P.N.K. Chandrasiri
Microbial Biotechnology Unit
National Institute of Fundamental Studies (NIFS)
Kandy, Sri Lanka

Himani Chaturvedi
Department of Microbiology
Barkatullah University
Bhopal, India

Inho Choi
Department of Medical Biotechnology
Yeungnam University
Republic of Korea

Puneet Singh Chauhan
Division of Plant Microbe Interactions
CSIR-National Botanical Research Institute
Lucknow, India

G.C. Daniels
Alberta Agriculture and Forestry
Crop Diversification Centre South
Brooks, Canada

Deepak Dwivedi
Department of Microbiology
Barkatullah University
Bhopal, India

Mohammad Faizan
Department of Botany
Aligarh Muslim University
Aligarh, India

Ahmad Faraz
Department of Botany
Aligarh Muslim University
Aligarh, India

Rodolfo García-Contreras
Universidad Nacional Autónoma de México
Department of Microbiology and Parasitology
Faculty of Medicine
Mexico City, Mexico

Elisabeth Grohmann
Department of Life Sciences and Technology
Beuth University of Applied Sciences Berlin
Berlin, Germany

Govind Gupta
Department of Microbiology
Barkatullah University
Bhopal, India

Michael W. Harding
Alberta Agriculture and Forestry
Crop Diversification Centre South
Brooks, Canada

Kusum Harjai
Department of Microbiology
Panjab University
Chandigarh, India

Shamsul Hayat
Department of Botany
Aligarh Muslim University
Aligarh, India

Fohad Mabood Husain
Department of Food Science and Nutrition
College of Food and Agriculture Sciences
King Saud University
Riyadh, Saudi Arabia

Huma Jafri
Department of Agricultural Microbiology
Aligarh Muslim University
Aligarh, India

Rajendran Janani
Rhizosphere Biology Laboratory
Department of Microbiology
Bharathidasan University
Tiruchirappalli, India

Sadaf Kalam
Molecular Plant-Microbe Interaction Lab
Department of Plant Sciences
School of Life Sciences
University of Hyderabad
Hyderabad, India

Ashraf A. Khan
Division of Microbiology (HFT-250)
National Center for Toxicological Research
U.S. Food and Drug Administration
Jefferson, Arkansas, USA

Masudulla Khan
Division of Plant pathology and Biocontrol
Department of Botany
Aligarh Muslim University
Aligarh, India

M.E. Khan
School of Chemical Engineering
Yeungnam University
Republic of Korea

Mohammad Shavez Khan
Department of Agricultural Microbiology
Faculty of Agricultural Sciences
Aligarh Muslim University
Aligarh, India

Mohd Sajjad Ahmad Khan
Department of Biology
College of Medicine
Imam Abdulrahman Bin Faisal University, Dammam
Saudi Arabia

Mayuri Khare
Department of Microbiology
Barkatullah University
Bhopal, India

Meenu Mahaeshewari
Department of Agricultural Microbiology
Faculty of Agricultural Sciences
Aligarh Muslim University
Aligarh, India

Lyriam L.R. Marques
MicroBio SMARTS
Calgary, Canada

Alexander Cardoso Machado
Environmental Biotechnology Laboratory
West Zone State University – Uezo
RJ, Brazil

Mohd. Oves
Faulty of Science
Centre of Excellence in Environmental Studies
King Abdul Aziz University
Jeddah, Saudi Arabia

Daniel Padilla-Chacón
Colegio de Postgraduados
Campus Montecillo
Posgrado de Botánica
Montecillo, Mexico City, Mexico

Alwar Ramanujam Padmavathi
Department of Biotechnology
Alagappa University
Karaikudi, India

and

Nanotec-PSU Center of Excellence on Drug Delivery System and Department of Pharmaceutical Technology
Faculty of Pharmaceutical Sciences
Prince of Songkla University
Hat Yai, Songkhla, Thailand

Shunmugiah Karutha Pandian
Department of Biotechnology
Alagappa University
Karaikudi, India

Vânia Lúcia Muniz de Pádua
Environmental Biotechnology Laboratory
West Zone State University – Uezo
RJ, Brazil

John Pichtel
Ball State University
Natural Resources and Environmental Management, Muncie, USA

Faizan Abul Qais
Department of Agricultural Microbiology
Aligarh Muslim University
Aligarh, India

Velu Rajesh Kannan
Rhizosphere Biology Laboratory
Department of Microbiology
Bharathidasan University
Tiruchirappalli, India

Kendra P. Rumbaugh
Department of Surgery
Texas Tech University Health Sciences Center
Lubbock, Texas, USA

Neha Sabharwal
Department of Microbiology
Panjab University
Chandigarh, India

Gamini Seneviratne
Microbial Biotechnology Unit
National Institute of Fundamental Studies (NIFS)
Kandy, Sri Lanka

Byron Shore
Nautilus Environmental
Calgary, Canada

Zaki Anwar Siddiqui
Division of Plant Pathology and Biocontrol,
Department of Botany
Aligarh Muslim University
Aligarh, India

Arpita Singh
Division of Plant Microbe Interactions
CSIR-National Botanical Research Institute
Lucknow, India

Vinod Singh
Department of Microbiology
Barkatullah University
Bhopal, India

Sunil Kumar Snehi
Department of Microbiology
Barkatullah University
Bhopal, India

John B. Sutherland
Division of Microbiology (HFT-250)
National Center for Toxicological Research
U.S. Food and Drug Administration
Jefferson, Arkansas, USA

Ankita Vaishampayan
Beuth University of Applied Sciences Berlin
Department of Life Sciences and Technology
Berlin, Germany

P.C. Wijepala
Microbial Biotechnology Unit
National Institute of Fundamental Studies (NIFS)
Kandy, Sri Lanka

1

Biofilms: An Overview of Their Significance in Plant and Soil Health

Iqbal Ahmad[1], Mohammad Shavez Khan[1], Mohd Musheer Altaf[1], Faizan Abul Qais[1], Firoz Ahmad Ansari[1] and Kendra P. Rumbaugh[2]

[1] *Department of Agricultural Microbiology, Aligarh Muslim University, Aligarh, India*
[2] *Department of Surgery, Texas Tech University Health Sciences Center, Lubbock, Texas, USA*

1.1 Introduction

The green revolution has enhanced agricultural productivity to a great extent with the increased use of high-yielding crop varieties, heavy farm equipment, synthetic fertilizers, pesticide applications, improved irrigation, better soil management, and massive conversion of forest to agricultural lands [1, 2]. But there is a growing concern that intensive agricultural practices promote large-scale ecosystem degradation and loss of productivity. Adverse environmental effects include deforestation, soil degradation, large-scale greenhouse gas emissions, accumulation of pesticides and chemical fertilizers, pollution of groundwater, and decreased water table due to excessive irrigation [1, 3].

The world population is currently around 7 billion and is projected to approximately 8 billion by the year 2025 and 9 billion by 2050. Considering this population growth and the environmental damage due to ever-increasing industrialization, it is clear that feeding the world's population will be a daunting task over the next 50 years. Therefore, there is a need for new strategies and approaches to improve agricultural productivity in a sustainable and environmentally friendly manner [4]. The effective use of beneficial microorganisms in agriculture in an integrated manner is an attractive technology to address these problems. The role of soil microorganisms in agriculture to improve the availability of plant nutrients and plant health is well known [5]. However, the ability of root-associated microbes to improve nutrient supply and plant protection has yet to be fully exploited [6].

The colonization of the adjacent volume of soil under the plant root is known as rhizosphere colonization. Rhizosphere colonization not only works as a fundamental step in the pathogenesis of soil microbes but also plays an important role in the employment of microorganisms for beneficial purposes [7]. Beneficial rhizobacteria normally promote plant growth by establishing themselves on plant roots and suppressing the colonization or eliminating the presence of pathogenic microorganisms [8]. The competitive exclusion of deleterious rhizosphere organisms is directly linked to the ability to successfully colonize a root surface. However, disease suppressive mechanisms were shown by plant growth–promoting rhizobacteria (PGPR) to be of no use until these microbes successfully colonized and established themselves on root surfaces [9, 10].

Bacterial root colonization is primarily influenced by the presence of the specific character of bacteria necessary for adherence and subsequent colonization. Moreover, several biotic and abiotic factors also play significant roles in bacterial-plant root interactions and colonization. When an organism colonizes a root, factors like water content, temperature, pH, soil characteristics, composition of root exudates, mineral contents, and other microorganisms may influence the process of root colonization. However, plants are the major determinant of microbial diversity [11]. Recent studies on the root-microbe interaction have indicated that rhizobacteria can colonize the root zone and form biofilm and biofilm-like structures. This phenomenon is considered to be a survival strategy by the rhizobacteria, which provides protection to the plant under stress conditions [12].

Traditionally, microbes have been characterized as freely suspended (planktonic) cells; although, many pioneering microbiologists recognized the surface-associated growth of microorganisms on tooth surfaces, aquatic environments, and other biotic and abiotic surfaces. However, a detailed examination of biofilms only became possible after observation under the electron microscope [13, 14]. Based on the observation of dental plaque and other sessile communities, in 1987 Costerton et al. put forth a theory on biofilms that explained the mechanisms of microbial adherence to living and nonliving material, and the benefits associated with this lifestyle. Since then, studies on biofilms in environmental, industrial, and ecological settings relevant to public health have increased significantly [15]. Much of the work on biofilms in the last few decades has demonstrated tremendous growth and understanding through the utilization of scanning electron microscopy, scanning confocal laser microscopy, and both standard microbiology cultural techniques and molecular-based investigation. The ultrastructures of biofilm, roles of various adhesins, genes, and regulatory pathways have all been explored in model organisms [16]. Our understanding of biofilms in natural settings has also substantially improved as new methods allow us to better distinguish different microbial species within complex communities [17–19].

According to Costerton, "the father of biofilm," a biofilm is defined as "a structural community of bacterial cells enclosed in a self-produced polymeric matrix and adherent to an inert or living surface" [20]. However, this definition was later modified to include other characteristics of biofilm such as irreversible cell attachment, altered phenotype with respect to growth rate, and characteristic changes in gene transcription [21]. The composition of the self-produced polymeric material is mainly exopolysaccharide, protein, lipid, and DNA [19]. (Chapter 9 provides details of EPS composition.)

Biofilm formation is a complex process involving various steps such as initial adsorption or reversible attachment, irreversible attachment and the formation of a microbial monolayer on the substrate, early development of microcolonies, maturation of the biofilm structure, including the formation of characteristic architectural features, and lastly, the dispersion (or shedding) of planktonic cells from the biofilm [22]. Each of these stages is very distinct in their morphology and regulation [23]. The sessile growth of microorganisms has distinct phenotypes compared to planktonic cells and exhibits enhanced resistance to antimicrobial compounds and alterations in nutrient uptake [24].

Biofilms provide an important and fundamental strategy for adaptation and survival in the environment, as well as in the pathogenesis of various bacterial pathogens associated with humans, animals and plants [25]. Other applications of biofilms, which have been subsequently studied and are under active investigation, relate to the environmental sciences and food industry. However, in this chapter we will only address the roles of biofilm in plant and soil health, as well as briefly touch on their public health perspective.

1.2 Biofilm Associated with Plants

Biofilms are assemblages of microorganisms adhered to each other and/or to a surface and embedded in a matrix of exopolymers [26]. Biofilms are microniches, which are entirely different from their surrounding environment, and which allow microbes to work as a functional unit, accomplishing tasks not possible in their planktonic state or outside biofilms. The list of the possible effects of biofilms on bacterial ecology and biology, such as protection from desiccation, salinity, UV exposures, acid exposures, metal toxicity, predation and bactericides, and enhancement of genetic exchange and of synergistic interactions is impressive [22, 26]. Biofilms might also foster the expression of density-dependent phenotypes. Induction of the expression of certain bacterial genes, in a density-dependent manner, is known to require the accumulation of diffusible molecules such as acyl homoserine lactones, via a process called quorum sensing (QS) [27].

Research on microbial biofilms is proceeding extensively on many fronts in the medical, environmental, and food industries [22, 28]. Biofilm formation by bacteria on various biotic and abiotic surfaces such as mineral crystals, corrosion particles, clay, silt particles, living cells/tissues of human, animals, and plants has been extensively demonstrated. However, our understanding of plant-associated biofilms is still limited. This is probably due to the complexity of microbes in the soil-root association and difficulties in studying the mixed biofilm under natural/ simulated models [29]. However, over the last decade, many researchers have explored the beneficial association of biofilm with plants [26, 30], which can be exploited to enhance plant protection and promote growth even under stress conditions [31].

Plant-associated microbes can be distinguished as commensal, mutualistic, and pathogenic, and can interact with different parts of the plant such as leaves, stems, roots, seeds, and the vascular system. A number of well-studied, pathogenic, plant bacteria that form biofilms on leaves, vascular system and other plant parts are described in detail by various investigators, as well as in this book (see Chapters 20 and 21). For example, pathogenic bacteria such as *Pseudomonas syringae* colonize leaves and cause brown spot disease. Various studies have demonstrated the importance of surface colonization and aggregation for bacterial survival and competition on aerial plant surfaces [32, 33]. Similarly, vascular pathogens colonizing xylem are prevalent and of great economic importance. *Xyllela fastidosa* is an endophyte and cause of Pierce's disease on grapevines and citreous variegated chlorosis [34]. The gene expression profile of *Xyllela fastidosa* growing as a biofilm indicates the role of several genes likely involved in attachment, such as fimbrial proteins and surface proteins. Elevated expression of plasmid HGT has also been reported in biofilms [35]. In addition, *Xanthomonas campestris, Pantoea stewart,* and *Ralstonia solanacearum,* among others, have been studied in recent years for their capacity to form biofilm during the infection process [36–38].

Conversely, other bacterial species form mutualistic or beneficial biofilms within rhizospheres and on root surfaces. These relationships have been the subject of recent investigations, and some excellent review articles have been published [16, 39–41]. This topic is further discussed in Chapters 2 and 3.

The attachment and surface colonization of PGPR has been widely studied in agriculture and horticulture. Rhizobacteria was found to be effective in root colonization and plant growth promotion after inoculation into wheat or rice and other seedlings [42]. Competitive root colonization by PGPR is considered to be one of the major prerequisites for sustained crop productivity. In many cases, attachment by bacteria leading to root surface colonization also results in biofilm formation [43]. Studies conducted on various PGPR, such as *Bacillus*, *Pseudomonas*, *Rhizobium*, and *Azotobacter*, have demonstrated successful biofilm formation and root colonization [44–47], although both are also influenced by biotic and abiotic factors [48]. Therefore, it is now considered that biofilm-forming PGPR will be more effectively colonized on the plant roots when inoculated and will thus be able to sufficiently withstand the fluctuating conditions of the soil environment to perform its plant growth–promotion activity.

Another aspect of biofilm research that has received increased attention is the association of human pathogens with plants [49]. Biofilm on seeds and sprouts and salad crops for human consumption are a potential health concern, as they may harbor pathogenic or opportunistic pathogens. Plant-associated pathogens such as *Salmonella*, *E. coli*, *Enterococcus faecalis*, and *Pseudomonas aeruginosa* [50–53] may form biofilms, with maximum thicknesses ranging from 5 to 12 μm [54] on plants. Such bacterial species may come into contact with humans, causing sickness, as well as poor food quality and other hygiene concerns. Details on this aspect are also discussed in Chapter 23 of this book.

1.3 Biofilm Formation Mechanisms: Recent Update on Key Factors

Biofilm formation by several human, animal, and plant pathogenic bacteria has been well studied and reported. A general layout of the biofilm formation process and key regulatory factors is depicted in Figure 1.1.

The mechanism of biofilm formation is a highly regulated process. Each species responds to its own set of environmental conditions via distinct molecular mechanisms. However, several stimuli are generally important for plant-associated biofilms. Recently, new insights on the molecular regulation of biofilm formation in plant-associated bacteria have been described by Castiblanco and Sundin [55]. The authors also review progress in understanding the role of cyclic diGMP, cyclic GMP, and small RNAs during the regulation of biofilm formation by plant pathogens such as *Erwina amylovora*, *Agrobacterium tumefaciens*, and *Xanthomonas* spp. [55]. This topic is also discussed in Chapters 20 and 21 of this book.

However, the exact mechanisms of regulation have yet to be discovered and understood in many plant-associated biofilms, especially in regard to rhizospheres and mixed-species biofilms. We are just beginning to understand how various components of complex soil impacts biofilm establishment in the rhizosphere and functions under natural conditions. Novel mechanisms of genetic regulation in biofilms by pathogenic

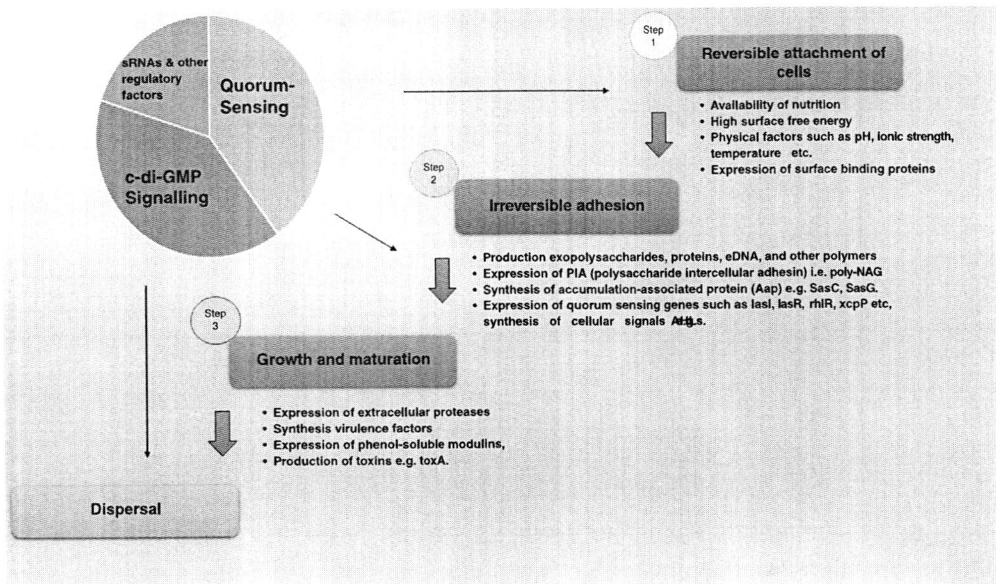

Figure 1.1 Key steps involved in biofilm formation and the role of regulatory factors.

plant bacteria include the secondary messenger molecule cyclic diGMP in *E. amylovora* [56], *A. tumefaciens* [57], and *P. syrengie* [58]. Similarly, the role of cyclic GMP in biofilm formation was also reported by An et al. [59]. Small RNA and RNA-binding proteins in pathogenic plant bacteria have also been shown to activate biofilm formation *in vitro* by other investigators [60–62].

Excellent review articles on the molecular mechanisms and genes involved in biofilm formation by various bacteria have been published recently [26]. A number of regulatory pathways that control biofilm formation are explained, for example, in *Bacillus subtilis*. SpoOA is a central transcriptional regulator that controls more than 100 genes, including those necessary for biofilm matrix gene expression and sporulation [63, 64]. Similarly, SlrR/SlrA controls the initiation of biofilm formation in *Bacillus subtilis* [65, 66]. The molecular mechanisms involved in biofilm formation by various Gram-negative bacteria, especially in *Pseudomonas* spp., have been well documented [67, 68].

The role of plant exudates in attracting bacteria toward point of release from root surfaces and providing optimal nutrient availability is well known. Some bacteria like *A. brasilense* may use aerotaxics to identify an optimal O_2 concentration that will allow, but not inhibit, nitrogen fixation [69]. These processes are helpful in root colonization. Motility is also involved in the biofilm formation of several bacteria, although its importance is often conditional. Many reports are contradictory because the importance of motility in biofilm formation seems to differ from strain to strain and in different conditions. Bacterial fimbrial and afimbrial adhesins are widely known for their role in bacterial–host and solid-surface adhesions [70]. Various types of fimbrial adhesins are also known, and play important roles in adhesion and biofilm formation on solid surfaces, including the type I fimbrial structure and type IV pili (T4P) proteins in pathogenic, plant bacteria [71]. Afimbrial adhesins in Gram-negative bacteria are secreted by type V secretion systems (T5SS) or autotransporters. These adhesins are surface proteins and are used by bacteria for adhesion [72].

Newly described attachment strategies by plant pathogenic bacteria in biofilm formation include other secretion systems, which play an important role the secretion of different types of proteins and nucleic acids to the extracellular environment or direct translocation into adjacent eukaryotic or prokaryotic cells [73]. The type III secretion system (T3SS) is the most widely studied, and its role in biofilm formation has been described. Another secretion system described includes the type VI (T6SS) [74]. These secretion systems appear to play important roles in virulence and pathogenicity of the plant pathogenic bacteria and biofilm formation, as some of the proteins they secrete are essential for the attachment process. Other attachment strategies have been reported in plant-bacterial interactions and biofilm formation for example, *Bacillus amyloliquifaciens* FZB42, a PGPR, utilizes a collagen-like protein (CLPs), which was found to be an ECM component of biofilms present in the root of *A. thaliana* [75].

A number of signals and/or conditions have been identified that trigger biofilm formation by plant pathogenic bacteria on plant surfaces, including: (i) nutritional status of the plant; (ii) plant tissue; (iii) QS signals; and (iv) O_2 concentration and iron availability [76]. The role of calcium [77] has been explored in *X. fastidiosa*. Similarly, other metals (e.g., Cu, Zn, Mn) also differently influence biofilm and cell aggregation or inhibition. Recently, Nagar and Schwarz [78] published an interesting review on self-inhibition of biofilm development and highlighted that the transition between the planktonic and biofilm modes of growth is a highly regulated developmental shift that has a significant impact on cell fate. There are three mechanisms involved in the self-inhibition process:

1) The process is regulated by secreted small molecules and decrease in biofilm development.
2) Physiochemical properties of the substratum are modified by extracellular polysaccharides.
3) eDNA masks an adhesive structure.

1.4 Biofilm in Soil and Rhizospheres

Bacteria are not evenly distributed in the soil environment. Microbial populations are found in higher density near/in a rhizosphere, or in decaying organic matter compared to bulk soil. The microcolonies of bacteria, which are found in bulk soil, are often composed of different bacterial species [79, 80]. However, in the presence of a nutrient source, these microcolonies can develop into a multispecies biofilm. It is thought that due to dramatic changes in physical and chemical conditions in soil, bacteria periodically adapt the biofilm mode of growth for self-protection. For example, production of EPS by soil bacteria provides protection to water stress [81, 82]. Another stress condition is caused by the antibiotics produced naturally by microorganisms in the soil. Similarly, heavy metal contamination or soil pollution with organic compounds can result in a stressful environment. Bacteria present in biofilms become more tolerant to antibiotics and other toxic pollutants [83]. As these soil bacterial biofilms also consist of a variety of species interacting metabolically and socially, they can also convey selective advantages to their inhabitants [84].

1.5 Genetic Exchange in Biofilms

Genetic exchange by bacteria under *in vitro* and *in situ* conditions has been widely studied since the discovery of transferable plasmids in Japan [85]. Similarly, biofilms have long been a subject of interest in environmental, medical, and industrial microbiology. The interconnection between biofilm formation and horizontal gene transfer (HGT) has been the topic of recent attention. There is an enhanced rate of horizontal gene transfer due to high density of bacteria in biofilms [86]. Thus, the chances of acquiring new genes relevant to tolerance and adaptability are increased and therefore the environmental or ecological fitness of bacteria is increased [87]. The roles of QS signals in this process have also been demonstrated by several investigators [88, 89]. An excellent review on the literature of the above areas by Madsen et al. [90] reached the following conclusions:

1) HGT rates are typically higher in biofilm communities compared to planktonic cells.
2) Biofilms promote plasmid stability and may improve host range.
3) Plasmids are well-suited to promote the evolution of social traits such as biofilm formation.
4) This exchange may result in overall interconnectedness between HGT, mobile genetic elements, and social evolution of bacteria.

1.6 Diversity and Function of Soil Biofilms

Succession of microbial populations on any surface, but especially in soil or rhizospheres, is a complex process and influenced by a number of physical, chemical and biological factors. Most of the biofilms in nature are mainly composed of bacteria attached to the surface of soil or water environments. But other microorganisms like fungi, algae, and protozoa also play important roles in the establishment of bacterial/polymicrobial biofilms [91].

Biofilms play a significant role in the degradation of organic material in the soil, since this degradation is chiefly dependent on the extracellular enzymes elaborated by soil microbes. The bacteria attached to soil or present near degradable sources such as roots and litters have advantages compared to planktonic bacteria for nutrient availability [82]. Various enzymes elaborated chiefly by heterotrophic populations like fungi and heterotrophic bacteria, degrade a variety of organic matter. Lignolytic and cellulolytic microorganisms also play a specific role [92, 93]. Such processes in biofilms are enhanced due to altered growth rate and physiological capabilities [94]. Thus nutrient turnover, mineralization, and soil fertility are directly under the control of microbial activity in biofilms. Another important function of biofilms in soil is their ability to bioremediate metal and organic pollutants, and the bioremediation potential of mixed microbial consortiums or multispecies soil biofilms has been evaluated by several investigators. For example, VonCanstein et al. [95] demonstrated the role of biofilm in bioremediation of mercury in different environmental conditions. This topic is also addressed in Chapter 18 of this book.

The degradation of various organic pollutants, such as polyaromatic hydrocarbon (PAH), fenamiphos, and toluene, by soil bacteria have been demonstrated to be more efficient when they are in the biofilm mode of growth [96–98]. Similarly, microorganisms pathogenic to humans, animals, and plants, may survive in the soil for extended periods of time and become active pathogens once they reach a susceptible host. Many opportunistic pathogens such as *Pseudomonas aeruginosa*, species of *Salmonella*, *E. coli*, *Listeria*, and *Campylobacter* are naturally occurring in the soil environment [99–101]. Soil conditions may select for biofilm formation and attachment, which may also enhance their protection, survival, and pathogenicity [94].

1.7 The Role of Biofilms in Competitive Colonization by PGPR

Studies of soil biofilms are complex and require relevant models closely resembling natural soil environments. However, common techniques based on biofilm formation on glass surfaces or in flow models, and monitored through confocal and SEM analysis, are used more frequently. Thus, further progress in our understanding of complex microbial interactions in soil and rhizosphere environments awaits improvements in soil biofilm models.

Competitive root/rhizosphere colonization by plant pathogenic and root symbiotic or PGPR bacteria has been investigated, and their role is established. However, biofilm formation as a means of biocontrol has been studied for *Pseudomonas fluorescens* and wheat, *Bacillus subtilis* and several plants, and *Penibacillus polymyxa* and peanut plants

[102–106]. The role of QS in biofilm formation and the release of antifungal compounds, such as phenazine, has been well documented for wheat rhizospheres. In addition, surface chemistry on roots, and the production of surfactin was also found to influence biofilm formation and root colonization. Despite these advances, further investigation is needed to elucidate the exact mechanisms of interaction and the contribution of various microbial factors in rhizosphere competence [107].

1.8 Biofilm Synergy in Soil and Environmental Microbes

Under natural conditions, microbial biofilms are found on essentially any moist living or nonliving substrate, including plants and soil. Microbial communities within the biofilm and neighboring cells can influence the outcome of this interaction, which may alter community productivity. Interactions may be antagonistic such as competition, parasitism, predation, or social cheating, which can adversely affect community productivity. On the other hand, positive interactions among species such as cooperation, synergistic metabolism, construction of new niches, and can increase productivity. In a well-defined ecological succession, long-term evolved biofilm communities display synergistic interactions among the species within the biofilm [108]. For example, Poltak and Cooper [109] demonstrated long-term ecological succession experimentally by evaluating *Burkholderia cenocepacia* biofilms over 1,500 generations and concluded:

1) There is a successive adaptive diversification.
2) Mixed population is more productive due to complementary interactions.
3) Spatial partitioning and cross feeding generate community synergy.

Soil is typically a reservoir habitat for almost all types of microbes with varying metabolic capabilities. Soil provides a good setting for multispecies biofilm formation [110]. These microbes can interact positively or negatively through various microbial interaction mechanisms [29, 111]. *In vitro* interactions between two strains of the same species and interspecific interactions in biofilm formation have been reported in the literature [112]. Members of different species may also interact negatively through resource competition or by producing inhibitory compounds [113]. For example, Burmolle et al. [89] demonstrated a strong synergy among four epiphytic isolates from marine origin. Similar, synergistic interactions between *Candida albicans* and *S. aureus* or Streptococci have also been observed on abiotic surfaces and an oral mucosal analog [114]. Additionally, many investigators have reported the inability of single strains to form biofilms independently but can promote the formation of mixed-species biofilm [115–117].

The role of multispecies biofilms is evident in maintaining ecological balance in soil [40]. Many benefits are associated with the biofilm mode of growth, compared to the planktonic mode such as (i) protection from desiccation; (ii) protection from protozoan predation; (iii) increased resistance to antibacterial compounds; and (iv) an enhanced rate of genetic exchange [118, 119].

For example, it was recently demonstrated that there was a synergistic effect among seven different isolates co-cultured in combinations of four species, which included *Stenotrophomonas rhizophila, Xanthomonas retroflexes, Paenibacillus amylolyticus* and *Microbacterium oxydans*. The findings concluded that a high prevalence of synergy in multispecies biofilms indicated interspecific cooperation under natural conditions [120].

1.9 Biofilms in Drought Stress Management

Global climate change is considered to be one of the most serious threats to agricultural productivity worldwide in recent years. Sustainability in agricultural production implies high yield that can be maintained, even in the face of climate change. It is expected that global water shortage will be the key challenge for food security in the near future [121]. Rhizobacteria can alleviate plant drought stress. In this direction, the first report on drought tolerance enriched by PGPR was published by investigators in Sweden [122] and then followed by those in Canada [123].

Plants cope with drought through drought escape, dehydration avoidance, and dehydration tolerance. Tolerance mechanisms such as osmoprotection, detoxification, ion transport, or chaperone functions take over when tissues are no longer protected by avoidance mechanisms [121]. Drought tolerance enhancement by PGPR through 1-aminocyclopropane 1-carboxylate deaminase (ACC) can provide significant protection from drought and heat. The role of ACC deaminase in providing a plant tolerance mechanism is widely acknowledged [124, 125].

1.10 Plant Health and Biofilm

The health of green plants is of pivotal importance to everyone [126]. The term *plant health* is frequently used in two overlapping contexts: (i) the scientific and regulatory framework of checking plant imports for the presence of potential pathogens and pests [127, 128], and (ii) a less specific concept that touches on all areas of plant protection and is sometimes referred to as *plant health protection*.

For agricultural crops and the production and cultivation of medicinal plants, the major focus of plant health is centered on their protection from pests and pathogens, while enhancing productivity. Another term loosely associated with plant health is *plant growth promotion*. Although plant health cannot be directly compared with human and animal health, there are four similarities between human and plant health issues [129]:

1) Health variations between individuals, such as those due to age differences.
2) Health or disease is a dynamic process.
3) Health is subject to the geographical occurrences of pathogens.
4) Pathogens can develop resistance to treatment.

In the following section, our aim is to address the role of microbial biofilms in plant health. Many microbes deteriorate plant health and cause disease; however, several other microbes protect and promote plant health and control pathogens and pests.

1.11 How Microbial Biofilms Influence Plant Health?

Plant-associated biofilms above and below the ground can interact positively or negatively with plants, depending on the microorganisms involved. Various phytopathogenic bacteria form biofilms that cause plant diseases. This topic is the focus of Chapter 21. On the other hand, positively interacting rhizobacteria, known as PGPR, as well as certain bacterial biocontrol agents, also colonize plant roots, and this involves microcolonies and biofilm formation. Extensive literature is available on *Bacillus*, fluorescent *Pseudomonas*, and other PGPR, as discussed in Chapters 4 and 5. The overall interaction is depicted in Figure 1.2.

As our understanding about the microbial worlds associated with plant, human, animal and environments increases, we are beginning to see that the role of the microbiome on influencing human health is enormous, acting both positively and, in some cases, negatively [130]. Similarly, the rhizosphere microbiome has a direct impact on plant health. The rhizosphere microbiome exists mainly in biofilm mode. The collective genome of the rhizosphere microbial community is much larger than that of the plant and is often referred to as the plant's second genome [131].

In humans, the role of the intestinal microflora on health is now more understood, and similar functions can be ascribed to the human gut as to the plant rhizosphere. Root microbiomes are now under scrutiny for their exact role in plant health. The root microbiome can (i) influence disease suppressive soil and (ii) modulate the host immune system by beneficial microbes in the rhizosphere [131]. Plants actively shape their root microbiome through the secretion of active compounds that modulate bacterial QS, and influence the recruitment of beneficial microbes. The details of these mechanisms may be obtained in several published review articles [131–134].

Understanding the complex interactions between plants and microbes in the rhizosphere is still in its infancy [83]. Recent metagenomic studies on root microbes still provide very limited information. Nevertheless, complex interactions among various groups of microorganisms inhabiting rhizospheres, and their role in plant–microbe interactions in

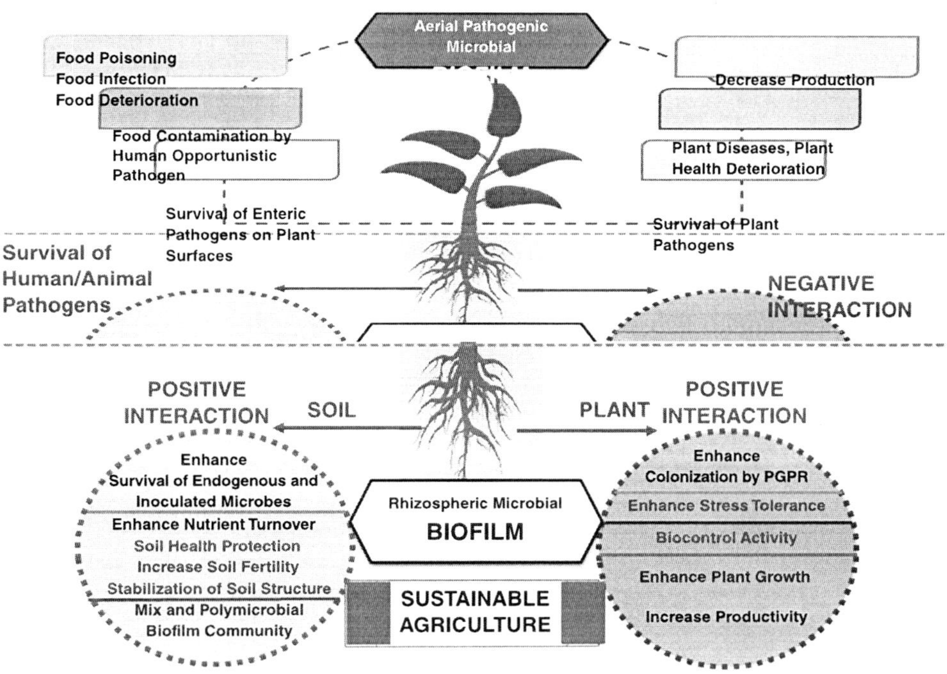

Figure 1.2 Biofilm interactions with plant and soil, and their significance.

biofilms must be explored to gain a better understanding, and for improved ecological niche engineering of plant and soil health in sustainable agriculture.

1.12 Soil Health and Biofilms

Biofilms grown on or around plant surfaces, tissues, soil, rhizosphere, and other habitats interact synergistically or antagonistically, both *in vitro* and in natural environments (Figure 1.2). Extensive research on biofilm stages, phenotypic, physiological, and molecular mechanisms has led to a better understanding of biofilms and application strategies. However, research on understanding and exploiting stable mixed biofilms for crop productivity, bioremediation, and the improvement of plant health (crop protection and fertilization) and for soil fertility and health is just beginning to gain momentum. New technologies for analyzing plant and soil-associated biofilms and determining their role in root colonization, plant growth promotion, and soil health improvement are needed. Some of the issues concerning the importance of microbes and microbial biofilms in maintaining soil health, and therefore sustainable environments and agriculture, are described next.

Soil has been defined by various scientists in different ways, depending on their purpose of study. Biologists, such as agricultural scientists, define the soil as the uppermost layer of the earth's crust, which supports the growth of plants and directly influences plant productivity [135]. Various management practices have long been known and used by farmers with the available resources to maintain soil fertility. Various human industrial and agricultural activities such as agrochemical discharge, industrial pollution, and extensive exploitation of soil resources in intensive agricultural practices, has resulted in a huge deficit of plant nutrient availability in soil versus the amount of nutrients taken up by plants. This has been realized in several parts of the world by the decreased productivity of soil. Therefore, scientists have promoted the concepts of the integrated plant nutrient supply system (IPNS), integrated plant nutrient management (IPNM), or the concept of organic farming for sustainable practices [136]. The term *soil health* is widely used in the discussion of sustainable agriculture pertaining to overall conditions or the quality of soil resources. Major factors known to cause soil quality degradation include erosion, decrease in organic matter content, and increased salinity [137].

Soil is a very complex system consisting of both abiotic and biotic components. Several physical, chemical, biochemical, and biological components of soil interact and give rise to the characteristics of a particular soil [93]. In a review published in *Philosophical Transaction of the Royal Society,* Kibblewhite et al. [138] define "the soil health as an integrative property that reflects the capacity of soil to respond to agricultural intervention or continue to support agricultural production and other ecosystem services." Nevertheless, Doran and Jones [139] defined soil health as "the continued capacity of soil to function as vital living system, within ecosystem and land use boundaries, to sustain biological productivity, maintain the quality of air, water environments and promote plant, animal and human health." Although there is no rigid definition of soil health, it constitutes the "soil quality," aspects that are based on the measurement of various parameters specific to soil properties such as physical, chemical, and biological.

The terms *soil quality* and *soil health* are sometimes used interchangeably. Warkentin and Fletcher [140] were probably the first to introduce the concept of soil quality as an approach to improve land use planning. Soil quality is simply defined as "the capacity of soil to function" [141, 142]. However, soil quality describes quantitative soil properties and linkages between properties and functions. Many prefer *soil health*, a term that clearly conveys the idea of a living thing.

There are various biochemical indicators, such as the production of soil enzymes, that are used to evaluate soil health, and several excellent review articles have been published describing soil quality [143, 144]. Another relevant alternative approach to studying soil health is based on an integrated approach that assumes the health of soil is more likely to be influenced by the interaction between different processes and properties of soil components [145]. It has been proposed that soil health is dependent on the maintenance of four functions [138]:

1) Carbon transformation
2) Nutrient cycling
3) Soil structure
4) Regulation of pests and diseases

Each of the above functions is controlled and regulated through multicomponent and multifunctional systems—an array of biological processes provided by diverse interacting living organisms under the influence of the abiotic soil environment. The ecosystem services provided by the soil are driven by biological process. Soil constitutes an important habitat for organisms and their reactions. The major components of soil include soil, air, water, pH and nutrients, and organic matter. Soil pores and gasses contribute significantly by providing a suitable habitat for a variety of organisms and their interactions. Various factors are known to have roles in controlling soil health, including (i) soil type; (ii) organisms; and (iii) nutrients. Therefore, in order to maintain good soil health for sustainable agriculture productivity, various integrated approaches have been adopted such as integrated plant nutrient management, integrated pest and disease management, integrated water management, and integrated water and land use management under the concept of organic farming.

1.13 How to Assess Soil Health?

Considering the complexity of soil and agricultural systems in different climates, it is difficult to develop common guidelines to assess soil health for all agricultural systems. It is evident that a single indicator of soil health is not appropriate, and it will be not be possible to measure all parameters to assess soil health. Several national and international proposals for soil assessment are linked to legal frameworks for the protection of soil [146–148]. Various soil quality/health assessment criteria have been developed which are based on the soil's physical, chemical and biological characterization. Frequently used sensitive indicators recommended to determine soil quality are (i) Soil organic matter; (ii) major plant nutrients and other physiochemical characteristics of soil; (iii) top soil depth; (iv) filtration rate [149, 150]; and (v) levels of soil enzymes such as β-glucosidase [151].

1.14 Impact of Biofilms on Soil Health

Soil health is dependent on the maintenance of carbon transformation, nutrient cycling, soil structure, and the regulation of pests and diseases. Soil microorganisms, both indigenous and those specifically introduced into soil, are widely known for their overall contribution in the processes just discussed. The role of autotrophic organisms is to provide a carbon source, and soil algae, blue green algae Cyanobacteria, and other chemoautotrophic bacteria have been documented to serve this purpose [152]. Additionally, a major contribution to the degradation of organic matter comes from the combined activities of extracellular enzymes produced by heterotrophic bacteria and fungi such as cellulose, hemicellulose, pectinase, and ligninolytic enzymes. Nutrient cycling of C, N, S, P, Zn, and Mn, for example, is regulated by this diverse microbial population [153]. The formation of mixed-species biofilms by diverse microbes in soil and on plant roots and other related surfaces, significantly helps in the survival of these microbes, improving soil modification [134]. Thus, intentional inoculation of soil and/or plants with probiotic biofilms can also improve field conditions [154].

1.15 Biofilm EPS in Soil Health

The majority of soil microbes exist in biofilms [155]. The major components of the biofilm matrix are polysaccharides, glycoconjugates, and protein. Measurements of microbial biomass, levels of ATP, and other enzymatic processes can be used to study soil microbial communities and their functions. As the EPS constitutes approximately 80 percent of a soil biofilm's dry mass, the extraction and estimation of biofilm EPS is now considered to be an important indicator for soil function [156].

Competitive advantages for microbial life are known to be influenced by EPS production, which provides advantages to biofilm communities such as (i) improved QS, (ii) colony adhesion, (iii) syntrophy, (iv) tolerance to heavy metals and desiccation [157], (v) improved stability of soil enzymes [158], (vi) bacterial gliding, and (vii) soil health [159]. The EPS is also reported to contribute to the process of soil aggregation [160], tolerance of wheat seedlings to saline [161], and drought tolerance of sunflowers inoculated with EPS-producing *Pseudomonas* [162]. Fully understanding the role of the EPS to mixed-species biofilms is difficult due to the complex nature of soil. However, Redmile-Gordon et al. [163] recently described a new method of EPS extraction from soil biofilms, which may help expand our knowledge.

Understanding multispecies biofilms under natural conditions, especially in soil, is complex. It is not yet well established that biofilms are associated with some properties that increase microbial fitness under specific selective pressure. In natural environments, the microbial communities are heterogeneous and are composed of multiple bacterial species. A recent study demonstrated that bacteria that coexist in natural environments facilitate interspecific biofilm formations [164].

This study also reported a positive relation between biofilm induction and phylogenetic history, suggesting that an increase in biofilm formation is a common adaptive response to long-term coexistence. Understanding multispecies biofilms under *in vitro* conditions is much more commonplace, and various model systems have been adopted.

However, attention now must be directed toward the study of these biofilms in their natural environments or controlled microcosms [112].

Based on the previous discussion, biofilm interaction with plant surfaces and soil and their significance in plant, soil, and environment health through various activities and processes are presented in Figure 1.2.

1.16 Conclusions and Future Directions

An in-depth understanding of microbial biofilm communities in plant and soil is still elusive. What is known is that biofilms are associated with a number of properties that increase the ecological fitness of bacteria by various operative mechanisms, including increased antimicrobial tolerance and protection from host defense systems, evasion of protozoan grazing, better utilization of secreted compounds, and increased horizontal gene transfer. However, in natural environments, biofilms are heterogeneous. Coexistence facilitates interspecific biofilm formation, and Madsen et al. have recently demonstrated complex microbial communities [164]. However, difficulties in investigating interactions in polymicrobial biofilms under natural environmental conditions, and in the succession of microbial communities in response to environmental conditions, have sustained a poor understanding of the soil environment. Biofilms associated with plants are better understood in the context of pathogenesis rather than symbiosis or mutualism, with few exceptions. Thus, while the significance of biofilms has been well documented in environmental, food, and medical aspects, our understanding of soil and rhizosphere-associated biofilms is still weak and woefully incomplete.

Recent advances and progress has been made on the metagenomic analysis of soil through various microbial surveys, such as the Earth Microbiome Project (EMP) [165], Terra Genome [166], and China Soil Microbiome Initiate. These are all excellent resources to explore taxonomic and functional diversity, as described by Nesme et al. [167], but it is likely that the future will hold even more questions regarding the genomic diversity of polymicrobial biofilms in natural environments, especially in soil and rhizospheres. One certainty, based on current knowledge, is that biofilm research will certainly expand our understanding of microbe–microbe and microbe–plant interactions, and will help in the development of new strategies to solve problems related to plant productivity and protection, soil health, and sustainability of agriculture.

References

1 D. Tilman, K.G. Cassman, P.A. Matson, R. Naylor and S. Polasky, Agricultural sustainability and intensive production practices, *Nature*, 418, 671–677 (2002).
2 T. Gomiero, D. Pimentel and M.G. Paoletti, Environmental impact of different agricultural management practices: conventional vs. organic agriculture, *Crit. Rev. Plant Sci.*, 30, 95–124 (2011).
3 J.A. Foley, N. Ramankutty, K.A. Brauman, E.S. Cassidy, J.S. Gerber, M. Johnston, N.D. Mueller, C. O'Connell, D.K. Ray, P.C. West, C. Balzer, E.M. Bennett, S.R. Carpenter, J. Hill, C. Monfreda, S. Polasky, J. Rockström, J. Sheehan, S. Siebert, G.D. Tilman and D.P.M. Zaks Solutions for a cultivated planet, *Nature* 478, 337–342 (2011).

4 B.R. Glick, Bacteria with ACC deaminase can promote plant growth and help to feed the world, *Microbiol. Res.*, 169, 30–39 (2014).
5 B. Lugtenberg, Life of microbes in the rhizosphere, *in Principles of plant-microbe interactions*, B. Lugtenberg (ed) Springer, Switzerland, 7–15 (2015).
6 J.M. Raaijmakers and B.J.J. Lugtenberg, Perspectives for rhizosphere research, *In Molecular Microbial Ecology of the Rhizosphere*, vol 2. de Bruijn FJ (ed) Wiley Blackwell, USA, 1227–1232 (2013).
7 E. Benizeri, E. Baudoin and A. Guckert, Root colonization by inoculated plant growth promoting rhizobacteria, *Biocontrol. Sci. Technol.*, 11, 357–374 (2001).
8 A. Beneduzi, A. Ambrosini and L.M.P. Passaglia, Plant growth-promoting rhizobacteria (PGPR): their potential as antagonists and biocontrol agents, *Genet. Mol. Biol.*, 35, 1044–1051 (2012).
9 C.S. Nautiyal, Rhizosphere competence of *Pseudomonas* sp. NBRI 9926 and *Rhizobium* sp. NBRI 9513 involved in the suppression of chickpea (*Cicer arietinum* L.) pathogenic fungi, *FEMS Microbiol. Ecol.*, 23, 145–158 (1997).
10 B. Meena, Biological control of pest and diseases using fluorescent pseudomonads, *in Basic and applied aspects of biopesticides*, K. Sahayaraj (ed), Springer, India, 17–29 (2014).
11 L. Philippot, J.M. Raaijmakers, P. Lemanceau and W.H. van der Putten, Going back to the roots: the microbial ecology of the rhizosphere, *Nat. Rev. Microbiol.*, 11, 789–799 (2013).
12 S. Timmusk and E. Nevo Plant root associated biofilms, *in Bacteria in agrobiology*, vol 3: D.K. Maheshwari (ed), Plant nutrient management, Springer, Berlin, 285–300 (2011).
13 J.R. Lawrence, D.R. Korber, B.D. Hoyle, J.W. Costerton and D.E. Caldwell, Optical sectioning of microbial biofilms, *J bacterial.*, 173(20), 6558–6567 (1991).
14 N.A. Fujishige, N.N. Kapadia and A.M. Hirsch, A feeling for the micro-organism: structure on a small scale biofilms on plant roots, *Bot. J. Linn. Soc.*, 150(1), 79–88 (2006).
15 J.W. Costerton, K.J. Cheng, G.G. Geesey, T.I. Ladd, J.C. Nickel, M. Dasgupta and T.J. Marrie, Bacterial biofilms in nature and disease, *Annu. Rev. Microbiol.*, 41(1), 435–464 (1987).
16 T. Danhorn and C. Fuqua, Biofilm formation by plant-associated bacteria, *Annu. Rev. Microbiol.*, 61, 401–422 (2007).
17 P. Stoodley, K. Sauer, D.G. Davies and J.W. Costerton, Biofilms as complex differentiated communities, *Annu. Rev. Microbiol.*, 56(1) 187–209 (2002).
18 J.L. Kirk, L.A. Beaudette, M. Hart, P. Moutoglis, J.N. Klironomos, H. Lee and J.T. Trevors, Methods of studying soil microbial diversity, *J Microbiol. Meth.*, 58(2) 169–188 (2004).
19 H.C. Flemming and J. Wingender, The biofilm matrix, *Nat. Rev. Microbiol.*, 8(9) 623–633 (2010).
20 J.W. Costerton, P.S. Stewart and E.P. Greenberg, Bacterial biofilms: a common cause of persistent infections., *Science*, 284(5418) 1318–1322 (1999).
21 R.M. Donlan and J.W. Costerton, Biofilms: survival mechanisms of clinically relevant microorganisms, *Clin. Microbiol. Rev.*, 15(2), 167–193 (2002).
22 L. Hall-Stoodley, J.W. Costerton and P. Stoodley, Bacterial biofilms: from the natural environment to infectious diseases, *Nat. Rev. Microbiol.*, 2(2), 95–108 (2004).
23 K. Sauer, A.K. Camper, G.D. Ehrlich, J.W. Costerton and D.G. Davies, *Pseudomonas aeruginosa* displays multiple phenotypes during development as a biofilm, *J. Bacterial.*, 184(4), 1140–1154 (2002).

24 P.C. Bogino, M.D.L.M. Oliva, F.G. Sorroche and W. Giordano, The role of bacterial biofilms and surface components in plant-bacterial associations, *Int. J. Mol. Sci.*, 14(8), 15838–15859 (2013).
25 R. Singh, O.C. Stine, D.L. Smith, J.K. Spitznagel, M.E. Labib and H.N. Williams, Microbial diversity of biofilms in dental unit water systems, *Appl. Environ. Microb.*, 69(6), 3412–3420 (2003).
26 H. Vlamakis, Y. Chai, P. Beauregard, R. Losick and R. Kolter, Sticking together: building a biofilm the *Bacillus subtilis* way, *Nat. Rev. Microbiol.*, 11(3), 157–168 (2013).
27 Y.H. Li and X. Tian, Quorum sensing and bacterial social interactions in biofilms, *Sensors*, 12(3), 2519–2538 (2012).
28 R. Van Houdt and C.W. Michiels, Biofilm formation and the food industry, a focus on the bacterial outer surface, *J. Appl. Microbiol.*, 109(4), 1117–1131 (2010).
29 M. Burmølle, D. Ren, T. Bjarnsholt and S.J. Sørensen, Interactions in multispecies biofilms: do they actually matter? *Trends Microbiol.*, 22(2), 84–91(2014).
30 S. Timmusk, N. Grantcharova and E.G.H. Wagner, *Paenibacillus polymyxa* invades plant roots and forms biofilms, *Appl. Environ. Microb.*, 71(11), 7292–7300 (2005).
31 S. Timmusk, I.A.A. El-Daim, L. Copolovici, T. Tanilas, A. Kännaste, L. Behers, E. Nevo, G. Seisenbaeva, E. Stenström and Ü. Niinemets, Drought-tolerance of wheat improved by rhizosphere bacteria from harsh environments: enhanced biomass production and reduced emissions of stress volatiles, *PloS One*, 9(5), 96086 (2014).
32 J.M. Monier and S.E. Lindow, Aggregates of resident bacteria facilitate survival of immigrant bacteria on leaf surfaces, *Microb. Ecol.* 49, 343–352 (2005).
33 J.M. Monier and S.E. Lindow, Spatial organization of dual-species bacterial aggregates on leaf surfaces, *Appl. Environ. Microbiol.*, 71, 5484–5493 (2005).
34 A.H. Purcell and D.L. Hopkins, Fastidious xylem-limited bacterial plant pathogens, *Annu. Rev. Phytopathol.* 34, 131–51 (1996).
35 A.A, de Souza, M.A. Takita, H,D, Coletta, C, Caldana, G.M. Yanai, Gene expression profile of the plant pathogen *Xylella fastidiosa* during biofilm formation *in vitro*, *FEMS Microbiol. Lett.*, 237, 341–53 (2004).
36 H.J. Park, H.W. Jung and S.W. Han, Functional and proteomic analyses reveal that wxcB is involved in virulence, motility, detergent tolerance, and biofilm formation in *Xanthomonas campestris pv. Vesicatoria*, *Biochem. Bioph. Res. Co.*, 452(3), 389–394 (2014).
37 T. Weller-Stuart, I. Toth, de P. Maayer and T. Coutinho, Swimming and twitching motility are essential for attachment and virulence of *Pantoea ananatis* in onion seedlings, *Mol. Plant Pathol.*, 12432 (2016).
38 T. Minh Tran, A. MacIntyre, D. Khokhani, M. Hawes and C. Allen, Extracellular DNases of *Ralstonia solanacearum* modulate biofilms and facilitate bacterial wilt virulence, *Environ. Microbiol.*, 13446, 2016.
39 T. Rudrappa, M.L. Biedrzycki and H.P. Bais, Causes and consequences of plant-associated biofilms. *FEMS Microbiol. Ecol.*, 64(2), 153–166 (2008).
40 M. Burmølle, A.H. Kjøller and S.J. Sørensen, An invisible workforce: biofilms in the soil, *in Microbial Biofilm: Current Research and Applications*, G. Lear and G.D. Lewis (eds.), Horizon Scientific Press, 61–71(2012).
41 L. Philippot, J.M. Raaijmakers, P. Lemanceau and W.H. van der Putten, Going back to the roots: the microbial ecology of the rhizosphere, *Nat. Rev. Microbiol.*, 11(11), 789–799 (2013).

42 D. Egamberdieva, Growth response of wheat cultivars to bacterial inoculation in calcareous soil, *Plant soil environ.*, 56(12), 570–573 (2010).
43 M. Fletcher and D.C. Savage (eds.), *Bacterial adhesion: mechanisms and physiological significance*, Springer Science & Business Media (2013).
44 H.P. Bais, T.L. Weir, L.G. Perry, S. Gilroy and J.M. Vivanco, The role of root exudates in rhizosphere interactions with plants and other organisms, *Annu. Rev. Plant Biol.*, 57, 233–266. (2006).
45 J. Yuan, N. Zhang, Q. Huang, W. Raza, R. Li, J.M. Vivanco and Q. Shen, Organic acids from root exudates of banana help root colonization of PGPR strain *Bacillus amyloliquefaciens* NJN-6, *Sci.* Rep-UK, 5 (2015).
46 Musheer and Ahmad, 2016.
47 M. Kröber, B. Verwaaijen, D. Wibberg, A. Winkler, A. Pühler and A. Schlüter, Comparative transcriptome analysis of the biocontrol strain *Bacillus amyloliquefaciens* FZB42 as response to biofilm formation analyzed by RNA sequencing, *J Biotechnol.*, 231, 212–223 (2016).
48 L. Philippot, J.M. Raaijmakers, P. Lemanceau and W.H. van der Putten, Going back to the roots: the microbial ecology of the rhizosphere, *Nat. Rev. Microbiol.*, 11(11), 789–799 (2013).
49 C.E. Morris and J.M. Monier, The ecological significance of biofilm formation by plant-associated bacteria, *Annu. Rev. Phytopathol.*, 41, 429–453 (2003).
50 W.F. Fett and P.H. Cooke, Reduction of *Escherichia coli* O157:H7 and *Salmonella* on laboratory-inoculated alfalfa seed with commercial citrus-related products, *J. Food Prot.*, 66, 1158–1165 (2003).
51 J.X. He, R.L. Baldini, E. Deziel, M. Saucier, Q.H. Zhang, N.T. Liberati, D. Lee, J. Urbach, H.M. Goodman and L.G. Rahme, The broad host range pathogen *Pseudomonas aeruginosa* strain PA14 carries two pathogenicity islands harboring plant and animal virulence genes, *Proc. Natl. Acad. Sci. USA*, 101, 2530–2535 (2004).
52 B.A. Annous, E.B. Solomon, P.H. Cooke and A. Burke, Biofilm formation by *Salmonella* spp. on cantaloupe melons, *J. Food Saf.*, 25, 276–287 (2005).
53 A.K. Jha, H.P. Bais and J.M. Vivanco, *Enterococcus faecalis* mammalian virulence-related factors exhibit potent pathogenicity in the *Arabidopsis thaliana* plant model, *Infect. Immun.*, 73, 464–75 (2005).
54 W.F. Fett and P.H. Cooke, A survey of native microbial aggregates on alfalfa, clover and mung bean sprout cotyledons for thickness as determined by confocal scanning laser microscopy, *Food Microbiol.*, 22, 253–259 (2005).
55 L.F. Castiblanco and G.W. Sundin, New insights on molecular regulation of biofilm formation in plant-associated bacteria, *J. Integr. Plant Biol,*. 58(4), 362–372 (2015).
56 A.C. Edmunds, L.F. Castiblanco, G.W. Sundin and C.M. Waters, Cyclic di-GMP modulates the disease progression of *Erwinia amylovora*, *J Bacteriol.*, 195, 2155–2165 (2013).
57 J. Xu, J. Kim, B.J. Koestler, J.H. Choi, C.M. Waters and C. Fuqua, Genetic analysis of *Agrobacterium tumefaciens* unipolar polysaccharide production reveals complex integrated control of the motile-to-sessile switch, *Mol. Microbiol.*, 89, 929–948 (2013).
58 C. Engl, C.J. Waite, J.F. McKenna, M.H. Bennett, T. Hamann and M. Buck, Chp8, A diguanylate cyclase from *Pseudomonas syringae* pv. tomato DC3000, suppresses the pathogen-associated molecular pattern flagellin, increases extracellular polysaccharides, and promotes plant immune evasion, *MBio* 5, e01168-14 (2014).

59 S.Q. An, K.H. Chin, M. Febrer, Y. McCarthy, J.G. Yang, C.L. Liu, D. Swarbreck, J. Rogers, J. Maxwell Dow, S.H. Chou and R.P. Ryan A cyclic GMP-dependent signalling pathway regulates bacterial phytopathogenesis. *EMBO J* 32, 2430–2438 (2013).

60 P.L. Zhu, S. Zhao, J.L. Tang and J.X. Feng, The rsmA-like gene rsmAXoo of *Xanthomonas oryzae* pv. oryzae regulates bacterial virulence and production of diffusible signal factor, *Mol. Plant Pathol.*, 12, 227–237 (2011).

61 X.H. Lu, S.Q. An, D.J. Tang, Y McCarthy, J.L. Tang, J.M. Dow and R.P. Ryan, RsmA regulates biofilm formation in *Xanthomonas campestris* through a regulatory network involving cyclic di-GMP and the Clp transcription factor, *PLoS One* 7,e52646 (2012).

62 Q. Zeng, R.R. McNally and G.W. Sundin, Global small RNA chaperone Hfq and regulatory small RNAs are important virulence regulators in *Erwinia amylovora*, *J. Bacteriol.*, 195, 1706–1717 (2013).

63 V. Molle, M. Fujita, S.T. Jensen, P. Eichenberger, J.E. González-Pastor, J.S. Liu and R. Losick, The Spo0A regulon of *Bacillus subtilis*, *Mol. Microbiol.*, 50, 1683–1701 (2003).

64 M. Fujita, J. E. Gonzalez-Pastor and R. Losick, Highand low-threshold genes in the Spo0A regulon of *Bacillus subtilis*, *J. Bacteriol.* 187, 1357–1368 (2005).

65 K. Kobayashi, SlrR/SlrA controls the initiation of biofilm formation in *Bacillus subtilis*, *Mol. Microbiol.*, 69(6), 1399–1410 (2008).

66 P. Patrick, Epigenetic Switching: Bacteria Hedge Bets about Staying or Moving, *Curr. Biol.*, 20(11), 480–482 (2010).

67 M. Fazli, H. Almblad, M.L. Rybtke, M. Givskov, L. Eberl and T. Tolker-Nielsen, Regulation of biofilm formation in *Pseudomonas* and *Burkholderia* species, *Environ. Microbial.*, 16(7), 1961–1981(2014).

68 M.T. Cabeen, S.A. Leiman and R. Losick, Colony-morphology screening uncovers a role for the *Pseudomonas aeruginosa* nitrogen-related phosphotransferase system in biofilm formation. *Mol. Microbiol.*, 99(3), 557–570 (2016).

69 I.B. Zhulin, V.A. Bespalov, M.S. Johnson and B.L. Taylor, Oxygen taxis and proton motive force in *Azospirillum brasilense*, *J. bacterial.*, 178(17), 5199–5204 (1996).

70 P.A. Zaini, S. Burdman, M.M. Igo, J.K. Parker, L. de La Fuente, Fimbrial and afrimbrial adhesins involved in bacterial attachment to surfaces, in Virulence Mechanisms of Plant Pathogenic Bacteria, N. Wang, J.B. Jones, G.W. Sundin, F.F. White, S.A. Hogenhout, M.C. Roper, L. de La Fuente and J.H. Ham (eds.), APS Press, St. Paul, 93–106 (2015).

71 J.M. Koczan, M.J. McGrath, Y. Zhao and G.W. Sundin, Contribution of *Erwinia amylovora* exopolysaccharides amylovoran and levan to biofilm formation: Implications in pathogenicity, *Phytopathology*, 99, 1237–1244 (2009).

72 J.C. Leo, I. Grin and D. Linke, Type V secretion: Mechanism(s) of autotransport through the bacterial outer membrane, *Philos. Trans. R. Soc. London B*, 367, 1088–1101 (2012).

73 J.H. Chang, D. Desveaux and A.L. Creason, The ABCs and 123s of bacterial secretion systems in plant pathogenesis, *Annu. Rev. Phytopathol.*, 52, 317–345 (2014).

74 H.T. Brian, T.G. Dong and J.J. Mekalanos, A view to a kill: The bacterial type VI secretion system, *Cell Host Microbe*, 15, 9–21 (2014).

75 X. Zhao, Y. Wang, Q. Shang, Y. Li, H. Hao, Y. Zhang, Z. Guo, G. Yang, Z. Xie and R. Wang, Collagen-like proteins (ClpA, ClpB, ClpC, and ClpD) are required for biofilm formation and adhesion to plant roots by *Bacillus amyloliquefaciens* FZB42, *PLoS One* 10: e0117414 (2015).

76 L. Burbank, M. Mohammadi and M.C. Roper, Siderophore-mediated iron acquisition influences motility and is required for full virulence of the xylem-dwelling bacterial phytopathogen *Pantoea stewartii* subsp. *stewartii, Appl. Environ. Microbiol.,* 81: 139–148 (2015).
77 L.F. Cruz, P.A. Cobine and L. De La Fuente, Calcium increases *Xylella fastidiosa* surface attachment, biofilm formation, and twitching motility, *Appl. Environ. Microbiol.*, 78, 1321–1331 (2012).
78 E. Nagar and R. Schwarz, To be or not to be planktonic? Self-inhibition of biofilm development, *Environ. Microbial.*, 17(5), 1477–1486 (2015).
79 N. Nunan, K. Wu, I.M. Young, J.W. Crawford and K. Ritz, Spatial distribution of bacterial communities and their relationships with the micro-architecture of soil, *FEMS Microbiol. Ecol.*, 44(2), 203–215 (2003).
80 G.L. Grundmann, Spatial scales of soil bacterial diversity—the size of clone, *FEMS Microbiol. Ecol.*, 48, 119–127 (2004).
81 I.W. Sutherland, Biofilm exopolysaccharides: a strong and sticky framework, *Microbiology*, 147(1), 3–9 (2001).
82 J. Jass, S.K. Roberts and H.M. Lappin-Scott, Microbes and enzymes in biofilms, *in Enzymes in the Environment Activity, Ecology and Applications*, R.G . Burns and R.P. Dick (eds.) CRC Press, 307–326 (2002).
83 Å.R. Almas, J. Mulder and L.R. Bakken, Trace metal exposure of soil bacteria depends on their position in the soil matrix, *Environ. Sci. Technol.*, 39(16), 5927–5932 (2005).
84 K.K. Jefferson, What drives bacteria to produce a biofilm? *FEMS Microbiol. Lett.*, 236(2), 163–173 (2004).
85 R. Nakaya, A. Akamura and Y. Murata, Resistance transfer agents in *Shigella, Biochem. Biophys. Res. Commun.*, 3, 654–659 (1960).
86 S.J. Sørensen, M. Bailey, L.H. Hansen, N. Kroer and S. Wuertz, Studying plasmid horizontal transfer *in situ*: a critical review, *Nat. Rev. Microbiol.*, 3(9), 700–710 (2005).
87 M. Burmølle, M.I. Bahl, L.B. Jensen, S.J. Sørensen and L.H. Hansen, Type 3 fimbriae, encoded by the conjugative plasmid pOLA52, enhance biofilm formation and transfer frequencies in *Enterobacteriaceae* strains, *Microbiology*, 154(1), 187–195 (2008).
88 K.B. Xavier and B.L. Bassler, LuxS quorum sensing: more than just a numbers game, *Curr. Opin. Microbial.*, 6(2), 191–197 (2003).
89 M. Burmølle, J.S. Webb, D. Rao, L.H. Hansen, S.J. Sørensen and S. Kjelleberg, Enhanced biofilm formation and increased resistance to antimicrobial agents and bacterial invasion are caused by synergistic interactions in multispecies biofilms, *Appl. Environ. Microbiol.*, 72(6), 3916–3923 (2006).
90 J.S. Madsen, M. Burmølle, L.H. Hansen and S.J. Sørensen, The interconnection between biofilm formation and horizontal gene transfer, *FEMS Immunol. Med. Mic.*, 65(2), 183–195 (2012).
91 M. Burmølle, L.H. Hansen, and S.J. Sørensen, Establishment and early succession of a multispecies biofilm composed of soil bacteria, *Microb. Ecol.*, 54(2), 352–362 (2007).
92 P. Nannipieri, E. Kandeler and P. Ruggiero, Enzyme activities and microbiological and biochemical processes, *in Enzymes in the Environment Activity, Ecology and Applications*, R.G . Burns and R.P. Dick (eds.) CRC Press, 1–33 (2002).
93 M. Alexander, *Introduction to soil microbiology, Krieger Publishing Company; 2 edition, ISBN-10: 0894645129* (2011).

94 M. Burmølle, L.H. Hansen and S.J. Sørensen, Use of a whole-cell biosensor and flow cytometry to detect AHL production by an indigenous soil community during decomposition of litter, *Microb. Ecol.*, 50(2), 221–229 (2005).
95 H. Von Canstein, Y. Li, J. Leonhäuser, E. Haase, A. Felske, W.D. Deckwer and I. Wagner-Döbler, Spatially oscillating activity and microbial succession of mercury-reducing biofilms in a technical-scale bioremediation system, *Appl. Environ. Microbiol.*, 68(4), 1938–1946 (2002).
96 L.T. Ou and J.E. Thomas, Influence of soil organic matter and soil surfaces on a bacterial consortium that mineralizes fenamiphos, *Soil Sci. Soc. Am. J.*, 58(4), 1148–1153 (1994).
97 P.A. Holden, J.R. Hunt and M.K. Firestone, Toluene diffusion and reaction in unsaturated *Pseudomonas putida* biofilms, *Biotechnol. Bioeng.*, 56(6), 656 (1997).
98 D.G. Karamanev, C. Chavarie and R. Samson, Soil immobilization: new concept for biotreatment of soil contaminants, *Biotechnol. Bioeng.*, 57(4), 471–476 (1998).
99 M.D. Winfield and E.A. Groisman, Role of nonhost environments in the lifestyles of *Salmonella* and *Escherichia coli*, *Appl. Environ. Microbiol.*, 69(7), 3687–3694 (2003).
100 G. Berg, L. Eberl and A. Hartmann, The rhizosphere as a reservoir for opportunistic human pathogenic bacteria, *Environ. Microbial.*, 7(11), 1673–1685 (2005).
101 M.J. Gray, N.E. Freitag and K.J. Boor, How the bacterial pathogen *Listeria monocytogenes* mediates the switch from environmental Dr. Jekyll to pathogenic Mr. Hyde, *Infect. Immun.*, 74(5), 2505–2512 (2006).
102 H.P. Bais, R. Fall and J.M. Vivanco, Biocontrol of *Bacillus subtilis* against infection of Arabidopsis roots by *Pseudomonas syringae* is facilitated by biofilm formation and surfactin production, *Plant Physiol.*, 134, 307–319 (2004).
103 H.L. Wei and L.Q. Zhang, Quorum-sensing system influences root colonization and biological control ability in *Pseudomonas fluorescens* 2P24, *Anton. Leeuw. Int. J. G.*, 89, 267–280 (2006).
104 T. Rudrappa and H.P. Bais, *Arabidopsis thaliana* root surface chemistry regulates in planta biofilm formation of *Bacillus subtilis*, *Plant Signal Behav.*, 2, 349–350 (2007).
105 W.M. Haggag and S. Timmusk, Colonization of peanut roots by biofilm forming *Paenibacillus polymyxa* initiates biocontrol against crown rot disease, *J. Appl. Microbiol.*, 104, 961–969 (2008).
106 B. Lugtenberg and F. Kamilova, Plant-growth-promoting rhizobacteria, *Annu. Rev. Microbiol.*, 63, 541–556 (2009).
107 P.R. Hirsch, A.J. Miller and P.G. Dennis, Do root exudates exert more influence on rhizosphere bacterial community structure than other rhizodeposits? *in Molecular microbial ecology of the rhizosphere, vol 1 and 2*, F.J. de Bruijn (ed) Wiley, USA, 229–242 (2013).
108 K Gross and B.J. Cardinale, Does species richness drive community production or vice versa? Reconciling historical and contemporary paradigms in competitive communities, *Am. Nat.*, 170, 207–220 (2007).
109 S.R. Poltak and V.S. Cooper, Ecological succession in long-term experimentally evolved biofilms produces synergistic communities, *ISME J.*, 5(3), 369–378 (2011).
110 M. Burmølle, A.H. Kjøller and S.J. Sørensen, An invisible workforce: biofilms in the soil, *in Microbial Biofilm: Current Research and Applications*, G. Lear and G.D. Lewis (eds.), Caister Academic Press, 61–71 (2012).

111 K.W. Lee, S. Periasamy, M. Mukherjee, C. Xie, S. Kjelleberg and S.A. Rice, Biofilm development and enhanced stress resistance of a model, mixed-species community biofilm. *ISME J.*, 8(4), 894–907 (2014).

112 H.L. Røder, S.J. Sørensen and M. Burmølle, Studying Bacterial Multispecies Biofilms: Where to Start? *Trends Microbiol.*, 24(6), 503–513 (2016).

113 D. Rao, J.S. Webb and S. Kjelleberg, Competitive interactions in mixed-species biofilms containing the marine bacterium *Pseudoalteromonas tunicate, Appl. Environ. Microbiol.*, 71(4), 1729–1736 (2005).

114 P.I. Diaz, Z. Xie, T. Sobue, A. Thompson, B. Biyikoglu, A. Ricker, L. Ikonomou and A. Dongari-Bagtzoglou, Synergistic interaction between *Candida albicans* and commensal oral streptococci in a novel in vitro mucosal model, *Infect. Immune.*, 80(2), 620–632 (2012).

115 A. Sharma, S. Inagaki, W. Sigurdson and H.K. Kuramitsu, Synergy between *Tannerella forsythia* and *Fusobacterium nucleatum* in biofilm formation, *Oral Microbiol. Immunol.*, 20(1), 39–42 (2005).

116 M. Yamada, A. Ikegami and H.K. Kuramitsu, Synergistic biofilm formation by *Treponema denticola* and *Porphyromonas gingivalis, FEMS Microbiol. Lett.*, 250(2), 271–277 (2005).

117 B.J. Klayman, P.A. Volden, P.S. Stewart and A.K. Camper, *Escherichia coli* O157: H7 requires colonizing partner to adhere and persist in a capillary flow cell, *Environ. Sci. Technol.*, 43(6), 2105–2111 (2009).

118 M.E. Davey and G.A. O'toole, Microbial biofilms: from ecology to molecular genetics, *Microbiol. Mol. Boil. R.*, 64(4), 847–867 (2000).

119 S.J. Sørensen, M. Bailey, L.H. Hansen, N. Kroer and S. Wuertz, Studying plasmid horizontal transfer *in situ*: a critical review, *Nat. Rev. Microbiol.*, 3(9), 700–710 (2005).

120 D. Ren, J.S. Madsen, S.J. Sørensen, and M. Burmølle, High prevalence of biofilm synergy among bacterial soil isolates in cocultures indicates bacterial interspecific cooperation, *ISME J.*, 9(1), 81–89 (2015).

121 S. Timmusk, K. Timmusk and L. Behers, Rhizobacterial plant drought stress tolerance enhancement: towards sustainable water resource management and food security, *J. Food Security*, 1(1), 6–9 (2013).

122 S. Timmusk and E. G. Wagner, The plant-growth-promoting rhizobacterium *Paenibacillus polymyxa* induces changes in *Arabidopsis thaliana* gene expression: a possible connection between biotic and abiotic stress responses, *Mol. Plant Microbe. Interact.* 12, 951 (1999).

123 S. Mayak, T. Tirosh and B. R. Glick, Plant growth-promoting bacteria that confer resistance to water stress in tomatoes and peppers, *Plant Sci.*, 166, 525 (2004).

124 B.R. Glick, B. Todorovic, J. Czarny, Z. Cheng, J. Duan and B. McConkey, Promotion of plant growth by bacterial ACC deaminase, *Crit.l Rev. Plant Sci.*, 26(5–6), 227–242 (2007).

125 E. Ngumbi and J. Kloepper, Bacterial-mediated drought tolerance: Current and future prospects, *App.Soil Eco.*, 105, 109–125 (2016).

126 J.A. Lucas, Plant Pathology and Plant Pathogens. Oxford, UK: Blackwell Science. 1998.

127 D.L. Ebbels, Principles of Plant Health and Quarantine, *In Principles Of Plant Health And Quarantine*, D.L. Ebbels (ed), Wallingford, UK: CAB International, 1–8 (2003).

128 A. MacLeod, M. Pautasso, M.J. Jeger and R. Haines-Young, Evolution of the international regulation of plant pests and challenges for future plant health, *Food Security* 2, 49–70 (2010).

129 T.F. Döring, M. Pautasso, M.R. Finckh and M.S. Wolfe, Concepts of plant health–reviewing and challenging the foundations of plant protection. *Plant Pathol.*, 61(1), 1–15 (2012).

130 J.R. Marchesi, D.H. Adams, F. Fava, G.D. Hermes, G.M. Hirschfield, G. Hold, M.N. Quraishi, J. Kinross, H. Smidt, K.M. Tuohy and L.V. Thomas, The gut microbiota and host health: a new clinical frontier, *Gut*, 65(2), 330–339 (2016).

131 R.L. Berendsen, C.M. Pieterse and P.A. Bakker, The rhizosphere microbiome and plant health, *Trends Plant Sci.*, 17(8), 478–486 (2012).

132 M. Teplitski, J.B. Robinson and W.D. Bauer, Plants secrete substances that mimic bacterial N-acyl homoserine lactone signal activities and affect population density-dependent behaviors in associated bacteria, *Mol. Plant Microbe Interact.*, 13, 637–648 (2000).

133 P. Garbeva, J.A. van Veen and J.D. van Elsas, Microbial diversity in soil: selection of microbial populations by plant and soil type and implications for disease suppressiveness, *Annu. Rev. Phytopathol.*, 42, 243–270 (2004).

134 G. Berg and K. Smalla, Plant species and soil type cooperatively shape the structure and function of microbial communities in the rhizosphere, *FEMS Microbiol. Ecol.*, 68, 1–13 (2009).

135 A. E. Hartemink, Chapter Two—The definition of soil since the early 1800s, *Adv. Agron.*, 137, 73–126 (2016).

136 W. Zhu, S. Wang and C.D. Caldwell, Pathways of assessing agroecosystem health and agroecosystem management, *Acta ecologica sinica*, 32(1), 9–17 (2012).

137 European Communities (EUCom), Material Use in the European Union 1980–2000: indicators and analysis, Edition. – Office for Official Publications of the European Communities, 103, Luxembourg. (http://epp.eurostat.ec.europa.eu/cache/ITY_OFFPUB/KS-AO-02-005/EN/KS-AO-02-005-EN.PDF) (2002).

138 M. G. Kibblewhite, K. Ritz and M. J. Swift, Soil health in agricultural systems, *Phil. Trans. R. Soc. B*, 363, 685–701 (2008).

139 J.W. Doran, and A. J. Jones (eds.), *Methods for assessing soil quality, vol. 49. SSSA special publication*, Madison, WI: ASA (1996).

140 B.P. Warkentin and H.F. Fletcher, Soil quality for intensive agriculture, *in Intensive Agriculture Society of Science, Soil and Manure.* Proceedings of the International Seminar on Soil Environment and Fertilizer Management; National Institute of Agricultural Science: Tokyo, Japan, 594–598 (1977).

141 D.L. Karlen, M.J. Mausbach, J.W. Doran, R.G. Cline, R.F. Harris and G.E. Schuman, Soil quality: A concept, definition, and framework for evaluation, *Soil Sci. Soc. Am. J.*, 61, 4–10 (1997).

142 G.C. Daily, P.A. Matson and P.M. Vitousek, Ecosystem services supplied by soil, *in Nature's Services Societal Dependence On Natural Ecosystems*, G.C. Daily (ed) Island Press: Washington, DC, USA, 365–374 (1997).

143 J.W. Doran and T.B. Parkin, Quantitative indicators of soil quality: A minimum data set, *in Methods for Assessing Soil Quality*, J.W. Doran and A.D. Jones (eds), Soil Science Society of America: Madison, WI, USA, 25–37 (1996).

144 L. Van-Camp, B. Bujarrabal, A. R. Gentile, R. J. A. Jones, L. Montanarella, C. Olazabal and S.K. Selvaradjou, Reports of the technical working groups established under the thematic strategy for soil protection, vol. 5 Monitoring. EUR 21319 EN/5, pp. 653–718. Luxembourg, Europe: European Communities 2004.

145 R. F. Harris, D. L. Karlen and D. J. Mulla, A conceptual framework for assessment and management of soil quality and health, *in Methods for assessing soil quality*, J.W. Doran and A. J. Jones (eds) vol. 49, SSSA special publication, 61–82. Madison, WI: ASA (1996).

146 European Commission, *Communication of 16 April 2002 from the Commission to the Council, the European Parliament, the Economic and Social Committee and the Committee of the Regions: towards a thematic strategy for soil protection [COM (2002) 179 final]*. Brussels, Belgium: European Commission (2002).

147 Defra, The first soil action plan for England, London, UK, Department of Environment, Food and Rural Affairs (2004).

148 European Commission Directorate General for Research—Sustainable Development, Global Change and Ecosystems, Catalogue of projects funded during the Sixth Framework, pp. 362–363. Brussels, Belgium: European Commission (2007).

149 S.S. Andrews, D.L. Karlen and C.A. Cambardella, The soil management assessment framework: A quantitative soil quality evaluation method, *Soil Sci. Soc. Am. J.*, 68, 1945–1962. (2004).

150 B.J. Wienhold, D.L. Karlen, S.S. Andrews and D.E. Stott, Protocol for soil management assessment framework (SMAF) soil indicator scoring curve development, *Renew. Agric. Food Syst.*, 24, 260–266 (2009).

151 S.K. Das and A. Varma, Role of enzymes in maintaining soil health, *in Soil enzymology, soil biology*, chap 22, G. Shukla and A. Varma (eds), Springer, Berlin, 25–42 (2011).

152 M.T. Madigan, J.M. Martinko and J. Parker, *Brock biology of microorganisms*. Vol. 514. Upper Saddle River, NJ: Prentice Hall (1997).

153 V. Torsvik and L. Øvreås, Microbial diversity and function in soil: from genes to ecosystems. *Curr. Opin. Microbial.*, 5(3), 240–245 (2002).

154 E. Malusá, L. Sas-Paszt and J. Ciesielska, Technologies for beneficial microorganisms inocula used as biofertilizers, *Scientific World J.*, 2012 (2012).

155 B. Vu, M. Chen, R.J. Crawford and E.P. Ivanova, Bacterial extracellular polysaccharides involved in biofilm formation, *Molecules* 14, 2535–2554 (2009).

156 C. Chenu, Clay polysaccharide or sand polysaccharide associations as models for the interface between microorganisms and soil e water related properties and microstructure, *Geoderma*, 56, 143–156 (1993).

157 S. Yang, B.T. Ngwenya, I.B. Butler, H. Kurlanda and S.C. Elphick, Coupled interactions between metals and bacterial biofilms in porous media: implications for biofilm stability, fluid flow and metal transport, *Chem. Geol.*, 337–338, 20–29 (2013).

158 E. Pohlon, C. Matzig and J. Marxsen, Desiccation affects bacterial community structure and function in temperate stream sediments, *Fund. App. Limnol.*, 182, 123–134 (2013).

159 R.G. Burns, J.L. DeForest, J. Marxsen, R.L. Sinsabaugh, M.E. Stromberger, M.D. Wallenstein, M.N. Weintraub and A. Zoppini, Soil enzymes in a changing environment: current knowledge and future directions, *Soil Biol. Biochem.*, 58, 216–234 (2013).

160 M. Spohn and L. Giani, Water-stable aggregates, glomalin-related soil protein, and carbohydrates in a chronosequence of sandy hydromorphic soils, *Soil Biol. Biochem.*, 42, 1505–1511 (2010).

161 S.K. Upadhyay, J.S. Singh, D.P. Singh, Exopolysaccharide-producing plant growth-promoting rhizobacteria under salinity condition, *Pedosphere* 21, 214–222 (2011).

162 V. Sandhya, Sk. Z. Ali, M. Grover, G. Reddy, B. Venkateswarlu, Alleviation of drought stress effects in sunflower seedlings by the exopolysaccharides producing *Pseudomonas putida* strain GAP-P45, *Biol. Fert. Soils*, 46, 17–26 (2009).

163 M.A. Redmile-Gordon, P.C. Brookes, R.P. Evershed, K.W.T. Goulding and P.R. Hirsch, Measuring the soil-microbial interface: Extraction of extracellular polymeric substances (EPS) from soil biofilms, *Soil Biol. Biochem.*, 72, 163–171 (2014).

164 J. S. Madsen, H. L. Røder, J. Russel, H. Sørensen, M. Burmølle and S. J. Sørensen, Coexistence facilitates interspecific biofilm formation in complex microbial communities, *Environ. Microbiol.*, 18, 2565–2574 (2016).

165 J.A. Gilbert, J.K. Jansson and R. Knight, The Earth Microbiome project: successes and aspirations, *BMC biology*, 12(1), 1 (2014).

166 H. Vogel, R.O. Musser and M.L. Celorio-Mancera, Transcriptome responses in herbivorous insects towards host plant and toxin feeding, *Ann. Plant Rev.*, 47, 197–233 (2014).

167 J. Nesme, W. Achouak, S.N. Agathos, M. Bailey, P. Baldrian, D. Brunel, Å. Frostegård, T. Heulin, J.K. Jansson, E. Jurkevitch and K.L. Kruus, Back to the future of soil metagenomics, *Fron. Microbial.*, 7 (2016).

2

Role of PGPR in Biofilm Formations and Its Importance in Plant Health

Govind Gupta, Sunil Kumar Snehi and Vinod Singh

Department of Microbiology, Barkatullah University, Bhopal, India

2.1 Introduction

In modern agricultural processes, indiscriminate use of fertilizers, particularly the nitrogenous and phosphorus, has led to substantial pollution of soil, air, and water. Excessive uses of these chemicals have several negative impacts on the environment and exert deleterious effects on soil microorganism, affect the fertility status of soil, and also pollute the environment [1]. These chemicals and fertilizers are often expensive for farmers, and they also have bad effects on agricultural field as well as human beings. To control this problem, agricultural practice is moving to a more sustainable and environmentally friendly approach worldwide. The search for microorganisms that improve soil fertility and enhance plant nutrition has continued to attract attention due to the increasing cost of fertilizers and some of their negative environmental impacts.

2.2 Rhizosphere: A Unique Source of Microorganisms for Plant Growth Promotion

The rhizosphere probably represents the most dynamic habitat on Earth and certainly is the most important zone in terms of defining the quality and quantity of the human terrestrial food resource. The pleasant environment of microorganisms around plant root called rhizosphere. The term *rhizosphere* was introduced for the first time by Hiltner [2]. In the rhizosphere, very important interactions takes place between the plant, soil, microorganisms, and soil microfauna, influenced by compounds exuded by the root and by microorganisms feeding on these compounds [3]. The rhizosphere region can be classified into three factions: (i) rhizospheric soil that adheres to the root when the root system is shaken manually, (ii) endorhizosphere (interior of the root), and (iii) rhizoplane (surface of the root) [4]. The rhizosphere is the front line between plant roots and soil-borne pests. Therefore, it seems logical that microorganisms that colonize the same niche could be the ideal candidates for sustainable agriculture [5].

A number of beneficial microorganisms are associated with the root system of higher plants, which depend on the exudates of these roots for their survival [6]. Many varieties of compounds present in root exudates, including polysaccharides and proteins, enable the bacteria to colonize plant roots [7, 8]. Most of the bacteria, fungi, viruses are found in rhizosphere. The importance of rhizobacterial population has been recognized and well-documented for maintenance of root health, nutrient uptake, tolerance of environmental stress, and crop responses. In the rhizosphere, plant–microbe interactions are responsible for a number of inherent processes such as carbon sequestration, ecosystem functioning, and nutrient cycling [9]. The rhizodeposition of root exudates, composed of small molecular weight metabolites, amino acids, mucilage, secreted enzymes and cell lysates, can range from less than 10 percent of the net carbon assimilation by a plant to as much as 44 percent of a nutrient-stressed plant's total carbon [10]. Soil microbes utilize this abundant carbon source, thereby implying that selective secretion of specific compounds may encourage beneficial symbiotic and protective relationships, whereas secretion of other compounds inhibits pathogenic associations [11].

2.3 Plant Growth–Promoting Rhizobacteria

Bacteria are the most abundant microorganisms in the rhizosphere [3]. Some of the rhizosphere bacteria can effect plant growth often referred to as a *plant growth–promoting rhizobacteria* (PGPR) [12]. According to Kloepper and Schroth [13], PGPR mediated plant growth promotion by the alteration of the whole microbial community in rhizosphere through the production of various substances. During the past decades, the use of PGPR for sustainable agriculture has increased tremendously in various parts of the world. PGPR promote plant growth, either directly by affecting their nutrient supply (nitrogen, phosphorus, potassium, and essential minerals) or modulating plant hormone levels, or indirectly by decreasing the inhibitory effects of various pathogens on plant growth and development in the forms of biocontrol agents, root colonizers, and environmental protectors [14–16].

PGPR can be classified into extracellular plant growth–promoting rhizobacteria (ePGPR) and intracellular plant growth–promoting rhizobacteria (iPGPR) [1]. The ePGPRs may exist in the rhizosphere, on the rhizoplane, or in the spaces between the cells of root cortex; on the other hand, iPGPRs locate generally inside the specialized nodular structures of root cells. The bacterial genera such as *Agrobacterium, Arthrobacter, Azotobacter, Azospirillum, Bacillus, Burkholderia, Caulobacter, Chromobacterium, Erwinia, Flavobacterium, Micrococcous, Pseudomonas,* and *Serratia* belong to ePGPR [17]. The iPGPR includes the endophytes and *Frankia* species, both of which can symbiotically fix atmospheric N_2 with the higher plants [18]. Endophyte includes a wide range of soil bacterial genera such as *Allorhizobium, Azorhizobium, Bradyrhizobium, Mesorhizobium,* and *Rhizobium* of the family Rhizobiaceae that generally invade the root systems in crop plants to form nodules [19, 20]. In addition, several actinomycetes, one of the major components of rhizosphere microbial populations, are also useful because of their significant ecological roles in soil nutrient cycling [21], as well in plant growth–promoting activities. Numbers of reports are available on the potential of actinomycetes as plant growth–promoting agents. Actinomycetes belonging to the genera *Actinoplanes, Actinomadura, Micromonospora, Streptomyces,*

Streptosporangium, *Streptoverticillium*, and *Spirillospora* were best used to colonize the plant rhizosphere, showing an immense potentiality as biocontrol agent against a range of root pathogenic fungi [22]. PGPR can promote plant growth directly or indirectly.

2.3.1 Direct Impact of Plant Growth–Promoting Rhizobacteria on Plant Nutrition

In agriculture, one of the limiting factors is providing plant nutrients, particularly nitrogen, phosphorus, and potassium. PGPR directly facilitate nutrient availability, solubilization of mineral nutrients, mineralize organic compounds, and production of phytohormones (Figure 2.1).

2.3.1.1 Nitrogen Fixation

Nitrogen is an essential element for all forms of life, and it is the most vital nutrient for plant growth and productivity. Although nitrogen makes up 78 percent of the atmosphere, it remains unavailable to plants. Actually no plant species is capable of transforming atmospheric dinitrogen into ammonia and expending it directly for its growth. Thus, the atmospheric nitrogen is converted into plant utilizable forms by biological nitrogen fixation (BNF), which changes nitrogen to ammonia by nitrogen-fixing microorganisms using a complex enzyme system known as nitrogenase. Various species of *Bacillus* strain such as *Bacillus brevis, Bacillus cereus, Bacillus circulans, Bacillus firmus Bacillus licheniformis, Bacillus megaterium, Bacillus pumilus,* and *B. subtilis* were showed biological nitrogen fixation on the basis of nitrogenase activity.

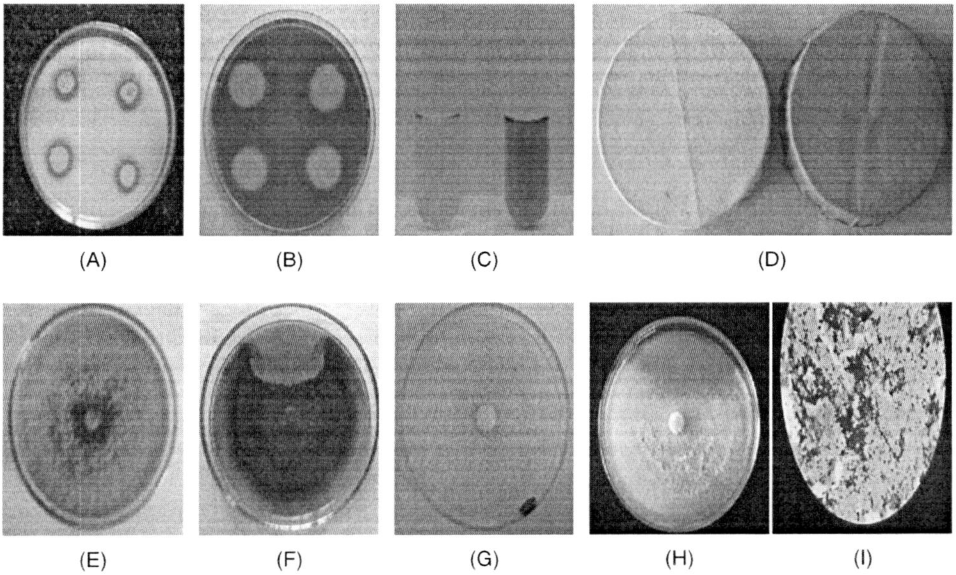

Figure 2.1 Schematic diagram showing various plant growth–promoting Attributes. (A) Phosphate solubilization; (B) Siderophore Production; (C) Indole acetic production; (D) HCN Production; (E) Control (*M. phaseolina*); (F) Inhibition of *M. phaseolina*; (G) Control (*F. solani*); (H) Inhibition of *F. solani*; (I) Biofilm formation. (*See color plate section for the color representation of this figure.*)

PGPR have the ability to fix atmospheric nitrogen and provide it to plants by two mechanisms: symbiotic and nonsymbiotic. Symbiotic nitrogen fixation is a mutualistic relationship between a microbe and the plant. The PGPR widely presented as symbionts are *Rhizobium*, *Bradyrhizobium*, *Sinorhizobium*, and *Mesorhizobium* with leguminous plants, *Frankia*, with nonleguminous trees and shrubs [23, 24]. On the other hand, nonsymbiotic nitrogen fixation is carried out by free living diazotrophs, and this can stimulate nonlegume plants growth such as radish and rice. Nonsymbiotic nitrogen-fixing rhizospheric bacteria belong to genera including *Azoarcus*, *Azotobacter*, *Acetobacter*, *Azospirillum*, *Burkholderia*, *Diazotrophicus*, *Enterobacter*, *Gluconacetobacter*, *Pseudomonas*, and cyanobacteria (*Anabaena*, *Nostoc*) [25].

2.3.1.2 Phosphorus Solubilization

Phosphorus (P) is an essential plant nutrient that has low availability in many agricultural soils. It is one of the most important elements after nitrogen which is present for plant growth [26]. To obviate this problem, farmers use phosphorus fertilizers, which is an expensive, less effective, and environmentally unsafe method. The eco-friendly approaches inspire a wide range of exploitation of beneficial PGPR led to improved phosphorus uptake from unavailable to available forms to the plants [27]. The main phosphate solubilization mechanisms employed by PGPR include: (1) release of complexing or mineral-dissolving compounds (e.g., organic acid anions, protons, hydroxyl ions, CO_2), (2) liberation of extracellular enzymes (biochemical phosphate mineralization), and (3) the release of phosphate during substrate degradation (biological phosphate mineralization) [28]. Phosphate solubilizing PGPR included in the genera *Bacillus*, *Beijerinckia*, *Pseudomonas*, *Arthrobacter*, *Burkholderia*, *Enterobacter*, *Erwinia*, *Flavobacterium*, *Microbacterium*, *Rhizobium*, *Rhodococcus*, and *Serratia* have ability by their capacity to solubilize precipitated forms of phosphorus to available forms for plant growth and yield [29].

2.3.1.3 Potassium Solubilization

In nutrient-deficient ecosystems, potassium (K) is the third major macronutrient that is necessary for obtaining the high yield of field crops [30]. Without adequate potassium, the plants will have poorly developed roots, grow slowly, produce small seeds, and have lower yields. The concentrations of soluble potassium in the soil are usually very low, and more than 90 percent of potassium in the soil exists in the form of insoluble rocks and silicate minerals [31]. Sheng [32] studied the ability of a potassium-releasing bacterial strain *Bacillus edaphicus* to promote cotton and rape growth in K-deficient soil. The strain was able to mobilize potassium as well as promote root and shoot growth. Similarly, inoculation of maize and wheat plants with *Bacillus mucilaginosus*, *Azotobacter chroococcum*, and *Rhizobium* resulted in significantly higher mobilization of potassium from waste mica, which, in turn, acted as a source of potassium for plant growth [9]. Potassium-solubilizing PGPR could be exploited to reduce the use of agrochemicals, lead to improved nutrient uptake ability, and play a major role in improving soil fertility under field conditions.

2.3.1.4 Siderophore Production

Iron (Fe) is the most abundant element on Earth's crust, found in the form of ferromagnesium silicates. Most of the iron in the soil is found in silicate minerals or iron oxides and hydroxides, forms that are not readily utilizable by microorganisms and plants. PGPR have evolved specialized mechanisms for the assimilation of iron, including the production of low molecular weight iron-chelating compounds known as siderophores, which transport this element into their cells [33, 34]. Siderophores are divided into three main families, depending on the characteristic functional group (i.e., hydroxamates, catecholates and carboxylates). More than 500 different types of siderophores are known, of which 270 have been structurally characterized [35]. Microorganisms produce a wide range of siderophores. Most of the bacterial siderophores are catecholates (i.e., enterobactin), and some are carboxylates (i.e., rhizobactin) and hydroxamates (i.e., ferrioxamine B). However, there are also certain types of bacterial siderophores that contain a mix of the main functional groups (i.e., pyoverdine) [36]. The benefits of bacterial siderophores on the growth of plants have been demonstrated by using radiolabeled ferricsiderophores, as a sole source of iron showed that plants are able to take up the labeled iron by a large number of PGPR including *Aeromonas, Azadirachta, Azotobacter, Bacillus, Burkholderia, Pseudomonas, Rhizobium, Serratia,* and *Streptomyces* sp. [37].

2.3.1.5 Phytohormone Production

One of the direct growth–promotion mechanism used by PGPR is the production of phytohormones, including indole acetic acid, gibberellins, cytokinins, ethylene, and abscisic acid [38].

2.3.1.6 Indole Acetic Acid (IAA) Production

The ability to synthesize phytohormones is widely distributed among plant-associated bacteria, and 80 percent of the bacteria associated with plants are able to produce indole acetic acid [39]. IAA is the most common and best-characterized phytohormone, which stimulates root elongation and increased density of both root hairs and lateral roots [17]. In addition to IAA, PGPR also release other compounds in the rhizosphere, like indole-3-butyric acid (IBA), Trp, and tryptophol or indole-3-ethanol (TOL) that can contribute to plant growth promotion [40]. Phytohormones such as IAA produced by various PGPR such as *Agrobacterium, Azospirillum, Alcaligenes faecalis, Klebsiella, Enterobacter, Bradyrhizobium, Rhizobium, Pseudomonas, Enterobacter, Acetobacter diazotrophicus,* and *Herbaspirillum seropedicae* [41, 42]. Tryptophan is an amino acid commonly found in root exudates, has been identified as main precursor molecule for biosynthesis of IAA in PGPR [43].

2.3.1.7 Gibberellins and Cytokinins Production

Gibberellins is another phytohormone that is synthesized by higher plants, fungi, and bacteria. They affect cell division and elongation and are involved in several plants developmental processes, including seed germination, stem elongation, flowering, fruit setting, and delay of senescence in many organs of a range of plant species [44]. According to MacMillan [45], up to 136 different gibberellins have been identified and characterized. Some strains of phytopathogens can also synthesize cytokinins. Cytokinins are N6-substituted aminopurines that play a key role in a wide range of physiological processes such as plant cell division, interruption of the quiescence of dormant buds, activation of seed germination, promotion of branching, root growth, accumulation of chlorophyll, leaf expansion, and delay of senescence [46]. However, it appears that PGPR produce lower cytokinin levels compared to phytopathogens, so that the effect of the PGPR on plant growth is stimulatory while the effect of the cytokinins from pathogens is inhibitory. Several PGPR such as *Agrobacterium*, *Arthrobacter*, *Azospirillum*, *Azotobacter*, *Bacillus*, *Burkholderia*, *Clostridium*, *Flavobacterium*, *Micrococcus*, *Pseudomonas fluorescens*, *Rhizobium* sp., *Xanthomonas* can produce gibberellins and *Arthrobacter*, *Azospirillum*, *Paenibacillus polymyxa*, *Pseudomonas fluorescens*, *Rhodospirillum rubrum* can produce cytokinins respectively [44, 47].

2.3.2 In Direct Impact of Plant Growth–Promoting Rhizobacteria on Plant Nutrition

Numerous PGPR is a possible alternative to reducing the chemical input in agriculture to obtain environment friendly sustainable fertility of soil by indirectly. This include by antibiotic production, lytic enzyme production, induced systemic resistance, hydrogen cyanide production, and exopolysaccharides production or biofilm formation (Figure 2.1).

2.3.2.1 Antibiotic Production

Antibiotics are known to possess antiviral, antimicrobial, antihelminthic, phytotoxic, antioxidant, cytotoxic, antitumor, and plant growth–promoting activities [48]. The antibiotics disrupt the functions of ribosomes and inhibit the formation of initiation complex on small subunit of ribosomes. PGPR producing antibiotics are said to be effective against Gram-positive and Gram-negative bacteria and pathogenic fungi [49, 50]. A variety of antibiotics have been identified, including compounds such as amphisin, 2,4-diacetylphloroglucinol (DAPG), oomycin A, phenazine, pyoluteorin, pyrrolnitrin, tensin, tropolone, and cyclic lipopeptides produced by pseudomonads [51] and oligomycin A, kanosamine, zwittermicin A, and xanthobaccin produced by *Bacillus*, *Streptomyces*, and *Stenotrophomonas* sp. to prevent the proliferation of plant pathogens (generally fungi) and promote the growth indirectly [16, 52].

2.3.2.2 Enzyme Production

PGPR produce extracellular enzymes within the rhizosphere, which is necessary for degradation of both plant and microbial biomass and the degradation and utilization of exudates released by the plants. Enzymes commonly found in the rhizosphere are those that degrade biopolymers present in plant and microbial biomass, such as amylases, cellulases, mannanases, pectinases, and xylanases that degrade starch, cellulose, mannose, pectin and xylan, respectively. Plant growth–promoting rhizobacterial strains can produce certain lytic enzymes such as chitinases, dehydrogenase, β-glucanase, lipases, phosphatases, and proteases [53, 54], attacking pathogens by excreting cell wall hydrolases, which protect them from biotic and abiotic stresses by suppression of pathogenic fungi [55]. In recent years, *Pseudomonas fluorescens* has been suggested as a potential biological control agent due to its ability to colonize rhizosphere and protect plants against a wide range of important agronomic fungal diseases such as black root rot of tobacco, root rot of mustard, and damping-off of sugar beet in field condition [55].

2.3.2.3 Induced Systemic Resistance

Another indirect positive benefit of PGPR is the activation of a defense mechanism called induced systemic resistance (ISR). ISR is a new approach for sustainable agriculture against pathogens [56]. Exudates produced by PGPR are able to stimulate ISR by activating components such as lipoxygenases, lipid peroxidases, and reactive oxygen species protect against diseases caused by different organisms [50, 57]. Hence, multiple genes are performing multiple functions to enhance ISR as a powerful tool for plant growth promotion and disease suppression. Bacteria-produced salicylic acid (SA) contributes to the induction of systemic resistance. Several *Pseudomonas* spp. are able to induce ISR in a wide range of plants against different pathogens. ISR elicitation is a widespread phenomenon not only for fluorescent pseudomonads but also for a variety of nonpathogenic microorganisms and biological control agents [58].

2.3.2.4 Hydrogen Cyanide Production

Phytopathogenic microorganisms are a major and chronic threat to sustainable agriculture and ecosystem stability worldwide. They subvert the soil ecology, disrupt environment, degrade soil fertility, and consequently show harmful effects on human health, along with contaminating ground water. Hydrogen cyanide (HCN) is the most potent volatile compound produced by many plant growths–promoting rhizobacteria. The HCN produced by antagonistic fluorescent pseudomonads have exemplary antifungal activity against phytopathogens. Hydrogen cynide is generally produced by the genus *Bacillus,* including *B. amyloliquefaciens, B. cereus, B. mycoides, B. pumilus, B. sphaericus,* and *B. subtilis* [59]. Recently, some authors demonstrated that some PGPRs can produce VOCs as signals that stimulate the growth of plants [60].

2.3.2.5 Exopolysaccharides Production or Biofilm Formation

Certain PGPR synthesize a wide spectrum of multifunctional polysaccharides, including extracellular polysaccharides, intracellular polysaccharides, and structural polysaccharides. The key functions of exopolysaccharides comprise the mediation of the initial attachment of cells to different substrata and protection against environmental stress and dehydration. Production of exopolysaccharides is generally important in biofilm formation. Biofilms are structurally complex communities of microbial cells that adhere to a surface and are surrounded by an extracellular polymeric matrix. It is defined as a complex microbial aggregate embedded in a hydrated extracellular polymeric substance (EPS) matrix.

The term *biofilm* was coined and described in 1978 [61]. Growth of microbial aggregates attached to any solid substratum is the characteristic nature of biofilms. Biofilm formation causes severe complications in healthcare, agriculture, and industries, which results in recurrent infections, plant diseases, and biocorrosion, respectively [62]. Beside from polysaccharides, biofilms also consist of proteins, nucleic acids, lipids and humic substances [63]. Biofilm formation begins with an initial, reversible attachment, when planktonic bacteria make contact with the substrate and become temporarily fixed, with some cells being able to detach. In this process, an important role is played by flagella. This stage involves the work of nonspecific physicochemical forces of interaction between the molecules and structures on the surfaces of the microorganism and the solid substrate [64]. The start of biofilm formation may be signaled by osmolarity, the pH of the medium, soil content of metals, oxygen supply, temperature, and other factors. Effective colonization of plant roots by EPS-producing microbes helps to hold the free phosphorous from the insoluble one in soils and circulating essential nutrient to the plant for proper growth and development and protecting it from the attack of foreign pathogens [65].

2.4 Biofilm Producing Plant Growth–Promoting Rhizobacteria

Generally, PGPR are recognized in four stages in the process of biofilm formation:

1) Initial reversible attachment of an individual bacterium to a surface, leading to a stronger, irreversible attachment by bacteria
2) Microcolony formation with early stage development of biofilm architecture
3) Biofilm maturation
4) Eventual dispersal of single cells from the biofilm [66]

PGPR are often attached to a surface and coated with a polymeric substance, characteristic of a biofilm [67]. *Pseudomonas* species produce a variety of organic molecules, including polysaccharides, nucleic acids, and proteins that are used to form biofilm matrices. EPS released by the *Rhizobium* assist in biofilm formation that enhances plant growth and provides protection from pathogens [33]. Beneficial biofilms that can be developed in vitro and be used as biofertilizers are called biofilmed biofertilizers. Beneficial biofilms attached to the plant roots of some crops may help cycle nutrients as well as biocontrol of pests and diseases and consequently improve productivity and crop yields (Table 2.1). Additionally, rhizobial polysaccharides are highly important in promoting plant growth, work as an active signal molecule during beneficial interactions, and provide defense response during the infection process [68]. Some PGPR-EPS can also bind cations, including Na^+, suggesting a role in mitigation of salinity stress by reducing the content of Na^+ available for plant uptake [65]. The co-inoculation of biofilmed inocula can be used for successful establishment of introduced beneficial microorganisms in plants for biocontrol of diseases [69]. When two (rather than one) plant

Table 2.1 Some Known Rhizospheric Bacteria Used as a Biofilm Promoting Plant Growth.

S.N.	Bacteria	Characteristics	References
1.	*Azorhizobium caulinodans*	Root colonization of rice	[77]
2.	*Azorhizobium brasilense*	Root colonization of wheat	[78]
3.	*Acinetobacter calcoaceticus* P23	Root colonization of duckweed	[79]
4.	*Bacillus amyloliquefaciens* S499	Root colonization of tomato, maize, and Arabidopsis thaliana	[80]
5.	*Bacillus polymyxa*	Root colonization of cucumber	[81]
6.	*Cyanobacteria* spp	Enhanced mixed-species biofilm formation with *Rhizobium*, *Azotobacter* and *Pseudomonas* spp.	[82]
7.	*Klebsiella pneumoniae*	Root colonization of wheat	[83]
8.	*Pantoea agglomerans*	Root colonization of chickpea and wheat	[84,85]
9.	*Rhizobium leguminosarum* bv. viciae 3841	Root colonization of various legumes	[86–88]
10.	*Rhizobium (Sinorhizobium)* sp. strain NGR234	Root colonization of various legumes and competitive colonization in the rhizosphere of cowpea	[89]

growth–promoting rhizobacterial biofilm biofertilizers were applied, there was a propensity for increased photosynthesis. The co-inoculation of 50% of the recommended fertilizers with PGPR biofilm biofertilizers helped increase leaf growth compared to 100% of the recommended fertilizer application with PGPR biofilm biofertilizers respectively [70]. This integrated approach for growth promotion of plants has to achieve more sustainable agriculture as well as fertility of soil.

2.5 Role of PGPR in Biofilm Formations

Biofilm acts as a protective barrier that prevents the penetration of drug molecules and also secretes various inactivating and modifying enzymes [71]. To combat biofilm-related problems in any industry, including agriculture, antibiofilm agents are the wisest choice. PGPR have evolved sophisticated mechanisms to coordinate gene expression at population and community levels via the synthesis and perception of diffusible molecules.

Because the concentration of the emitted signal in a confined environment reflects the bacterial cell number per volume unit, such a regulatory pathway is termed *quorum sensing* (QS) [72]. Bacterial quorum sensing mechanism is based on two groups of signal molecules: peptide derivatives typical for Gram-positive bacteria and fatty acid derivatives exploited by Gram-negative bacteria [15, 44]. Quorum sensing [QS], their regulation, interaction between bacteria and host plant, mechanism and application of bioinformatics in development of Anti-QS is described in chapter 3, 6, 7, 16, 20, 22 and 25 in detail of this book.

The most intensively investigated signal molecules in PGP Gram-negative bacteria especially *Pseudomonas* are N-Acyl-L-homoserine Lactones (N-AHLs). The N-Acyl-L-homoserine Lactones are responsible for controlling various activities such as biofilm formation, antibiotic production, resistance, conjugation, replication, exoenzyme synthesis, swarming, and bioluminescence [73]. The PGPR *Pseudomonas aureofaciens* competes with soil fungi of the genus *Fusarium* by quorum sensing–dependent and AHL-based production of an antifungal antibiotic phenazine, which suppresses *Fusarium* growth [74]. Another signal molecule in PGPR are diffusible signal molecules (DSM), Quinolone, Autoinducer-2, and Diketopiperazines, which is used for microbe–microbe communication for plant development and productivity.

2.6 Future Research and Development Strategies for Biofilm Producing Sustainable Technology

Using PGPR is a promising and environmentally friendly approach to obtaining sustainable fertility of the soil and plant growth. The eco-friendly approach has the potential for high output yield and enhanced crop production as well as soil fertility through biofilm-producing biofertilizers. However, inconsistency in field application of such microbial inocula has limited its widespread commercial application, probably due to the incapability of such inocula to successfully compete with indigenous microflora in establishing themselves in the rhizosphere [75].

Future research in rhizosphere biology will rely on the development of biofilm-forming plant growth–promoting rhizobacterial inoculants, which can protect against adverse environmental conditions such as high salinity, tannin concentrations, low pH, and heavy metals contamination. In the future it will also be helpful to identify the compounds and correspondence in complex mixtures of biofilm-producing PGPR. Future research will be also helpful for optimizing the inoculants of mono as well as multistrain biofilm-forming PGPR for reducing the harmful impact of stress on plant growth, reducing the need for agrochemicals (fertilizers and pesticides) for improve soil fertility [76]. However, future research has to be focused on the new concept of biofilm engineering in agro-ecological systems regarding selection of combinations of microbes for highest efficiency, simultaneous biofertilizing and biocontrolling activities is a key in future research in this technology. Thus, more research should be done under laboratory and field conditions in order to optimize biofilmed inocula for various crops. Apart from that, increased shelf life of biofilm-forming PGPR products, not phytotoxic to crop plants, would improve tolerance for adverse environmental conditions, improve yields, and make biofilm-forming PGPR products more cost-effective for agricultural farmers.

2.7 Conclusions

The studies here demonstrate the importance of PGPR and biofilm formation, which is an important component of plant–microbe interactions, its effect on plant health and soil fertility. The applicability of PGPR is a promising sustainable and environmentally friendly approach to obtain sustainable fertility of the soil and plant growth. This approach inspires a wide range of exploitation of PGPR, which could lead to reducing the need for agrochemicals (fertilizers and pesticides) for improved soil fertility by a variety of mechanisms that involve soil structure formation, decomposition of organic matter, recycling of essential elements, solubilization of mineral nutrients, production of numerous plant growth regulators, degradation of organic pollutants, stimulation of root growth, crucial for soil fertility, biocontrol of soil and seedborne plant pathogens, and promotion of changes in vegetation. Plant growth–promoting rhizobacterial biofilms have been shown to play a fundamental role in futuristic agricultural approaches such as biofertilizers, plant growth promoters, and biocontrolling agents for plant health.

Acknowledgments

Authors are grateful to Director, Madhya Pradesh Council of Science & Technology, Bhopal, Madhya Pradesh and Vice Chancellor of Barkatullah University, Bhopal, Madhya Pradesh, India, for their financial support and necessary facilities, respectively.

References

1 G. Gupta, S.S. Parihar, N.K. Ahirwar, S.K. Snehi and V. Singh, Plant growth promoting rhizobacteria (PGPR): Current and future prospects for development of sustainable agriculture, *J Microb Biochem Technol*, 7, 096–102 (2015).

2 L. Hiltner, Iiber neuere Erfahrungen und unter besonderer Berucksichtingung der Grundunging und Brache, *Arb Dtsch Landw Ges*, 98, 59–78 (1904).
3 H. Antoun and D. Prévost, Ecology of plant growth promoting rhizobacteria, Z.A. Siddiqui (Eds), *PGPR: biocontrol and biofertilization*, Springer, Dordrecht, pp 1–38 (2005).
4 J.M. Gobat, M. Aragno and W. Matthey, *The living soil. Fundamentals of soil science and soil biology, Lausanne*, Suisse: science publisher Inc., USA (2004).
5 D.M. Weller, *Pseudomonas* biocontrol agents of soilborne pathogens: looking back over 30 years, *Phytopathology*, 97, 250–256 (2007).
6 J.M. Whipps, Microbiol Interaction and biocontrol in the rhizosphere, *J Exp Bot* 52, 487–511 (2001).
7 E. Somers, J. Vanderleyden and M. Srinivasan, Rhizosphere bacterial signalling: A love parade beneath our feet, *Crit Rev Microbiol*, 30, 205–240 (2004).
8 D.N. Rodriguez–Navarro, M.S. Dardanelli and J.E. Ruiz-Sainz, Attachment of bacteria to the roots of higher plants, *FEMS Microbiol Lett*, 272, 127–136 (2007).
9 G. Singh, D.R. Biswas and T.S. Marwah , Mobilization of potassium from waste mica by plant growth promoting rhizobacteria and its assimilation by maize (*Zea mays*) and wheat (*Triticum aestivum* L.), *Journal of Plant Nutrition*, 33, 1236–1251 (2010).
10 E. Patterson and A. Sims, Effect of nitrogen supply and defoliation on loss of organic compounds from roots of *Festuca rubra*, *J Exp Bot*, 51, 1449–1457 (2000).
11 H.P. Bais, B. Prithiviraj, A.K. Jha, F.M. Ausubel and J.M. Vivanco, Mediation of pathogen resistance by exudation of antimicrobials from roots, *Nature* 434, 217–221 (2005).
12 J.W. Kloepper, J. Leong, M. Teintze and M.N. Scroth, Enhancement of plant growth by siderophores produced by plant growth promoting rhizobacteria, *Nature (London)*, 286, 885–886 (1980).
13 J.W. Kloepper and M.N. Schroth, Plant growth promoting rhizobacteria on radishes. In: Proceedings of the 4th international conference on plant pathogenic bacteria, *Angers*, 879–882 (1978).
14 N.K. Ahirwar, G. Gupta, V. Singh, R.K. Rawlley and S. Ramana, Influence on growth and fruit yield of tomato (*Lycopersicon esculentum* Mill.) plants by inoculation with *Pseudomonas fluorescence* (SS5): Possible role of plant growth promotion, *Int J. Curr Microbiol App Sci*, 4(2), 720–730 (2015).
15 F. Ahmad, I. Ahmad and M.S. Khan, *Screening of free-living rhizospheric bacteria for their multiple plant growth promoting activities, Microbiological Research, ISSN 0944-5013* (2008).
16 E. Khare and N.K. Arora, Dual activity of pyocyanin from *Pseudomonas aeruginosa*: Antibiotic against phytopathogen and signal molecule for biofilm development by rhizobia, *Canadian Journal of Microbiology* 57(9), 708–713 (2011).
17 E.J. Gray and D.L. Smith, Intracellular and extracellular PGPR: commonalities and distinctions in the plant-bacterium signaling processes, *Soil Biol Biochem* 37, 395–412 (2005).
18 J.P. Verma, J. Yadav and K.N. Tiwari, Enhancement of nodulation and yield of chickpea by coinoculation of indigenous *Mesorhizobium* sp. and plant growth–promoting rhizobacteria in eastern Uttar Pradesh, *Commun Soil Sci Plant Anal*, 43, 605–621 (2012).

19 G. Singh, H. Sekhon and P. Sharma, Effect of irrigation and biofertilizer on water use, nodulation, growth and yield of chickpea (*Cicer arietinum* L), *Arch Agron Soil Sci*, 57, 715–726 (2011).

20 I. Ahmad, S.A. Khan, F. Aqil and M. Singh, Microbial applications in agriculture and the environment: A broad perspective, I. Ahmad et al. (Eds.), *Microbes and Microbial Technology: Agricultural and Environmental Applications*, DOI 10.1007/978-1-4419-7931-5_1, (pp 01–28) © Springer, USA (2011).

21 A.K. Halder, A.K. Misra and P.K. Chakrabarty, Solubilization of inorganic phosphates by *Bradyrhizobium*, *Indian J Exp Biol*, 29, 28–31 (1991).

22 C. Franco, P. Michelsen, N. Percy, V. Conn, E. Listiana, S. Moll, R. Loria and J. Coombs, Actinobacterial endophytes for improved crop performance, *Australas Plant Pathol*, 36, 524–531 (2007).

23 M.A. Iqbal, M. Khalid, S.M. Shahzad, M. Ahmad, N. Soleman and N. Akhtar, Integrated use of *Rhizobium leguminosarum*, plant growth promoting rhizobacteria and enriched compost for improving growth, nodulation and yield of lentil (*Lens culinaris* Medik), *Chilean J Agric Res*, 72, 104–110 (2012).

24 H.H. Zahran, Rhizobia from wild legumes: diversity, taxonomy, ecology, nitrogen fixation and biotechnology, *J Biotechnol*, 91, 143–153 (2001).

25 J.K. Vessey, Plant growth promoting rhizobacteria as biofertilizers, *Plant Soil*, 255, 571–586 (2003).

26 L.A. Fernandez, P. Zalba, M.A. Gomez and M.A. Sagardoy, Phosphate-solubilization activity of bacterial strains in soil and their effect on soybean growth under green house conditions, *Biol Fertil Soils*, 43, 803–805 (2007).

27 S. Singh, G. Gupta, E. Khare, K.K. Behal and N.K. Arora, Effect of enrichment material on the shelf life and field efficiency of bioformulation of *Rhizobium* sp. and P-solubilizing *Pseudomonas fluorescens*, *Science Research Reporter*, 4(1), 44–50 (2014).

28 S.B. Sharma, R.Z. Sayyed, M.H. Trivedi and T.A. Gobi, Phosphate solubilizing microbes: sustainable approach for managing phosphorus deficiency in agricultural soils, *Springer plus* 2, 587 (2013).

29 P.N. Bhattacharyya and D.K. Jha, Plant growth–promoting rhizobacteria (PGPR): emergence in agriculture, *World J Microbiol Biotechnol*, 28, 1327–1350 (2012).

30 H.S. Han and K.D. Lee, Phosphate and potassium solubilizing bacteria effect on mineral uptake, soil availability and growth of eggplant, *Res J Agric Biol Sci*, 1, 176–180 (2005).

31 P. Parmar and S.S. Sindhu, Potassium solubilization by rhizosphere bacteria: influence of nutritional and environmental conditions, *J Microbiol Res*, 3, 25–31 (2013).

32 X.F. Sheng, Growth promotion and increased potassium uptake of cotton and rape by a potassium releasing strain of *Bacillus edaphicus*, *Soil Biol Biochem*, 37, 1918–1922 (2005).

33 N.K. Arora, S. Tewari and R. Singh, Multifaceted plant-associated microbes and their mechanisms diminish the concept of direct and indirect PGPRS, N.K. Arora (Eds.), *Plant Microbe Symbiosis: Fundamentals and Advances*, pp 411–449 (2013).

34 B. Schwyn and J.B. Neilands, Universal chemical assay for the detection and determination of siderophores, *Anal Biochem*, 160, 47–56 (1987).

35 H. Boukhalfa, J. Lack, S.D. Reilly, L. Hersman and M.P. Neu, Siderophore production and facilitated uptake of iron and plutonium in *P. putida*, AIP Conf Proc, 673, 343–344 (2003).

36 P. Cornelis, Iron uptake and metabolism in pseudomonads, *Appl Microbiol Biotechnol* 86, 1637–1645 (2010).
37 N. Sujatha and K. Ammani, Siderophore production by the isolates of fluorescent Pseudomonads, *Int J Cur Res Rev*, 5, 1–7.
38 S. Desai, M. Grover, E.L. Amalraj, G.P. Kumar and S.K.M. Ahmed, Exploiting plant growth promoting rhizomicroorganisms for enhanced crop productivity, T. Satyanarayana, B.N. Johri, A. Prakash (Eds), *Microorganisms in sustainable agriculture and biotechnology*, Springer, New York, pp 227–241 (2012).
39 P. Cheryl and B. Glick, Bacterial biosynthesis of indole-3-acetic acid, *Can J Microbiol* 42, 207–220 (1996).
40 H. El-Khawas and K. Adachi, Identification and quantification of auxins in culture media of *Azospirillum* and *Klebsiella* and their effect on rice roots, *Biol Fertil Soils*, 28,377–381 (1999).
41 E. Khare and N.K. Arora, Effect of Indole-3-Acetic Acid (IAA) Produced by *Pseudomonas aeruginosa* in Suppression of Charcoal Rot Disease of Chickpea, *Curr Microbiol*, 61, 64–68 (2010).
42 A. Costacurta and J. Vanderleyden, Synthesis of phytohormones by plant-associated bacteria, *Crit Rev Microbiol* 21, 1–18 (1995).
43 H.A. Etesami, H.A. Alikhani and A. Akbari, Evaluation of plant growth hormones production (IAA) ability by Iranian soils rhizobial strains and effects of superior strains application on wheat growth indexes, *World Appl Sci J*, 6, 1576–1584 (2009).
44 S. Tewari and N.K. Arora, Transactions among microorganisms and plant in the composite rhizosphere, N.K. Arora (Eds.), *Plant Microbe Symbiosis: Fundamentals and Advances*, pp 1–50 (2013).
45 J. MacMillan, Occurrence of gibberellins in vascular plants, fungi and bacteria, *J Plant Growth Regul*, 20, 387–442 (2002).
46 B.F. Salisbury and C.W. Ross, *Plant physiology*, 4th Eds. Wadsworth Publishing Company, Belmont, ISBN 10:0-534-15162-0 (1992).
47 S.M. Kang, M. Waqas, A.L. Khan and I. J. Lee, *Plant-growth–promoting rhizobacteria: potential candidates for gibberellins production and crop growth promotion, Use of microbes for the alleviation of soil stresses*, Springer, New York, 1, 1–19 (2014).
48 S.D. Kim, Colonizing ability of *Pseudomonas fluorescens* 2112, among collections of 2, 4-diacetylphloroglucinol-producing *Pseudomonas fluorescens* spp. in pea rhizosphere, *J Microbiol Biotechnol*, 22, 763–770 (2012).
49 V. Singh, S.S. Parihar, G. Gupta and M. Khare, Molecular characterization of *Ganoderma lucidum* and analysis of few secondary metabolites use as antimicrobial activity, *International Journal of Pharma & Bio Sciences*, Spl Ed. (Int-BIONANO), 83–90 (2016).
50 I.V. Maksimov, R.R. Abizgil'Dina and L.I. Pusenkova, Plant growth promoting rhizobacteria as alternative to chemical crop protectors from pathogens (review), *Appl Biochem Microbiol*, 47, 333–345 (2011).
51 J.E. Loper and H. Gross, Genomic analysis of antifungal metabolite production by *Pseudomonas fluorescens* Pf-5, *Eur J Plant Pathol*, 119, 265–278 (2007).
52 K. Saharana, M.V.R.K. Sharma, A. Prakash, B.N. Johri, V.S. Bisariaa and V. Sahai,. Shelf-life enhancement of bio-inoculant formulation by optimizing the trace metals ions in the culture medium for production of DAPG using fluorescent pseudomonad R62, *Enzyme Microbiol Technol*, 48(1), 33–38 (2011).

53 M. Joshi, R. Shrivastava, A.K. Sharma and A. Prakash, Screening of resistant verities and antagonistic *Fusarium oxysporum* for biocontrol of *Fusarium* Wilt of Chilli, *Plant Pathol Microbiol* 3, 134 (2012).
54 R. Hayat, S, Ali, U. Amara, R. Khalid and I. Ahmed, Soil beneficial bacteria and their role in plant growth promotion: A review, *Ann Microbiol*, 60, 579–598 (2010).
55 N.K. Arora, E. Khare, A. Verma and R.K. Sahu, In vivo control of *Macrophomina phaseolina* by a chitinase and ß-1,3 glucanse producing pseudomonad NDN1, *Symbiosis* 46, 129–135 (2008).
56 L.C. van Loon, Plant responses to plant growth–promoting rhizobacteria, *Eur J Plant Pathol*, 119, 243–254 (2007).
57 B. Lugtenberg and F. Kamilova, Plant-growth–promoting rhizobacteria, *Annu Rev Microbiol*, 63, 541–556 (2009).
58 P. Bakker, C. Pieterse and L. van Loon, Induced systemic resistance by fluorescent *Pseudomonas* spp., *Phytopathol*, 97, 239–246 (2007).
59 F.M. Gutierrez-Luna, J. López-Bucio, J. Altamirano-Hernández, E. Valencia-Cantero, H.R. Cruz and L. Macías-Rodríguez, Plant growth–promoting rhizobacteria modulate root-system architecture in *Arabidopsis thaliana* through volatile organic compound emission, *Symbiosis*, 51, 75–83 (2010).
60 H.A. Naznin, M. Kimura, M. Miyazawa, M. Hyakumachi, Analysis of volatile organic compounds emitted by plant growth promoting fungus *phoma* sp. GS8-3 for growth promotion effects on tobacco, *Microbe Environ*, 28(1),42–49 (2012).
61 J.W. Costerton, G.G. Geesey and K.J. Cheng, How bacteria stick, *Sci Am* 238, 86-95 (1978).
62 M.S. Aparna and S. Yadav, Biofilms: microbes and disease, *Braz J Infect Dis*, 12, 526–530 (2008).
63 V. Barbara, C. Miao, J. Russell, Crawford and P.I. Elena, Bacterial extracellular polysaccharides involved in biofilm formation, *Molecules*, 14, 2535–2554 (2009).
64 T.S. Ilyina, Y.M. Romanova and A.L. Gintsburg, Biofilms as a mode of existence of bacteria in external environment and host body: the phenomenon, genetic control, and regulation systems of development, *Russ J Genet*, 40, 1189–1198 (2004).
65 S.K. Upadhyay, J.S. Singh and D.P. Singh, Exopolysaccharide-producing plant growth–promoting rhizobacteria under salinity condition, *Pedosphere*, 21, 214–222 (2011).
66 J.J. Schuster and G.H. Markx, Biofilm architecture, *Adv Biochem Eng Biotechnol*, 146, 77–96 (2014).
67 E.E. Mann and D.J. Wozniak, *Pseudomonas* biofilm matrix composition and niche biology, *FEMS Microbiol Rev*, 36: 893–916 (2012).
68 M. Parada, J. Vinardell, F. Ollero, A. Hidalgo, R. Gutiérrez, A. Buendía-Clavería, W. Lei, I. Margaret, F. López-Baena, A. Gil-Serrano, M. Rodríguez-Carvajal, J. Moreno and J. Ruiz-Sainz, *Sinorhizobium fredii* HH103 mutants affected in capsular polysaccharide (KPS) are impaired for nodulation with soybean and *Cajanus cajan*, *Mol Plant Microbe Interact*, 19, 43–52 (2006).
69 H.S. Jayasinghearachchi and G. Seneviratne, A mushroom-fungus helps improve endophytic colonization of tomato by *Pseudomonas fluorescence* through biofilm formation, *Res J Microbiol*, 1: 83–89 (2006b).
70 G. Seneviratne, R.M.M.S. Thilakaratne, A.P.D.A. Jayasekara, K.A.C.N. Seneviratne, K.R.E. Padmathilake and M.S.D.L. De silva, Developing beneficial microbial biofilms on roots of non-legumes: A novel biofertilizing technique, M.S. Khan, A. Zaid and J. Musarrat (Eds) *Microbial strategy for crop improvement*, Springer, Berlin (2009).

71 H.C. Flemming and J. Wingender, The biofilm matrix, *Nat Rev Microbiol*, 8, 623–633 (2010).

72 W.C. Fuqua, S.C. Winans and E.P. Greenberg, Quorum sensing in bacteria, the LuxR-LuxI family of cell density-responsive transcriptional regulators, *J Bacteriol*, 176, 269–275 (1994).

73 J.P. Pearson, K.M. Gray, L, Passador, K.D. Tucker, A. Eberhard, B.H. Iglewski and E.P. Greenberg, Structure of auto inducer required for the expression of *Pseudomonas aeruginosa* virulence gene, *Proc Natl Acad Sci USA*, 91, 197–201 (1994).

74 M. de Boer, Combining *Pseudomonas* strains to improve biological control of *Fusarium* wilt in radish, *Ph.D. thesis*, Utrecht University, Utrecht (2000).

75 E. Bent and C.P. Chanway, The growth-promoting effects of a bacterial endophyte on lodgepole pine are partially inhibited by the presence of other rhizobacteria, *Can J Microbiol* 44, 980–988 (1998).

76 M. Kumar, S. Mishra, V. Dixit, M. Kumar, L. Agarwal, P.S. Chauhan and C.S. Nautiyal, Synergistic effect of *Pseudomonas putida* and *Bacillus amyloliquefaciens* ameliorates drought stress in chickpea (*Cicer arietinum* L.), *Plant Sign Beha*, 11(1), e1071004 (2016).

77 C. Van Nieuwenhove, V. Holm, S.A. Kulasooriya, K. Vlassak, Establishment of *Azorhizobium caulinodans* in the rhizosphere of wetland rice (*Oryza sativa* L.), *Biol Fertil Soils*, 31, 143–149 (2000).

78 C. Kim, M.L. Kecskes, R.J. Deaker, K. Gilchrist, P.B. New, I.R. Kennedy, Wheat root colonization and nitrogenase activity by *Azospirillum* isolates from crop plants in Korea, *Can J Microbiol*, 51, 948–956 (2005).

79 F. Yamaga, K. Washio, M. Morikawa, Sustainable biodegradation of phenol by *Acinetobacter calcoaceticus* P23 isolated from the rhizosphere of duckweed *Lemna aoukikusa*, *Environ Sci Technol*, 44, 6470–6474 (2010).

80 B. Fan, X. H. Chen, A. Budiharjo, W. Bleiss, J. Vater and R. Borriss, Efficient colonizationof plant roots by the plant growth promoting bacterium *Bacillus amyloliquefaciens* FZB42, engineered to express green florescent protein, *J of Biotechnol*, 151, 303–311 (2011).

81 J. Yang, P.D. Kharbanda, M. Mirza, Evaluation of *Paenibacillus polymyxa* PKB1 for biocontrol of Pythium disease of cucumber in a hydroponic system, *Acta Hortic (ISHS)*, 635, 59–6 (2004).

82 R. Prasanna, S. Pattnaik, T.C. Sugitha, L. Nain, A.K. Saxena, Development of cyanobacterium-based biofilms and their in vitro evaluation for agriculturally useful traits, *Folia Microbiol (Praha)*, 56, 49–58 (2011).

83 A.L. Iniguez, Y. Dong, E.W. Triplett, Nitrogen fixation in wheat provided by *Klebsiella pneumoniae* 342, *Mol Plant-Microbe Interact*, 17,1078–1085 (2004).

84 P.S. Chauhan and C.S. Nautiyal, The *pur*B gene controls rhizosphere colonization by *Pantoea agglomerans*, *Letters in Applied Microbiology*, 50, 205–210 (2010).

85 N. Amellal, G. Burtin, F. Bartoli, T. Heulin, Colonization of wheat roots by an exopolysaccharide-producing *Pantoea agglomerans* strain and its effect on rhizosphere soil aggregation, *Appl Environ Microbiol*, 64, 3740–3747 (1998).

86 E.M. Vanderlinde, A. Muszynski, J.J. Harrison, S.F. Koval, D.L. Foreman, H. Ceri, *Rhizobium leguminosarum* biovar viciae 3841, deficient in 27-hydroxyoctacosanoate-modified lipopolysaccharide, is impaired in desiccation tolerance, biofilm formation and motility, *Microbiol*, 155, 3055–3069 (2009).

87 M. Janczarek, A. Skorupska, Modulation of rosR expression and exopolysaccharide production in *Rhizobium leguminosarum* bv. *Trifolii by phosphate and clover root exudates, Int J Mol Sci*, 12, 4132–4155 (2011).
88 A. Williams, A. Wilkinson, M. Krehenbrink, D.M. Russo, A. Zorreguieta, J.A. Downie, Glucomannan-mediated attachment of *Rhizobium leguminosarum* to pea root hairs is required for competitive nodule infection, *J Bacteriol*, 190, 4706–4715 (2008).
89 D. Krysciak, C. Schmeisser, S. Preuss, J. Riethausen, M. Quitschau, S. Grond, Involvement of multiple loci in quorum quenching of autoinducer I molecules in the nitrogen-fixing symbiont *Rhizobium* (*Sinorhizobium*) sp. strain NGR234, *Appl Environ Microbiol*, 77, 5089–5099 (2011).

3

Concept of Mono and Mixed Biofilms and Their Role in Soil and in Plant Association

Janaína J. de V. Cavalcante[1], *Alexander M. Cardoso*[2] *and Vânia L. Muniz de Pádua*[2]

[1] *National Institute of Metrology, Quality and Technology, Av. Nossa Senhora das Graças, 50, 25250-020, Duque de Caxias, RJ, Brazil*
[2] *Centro Universitário Estadual da Zona Oeste, Av. Manuel Caldeira de Alvarenga, 1203, 23070-200, Rio de Janeiro, RJ, Brazil*

3.1 Introduction

Biofilm is a society of sessile microorganisms (Figure 3.1) assembled as microcolonies with distinct community properties surrounded by an extrapolymeric substance (EPS) matrix. Indeed, biofilms are the dominant lifestyle of microorganisms in all environments. While microorganisms have traditionally been studied as planktonic cells, most are found attached to surfaces in multicellular assemblies forming biofilms, a mode of growth that allows them better survival in hostile environments, resulting in adaptive advantages such as protection, nutrient availability, metabolic cooperativity, gene expression regulation, and acquisition of new genetic traits [13]. Although biofilm is mainly of prokaryotic origin, every species of microorganism, Gram-negative and Gram-positive bacteria, archaea, protozoa, and fungi have mechanisms by which they can adhere to substrate, and to each other, allowing them to construct and inhabit a biofilm community.

Biofilm is a ubiquitous structure that can be formed on nearly every biotic or abiotic substrate. Nevertheless, biofilms on different environments present unique properties, reflecting the conditions of the colonized site. On plants surfaces, they are formed mostly by bacteria in pathogenic, mutualistic, or symbiotic association on leaves, roots, and in the soil. These microbes have a profound influence on soil composition, plant health, and productivity. There is growing appreciation that the intensity, duration, and effect on plant–microbe interactions are significantly influenced by the interaction with its substrate and cell-to-cell communication between colonizing microbes [12, 13].

Natural biofilms are found predominantly as a diversified community of microorganisms, called mixed biofilms [21]. Biofilms formed by mixed-species, compared with that formed by monospecies, are highly resistant to antimicrobials, possibly primarily due to higher exopolysaccharide (EPS) production, better nutrient utilization, and interspecific protection. There is a perception that the end result of microbe collaborations is considerably affected by physical interactions as occur in the biofilms, impacting plant

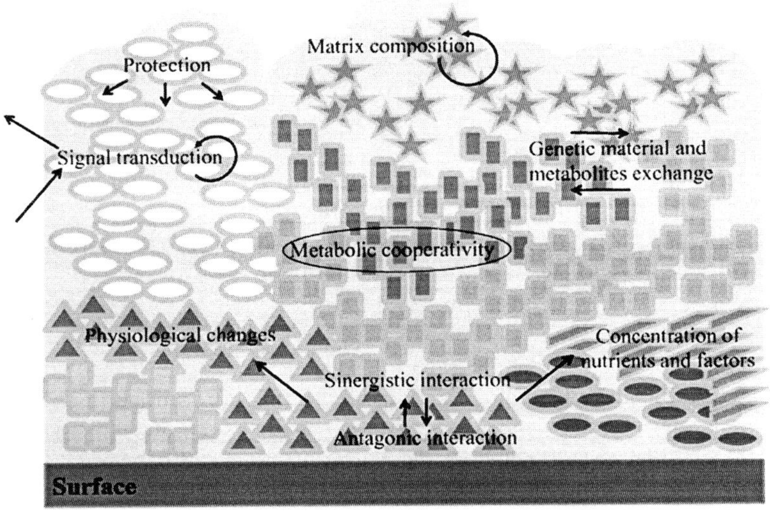

Figure 3.1 Biofilm in a community of mixed species. Most of the natural biofilms contain cells with different genotypes and phenotypes that express distinct metabolic pathways, stress responses, and other biological activities. The cells are embedded in a self-produced matrix that allows concentrate nutrients, besides the intercellular exchange of metabolites, genetic material, and signaling molecules. The interaction can be synergistic or antagonic, cooperative, or competitive, and result in phenotypic changes that confer survival advantages.

development and the soil conditions [13]. It is well-known that biofilm increases resistance to adverse conditions, as well as to certain environmental stresses, antimicrobial substances, protozoan predation, and resistance to pathogens. The success of a biofilm niche is allied to microbial communication, as well as competition for nutrients and physical resources, constituting a layer of complexity that strongly influences the whole biofilm community [12].

The fitness of mixed-species colonizers of plant, rhizosphere, or soil biofilms include soil composition and different types of biological associations, such as cooperative, competitive, or neutral, based on the genetic background of the involved species, physical, chemical, and biological processes, and also the available resources. Mixed biofilm growth mode may promote benefits for those involved due to the consortial metabolism, or opportunity of acquired genes by conferring useful functions through horizontal gene transfer [3].

Plants are deeply involved in this respect, since nutrients and plant defense compounds are released and biofilms formation is strongly influenced by the exuded and/or sequestrated molecules. Into plant vascular vases, biofilm can be an obstruction that causes tissue damage. However, on the phyllosphere, biofilm can be found in microniches with differentiated abundance of nutrients and moisture. Under these circumstances, physiological conditions can be shaped in a particular way, affecting the competition and other biological association or virulent responses on these sites [12].

More research aiming to better understand the multi-species biofilms panorama and the effects of such interactions on the development, nature, and survival of the biofilm community and its consequences is needed. This chapter summarizes the current knowledge concerning the biofilm role in soil and in plant association focusing on symbiotic association, nutrition, signaling, and cell communication related to mono and mixed-species biofilms.

3.2 Soil- and Plant-Associated Biofilms

Soil is by far the most diverse habitat on the planet, containing a myriad of novel microorganisms that make up the soil microbiome. This is evidenced by the significant biomass and activities displayed by microbial biofilms in soils and plants [24, 30]. Biofilms offer huge potential for offering protection against pathogens and nutrients to plants, bioremediation of hazardous compounds, soil erosion prevention, and formation of biobarriers to protect soil and groundwater from contamination [13]. Soils comprise a heterogeneous location that can affect microbial growth and its survival as biofilm [42].

Bacterial and fungal biofilm formation in soil have been well documented [4, 23, 27, 40]. On the other hand, there are only a few reports of archaeal biofilms [41]. A closer look at the properties of Archaea illustrates that they display a variety of features that are either shared only with Eukarya or only with Bacteria, placing them in the middle of these two domains [1, 10]. Archaea dominate among ammonia-oxidizing prokaryotes in soils, suggesting that they may be the most abundant ammonia-oxidizing organisms in soil ecosystems, supplying to enhance nitrate availability in soils for plant nutrition [32].

The root surface (rhizoplane) and the region immediately surrounding a root (rhizosphere) constitute an ecological niche in soil where both nutrients and organic material enhances microbial growth and drive the structuring of the microbial communities in the rhizosphere [9]. Interactions between the plant and the surrounding microorganisms select for the establishment of certain microbial populations that will benefit the plant, suggesting that a highly evolved association may exist between the plants and the microbe residing in soil. It has been described a variety of cell components, such as outer membrane proteins, wall polysaccharides (capsules), lipopolysaccharides, and cell surface agglutinin for microbial attachment to roots [39] and exopolysaccharide production [2] in the rhizosphere.

Many bacterial pathogens, symbionts, and commensals are especially creative in their regulatory processes to adapt to sudden changes in nutrient availability, and the response to host secondary defenses. A particularly important example of bacterial adaptation is the ability to sessile grown as biofilm, forming an optimized niche embedded in a gel-like extracellular polymer network of microbial origin. The bacterium *Gluconacetobacter diazotrophicus* (GDI) is found mainly in the intercellular spaces and vascular tissues of plants from different families, especially sugarcane, resulting in substantial benefits to culture. The bacterium must adapt to the interaction with insects, spores of fungi and, particularly, to different plant species and in different circumstances [6].

GDI is a nitrogen-fixing bacterium with an endophytic lifestyle that colonizes several plant species, promoting plant growth without disease symptoms. GDI was observed to form assemblages referred to as biofilms on root surfaces and within intercellular spaces. Biofilms are purported to form microniches distinct from the ambient environment and permit bacteria to generate functions not possible outside them, some of which are fundamental for biological associations. Biofilm formed under glass wool soaked in culture medium and co-cultivated with sugarcane plantlets have revealed that carbohydrate metabolism was the most represented functional category among the differential expressed proteins. Functions related to secretion pathway, virulence determinant, and fatty acid synthesis were also represented. Components of protein folding process were identified and suggest a dependency of such mechanism for particular structures required during biofilm formation. Some of the proteins found can be related to others plant–microbe interactions studies, highlighting the biofilm significance in initiating and sustaining contact with the plant host [49].

It has been shown that a flagellar mutant of GDI was nonmotile and displayed reduced capacity to form biofilm, suggesting that flagella play an important role in motility, adherence, biofilm formation, and interaction with the plant and host colonization [46]. Also, levansucrase and exopolysaccharide production are required for biofilm formation and plant colonization by this nitrogen-fixing endophyte [38, 56].

3.3 Microbial Signaling, Regulation, and Quorum Sensing

Microbial biofilm formation is a highly regulated and complex process, which requires the coordinated expression of many specific genes. Each species responds to its own set of environmental signal via a distinct set of molecular devices. Despite this mechanistic diversity, there are several stimuli that are generally important for plant-associated bacteria and proceed through conserved regulatory systems [57].

The alteration between the planktonic form and multicellular surface-bound aggregates, or biofilms, is controlled by a cell-to-cell communication system known as quorum sensing (QS). Plant-associated bacteria frequently employ this signaling mechanism to modulate and coordinate its interactions with plants, including control of tissue maceration, antibiotic production, toxin release, and horizontal gene transfer [58]. Plants live in association with bacteria, but also with other microbes as fungi and archaea, and it is believed that plant evolution was influenced by the presence of these associated microorganisms. During this evolution, diverse signaling systems emerged that permitted mutual plant-microorganism sensing [57].

QS is a population-based regulatory strategy that permits microbes to sense and respond to their environment, resulting in cooperative activity [22]. Traits regulated by QS include surface attachment and biofilm formation, toxin and exopolysaccharide production, virulence, bioluminescence, sporulation and genetic competence, among others, which usually are essential for successful establishment of a symbiotic or a pathogenic relationship with the eukaryotic hosts [14, 16, 25, 35, 36, 50]. It is based on synthesis, detection, and response to signal molecules named QS signals (QSS) or autoinducers. As cell density increases (i.e., as a *quorum* is reached), QSS accumulate in the surroundings and are sensed by bacterial proteins called QS regulators, which in turn control the expression of specific genes. The products of these genes express activities that are useful when executed by groups of bacteria acting in synchrony by regulating various physiological activities [47].

Structurally, QSS have a low molecular weight and belong to a wide range of chemical classes, including acyl homoserine lactones (AHLs), quorum sensing peptides (QSPs) diketopiperazines (DKPs), cis-unsaturated fatty acids, or diffusible signal factors (DSF family signals), 4-hydroxy-2-alkylquinolines (HAQs), furanosyl borate diesters (autoinducer-2 or AI-2), and others [44]. Diverse polysaccharides and QSS production in different strains of organisms allow for variations among biofilm structures. Among all the known QS signaling systems, AHLs are the most prevalent molecules predominantly found in Gram-negative bacteria [64]. Gram-positive bacteria usually use autoinducing peptides (AIPs) for signaling. However, there are some signaling molecules, like autoinducer-2 (AI-2), produced by both groups of bacteria that also facilitate interspecies communication [20].

Survey studies suggest that greater numbers of AHL-producing bacteria live associated with plants than within bulk soil populations [18]. Corroborating with this idea, recently it was found that GDI PAL5 produces eight molecules from the AHL family in an active QS system modulated by environmental factors [7]. Interestingly, *Escherichia* and *Salmonella* have a receptor for AHL (SdiA), despite not producing AHLs, enabling them to detect and respond to nearby AHL-producing bacteria [31]. Therefore, the ability to detect and respond to the QS signals of other species may constitute an important aspect of interpopulation communication and community structuring in natural environments. Several soil and rhizosphere bacteria degrade AHLs, and measurement of the total degradative activity in soils suggests that this activity can significantly affect signaling [62]. These studies suggest that QS signaling molecules have the potential to reach unexpectedly long distances, supporting communication at specific sites.

QS is recognized to control the production of extracellular substances within a biofilm. It has been observed that quorum sensing controls biofilm formation in *Vibrio cholerae* [25, 61], *Pseudomonas aeruginosa* [14], *Sinorhizobium meliloti* [36], as well as in some other types of bacteria [60] by regulating transcription of genes involved in the exopolysaccharide production. Moreover, QS molecules produced in the rhizosphere, changed gene expression in *Arabidopsis thaliana* [59]. Similarly, a proteomics study showed that AHLs modulate the expression of a large number of genes in the legume *Medicago truncatula* [37].

In addition, plants seem to actively participate during the process of QS, since different plants produce several unidentified compounds that may interfere with the bacterial QS machinery, suggesting that these plant compounds perform as AIPs or AHL mimics and bind to the autoinducer-binding domain of QS regulators. Recently, rosmarinic acid was identified as a plant-derived compound that acts as an AHL mimic and directly interacts with a bacterial QS regulator in *Pseudomonas aeruginosa* PAO1. Rosmarinic acid induced quorum sensing–dependent gene expression, increased biofilm formation and the production of the virulence factors, indicating that rosmarinic acid secretion is a plant defense mechanism to stimulate a premature quorum sensing response [11]. However, there is a lack of information available regarding the molecular identity of more active compounds that are responsible for this kind interference [54, 57].

Remarkable evidences have shown that in addition to the well-documented quorum sensing systems other molecules may work as signal molecules. At sub-inhibitory concentrations, various antibiotics may act as signals, modulating bacterial functions in subtle ways. Unexpectedly, some antibiotics at low concentrations may even be beneficial to the bacteria in natural environments [15]. In addition to antibiotics, several other secondary metabolites are known to be involved in microbial signaling [52]. The role of 3', 5'-cyclic diguanylic acid (cdG) in the regulation of motility, biosynthesis of exopolysaccharides, biofilm formation, quorum sensing, and symbiosis in the plant-symbiotic

Sinorhizobium meliloti has been recently revealed [51], proving that enhanced levels of cdG promote sessility. High intracellular cdG concentrations have been shown to favor production of exopolysaccharides (EPS), fimbriae, pili, and adhesion, which contribute to biofilm formation. The cdG is a ubiquitous second messenger in bacteria that is known to promote the transition from a planktonic motile to a sessile, frequently biofilm-associated lifestyle. Its levels are controlled by diguanylate cyclases (DGCs) that synthesize cdG from two GTP molecules and by phosphodiesterases (PDEs) that degrade it. The transition from a motile to a sessile lifestyle also includes inhibition of flagellar motility by cdG [45, 51, 52].

A variety of other cellular functions, not directly involved in the switch between motile and sessile form, is also subject to regulation by cdG, such as virulence-associated processes, quorum sensing, cell cycle control, and cell differentiation [45, 54]. Although these results seemed to be very promising, no one could dismiss the possibility of finding novel molecules involved in the microbial biofilm regulation.

3.4 Biotechnology

The biotechnology sector is just beginning to better elucidate the questions and possible applications about plant and soil biofilms. For instance, bioelectrochemical systems have been proposed to promote sustainable agriculture [65] and remediation of contaminated soil [33, 34]. These efforts can be improved by using procedures associated with biofilm, as suggested by a research focused on the observation/description of the effect of developed microbial biofilms on the restoration of soils deteriorated by conventional agricultural practices [53]. Another possible application is suggested in metagenomics studies of acid mine drainage biofilm that assessed possible influences over environmental geochemistry from the microbial tolerance mechanisms to the extremely acidic environment [55].

Moreover, as an important part of infection strategy, many bacterial plant pathogens produce extracellular carbohydrates [5,8]. Microbial enzymes that degrade extracellular carbohydrates are effective at degrading or inhibiting biofilm formation [8, 29] and several enzymes are used to form protective biofilms on farm equipment, greenhouse surfaces or plant debris as well as can be developed as antimicrobial agents with novel mechanisms of action. Approaches against pathogens might include the interference with the quorum sensing system by the quorum quenching enzymes and be used to develop a plan of disease management, with applications in many fields, including agriculture [17, 26, 63]. Microbial consortium is related to the biofilm mode of growth and has been correlated to improve plant health and tolerance to adverse conditions [19, 28], suggesting future directions for engineering biofilm with a microbial consortium of interest.

Finally, biosurfactants are considered to be less toxic, and they have the potential to be commercially produced for many uses, including agriculture industries, such as for plant pathogen elimination and for increasing the bioavailability of nutrient. Interestingly, biosurfactants play vital roles on bacterial motility, signaling and biofilm formation, and are produced by several rhizosphere and plant-associated microbes. The use of biosurfactant-based strategies can be applied to develop a sustainable agriculture, improving soil and food quality by soil remediation [43, 48].

3.5 Outlook

Studies on biofilm have led to an exciting expansion of our knowledge of the cell to cell communication, signaling, and quorum sensing. It also shed light on the molecular mechanisms underlying microbial biofilms formation and performance; further studies with archaea, rhizobacteria, and recently discovered microorganisms may uncover more surprises. The emerging concept of mono and mixed biofilms as social organisms is a disciplinary landmark and reflects the convergence of microbiology and ecology. Biofilms are not just an unintentional mixture of different types of microorganisms; rather, they are a highly coordinated society that promotes a fundamental role in soil and in plant association.

Acknowledgments

Work in the authors' laboratory was supported by grants from the Fundação Carlos Chagas Filho de Amparo à Pesquisa do Estado do Rio de Janeiro (FAPERJ) and Conselho Nacional de Desenvolvimento Científico e Tecnológico (CNPq).

References

1 Alqueres, S.M.C., Almeida, R.V., Clementino, M.M., Vieira, R.P., Almeida, W.I., Cardoso, A.M., et al. Exploring the Biotechnological Applications in the Archaeal Domain. *Braz J. Microbiol.* 2007; 36:398–405.
2 Amellal, N., Burtin, G., Bartoli, F., and Heulin T. Colonization of wheat roots by an exopolysaccharide-producing Pantoea agglomerans strain and its effect on rhizosphere soil aggregation. *Appl. Environ. Microbiol.* 1998; 64(10):3740–3747.
3 Aminov, R.I. Horizontal gene exchange in environmental microbiota. *Front Microbiol.* 2011; 2:158.
4 Bach, E., Seger, G.D.S., Fernandes, G.C., Lisboa, B.B., and Passaglia, L.M.P. Evaluation of biological control and rhizosphere competence of plant growth promoting bacteria. *Appl. Soil. Ecol.* 2016; 99:141–149.
5 Beauregard, P.B., Chai, Y., Vlamakis, H., Losick, R., and Kolter, R. 2013. Bacillus subtilis biofilm induction by plant polysaccharides. *Proc. Natl Acad. Sci. USA.* 2013; 110(17):E1621–1630.
6 Bertalan, M., Albano, R., Pádua, V.L., Rouws, L.F., Rojas, C., Hemerly, A., et al. Complete genome sequence of the sugarcane nitrogen-fixing endophyte Gluconacetobacter diazotrophicus Pal5. *BMC Genomics.* 2009; 10:450–467.
7 Bertini, E.V., Nieto Peñalver, C.G., Leguina, A.C., Irazusta, V.P., and de Figueroa, L.I. Gluconacetobacter diazotrophicus PAL5 possesses an active quorum sensing regulatory system. *Antonie van Leeuwenhoek.* 2014; 106(3):497–506.
8 Burns, R.G., De Forestb, J.L., Marxsenc, J., Sinsabaughd, R.L., Strombergere, M.E., Wallensteinf, M.D., et al. Soil enzymes in a changing environment: Current knowledge and future directions. *Soil Biol. and Biochemistry.* 2013; 58:216–234.
9 Campbell, R., and Greaves, M.P. Anatomy and community structure of the rhizosphere. In: J. M. Lynch (ed.), *The rhizosphere.* West Sussex, U.K: John Wiley & Sons, 1990, pp. 11–34.

10 Cavicchioli, R., Curmi, P.M., Saunder, N., and Thomas, T. Pathogenic archaea: do they exist? *Bioessays*. 2003; 25(11):1119–1128.

11 Corral-Lugo, A., Daddaoua, A., Ortega, A., Espinosa-Urgel, M., and Krell, T. Rosmarinic acid is a homoserine lactone mimic produced by plants that activates a bacterial quorum-sensing regulator. *Sci. Signal.* 2016; 9(409):ra1.

12 Danhorn, T., and Fuqua, C. Biofilm formation by plant-associated bacteria. *Annu. Rev. Microbiol.* 2007; 61:401–422.

13 Davey, M.E., and O'Toole, G.A. Microbial Biofilms: from Ecology to Molecular Genetics. *Microbiol. Mol. Biol. Rev.* 2000; 64(4):847–867.

14 Davies, D.G., Parsek, M.R., Pearson, J.P., Iglewski, B.H., Costerton, J.W., and Greenberg, E.P. The involvement of cell-to-cell signals in the development of a bacterial biofilm. *Science*. 1998; 280(5361):295–298.

15 Davies, J. Everything depends on everything else. *Clin. Microbiol. Infect.* 2009 Jan;15 (Suppl 1):1–4.

16 Dunne, M. Bacterial adhesion: seen any good biofilms lately? *Clin. Microbiol. Rev.* 2002; 15(2):155–166.

17 Eberl, L., Von Bodman, S.B., and Fuqua, C. Biofilms on Plant Surfaces. In: S. Kjelleberg, M. Givskov (eds.), *The Biofilm Mode of Life: Mechanisms and Adaptations*. United Kingdom: Horizon Bioscience, 2007, pp. 215–233.

18 Elasri, M., Delorme, S., Lemanceau, P., Stewart, G., Laue, B., Glickmann, E., et al. Acyl-homoserine lactone production is more common among plant-associated *Pseudomonas* spp. than among soilborne *Pseudomonas* spp. *Appl. Environ. Microbiol.* 2001; 67(3):1198–1209.

19 Fatima, K., Imran, A., Amin, I., Khan, Q.M., and Afzal, M. Plant species affect colonization patterns and metabolic activity of associated endophytes during phytoremediation of crude oil-contaminated soil. *Environ. Sci. Pollut. Res. Int.* 2016; 23(7):6188–6196.

20 Federle, M.J., and Bassler, B.L. Interspecies communication in bacteria. *J. Clin. Invest.* 2003; 112(9):1291–1299.

21 Fett, W.F. Naturally occurring biofilms on alfalfa and other types of sprouts. *J. Food Prot.* 2000; 63(5):625–632.

22 Fuqua, W.C., Winans, S.C., and Greenberg, E.P. Quorum sensing in bacteria: the LuxR-LuxI family of cell density-responsive transcriptional regulators. *J. Bacteriol.* 1994; 176(2):269–275.

23 Gao, T., Foulston, L., Chai, Y., Wang, Q., and Losick R. Alternative modes of biofilm formation by plant-associated *Bacillus cereus*. *Microbiologyopen*. 2015; 4(3):452–464.

24 Garbeva, P., van Veen, J.A., and van Elsas, J.D. Microbial diversity in soil: selection microbial populations by plant and soil type and implications for disease suppressiveness. *Annu. Rev. Phytopathol.* 2004; 42:243–270.

25 Hammer, B.K., and Bassler, B.L. Quorum sensing controls biofilm formation in Vibrio cholerae. *Mol. Microbiol.* 2003; 50(1):101–114.

26 Hentzer, M., and Givskov, M. Pharmacological inhibition of quorum sensing for the treatment of chronic bacterial infections. *J. Clin. Invest.* 2003; 112(9):1300–1307.

27 Herath, H.M.L.I., Menikdiwela, K.R., Igalavithana, D., and Seneviratne, G. Developed fungal-bacterial biofilms having nitrogen fixers: universal biofertilizers for legumes and non-legumes. In: Frans J. de Bruijn (eds.), *Biological Nitrogen Fixation*. Wiley-Blackwell; 2015, pp.1041–1046.

28 Kakar, K.U., Ren, X.L., Nawaz, Z., Cui, Z., Li, B., Xie, G., et al. Consortium of rhizobacterial strains and biochemical growth elicitors improve cold and drought stress tolerance in rice (*Oryza sativa L.*). *Plant Biol.* 2016; 18(3):471–483.

29 Kaplan, J.B. Biofilm matrix-degrading enzymes. In: Gianfranco Donelli (ed.), *Microbial Biofilms*. New York: Springer, 2014, pp. 203–213.

30 Lakshmanan, V., Selvaraj, G., and Bais, H.P. Functional soil microbiome: belowground solutions to an aboveground problem. *Plant Physiol.* 2014; 166(2):689–700.

31 Lindsay, A., and Ahmer, B.M. Effect of sdiA on biosensors of N-acylhomoserine lactones. *J. Bacteriol.* 2005; 187(14):5054–5058.

32 Leininger, S., Urich, T., Schloter, M., Schwark, L., Qi, J., Nicol, G.W., et al. Archaea predominate among ammonia-oxidizing prokaryotes in soils. *Nature.* 2006; 442(7104):806–809.

33 Lu, L., Huggins, T., Jin, S., Zuo, Y., and Ren, Z.J. Microbial metabolism and community structure in response to bioelectrochemically enhanced remediation of petroleum hydrocarbon-contaminated soil. *Environ. Sci. Technol.* 2014; 48(7):4021–4029.

34 Lu, L., Yazdi, H., Jin, S., Zuo, Y., Fallgren, P.H., and Ren, Z.J. Enhanced bioremediation of hydrocarbon-contaminated soil using pilot-scale bioelectrochemical systems. *J. Hazard. Mater.* 2014; 274:8–15.

35 Mangwani, N., Dash, H.R., Chauhan, A., and Das, S. Bacterial Quorum sensing: functional features and potential applications in biotechnology. *J. Mol. Microbiol. Biotechnol.* 2012; 22(4):215–227.

36 Marketon, M.M., Glenn, S.A., Eberhard, A., and Gonzalez, J.E. Quorum sensing controls exopolysaccharide production in *Sinorhizobium melioti*. *J. Bacteriol.* 2003; 185(1):325–331.

37 Mathesius, U., Mulders, S., Gao, M., Teplitski, M., Caetano-Anollés, G., Rolfe, B.G., and Bauer, W.D. Extensive and specific responses of a eukaryote to bacterial quorum sensing signals. *Proc. Natl Acad. Sci USA.* 2003; 100(3):1444–1449.

38 Meneses, C.H.S.G., Rouws, L.F., Simões-Araújo, J.L., Vidal, M.S., and Baldani, J.I. Exopolysaccharide production is required for biofilm formation and plant colonization by the nitrogen-fixing endophyte *Gluconacetobacter diazotrophicus*. *Mol. Plant Microbe. Interact.* 2011; 24(12):1448–1458.

39 Michiels, K.W., Croes, C.L., and Vanderleyden, J. Two different modes of attachment of *Azospirillum brasilense* Sp7 to wheat roots. *J. Gen. Microbiol.* 1991; 137:2241–2246.

40 Mori, Y., Inoue, K., Ikeda, K., Nakayashiki, H., Higashimoto, C., Ohnishi, K., et al. The vascular plant pathogenic bacterium *Ralstonia solanacearum* produces biofilms required for its virulence on the surfaces of tomato cells adjacent to intercellular spaces. *Mol. Plant Pathol.* 2016; 17(6):890–902.

41 Orell, A., Frols, S., and Albers, S.V. Archaeal Biofilms: The great unexplored. *Ann. Rev. Microbiol.* 2013; 67:337–354.

42 Paul, E.A., and Clark, F.E. *Soil Microbiology and Biochemistry*. Academic Press: San Diego, Calif.; 1989.

43 Primo, E.D., Ruiz, F., Masciarelli, O., and Giordano, W. Biofilm formation and biosurfactant activity in plant-associated bacteria, p. 337–349. In: Dinesh K. Maheshwari (ed), *Bacterial Metabolites in Sustainable Agroecosystem*. Volume 12 of Sustainable Development and Biodiversity. Springer International Publishing: Switzerland; 2015.

44 Rajput, A., Kaur, K., and Kumar, M. SigMol: repertoire of quorum sensing signaling molecules in prokaryotes. *Nucleic Acids Res.* 2016; 44(D1):D634–639.

45 Römling, U., Galperin, M.Y., and Gomelsky, M. Cyclic di-GMP: the first 25 years of a universal bacterial second messenger. *Microbiol. Mol. Biol. Rev.* 2013; 77(1):1–52.

46 Rouws, L.F., Simões-Aráujo, J.L., Hemerly, A.S., and Baldani, J.I. Validation of a Tn5 transposon mutagenesis system for *Gluconacetobacter diazotrophicus* through characterization of a flagellar mutant. *Arch. Microbiol.* 2008; 189:397–405.

47 Rutherford, S.T., and Bassler, B.L. Bacterial quorum sensing: Its role in virulence and possibilities for its control. *Cold Spring Harb Perspect. Med.* 2012; 2(11). pii: a012427.

48 Sachdev, D.P., and Cameotra, S.S. Biosurfactants in agriculture. *Appl. Microbiol. Biotechnol.* 2013; 97(3):1005–1016.

49 Santos, M.F., Padua, V.L.M., and Nogueira, E.M., Hemerly A.S., Domont G.B. Proteome of *Gluconacetobacter diazotrophicus* co-cultivated with sugarcane plantlets. *J. Proteomics.* 2010; 73(5):917–931.

50 Sakuragi, Y., and Kolter, R. Quorum-sensing regulation of the biofilm matrix genes (pel) of *Pseudomonas aeruginosa. J. Bacteriol.* 2007; 189(14):5383–5386.

51 Schäper, S., Krol, E., Skotnicka, D., Kaever, V., Hilker, R., Søgaard-Andersen, L., and Becker, A. Cyclic di-GMP regulates multiple cellular functions in the symbiotic *Alphaproteobacterium sinorhizobium* meliloti. *J. Bacteriol.* 2015; 198(3):521–535.

52 Shank, E.A., and Kolter, R. New developments in microbial interspecies signaling. *Curr. Opin. Microbiol.* 2009; 12(2):205–214.

53 Seneviratne, G., Jayasekara, A.P.D.A., De Silva, M.S.D.L., and Abeysekera, U.P. Developed microbial biofilms can restore deteriorated conventional agricultural soils. *Soil Biol. Biochem.* 2011; 43(5): 1059–1062.

54 Ta, C.A.K., and Arnason, J.T. Mini review of phytochemicals and plant taxa with activity as microbial biofilm and quorum sensing inhibitors. *Molecules.* 2015; 21(1):E29.

55 Tyson, G.W., Chapman, J., and Hugenholtz, P. Community structure and metabolism through reconstruction of microbial genomes from the environment. *Nature.* 2004; 428(6978): 37–43.

56 Velázquez-Hernández, M.L., Baizabal-Aguirre, V.M., Cruz-Vázquez, F., Trejo-Contreras, M.J., Fuentes-Ramírez, L.E., Bravo-Patiño A, et al. Gluconacetobacter diazotrophicus levansucrase is involved in tolerance to NaCl, sucrose and desiccation, and in biofilm formation. *Arch. Microbiol.* 2011; 193(2):137–149.

57 Venturi, V., and Fuqua C. Chemical signaling between plants and plant-pathogenic bacteria. *Annu. Rev. Phytopathol.* 2013; 51:17–37.

58 von Bodman, S.B., Bauer, W.D., and Coplin, D.L. Quorum sensing in plant-pathogenic bacteria. *Annu. Rev. Phytopathol.* 2003; 41:455–482.

59 von Rad, U., Klein, I., Dobrev, P.I., Kottova, J., Zazimalova, E., Fekete, A., et al. Response of *Arabidopsis thaliana* to N-hexanoyl-DL-homoserine-lactone, a bacterial quorum sensing molecule produced in the rhizosphere. *Planta.* 2008; 229(1):73–85.

60 Vuong, C., Gerke, C., Somerville, G.A., Fischer, E.R., and Otto, M. Quorum-sensing control of biofilm factors in *Staphylococcus epidermidis. J. Infect. Dis.* 2003; 188(5):706–718.

61 Waters, C.M., Lu, W., Rabinowitz, J.D., and Bassler, B.L. Quorum sensing controls biofilm formation in Vibrio cholerae through modulation of cyclic di-GMP levels and repression of vpsT. *J. Bacteriol.* 2008; 190(7):2527–2536.

62 Wang, Y.J., and Leadbetter, J.R. Rapid acyl-homoserine lactone quorum signal biodegradation in diverse soils. *Appl. Environ. Microbiol.* 71(3):1291–1299.
63 Webb, J.S., Givskov, M., and Kjelleberg, S. Bacterial biofilms: prokaryotic adventures in multicellularity. *Curr. Opin. Microbiol.* 2003; 6(6):578–585.
64 Whitehead, N.A., Barnard, A.M., Slater, H., Simpson, N.J., and Salmond, G.P. Quorum-sensing in Gram-negative bacteria. *FEMS Microbiol. Rev.* 2001; 25(4):365–404.
65 Xiaojin, L., Abu-Reesh, I.M., and He, Z. Development of bioelectrochemical systems to promote sustainable agriculture. *Agriculture.* 2015; 5:367–388.

4

Bacillus Biofilms and Their Role in Plant Health

Mohd Musheer Altaf[1], Iqbal Ahmad[1], Mohd Sajjad Ahmad Khan[2] and Elisabeth Grohmann[3]*

[1] Department of Agricultural Microbiology, Aligarh Muslim University, Aligarh, India
[2] Department of Biology, College of Medicine, Imam Abdulrahman Bin Faisal University, Dammam, Saudi Arabia
[3] Department of Life Sciences and Technology, Beuth University of Applied Sciences Berlin, Berlin, Germany

4.1 Introduction

Agriculture in the twenty-first century faces challenges associated with loss of soil fertility, increased use of chemical fertilizers, pesticides, drought, salinity, fluctuating climatic conditions, and growing attack on crops by pathogens and pests. The global necessity to increase agricultural production from a steadily decreasing and degrading land resource base had placed considerable burden on the fragile agroecosystem [1, 2]. World total population is currently around 7 billion, and this is projected to increase to approximately 8 billion people by 2025 and 9 billion by 2050. Considering the increase in worldwide population with the increase in environmental damage due to ever-increasing industrialization, it is clear that in the coming 50 years it will be a daunting task to feed the existing population, a problem that will increase with time. To feed the growing population, agricultural productivity will need to increase in a sustainable and environmentally friendly manner [3]. Technology based on integrating plant growth–promoting rhizobacteria (PGPR) in agriculture is an attractive approach to addressing these problems.

PGPR are beneficial bacteria that colonize the roots and enhance plant growth both directly and indirectly [4]. These microorganisms are associated with the rhizosphere and rhizoplane of dicots and monocots plants and are normally found in different environmental conditions. Microorganisms with PGP traits belong to genera such as *Azoarcus, Azospirillum, Azotobacter, Arthrobacter, Achromobacter, Bacillus, Bradyrhizobium, Burkholderia, Clostridium, Enterobacter, Gluconacetobacter, Paenibacillus, Pseudomonas*, and *Serratia* [5]. Among these, species of *Bacillus* and *Pseudomonas* are the most extensively studied and used as bioinoculant for enhanced crop production. Aerobic endospore-forming bacteria, typically *Bacillus* and its related species, comprise one of the major genera of soil microflora. *Bacillus* spp. is a well-studied bacterial group, omnipresent in nature with beneficial traits useful for environmental, industrial, and agricultural applications [6].

Several species of *Bacillus* are also found as endophytes (inner tissues) associated with different species of plants, including banana, wheat, rice, cassava, cucumber, grape, cotton, peas, spruce, sugar beet, and sweet corn, where they contribute significantly to plant growth promotion and biocontrol of phytopathogens [7–10]. In a study related to the diversity of endophytic *Bacillus* in the cotton plant (*Gossypium* sp. Dushanbe, Tajikistan), Reva et al. [11] reported 76 strains, the majority of which were identified as *B. amyloliquefaciens, B. licheniformis, B. megaterium, B. pumilus,* and *B. subtilis.* Similarly, Ceballos et al. [12] reported the cultivable endophytic population of aerobic endospore-forming bacteria from banana leaves that constitute 44 percent of the total population. Further, Raton et al. [13] isolated 18 aerobic endospore-forming bacteria from the rhizosphere of sugarcane and found that the majority of them belong to the genus *Bacillus.*

It is now commonly accepted that most microbes live in composite sessile communities denominated as biofilms under natural conditions. In biofilms, cells grow as multicellular colonies, enclosed in a self-produced extracellular matrix that normally consists of lipids, polysaccharides, proteins, and DNA [14–16]. Biofilms provide several advantages to microbes. For example, biofilms provide nutrients and protection from adverse environmental conditions. In addition, biofilms act as a main source of bacteria and other microorganisms in the environment [17]. Because of these characteristics, biofilms have been frequently exploited in industrial settings for bioremediation purposes [18]. Moreover, biofilms display the capability to be employed in agricultural practices as a biological alternative to chemical fertilizers [19, 20]. It has been reported that the majority of biofilms in the natural environment are polymicrobial in composition. Although research involving single-species biofilms has provided much knowledge related to biofilms, there is still a lot to be discovered from single-species biofilms.

Fundamental aspects and the molecular basis of biofilm formation by *B. subtilis,* a well-known PGPR have been well studied [21–23]. Definitely, this research has led to the discovery of several basic principles that support the biofilm structure, the role of the integration of complex signaling networks that occur in biofilms, how bacteria sense and respond to specific signals, and how genetically identical cells distinguish themselves into distinct cell lineages [22, 24]. Moreover, several details of the macromolecules that provide structure to the biofilms are now known [25–27].

Recent studies on microbial biofilms associated with plants have gained momentum in the last few years. Excellent review articles have detailed both the benefits and drawbacks of using bacterial biofilms in terms of agricultural, environmental, and ecological significance [20, 22, 28–30].

The areas around the roots are rich in nutrients because almost 40 percent of plant photosynthetic products are secreted by plants through the roots in the form of root exudates. Therefore, this area maintains a beneficial population of microbes [31]. Moreover, the functional diversity of rhizospheric microorganisms differs significantly from the functional diversity of bulk soil microorganisms [32]. The population size of bulk soil is bigger compared to the rhizospheric population, whereas in the rhizospheric soil the phylogenetic diversity is limited [9, 33]. The availability of easily metabolizable organic compounds in the rhizosphere helps in the survival of beneficial microbes that are more active and denser but with limited genetic diversity compared to bulk soil microorganisms [34].

The importance of polymicrobial biofilms and their role in plant health and disease have been recently reviewed [20]. However, the role of biofilm-forming rhizobacteria belonging to the genus *Bacillus* and their role in plant health is less explored. This chapter reviews the scientific literature available on the role of plant-associated biofilms of *Bacillus* spp. and their impact on plant growth and health.

4.2 Interaction of *Bacillus* within Plant Rhizosphere and Biofilm Development

The term *rhizosphere* was defined by Lorenz Hiltner as the soil section affected by plant roots [35, 36]. This region of plant–microbe communication has been in the research focus of scientists from various disciplines. Moreover, the aerial parts of the plant, called *phyllosphere,* are a habitat for different microbial communities that have been studied in detail [37]. The activity of different microbes in the rhizospheric region is vital for plant growth and development as it helps the host plants with nutrient acquisition and protection from phytopathogens [38]. PGPR make an important group of beneficial microbes in the rhizospheric area [39].

The success of PGPR by imparting benefits to plant in the form of phytohormone production and biocontrol activity can be achieved by effective colonization of rhizosphere and rhizoplane. The mode of action of PGPR is essential for successful plant–microbe interaction [40–42]. This creates the need for rhizosphere-competent (rhizobacteria that compete for nutrients and survive and colonize the rhizosphere) beneficial rhizobacteria [41, 43].

Colonization of plant roots by microbes is a step-by-step process. This includes recognition, adherence, invasion, colonization and growth, and many other approaches for the successful interactions. The plant roots communicate with the soil-dwelling bacterial population by releasing signals that are received by the microbes and in response the bacteria release signals that start the colonization process [44]. Among the genus *Bacillus,* the process of root colonization was studied by using *B. subtilis* as a model organism. Bais et al. [45] reported the role of surfactin produced by *B. subtilis* in colonization of *A. thaliana* roots and also demonstrated that a *B. subtilis* mutant ineffective in surfactin production lacks the ability to colonize the root. The colonization of *Bacillus* of the root surface of the host plant requires extracellular matrix. Beauregard et al. [46] reported the role of both transcriptional regulator Spo0A and the matrix antirepressor SinI in root colonization by *B. subtilis.*

The interaction of bacteria with plant tissues is mainly controlled by the host plant. Nutrient secretion by a plant in the form of root exudates, and defense responses, all together, directly control bacteria to interact/colonize with the host plant and build beneficial, harmful or neutral associations [47]. Moreover, plants directly influence the microbial communities that colonize their rhizosphere. This phenomenon, called *rhizosphere effect,* significantly controls the microbial population linked with the plant. This encourages the plant to select a small number of bacterial species while inhibiting the growth of others and not influencing other species [48].

Over the past decades, the genus *Bacillus* has been used to study the plant root associated biofilms [22]. A general schematic diagram showing biofilm formation by *Bacillus* on wheat root surface is shown in Figure 4.1. Inside the biofilm, the genetically similar

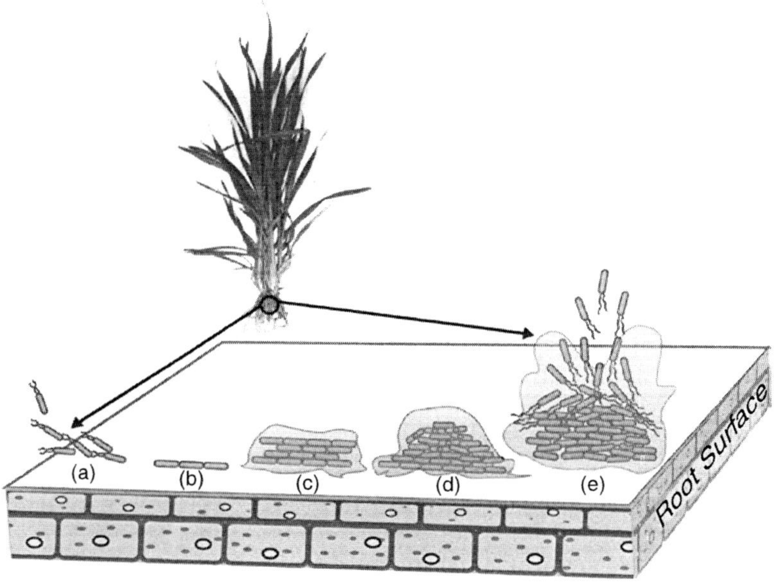

Figure 4.1 Schematic diagram showing the life cycle of *Bacillus* biofilm formation on the root surface of wheat (*Triticum aestivum*) plant. (a–b) Attachment of microorganism to sterile surface and suppression of motility. (c) Production of extracellular matrix, irreversible attachment and aggregation of planktonic cells. (d) Maturation and development of biofilm architecture. (e) Dispersion of planktonic cells from biofilm. (*See color plate section for the color representation of this figure.*)

bacterial cells give rise to subpopulations of phenotypically different cells, living together. This process of differentiation starts with the expression of genes, when bacterial cells receive stimulus from an external source like from the lipopeptide surfactin. In the beginning, the bacterial cells are short rods. As biofilm formation proceeds, the non-motile planktonic cells start forming long chains. At this stage the adherence is not reversible. This stage is also accompanied with the production of extracellular matrix, which plays an important role in maintaining the architecture of the biofilm and the sustenance of the biofilm community [49, 50].

Further, during the course of biofilm development and maturation, the division of bacterial cells takes place, which increase the size of the biofilm. The extracellular matrix provides integration and protection to the biofilm community from any adverse or harmful conditions. *Bacillus* biofilm at this stage consists of motile cells, spores along with extracellular matrix producers [22, 24].

The final stage in biofilm cycle is dispersion. Several reasons are responsible for the dispersion of biofilm. Limited resources and waste accumulation are some of the main reasons. Thus, it becomes essential for biofilm cells to disperse or to face death. Similarly, when the surrounding environment is not suitable for biofilm growth and development any more, or when there is an attack on the biofilm community, then it becomes a necessity for the planktonic cells to move to a safer site for their survival.

The biofilm dispersion process involves three steps. Initially, the stress condition inside biofilm signals the bacterial cells to alter their gene expression into dispersal mode. Second, the signaling surpasses the diffusion environment. Third, the most important steps, involve the disintegration of the extracellular matrix that works as a fortress for the biofilm community to enable the discharge of biofilm inhabitants into the surrounding environment [51, 52]. The *B. subtilis* biofilm dispersion takes place by the secretion of a mixture of D-amino acids (D-tyrosine, D-leucine, D-tryptophan and D-methionine), which dissolve the mature biofilm and stop the biofilm formation [52]. These amino acids are normally secreted by *B. subtilis* in the stationary phase of maturing biofilms [53]. These D-amino acids belong to the family of noncanonical D-amino acids. These D-amino acids prevent the biofilm formation by interrupting the attachment of protein amyloid fibers that normally facilitate binding of biofilm [54].

4.3 Multispecies Biofilms and Their Significance

Primarily, the microbial interactions are necessary for effective colonization, biofilm formation and continued symbiotic cooperation. Several studies revealed the role of quorum sensing in biofilm formation. Generally, the individual microbial cells use the chemical signals to coordinate prior to biofilm formation. Quorum sensing permits the microbes to determine the actual conditions of their population [20]. *N*-acyl-homoserine lactones (AHLs), autoinducer-2 (AI-2) and 2-heptyl-3-hydroxy-4-quinoline (PQS) are the main diffusible signals employed in bacterial conversation. *B. subtilis* biofilm formation is directly affected by the presence of other bacterial species, mainly by bacteria of the same genus [55].

Till date the majority of observations related to bacterial biofilms have been obtained in investigations related to single species biofilms. Whereas, the predominant form of microbial biofilms in natural habitats consist of numerous bacterial species and in several cases involves the interactions of fungi, algae, and protozoa [56]. Therefore, there is an utmost need to explore the structure and cooperation in mixed species biofilms. Moreover, multispecies biofilms depict the scenario where collaboration among different genera and species can potentially evolve, which benefits the microbes in different ways like increase tolerance to antibiotics [57]. Bacteria in multispecies biofilms interact with each other in several different ways. The main types of interactions are competition, cooperation, co-aggregation, and co-metabolism [58]. In case of co-metabolism the one community generates or eradicates a substrate, permitting the growth of another community. Under co-aggregation the attachment between the cells of different species takes place with specific surface associated components. Other interaction involves the transfer of plasmid DNA from one species to another [57].

Multispecies biofilms are highly significant in many areas, such as medicine, ecology and biotechnology (wastewater treatment, bioremediation, biodetoxification, protection from corrosion and agriculture) [59], but here we focus only on plant root associated biofilms formed by PGPR. The production of extracellular matrix by cyanobacterial and algal biofilms significantly enhanced the water retaining capacity of soils under drought conditions, which prevents soil erosion [60]. Prasanna et al. [61] reported the use of multispecies biofilm consisting of cyanobacteria (*Anabaena torulosa, Anabaena laxa* and *Calothrix elenkinii*) along with bacterial strains of *Azotobacter chroococcum, Bacillus subtilis, Pseudomonas fluorescens, Bradyrhizobium* sp., *Azotobacter* sp., and *Trichoderma viride* to increase the growth and yield of *Vigna radiata* and *Glycine max*.

4.4 Biofilm Detection and Characterization

Evaluation of complicated microbial communities is not an easy task, particularly if one wants to evaluate the individual components separately or jointly. However, the current developments in enhanced resolution in microscopy, lower costs of new techniques like next generation sequencing, pyrosequencing, advanced mass spectrometry, and high resolution microscopy, contributed to the generation of high quality data for the evaluation of whole microbial communities in mixed species biofilms [57, 62].

The key approaches used for studying single and multispecies biofilms are similar. However, the data obtained from the investigation of multispecies biofilm using these approaches are complex. This problem can be mitigated by the collective use of high throughput and high resolution methods. Presently several methods ranging from classical spectrophotometry to metagenomics, metatranscriptomics, metaproteomics, and metabolomics are being extensively applied for biofilm studies [59, 63]. Examples of some of these techniques are illustrated in Figures 4.2 to 4.4. Besides above mentioned methods there are some physicochemical approaches that are routinely used for the analysis of architecture of biofilms. These include Fourier transformation infrared spectroscopy (FTIR), nuclear magnetic resonance (NMR), mass spectrometry, such as matrix assisted laser desorption/ionization (MALDI) time of flight (TOF) mass spectrometry.

4.5 *Bacillus* Biofilm and Plant Health Promotion

Plants maintain different genera of bacteria on their surface like roots, stems and leaves and endophytes, which play a crucial role in plant health. Among all the plant-dwelling beneficial microbes, the aerobic endospore-forming *Bacillus* spp. are universally used in modern agricultural practices owing to their robust cell wall morphology, capability to form stress-resistant spores and to secrete several types of antimicrobial compounds. Capitalizing these properties the *Bacillus* spp. can easily inhabit the plant surface and can displace other microbes on the plant surface. Thus, the colonization sites of these

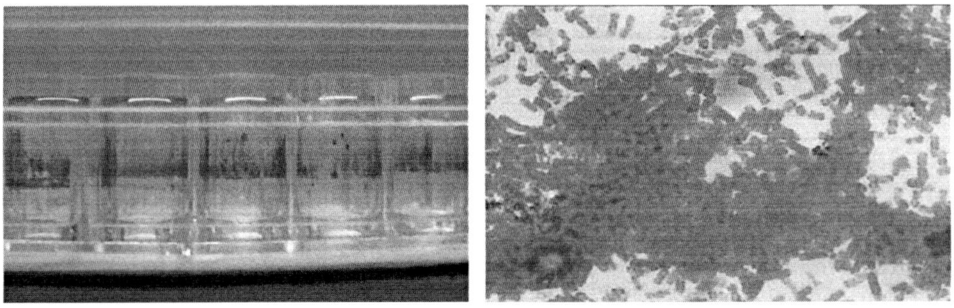

(a) (b)

Figure 4.2 Representative images of *Bacillus cereus* biofilm. (a) Side view of biofilm formed on air-liquid interface on 96 well polystyrene microtiter plates using crystal violet assay. (b) Biofilm on glass coverslips as viewed by light microscopy (Magnification 100x).

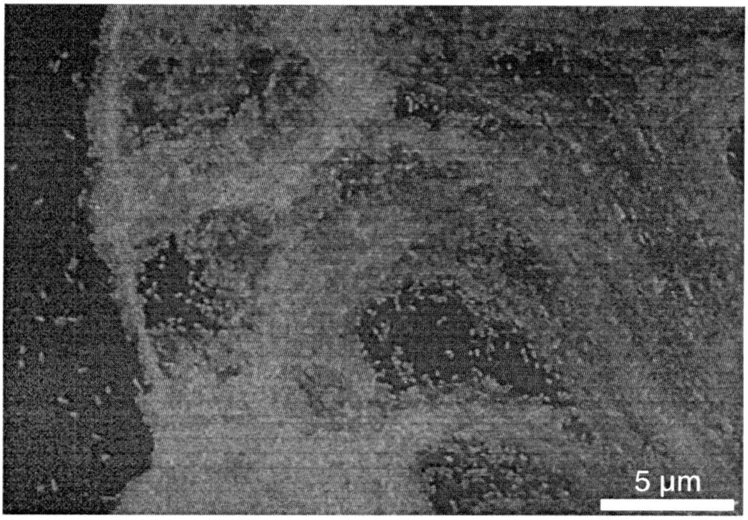

(a)

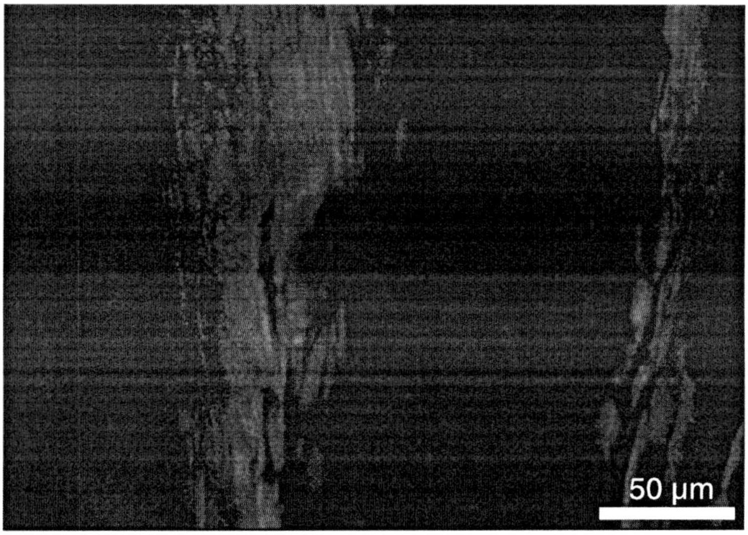

(b)

Figure 4.3 Confocal laser scanning microscopy images of *Bacillus subtilis* biofilm formed on glass coverslip (a), and chickpea root (b). Biofilms were stained with acridine orange. (*See color plate section for the color representation of this figure.*)

bacteria are consistent, which makes their application more appropriate in the management of agro-ecosystems [64].

Considerable amount of data have already been generated on the role of *Bacillus* as plant growth–promoting agent through its various plant growth–promoting activities and biocontrol properties [6, 65–67]. However, recent interest in biofilm and its role in

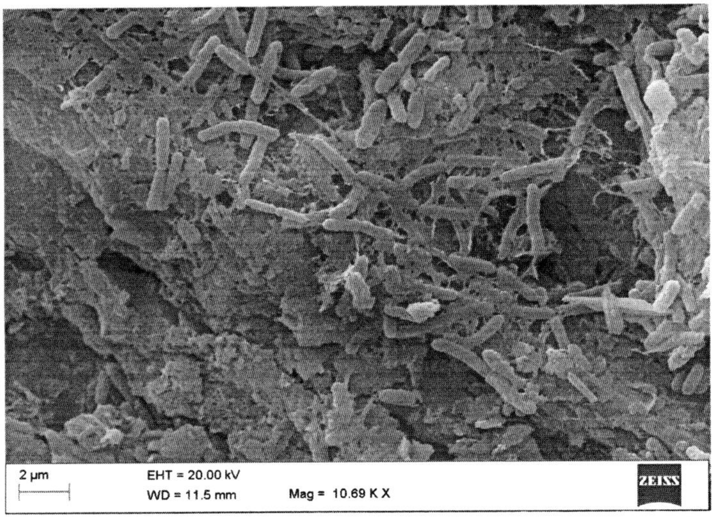

Figure 4.4 Scanning electron micrograph of *Bacillus subtilis* colonization and biofilm formation on root of *Cicer arietinum* (chickpea).

plant health have directed attention toward exploring the contribution of biofilm to above activities. Some of the relevant recent publications are presented here.

Chen et al. [68] reported that *B. subtilis* forms robust biofilm on plant root surfaces, which suppresses the wilt disease in tomato plants. Further, the study revealed that the biofilm-forming ability of *B. subtilis* enhanced the colonization potential as well as the local availability of antibiotics around the roots. These antibiotics work as signals, which consequently induce biofilm formation. Xu et al. [69] showed that bacillomycin D produced by *Bacillus amyloliquefaciens* SQR9 stimulates biofilm formation and also contributes to antagonistic activity against *Fusarium oxysporum*.

Zeriouh et al. [70] found that *B. subtilis* UMAF6614 secretes surfactin that triggers biofilm formation on melon phyllosphere, which promotes the stability and sufficient production of suppressive lipopeptides, bacillomycins and fengycins, which powerfully suppress the phytopathogens. Xu et al. [71] found that an endophytic strain of *Bacillus cereus* 0-9 efficiently colonized the wheat root and showed biofilm formation that participated in the biocontrol activity against the wheat sharp eyespot.

Other bacilli like *Paenibacillus* have also been implicated in plant growth promotion, biocontrol and drought tolerance through biofilm formation [72, 73].

4.6 Conclusion and Future Prospects

During the last decade, significant work has been carried out on plant–microbe interactions, emphasizing the role of biofilm formation. Biofilm formation is a universal phenomenon normally displayed by microbes in natural environments, including plant surfaces. Microbial populations gain several advantages as biofilms. Biofilm acts as a fortress for microbes that provides protection to them from various forms of stress and increases their survivability. The biofilm also provides better nutrition in the rhizosphere and rhizoplane available in the form of rhizodeposits and root exudates. The PGPR, especially bacteria belonging to the genus of free-living endospore-forming *Bacillus*, provide an opportunity for efficient use as bioinoculant as they have the ability to survive better even under stress conditions compared to other PGPR. Moreover, a combination of high-throughput techniques will explain the molecular basis involved in bacterial biofilm formation, and research continues to uncover additional mechanisms that can be employed in resistance to drought, salinity, and other stress conditions. These types of investigations are necessary for exploiting biofilm for increasing plant growth and productivity, thereby decreasing the dependence on chemical fertilizers. Further, more detailed studies are needed on multispecies biofilms, because these represent the actual conditions in natural environments.

References

1 K.V.B.R. Tilak, N. Ranganayaki, K.K. Pal, R. De, A.K. Saxena, C.S. Nautiyal, S. Mittal, A.K. Tripathi, B.N. Johri, Diversity of plant growth and soil health supporting bacteria, *Curr. Sci.*, 89, 136–150 (2005).

2 T. Gomiero, D. Pimentel, M. G. Paoletti, Environmental impact of different agricultural management practices: conventional vs. organic agriculture, *Crit. Rev. Plant Sci.*, 30, 95–124 (2011).

3 B.R. Glick, Bacteria with ACC deaminase can promote plant growth and help to feed the world, *Microbiol. Res.*, 169, 30–39 (2014).

4 J.M. Barea, M.J. Pozo, R. Azcón, C. Azcón-Aguilar, Microbial co-operation in the rhizosphere, *J. Exp. Bot.* 56, 1761–1778 (2005).

5 P.N. Bhattacharyya and D.K. Jha, Plant growth-promoting rhizobacteria (PGPR): emergence in agriculture, *World J. Microbiol. Biotechnol.*, 28, 1327–1350 (2012).

6 A. Kumar, A. Prakash, B.N. Johri, *Bacillus* as PGPR in crop ecosystem, *in Bacteria in Agrobiology: Crop Ecosystems*, D.K. Maheshwari,(ed), Springer, Berlin, 2011.

7 C.R. Bell, G.A. Dickie, W.L.G. Harvey, J.W.Y.F. Chan, Endophytic bacteria in grapevine, *Can. J. Microbiol.*, 41, 46–53 (1995).

8 M. Shishido, C. Breuil, C.P. Chanway, Endophytic colonization of spruce by plant growth–promoting rhizobacteria, *FEMS Microbiol. Ecol.*, 29,191–196 (1999).

9 G. Berg, C. Zachow, J. Lottmann, M. Gotz, R. Costa, K. Smalla, Impact of plant species and site on rhizosphere-associated fungi antagonistic to *Verticillium dahliae Kleb*, *Appl. Environ. Microbiol.*, 71, 4203–4213 (2005).

10 S.H. Ji, M.A. Gururani, S-C, Chun, Isolation and characterization of plant growth–promoting endophytic diazotrophic bacteria from Korean rice cultivars, *Microbiol. Res.*, 169, 83–98 (2014).

11 O.N. Reva, V.V. Smirnov, B. Pattersson, F.G. Priest, *Bacillus endophyticus* spp. nov., isolated from the inner tissues of cotton plants (*Gossypium* sp.), *Int. J. Syst. Evol. Microbiol.*, 52,101–107(2002).

12 I. Ceballos, S. Mosquera, M. Angulo, J. Mira, L. Argel, D. Uribe-Velez, M. Romero-abarez, S. Orduz-Peralta, V. Villegas, Cultivable bacteria populations associated with leaves of banana and plantain plants and their antagonistic activity against *Mycosphaerella fijiensis*, *Microb. Ecol.*, 64(3), 641–653 (2012).

13 T.D.L.M.O. Raton, R. Yano, O.R. Gamez, E.I.S. Floh, M.D.J.S. Dıaz, H.R. Barbosa, Isolation and characterisation of aerobic endospore forming Bacilli from sugarcane rhizosphere for the selection of strains with agriculture potentialities, *World J. Microbiol.Biotechnol.*, 28, 1593–1603 (2012).
14 S.S. Branda, S. Vik, L. Friedman, R. Kolter, Biofilms: The matrix revisited, *Trends Microbiol.*, 13, 20–26 (2005).
15 L. Hall-Stoodley and P. Stoodley, Evolving concepts in biofilm infections, *Cell Microbiol.*, 11, 1034–1043 (2009).
16 L. Hobley, C. Harkins, C.E. MacPhee, N.R. Stanley-Wall, Giving structure to the biofilm matrix: an overview of individual strategies and emerging common themes, *FEMS Microbiol. Rev.*,39(5), 649–669 (2015).
17 J.W. Costerton, Overview of microbial biofilms, *J.Ind. Microbiol.*, 15,137-140 (1995).
18 B. Halan, K. Buehler, A. Schmid, Biofilms as living catalysts in continuous chemical syntheses, *Trends Biotechnol.*, 30, 453–446 (2012).
19 G. Seneviratne, A. Jayasekara, M.S.D.L. De Silva, U.P. Abeysekera, Developed microbial biofilms can restore deteriorated conventional agricultural soils, *Soil Biol. Biochem.*, 43, 1059–1062 (2011).
20 A.A. Angus and A.M. Hirsch, Biofilm formation in the rhizosphere: multispecies interactions and implications for plant growth, *in Molecular microbial ecology of the rhizosphere*, de Bruijn, FJ (ed.), Hoboken, NJ: John Wiley & Sons, 2013.
21 S.S. Branda, J.E. Gonza´lez-Pastor, S. Ben-Yehuda, R. Losick, R. Kolter, Fruiting body formation by *Bacillus subtilis, Proc. Natl. Acad. Sci. USA.*, 98,11621–11626 (2001).
22 H. Vlamakis, Y. Chai, P. Beauregard, R. Losick, R. Kolter, Sticking together: building a biofilm the Bacillus subtilis way, *Nat. Rev. Microbiol.*, 3,157–168 (2013).
23 L.S. Cairns, L. Hobley, N.R. Stanley-Wall, Biofilm formation by *Bacillus subtilis*: new insights into regulatory strategies and assembly mechanisms, *Mol. Microbiol.*, 93, 587–598 (2014).
24 D. López, H. Vlamakis, R. Kolter, Biofilms, *Cold Spring Harb. Perspect. Biol.*, 2, a000398 (2010).
25 D. Romero, H. Vlamakis, R. Losick, R. Kolter, An accessory protein required for anchoring and assembly of amyloid fibres in *B. subtilis biofilms, Mol. Microbiol.*, 80,1155–1168 (2011).
26 K. Kobayashi and M. Iwano, BslA (YuaB) forms a hydrophobic layer on the surface of *Bacillus subtilis* biofilms, *Mol. Microbiol.*, 85, 51–66 (2012).
27 L. Hobley, A. Ostrowski, F.V. Rao, K.M. Bromley, M. Porter, A.R. Prescott, C.E. MacPhee, D.M. van Aalten, N.R. Stanley-Wall, BslA is a self-assembling bacterial hydrophobin that coats the *Bacillus subtilis* biofilm, *Proc. Natl. Acad. Sci. USA.*, 110, 13600–13605 (2013).
28 M.E. Davey, and G.A. O'Toole, Microbial biofilms: from ecology to molecular genetics, *Microbiol. Mol. Biol. Rev.*, 64, 847–867(2000).
29 C.E. Morris and J.M. Monier, The ecological significance of biofilm formation by plant-associated bacteria, *Ann. Rev. Phytopathol.*, 41,429–453 (2003).
30 T. Danhorn and C. Fuqua, Biofilm formation by plant associated bacteria, *Ann. Rev. Microbiol.*, 61, 401–422 (2007).
31 L. Ping and W. Boland, Signals from the underground: bacterial volatiles promote growth in *Arabidopsis, Trends Plant Sci.*, 9(6), 263–266 (2004).

32 P. Hinsinger, A. G. Bengough, D. Vetterlein, I. M. Young, Rhizosphere: biophysics, biogeochemistry and ecological relevance, *Plant Soil.*, 321 (1), 117–152 (2009).
33 L. Marilley and M. Aragno, Phylogenetic diversity of bacterial communities differing in degree of proximity of *Lolium perenne* and *Trifolium repens* roots, *Appl. Soil Ecol.*, 13, 127–136 (1999).
34 E. Inbar, S.J. Green, Y. Hadar, D. Minz, Competing factors of compost concentration and proximity to root affect the distribution of Streptomycetes. *Microbiol. Ecol.*, 50, 73–81 (2005).
35 L. Hiltner, Ueber neue Erfahrungen und Probleme auf dem Gebiete der Bodenbakteriologie. Arbeiten der Deutschen Landwirtschaft Gesellschaft, 98, 59–78 (1904).
36 A. Hartmann, M. Rothballer, M. Schmid, Lorenz Hiltner, a pioneer in rhizosphere microbial ecology and soil bacteriology research. *Plant Soil.*, 312, 7–14 (2008).
37 J.A. Vorholt, Microbial life in the phyllosphere. *Nat. Rev. Microbiol.*, 10, 828–840 (2012).
38 R.L. Berendsen, C.M.J. Pieterse, P.A.H. Bakker, The rhizosphere microbiome and plant health, *Trends Plant Sci.*, 17, 478–486 (2012).
39 B. Lugtenberg and F. Kamilova, Plant-growth-promoting rhizobacteria, *Annu. Rev. Microbiol.*, 63, 541–556 (2009).
40 B.J.J. Lugtenberg, L. Dekkers, G.V. Bloemberg, Molecular determinants of rhizosphere colonization by Pseudomonas, *Ann. Rev. Phytopathol.*, 39, 461–490 (2001).
41 F. Kamilova, S. Validov, T. Azarova, I. Mulders, B. Lugtenberg, Enrichment for enhanced competitive root tip colonizers selects for a new class of biocontrol bacteria, *Environ. Microbiol.*, 7, 1809–1817 (2005).
42 M. Ongena and P. Jacques, *Bacillus* lipopeptides: versatile weapons for plant disease biocontrol, *Trends Microbiol.*, 16(3), 115–125 (2008).
43 A. Beneduzi, A. Ambrosini, L.M.P. Passaglia, Plant growth-promoting rhizobacteria (PGPR): their potential as antagonists and biocontrol agents, *Genet. Mol. Biol.*, 35, 1044–1051 (2012).
44 H.P. Bais, T.L. Weir, L.G. Perry, S. Gilroy, J.M. Vivanco, The role of root exudates in rhizosphere interactions with plants and other organisms, *Ann. Rev. Plant Biol.*, 57, 233–266 (2006).
45 H.P. Bais, R. Fall, J.M. Vivanco, Biocontrol of Bacillus subtilis against infection of Arabidopsis roots by Pseudomonas syringae is facilitated by biofilm formation and surfactin production. *Plant physiol.*, 134, 307–319 (2004).
46 P.B. Beauregard, Y. Chai, H. Vlamakis, R. Losick, R. Kolter, *Bacillus subtilis* biofilm induction by plant polysaccharides, *Proc. Natl. Acad. Sci. USA.*, 110, 1621–1630 (2013).
47 X.F. Huang, J.M. Chaparro, K.F. Reardon, R. Zhang, Q. Shen, J.M. Vivanco, Rhizosphere interactions: root exudates, microbes, and microbial communities, *Botany*, 92, 267–275 (2014).
48 S.J. Green, F.C. Michel, Y. Hadar, D. Minz, Contrasting patterns of seed and root colonization by bacteria from the genus *Chryseobacterium* and from the family Oxalobacteriaceae. *ISME J.*,1, 291–299 (2007).
49 M. Marvasi, P.T. Visscher, L. Casillas Martinez, Exopolymeric substances (EPS) from *Bacillus subtilis*: Polymers and genes encoding their synthesis. *FEMS Microbiol. Lett.*, 313(1), 1–9 (2010).

50 H.C. Flemming and J. Wingender, The biofilm matrix, *Nat. Rev. Microbiol.*, 8, 623–633(2010).

51 E. Karatan and P. Watnick, Signals, regulatory networks, and materials that build and break bacterial biofilms, *Microbiol. Mol. Biol. Rev.*, 73, 310–347(2009).

52 I. Kolodkin-Gal, D. Romero, S. Cao, J. Clardy, R. Kolter, R. Losick, D-amino acids trigger biofilm disassembly, *Science*, 328, 627–629 (2010).

53 H. Lam, D.C. Oh, F. Cava, C.N. Takacs, J. Clardy, M.A. de Pedro, M.K. Waldor, D-amino acids govern stationary phase cell wall remodeling in bacteria, *Science*, 325,1552–1555 (2009).

54 F. Cava, M.A. de Pedro, H. Lam, B.M. Davis, M.K. Waldor, Distinct pathways for modification of the bacterial cell wall by non-canonical D-amino acids, *EMBO J.*, 30, 3442–3453 (2011).

55 E.A. Shank, V. Klepac-Ceraj, L. Collado-Torres, G.E. Powers, R. Losick, R. Kolter, Interspecies interactions that result in *Bacillus subtilis* forming biofilms are mediated mainly by members of its own genus, *Proc. Natl. Acad. Sci. U S A.*, 108, 1236–1243 (2011).

56 J. Jass, S.K. Roberts, H.M. Lappin-Scott, Microbes and enzymes in biofilms, *in Enzymes in the Environment. Activity, Ecology and Applications*, R.G. Burns and R.D. Dick (Eds), Marcel Dekker, New York, 2002.

57 M. Burmolle, D.W. Ren, T. Bjarnsholt, S.J. Sorensen, Interactions in multispecies biofilms: do they actually matter? *Trends Microbiol.*, 22, 84–91 (2014).

58 S. Elias and E. Banin, Multi-species biofilms: living with friendly neighbours, *FEMS Microbiol. Rev.*, 36(5), 990–1004 (2012).

59 N.A. Nozhevnikova, E. A. Botchkoval, V. K. Plakunov, Multi-species biofilms in ecology, medicine, and biotechnology, *Microbiol.*, 84 (6), 731–750 (2015).

60 G. Roeselers, M.C.M. van Loosdrecht, G. Muyzer, Phototrophic biofilms and their potential applications, *J. Appl. Phycol.*, 20, 227–235 (2008).

61 R. Prasanna, S. Triveni, N. Bidyarani, S. Babu, K. Yadav, A. Adak, S. Khetarpal, M. Pal, Y.S. Shivay, A.K. Saxena, Evaluating the efficacy of cyanobacterial formulations and biofilmed inoculants for leguminous crops, Arch. Agron. Soil Sci., 60, 349–366 (2014).

62 D. Medini, D. Serruto, J. Parkhill, D. A. Relman, C. Donati, R. Moxon, S. Falkow, R. Rappuoli, Microbiology in the post-genomic era, *Nat. Rev. Microbiol.*, 6, 419–430 (2008).

63 F. Abram, Systems-based approaches to unravel multi-species microbial community functioning, *Comput. Structur. Biotechnol. J.*, 13, 24–32 (2015).

64 S. Timmusk and E. Nevo, Plant root associated biofilms, *in Bacteria in Agrobiology, Plant Nutrient Management*, D. K. Maheshwari (Ed), Springer, Berlin, 2011.

65 D.K. Choudhary and B.N Johri, Interactions of *Bacillus* spp. and plants with special reference to induced systemic resistance (ISR), *Microbiol. Res.*, 164, 493–513 (2008).

66 K. Sunar, P. Dey, U. Chakraborty, B. Chakraborty, Biocontrol efficacy and plant growth–promoting activity of *Bacillus altitudinis* isolated from Darjeeling hills, India, *J. Basic Microbiol.*, 53, 1–14 (2013).

67 S.P. Chowdhury, A. Hartmann, X.W. Gao, R. Borriss, Biocontrol mechanisms by root-associated *Bacillus amyloliquefaciens* FZB42-a review, *Front. Microbiol.*,6, 780 (2015).

68 Y. Chen, F. Yan, Y. Chai, H. Liu, R. Kolter, R. Losick, J.H. Guo, Biocontrol of tomato wilt disease by *Bacillus subtilis* isolates from natural environments depends on conserved genes mediating biofilm formation, *Environ. Microbiol.*, 15, 848–864 (2013).

69 Z. Xu, J. Shao, B. Li, X. Yan, Q. Shen, R. Zhang, Contribution of bacillomycin D in *Bacillus amyloliquefaciens* SQR9 to antifungal activity and biofilm formation. *Appl. Environ.Microbiol.*, 79, 808–815 (2013).

70 H. Zeriouh, A. de Vicente, A. Perez-Garcia, D. Romero, Surfactin triggers biofilm formation of *Bacillus subtilis* in melon phylloplane and contributes to the biocontrol activity, *Environ. Microbiol.*, 16, 2196–2211 (2014).

71 Y.B. Xu, M. Chen, Y. Zhang, M. Wang, Y. Wang, Q. B. Huang, X. Wang, G. Wang, The phosphotransferase system gene ptsI in the endophytic bacterium *Bacillus cereus* is required for biofilm formation, colonization, and biocontrol against wheat sharp eyespot, *FEMS Microbiol. Lett.*, 354,142–152 (2014).

72 W.M. Haggag and S. Timmusk, Colonization of peanut roots by biofilm forming *Paenibacillus polymyxa* initiates biocontrol against crown rot disease, *J. App. Microbiol.*, 104(4), 961–969 (2008).

73 S. Timmusk, S-B. Kim, E. Nevo, I. Abd El Daim, B. Ek, J. Bergquist, L. Behers, Sfp-type PPTase inactivation promotes bacterial biofilm formation and ability to enhance wheat drought tolerance, *Front. Microbiol.*,6, 387 (2015).

5

Biofilm Formation by *Pseudomonas* spp. and Their Significance as a Biocontrol Agent

Zaki A. Siddiqui and Masudulla Khan

Section of Plant Pathology and Nematology, Department of Botany, Aligarh Muslim University, Aligarh, India

5.1 Introduction

Pseudomonas Migula 1894 is one of the most ubiquitous bacterial genera worldwide; different species have been isolated from highly diverse ecological niches [1]. Genus *Pseudomonas* currently includes 202 species, using classification methods, including a combination of 16S rRNA, analysis of cellular fatty acids, and differentiation via classical physiological and biochemical tests [2]. The genus consists of a group of medically and biotechnologically important bacteria having vast metabolic versatility [3, 4]. They are nonsporulating, aerobic, Gram-negative rods that are found in both biofilms and planktonic forms. The worldwide distribution of *Pseudomonas* is presumably due to high genetic and physiological adaptability [5], and the presence of numerous genomic islands in the genus is a key to the adaptability traits for diverse environments [6].

Members of the genus *Pseudomonas* (*sensu stricto*) belong to Palleroni's RNA group I, in the *Gammaproteobacteria*. Gene trees of the *Pseudomonas* genus allow for the discrimination of two lineages or intrageneric groups (IG), termed IG *P. aeruginosa* and IG *P. fluorescens* [7]. The IG *P. aeruginosa* is divided into three main groups, represented by the species *P. aeruginosa*, *P. stutzeri*, and *P. oleovorans*. The IG *P. fluorescens* is divided into six groups, represented by the species *P. fluorescens*, *P. syringae*, *P. lutea*, *P. putida*, *P. anguilliseptica*, and *P. straminea*. The *P. fluorescens* group is the most complex and includes nine subgroups, represented by the species *P. fluorescens*, *P. gessardi*, *P. fragi*, *P. mandelii*, *P. jesseni*, *P. koreensis*, *P. corrugata*, *P. chlororaphis*, and *P. asplenii*. *Pseudomonas rhizospherae* is affiliated with *P. fluorescens* IG in the phylogenetic analysis but is independent of any group. Some species are located on phylogenetic branches that are distant from defined clusters, such as those represented by the *P. oryzihabitans* group and the type strains *P. pachastrellae*, *P. pertucinogena*, and *P. luteola*. Additionally, 17 strains of *P. aeruginosa*, *P. entomophila*, *P. fluorescens*, *P. putida*, *P. syringae*, and *P. stutzeri*, for which genome sequences have been determined, have been included in the *P. fluorescens* IG in order to obtain a comprehensive view regarding the phylogenetic relationships within the *Pseudomonas* genus [7].

Pseudomonas spp. has been applied in engineered systems for biocontrol, plant growth promotion and bioremediation. The antibiotic 2,4-diacetylphloroglucinol

(DAPG)-producing strains are predominant among biocontrol microorganisms [8]. Soil-borne *Pseudomonas* spp. have received much attention as biocontrol agents of plant pathogens because of their catabolic versatility, excellent root-colonizing abilities, and capacity to produce a wide range of antifungal metabolites. In addition, some *Pseudomonas* have been shown to elicit a disease-resistance response in crop species, a phenomenon known as induced systemic resistance (ISR) [9, 10]. This dual activity of *Pseudomonas* further highlights their potential as biocontrol agents of plant pathogens [11, 12]. *Pseudomonas* spp. have been used successfully for the control of nematode (Table 5.1), fungal (Table 5.2), and bacteterial (Table 5.3) diseases of plants in different parts of world.

Table 5.1 Effects of *Pseudomonas* spp. on Nematode Diseases of Plants.

Nematode	*Pseudomonas* sp.	Effect	Reference
M. incognita	354 isolates	*P. fluorescens* (strains JOB204, JOB 209) and *Bacillus* (JOB203) were most effective, and clover plants treated with these bacteria had fewer galls and large roots.	[13]
M. incognita	*P. fluorescens*	Reduced galling and nematode populations on tomato.	[14–17]
M. incognita	*P. fluorescens*	Seed treatment significantly reduced galling on okra.	[18]
M. incognita	*P. fluorescens*	Management of *M. incognita* on brinjal was obtained with *P. fluorescens*.	[19]
M. incognita	*P. aeruginosa*	Reduced galling on tomato.	[20]
M. incognita	*P. aeruginosa*	*P. aeruginosa* caused a significant reduction in galling and nematode multiplication on chickpea.	[21]
M. incognita	*P. straita*	Reduced reproduction of *M. incognita* on pea.	[22]
M. incognita	*P. straita*, *Rhizobium* sp.	Inoculation of *Rhizobium* resulted in greater increase in chickpea growth than caused by *P. straita*.	[23]
M. incognita, *H. cajani*	Fluorescent pseudomonads and *Bacillus* isolates	Four isolates of *Pseudomonas* and 2 of *Bacillus* (Pa70, Pf18, Pa116, Pa324, B18 and B160) were considered potentially useful for the biocontrol of nematodes.	[24]
M. incognita	20 isolates of fluorescent pseudomonads	Isolates Pf604, Pf605, Pf611, and Pa616 have an inhibitory effect on the hatching and penetration of nematodes.	[25]
M. incognita	*Bacillus* and *Pseudomonas* isolates	Out of 18 isolates, B 28 was best in improving tomato growth of *M. incognita* inoculated plants.	[26]
M. incognita, *H. cajani*	*Bacillus* and *Pseudomonas* isolates	Combined use of Pa324 with B18 provided better biocontrol of nematodes on pigeon pea.	[27]

Table 5.1 (Continued)

Nematode	*Pseudomonas* sp.	Effect	Reference
M. incognita	*P. putida*, *Paenibacillus polymyxa*	*P. putida* most effectively reduces galling and nematode multiplication on chickpea.	[28]
M. incognita	*P. alcaligenes*, *B. pumilus*	*P. alcaligenes* resulted in a greater increase in shoot dry weight of chickpea plants with nematodes.	[29]
M. incognita	*P. putida*	*P. putida* caused significant reduction in galling and nematode multiplication on tomato.	[30,31]
M. incognita	*P. putida*, *P. alcaligenes*, *Pseudomonas* isolate Pa 28	*P. putida* caused greatest reduction in galling and nematode multiplication followed by *P. alcaligenes* and Pa 28.	[32]
M. incognita	*Pseudomonas* and *Bacillus* isolates	*Pseudomonas* isolates were more effective in reducing galling and nematode multiplication in pea.	[33]
M. incognita	*B. thuringiensis*, *P. fluorescens*, *R. leguminosarum*	Results of *in vitro* and *in vivo* experiments showed that all tested bacteria significantly suppressed nematodes.	[34]
M. incognita	*Pseudomonas* sp.	Isolates Pa 8, Pa 9 have the potential for biocontrol of nematodes on tomato.	[35]
M. incognita	*P. fluorescens*, *Rhizobium* sp.	Use of *Rhizobium* plus *P. fluorescens* caused greater reduction in galling and nematode multiplication on pea.	[36]
M. incognita	*Bacillus megaterium* *P. fluorescens* *Trichoderma viride* *Paecilomyces lilacinus*	All rhizospheric microbes displayed nematicidal potential via ovicidal and larvicidal actions *in vitro* and resulted in significant improvement in plant growth parameters.	[37]
M. javanica			
M. javanica	*P. fluorescens*	Reduced nematode multiplication and morphometrics of *M. javanica* females on tomato in different soils.	[38]
M. javanica	*P. aeruginosa*	Reduced galling and nematode populations on tomato.	[39]
M. javanica	*P. fluorescens*	*P. fluorescens* with *Glomus mosseae* was better at improving chickpea growth and reducing galling and nematode multiplication.	[40]
M. javanica	*P. aeruginosa*, *P. fluorescens*	Bare root dip or soil drench treatment reduced nematode penetration into tomato roots.	[41]

(Continued)

Table 5.1 (Continued)

Nematode	*Pseudomonas* sp.	Effect	Reference
M. javanica	*P. fluorescens* CHA0	*P. fluorescens* with ammonium molybdate reduced nematode penetration in mung bean.	[42]
M. javanica	*P. putida*, *P. alcaligenes*, *P. polymyxa*, *B. pumilus*	*P. putida* was best in reducing galling and nematode reproduction on lentil.	[43]
M. javanica	*P. fluorescens* EPS291 and EPS817	Both isolates significantly increased plant growth and reduced nematode reproduction in micropropagated banana.	[44]
M. javanica	*P. putida* *P. alcaligenes*	All PGPR strains significantly reduced disease severity in chickpea.	[45]
M. javanica	*Pseudomonas* sp., *Bacillus* sp.	Out of 15 strains, Rh37 showed maximum mortality (53%) after 48 hr as compared to control on tomato.	[46]
M. javanica	*P. fluorescens* *Bacillus subtilis*, and *Aspergillus awamori*	*P. fluorescens* caused greatest reduction in galling and nematode reproduction followed by *B. subtilis* and *A. awamori*.	[47]
Heterodera cajani			
M. incognita, *H. cajani*, *H. zeae*, *H. avenae*	*Bacillus subtilis*, *B. cereus*, *B. pumilus*, *Pseudomonas* spp.	*B. subtilis* and *B. pumilus* were most effective against all tested species. The noncellular extract exhibited larvicidal properties.	[48]
H. cajani	*P. fluorescens*	Reduced multiplication of *H. cajani* on pigeon pea.	[49]
Heterodera schachtii			
H. schachtii	290 isolates	Eight isolates were antagonistic to *H. schachtii*; three isolates were identified as *P. fluorescens*.	[50]
H. schachtii	8 isolates	Nematode penetration was reduced by 6 of 8 isolates tested.	[51]
Other nematodes species			
Criconemella xenoplax	*P. aureofaciens*	Bacteria inhibited nematode multiplication in greenhouse tests.	[52]
Panagrellus sp.	*P. fluorescens*	Bacteria cultivated on plate count broth reduced nematodes by up to 57.4%.	[53]
R. reniformis	*P. solanacearum*	Slight inhibition of nematode activity on aubergine roots.	[54]
R. similis, *Meloidogyne* spp.	*P. putida*, *P. fluorescens*	Inhibited invasion of *R. similis* and *Meloidogyne* spp. in banana, maize, and tomato.	[55]
Heterodera cruciferae	Fluorescent pseudomonads	Growth and hatching of nematode eggs were inhibited.	[56]

Table 5.2 Effects of *Pseudomonas* spp. on Fungal Diseases of Plants.

Fungus	*Pseudomonas* sp.	Effect	Reference
***Gaeumannomyes* sp.**			
G. graminis var. *tritici*	*P. fluorescens*	Strains of *P. fluorescens* may be involved in the suppression of *G. graminis var. tritici*.	[57]
G. graminis	*Pseudomonas* spp. (fluorescent strains)	27% yield increase due to biocontrol of bacteria in winter wheat under field conditions.	[58]
G. graminis var. *tritici*	*P. aureofaciens* Q2-87	Inhibition of fungus was demonstrated both *in vitro* and *in vivo*.	[59]
G. graminis *Rhizoctonia solani*	*Bacillus subtilis*, *B. cereus* isolates, *P. corrugata*	*Bacillus* isolate A47 and *B. subtilis* B908 reduced take-all disease in sodic acid soil while *B. subtilis* B931 was more effective in reducing Rhizoctonia root-rot in wheat grown on a calcareous sandy loam.	[60]
***Pythium* spp.**			
Pythium sp.	*P. fluorescens*	In *Pythium*-contaminated sites, significant increases were observed in plant height, number of heads, and grain yield of winter wheat.	[61]
P. ultimum P17	Fluorescent pseudomonads	Significantly suppressed root-rot disease on tulip.	[62]
P. aphanidermatum OP4	Fluorescent Pseudomonas (CH31, CH1)	Suppressed root-rot disease on cucumber.	[63]
P. aphanidermatum	*P. corrugate*, *P. aureofaciens*	Induced systemic resistance in cucumber roots.	[64]
Pythium sp.	*P. putida*, *B. subtilis*, *E. aerogenes*, *E.agglomerans*, *B. cereus*	Most strains increased root length of cucumber in *Pythium*-infected plants *in vitro*.	[65]
P. aphanidermatum f. sp. *cucurbitacearum*	*B. subtilis* *P. putida*	Growth and yield of lettuce and cucumber were increased and disease severity reduced.	[66]
P. aphanidermatum, *Aspergillus* sp., *F. oxysporum* f. sp. *ciceri*, *Rhizoctonia solani*	*Pseudomonas* isolates	Two strains, i.e., MRS23 and CRP55P, have shown antifungal activity.	[67]

(Continued)

Table 5.2 (Continued)

Fungus	*Pseudomonas* sp.	Effect	Reference
P. aphanidermatum	*P. fluorescens*, *P. putida*	*P. fluorescens* isolate Pf1 was effective in reducing the damping-off incidence in tomato and hot pepper.	[68]
P. aphanidermatum OP4	Fluorescent Pseudomonas (CH31, CH1)	Suppressed root-rot disease on cucumber.	[63]
***Fusarium* spp.**			
Fusarium sp.	*P. fluorescens*	Observed induced resistance and phytoalexin accumulation in carnation.	[69]
F. oxysporum f.sp. *raphani* *A. brassicicola*, *F. oxysporum*	*P. fluorescens*	Protected radish plants through induction of systemic resistance against these pathogens.	[70]
F. culmorum	*P. chlororaphis* *2E3,O6*	Strong inhibition of the fungus on spring wheat in the field.	[71]
F. udum	*P. fluorescens*	Wilt incidence was reduced in pigeon pea.	[49]
F. oxysporum f. sp. *lycopersici*	*P. fluorescens* PRS9, *B. polymyxa*	Reduced the wilting index and rhizosphere population of fungus on tomato.	[14]
F. oxysporum *M. phaseolina* *F. solani* *R. solani*	*P. aeruginosa*	Significantly suppressed growth of root-infecting fungi on tomato.	[39]
F. oxysporum f. sp. *lycopersici*	*B. subtilis* *P. fluorescens* *Aspergillus awamori*, *A. niger*, *P. digitatum*	Use of all the PSM increased yields and also reduced the rhizospheric population of wilt fungus by 23%–49% on tomato.	[72]
F. oxysporum	Flourescent Pseudomonads isolates	All 5 isolates have shown antifungal activity against *Fusarium*.	[73]
F. moniliformae, *F. graminearum*, *M. phaseolina*	*Pseodomonas* sp. EM85 *Bacillus* sp. (MR-11(2), MRF)	All isolates had the ability to suppress the diseases caused by *F. moniliforme*, *F. graminearum* and *M. phaseolina* on maize.	[74]
F. oxysporum f. sp. *ciceri* *Aspergillus* sp. *P. aphanidermatum* *R. solani*	*Pseudomonas* isolates	Two strains, MRS23 and CRP55P, have shown antifungal activity.	[67]

Table 5.2 (Continued)

Fungus	*Pseudomonas* sp.	Effect	Reference
F. oxysporum, *R. solani*	*P. fluorescens*	Out of 40 strains, 18 showed strong antifungal activity.	[75]
F. oxysporum f. sp. *lycopersici*	*P. fluorescens* Pf1	Pf1 protected tomato plants from wilt disease.	[76]
F. udum, *F. oxysporum* f. sp. *ciceris*	*P. aeruginosa PNA 1*	*P. aeruginosa* protected pigeon pea and chickpea from Fusarium wilt.	[77]
F. chlamydosporium	*P. fluorescens*	Reduced the severity of disease on *Coleus*.	[78]
F. oxysporum f. sp. *melonis*	*P. putida*	Control on muskmelon achieved by seed treatment of *P. putida*.	[79]
F. udum	*Bacillus* and fluorescent Pseudomonads isolates	Four isolates, namely Pa116, P324, B18, and B160, have shown antifungal activity.	[24]
F. udum	Fluorescent Pseudomonads	Four isolates have shown antifungal activity; isolate Pf605 reduced the wilt disease index of pigeon pea under greenhouse conditions.	[25]
F. oxysporum f. sp. *lycopersici*	Fluorescent Pseudomonads	Significantly reduced disease severity on tomato.	[80]
F. oxysporum f. sp. *lycopersici*	*P. fluorescens*	Bacteria in culture medium inhibited fungal growth by about 50%.	[81]
F. oxysporum	*B. subtilis* and *P. fluorescens*	Both bacteria reduced the incidence of *Fusarium* wilt in tomato significantly.	[82]
***Rhizoctonia* spp.**			
R. solani	*P. cepacia* R55, R85 *P. putida* R104	Increase of 62%–78% of dry weight of winter wheat grown in *R. solani*-infected soil.	[83]
R. solani	*P. fluorescens*	Mixture of 3 strains reduced disease and promoted growth of rice.	[84]
R. solani	*Bacillus subtilis* *Burkholderia cepacia*	Combination of *B. subtilis* RB14-C with *B. cepacia* BY can lead to greater damping-off suppression than by these strains separately.	[85]
R. solani	*P. fluorescens* A6RI	Inoculation of A6RI increased growth of pathogen-inoculated plants.	[86]

(Continued)

Table 5.2 (Continued)

Fungus	*Pseudomonas* sp.	Effect	Reference
R. solani	Fluorescent pseudomolnads	Out of 103 isolates, only 52 showed antifungal activity against *R. solani in vitro*.	[87]
Macrophomina phaseolina			
M. phaseolina	*P. fluorescens* 4-92	*P. fluorescens* increased disease resistance by 33% in chickpea.	[88]
M. phaseolina	Fluorescent *Pseudomonas* GRC_2	GRC_2 strain reduced charcoal rot disease of peanut in *M. phaseolina*-infested soil.	[89]
M. phaseolina	*P. fluorescens*	Seed treatment with *P. fluorescens* and neem cake as soil application reduced root rot indices on green gram.	[90]
M. phaseolina	*P. alcaligenes* *Bacillus pumilus*	*P. alcaligenes* caused a greater reduction against root-rot than did *B. pumilus* on chickpea.	[29]
M. *phaseolina*	*P. aeruginosa*	*P. aeruginosa* PGPR2 has several potential applications as fungicidal agents in agriculture.	[91]
M. phaseolina	*P. fluorescens*	Stem cutting and soil application of a talc-based formulation of Pf1 significantly reduced root-rot incidence.	[92]
M. phaseolina *R. solani*	*P. aeruginosa*	Strains PGPR-3 and PGPR-13 significantly controlled the infection of mungbean	[93]
Other fungi			
Sclerotium rolfsii, *Fusarium* sp.	*P. putida* *P. fluorescens* *P. alcaligenes*	Reduced incidence of disease caused by *S. rolfsii* in bean, and fusarium wilt of cotton and tomato.	[94]
Colletotrichum orbiculare	*P. putida*, *S. marcescens*, *Flavomonas oryzihabitans*, *B. pumilus*	PGPR-mediated ISR was operative under field conditions against naturally occurring anthracnose of cucumber.	[95]
Verticillium dahliae	*Pseudomonas PsJN*	Reduced disease incidence in tomato.	[96]
Colletotrichum capsici	*Pseudomonas fluorescens*	Increased accumulation of enzymes involved in phenyl propanoid pathway and PR-proteins in hot pepper.	[97]
Botrytis cineria	*Pseudomonas* PsJn	Ps Jn inhibits growth of *B. cineria* by disrupting cellular membranes and causing cell death.	[98]

Table 5.2 (Continued)

Fungus	*Pseudomonas* sp.	Effect	Reference
Colletotrichum lindemuthianum	*P. aeruginosa*, *P. fluorescens*	*P. aeruginosa* induced resistance only in resistant interactions while *P. fluorescens* induced resistance in susceptible and moderately resistant interactions on bean.	[99]
Cnaphalocrocis medinalis	*P. fluorescens* strains Pf1, FP7	Mixture of two strains performed better than the individual strains in reducing sheath blight of rice.	[100]
Colletotrichum falactum	*P. fluorescens*	Induced systemic resistance against red rot of sugarcane.	[101]
Phytophthora infestans	*P. fluorescens* 89B61 *B. pumilus* SE34	Elicited systemic protection against late blight of tomato and reduced disease severity.	[102]
Sclerospora graminicola	*P. fluorescens*	Isolates offered protection ranging from 20% to 75% against downy mildew to pearl millet.	[103]
Colletotrichum gloeosporioides	*P. fluorescens* FP7	Suppressed anthracnose pathogen on mango leading to improved yield attributes.	[104]
Alternaria triticina	*P. fluorescens* *A. chroococcum*	*P. fluorescens* caused greater reduction in *A. triticina*-infected leaf area than *A. chroococcum*.	[105]
Alternaria triticina	*Bacillus* and Fluorescent Pseudomonads	B28 was found best in improving plant growth and also caused reduction in percent leaf infected area of wheat.	[106]
Exobasidium vexans	*Pseudomonas* and *Bacillus*	Seed treatment with PGPR strains reduced disease severity on tea under field conditions.	[107]
Sclerotina sclerotiorum	*P. aeruginosa*	*Pseudomonas* strains effectively reduced growth of *S. sclerotiorum in vitro*.	[108]
Verticillium dahliae	*P. fluorescens* PICF7	Root colonization by this bacterium offers a wide array of systemic defense responses in distant tissues (stems, leaves).	[109]
Athelia rolfsii	*P. fluorescens* strain ALEB 7B	Pre-inoculating *Atractylodes lancea* seedlings significantly reduces southern blight morbidity rate.	[110]

Table 5.3 Effects of *Pseudomonas* spp. on Bacterial Diseases of Plants

Pathogenic bacteria	*Pseudomonas* sp.	Effect	Reference
Xanthomonas compestris pv. *citri*	*P. fluorescens*	Control of citrus canker by siderophore production.	[111]
Erwinia carotovora	*P. putida* W4P63	Increased yields of Rosset Burbank potato and suppressed soft rot potential of tubers.	[112]
E. amylovora	*P. fluorescens* A506	Reduction in population size of *E. amylovora* in pear flowers.	[113]
P. syringae pv. *tomato*	*P. fluorescens* WCS417	*P. fluorescens* protected radish via induction of systemic resistance.	[70]
P. solanacearum	*P. fluorescens* M29 and M40	Isolate M40 significantly reduced tomato wilt.	[114]
P. syringae pv. *lachrymans*	*P. putida*, *S. marcescens*, *Flavomonas oryzihabitans*, *B. pumilus*	PGPR strains caused significant protection against pathogens on cucumber.	[95]
E. amylovora	*P. fluorescens* A506	Strain A506 and antibiotics acted additively in the control of frost and fire blight disease.	[115]
Ralstonia solanacearum	Fluorescent pseudomonads	All 3 strains suppressed wilt of tomato and increased yields.	[116]
Xanthomonas oryzae pv. *oryzae*	*P. fluorescens*	Imparted resistance to the rice bacterial blight pathogen.	[117]
Ralstonia solanacearum	*Serratia* J2, *Pseudomonas*, *Bacillus* BB11	All 3 strains suppressed wilt of tomato and increased yields.	[118]
Ralstonia solanacearum	*Bacillis* and *Pseudomonas* isolates	Out of 120 isolates, 6 showed antagonistic activity against bacterial wilt on potato *in vitro*.	[119]
R. solanacearum	*P. fluorescens*	Bacterial wilt of potato was significantly reduced by 59.83%.	[120]
R. solanacearum	*P. fluorescens*	Soil disinfection with lime one month before transplantation and the use of *P. fluorescens* were effective in minimizing wilt incidence in tomato.	[121]
R. solanacearum	*Pseudomonas* *Enterobacter* isolates	Plants treated with *Pseudomonas*, *Enterobacter* and *Bacillus* isolates reduced wilt of eggplant by more than 70%.	[122]
R. solanacearum	*P. fluorescens*	Disease incidence was significantly reduced in plants raised from *P. fluorescens*-treated seeds.	[123]

Table 5.3 (Continued)

Pathogenic bacteria	*Pseudomonas* sp.	Effect	Reference
R. solanacearum	*P. putida*	Bacterial suspension of *P. putida* (Pf-20) was able to suppress disease and prolong the incubation period.	[124]
Xanthomonas translucens pv. *pistaciae*	*B. subtilis*, *P. fluorescens*	Both bacteria inhibited dieback disease of pistachio.	[125]
R. solanacearum	*P. fluorescens*	Suppressed 83.3% of bacterial wilt of brinjal in field experiments.	[126]
Ralstonia solanacearum	*P. fluorescens*, *P. putida*, *Bacillus subtilis*, *Enterobacter aerogenes*	*P. fluorescens* exhibited the highest disease reduction of tomato bacterial wilt disease followed by *P. putida*, *B. subtilis* and *E. aerogenes*.	[127]
Clavibacter michiganensis 1–07	*P. chlororaphis* Strain UFB2	Strain UFB2 was highly efficient in inhibiting bacterial canker of tomato.	[128]

5.2 Biofilms

The term *biofilm* was coined and described by Costerton et al. [129]. Clusters of varied microbial populations occur in almost all moist environments where nutrient flow is available and surface attachment is possible [130]. Mature biofilms can be described as complex three-dimensional structures where cells are embedded in a thick matrix of extracellular polymeric substances (EPS). These structures are crossed by fluid-filled channels that enable nutrient transport to interior regions of the biofilm and elimination of waste products [131, 132].

The processes involved in biofilm development are relevant to both bacterial survival and host plant colonization [133]. Bacteria gain advantages from congregating in biofilms, including protection from predation, desiccation, and exposure to antibacterial substances, and improved acquisition of nutrients released in the plant environment. Biofilms provide survival sites for both beneficial and opportunistic pathogenic bacteria by providing protection and increasing the potential of bacterial survival and evolution in the plant environment. Biofilms have been shown to enhance (i) the fitness of individual bacteria and (ii) more general plant health and productivity as a result of the cumulative selective advantage of the individual bacteria [133].

Bacteria in natural environments persist by forming biofilms [134]. Effective colonization of plant roots by biocontrol agents plays an important role in growth promotion by production of metabolites, effects on nutrient uptake, and induced plant resistance and antibiotics production. Biofilm formation is a dynamic process, and extracellular polymeric substances play an important role in the attachment and colonization of microorganisms to contact surfaces. Biofilms are highly structured, surface-attached communities of cells encased within a self-produced extracellular polymeric substance matrix [135, 136]. Bacterial biofilms established on plant roots could protect and colonization sites and act as a sink for nutrients in the rhizosphere, thus reducing the availability of root exudates and nutritional elements for pathogen stimulation or subsequent colonization on the root [137].

Biofilms confer resistance to many antimicrobials, protection from protozoan grazing, and also protect against host defenses [138–140]. The increased resistance to environmental stresses observed in biofilm cells appears to be a function of an increase in the proportion of persister cells within the biofilm [141]. Persister cells are resistant to many antibiotics and are nondividing, despite being genetically identical to the population as a whole. Persister cells have been proposed to be protected from the action of antibiotics because they express toxin–antitoxin systems where the target of the antibiotics is blocked by the toxin modules [141]. In addition to an increase in persisters, the presence of an extracellular matrix protects constituents cells from external aggressions. Extracellular matrices also act as a diffusion barrier to small molecules [140, 142]. The diffusion of nutrients, vitamins, or cofactors in biofilms is slower, resulting in a bacterial community in which some of cells are metabolically inactive. The rate of bacterial growth is influenced by the fact that cells within a biofilm are confined to a limited space [143]. Hence, biofilm formation represents, in a way, the natural stationary phase of bacterial growth. During the stationary phase, bacteria profoundly change their physiology by increasing production of secondary metabolites such as antibiotics, pigments, and other small molecules [144]. These secondary metabolites also function as signaling molecules to initiate the process of biofilm formation or to inhibit biofilm formation by other organisms that occur in the same habitat [145].

Bacteria are protected from the inhibitory effects of antimicrobial compounds, biocides, chemical stresses (such as pH and oxygen), and physical stresses (like pressure, heat, and freezing). The polymeric matrix increases the binding of water and therefore decreases the possibility of dehydration of the bacterial cells—a stress that planktonic cells are subject to. The proximity of the microorganisms in biofilms allows nutrients, metabolites, and genetic material to be readily exchanged [134, 146, 147]. Cell division is uncommon in a mature biofilm, and energy is used to produce exopolysaccharides, which the biofilm cells can use as nutrients [148]. Jefferson [149] stated that biofilms are the default mode of growth for some bacterial species, whereas planktonic growth is an *in vitro* artifact.

Pseudomonads are well known for their plant growth–promoting potential. The traits involved in bacterial rhizosphere competence that might play an important role include the ability to form biofilms and cell-to-cell communication systems (quorum sensing) that enable bacteria to coordinate the expression of special phenotypes in a cell density-dependent manner. Many root-colonizing bacteria, including *P. putida*, employ QS systems that rely on *N*-acyl homoserine lactone (AHL) signal molecules to express certain phenotypic traits in a cell density–dependent manner [150–152]. Steidle et al. [153] demonstrated that the tomato rhizosphere isolate *P. putida* IsoF produces a wide spectrum of AHLs, including 3-oxo-dodecanoyl-homoserine lactone (3-oxo-C12-HSL) and 3-oxo-decanoyl-homoserine lactone (3-oxo-C10-HSL) and, as minor products, 3-oxo-octanoyl-homoserine lactone (3-oxo-C8-HSL) and 3-oxo-hexanoyl-homoserine lactone (3-oxo-C6-HSL). The QS system of this strain consists of PpuI, which directs the synthesis of AHL signal molecules, and PpuR, which binds to the AHLs and regulates the *ppuI* expression in a positive feedback loop. The wild-type *P. putida* IsoF formed homogenous unstructured biofilms that uniformly covered the surface, and a *ppuI* mutant-formed structured biofilms with characteristic microcolonies and water-filled channels. When the medium was supplemented with AHL signal molecules, the mutant biofilm lost its structure and converted to an unstructured biofilm similar to that formed by the wild-type IsoF. These results suggest that the QS system influences biofilm structural development [154].

Comparison of the protein or transcription profiles of *Pseudomonas* spp. and *Escherichia coli* grown planktonically with those of their sessile counterparts suggest that a large number of genes could be differentially regulated during biofilm formation [155–160]. Sauer and Camper [158] detected 45 differentially expressed proteins in biofilms of *P. putida* ATCC 39168 following six hours of attachment, indicating that this strain undergoes a global change in gene expression after surface adherence. Furthermore, the expression rate of 16 proteins was changed when planktonic cells were grown in medium supplemented with 3-oxo-C12-HSL. Only one protein, the periplasmatic putrescine binding protein PotF, was found to be down regulated in biofilm cells as well as in the presence of AHL signal molecules. It was suggested that QS does not play an important role in the initial attachment process.

Sauer and Camper [158] studied global changes in the expression profile of intracellular proteins during the initial stage of biofilm formation of an AHL-negative *P. putida* strain. The impact of QS and surface growth on gene expression was analyzed in the rhizosphere isolate *P. putida* IsoF, which possesses an AHL-dependent cell-to-cell-communication system. Analyses focused on mature biofilms as the QS regulatory system is induced at high cell densities. It is likely that proteins on the surface of the bacterial cell play a particularly important role in surface colonization. The protein profiles of the *P. putida* wild-type IsoF and the *ppuI* mutant F117, grown either as biofilm or in suspension, were compared by two-dimensional gel electrophoresis (2-DE). QS-controlled genes were identified by comparative proteome analyses of planktonically and surface-adhered cultures of the IsoF wild type and the *ppuI* mutant F117. Differentially expressed spots were identified by matrix-assisted time of flight mass spectrometry (MALDI-TOF MS) plus a database search in the recently completed genome sequence of *P. putida* KT2440 [161].

5.3 Mechanisms of Biofilm Formation

Bacterial biofilm formation is a highly regulated process. Each bacterial species responds to its environmental conditions via a distinct set of molecular mechanisms.

5.3.1 Quorum Sensing

Biofilms are often the site for quorum sensing, and this process can influence their formation [162]. Plant-associated bacteria frequently employ this signaling mechanism to modulate and coordinate their interactions with plants, including control of tissue maceration, antibiotic production, toxin release, and horizontal gene transfer (HGT) [163]. Several different signals function in plant-associated bacteria, including (i) acylated homoserine lactones (AHLs) among proteobacteria; (ii) gamma-butyrolactones in *Streptomyces* species; (iii) *cis*-11-methyl-2-dodecanoic acid (also called DSF) in species of *Xanthomonas, Xylella,* and their relatives; and (iv) oligopeptides among Gram-positive microbes. It is virtually certain that there are additional diffusible signals that promote or control plant colonization.

Greater numbers of AHL-producing bacteria live in association with plants as compared to bulk soil populations [151]. *P. fluorescens* 2P24 requires AHLs for biofilm formation and controls take-all disease on wheat [164]. Microscopic analysis and geostatistical modeling of AHL signaling by *P. putida* on tomato and wheat roots revealed the requisite cell density to be very low and the calling distance between cells to be as great as 78 μm along the root [165]. Several soil and rhizosphere bacteria degrade AHLs, and measurement of the total degradative activity in soils suggests that this can significantly affect AHL signaling [166]. AHL signaling has the potential to reach over surprisingly long distances within and between biofilms, but quorum-quenching mechanisms may act to insulate communication at specific sites.

5.3.2 Regulation in Response to Phosphorus Starvation

Controlling aggregation and surface attachment in response to availability of phosphorus is an important fitness factor for growth of plant-associated bacteria [167]. In many bacteria, the PhoR-PhoB two-component system, similar to that identified in *E. coli*, activates the response to phosphorus limitation. The impact of limiting phosphorus on biofilms depends on the organism and its role in the ecosystem. *Pseudomonas aureofaciens* PA147-2 exhibits a *phoB*-dependent inhibition of biofilm formation under phosphate starvation [168].

5.3.3 Phase Variation

Phase variation allows for drastic phenotypic changes via a small reversible change of the genotype. *Pseudomonas brassicacearum* NFM421 use a flagellin-over producing, hypermotile phase variant to colonize *Arabidopsis thaliana* root tips [169]. The low probability of relatively rare genetic events such as phase variation is offset by the large populations harbored within biofilms.

5.3.4 Motility and Chemotaxis

Plant exudates are released from specific sites, and many bacteria preferentially chemotax toward and colonize positions on the plant surface with optimal nutrient availability [167]. Bacteria move by different mechanisms, including flagellar swimming, swarming, twitching, and gliding motility. Flagellar motility provides access to attachment sites on the plant, and possibly drives initial attachment, but is also involved in biofilm maturation. Twitching motility conferred by retractable-type IV pili is used by *Pseudomonas* spp. to spread across the surface and form mushroom structures in biofilms [170].

The rate of chemotaxis by different bacteria toward different nutrients is highly dependent on the species and strain. Bacterial strains show a stronger reaction to their native host plants, and endophytes often respond more strongly than do rhizobacteria [171]. Motility is involved not only in the initiation and development of biofilms but also in their dispersal—it is an essential mechanism for spreading and colonizing new habitats [157].

5.3.5 Surface Adhesins

Bacterial surface-associated structures, including polysaccharides and proteinaceous pili or fimbriae, are often involved in attachment to inert surfaces and plants and other bacterial cells [167]. The interplay of polysaccharides and proteins that specifically bind to them is a common theme in surface attachment and aggregation of several plant-associated microbes. Large adhesion protein (LapA) is an important adhesin in certain *Pseudomonas* species. In *P. putida* it is involved in attachment to abiotic surfaces as well as plant seeds [172, 173]. A smaller surface protein, the outer membrane porin OmpA, is required for binding of *E. coli* O157:H7 to alfalfa sprouts and seed coats. Other adhesin and fimbrial genes of O157:H7 were sufficient for conferring plant attachment competence to *E. coli* K12 but could be mutated without effect in O157:H7, suggesting redundancy among adhesion systems [174].

5.3.6 Biofilm Matrix Components

The biofilm matrix connects the cells and imparts many key features to the biofilm, including protecting the cells from desiccation and other stresses [167]. The matrix usually consists of extracellular polysaccharides, but it can also include proteins and even DNA [135]. Bacteria can vary their matrix composition and use different adhesins in changing environmental conditions. Different strains of the same species may employ different matrix components [167].

Production of copious extracellular polysaccharides can increase biofilm formation and promote colonization of plants. This is exemplified by a regulatory mutant of *P. fluorescens* CHA0 that is hypermucoid and shows dramatically increased attachment to roots as well as to the mycelium of arbuscular mycorrhizal fungi [175]. Such mutants of plant growth–promoting bacterial species may have commercial relevance as agricultural inoculants. Abundance of a polysaccharide, however, does not necessarily correlate with its role in biofilm formation. In *P. aeruginosa*, for example, the exopolysaccharide alginate can influence biofilm architecture when overproduced, but it is not required and could not be detected in the matrix of wild-type biofilms [164].

A survey of plant-associated *Pseudomonas* isolates found that the majority formed pellicles with a wide variety of structures, and one-fifth produced a β-linked polysaccharide presumed to be cellulose [176]. Acetylated cellulose is an essential matrix component of pellicles formed by the wrinkly spreader mutant of *P. fluorescens* strain SBW25 [177]. Compared with the wild type, the mutant overproduces cellulose, which is thought to interact with membrane lipopolysaccharides in pellicle formation. Introduction of the wrinkly spreader mutation elevated cellulose production, surface attachment, and pellicle formation in several *Pseudomonas* strains [176].

The effect of desiccation on *P. putida* biofilms grown in air has been studied. Growth of *P. putida* under dehydrating conditions resulted in taller and more porous biofilms with a thick polysaccharide layer at the air interface [178]. Identification of genes activated in response to desiccation included those for alginate synthesis [179].

5.4 Metabolites Affecting Biofilm Formation

5.4.1 Plant Defense Compounds

Plants produce a number of compounds in response to microbial colonizers. The systemic plant defense signaling molecule salicylic acid acts to directly suppress virulence genes in *P. aeruginosa*, inhibiting attachment and biofilm formation on *Arabidopsis thaliana* roots [180]. On the other hand, rosmarinic acid is effective against free-swimming *P. aeruginosa* but far less effective on biofilms [181]. These studies demonstrate the complex interplay between plant defense, biofilms, and infection.

5.4.2 Phenazine

Phenazine is produced by *P. chlororaphis* strain 30–84, which is regulated by the PhzR-PhzI quorum-sensing system. Both *phzR* and *phzI* regulatory mutants and *phzB* phenazine biosynthetic mutants formed significantly reduced and immature biofilms *in vitro* but were rescued by heterologous *phzB* expression for exogenous phenazine. The *phzR/I* mutants were also deficient in colonization of wheat seeds and roots [182]. Phenazine apparently also contributes significantly to biofilm maturation *in vitro* and on plants.

5.4.3 Surfactants

Rhamnolipids shape *P. aeruginosa* biofilm architecture by maintaining open fluid channels between cell clusters [183]. The role of surfactants in biofilm formation in other species has also been examined. Cyclic lipopeptides (CLPs) are another class of surfactants produced by several pseudomonads [184]. CLPs can increase virulence of plant pathogens and are of great importance for biocontrol species because of their antimicrobial properties and their role in motility and biofilm formation. *P. putida* strain PCL1445 forms two CLP surfactants, both of which inhibit biofilm formation of several *Pseudomonas* species and can even dissolve existing biofilms. The CLPs appear to have a similar function in PCL1445, because CLP-less mutant shows increased for aggregation and forms thicker biofilms [185].

5.5 Biofilm Formation and Biological Control of Plant Diseases

Bacteria develop biofilms on a range of surfaces [146, 186]. Biofilm formation is commonly considered to occur in four main stages:

1) Bacterial attachment to a surface
2) Microcolony formation
3) Biofilm maturation
4) Detachment (also termed dispersal) of bacteria, which may then colonize new areas [*187*]

The dramatic phenotypical changes that occur during these processes must involve a coordinated series of molecular events. *P. fluorescens* forms a biofilm under all conditions tested [188]. A genetic analysis of biofilm formation by this organism revealed that it utilizes multiple genetic pathways to initiate biofilm development. For example, mutants unable to form a biofilm when grown on glucose overcame this defect by growth on citrate, suggesting an alternative citrate-dependent pathway for biofilm formation [188]. The ability to organize structurally and to distribute metabolic activities between the different bacteria demands, consequently, a high degree of coordinated cell–cell interaction. For example, N-Acyl homoserine lactone (AHL)-mediated cell-to-cell signalling is an important regulatory factor that directly controls development of biofilms [189, 190] in various Gram-negative bacteria, including *Burkholderia cepacia* [191, 192] and *Pseudomonas aeruginosa* [193–196].

Beneficial rhizobacteria are associated with the surfaces of plant roots and may increase plant yield by mechanisms that impart improved mineral nutrient uptake, disease suppression, or phytohormone production. Beneficial rhizobacteria can promote plant growth, protect against pathogen attack, and play a role in the degradation of organic polymers in the soil [197]. Bacteria form adhering biofilms on inert surfaces under the control of a variety of transcription factors [198, 199]. Kinsinger et al. [200] and Bais et al. [201] demonstrated the biocontrol ability of a wild-type bacterial strain against plant pathogens. Arabidopsis root surfaces treated with bacteria were analyzed with confocal scanning laser microscopy to reveal a three-dimensional biofilm. Root-associated pseudomonads promote the growth of host plants or are used as biocontrol agents [202]. *Pseudomonas putida* can respond rapidly to the presence of root exudates in soils, converging at root colonization sites and establishing stable biofilms networks [203]. Haggag and Timmusk [204] investigated the role of biofilm-forming *Paenibacillus polymyxa* strains in controlling *Aspergillus niger* and highlighted the importance of biofilms in biocontrol initiation. Both strains were able to suppress the pathogen, but the superior biofilm former offers significantly better protection against *A. niger* crown rot. Successful colonization with the biocontrol agent is necessary to make biocontrol effective and reproducible.

Microorganisms often inhabit the plant leaf surface in biofilms. *Burkholderia* sp. FP62 is a biocontrol agent of *B. cinerea* in geraniums and forms extensive biofilms in the phyllosphere [205]. Scanning electron micrographs demonstrate extensive phyllosphere colonization (60 to 70 percent of the leaf surface). FP62 biofilms appeared to be many cell layers thick and enveloped in a polymer-like matrix. The biofilm phenotype of this strain is related to biocontrol. Isolation of transposon mutants that are deficient in biofilm formation in an *in vitro* biofilm assay also lacked the capacity to control *B. cinerea* when applied to geranium leaves [205]. The biofilm mutants are less efficient in phyllosphere colonization, lacking many of the characteristics of wild-type biofilms. The biofilm mutation and biocontrol capacity could be restored through the addition of exogenous polymers to the biocontrol formulation of the mutants. The addition of polymers to the formulation of other biocontrol agents also improved their biocontrol capacity, suggesting that biofilms contribute to biocontrol efficacy and are an important aspect of phyllosphere competence.

5.6 Conclusion

The formation of bacterial biofilms, or surface-attached communities, is important in agriculture. Most bacteria in natural environments are found predominately attached to soil particles or plant roots. Association of bacteria such as *Pseudomonas* sp. with plant roots can protect plants from pathogens present in soil. Bacteria living in the rhizosphere can occupy three niches: as a biofilm on soil colloids; as a biofilm on plant surfaces; or as planktonic cells. The ability to interact with plant and soil surfaces, and to transition between a planktonic and a sessile lifestyle, affects protection of plants and survival of the bacteria.

The role of biofilm formation in bacterial colonization of plant roots, a process thought to contribute to biological control, is not well understood. A better understanding of the role of genes for biofilm formation in biological control, root colonization, attachment to soil, and bacterial survival are needed. Further elucidation of the mechanisms involved in the various stages of biofilm formation will improve our understanding of microbial adaptations to this mode of life in a wide range of environments and the interactions of bacteria, host plant, and pathogen. Multidisciplinary studies using new approaches will clarify the ways in which bacteria travel and interact in a variety of surface microenvironments during biofilm development. Such knowledge will enhance our understanding of biofilm formation on plant surfaces and their proper use in the biocontrol of plant pathogens.

References

1 A. Peix, M.H. Ramírez-Bahena and E. Velázquez, Historical evolution and current status of the taxonomy of genus *Pseudomonas*, *Infect Genet Evol.*, 9, 1132–1147 (2009).
2 B.J. Tindall, P. Kampfer, J.P. Euze′ by and A. Oren, Valid publication of names of prokaryotes according to the rules of nomenclature: Past history and current practice, *Int. J. Syst. Evol. Microbiol.*, 56, 2715–2720 (2006).
3 J. He, R.L. Baldini, E. Deziel, M. Saucier and Q.H. Zhang, et al., The broad host range pathogen *Pseudomonas aeruginosa* strain PA14 carries two pathogenicity islands harboring plant and animal virulence genes, *Proc. Natl. Acad. Sci. USA*, 101, 2530–2535 (2004).
4 N. Høiby, T. Bjarnsholt, M. Givskov, S. Molin, and O. Ciofu, Antibiotic resistance of bacterial biofilms, *Int. J. Antimicrob. Agents.*, 35, 322–332 (2010).
5 A.J. Spiers, A. Buckling and P. B. Rainey, *The causes of Pseudomonas diversity*, 146, 2345–2350 (2000).
6 S.E. Battle, J. Rello, and A.R. Hauser, Genomic islands of *Pseudomonas aeruginosa*, *FEMS Microbiol.Lett.*, 290, 70–78 (2009).
7 M. Mulet, J. Lalucat, and E. García-Valdés, DNA sequence-based analysis of the *Pseudomonas* species, *Environ. Microbiol.*, 12, 1513–1530 (2010).
8 X. H. J.J. X. Liu, and Z.R. Pyruvate kinase M2 regulates gene transcription by acting as a protein kinase, *Mol. Cell.*, 45(5), 598–609 (2012).
9 L.C. Van Loon, P.A.H.M. Bakker and C.M.J. Pieterse, Systemic resistance induced by rhizosphere bacteria, *Annu. Rev. Phytopathol.*, 36, 453–483(1998).
10 C.M.J. Pieterse, and L.C. van Loon, Salicylic acid-independent plant defence pathways, *Trends Plant Sci.*, 4, 52–58 (1999).
11 Z.A. Siddiqui, PGPR: Prospective biocontrol agents of plant pathogens. In *PGPR: Biocontrol and Biofertilization*, Z.A. Siddiqui (eds.) pp 111–142, The Netherlands: Springer (2006), 318 pp.

12 M.S. Akhtar and Z.A. Siddiqui, Role of plant growth promoting rhizobacteria in biocontrol of plant diseases and sustainable agriculture, pp. 157–195. In *Microbiology Monograph 18: Plant growth and health promoting bacteria* (ed.) Maheshwari, D.K. Springer-Verlag Berlin Heidelberg. ISBN9783642136115 (2010).
13 J.O. Becker, E. Zavaleta-Mejia, S.F. Colbert, M.N. Schroth, A.R. Weinhold, J.G. Hancock and S.D. VanGundy, Effects of rhizobacteria on root-knot nematodes and gall formation, *Phytopathology*, 78, 1466–1469 (1988).
14 M.R. Khan and M. Akram, Effects of certain antagonistic fungi and rhizobacteria on wilt disease complex of tomato caused by *Meloidogyne incognita and Fusarium oxysporum f.* sp. *Lycopersici, Nematol. Medit.*, 28, 139–144 (2000).
15 Z.A. Siddiqui, A. Iqbal and I. Mahmood, Effects of *Pseudomonas fluorescens* and fertilizers on the reproduction of *Meloidogyne incognita* and growth of tomato, *Appl. Soil Ecol.*, 16, 179–185 (2001).
16 Z.A. Siddiqui and I. Mahmood, Effects of plant straws and plant growth promoting bacteria on the reproduction of *Meloidogyne incognita* and growth of tomato, *Boil. Agri. Hort.*, 21, 53–62 (2003).
17 Z.A. Siddiqui, Effects of plant growth promoting bacteria and composted organic fertilizers on the reproduction of *Meloidogyne incognita* and tomato growth, *Bioresource Technol.*, 95, 223–227 (2004).
18 S.L. Devi and U. Dutta, Effect of *Pseudomonas fluorescens* on root-knot nematode (*Meloidogyne incognita*) of okra plant, *Indian J. Nematol.*, 32, 215–216 (2002).
19 Z.A. Siddiqui, S. Khan and I. Mahmood, Use of Rhizobacteria for the management of *Meloidogyne incognita* on *Solanum melongena, Thai. J. Agri. Sci.*, 35, 1–8 (2002).
20 I.A. Siddiqui and S.E. Haque, Suppression of the root rot-root knot disease complex by *Pseudomonas aeruginosa* in tomato: The influence of inoculums density, nematode populations, moisture and other plant associated bacteria, *Plant Soil*, 237, 81–89 (2001).
21 Z.A. Siddiqui and M.S. Akhtar, Biocontrol of a chickpea root-rot disease complex with phosphate solubilizing microorganisms, *J. Plant Pathol.*, 89, 31–42 (2007).
22 Z.A. Siddiqui and L.P. Singh, Effects of fly ash, *Pseudomonas striata* and *Rhizobium* sp. on the reproduction of nematode *Meloidogyne incognita* and on the growth and transpiration of pea, *J. Environ. Biol.*, 26, 117–122 (2005).
23 M.S. Akhtar and Z.A. Siddiqui, Biocontrol of a root-rot disease complex of chickpea by *Glomus intraradices, Rhizobium* sp. *and Pseudomonas straita, Crop Protec.*, 23, 410–417 (2008).
24 S. Siddiqui, Z.A. Siddiqui and A. Iqbal, Evaluation of fluorescent pseudomonads and *Bacillus* isolates for the biocontrol of wilt disease complex of pigeonpea, *World J. Microbiol. Biotechnol.*, 21, 729–732 (2005).
25 Z.A. Siddiqui and U. Shakeel, Use of fluorescent pseudomonads isolates for the biocontrol of wilt disease complex in pigeonpea in green house and under pot condition, *Plant Pathol. J.*, 5, 99–105 (2006).
26 Z.A. Siddiqui, B. Sharma and S. Siddiqui, Evaluation of *Bacillus* and *Pseudomonas* isolates for the biocontrol of *Meloidogyne incognita* on tomato, *Acta Phytopath. et Entom. Hung.*, 42, 25–34 (2007).
27 Z.A. Siddiqui, U. Shakeel and S. Siddiqui, Biocontrol of wilt disease complex of pigeonpea by fluorescent pseudomonads and *Bacillus* spp. under pot and field conditions. *Acta Phytopath. et Entom. Hung.*, 43, 79–94 (2008).

28 M.S. Akhtar and Z.A. Siddiqui, Biocontrol of a chickpea root-rot disease complex with *Glomus intraradices, Pseudomonas putida and Paenibacillus polymyxa. Australa. Plant Pathol.*, 36, 175–180 (2007).
29 M.S. Akhtar and Z.A. Siddiqui, *Glomus intraradices, Pseudomonas alcaligenes, Bacillus pumilus* as effective biocontrol agents for the root-rot disease complex of chickpea (*Cicer arietinum* L.), *J. General Plant Pathol.*, 74, 53–60 (2008).
30 Z.A. Siddiqui and M.S. Akhtar, Effects of organic wastes, *Glomus intraradices and Pseudomonas putida on the growth of tomato and on the reproduction of root-knot nematode Meloidogyne incognita. Phytoparasitica*, 36, 460–471 (2008).
31 Z.A. Siddiqui and K. Futai, Biocontrol of *Meloidogyne incognita* using antagonistic fungi, plant growth-promoting rhizobacteria and cattle manure on tomato, *Pest Manag. Sci.*, 65, 943–948 (2009).
32 M.S. Akhtar and Z.A. Siddiqui, Use of plant growth promoting rhizobacteria for the biocontrol of root-rot disease complex of chickpea, *Australa. Plant Pathol.*, 38, 44–50 (2009).
33 Z.A. Siddiqui, A. Qureshi and M.S. Akhtar, Biocontrol of root-knot nematode *Meloidogyne incognita* by *Pseudomonas* and *Bacillus* isolates on *Pisum sativum, Arch. Phytopathol. Plant Protec.*, 42, 1154–1164 (2009).
34 A.H. Ashoub and M.T. Amara, Biocontrol activity of some bacterial genera against root-knot nematode, *Meloidogyne incognita, J. Americ. Sci.*, 6, 321–328 (2010).
35 P. Singh and Z.A. Siddiqui, Biocontrol of root-knot nematode Meloidogyne incognita by the isolates of Pseudomonas on tomato, *Arch. Phytopathol. Plant Protec.*, 43, 1423–1434 (2010).
36 Z.A. Siddiqui and N. Zehra, Biocontrol of wilt disease complex of pea using Pseudomonas fluorescens and *Rhizobium* sp. *Arch. Phytopathol. Plant Protec.*, 45, 2340–2346 (2012).
37 S.K. Saikia, S. Tiwari and R. Pandey, Rhizospheric biological weapons for growth enhancement and *Meloidogyne incognita* management in *Withania somnifera cv. Poshita, Biol. Control*, 65, 225–234 (2013).
38 Z.A. Siddiqui and I. Mahmood, Effect of a plant growth promoting bacterium, an AM fungus and soil types on the morphometrics and reproduction of *Meloidogyne javanica* on tomato, *Appl. Soil Ecol.*, 8, 77–84 (1998).
39 I.A. Siddiqui, S.A. Qureshi, V. Sultana, S.E. Haque and A. Ghaffar, Biological control of root rot-root knot disease complex in tomato, *Plant Soil*, 227, 163–169 (2000).
40 Z.A. Siddiqui and I. Mahmood, Effects of rhizobacteria and root symbionts on the reproduction of *Meloidogyne javanica* and growth of chickpea, *Bioresource Technol.*, 79, 41–45 (2001).
41 I.A. Siddiqui and S. S. Shaukat, Rhizobacteria-mediated induction of systemic resistance (ISR) in tomato against *Meloidogyne javanica, J. Phytopathol.*, 150, 469–473 (2002).
42 M. Hamid, I.A. Siddiqui and S.S. Siddiqui, Improvement of *Pseudomonas fluorescens* CHA0 biocontrol activity against root-knot nematode by the addition of ammonium molybdate, *Lett. Appl. Microbiol.*, 36, 239–244 (2003).
43 Z.A. Siddiqui, G. Baghel and M.S. Akhtar, Biocontrol of *Meloidogyne javanica* by *Rhizobium* and plant growth-promoting rhizobacteria on lentil, *World J. Microbiol. Biotechnol.*, 23, 435–441 (2007).
44 A.S. Rodriguez-Romero, E. Badosa, E. Montesinos and M.C. Growth promotion and biological control of root-knot nematodes in micropropagated banana during the

nursery stage by treatment with specific bacterial strains, *Ann. Appl. Biol.*, 152, 41–48 (2008).

45 Z.A. Siddiqui and M.S. Akhtar, Effect of plant growth promoting rhizobacteria, nematode parasitic fungi and root-nodule bacterium on root-knot nematodes *Meloidogyne javanica* and growth of chickpea, *Biocont. Sci. Technol.*, 19, 511–521 (2009).

46 J.A. Tariq, M.I. Haq, F.Y. Hafeez, S.T. Sahi and M. Khan, Potential of rhizobacteria for the biocontrol of *Meloidogyne javanica, Pak. J. Agri.Sci. Agri. Engg. Vet. Sci.*, 27, 66–72 (2011).

47 N. Singh and Z.A. Siddiqui, Effects of *Bacillus subtilis, Pseudomonas fluorescens* and *Aspergillus awamori* on the wilt-leaf spot disease complex of tomato, *Phytoparasitica*, 43, 61–75 (2015).

48 N. Gokte and G. Swarup, On the potential of some bacterial biocides against root-knot and cyst nematodes, *Indian J. Nematol.*, 18, 152–153 (1988).

49 Z.A. Siddiqui, I. Mahmood and S. Hayat, Biocontrol of *Heterodera cajani* and *Fusarium udum* on pigeonpea using *Glomus mosseae, Paecilomyces lilacinus* and *Pseudomonas fluorescens, Thai J. Agri. Sci.*, 31, 310–321 (1998).

50 M. Oostendorp and R.A. Sikora, Seed treatment with antagonistic rhizobacteria for the suppression of *Heterodera schachtii* early root infection of sugarbeet, *Rev. Nematol.*, 12, 77–83 (1989).

51 M. Oostendorp and R.A. Sikora, *In-vitro* relationships between rhizosphere bacteria and *Heterodera schachtii, Rev. Nematol.*, 13, 269–274 (1989).

52 S.W. Westcott and D.A. Kluepfel, Inhibition of *Criconemella xenoplax* egg hatch by a strain of *Pseudomonas aureofaciens, J. Nematol.*, 24, 626 (1992).

53 M. Weidenborner and B. Kunz, Infuence of fermentation conditions on nematicidal activity of *Pseudomonas fluorescens. Zeitschrift fur Pfleanzenkrankheiten und pflanzenschuts.*, 100, 90–94 (1993).

54 A. Kermarrec, G. Jacqua and J. Anais, Effect of *Fusarium solani* and *Pseudomonas solanacearum* on the infestation of aubergine with the plant parasitic nematode, *Rotylenchulus reniformis, Nematologica*, 40, 152–154 (1994).

55 P.M. Aalten, D. Vitour, D. Blanvillain, S.R. Gowen and L. Sutra, Effect of rhizosphere fluorescent *Pseudomonas* strains on plant parasitic nematodes *Radopholus similis* and *Meloidogyne* spp., *Lett. Appl. Microbiol.*, 27, 357–361 (1998).

56 H.M. Aksoy and S. Mennan, Biological control of *Heterodera cruciferae* (Tylenchida: Heteroderidae) Franklin 1945 with fluorescent *Pseudomonas* spp., *J. Phytopathol.*, 152, 514–518 (2004).

57 R.J. Cook and A.D. Rovira, Role of bacteria in the biological control of *Gaeumannomyces graminis* by suppressive soils, *Soil Biol. Biochem.*, 8, 269–273 (1976).

58 J.R. De Freitas and J.J. Germida, Plant growth promoting rhizobacteria for winter wheat. *Can. J. Microbiol.*, 36, 265–272 (1990).

59 L.A. Harrison, L. Letendre, P. Kovacevich, E. Piersn and D. M. Weller, Purification of an antibiotic effective against *Gaeumannomyces graminis* var. *tritici* produced by a biocontrol agent *Pseudomonas aureofaciens, Soil Biol. Biochem.*, 25, 215–221 (1993).

60 H.R. Maarten, Y. Zhinong, E.T. Teri, D. Albert, W.T. Rovira, and L.C. Raymond, Use of strains of *Bacillus* isolates in china to suppress take all and rhizoctonia root rot and promote seedling growth of glasshouse grown wheat in Australia soils. *Soil Biol. Biochem.*, 1, 19–29 (1998).

61 D.M. Weller, and R.J. Cook, Increased growth of wheat by seed treatment with fluorescent pseudomonads, and implications of *Pythium* control, *Can. J. Plant Pathol.*, 8, 328–334 (1986).
62 A. Emma, Flourescent *Pseudomonas* isolate E11.3 a biocontrol agent for Pythium root-rot in tulips, *Eur. J. Plant Pathol.*, 96, 261–272 (1990).
63 F. Moulin, P. Lemanceau and C. Alabouvette, Suppression of Pythium root-rot of cucumber by a fluorescent *Pseudomonas* is related to reduced root colonization by *Pythium aphanidermatum*, *J. Phytopathol.*, 144, 125–129 (1996).
64 C. Chen, R.R. Belanger, N. Benhamou, and T.C. Paulitz, Role of salicyclic acid in systemic resistance induced by *Pseudomonas* spp. against *Pythium aphanidermatum* in cucumber roots, *Eur. J. Plant Pathol.*, 105, 477–486 (1999).
65 R.S. Uthede, C.A. Koch, and J.G. Menzies, Rhizobacterial growth and yield promotion of cucumber plants inoculated with *Pythium aphanidermatum*, *Can. J. Plant Pathol.*, 21, 265–271 (1999).
66 G.A. Amer, and R.S. Utkhede, Development of formulations of biological agents for management of root rots of lettuce and cucumber. *Can. J. Microbiol.*, 46, 809–816 (2000).
67 A. Goel, S. Sindhu and K. Dadarwal, Stimulation of nodulation and plant growth of chickpea (*Cicer arietinum* L.) by *Pseudomonas* spp. antagonistic to fungal pathogens. *Biol. Fert. Soils*, 36, 391–396 (2002).
68 V. Ramamoorthy, T. Raghuchander and R. Samiyappan, Enhancing resistance of tomato and hot pepper to *Pythium* diseases by seed treatment with fluorescent pseudomonads, *Eur. J. Plant Pathol.*, 108, 429–441 (2002).
69 R. Van Peer, G.J. Niemann and B. Schippers, Induced resistance and phytoalexin accumulation in biological control of Fusarium wilt of carnation by *Pseudomonas* sp. strain WCS417r. *Phytopathology*, 81, 728–733 (1991).
70 E. Hoffland, J. Hakulinen and J.A. van Pelt, Comprison of systemic resistance induced by avirulent and nonpathogenic *Pseudomonas* species, *Phytopathology*, 86, 757–762 (1996).
71 B.R. Kropp, E. Thomas, J.I. Pounder and A.J. Anderson, Increased emergence of spring wheat after inoculation with *Pseudomonas chlororaphis* isolate 2E3 under field and laboratory conditions. *Biol. Fert. Soils*, 23, 200–206 (1996).
72 M.R. Khan and S.M. Khan, Biomanagement of Fusarium wilt of tomato by the soil application of certain phosphate solubilizing microorganisms, *Int. Pest Manag.*, 47, 227–231 (2001).
73 B.S.D. Kumar, I. Berggren and A.M. Martensson, Potential for improving pea production by co-inoculation with fluorescent *Pseudomonas* and *Rhizobium*. *Plant Soil*, 229, 25–34 (2001).
74 K.K. Pal, K.V.B.R. Tilak, A.K. Saxena, R. Dey and C.S. Singh, Suppression of maize root diseases caused by *Macrophomina phaseolina*, *Fusarium moniliformae* and *Fusarium graminearum* by growth promoting rhizobacteria. *Microbiol. Res.*, 56, 209–223 (2001).
75 N.R. Kumar, V.T. Arasu and P. Gunasekaran, Genotypeing of antifungal compounds producing plant growth-promoting rhizobacteria, *Pseudomonas fluorescens*. *Curr. Sci.*, 82, 1463–1466 (2002).
76 V. Ramamoorthy, T. Raghuchander and R. Samiyappan, Induction of defense related proteins in tomato roots treated with *Pseudomonas fluorescens* Pf1 and *Fusarium oxysporum* f.sp. *lycopersici*. *Plant Soil*, 239, 55–68 (2002).

77 V. Anjaiah, P. Cornelis and N. Koedam, Effect of genotype and root colonization in biological control of Fusarium wilts in pigeonpea and chickpea by *Pseudomonas aeruginosa* PNA1. *Can. J. Microbiol.*, 49, 85–91 (2003).

78 V.U. Boby and D. J. Bagyaraj, Biological control of root-rot of *Coleus forskohlii* Briq. using microbial inoculants. *World J. Microbiol. Biotechnol.*, 19, 175–180 (2003).

79 T. Bora, H. Ozaktan, E. Gore and E. Aslan, Biological control of *Fusarium oxysporum* f. sp. *melonis* by wettable powder formulations of two strains of *Pseudomonas putida. J. Phytopathol.*, 152, 471–475 (2004).

80 F. Yigit and M. Dikilitas, Control of Fusarium wilt of tomato by combination of fluorescent *Pseudomonas*, non pathogenic *Fusarium* and *Trichoderma harzianum* T-22 in green house condition, *Plant Pathol. J.*, 6, 159–163 (2007).

81 Z.O. Ardebili, N.O. Ardebili and S.M. Mahdi Hamdi, Physiological effects of '*Pseudomonas fluorescens*' CHA0 on tomato ('*Lycopersicon esculentum*' Mill.) plants and its possible impact on *Fusarium oxysporum* f. sp. 'lycopersici,' *Austral. J. Crop Sci.*, 5, 1631–1638 (2011).

82 G. Selvakumar, R. Reetha, and P. Thamizhiniyan, The Pgpr as elicitors of plant defence mechanisms and growth stimulants on tomato (*Lycopersicum esculentum* Mill.) *Bot. Res. Intern.*, 6(2), 47–55 (2013).

83 J.R. De Freitasand and J.J. Germida, *Pseudomonas cepacia* and *Pseudomonas putida* as winter wheat inoculants for biocontrol of *Rhizoctonia solani. Can. J. Microbiol.*, 37, 780–784 (1991).

84 R. Nandakumar, R. Viswanathan, S. Babu, J. Shella, , T. Raghuchander, and R. Samiyappan, A new bio-formulation containing plant growth promoting rhizobacterial mixture for the management of sheath blight and enhanced grain yield in rice, *Biocontrol*, 46, 493–510 (2001).

85 M. Szczech, and M. Shoda, Biocontrol of Rhizoctonia damping-off of tomato by *Bacillus subtilis* combined with *Bukholderia cepacia. J. Phytopathol.*, 152, 549–556 (2004).

86 G. Berta, S. Sampo, E. Gamalero, N. Musasa, and P. Lemanceau, Suppression of Rhizoctonia root-rot of tomato by *Glomus mosseae* BEG 12 and *Pseudomonas fluorescens* A6RI is associated with their effect on the pathogen growth and on the root morphogenesis. *Eur. J. Plant Pathol.*, 111, 279–288 (2005).

87 M. Ahmadzadeh and A.S. Tehrani, Evaluation of fluorescent pseudomonads for plant growth promotion, antifungal activity against *Rhizoctonia solani* on common bean, and biocontrol potential, *Biol. Control*, 48, 101–107 (2009).

88 A.K. Srivastava, T. Singh, T.K. Jana, and D.K. Arora, Induced resistance and control of charcoal rot in *Cicer arietinum* (chickpea) by *Pseudomonas fluorescens, Can. J. Bot.*, 79, 787–795 (2001).

89 C.P. Gupta, R.C. Dubey and D.K. Maheshwari, Plant growth enhancement and suppression of *Macrophomina phaseolina* causing charcoal rot of peanut by fluorescent *Pseudomonas. Biol. Fert. Soils*, 35, 399–405 (2002).

90 M.Z. Begum, and M.S. Kumar, Management of disease complex involving *Heterodera cajani* Koshy, 1967 and *Macrophomina phaseolina* (Tassi) Goid on green gram (*Vigna radiata* L. Wilczek), *Indian J. Nematol.*, 35, 192–194 (2005).

91 D. Illakkiam, P. Ponraj, M. Shankar, S. Muthusubramanian, J. Rajendhran, and P. Gunasekaran, Identification and structure elucidation of a novel antifungal compound produced by *Pseudomonas aeruginosa*PGPR2 against *Macrophomina phaseolina, Appl. Biochem. Biotechnol.*, 171, 2176–2185 (2013).

92 S. Vanitha and R. Ramjegathesh, Biocontrol potential of *Pseudomonas fluorescens* against coleus root-rot disease. *J. Plant Pathol. Microb.*, 5, 216 (2014).

93 N. Khatoon, H.A. Shafique, R. Noreen, V. Sultana, R. Badar and S.E. Haque, Role of endophytic and rhizospheric fluorescent pseudomonas associated with mungbean in suppressing the root rotting fungi of mungbeaan. *Int. J. Biol. Res.*, 2, 115–123 (2014).

94 A. Gamliel, and J. Katan, Suppression of major and minor pathogens by fluorescent pseudomonads in solarized and nonsolarized soils, *Phytopathology*, 83, 68–75 (1993).

95 G. Wei, J.W. Kloepper and S. Tuzun, Induced systemic resistance to cucumber diseases and increased plant growth by plant growth promoting rhizobacteria under field conditions. *Phytopathology*, 86, 221–224 (1996).

96 V. Sharma and J. Nowak, Enhancement of Verticillium wilt resistance in tomato transplants by *in vitro* co-culture of seedlings with a plant growth promoting rhizobacterium (*Pseudomonas* sp. Strain PsJN). *Can. J. Microbiol.*, 44, 528–536 (1998).

97 V. Ramamoorthy and R. Samiyappan, Induction of defense related genes in *Pseudomonas fluorescens* treated chili plants in response to infection by *Colletotrichum capsici*, *J. Mycol. Plant Pathol.*, 31, 146–155 (2001).

98 E.A. Barka, S. Gognies, J. Nowak, J. Audran, and A. Belarbi, Inhibitory effect of endophyte bacteria on *Botrytis cineria* and its influence to promote the grapevine growth, *Biol. Control*, 24, 135–142 (2002).

99 J. Bigirimana and M. Hofte, Induction of systemic resistance to *Colletotrichum lindemuthianum* in bean by a benzothiadiazole derivative and rhizobacteria, *Phytoparasitica*, 30, 159–168 (2002).

100 R. Radja Commare, R. Nandkumar, A. Kandan, S. Suresh, M. Bharathi, T. Raguchander and R. Samiyappan, *Pseudomonas fluorescens* based bioformulation for the management of sheath blight disease and leaffolder insect in rice, *Crop Protect.*, 21, 671–677 (2002).

101 R. Viswanathan and R. Samiyappan, Induced systemic resistance by fluorescent pseudomonas against red rot disease of sugarcane caused by *Colletotrichum falcatum*, *Crop Protec.*, 21, 1–10 (2002).

102 Z. Yan, M.S. Reddy, C.M. Ryu, J.A. Mc Inroy, M.A. Wilson and J.W. Kloepper, Induced systemic protection against tomato late blight elicited by plant growth-promoting rhizobacteria, *Phytopathology*, 92, 1329–1333 (2002).

103 S. Niranjan-Raj, N.P. Shetty and H.S. Shetty, Seed bio-priming with *Pseudomonas fluorescens* isolates enhances growth of pearl millet plants and induces resistance against downy mildew, *Intern. J. Pest Manag.*, 50, 41–48 (2004).

104 R. Vivekananthan, M. Ravi, A. Ramanathan and R. Samiyappan, Lytic enzymes induced by *Pseudomonas fluorescens* and other biocontrol organisms mediate defence against the anthracnose pathogen in mango. *World J. Microbiol. Biotechnol.*, 20, 235–244 (2004).

105 Z.A. Siddiqui and L.P. Singh, Effects of fly ash and soil microorganisms on the plant growth, photosynthetic pigments and leaf blight of wheat, *J. Plant Dis. Protect.*, 112, 146–155 (2005).

106 Z.A. Siddiqui, Biocontrol of *Alternaria triticina* by plant growth promoting rhizobacteria on wheat, *Arch. Phytopathol. Plant Protec.*, 40, 301–308 (2007).

107 D. Saravanakumar, C. Vijayakumar, N. Kumar and R. Samiyappan, PGPR-induced defense responses in the tea plant against blister blight disease, *Crop Protec.*, 26, 556–565 (2007).

108 V.K. Deshwal, *Pseudomonas aeruginosa* as biological control agent against plant pathogenic fungus *Sclerotinia sclerotiorum, Intern. J. Plant Anim. Environ. Sci.*, 2(1), 14–17 (2012).

109 C.G.L. Cabanás, E. Schilirò, A. Valverde-Corredor and J. Mercado-Blanco, The biocontrol endophytic bacterium *Pseudomonas fluorescens* PICF7 induces systemic defense responses in aerial tissues upon colonization of olive roots, *Front. Microbiol.*, 5, 427 (2014).

110 J.Y. Zhou, X.Y. Zhao and C.C. Dai, Antagonistic mechanisms of endophytic *Pseudomonas fluorescens* against *Athelia rolfsii. J. Appl. Microbiol.*, 117, 1144–1158 (2014).

111 N. Unnamalai and S.S. Gnanamanickam, *Pseudomonas fluorescens* is an antagonist to *Xanthomonas citri* (Hasse) Dye, the incitant of citrus canker. *Curr. Sci.*, 53, 703–704 (1984).

112 G.W. Xu and D.C. Gross, Field evaluation of the interactions among fluorescent pseudomonads, *Erwinia carotovora* and potato yields, *Phytopathology*, 76, 423–430 (1986).

113 M. Wilson and S.E. Lindow, Interactions between the biological control agent *Pseudomonas fluorescens* A506 and *Erwinia amylovora* in pear blossoms, *Phytopathology*, 83, 117–123 (1993).

114 D.H. Kim, and I.J. Misaghi, Biocontrol performance of two isolates of *Pseudomonas fluorescens* in modified soil atmosphere, *Phytopathology*, 86, 1238–1241 (1996).

115 S.E. Lindow, G. Mc Gourty and R. Elkins, Interaction of antibiotics with *Pseudomonas fluorescens* strain A506 in the control of fire blight and frost injury to pear, *Phytopathology*, 86, 841–849 (1996).

116 K.S. Jagadeesh, J.H. Kulkarni and P.U. Krisharaj, Evaluation of role of fluorescent siderophore in the biological control of bacterial wilt in tomato using Tn^5 mutants of fluorescent *Pseudomonas* sp. *Curr. Sci.*, 81, 882–883 (2001).

117 P. Vidhyasekaran, N. Kamala, A. Ramanathan, K. Rajappan, V. Paranidharan and R. Velazhahan, Induction of systemic resistance by *Pseudomonas fluorescens* Pf1 against *Xanthomonas oryzae* pv. *oryzae* in rice leaves. *Phytoparasitica*, 29, 155–166 (2001).

118 J. Guo, H. Ying Qi, Y. Guo, H. Ge, L. Gong, L. Zhang and P. Sun, Biocontrol of tomato wilt by plant growth promoting rhizobacteria, *Biol. Control*, 29, 66–72 (2004).

119 N. Aliye, C. Fininsa and Y. Hiskias, Evaluation of rhizosphere bacterial antagonists for their potential to bioprotect potato (*Solanum tuberosum*) against bacterial wilt (*Ralstonia solanacearum*), *Biol. Control*, 47, 282–288 (2008).

120 H. Kurabachew, F. Assefa and Y. Hiskias, Evaluation of Ethhiopian isolates of *Pseudomonas fluorescens* as biocontrol agent against potato bacterial wilt caused by *Ralstonia (Pseudomonas) solanacearum, Acta Agric. Slovanica*, 90, 125–135 (2007).

121 S. Biswas and N.P. Singh, Integrated management of wilt of tomato caused by *Ralstonia solanacearum, Indian J. Mycol. Pl. Pathol.*, 38 (2), 182–184 (2008).

122 R. Ramesh, A. Joshi and M. Ghanekar, Pseudomonads: major antagonistic endophytic bacteria to suppress bacterial wilt pathogen, *Ralstonia solanacearum* in the eggplant (*Solanum melongena* L.), *World J. Microbiol. Biotechnol.*, 25, 47–55 (2009).

123 S.C. Vanitha, S.R. Niranjana, C.N. Mortensen and S. Umesha, Bacterial wilt of tomato in Karnataka and its management by *Pseudomonas fluorescens. Biocontrol*, 54, 685–695 (2009).

124 T. Arwiyanto, Y.S.Y.S. Maryudani and S.D. Nurcahyanti, Protection of eggplant and chilli from bacterial wilt (*Ralstonia solanacearum*) with antagonistic bacteria, ISHS Acta Horticulturae 933: XXVIII-IHC2010: International Symposium on Organic Horticulture: Productivity and Sustainability. 10.17660/ActaHortic.2012.933.54 (2010).

125 A. Salowi, Towards biological control of Pistachio dieback. Thesis (M.Sc. Ag.) University of Adelaide, School of Agriculture, Food and Wine. http://hdl.handle.net/2440/63478 (2010).

126 G. Chakravarty and M.C. Kalita, Comparative evaluation of organic formulations of *Pseudomonas fluorescens* based biopesticides and their application in the management of bacterial wilt of brinjal (*Solanum melongena* L.). *Afr. J. Biotechnol.*, 10, 7174–7182 (2011).

127 M.A. Seleim, F.A. Saead, K.M.H. Abd-El-Moneem and K.A.M. Abo-elyousr, Biological control of bacterial wilt of tomato by plant growth promoting rhizobacteria. *Plant Pathol. J.*, 10(4), 146–153 (2011).

128 P. Deng, , and , Complete genome of *Pseudomonas chlororaphis* strain UFB2, a soil bacterium with antibacterial activity against bacterial canker pathogen of tomato, *Stand Genomic Sci.*, 10, 117 (2015).

129 J.W. Costerton, G.G. Geesey and K.J. Cheng, How bacteria stick, *Sci. Amer.*, 238, 86–95 (1978).

130 R., Singh, D. Paul and R.K. Jain, Biofilm: implication and bioremediation, *Trends Microbiol.*, 14, 389–397 (2006).

131 G.O. Toole, H.B. Kaplan and R. Kolter, Biofilm formation as microbial development, *Annu. Rev. Microbiol.*, 54, 49–79 (2000).

132 L. Hall-Stoodley, J.W. Costerton and P. Stoodley, Bacterial biofilms: from the natural environment to infectious diseases, *Nat. Rev. Microbiol.*, 2, 95–108 (2004).

133 P.C. Bogino, M.M. Oliva, and W. Giordano, The role of bacterial biofilms and surface components in plant-bacterial associations, *Int. J. Mol. Sci.*, 14, 15838–15859 (2013).

134 M.E. Davey, and A.G. O'Toole, Microbial biofilms: from ecology to molecular genetics, *Microbiol. Mol. Biol. Rev.*, 64, 847–867 (2000).

135 S.S. Branda, S. Vik, L. Friedman and R. Kolter, Biofilms: the matrix revisited, *Trends Microbiol.*, 13, 20–26 (2005).

136 R. Kolter, and E.P. Greenberg, Microbial sciences: the superficial life of microbes, *Nature* 441, 300–302 (2006).

137 D.M. Weller and L.S. Thomashow, Current challenges in introducing beneficial microorganisms into the rhizosphere, In *Molecular Ecology of Rhizosphere Microorganisms* ed. O'Gara, F., Dowling, D.N. and Boesten, B. New York: Academic Press (1994), pp. 1–18.

138 T.F. Mah and G.A. O'Toole, Mechanisms of biofilm resistance to antimicrobial agents, *Trends Microbiol.*, 9, 34–39 (2001).

139 C. Matz and S. Kjelleberg, Off the hook–how bacteria survive protozoan grazing, *Trends Microbiol.*, 13, 302–307 (2005).

140 G.G. Anderson, and G.A. O'Toole, Innate and induced resistance mechanisms of bacterial biofilms. *Curr. Top. Microbiol. Immunol.*, 322, 85–105 (2008).

141 K. Lewis, Persister cells and the riddle of biofilm survival, *Biochemistry (Mosc)* 70, 267–274 (2005).

142 L. Hall-Stoodley and P. Stoodley, Evolving concepts in biofilm infections, *Cell Microbiol.* 11, 1034–1043 (2009).

143 P.S. Stewart and M.J. Franklin, Physiological heterogeneity in biofilms, *Nat. Rev. Microbiol.*, 6, 199–210 (2008).

144 M.F. Martin and P. Liras, Organization and expression of genes involved in the biosynthesis of antibiotics and other secondary metabolites, *Annu. Rev. Microbiol.*, 43, 173–206 (1989).

145 D. Lopez and R. Kolter, Extracellular signals that define distinct and coexisting cell fates in *Bacillus subtilis, FEMS Microbiol. Rev.*, 34, 134–49 (2009).

146 R.M. Donlan, Biofilms: microbial life on surfaces, *Emerg. Infect. Dis.*, 8, 881–890 (2002).

147 N. Trachoo, Biofilms and the food industry, *Songklanakarin J. Sci. Technol.*, 25, 807–815 (2003).

148 P.I. Watnick and R. Kolter, Biofilm, city of microbes, *J. Bacteriol.*, 182, 2675–2679 (2000).

149 K.K. Jefferson, What drives bacteria to produce a biofilm? *FEMS Microbiol. Lett.*, 236, 163–173 (2004).

150 G. Berg, N. A. L. A. Zock and K. Smalla, Plant-dependent genotypic and phenotypic diversity of antagonistic rhizobacteria isolated from different *Verticillium* host plants, *Appl. Environ. Microbiol.*, 68, 3328–3338 (2002).

151 M. Elasri, S. Delorme, P. Lemanceau, G. Stewart, B. Laue, et al., Acyl-homoserine lactone production is more common among plant-associated *Pseudomonas* spp. than among soilborne *Pseudomonas* spp. *Appl. Environ. Microbiol.*, 67, 1198–1209 (2001).

152 E.A. Pierson, D.W. Wood, J.A. Cannonand, F.M. Blachere, Interpopulation signaling via N-acyl-homoserine lactones among bacteria in the wheat rhizosphere. *Mol. Plant-Microb. Interact.*, 11, 1078–1084 (1998).

153 A. Steidle, M. Allesen-Holm, K. Riedel, G. Berg, M. Givskov, S. Molin, and L. Eberl, Identification and Characterization of an *N*-Acylhomoserine Lactone-Dependent Quorum-Sensing System in *Pseudomonas putida* Strain IsoF, *Appl. Environ. Microbiol.*, 68, 6371–6382 (2002).

154 A. Steidle, K. Sigl, R. Schuhegger, A. Ihring, M. Schmid, S. Gantner, M. Stoffels, K Riedel, M. Givskov, A. Hartmann, C. Langebartels, and L. Eberl. Visualization of *N*-acylhomoserine lactone-mediated cell–cell communication between bacteria colonizing the tomato rhizosphere. *Appl. Environ. Microbiol.*, 67, 5761–5770 (2001).

155 C. Beloin, J. Valle, P. Latour-Lambert, P. Faure, M. Kzreminski, D. Balestrino, J.A. Haagensen, S. Molin, G. Prensier, B. Arbeille, and J.M. Ghigo, Global impact of mature biofilm lifestyle on *Escherichia coli* K-12 gene expression. *Mol. Microbiol.*, 51, 659–674 (2004).

156 M. Hentzer, H. Wu, J. B. Andersen, N. Bagge, Z. Song, P. Steinberg, and M. Givskov, Attenuation of *Pseudomonas aeruginosa* virulence by quorum sensing inhibitors, *EMBO J.*, 22, 3803–3815 (2003).

157 K. Sauer, A.K. Camper, G.D. Ehrlich, J.W. Costerton, and D.G. Davies, *Pseudomonas aeruginosa* displays multiple phenotypes during development as a biofilm, *J. Bacteriol.*, 184, 1140–1154 (2002).

158 K. Sauer, and A. K. Camper, Characterization of phenotypic changes in *Pseudomonas putida* in response to surface-associated growth. *J. Bacteriol.*, 183, 6579–6589 (2001).

159 M.A. Schembri, K. Kjaergaard, and P. Klemm, Global gene expression in *Escherichia coli* biofilms, *Mol. Microbiol.*, 48, 253–267 (2003).

160 M. Whiteley, M.G. Bangera, R.E. Bumgarner, M.R. Parsek, G.M. Teitzel, S. Lory *and* E.P. Greenberg, Gene expression in *Pseudomonas aeruginosa* biofilms, *Nature* 413, 860–864 (2001).

161 K.E. Nelson, C. Weinel, I.T. Paulsen, R.J. Dodson, H. Hilbert, V.A.P. Martins dos Santos, *et al.* Complete genome sequence and comparative analysis of the metabolically versatile *Pseudomonas putida* KT2440, *Environ. Microbiol.*, 4, 799– 808 (2002).

162 C. Fuqua, and E.P. Greenberg, Listening in on bacteria: acyl-homoserine lactone signaling, *Nat. Rev. Mol. Cell Biol.*, 3, 685–695 (2002).

163 S.B. von Bodman, W.D. Bauer, and D.L. Coplin, Quorum sensing in plant-pathogenic bacteria, *Annu. Rev. Phytopathol.*, 41, 455–482 (2003).

164 H.L. Wei, and L.Q. Zhang, Quorum-sensing system influences root colonization and biological control ability in *Pseudomonas fluorescens* 2P24, *Antonie van Leeuwenhoek*, 89, 267–280 (2006).

165 S. Gantner, M. Schmid, C. Dürr, R. Schuhegge, A. Steidle, et al. In situ quantitation of the spatial scale of calling distances and population density-independent *N*-acylhomoserine lactone-mediated communication by rhizobacteria colonized on plant roots, *FEMS Microbiol. Ecol.*, 56, 188–194 (2006).

166 Y.J. Wang, and J.R. Leadbetter, Rapid acyl-homoserine lactone quorum signal biodegradation in diverse soils, *Appl. Environ. Microbiol.*, 71, 1291–1299 (2005).

167 T. Danhorn and C. Fuqua, Biofilm formation by plant-associated bacteria, *Annu. Rev. Microbiol.*, 61, 401–422 (2007).

168 W. Achouak, S. Conrod, V. Cohen and T. Heulin, Phenotypic variation of *Pseudomonas brassicacearum* as a plant root-colonization strategy. *Mol. Plant-Microb. Interact.*, 17, 872–879 (2004).

169 R.D. Monds, M.W. Silby and H.K. Mahanty, Expression of the Pho regulon negatively regulates biofilm formation by *Pseudomonas aureofaciens* PA147-2, *Mol. Microbiol.* 42, 415–426 (2001).

170 M. Klausen, A. Heydorn, P. Ragas, L. Lambertsen, A. Aaes-Jørgensen, et al., Biofilm formation by *Pseudomonas aeruginosa* wild type, flagella and type IV pili mutants, *Mol. Microbiol.*, 48, 1511–1524 (2003).

171 M. Bacilio-Jiménez, S. Aguilar-Flores, E. Ventura-Zapata, E. Pérez-Campos, S. Bouquelet, et al., Chemical characterization of root exudates from rice (*Oryza sativa*) and their effects on the chemotactic response of endophytic bacteria, *Plant Soil*, 249, 271–277 (2003).

172 M. Espinosa-Urgel, A. Salido and J.L. Ramos, Genetic analysis of functions involved in adhesion of *Pseudomonas putida* to seeds, *J. Bacteriol.* 182, 2363–2369 (2000).

173 S.M. Hinsa, M. Espinosa-Urgel, J.L. Ramos and G.A. O'Toole, Transition from reversible to irreversible attachment during biofilm formation by *Pseudomonas fluorescens* WCS365 requires an ABC transporter and a large secreted protein, *Mol. Microbiol.* 49, 905–918 (2003).

174 A.G. Torres, C. Jeter, W. Langley and A.G. Matthysse, Differential binding of *Escherichia coli* O157, H7 to alfalfa, human epithelial cells, and plastic is mediated by a variety of surface structures. *Appl. Environ. Microbiol.*, 71, 8008–8015 (2005).

175 V. Bianciotto, S. Andreotti, R. Balestrini, P. Bonfante and S. Perotto, Extracellular polysaccharides are involved in the attachment *of Azospirillum brasilense* and *Rhizobium leguminosarum* to arbuscular mycorrhizal structures, *Eur. J. Histochem.*, 45, 39–49 (2001).

176 S. Ude, D.L. Arnold, C.D. Moon, T. Timms-Wilson and A.J. Spiers, Biofilm formation and cellulose expression among diverse environmental *Pseudomonas* isolates, *Environ. Microbiol.*, 8, 1997–2011 (2006).

177 A.J. Spiers and P.B. Rainey, The *Pseudomonas fluorescens* SBW25 wrinkly spreader biofilm requires attachment factor, cellulose fibre and LPS interactions to maintain strength and integrity, *Microbiol.*, 151, 2829–2839 (2005).

178 W.S. Chang and L.J. Halverson, Reduced water availability influences the dynamics, development, and ultrastructural properties of *Pseudomonas putida* biofilms, *J. Bacteriol.*, 185, 6199–6204 (2003).

179 M. vande Mortel and L.J. Halverson, Cell envelope components contributing to biofilm growth and survival of *Pseudomonas putida* in low-water-content habitats, *Mol. Microbiol.*, 52, 735–750 (2004).

180 B. Prithiviraj, H.P. Bais, T. Weir, B. Suresh, E.H. Najarro, et al. Down regulation of virulence factors of *Pseudomonas aeruginosa* by salicylic acid attenuates its virulence on *Arabidopsis thaliana* and *Caenorhabditis elegans*, *Infect. Immun.*, 73, 5319–5328 (2005).

181 T.S. Walker, H.P. Bais, E. Deziel, H.P. Schweizer, L.G. Rahme, et al. *Pseudomonas aeruginosa*-plant root interactions. Pathogenicity, biofilm formation, and root exudation, *Plant Physiol.*, 134, 320–331 (2004).

182 V. Maddula, Z. Zhang, E.A. Pierson and L.S. Pierson, Quorum sensing and phenazines are involved in biofilm formation by *Pseudomonas chlororaphis* (*aureofaciens*) strain 30–84, *Microb. Ecol.*, 52, 289–301 (2006).

183 M.E. Davey, N.C. Caiazza, and G.A. O'Toole, Rhamnolipid surfactant production affects biofilm architecture in *Pseudomonas aeruginosa* PAO1. *J. Bacteriol.*, 185, 1027–1036 (2003).

184 J.M. Raaijmakers, I. de Bruijn and M.J.D. de Kock, Cyclic lipopeptide production by plant-associated *Pseudomonas* spp.: diversity, activity, biosynthesis and regulation, *Mol. Plant-Microbe Interact.*, 19, 699–710 (2006).

185 I. Kuiper, E.L. Lagendijk, R. Pickford, J.P. Derrick, G.E. Lamers, et al. Characterization of two *Pseudomonas putida* lipopeptide biosurfactants, putisolvin I and II, which inhibit biofilm formation and break down existing biofilms, *Mol. Microbiol.*, 51, 97–113 (2004).

186 H.C. and J. Relevance of microbial extracellular polymeric substances (EPSs)--Part I: Structural and ecological aspects, *Water Sci. Technol.*, 43, 1–8 (2001).

187 P. D. J.G. and R. Nijland, Molecular mechanisms of compounds affecting bacterial biofilm formation and dispersal, *Appl. Microbiol. Biotechnol.*, 86, 813–823 (2010).

188 G.A. O'Toole, and R. Kolter, The initiation of biofilm formation in *Pseudomonas fluorescens* WCS365 proceeds via multiple, convergent signaling pathways: a genetic analysis. *Mol. Microbiol.*, 28, 449–461 (1998).

189 M. Hentzer, L. Eberl *and* M. Givskov, Quorum sensing in Biofilms: gossip in the slime world? Microbial Biofilms (Ghannoum M. and G. O'Toole, eds), pp. 118–140. ASM Press, Washington, DC (2004).

190 S. Kjelleberg and S. Molin, Is there a role for quorum sensing signals in bacterial biofilms? *Curr. Opinion Microbiol.*, 5, 254–258 (2002).

191 B. Huber, K. Riedel, M. Hentzer, A. Heydorn, A. Gotschlich, et al., The cep quorum-sensing system of *Burkholderia cepacia* H111 controls biofilm formation and swarming motility, *Microbiol.*, 147, 2517–2528 (2001).

192 B. Huber, K. Riedel, M. Köthe, M. Givskov, S. Molin, and L. Eberl, Genetic Analysis of functions involved in the late stages of biofilm development in *Burkholderia cepacia* H111. *Mol. Microbiol.*, 46, 411–426 (2002).

193 D.G. M.R. J.P. B.H. J.W. Costerton and E.P. Greenberg . The involvement of cell-to-cell signals in the development of a bacterial biofilm, *Science*, 280, 295–298 (1998).

194 A. Heydorn, B. Ersbøll, J. Kato, , M.R. Parsek, T. Tolker-Nielsen, M. Givskov and S. Molin, Statistical Analysis of *Pseudomonas aeruginosa* Biofilm Development: Impact of Mutations in Genes Involved in Twitching Motility, Cell-to-Cell Signaling, and Stationary-Phase Sigma Factor Expression. *Appl. Environ. Microbiol.*, 68, 2008–2017 (2002).

195 P. Stoodley, Z. Lewandowski, J.D. Boyle, and H.M. Lappin-Scott, Structural deformation of bacterial biofilms caused by short-term fluctuations in fluid shear: an in situ investigation of biofilm rheology, *Biotechnol. Bioeng.*, 65, 83–92 (1999).

196 S.S. Yoon, R.F. Hennigan, G.M. Hilliard, U.A. Ochsner, K. Parvatiyar, , M.C. Kamani, H.L. Allen, T.R. DeKievit, P.R. Gardner, U. Schwab, *Pseudomonas aeruginosa* anaerobic respiration in biofilms: relationships to cystic fibrosis pathogenesis, *Dev. Cell*, 3, 593–603 (2002).

197 E.A.B. Emmert and J. Handelsman, Biocontrol of plant disease: a Gram-positive perspective, *FEMS Microbiol. Lett.*, 171, 1–9 (1999).

198 M.A Hamon and B.A. The sporulation transcription factor Spo0A is required for biofilm development in *Bacillus subtilis, Mol. Microbiol.*, 42, 1199–1209 (2001).

199 N.R. Stanley, R.A. Britton, A.D. Grossman and B.A. Lazazzera, Identification of catabolite repression as a physiological regulator of biofilm formation by *Bacillus subtilis* by use of DNA microarrays. *J. Bacteriol.*, 185, 1951–1957 (2003).

200 R.F. Kinsinger, M.C. Shirk, and R. Fall, Rapid surface motility and biofilm formation in *Bacillus subtilis* is dependent on extracellular surfactin and potassium ion, *J. Bacteriol.*, 185, 5627–5631 (2003).

201 H.P. Bais, R. Fall and J.M. Vivanco, Biocontrol of *Bacillus subtilis* against infection of *Arabidopsis* roots by *Pseudomonas syringae* is facilitated by biofilm formation and surfactin production, *Plant Physiol.*, 134, 307–319 (2004).

202 B.J.J. Lugtenberg, L. Dekkers, and G.V. Bloemberg, Molecular determinants of rhizosphere colonization by *Pseudomonas. Annu. Rev. Phytopathol.*, 39, 461–490 (2001).

203 M. Espinosa-Urgel, R. Kolter, and J.L. Ramos, Root colonization by *Pseudomonas putida*: love at first site. *Microbiol.*, 148, 341–43 (2002).

204 W.M. Haggag, and S. Timmusk, Colonization of peanut roots by biofilm forming *Paenibacillus polymyxa* initiates biocontrol against crown rot disease, *J. Appl. Microbiol.*, 104, 961–969 (2008).

205 W.M. Haggag, *The role of biofilm exopolysaccharides on biocontrol of plant diseases.* Intech, Cape Town, ISBN: 978-953-307-109-1, p. 612 (2010).

6

Quorum Sensing Mechanisms in Rhizosphere Biofilms

Jorge Barriuso

Centro de Investigaciones Biológicas, Consejo Superior de Investigaciones Científicas, Madrid, Spain

6.1 Background

Since the 1960s, it has been known that unicellular microorganisms do not live as independent cells isolated in the environment. There is a communication between single cells that allows the population to coordinate their behavior [1]. This cell-to-cell signaling system is known as *quorum sensing* (QS) and is mediated by small, diffusible, chemical signal molecules. The QS molecules (QSM) are secreted by the microbial cells and accumulate in the environment as the population density grows. Each individual cell it is able to sense the amount of QSM present in the medium, and when the concentration reaches a threshold, the gene expression in the population is altered (induced or repressed) in a coordinated manner [2]. Hence, the QS phenomena depend on cell population density. These molecules are autoinducible, which means that the sensing of the QSM stimulates the biosynthesis of more QSM [3].

The QS mechanisms allow the coordination of diverse processes in microbial populations and communities, and suppose a great evolutionary advantage for the adaptation to the rapid changing conditions of the environment. Microorganisms utilize QS mechanisms to adapt to unfavorable conditions. Furthermore, some authors propose the role of these processes as neo-Darwinian mechanisms of evolution, with an important participation in the apparition of the first multicellular organisms [4, 5].

QS has extensively been studied in bacteria during the past 50 years [4, 6] and it is known to be involved in the regulation of processes such as pathogenesis, symbiosis, competence, conjugation, nutrients uptake, morphological differentiation, secondary metabolites biosynthesis, or biofilms production [7]. In one study using Gram-negative bacteria, it was discovered that QSM was produced by a two-component system: the enzyme synthase LuxI that produces N-acyl-homoserine-lactone (AHLs) and a sensor receptor (LuxR) responsible of the detection of the molecule in the environment [8, 9]. Later on, quinolones were found as QSM in other Gram-negative bacteria such as *Pseudomonas aeuroginosa* [10]. In Gram-positive bacteria, QS mechanisms were discovered mediated by peptides, and the autoinducer 2 (AI-2) was also later described in

Gram-negative bacteria [4]. In the following years, ɣ-butirolactone was found as the QSM used by species of the genus *Streptomyces* [7].

More recently, this phenomenon has been described in eukaryotes [11–13], particularly in fungi [14]. QS mechanisms have been extensively studied in the dimorphic yeast *Candida albicans* due to its clinical importance in humans. In this fungus, the sesquiterpene alcohol farnesol (1-hydroxy-3,7,11-trimethyl-2,6,10-dodecatriene) was described as the QSM [15].

QS communication can occur between different species, and even between organisms from different kingdoms. The populations are able to communicate with the populations from another species; this cross-talking has been described between bacteria and fungi, and even between bacteria or fungi with their hosts, such as plants [16] or mammalians [17]. For example, opaque cells from *C. albicans* are able to attract leukocytes to the site of infection [18], or certain plants are able to produce bacterial QSM [19]. In this sense, the fungal QSM farnesol has been proved to be effective in the disruption of bacterial biofilms [20]. On the other hand, *Bacillus subtilis* biofilm formation is influenced by the QSM from another species [21]. This may be particularly important in the context of sustainable agricultural practices with the aim of enhance the effectiveness of bioinoculants.

Although microorganisms have historically been studied as planktonic (or free swimming) cells, most of them live forming structures attached to biofilm surfaces [22]. Biofilms are a group of matrix-enclosed microbial accretions that adhere to biological or non-biological surfaces. These adherent cells are frequently embedded within a self-produced extracellular polymeric matrix. This structured microbial community is a protected growth modality that allows bacteria to survive in hostile environments. They present enhanced resistance to toxic compounds and, in the case of pathogens, increased virulence [23]. Adhesion of microbial cells, subsequent biofilm formation, and maturation of its structure are mediated by changes in cell phenotype and physiology. These changes are regulated by QS among a variety of mechanisms.

Moreover, some organisms have evolved strategies to disrupt the proper QS signaling of other microorganisms. This phenomenon is called *quorum quenching* [24] and usually leads to inactivation/degradation of the QSM [25]. This strategy prevents the sufficient accumulation of QSM in the surrounding environment of the microorganism cells, preventing the initiation of the QS mediated processes [26]. In the case of the degradation of AHLs, the most commonly widespread QSM, two major enzymes are responsible: Lactonases break the lactone ring and acylases break the acyl side chain [27]. This is the case of the actinobacterium *Rhodococcus erythropolis*, which exhibits the ability to degrade complex compounds such as AHLs [28].

Bacterial biofilms are most often discussed in the context of chronic mammalian infections, as they are frequently involved in the resistance to the host immune system and antibiotic therapy [29]. QS allows the microorganisms to form biofilms and resist harsh conditions such as the treatment with antibiotic. Hence, quorum quenching is one of the strategies used to control the development of drug resistance in microbes.

It is worth mentioning the danger of infections resulting from biofilm formation on medical devices—for example, the colonization of *Pseudomonas aeruginosa* on catheters used in cystic-fibrosis-affected patients is a big threat. In the pathogen *P. aeruginosa*, the *lasI* gene is involved in the development of biofilms [30]. Microbial biofilms are prevalent in many environments in nature; they are found attached to any surface and provide a niche for individuals to act as a group.

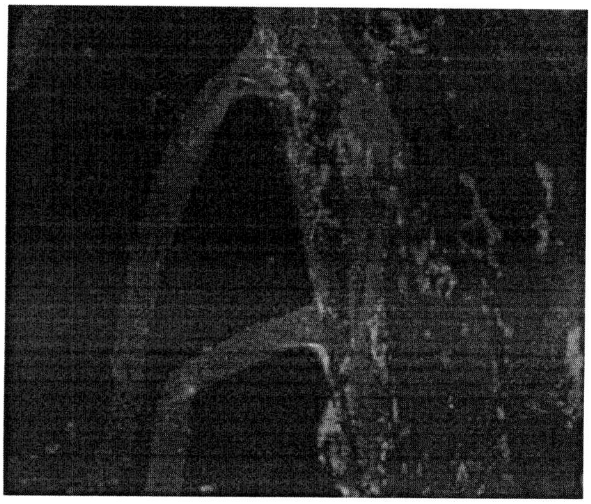

Figure 6.1 Confocal laser scanning micrograph of bacterium *Serratia liquefaciens* MG44 labeled with red probe Eub-338-I-Cy3, forming biofilm on the roots of *Arabidopsis thaliana*. (*See color plate section for the color representation of this figure.*)

Although it is well understood that in animals, biofilms can contribute to chronic diseases by allowing bacteria to persist, we do not fully understand the contributions of biofilm formation on the plant roots (Figure 6.1) and the implications to plant fitness and productivity [31]. Furthermore, soil fertility is fundamental for crops sustainability worldwide.

Human population growth and climate change concerns have resulted in an upsurge in attention to the anthropogenic activities that affect one of our most important resources, the soil. The soil environment is inherently complex and diverse, and its fertility is dependent on the microorganisms inhabiting this area, which are implied in the biogeochemical cycles of nature and in charge of the recycling of the biosphere nutrients [32]. Preserving biodiversity is essential for maintaining soil fertility, but also the organization of these microorganisms, and the formation of biofilms, may be fundamental to the correct functioning of the rhizosphere. Some authors have studied the survival mechanisms for rhizobacteria subjected to stress by studying the formation of biofilms and describing many factors that mediate biofilm formation. This structure provides protection from fluctuating severe conditions in the rhizosphere such as desiccation, extreme pH levels, temperature, salt, and nutrient availability [33].

6.2 QS in Biofilms Formation

Numerous studies have connected QS to biofilm formation in a number of species. Signaling compounds seem to be important in biofilm formation and exopolymer synthesis [30]. Before forming biofilms, planktonic cells coordinate their efforts by means of chemical signaling to colonize the root surface. QS allows bacteria to assess the size and status of their population, and coordinate the changes in gene expression required for the first steps in biofilm formation [31].

The rhizosphere is a critical interface between plant roots and soil, where beneficial and harmful interactions between plants and microorganisms take place. This zone of soil immediately surrounding plant roots, where complex biological and ecological processes occur, forms an environment that fulfills the requirements for biofilm formation, including sufficient moisture and supply of nutrients, which are provided by the plant. Besides biocontrol, certain bacteria such as plant growth–promoting rhizobacteria (PGPR) not only induce plant growth but also protect plants from soil-borne pathogens in a process known as induction of resistance, stimulating the defense systems of the plant [34]. Contrastingly, other rhizobacteria in a biofilm matrix may cause pathogenesis in plants. Research suggests that biofilm formation on plants is associated both with biological control and pathogenic response [35].

The rhizosphere is an area of high competence due to the great concentration of microorganisms struggling for the resources exudated by the plant. QS mechanisms could be an effective survival strategy for the efficient colonization and competition with other species (Figure 6.2). The successful colonization of the rhizosphere by the biocontrol organism defends the host roots from disease by pathogens. QS-regulated biofilm formation in *P. fluorescens* allows the organism to create a niche, in which antifungal compounds are secreted [36]. Studies with *P. polymyxa* strains revealed that the more efficient biofilm former was also more efficient against fungal pathogens, both *in vitro* and *in vivo* [37].

6.2.1 Positive Interactions

6.2.1.1 Plant Growth–Promoting Rhizobacteria (PGPR)

Plants benefit extensively by harboring microbes. They promote plant growth and confer enhanced resistance to various pathogens. PGPR may exert their beneficial effect on

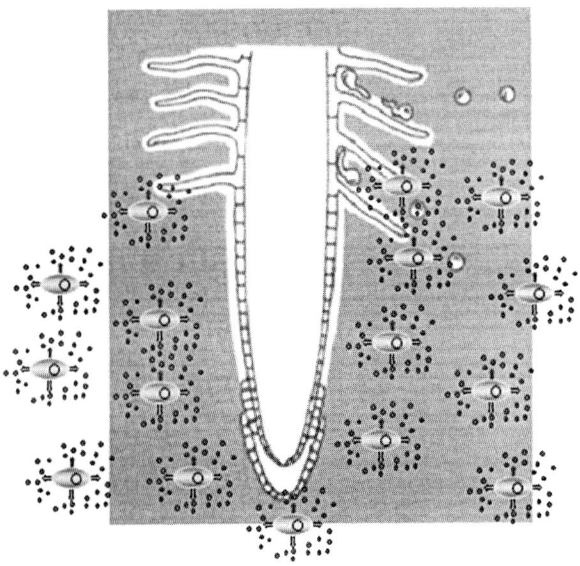

Figure 6.2 Representation of bacteria cells secreting a QSM in the rhizosphere.

the plant through a plethora of direct and indirect mechanisms, including nutrients uptake, hormone production, biocontrol, and stimulation of the plant defense system, for example. To be effective in the growth and fitness promotion of the plant, it is imperative the colonization and persistence of the bacteria in the rhizosphere [34].

The specie *Azospirillum amazonense* found in association with several crops of economic importance is described as PGPR; however, there was a lack of information on its physiology. Sant'Anna et al. [38] realized a comprehensive analysis of the genomic features of this bacterium; they identified the genes related to nitrogen/carbon metabolism, energy productions, phytohormone production, transport, QS, antibiotic resistance, chemotaxis/motility and bacteriophytochrome biosynthesis. Nitrogen fixation genes should be directly implicated in plant growth promotion; moreover, a LuxIR system was identified in the genome, suggesting the ability to produce and sense the QSM AHLs. The QS phenomenon in *Azospirillum* seems to regulate functions linked to rhizosphere competence and adaptation to plant roots. The genome of *A. amazonense* also presented a *Klebsiella pneumoniae ahlK* homolog [39], a predicted gene that codes for a putative lactonase implicated in AHL degradation. Since bacterial plant pathogens rely on QS mechanisms to infect plants, *A. amazonense* homoserine lactonase may reduce the deleterious activities of these pathogens in the rhizosphere.

Enterobacter cloacae GS1 is another PGPR that colonizes rice (*Oryza sativa*) roots. In the rhizosphere environment, *E. cloacae* GS1 is unable to synthesize AHL, but other microbial inhabitants do. Despite the fact that this bacterium possesses the AHL-dependent transcriptional regulator-encoding gene *sdiA*. Shankar et al. [40] studied the AHL-dependent rice root colonization by *E. cloacae* GS1. The effects of *sdiA* inactivation included increased root colonization and biofilm formation, suggesting a negative role for SdiA in bacterial adhesion. Exogenous addition of AHLs had a negative effect on root colonization and biofilm formation suggesting that AHLs produced by both plant and microbes in the rhizosphere act as interspecies signals to negatively impact cellular adhesion and, hence, root colonization in *E. cloacae* GS1.

Endophytes isolated from rice (bacteria and fungi) produced two types of interactions—biofilms and mixed cultures with no such attachments. Acidity in cultures with biofilms was higher than that of fungi alone, bacteria alone, or the mixed cultures. This lower pH measurement in the biofilms was due to the higher production of the plant hormone indoleacetic acid (IAA). IAA is an important hormone to promote the plant growth; moreover, microbial acid production is important for suppressing plant pathogens. Thus, the biofilm formation and QS mechanisms in endophytic environment seems to be very important for healthy and improved plant growth [41]. Application of beneficial inocula of bacteria to the rhizosphere is important for improved plant production in any agro-ecosystem. However, the conventional practice of inoculation with monocultures or mixed cultures of PGPR may not give the optimal effect, which may only be achieved by biofilm formation.

The rhizosphere-associated bacterium *Serratia plymuthica* HRO-C48 is able not only to stimulate plant growth but also to suppress symptoms caused by soil-borne pathogens. Investigations on the mechanisms regulating these PGPR traits show promising opportunities for application in biocontrol strategies. The presence of a QS mechanism mediated by AHL molecules in this bacterium induced to think that may be implied in its PGPR effect. To probe that, the AHL-degrading lactonase AiiA was heterologously expressed in the strain, resulting in abolished AHL production [42]. The transformed strain was not able to produce biocontrol in the pathosystem *Verticillium dahliae*–oilseed rape, while the wild type was completely effective in preventing the symptoms of the disease. HRO-C48 emits a broad spectrum of volatile organic compounds that are involved in antifungal activity and, interestingly, whose relative abundances are influenced by QS. In addition, it was shown affected the regulatory function of AHLs in the synthesis of IAA, swimming motility and production of extracellular hydrolytic enzymes. However, the ability to form biofilms on polystyrene surfaces was not affected in the transformed strain.

The endophytes present in *Pterocarpus santalinus* plants were screened for the presence of AHLs degrading strains. *Bacillus firmus* PT18 and *Enterobacter asburiae* PT39 exhibited potent AHL degrading ability by inhibiting 80 percent violacein production in biosensor strain *Chromobacterium violaceum*. When cell-free lysate of those bacteria were applied to pathogenic *P. aeruginosa* PAO1 and PAO1-JP2 biofilm, it resulted in significant inhibition of biofilm formation [43]. The biofilm inhibition was confirmed by visualization of biofilm slides under fluorescence microscopy. Amplification of the gene (aiiA) indicated the presence of AHL lactonase in cell-free lysate. Therefore, the study shows the potential of endophyte bacteria to control the pathogenesis of *P. aeruginosa* using quorum quenching mechanisms [43].

6.2.1.2 Rhizobia

Biofilm formation on plant roots appears to be associated with symbiotic interactions. These structures help to create protective niches for rhizobia, where there is a highly ordered, three-dimensional organization of the extracellular matrix. However, the role of biofilms in rhizobial-legume symbiosis is still not clear, the mechanisms involved in bacterial biofilm formation and attachment on plant roots, as well as the relation of these mechanisms to rhizobial function, needs to be further investigated [44].

Biofilm formation allows soil bacteria to colonize their surrounding habitat and survive environmental stresses, this mode of life is often essential for survival in bacteria of the genera *Mesorhizobium*, *Sinorhizobium*, *Bradyrhizobium*, and *Rhizobium*. QS mechanisms control numerous functions, including exopolysaccharide production [45], motility, nitrogen fixation, and nodulation [46], all of them related to symbiosis. Many studies show clearly that *Mesorhizobium* is one of the genera of bacteria in which QS plays an important role in biofilm formation, attachment, colonization, and nodulation of legumes. In the case of *S. fredii*, *S. meliloti* and *B. japonicum* the role of biofilm formation in symbiosis has been investigated [44]. Biofilm formation is crucial for an optimal root colonization and symbiosis between *S. fredii* SMH12 and *Glycine max* cv Osumi. In this bacterium, nod-gene-inducing flavonoids and the NodD1 protein are required for the transition to the biofilm form. In addition, QS mechanisms are required for the full development of this structure. This rhizobia strain transformed with a lactonase enzyme is defective in soybean root colonization, leading to a decrease in the symbiotic parameters [47]. On the other hand, unlike what has been described in other rhizobial species, disruption of the CinIR QS system in *R. leguminosarum* led to an increase in biofilm formation. This effect seemed to be mediated by a transcriptional regulator. Mutants in some components of the QS system presented an enhanced attachment to glass and altered expression of the exopolysaccharide glycanase PlyB, responsible for the cleavage of the acidic exopolysaccharide [48]. Moreover, overexpression of the *Agrobacterium tumefaciens* quorum regulator TraR in *M. huakuii* Mh93 interfered with the endogenous QS system of the rhizobia, forming thinner biofilms than the control strain, suggesting that quorum sensing positively regulates biofilm formation *in M. huakuii* [49].

6.2.2 Negative Interactions

Species from the genus *Pseudomonas* are well-known human and plant pathogens. These species easily form biofilms on various types of surfaces, providing increased resistance to environmental influences, including resistance to disinfectants. A strategy to fight against these infections may be the destruction of pseudomonas biofilms. The formation of these structures in pseudomonads is similar to that in other microorganisms, a sophisticated process with many regulatory elements, including QS mechanisms [50].

Particularly, *P. aeruginosa* is an opportunistic human pathogen capable of forming biofilm. Pathogenic *P. aeruginosa* strains PAO1 and PA14 are able of infecting the roots of *Arabidopsis* and sweet basil (*Ocimum basilicum*), causing plant mortality. The colonization of the plant roots is mediated by the formation of biofilm as observed by diverse microscopy techniques [51]. Sweet basil plants responds to the infection-secreting rosmarinic acid through the roots, with *in vitro* antibacterial activity against planktonic cells of both *P. aeruginosa* strains, while *Arabidopsis* does not naturally secrete rosmarinic acid as a root exudate. Biofilm resisted the microbicidal effects of the acid *in vivo* and *in vitro*. However, induction of rosmarinic acid secretion by sweet basil roots and exogenous supplementation of *Arabidopsis* root exudates with rosmarinic acid before infection, conferred resistance to *P. aeruginosa*. Studies with QS mutants PAO210 (Δ*rhlI*) and PAO216 (Δ*lasI* Δ*rhlI*) demonstrated that all of the strains were pathogenic to *Arabidopsis*, while both mutants were not pathogenic toward sweet basil. These mutants were affected in their capacity to form biofilm, suggesting that the QS mechanisms are involved in the biofilm formation, pathogenesis, and in the resistance of these bacteria to the biocides produced by the plant [51].

6.2.3 Cross-Communication

Eukaryotes, including plants, can interfere with bacterial QS systems by synthesizing molecules that disturb these mechanisms. It is known that *O. sativa* (rice) and *Phaseolus vulgaris* (bean) plants can produce certain type of molecules that mimic the structure and effect of the bacterial QSM AHLs. Using a lactonase enzyme, it was shown that bean and rice seed-extract contain AHL-mimic molecules that lack the typical lactone ring. Interestingly, these molecules specifically alter the QS-regulated biofilm formation of two plant-associated bacteria, *Sinorhizobium fredii* SMH12 and *Pantoea ananatis* AMG501, suggesting that these plants are able to enhance or to inhibit the colonization performance of the bacterial strains [19]. In another example, exudates from *Pisum sativum* (pea) and various other species of higher plants were found to contain activities that mimicked AHL signals, stimulating AHL-regulated behaviors in some bacterial strains while inhibiting such behaviors in others. The AHL signal-mimic compounds could be important in determining the interactions between plants and diverse pathogenic, symbiotic, and saprophytic bacteria [52].

On the other hand, two *Burkholderia graminis* strains (M12 and M14), which presented PGPR activities, were shown to produce a variety of AHLs as QS signaling

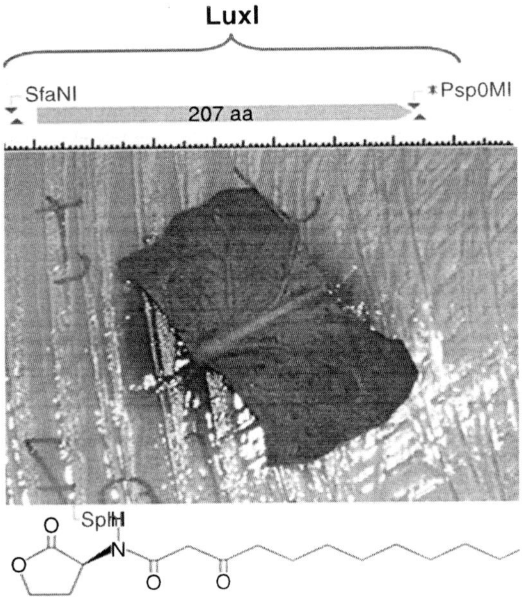

Figure 6.3 Tobacco transgenic plant leaf transformed with the AHL-synthase LuxI from *Burkholderia graminis*. Production of violacein (purple color) in the agar plate is due to the detection of the QSM by the strain CV026 of *Chromobacterium violaceum*. (*See color plate section for the color representation of this figure.*)

molecules [16]. Two tomato transgenic lines were engineered to exudate by the roots the main AHL QS signal produced by each *B. graminis* strain (Figure 6.3). To determine whether plant growth promotion and protection against salt stress were mediated by QS, these PGPRs were inoculated on wild-type plants, as well as their corresponding transgenic lines expressing YenI (short-chain AHL producer) and LasI (long-chain AHL producer). The growth promotion effect disappeared in both transgenic lines when inoculated with M12. However, while M14 did not promote growth in wild-type tomatoes, it was effective in the LasI transgenic line, revealing that AHL QS signal molecules mediate the ability of both PGPR strains to promote plant growth and to induce protection against salt stress [16].

6.3 Conclusions

The study of the QS mechanisms underlaying the formation of biofilms in the rhizosphere is of great importance since these structures are vital in the root colonization processes, and for the persistence and survival in the environment of diverse microorganisms, some of them with a positive effect on plant health and fitness and others with a negative effect, such as pathogens. These interactions are of great importance to conserve soil quality and fertility, and eventually to improve crops productivity through the management of the rhizosphere populations.

References

1 Nealson KH, Platt T, Hastings JW. Cellular control of the synthesis and activity of the bacterial luminescent system. *J. Bacteriol.* 1970;104(1):313–322.
2 Lu Y, Su C, Unoje O, Liu H. Quorum sensing controls hyphal initiation in *Candida albicans* through Ubr1-mediated protein degradation. *Proc. Natl. Acad. Sci. USA.* 2014;111(5):1975–1980.
3 Williams HE, Steele JC, Clements MO, Keshavarz T. Gamma-Heptalactone is an endogenously produced quorum-sensing molecule regulating growth and secondary metabolite production by Aspergillus nidulans. *App. Microbiol. Biotech.* 2012;96:773–781.
4 Miller MB, Bassler BL. Quorum sensing in bacteria. *Annu. Rev. Microbiol.* 2001;55:165–199.
5 Wuster A, Babu MM. Transcriptional control of the quorum sensing response in yeast. *Mol. Biosyst.* 2010;6(1):134–141.
6 Fuqua WC, Winans SC, Greenberg EP. Quorum sensing in bacteria—The LuxR-LuxI family of cell density-responsive transcriptional regulators. *J. Bacteriol.* 1994;176:269–275.
7 Bandara HMHN, Lam OLT, Jin LJ, Samaranayake L. Microbial chemical signaling: a current perspective. *Cr. Rev. Microbiol.* 2012;38:217–249.
8 Puskas A, Greenberg EP, Kaplan S, Schaefer AL. A quorum-sensing system in the free-living photosynthetic bacterium *Rhodobacter sphaeroides*. *J. Bacteriol.* 1997;179(23):7530–7537.
9 Bassler BL. How bacteria talk to each other: regulation of gene expression by quorum sensing. *Curr Opin Microbiol.* 1999;2(6):582–587.
10 Pesci EC, Milbank JB, Pearson JP, McKnight S, Kende AS, Greenberg EP, Iglewski BH. Quinolone signaling in the cell-to-cell communication system of *Pseudomonas aeruginosa*. *Proc. Natl. Acad. Sci. USA*. 1999;28;96(20):11229–11234.
11 Nickerson KW, Atkin AL, Hornby JM. Quorum sensing in dimorphic fungi: farnesol and beyond. *Appl. Environ. Microbiol.* 2006;72(6):3805–3813.
12 de Salas F, Martínez MJ, Barriuso J. Quorum-sensing mechanisms mediated by farnesol in *Ophiostoma piceae*: effect on secretion of sterol esterase. *Appl. Environ. Microbiol.* 2015;81(13):4351–4357.
13 Raina S, Odell M, Keshavarz T. Quorum sensing as a method for improving sclerotiorin production in *Penicillium sclerotiorum*. *J. Biotechnol.* 2010;148(2–3):91–98.
14 Hornby JM, Jensen EC, Lisec AD, et al. Quorum sensing in the dimorphic fungus *Candida albicans* is mediated by farnesol. *App. Environ. Microbiol.* 2001;67(7):2982–2992.
15 Hogan DA. Talking to themselves: Autoregulation and quorum sensing in fungi. *Eukaryot. Cell.* 2006;5(4):613–619.
16 Barriuso J, Ramos Solano B, Fray RG, Cámara M, Hartmann A, Gutiérrez Mañero FJ. Transgenic tomato plants alter quorum sensing in plant growth–promoting rhizobacteria. *Plant Biotechnol. J.* 2008;6(5):442–452.
17 Camara M, Williams P, Hardman A. Controlling infection by tuning in and turning down the volume of bacterial small-talk. *Lancet Infect. Dis.* 2002;2:667–676.

18 Geiger J, Wessels D, Lockhart SR, Soll DR. Release of a potent polymorphonuclear leukocyte chemoattractant is regulated by white-opaque switching in *Candida albicans*. *Infect. Immun.* 2004;72(2):667–677.
19 Pérez-Montaño F, Jiménez-Guerrero I, Contreras Sánchez-Matamoros R, López-Baena FJ, Ollero FJ, Rodríguez-Carvajal MA, Bellogín RA, Espuny MR. Rice and bean AHL-mimic quorum-sensing signals specifically interfere with the capacity to form biofilms by plant-associated bacteria. *Res Microbiol.* 2013; 164(7):749–60.
20 Madhani HD. Quorum sensing in fungi: Q&A. *PLoS Pathog.* 2011; 7(10):e1002301.
21 Shank EA, Klepac-Ceraj V, Collado-Torres L, Powers GE, Losick R, Kolter R. Interspecies interactions that result in *Bacillus subtilis* forming biofilms are mediated mainly by members of its own genus. *Proc. Natl. Acad. Sci. USA.* 2011; 108:E1236–1243.
22 Rudrappa T, Biedrzycki ML, Bais HP. Causes and consequences of plant-associated biofilms. *FEMS Microbiol. Ecol.* 2008;64(2):153–166.
23 Watnick P and Kolter R. Biofilm, city of microbes. *J. Bacteriol.* 2000;182: 2675–2679.
24 Barrios AFG, Covo V, Medina LM, Vives-Florez M, Achenie L. Quorum quenching analysis in *Pseudomonas aeruginosa* and *Escherichia coli*: network topology and inhibition mechanism effect on the optimized inhibitor dose. *Bioprocess Biosyst. Eng.* 2009;32:545–556.
25 Zhu J, Kaufmann GF. Quo vadis quorum quenching? *Curr. Opin. Pharmacol.* 2013;13(5):688–698.
26 Medina-Martinez MS, Uyttendaele M, Rajkovic A, Nadal P, Debevere J. Degradationof N-acyl-L-homoserine lactones by *Bacillus cereus* in culture media and pork extract. *Appl. Environ. Microbiol.* 2007;73:2329–2332.
27 Rashid R, Moroshoshi T, Someya N, Ikeda T. Degradation of N-acylhomoserine lactone quorum sensing signaling molecules by potato Root surface-associated *Chryseobacterium* strains. *Microbes. Environ.* 2011;26:144–148.
28 Kwasiborski A, Mondy S, Chong TM, Chan KG, Beury-Cirou A, Faure D. Core genome and plasmidome of the quorum-quenching bacterium *Rhodococcus erythropolis*. *Genetica.* 2015;143(2):253–261.
29 Hall-Stoodley L, Costerton JW, Stoodley P. Bacterial biofilms: from the natural environment to infectious diseases. *Nat. Rev. Microbiol.* 2004;2:95–108.
30 Davies DG, Parsek MR, Pearson JP, Iglewski BH, Costerton JW, Greenberg EP. The involvement of cell-to-cell signals in the development of a bacterial biofilm. *Science.* 1998;10;280(5361):295–298.
31 Angus A, Hirsch A. Molecular microbial ecology of the rhizosphere, Vol. 2, Chapter 66. *Frans de Bruijn*, ed. Hoboken, NJ: John Wiley and Sons; 2013. pp. 703–712.
32 Barriuso J, Marín S, Mellado RP. Potential accumulative effect of the herbicide glyphosate on glyphosate-tolerant maize rhizobacterial communities over a three-year cultivation period. *PLoS ONE.* 2011; 6(11):e27558.
33 Hirsch AM. How rhizobia survive in the absence of a legume host, a stressful world indeed. In: Seckbach J, Grube M, editors. *Symbiosis and Stress: Joint Ventures in Biology, Cellular Origin, Life in Extreme Habitats and Astrobiology.* Volume 17. Springer; New York; 2010. pp 375–391.
34 Ramos Solano B, Barriuso Maicas J, Pereyra de la Iglesia MT, Domenech J, Gutiérrez Mañero FJ. Systemic disease protection elicited by plant growth promoting

rhizobacteria strains: Relationship between metabolic responses, systemic disease protection, and biotic elicitors. *Phytopathology*. 2008;98(4):451–457.

35 Matthysse AG, Marry M, Krall L, Kaye M, Ramey BE, Fuqua C, White AR. The effect of cellulose overproduction on binding and biofilm formation on roots by *Agrobacterium tumefaciens*. *Mol. Plant. Microbe. Interact.* 2005;18(9):1002–1010.

36 Wei HL, Zhang LQ. Quorum-sensing system influences root colonizationand biological control ability in *Pseudomonas fluorescens* 2P24. *Antonie Van Leeuwenhoek*. 2006;89:267–280.

37 Haggag WM, Timmusk S. Colonization of peanut roots by biofilmforming *Paenibacillus polymyxa* initiates biocontrol against crown rot disease. *J. Appl. Microbiol.* 2008;104:961–969.

38 Sant'Anna FH, Almeida LG, Cecagno R, Reolon LA, Siqueira FM, Machado MR, Vasconcelos AT, Schrank IS. Genomic insights into the versatility of the plant growth-promoting bacterium *Azospirillum amazonense*. *BMC Genomics*. 2011;12(12):409.

39 Park SY, Lee SJ, Oh TK, Oh JW, Koo BT, Yum DY, Lee JK. AhlD, an N-acylhomoserine lactonase in *Arthrobacter* sp., and predicted homologues in other bacteria. *Microbiology*. 2003;149:1541–1550.

40 Shankar M, Ponraj P, Illakkiam D, Rajendhran J, Gunasekaran P. Inactivation of the transcriptional regulator-encoding gene *sdiA* enhances rice root colonization and biofilm formation in *Enterobacter cloacae* GS1. *J. Bacteriol.* 2013;195(1):39–45. doi:10.1128/JB.01236-12.

41 Bandara WM, Seneviratne G, Kulasooriya SA. Interactions among endophytic bacteria and fungi: effects and potentials. *J. Biosci.* 2006;31(5):645–650.

42 Müller H, Westendorf C, Leitner E, Chernin L, Riedel K, Schmidt S, Eberl L, Berg G. Quorum-sensing effects in the antagonistic rhizosphere bacterium *Serratia plymuthica* HRO-C48. *FEMS Microbiol. Ecol.* 2009;67(3):468–478.

43 Rajesh PS, Ravishankar Rai V. Quorum quenching activity in cell-free lysate of endophytic bacteria isolated from *Pterocarpus santalinus* Linn., and its effect on quorum sensing regulated biofilm in *Pseudomonas aeruginosa* PAO1. *Microbiol. Res.* 2014;169(7–8):561–569.

44 Rinaudi LV, Giordano W. An integrated view of biofilm formation in rhizobia. *FEMS Microbiol. Lett.* 2010;304(1):1–11.

45 Glenn SA, Gurich N, Feeney MA &Gonz´alez JE. The ExpR/Sin quorum-sensing system controls succinoglycan production in *Sinorhizobium meliloti*. *J. Bacteriol.* 2007;189:7077–7088.

46 Hoang HH, Becker A and Gonzalez JE. The LuxR homolog ExpR, in combination with the Sin quorum sensing system, plays a central role in *Sinorhizobium meliloti* gene expression. *J. Bacteriol.* 2004;186:5460–5472.

47 Pérez-Montaño F, Jiménez-Guerrero I, Del Cerro P, Baena-Ropero I, López-Baena FJ, Ollero FJ, Bellogín R, Lloret J, Espuny R. The symbiotic biofilm of *Sinorhizobium fredii* SMH12, necessary for successful colonization and symbiosis of *Glycine max* cv Osumi, is regulated by Quorum Sensing systems and inducing flavonoids via NodD1. *PLoS One*. 2014;28;9(8):e105901.

48 Edwards A, Frederix M,Wisniewski-Dy´e F, Jones J, Zorreguieta A & Downie JA. The cin and rai quorum-sensing regulatory systems in *Rhizobium leguminosarum* are coordinated by ExpR and CinS, a small regulatory protein coexpressed with CinI. *J. Bacteriol.* 2009;191:3059–3067.

49 Wang H, Zhong Z, Cai T, Li S and Zhu J. Heterologous overexpression of quorum-sensing regulators to study celldensity-dependent phenotypes in a symbiotic plant bacterium *Mesorhizobium huakuii*. *Arch. Microbiol.* 2004;182:520–525.
50 Masák J, Čejková A, Schreiberová O, Rezanka T. 2014. Pseudomonas biofilms: possibilities of their control. *FEMS Microbiol. Ecol.* 89(1):1–14.
51 Walker TS, Bais HP, Déziel E, et al. *Pseudomonas aeruginosa*—Plant root interactions. Pathogenicity, biofilm formation, and root exudation. *Plant Physiol.* 2004;134(1):320–331. doi:10.1104/pp.103.027888.
52 Teplitski M, Robinson JB, Bauer WD. Plants secrete substances that mimic bacterial N-acyl homoserine lactone signal activities and affect population density-dependent behaviors in associated bacteria. *Mol. Plant Microbe. Interact.* 2000;13(6):637–648.

7

Biofilm Formation and Quorum Sensing in Rhizosphere

Kusum Harjai and Neha Sabharwal

Department of Microbiology, Panjab University, Chandigarh, India

7.1 Introduction

The term *rhizosphere* originating from a Greek word *rhiza,* meaning root, was coined in 1904 by the German agronomist and plant physiologist Lorenz Hiltner [1]. The rhizosphere constitutes the area surrounding the plant root with characteristic microbiota whose inhabitancy is associated with the respective plant root secretions. The sloughed-off plant cells nourish rhizospheric microbiota, and this phenomenon is termed *rhizodeposition*. Predators like protozoa and nematodes are also prevalent in the rhizosphere. Thus, the rhizosphere acts as the metabolically and immunologically busiest spot of the plant [2].

Over the years, the rhizosphere has been extensively studied. Based on the distance and access from the root, it constitutes three zones: The *endorhizosphere* includes portions of the cortex and endodermis in which microbes and cations occupy the "free space" between cells (apoplastic space). The *rhizoplane* is the medial zone directly adjacent to the root including the root epidermis, mucilage, and outer cortex. Here, soil particles, bacteria, and fungal hyphae adhere. Rhizoplane microorganisms tend to be found on older rather than younger roots [3, 4, 4a]. The outermost zone is the *ectorhizosphere,* which extends from the rhizoplane out into the bulk soil (soil that is not part of the rhizosphere). Rhizosphere does not possess definite shape or size due to its intricacy and diversity of plant root systems. The chemical, physical, and biological properties of rhizosphere often change both longitudinally and radially along the root [5].

7.2 Importance of Rhizosphere

Rhizosphere is considered as the most active portion of soil. The biogeochemical reactions in rhizosphere can influence geographically and a range of global scale processes. It is the need of the hour to have a better vision of these processes for maintaining the sustainability of our planet [6]. The efforts toward this is progressing very slowly, as the majority is focused on just meeting the demand of staple food crops with respect to doubling need in the next 50 years [7, 8]. Rhizosphere microorganisms strongly influence nutrient uptake of plants by altering nutrient availability. They accelerate soil remediation by immunizing the plants toward abiotic stress and by improving plant growth [9]. Thereby rhizosphere plays a prominent role with respect to global climatic changes and population explosion that require heightened output of food, feed, and fiber on less productive (and often infertile) lands—the current scenario in most of the developing countries [10]. Thus, the scientific community is looking forward to unraveling the untold inner secrets of rhizosphere processes.

7.3 Constituents of Rhizosphere

7.3.1 Physical/Chemical

The rhizodeposits including sloughed-off root cap and border cells, mucilage, and exudates are the major constituents of rhizosphere. Mucilage is secreted by root cap and epidermal cells. It is a viscous, insoluble, high molecular weight, polysaccharide-rich material that assists in nutrient acquisition, sequestration of toxic metals, and protecting plants from desiccation [11].

The organic compounds in the rhizodeposits can be of high or low molecular weight. The former include mucilage and cellulose, which are unable to be processed by microbes, whereas the latter include organic acids, amino acids, proteins, sugar, phenolics, and other secondary metabolites that can be easily consumed by microbes.

7.3.2 Rhizosphere—A Hot Niche of Microbial Activity

The occurrence of high microbial density is the characteristic feature of rhizhosphere. Rhizodeposits provide the nourishment for microbiota to proliferate in the rhizosphere. These chemical cocktails also serve as (i) invasive agents (i.e. allelopathy), (ii) chemotactic agents (like in rhizobia and legumes), and (iii) agents that promote proliferation of beneficial microbes on root surfaces (e.g. *Bacillus subtilis, Pseudomonas florescence*) [12]. Among these, acidic soils favor fungal growth, whereas bacteria survive in a broad pH spectrum. Thus, pH imparts a significant role in selecting the diversity of micorbiota of the rhizosphere.

7.3.2.1 Bacteria

Rhizospheric bacteria are predominantly Gram-negative, rod-shaped, nonsporulating bacteria that thrive in root exudates (e.g., *Pseudomonas, Agrobacterium* etc.). Gram-positive rods, cocci, and aerobic spore forming bacteria like *Bacillus* and *Clostridium* are rarely seen in the rhizosphere, whereas *Pseudomonas, Arthrobacter, Agrobacterium, Alcaligenes, Azotobacter, Mycobacterium, Flavobacter, Cellulomonas* and *Micrococcus* are the most common genera of bacteria found in the rhizosphere. The root respiration limits the oxygen level and hence survival of aerobic bacteria.

Among the Gram-negative flora in the rhizosphere, the overwhelming literature pertains to abundance of pseudomonads, including *Pseudomonas* species like *P. aeruginosa, P. aureofaciens, P. fluorescens,* and *P. syringae*. They have adapted in a better way, making them the commonest rhizosphere organisms [13, 14]. These bacteria produce antibiotics, particularly Phenazine-1-carboxylic acid (PCA), which operates via a two-component system constituting an environmental sensor (presumably a membrane protein) and a cytoplasmic response factor [15]. Another feature of these bacteria is the production of iron (Fe^{3+}) chelators or siderophores under iron scarcity. This process of iron sequestera-tion also helps in avoiding the presence of pathogenic fungi, which require iron to grow [16, 17]. Pseudomonads produce pyoverdine as the major siderphore and pyochelin and its precursor salicylic acid in a limited manner. It is assumed that pyochelin produced by *Pseudomonas aeruginosa* Migula 7NSK2 plays a significant role in protecting tomato plants from *Pythium* [18].

7.3.2.2 Fungi

Plants roots hardly support the growth of fungi in the rhizosphere as compared with the support provided to bacteria. However, *Fusarium, Verticillium, Aspergillus,* and *Penicillium* fungal genera are still present in the rhizosphere. *Phytophthora, Pythium, and Aphanomyces* are some other lower fungi that thrive on root exudates. The *phytohormones* produced by fungi such as *Gibberella* and *Fujikurio* influence the plant growth. The majority of all terrestrial plant roots share a mutualistic relationship with one or more mycorrhizal fungi—for example: *Boletus betulicola, Pisolithus tinctorius, Glomus tenue,* and *Scutellospora* [3]. Antibiotic production by fungi, which has been used in disease biocontrol activity, is reported by *Trichoderma/Gliocladium* [19] and *Talaromyces flavus* isolates [20, 21].

7.3.2.3 Actinomycetes and Protozoa

Actinomycetes are bacteria known to constitute a large part of the rhizosphere microbiota. Evidence indicates that actinomycetes, being less supported by roots, as compared to bacteria, enhance plant growth and protect plant roots against fungal pathogens [22]. Among them, *Nocardia* and *Streptomyces* sp, the phosphate solublizers, are able to suppress the growth of many fungal pathogens. Actinomycetes producing antifungal compounds have been isolated from plant rhizosphere such as *Vitis vinifera, Argania spinosa,* sagebrush (*Artemisia tridentata*), and *Zingiber officinale* [23]. Additionally, actinomycetes are source of important therapeutic products and can produce valuable pharmaceutical substances of biotechnological interest [4].

The micro-fauna (protozoa and nematodes) in soil also play an indespensable role in the rhizosphere. They aid in providing nutrients to the plant, enrichment, and stabilization of soil organic carbon. In addition, they also exert significant hormonal effects on roots, microbial diversity, functional stability, and soil remediation. The specialized metabolism performed by protozoa and nematodes not only benefit the rhizosphere ecological functions, but also have an impact on whole soil and the terrestrial organisms [24]. Here, dominant species are flagellates and amoebae while ciliates represent a rare community.

7.4 Communication in Rhizosphere

There are molecular communications in the rhizosphere, which have been observed between plants and bacteria and within bacterial genera. Bacteria use signaling pathways to coordinate gene expression within the population. In turn, plants can respond to bacterial infections.

It has already been discussed that under appropriate cell density, bacteria can modify their behavior to act as multicellular entities. In natural ecosystems, bacteria are adapted to live in communities by exploiting the intercellular communication network. Such microbial cell-to-cell communications, which operate via small signaling molecules, have recently been unraveled [25]. Among various cell-density-dependent signaling pathways, the major regulating mechanism is identified as quorum sensing (QS). The term QS was coined by Fuqua and co-workers [26]. It is an excellent mechanism of inter-talk between bacteria and a wonderful cell-density-dependent regulatory network. The chemical signals called autoinducers form the building blocks of QS. According to the population explosion in bacteria, the extracellular concentration of these autoinducers increases. Upon reaching a threshold concentration, individual cells sense the surrounding cell population. When there are enough bacteria—a *quorum*—they initiate the regulation and thereby change the gene expression [27].

Chemically, these autoinducers are N-acyl-homoserine lactones (AHLs), constituting a conserved lactone moiety and an acyl chain. The length, degree of oxidation and saturation of carbon chain, of the acyl side chain may vary in different signals [28]. QS system is integrated into a complex, multilayered signal transduction network which regulates many bacterial physiological activities including symbiosis, virulence, biofilm formation competence, conjugation, motility, sporulation and antibiotic production [29–31]. Thus, QS optimizes the metabolic and behavioral activities of a community of bacteria. Nowadays it has been attributed to socio-microbiology of bacteria, the understanding of which would open many gates.

Gene expression is regulated by QS. In this process, the AHLs accumulate in the extracellular environment, attain a critical threshold concentration, and trigger the response that leads to gene expression. Evidences ascertain that the diversity in colonization of microflora in rhizosphere favors the possibility of QS signaling between the rhizospheric bacterial populations. It is estimated that 10 to 20 percent of the cultivable bacteria in soil and rhizospheric environments are AHL-producing [32]. Diverse bacterial populations produce the same AHLs or AHLs with similar structures and properties, indicating existence of QS.

AHLs consist of an HSL head group attached to a variable acyl side chain. The amphipathy of the AHL molecule seems to be a balance between the hydrophobic side chain and the hydrophilic HSL ring. These characteristics presumably allow the AHLs to traverse the phospholipid bilayer of the cell membrane and to navigate the aqueous intracellular and extracellular environments [33]. The acyl chain varies in length, from 4 to 18 carbons in the AHLs identified so far [34, 35]. Variability also exists in the third carbon position of the acyl chain, where there can be a hydrogen, hydroxyl, or oxo substitution. The overall length of the side chain and the chemical modification at the third carbon position provide the specificity to quorum-sensing signals.

It is evident that quorum sensing via AHLs is more common among plant-associated bacteria than the general population of soil bacteria [32]. Other than inter-bacterial communication, the functions controlled by QS includes horizontal transfer of plasmids, regulation of rhizospheric competence factors such as antibiotics as well as production of virulence factors for better plant associations [25]. A better knowledge of role of QS in the rhizosphere will facilitate sustained exploitation of bioinoculants in soil health, plant output, and bioremediation strategies that often determine the fate of a microorganism introduced in the natural ecosystems [27]

Broadly, microbially derived signaling molecules are of two main categories: (i) Gram-positive bacteria–generated amino acids and short peptide pheromones [36] and

(ii) Gram-negative bacteria generate fatty-acid derivatives (AHLs) [37]. *Burkholderia cepacia, Pseudomonas chlororaphis, P. fluorescens, Rhizobium elti, R. leguminosarum, Sinorhizobium meliloti,* plant pathogens, *Agrobacterium rhizogenes, A. tumefaciens, Erwinia carotovora, E. chrysanthemi, E. stewartii* and *Pseudomonas syringae,* and saprophytes, *Chromobacter violaceum, Nitrosomonas europaea, Pseudomonas corrugata,* and *Pseudomonas putida* exhibit QS systems based on acyl homoserine lactones (AHLs) to communicate in the rhizosphere [27].

7.5 Quorum Sensing in Rhizobia

Rhizobia are Gram-negative, motile, nonsporulating soil bacterial rods that fix nitrogen in root nodules of legumes. They are in symbiotic relationship with plant host in order to fix nitrogen. Nitrogen fixed from the air into ammonia acts as a natural fertilizer for the plants. Rhizobia fall into two classes of proteobacteria—alpha and beta proteobacteria.

Alpha proteobacteria includes:

- Bradyrhizobium
- Rhizobium
- Brucella
- Methylobacterium
- Aminobacter
- Mesorhizobium
- Phyllobacterium
- Shinella
- Sinorhizobium

Beta proteobacteria includes:

- Burkholderia

The most commonly studied bacteria in Rhizobia are: Rhizobium, Sinorhizobium, and Mesorhizobium.

7.5.1 Quorum Sensing in Rhizobium

Most of the identified rhizobial quorum-sensing regulation systems seem to be based on AHL synthesis. Rhizobia lack autoinducer-2 (AI-2) (LuxS-dependent) biosynthetic pathway [38]. QS in *Rhizobium elti* and *R. leguminosarum* has been well characterized. Production of AHLs by rhizobial strains has also been identified [39, 40]. There are three different biovars of *R. leguminosarum*: bv. *viciae* (which nodulates peas, vetch and lentils), bv. *trifolii* (which nodulates clover), and bv. *phaseoli* (which nodulates *Phaseolus* beans).

Most research has been done on *R. leguminosarum* bv. *viciae*. Four different LuxI-type AHL synthase genes have been identified in different isolates of *R. leguminosarum* bv. *viciae* [41]. Each of the following AHL synthase genes has a closely linked dedicated regulator encoded by a gene.

7.5.1.1 *cinI* and *cinR*

Many of the gene products required for legume rhizobia symbiotic relationship are encoded by symbiosis (Sym) plasmids. The QS system of *R. leguminosarum* is regulated by chromosomally encoded cinRI proteins [42]. The products of *cinR* and *cinI* genes were previously thought to be a bacteriocin named *small* [43]. However, the purified *small* bacteriocin molecule, which turned out to be an AHL, was identical to 3OH-C14:1-HSL in structure [42]. The product of *cinI* gene is responsible for synthesis of *N*-(3-hydroxy-7-*cis*-tetrade-cenoyl)-L-homoserine lactone (htde-DHL), whose expression is dependent on Htde DHL-activated cinR. CinR is a LuxR-type regulator which positively regulates *cinI* expression in response to htdeDHL (3OH-C14:1-HSL). The products of *cinI* and *cinR* genes additionally regulate the expression of at least one another chromosomally encoded AHL synthase, whereas two other HSL synthases are encoded on the Sym plasmid (pRL1J1) of *R. leguminosarum*. Other rhizobial species such as *R. meliloti* produce an array of compounds with AHL-like activity, but *R. fredii* produces just one strongly nonpolar compound with such activity [39, 40]. Mutations in *cinI* or *cinR* genes reduce the expression of all other AHL synthase genes [42, 44]. It appears that the *cinI*/*cinR* system acts as an overall switch, potentially influencing many aspects of rhizobial physiology [41].

3-OH-C14:1-HSL act as survival advantage during stationary phase. Reports indicated that the addition of exogenous 3-OH-C14:1-HSL promoted starvation survival in cultures of *R. leguminosarum bv. phaseoli*, entering stationary phase at minimal population density [45]. Likewise, in *R. leguminosarum*, small bacteriocin-associated growth inhibition is related to 3-OH-C14:1-HSL [34], which is produced by *cinI* [42].

In the complex cascade of quorum-sensing loops, cinMR locus appears to be the master control for three other AHL-dependent QS control systems described as follows.

7.5.1.2 *raiI* and *raiR*

The *raiI* and *raiR* genes are located on a large (nonsymbiotic) plasmid in *R. leguminosarum* bv. *phaseoli*. They are absent from the genome of *R. leguminosarum* bv. *viciae* and some other analyzed strains of *R. leguminosarum* bv. *viciae* [42]. However, the phenotype of *R. leguminosarum* is least affected by these genes. RaiR regulates the expression of *raiI* in response to the RaiI-made AHLs 3-OH-C8-HSL and C8-HSL [41]. Other genes regulated by RaiR are not yet identified.

7.5.1.3 *rhiI* and *rhiR*

The *rhiR* gene was originally identified because it was close to the genes (*nod*) required for legume nodulation and is required for the expression of the *rhiA* gene, which is highly expressed in the rhizosphere [46]. RhiR regulates the expression of *rhiI* and *rhiABC* operon (a unit made up of linked genes thought to regulate other genes responsible for protein synthesis) in response to RhiI-made C6-HSL, C7-HSL and C8-HSL [47].

The observations that the *rhi* genes are closely linked to nodulation and nitrogen fixation genes and are found only in bv. *viciae* but not in other biovars of *R. leguminosarum* suggest their probable role in growth and/or survival in association with specific legume hosts. Evidence indicates that mutations in *rhiA* or *rhiR* cause a significant reduction in nodulation in strains that are already compromised for nodulation ability [46].

7.5.1.4 *traI* and *traR*

TraR induces *traI* gene in response to TraI-made 3-oxo-C8-HSL. These genes are located on the symbiosis plasmid (pRL1JI). Together with *bisR* (encoding another LuxR-type regulator), both are required to induce the plasmid transfer genes.

Other rhizobial strains share some of these QS loci described above, but not all loci are found in all strains. In conclusion, we may say that *R. leguminosarum* harbors four known QS systems. The *cinRI* system resides on the chromosome and produces 3-OH-C14:1-HSL, which positively influences the *tra* and *rai* systems. BisR plays a dual role in activating *traR* and repressing *cinR* in response to 3-OH-C14:1-HSL, thereby linking the *cin* and *tra* systems. pRL1JI harbors both the *tra* and *rhi* systems, as well as the genes that confer growth sensitivity in response to 3-OH-C14:1-HSL. The *tra* system is responsible for the production of 3-oxo-C8-HSL and controls conjugal plasmid transfer, while the *rhi* system produces several short-chain AHLs and influences nodulation efficiency. The *raiRI* locus resides on pIJ9001 and also produces several short-chain AHLs [48].

In addition to the above, small bacteriocin along with the other N-acyl-homoserine lactones produced by these three AHL-based control systems regulate (i) growth inhibition of sensitive strains, (ii) transfer of the symbiotic plasmid pRL1JI, and (iii) expression of the rhizosphere-expressed (rhi) genes that influence nodulation [41].

7.5.2 Quorum Sensing in Sinorhizobium

Sinorhizobium is a genus of nitrogen-fixing bacteria (rhizobia), three of which (*Sinorhizobium meliloti, Sinorhizobium medicae,* and *Sinorhizobium fredii*) have been sequenced. QS was observed in genomically sequenced *Sinorhizobium meliloti* 1021, the Nitrogen-fixing bacterial symbiont of alfalfa (*Medicago sativa*) host plants. Its QS involves half a dozen different *N*-acyl homoserine lactone (AHL) signals and equal number of AHL receptors. It is dependent on SinI, the AHL synthase, and/or on ExpR, one of the AHL receptors.

7.5.2.1 *sinI* and *sinR*

AHLs produced by the SinI synthase have long acyl side chains, including C12-HSL, C14-HSL, 3-oxo-C14-HSL, C16-HSL, C16:1-HSL, 3-oxo-C16:1-HSL, and C18-HSL. Short-chain AHLs, including C6-HSL, 3-oxo-C6-HSL, and C8-HSL, are also produced by a second, unidentified AHL synthase [49].

Two classical LuxR AHL receptors, SinR and ExpR, have been identified in *S. meliloti* [50]. *sinR* is adjacent to *sinI* on the chromosome, and SinR induces *sinI* expression in response to SinI-made AHLs, thereby enhancing AHL production. Slightly delayed nodulation has been reported in *sinR* mutant. In microarray studies, three *sinI*-dependent genes had altered expression in the *sinR* mutant, which was consistent with SinI–AHL-dependent regulation via SinR [51].

7.5.2.2 *expR*

Functional ExpR is generated by spontaneous excision of the insertion sequence. It regulates the expression of many genes, including genes in the *exp* operon, required for the synthesis of EPSII in strain *S. meliloti* 1021 [50–52]. EPSII is one of the *S. meliloti* exopolysaccharides that is capable of eliciting responses in the host [53]. On the contrary, most wild-type strains of *S. meliloti* have an uninterrupted *expR* gene [50]. There is overexpression of more than 50 proteins and the expression of more than 80 genes in the laboratory cultures of *expR* mutants. It does not have any obvious nodulation phenotype [51, 52]. These genes and proteins carry out broad ranges of functions such as central metabolism, regulation, transport, motility, and symbiotically related behaviors.

7.5.2.3 *traI, traR* and *melI*

Strain *S. meliloti* 1021 lacks *traI* and *traR*. Strain Rm41 harbors *traI* and *traR* next to plasmid transfer genes. Strain Rm41 has an AHL synthase encoded by *melI,* which produces AHLs with short acyl side chains as well as *sinI*-synthesized long-chain AHLs [49].

7.5.3 Quorum Sensing in *Mesorhizobium*

Mesorhizobium tianshanense is a moderate-growth *Rhizobium* that forms nodules on a number of licorice plants. It produces multiple N-acyl homoserine lactone (AHL)-like molecules [54]. In the genome of *Mesorhizobium loti,* two LuxI-type AHL synthase genes are present adjacent to a LuxR-type regulator gene and other genes that might be associated with conjugal transfer. A strain of *Mesorhizobium huakuii* has also been shown to produce AHLs [55].

With regard to other rhizobia, although detectable amounts of AHLs have been tested, not much information is available regarding AHL synthesis genes and genes regulated by AHLs.

7.6 Quorum Sensing in Pseudomonads

7.6.1 Quorum Sensing in *Pseudomonas aeruginosa*

Rhizosphere is enriched with psedomonads. Quorum sensing in *Pseudomonas aeruginosa* has been extensively studied. There are two major QS systems operating, *las* and the *rhl* system. Each system constitutes a transcriptional activator and an autoinducer synthase. The autoinducers, PAI-1 and PAI-2, bind to specific target proteins, the transcriptional activators, and these complexes activate a large number of virulence factors. Two QS systems, *lasR*/*lasI* and *rhlR*/*rhlI,* are organized into a complex hierarchy in *P. aeruginosa,* which are united to regulate transcription of genes required for virulence. Recently, other than homoserine lactone molecules, a second intercellular signal has been identified as 2-heptyl-3-hydroxy-4-quinolone and designated the *Pseudomonas* quinolone signal (PQS).

7.6.1.1 *Las* System

The Las system consists of the transcriptional activator, *LasR,* which is homologous to the LuxR product in the QS system of *Vibrio fischeri.* This protein binds to the autoinducer, N-(3-oxododecanoyl)-L-homoserine lactone (PAI-1, C12-HSL or OdDHL) at high cell density and regulates the expression of *lasA* (LasA protease), *apr* (alkaline protease A), *toxA* (exotoxin A), *lasI* (the PAI-1 synthase), and *lasB* (elastase) [29, 56]. The *las* cell-to-cell signaling system is positively controlled by GacA as well as by *vfr,* which is required for the transcription of *lasR.*

7.6.1.2 *Rhl* System

The Rhl system consists of the transcriptional activator, RhlR and an autoinducer synthase, RhlI. RhlI directs the synthesis of N-butyryl-1-homoserine lactone (PAI-2, C4-HSL or BHL). PAI-2 binds to RhlR, and this complex activates the transcription of *rhlI, rhlA* and *rhlB,* an operon coding for rhamnosyltransferase (which is required for rhamnolipid production) and rpoS (a stationary-phase sigma factor) [56, 57]. Like the *las* cell-to-cell signaling, the *rhl*, also known as *vsm* (*v*irulence *s*econdary *m*etabolites), regulates the expression of various extracellular virulence factors of *P. aeruginosa* [27].

To initiate the second signaling cascade, the LasR–autoinducer complex activates *rhlR* expression. However, simultaneously, the LasR-dependent autoinducer, OdDHL prevents binding of the RhlI-dependent autoinducer, BHL to its cognate regulator. This second level control of RhlI/RhlR autoinduction by the LasI/LasR system ensures that the two systems initiate their cascades sequentially [58]. *P. aeruginosa* AHL systems function in a hierarchical manner, as the 3-oxo-C12-AHL-LasR complex positively regulates *rhlI*, *rhlR* and *mvfR* expression as well as *lasI*.

7.6.1.3 PQS System

The third interconnecting system in the QS network, the PQS system, is regulated by Las and Rhl QS systems, respectively [59]. PQS system is involved in the regulatory mechanisms along with Las and Rhl systems like partially controlling the expression of *lasB*. Expression of PQS essentially requires LasR, and it, in turn, induces transcription of *rhlI*. Thus, PQS serves as an additional link between the Las and Rhl circuits (Figure 7.1). It is believed that PQS initiates the Rhl cascade by allowing production of RhlI directed autoinducer only after the establishment of LasI/LasR signaling cascade [60]. PQS is packaged within outer membrane vesicles and transported to recipient cells.

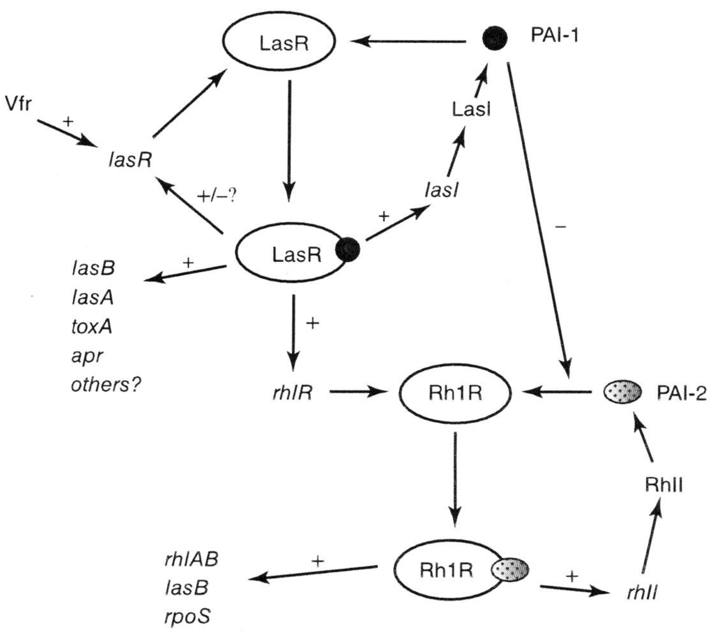

Figure 7.1 Model of the *P. aeruginosa* QS circuitry. LasR and RhlR are symbolized by circles. Plus (+) symbols indicate transcriptional activation of the gene. As culture density increases, *lasR* is activated by Vfr and PAI-1, reaches a threshold concentration and binds to LasR. Once RhlR associates with PAI-2, autoinduction of *rhlI* occurs and the remainder of the RhlR–PAI-2-controlled genes are activated. Las system also positively controls the expression of PQS, another system known to be in between the hierarchical level of Las and Rhl [58].

7.6.2 Quorum Sensing in Other Pseudomonads

The best-characterized QS system in Pseudomonadaceae is that of *P. aureofaciens*. This rhizospheric fluorescent pseudomonad produces three phenazine antibiotics, phenazine 1-carboxylic acid, 2-hydroxy-phenazine 1-carboxylic acid, and 2-hydroxy-phenazine. This specie inhibits phenazine-sensitive populations (like fungal pathogens) and thereby maintains better community profile in rhizosphere and in the surrounding soil environment. Regulation of phenazine production in *P. aureofaciens* is dependent on QS, involving the LuxRI homologue, PhzRI. The two genes encoding these regulators are placed close to each other on the chromosome and are transcribed convergently. PhzI is responsible for the synthesis of *N*-hexanoyl-L-homoserine lactone (HHC), the cognate AHL sensed by PhzR. Certain isolates of *P. fluorescens* also produce phenazines, wherein presence of *luxRI* homologues (also termed *phzRI*) has been confirmed. In addition, several phytopathogenic pseudomonads such as strains of *P. corrugata*, *P. savastanoi*, and five different pathovars of *P. syringae* (pvs. *syringae*, *tomato*, *angulata*, *coronofaciens*, and *tabaci*) produce various AHLs [61].

7.7 Biofilm Formation in Rhizosphere

Bacterial biomass exists in a statically attached form on a substratum called biofilm, which encapsulates the cells within a complex matrix comprising polysaccharides and proteins [62]. The bacteria in order to form a biofilm produce an exopolymer matrix and later embed themselves into it through external appendages (pili, flagella). The initial stage of any biofilm formation is attachment, followed by creation of monolayer, development of microcolonies, biofilm maturation, and dispersal [63]. Individual cells within the mature biofilms utilize open channels for metabolic needs. The genomic and environmental factors influence the ultimate biofilm assemblage [64].

Bacterial biofilms are mostly associated with chronic mammalian infections as they frequently cause disease by circumventing innate immune responses and antibiotic therapy [65]. Studies regarding biofilm have been centered on the structural and developmental dynamics of biofilms and gene expression in microbial community, primarily focusing on *P. aeruginosa* as a model [66].

It was generally believed that plants were immune to human pathogens. This view changed when *Pseudomonas aeruginosa* PA14, a human pathogen that causes infection in burn and cystic fibrosis patients, was recognized as a potent foliar pathogen in a variety of plants, including the model plant *Arabidopsis thaliana*. Many of the same *P. aeruginosa* virulence factors required for animal pathogenesis [67, 68] are also essential for plant pathogenesis [69]. Among these virulence factors, the ability to form biofilm in *P. aeruginosa* is crucial for infection of multiple hosts. Interestingly, the same mechanisms that allow bacteria to form biofilm within and on the human host also work in the plant's environment, especially in the rhizosphere.

Planktonic cells in rhizosphere also exist in communities of single or mixed species of bacteria; multicellular assemblies/biofilm, especially non–spore formers such as rhizobia. This phenotype is considered as an adaptation from the fluctuating conditions in the rhizosphere, such as desiccation, extreme pH levels, temperature, salt, and nutrient availability. In rhizosphere, there is generally multiple specie biofilms composed of either synergistically acting bacteria or antagonistic species that compete for root colonization which contribute toward plant health, growth, or development [64].

Quorum sensing has been connected to biofilm in rhizosphere. To initiate this, planktonic cells are engaged in chemical signaling to coordinate the gene expression. Before "settling down" on any surface to form biofilm, QS in bacteria estimates the size and status of their population. Intracellular communication by QS regulates the formation of biofilm as well as detachment of cells from biofilm through the collective expression of genes after bacterial population reach a certain threshold level. Mutations in either *lasI* or *rhlI* show reduced ability to adhere to or proliferate on the surface as compared to biofilm of wild-type bacteria, suggesting the importance of QS system [70].

Hydration is a critical trait of all microbes, including those associated with terrestrial plants. Unlike leaves, which can become very dry because of exposure to air and sunlight, roots are the site of constant moisture for rhizospheric microbiota, as they maintain a water film on their surface and are site for water uptake by plant. In addition, plant root exudation maintains a nutrient-rich condition, which in turn provide a target for chemotaxis of microbes [71]. Further, motility and chemotaxis provide a competitive advantage and allow soil-borne microbes to effectively colonize rhizospheric area [72]. Apart from intra-rhizospheric biofilm, the water flow around root systems can also act to inoculate cells and spores from some other site onto below ground tissues.

Within rhizobial biofilm, beneficial and harmful interactions occur between plant roots and microorganisms. Certain bacterial biofilm, such as that of plant growth–promoting rhizobacteria (PGPRs), not only induce plant growth but also protect plants from soil-borne pathogens (the process known as biocontrol). Nonpathogenic PGPRs associated with plant root surfaces thus contribute toward increase in plant yield by mechanisms such as improved mineral uptake, phytohormone production [73, 74], competitive suppression of pathogens by production of antibiotics [75], and induction of secondary metabolite-mediated systemic resistance [73]. In contrast, other rhizobacteria in a biofilm matrix may be pathogenic in plants. Various microorganisms are known to colonize the plant roots having diverse effects:

- *Biocontrol* has been observed in crop plants by bacteria like *Bacillus subtilis, Pseudomonas fluorescens, Pseudomonas putida, Pseudomonas chlororaphis, Microsphaeropsis sp, Enterobacter agglomerans, Pseudomonas aureofaciens, Bacillus polymyxa.*
- *Azorhizobium caulinodans, R. leguminosarum* and *Azospirillum brasilense, Klebsiella pneumonia* exert *beneficial* effects in rice and wheat plant, respectively.
- *Sinorhizobium* show *symbiosis* in legumes.
- *Bacillus cereus* exert *commensal* effect on many plants.
- *Pathogenic* effect has been observed by *Agrobacterium tumefaciens* on pea plant, diverse dicots and by *Burkholderia cepacia* on wheat and onion plant [76].

7.7.1 Beneficial Root Biofilm

Symbiotic rhizobia stands as the best model for studying beneficial plant-microbe interactions which is associated with nitrogen fixation in legumes. Rhizobia feed on root hair flavanoids released by legume root hair. Rhizobia attach to specific host plants via polysaccharides, which bind to plant lectins, mediated via calcium-binding proteins (initially called rhicadhesin), more recently called collectively as Rap adhesins [77, 78].

After attachment, rhizobia produces nod factors that stimulate the root hair curling and accumulation within the root hair, where they proceed to migrate down an infection thread through which they can invade cells in the root cortex and establish endosymbiosis [79]. Adherent cells form multicellular assemblies at the site of attachment. Both the populations of rhizobia associated with root hair and even the bacterial growth down the infection thread are considered as specialized biofilm. Rhizobial biofilm can also form on abiotic surfaces [74, 78].

Another example of beneficial biofilm formation in rhizosphere is a plant growth–promoting rhizobacterium, *Azospirillum brasilense*. It is α-proteobacterium that is linked as a symbiont with the root systems of cereals [80]. Dense biofilm are formed on root hair and root elongation zone by this organism [81]. Exopolysaccharides, flagellar motility (swimming and swarming), and specific outer-membrane proteins help in its effective colonization of cereal root systems. Once the adherent *A. brasilense* biofilm is formed, it promotes plant health by release of a number of different bioactive compounds that stimulate root hair proliferation and lateral root formation [82].

Some organisms can be used as biocontrol agents. They occupy the available surface areas, which are prone to pathogenic attack and thereby prevent disease initiation [83]. Hence, the colonization of biocontrol organism in the rhizosphere defends the host roots from pathogens. Biofilm formation as a means of biocontrol has been well documented for *Pseudomonas fluorescens* in wheat. *B. subtilis* is also used as biocontrol agent for various plants [84].

In wheat roots, natural pseudomonad communities constitute a predominant proportion of the microbiota that exists in the form of biofilm [85]. Other than natural population of pseudomonads, plant growth–promoting pseudomonads can also be used as biocontrol agents, where roots are colonized along surface fissures. The release of plant exudates into soil stimulates rapid mobilization and chemotaxis of pseudomonads toward root systems [86]. Flagellar motility and twitching motility play a significant role in root colonization. The LapA cell surface protein (large adhesion protein A) identified in *P. fluorescens* is required for the transition from reversible to irreversible attachment on abiotic surfaces [87]. Quorum-sensing regulated biofilm formation of *P. fluorescens* allows the organism to create a niche, into which it then secretes antifungal compounds, particularly the antibiotic phenazine, protecting the wheat rhizosphere. When *Pseudomonas putida* mutants were screened for deficiencies in seed binding, a *lapA* homologue was identified in this microbe, and its deficiency led to a deficiency in binding to roots as well [88]. These findings suggested that the LapA protein may function as an adhesin during plant attachment and biofilm formation of pseudomonads.

Gram-positive microbes, specifically *Bacillus* species, can also be effective biocontrol agents [89]. Bacillus species are abundantly associated with terrestrial plants. *Bacillus subtilis* develops dense surface-associated populations and secrete antimicrobial compounds, which make these adherent structures as effective biocontrol agents [12]. *B. subtilis* is sold commercially as biological control agent for agriculture, as it can promote growth and protect plants from infections by pathogenic bacteria, fungi, and even nematodes. Secretion of antimicrobial compounds by *B. subtilis* coupled with induced systemic resistance in the plant (in response to *B. subtilis*), enhances the capacity of the plant to resist various pathogens. For example, to recruit *B. subtilis*, plants secrete small molecules. When *A. thaliana* is infected with *P. syringae*, the plant secretes malic acid, which enhances *B. subtilis* biofilm formation on the root [90].

7.7.2 Pathogenic Root Biofilm

The association of pathogens and PGPR with root systems are quite similar. The pathogenic rhizospheric microorganisms produce an enormous amount of exopolysaccharide, which helps in biofilm formation and consequent infection. Production of exopolysaccharide can affect the interaction of microbes with roots and root appendages [91]. Exopolysaccharides are generally composed of monosaccharides and some noncarbohydrate substituents (such as acetate, pyruvate, succinate, and phosphate). The best-studied example is that of alginate from *Pseudomonas aeruginosa.* Pathogenic pseudomonads have been reported to form thicker, more confluent biofilm on root tissues, in contrast to the more heterogeneous colonization by beneficial pseudomonads [22, 92]. This difference reflects the interactions that lead to infection, rather than simple interaction.

Pathogenic *P. aeruginosa* strains PAO1 and PA14 are capable of infecting the roots of *Arabidopsis* and sweet basil (*Ocimum basilicum*), *in vitro* and in the soil. These organisms colonize the roots of these plants and form thick biofilm. They are capable of causing plant mortality in just seven days of post-inoculation. Upon *P. aeruginosa* infection, sweet basil roots secrete rosmarinic acid (RA), a multifunctional caffeic acid ester that exhibits *in vitro* antibacterial activity against planktonic cells of both *P. aeruginosa* strains. Despite of secretion of RA, it does not attain minimum inhibitory concentration level in sweet basil's root exudates before formation of *P. aeruginosa* biofilm, thus *P. aeruginosa* resists the microbicidal effects of RA and ultimately can cause plant mortality [93].

Another important plant root disease called *crown gall* is ubiquitous in plants. It is caused by *Agrobacterium tumefaciens,* an α-proteobacterium, and a close relative of symbiotic rhizobia [94]. Crown is the area where stem and root tissue converges. Infection occurs at wound sites along roots and crown. The mechanism of plant attachment is mediated initially by a yet-unidentified adhesin, followed by firmer attachment via cellulose fibril production. Once attached to root tissues, *A. tumefaciens* can form dense, structurally complex biofilm, extensively coating the epidermis and root hairs [95, 96]. It leads to a horizontal gene transfer from *A. tumefaciens* to the plant, directing uncontrolled proliferation of the tissue (the gall) and production of nutrients specific for the pathogen. The related species, including *A. rhizogenes* and *A. vitis,* cause hairy root disease and grape-specific necrosis, respectively. On abiotic surfaces, comparable biofilm are formed, similar to plant root surfaces [95–97].

A very important factor in pathogenic biofilm formation is oxygen limitation, which is a common condition in the rhizosphere and also within the biofilm. In *A. tumefaciens,* SinR regulator is a component of oxygen-limitation response pathway. Mutants disrupted for *sinR* develop sparse, patchy biofilm on plant roots and abiotic surfaces [96]. This clearly suggests the relation between oxygen levels and biofilm structure. Similarly, limiting phosphorous is common in the rhizosphere due to plant sequestration. Limiting phosphorous enhances biofilm formation by *A. tumefaciens*; in contrast, it decreases biofilm formation by *Pseudomonas aureofaciens* [97, 98].

7.7.3 Mixed-Species Biofilm

Multispecies biofilm in rhizosphere having multiple interactions (multitrophic interactions) are economically important for several agricultural crops. Such biofilm can be beneficial for many organisms. Biofilm formation in *Bradyrhizobium elkanii* SEMIA 5019 and *Penicillium* sp. significantly increases nodulation and nitrogen accumulation in soybeans roots as compared to their planktonic counterparts [99]. Similarly, co-inoculation of *Azospirillum* with nitrogen fixing rhizobia provides enhanced benefits to plant production, suggesting possible synergism within mixed communities of these microbes [100].

7.8 Conclusions

Bacteria interact extensively with plant roots and develop into complex multicellular populations, the biofilm, with the help of QS circuitry. These diverse populations are physically and metabolically linked in rhizosphere and to each other, and are also linked in communication via a diversity of molecular signals. Understanding the complexities and diversity of the biofilm formed in rhizosphere will not only contribute toward better understanding of the biology of root-biofilm interactions but also contribute to the field of rhizosphere biology. The abundance of natural beneficial strains in the vicinity of plant roots suppress plant pathogens without producing lasting effects on the rest of the soil microbial and plant communities. Moreover, the diversity of microbial communities in rhizosphere provides a rich source of potential biocontrol agents to control plant diseases. It offers an attractive alternative to the use of synthetic chemicals. In conclusion, beneficial rhizospheric biofilm driven by quorum sensing serves as a new model system for the study of rhizospheric microbial development, which can provide insights into microbial ecology and help in meeting the increasing crop yield demands.

References

1 Hartmann A, Rothballer M, and Schmid M. Lorenz Hiltner, a pioneer in rhizosphere microbial ecology and soil bacteriology research. *Plant Soil.* 2008;312:7–14.

2 Giri B, Giang PH, Kumari R, Prasad R, and Verma A. Microbial diversity in soils. In: Buscot F and Varma A, editors. *Microorganisms in Soils: Roles in Genesis and Functions.* New York: Springer; 2005, pp. 19–55.

3 Sylvia D, Fuhrmann J, Hartel P, and Zuberer D. *Principles and Applications of Soil Microbiology.* Pearson Education Inc. New Jersey; 2005.

4 Singer MJ, and Munns DN. *Soils: An introduction.* Englewood Cliffs, NJ: Pearson Education Inc.; 2006.

4a Strobel, G A and Long, D M. Endophytic microbes embody pharmaceutical potential. *ASM News.* 1998;64(5):263–268.

5 McNear Jr, DH. The Rhizosphere—Roots, Soil and Everything in Between. *Nature Education Knowledge.* 2013;4(3):1.

6 Morrisseey J, Dow J, Mark G, and O'Gara F. Are microbes at the root of a solution to world food production? *EMBO rept.* 2004;5(10) 922–926.

7 Gewin V. Food: An underground revolution. *Nature.* 2010;466:552–553.

8 Zhang F, Shen J, Zhang J, Zuo Y, Li L, and Chen X. Rhizosphere processes and management for improving nutrient use efficiency and crop productivity: implications for China. *Advances in Agronomy.* 2010;107:1-32.

9 Hrynkiewicz K, and Baum C The potential of rhizosphere microorganisms to promote the plant growth in disturbed soils. In: A Malik and E. Grohmann, eds. *Environmental Protection Strategies for Sustainable Development. New York: Springer*; 2011. pp. 35–64.

10 Tilman D, Cassman KG, Matson PA, Naylor R, and Polasky S. Agricultural sustainability and intensive production practices. *Nature.* 2002;418:671–677.

11 Hawes MC, Gunawardena U, Miyasaka S, et al. The role of root border cells in plant defense. *Trends in Plant Science.* 2000;5(3):128–133.

12 Bais HP, Fall R, and Vivanco JM. Biocontrol of Bacillus subtilis against infection of Arabidopsis roots by Pseudomonas syringae is facilitated by biofilm formation and surfactin production. *Plant Physiol.* 2004a;134: 307–319.

13 de Weger LA, van der Bij AJ, Dekkers LC, Simons M, Wijffelman CA, and Lugtenberg BJJ. Colonization of the rhizosphere of crop plants by plant-beneficial pseudomonads. *FEMS Microbiology Ecology.* 1995;17: 221–228.

14 Marilley L, and Aragno M. Phytogenetic diversity of bacterial communities differing in degree of proximity of Lolium perenne and Trifolium repens roots. *Applied Soil Ecology.* 1999;13:127–136.

15 Keel C, Défago G. Interactions between beneficial soil bacteria and root pathogens: mechanisms and ecological impact. In: Gange AC, Brown VK, eds. *Multitrophic Interactions in Terrestrial System.* Oxford: Blackwell Science; 1997, pp. 27–47.

16 O'Sullivan DJ, and O'Gara F. Traits of fluorescent *Pseudomonas* spp. involved in suppression of plant root pathogens. *Microbiological Reviews.* 1992;56:662–676.

17 Loper JE, and Henkels MD. Utilization of heterologous siderophores enhances levels of iron available to *Pseudomonas putida* in the rhizosphere. *Applied and Environmental Microbiology.* 1999;65:5357–5363.

18 Buysens S, Heungens K, Poppe J, and Höfte M. Involvement of pyochelin and pyoverdin in suppression of Pythium-induced damping-off of tomato by Pseudomonas aeruginosa 7NSK2. *Applied and Environmental Microbiology.* 1996;62:865–871.

19 Howell CR. The role of antibiosis in biocontrol. In: Harman GE, Kubicek CP, eds. *Trichoderma and Gliocladium, Vol. 2. Enzymes, biological control and commercial applications.* London: Taylor & Francis; 1998, pp. 173–184.

20 Kim KK, Fravel DR, Papavizas GC. Production, purification and properties of glucose oxidase from the biocontrol fungus *Talaromyces flavus. Canadian Journal of Microbiology.* 1990;36:199–205.

21 Fravel DR, and Roberts DP. In situ evidence for the role of glucose oxidase in the biocontrol of Verticillium wilt by *Talaromyces flavus. Biocontrol Science and Technology.* 1991;1:91–99.

22 Bais HP, Park SW, Weir TL, Callaway RM, and Vivanco JM. How plants communicate using the underground information superhighway. *Trends in Plant Science.* 2004b;9(1):26–32.

23 Alimuddin WJ, Asmara W, Mustofa. Antifungal production of a strain of *Actinomycetes* spp isolated from the rhizosphere of cajuput plant: Selection and detection of exhibiting activity against tested fungi. *Indonesian Journal of Biotechnology.* 2011;16(1):1–10.

24 Xiaoyun C, Manqiang L, Feng H, Xiaofang M, Huixin L. Contributions of soil micro-fauna (protozoa and nematodes) to rhizosphere ecological functions. *Acta Ecologica Sinica*, 2007;27(8):3132–3143.

25 Whitehead NA, Barnard AML, Slater H, L SNJ, Salmond GPC. Quorum-sensing in Gram-negative bacteria. *FEMS Microbiol Rev.* 2001;25:365–404.
26 Fuqua WC, Winans SC, and Greenberg EP. Quorum sensing in bacteria: the LuxR-LuxI family of cell density-responsive transcriptional regulators. *Bacteriol.* 1994;176:269–275.
27 Sharma A, Sahgal M, and Johri BN. Microbial communication in the rhizosphere: Operation of quorum sensing. *Current Science.* 2003;85(8).
28 Camilli A, and Bassler BL. Bacterial small-molecule signalling pathways. *Science.* 2006; 311:1113–1116.
29 Schuster M, Lostroh CP, Ogi T, and Greenberg EP. Identification, timing, and signal specificity of Pseudomonas aeruginosa quorum-controlled genes: a transcriptome analysis. *J Bacteriol.* 2003;185:2066–2079.
30 Wagner VE and Iglewski BH. P. aeruginosa biofilms in CF infection. *Clin Rev Allergy Immunol.* 2008;35(3):124–134.
31 de Kievit TR Quorum sensing in Pseudomonas aeruginosa biofilms. *Environ Microbiol.* 2009;11(2):279–288.
32 D'Angelo-Picard C, Faure D, Carlier A, Uroz S, Raffoux A, Fray R, and Dessaux Y. Bacterial populations in the rhizosphere of tobacco plants producing the quorumsensing signals hexanoyl-homoserine lactone and 3- oxo-hexanoyl-homoserine lactone. *FEMS Microbiol Ecol.* 2004;51:19–29.
33 Fuqua C, Parsek MR, Greenberg EP. Regulation of gene expression by cell-to-cell communication: acyl-homoserine lactone quorum sensing. *Annu Rev Genet.* 2001;35:439–468.
34 Gray KM, Pearson JP, Downie JA, Boboye BE and Greenberg EP. Cell-to-cell signaling in the symbiotic nitrogen-fixing bacterium Rhizobium leguminosarum: autoinduction of a stationary phase and rhizosphere-expressed genes. *J Bacteriol.* 1996;178(2):372–376.
35 Hardman AM, Stewart GS, and Williams P. Quorum sensing and the cell-cell communication dependent regulation of gene expression in pathogenic and non-pathogenic bacteria. *Antonie Van Leeuwenhoek.* 1998 Nov;74(4):199–210.
36 Shapiro J A. Thinking about bacterial populations as multicellular organisms. *Annu. Rev. Microbiol.* 1998;52:81–104.
37 Fuqua G, and Greenberg EP. Self perception in bacteria: quorum sensing with acylated homoserine lactones. *Curr. Opin. Microbiol.* 1998;1:183–189.
38 Winzer K, Hardie KR, Burgess N, Doherty N, et al. LuxS: Its role in central metabolism and the in vitro synthesis of 4-hydroxy-5-methyl-3(2H)-furanone. *Microbiology.* 2002;148:909–922.
39 Cha C, Gao P, Chen YC, Shaw PD, and Farrand SK. Production of acyl homoserine lactone quorum-sensing signals by gram-negative plant-associated bacteria. *Mol. Plant Microb. Interact.* 1998;11:1119–1129.
40 Shaw P D, Ping G, Daly S L, Cha C, Cronan Jr. J E, Rinehart K L, and Farrand S K. Detecting and characterizing N-acyl-homoserine lactone signal molecules by thin-layer chromatography. *Proc. Natl. Acad. Sci. USA.* 1997;94(12):6036–6041.
41 Wisniewski-Dye F, Downie J A. Quorum-sensing in Rhizobium. *Antonie Van Leeuwenhoek.* 2002;81:397–407.
42 Lithgow J K, Wilkinson A, Hardman A, Rodelas B, et al. The regulatory locus cinRI in Rhizobium leguminosarum controls a network of quorum-sensing loci. *Mol. Microbiol.* 2000;37:81–97.

43 Wijffelman CA, Pees E, Van Brussel AAN, and Hooykaas PJJ. Repression of Small bacteriocin excretion in Rhizobium leguminosarum and Rhizobium trifolii by transmissible plasmids. *Mol. Gen. Genet.* 1983;192:171–176.
44 Danino VE, Wilkinson A, Edwards A, and Downie JA. Recipient-induced transfer of the symbiotic plasmid pRL1JI in Rhizobium leguminosarum bv. viciae is regulated by a quorum-sensing relay. *Mol. Microbiol.* 2003;50:511–525.
45 Thorne S H, and Williams H D. Cell density-dependent starvation survival of Rhizobium leguminosarum bv. phaseoli: identification of the role of an N-acyl homoserine lactone in adaptation to stationary-phase survival. *J. Bacteriol.* 1999;181: 981–990.
46 Cubo MT, Economou A, Murphy G, Johnston AW, and Downie JA. Molecular characterization and regulation of the rhizosphere-expressed genes rhiABCR that can influence nodulation by Rhizobium leguminosarum biovar viciae. *J. Bacteriol.* 1992;174: 4026–4035.
47 Rodelas B, Lithgow JK, Wisniewski-Dye F, Hardman A, et al. Analysis of quorum-sensing-dependent control of rhizosphere-expressed (rhi) genes in Rhizobium leguminosarum bv. viciae. *J. Bacteriol.* 1999;181:3816–3823.
48 Gonzalez JE, and Marketon MM. Quorum sensing in nitrogen-fixing rhizobia. *Microbiol. Mol. Biol. Rev.* 2003;67(4):574–592.
49 Marketon MM, and Gonzalez J E. Identification of two quorum-sensing systems in Sinorhizobium meliloti. *J. Bacteriology* 2002;184:3466–3475.
50 Pellock, BJ, Teplitski M, Boinay RP, Bauer WD, and Walker GC. A LuxR homolog controls production of symbiotically active extracellular polysaccharide II by *Sinorhizobium meliloti. J. Bacteriol.* 2002;184:5067–5076.
51 Hoang HH, Becker A, Gonzalez JE (2004). The LuxR homolog ExpR, in combination with the Sin quorum sensing system, plays a central role in Sinorhizobium meliloti gene expression. *J. Bacteriol.* 186: 5460–5472.
52 Gao M, Chen H, Eberhard A, Gronquist MR, et al. sinI- and expR-dependent quorum sensing in *Sinorhizobium meliloti. J. Bacteriol.* 2005;187(23):7931–7944.
53 González JE, Reuhs BL, and Walker GC. Low molecular weight EPS II of *Rhizobium meliloti* allows nodule invasion in Medicago sativa. *Proc. Natl. Acad. Sci USA.* 1996 Aug 6;93(16):8636–8641.
54 Zheng H, Zhong Z, Lai X, Chen WX, Li S, and Zhu J. A LuxR/LuxI-Type quorum-sensing system in a plant bacterium, *Mesorhizobium tianshanense*, controls symbiotic nodulation. *J. Bacteriol.* 2006;188(5):1943–1949.
55 Zhu J, Chai Y, Zhong Z, Li S, and Winans SC. Agrobacterium bioassay strain for ultrasensitive detection of N-acylhomoserine lactone-type quorum-sensing molecules: detection of autoinducers in *Mesorhizobium huakuii. Appl. Environ. Microbiol.* 2003;69: 6949–6953.
56 Schuster M, and Greenberg EP. Early activation of quorum sensing in Pseudomonas aeruginosa reveals the architecture of a complex regulon. *BMC Genomics.* 2007;8:287.
57 Pearson JP, Pesci EC, and Iglewski BH. Roles of *Pseudomonas aeruginosa* las and rhl quorum sensing systems in control of elastase and rhamnolipid biosynthesis genes. *J. Bacteriol.* 1997;179:5756–5767.
58 Pesci EC, Pearson JP, Seed PC, and Iglewski BH. Regulation of Las and Rhl quorum sensing in *Pseudomonas aeruginosa. J. Bacteriol.* 1997;179(10):3127–3132.

59 McKnight SL, Iglewski BH, and Pesci EC. The Pseudomonas Quinolone Signal Regulates rhl Quorum Sensing in Pseudomonas aeruginosa. *J. Bacteriol.* 2000;182(10):2702–2708.
60 Gera C, and Srivastava S. Quorum-sensing: The phenomenon of microbial communication. *Curr. Sci.* 2006;90(5):666–676.
61 Wood DW, Gong F, Daykin MM, Williams P, and Pierson III, LS. N-acyl-homoserine lactone-mediated regulation of phenazine gene expression by *Pseudomonas aureofaciens* 30-84 in the wheat rhizosphere. *J. Bacteriol.* 1997;179:7663–7670.
62 Donlan RM, and Costerton JW. Biofilms: Survival mechanisms of clinically relevant microorganisms. *Clin. Microbiol. Rev.* 2002;15(2):167–183.
63 Kostakioti M, Hadjifrangiskou M, and Scott JH. Bacterial biofilms: Development, dispersal and therapeutic strategies in the dawn of the postantibiotic era. *Cold Spring Harb Perspect Med.* 2013;3(4):a010306.
64 Angus AA, Hirsch AM. Biofilm formation in the rhizosphere: Multispecies interactions and implications for plant growth. Chapter 66. In: Frans J. de Bruijn, editor. *Molecular Microbial Ecology of the Rhizosphere*, Volume 2. First Edition. Hoboken, NJ: John Wiley & Sons, Inc., 2013.
65 Hall-Stoodley L, Costerton JW, and Stoodley P. Bacterial biofilms: from the natural environment to infectious diseases. *Nat. Rev. Microbiol.* 2004;2:95–108.
66 O'Toole G, Kaplan HB, and Kolter R. Biofilm formation as microbial development. *Annu. Rev. Microbiol.* 2000;54:49–79.
67 Mittal R, Khandwaha RK, Gupta V, Mittal PK, and Harjai K. Phenotypic characters of urinary isolates of Pseudomonas aeruginosa and their association with mouse renal colonization. *Indian J. Med. Res.* 2006;123:67-72.
68 Sabharwal N, Dhall S, Chhibber S, and Harjai K. Molecular detection of virulence genes as markers in *Pseudomonas aeruginosa* isolated from urinary tract infections. *Int. J. Mol. Epidemiol. Genet.* 2014b;5(3):125–134.
69 Rahme LG, Stevens EJ, Wolfort SF, Shao J, Tompkins RG, and Ausubel FM. Common virulence factors for bacterial pathogenicity in plants and animals. *Science.* 1995;268:1899–1902.
70 Davies DG, Parsek MR, Pearson JP, Iglewski BH, Costerton JW, and Greenberg EP. The involvement of cell-to-cell signals in the development of a bacterial biofilm. *Science.* 1998;280:295–298.
71 Walker TS, Bais HP, Grotewold, E, and Vivanco, JM. Root exudation and rhizosphere biology. *Plant Physiol.* 2003;132:44–51.
72 Lugtenberg BJJ, Chin-A-Woeng TF, and Bloemberg GV. Microbe-plant interactions: principles and mechanisms. *Antonie Van Leeuwenhoek.* 2002;81:373–383.
73 Ryu CM, Farag MA, Hu CH, Reddy MS, Kloepper JW, and Paré PW. Bacterial volatiles induce systemic resistance in Arabidopsis. *Plant Physiol.* 2004;134:1017–1026.
74 Fujishige NA, Kapadia NN, De Hoff PL and Hirsch AM. Investigations of Rhizobium biofilm formation. *FEMS Microbiol. Ecol.* 2006;56:195–206.
75 Mazzola M, Cook RJ, Thomashow LS, Weller DM, and Pierson LS III. Contribution of phenazine antibiotic biosynthesis to the ecological competence of fluorescent pseudomonads in soil habitats. *Appl. Environ. Microbiol.* 1992;58:2616–2624.
76 Rudrappa T, Biedrzycki ML, and Bais HP. Causes and consequences of plant-associated biofilms. *FEMS Microbiol. Ecol.* 2008;64:153–166.
77 Laus MC, Logman TJ, Lamers GE, Van Brussel AA, Carlson RW, and Kijne JW. A novel polar surface polysaccharide from *Rhizobium leguminosarum* binds host plant lectin. *Mol. Microbiol.* 2006;59:1704–1713.

78 Russo DM, Williams A, Edwards A, Posadas DM, et al. Proteins exported via the PrsD-PrsE type I secretion system and the acidic exopolysaccharide are involved in biofilm formation by Rhizobium leguminosarum. *J. Bacteriol.* 2006;188:4474–4486.
79 Gage DJ. Infection and invasion of roots by symbiotic, nitrogen-fixing rhizobia during nodulation of temperate legumes. *Microbiol. Mol. Biol. Rev.* 2004;68:280–300.
80 Burdman S, Okon Y, and Jurkevitch E. Surface characteristics of *Azospirillum brasilense* in relation to cell aggregation and attachment to plant roots. *Crit. Rev. Microbiol.* 2000;26:91–110.
81 Assmus B, Hutzler P, Kirchhof G, Amann R, Lawrence JR and Hartmann A. In-situ localization of *Azospirillum brasilense* in the rhizosphere of wheat with fluorescently labeled, ribosomal-RNA-targeted oligonucleotide probes and scanning confocal laser microscopy. *Appl. Environ. Microbiol.* 1995;61:1013–1019.
82 Umali-Garcia M, Hubbell DH, Gaskins MH, and Dazzo FB. Association of Azospirillum with grass roots. *Appl. Environ. Microbiol.* 1980;39:219–226.
83 Lugtenberg B, and Kamilova F. Plant-growth-promoting rhizobacteria. *Annu. Rev. Microbiol.* 2009;63:541–556.
84 Rudrappa T, and Bais HP. Arabidopsis thaliana root surface chemistry regulates in plant biofilm formation of *Bacillus subtilis. Plant Signal Behav.* 2007;2:349–350.
85 Watt M, Hugenholtz P, White R, and Vinall K. Numbers and locations of native bacteria on field-grown wheat roots quantified by fluorescence in situ hybridization (FISH). *Environ. Microbiol.* 2006;8:871–884.
86 Espinosa-Urgel M, Kolter R, and Ramos JL. Root colonization by Pseudomonas putida: love at first site. *Microbiology.* 2002;148:341–343.
87 Hinsa SM, Espinosa-Urgel M, Ramos JL, and O'Toole GA. Transition from reversible to irreversible attachment during biofilm formation by Pseudomonas fluorescens WCS365 requires an ABC transporter and a large secreted protein. *Mol. Microbiol.* 2003;49:905–918.
88 Espinosa-Urgel M, Salido A, and Ramos JL. Genetic analysis of functions involved in adhesion of Pseudomonas putida to seeds. *J. Bacteriol.* 2000;182:2363–2369.
89 Ugoji EO, Laing MD, and Hunter CH. Colonization of *Bacillus* spp. on seeds and in plant rhizoplane. *J. Environ. Biol.* 2005;26:459–466.
90 Vlamakis H, Chai Y, Beauregard P, Losick R and Kolter R. Sticking together: building a biofilm the Bacillus subtilis way. *Nature Reviews Microbiology.* 2013;11:157–168.
91 Bianciotto V, Andreotti S, Balestrini R, Bonfante P, and Perotto S. Mucoid mutants of the biocontrol strain *Pseudomonas fluorescens* CHA0 show increased ability in biofilm formation. *Current Opinion in Microbiology.* 2004;7:602–609.
92 Walker TS, Bais HP, Deziel E, Schweizer HP, et al. Pseudomonas aeruginosa-plant root interactions. Pathogenicity, biofilm formation, and root exudation. *Plant Physiol.* 2004;134:20–331.
93 Travis SW, Pal BH, Éric D, Herbert PS et al. Pseudomonas aeruginosa-plant root interactions. Pathogenicity, biofilm formation, and root exudation. *Plant Physiology,* 2004;134(1):320–331.
94 Escobar MA, and Dandekar AM. Agrobacterium tumefaciens as an agent of disease. *Trends Plant Sci.* 2003;8:380–386.
95 Matthysse AG, Marry M, Krall L, Kaye M, et al. The effect of cellulose overproduction on binding and biofilm formation on roots by Agrobacterium tumefaciens. *Mol. Plant. Microbe Interact.* 2005;18:1002–1010.

96 Ramey BE, Matthysse AG, and Fuqua C. The FNR-type transcriptional regulator SinR controls maturation of *Agrobacterium tumefaciens* biofilms. *Mol. Microbiol.* 2004;52:1495–1511.
97 Danhorn T, Hentzer M, Givskov M, Parsek M, and Fuqua C. Phosphorous limitation enhances biofilm formation of the plant pathogen *Agrobacterium tumefaciens* through the PhoR-PhoB regulatory system. *J. Bacteriol.* 2004;186:4492–4501.
98 Monds RD, Silby MW, and Mahanty HK. Expression of the Pho regulon negatively regulates biofilm formation by *Pseudomonas aureofaciens* PA147-2. *Mol. Microbiol.* 2001;42:415–426.
99 Jayasinghearachchi HS, and Sereviratne G. A bradyrhizobial-Penicillium spp. biofilm with nitrogenase activity improves N_2 fixing symbiosis of soybean. *Biol. Fert. Soils.* 2004;40:432–434.
100 Burdman S, Jurkevitch E, Schwartsburd B, Hampel M, Okon Y. Aggregation in *Azospirillum brasilense*: Effects of chemical and physical factors and involvement of extracellular components. *Microbiology.* 1998;144(7):1989–1999.

8

The Significance of Fungal Biofilms in Association with Plants and Soils

Michael W. Harding[1], Lyriam L.R. Marques[2], Bryon Shore[3] and G.C. Daniels[1]

[1] *Alberta Agriculture and Forestry, Crop Diversification Centre South, Brooks, Canada*
[2] *MicroBio SMARTS, Calgary, Canada*
[3] *Nautilus Environmental, Calgary, Canada*

8.1 Introduction

Phytopathogenic microorganisms cause worldwide economic losses in all industries involving plant production such as agriculture, horticulture, floriculture, turf-grass, nursery crops, and forestry operations. In addition, pathogens attack plant materials in post-harvest storages. Global economic losses due to plant diseases were estimated at 10 to 15 percent, resulting in a cost of $76.1 billion between 1988 and 1990 [1, 2]. Many microorganisms that interact with plants can survive or thrive within soils. Soils are living ecosystems composed of heterogeneous physical, chemical, and biological environments interacting in a complex manner. Soil microorganisms, mainly bacteria and fungi, are the predominant component of the soil biota and strongly influence primary production through involvement in several key ecosystem functions, including: (i) nutrient recycling through decomposition of dead organic matter, (ii) biogeochemical cycles, (iii) soil formation, structure and moisture regime, through their contribution to the development of aggregates and pore spaces which influence water infiltration and movement in the soil, (iv) degradation of soil contaminants, and (v) promotion of plant health through symbiotic interactions that improve nutrition, promote growth, and offer protection against plant pathogens.

Microbiologists have documented that microbial populations in their natural environments form biofilms—complex communities attached to surfaces in a self-produced polymeric matrix [3–6]. The study of microbial biofilms has led to a shift in our understanding of how microorganisms grow, survive, adapt, and exploit hosts and resources. The bulk of biofilm research has been done on bacteria in aquatic or clinical settings, and a few examples on plants [7–9]. Fungal biofilms have also been studied, mainly in model yeast pathogens responsible for human and animal diseases [10–14]. Only recently have reports of filamentous fungal biofilms been gradually accumulating. The focus of this chapter is to review and integrate what is known about filamentous fungal biofilms with special attention to those formed on or within plant tissues or in soils.

8.2 What Is a Biofilm?

The linguistic meaning of the word *biofilm* literally translates to "biological material in a thin layer," but the more we learn about biofilms, the more difficult it is for all to agree on a single definition [15, 16]. A definition for filamentous biofilms becomes additionally challenging because of the unique morphology and tip growth of hyphal filaments when compared to the single-celled populations of bacteria or yeasts undergoing binary fission or budding. Filamentous fungi cannot always be cultured, measured, manipulated, or enumerated using the same experimental protocols as those used for the individual cells within bacteria and yeast biofilms. As a result, methods for describing and quantifying filamentous fungal biofilms are quite different from those established for bacteria and budding yeasts. Despite these challenges, it is important to define the term *biofilm* for the purposes of this chapter such that it is inclusive to bacteria, yeast, and filamentous microorganisms growing in natural environments or under laboratory conditions. Herein, we use *biofilm* to refer to a microbial community that grows attached to a surface in a self-produced extracellular matrix of polymeric substances [4, 6, 15]. Additionally, a biofilm will have some unique, measurable phenotype(s) distinct from its planktonic counterparts [17]. Finally, for submerged biofilms, channels within the biofilm allowing fluid to pass through are also a key component [18].

Biofilms frequently have detectable and/or measurable characteristics that are advantageous for survival or completion of the microbial life cycle. These benefits can include communication, attachment, colonization, or tolerance to extreme conditions or chemical challenges, enhanced virulence, genetic recombination or exchange, or exploitation of resources [3, 15, 19]. The benefits of biofilm formation are theorized to have undergone strong positive selective pressure very early in the evolution of microbes because fossils of microbial biofilms and stromatolites have been found on rocks dated at approximately 3.5 million years old [20], and virtually every microbial species evaluated shows capacity to form a biofilm [21].

8.3 Where Do We Find Filamentous Fungal Biofilms?

Filamentous fungal biofilms have been found in diverse environments. First, filamentous fungal biofilms in clinical infections are reported for species of *Aspergilllus, Zygomycetes, Coccidiodes, Trichosporon,* and *Cryptococcus* and represent a significant burden of hospital infections [22]. For *Aspergillus* species, adhesion, matrix components, gene expression, and drug resistance have all been studied in a biofilm context [16, 23, 24]. Second, filamentous fungi have been found growing as biofilms attached to pipes in hospital water-distribution systems, and have been implicated as a potential source of inoculum for hospital infections [25]. For example, *Aspergillus, Alternaria, Botrytis, Cladosporium, and Penicillium* species have been recovered from drinking water distribution systems [26, 27]. Third, environmental biologists have described fungal biofilms formed by wood decay and mycorrhizal fungal species. Biofilm-like structures, including abundant exopolymeric matrix, have been reported in beech vessels colonized by *Coriolus versicolor* [28], and mycorrhizal fungi on or near plant root surfaces [29, 30]. Fourth, filamentous fungal biofilms such as those formed by *Aspergillus* and *Trichoderma* have been found in industrial fermentation processes [16, 18, 31–34]. Fifth, biofilms on building materials and textiles have been described. A study by Barratt et al. [35] reported fungal biofilms of *Geomyces* sp and *Nectria* sp on polyurethane textiles. The biofilms were presented as a network of branched hyphae and spores, layered, and covered in exopolymeric matrix, which caused degradation of substrate buried in soil for 44 days. Rocks, minerals, and buildings, including concrete, stone, brick, plaster, wood, plastic, and painted surfaces are all colonized by fungal, algal, and bacterial biofilms, which accelerate their weathering and/or deterioration [36, 37].

Filamentous biofilms have also been found on plants. Filamentous biofilms on a field crop were first reported by Galiana et al. [38] when they published a brief report of the Oomycete pathogen *Phythophthora parasitica* forming biofilms in association with infection of tobacco leaves. They reported evidence for a QS mechanism that allowed solitary zoospores to aggregate and initiate an infection with biofilm characteristics. A more thorough characterization of *P. parasitica* biofilms was subsequently published showing differential gene expression during distinct phases of biofilm formation [39]. The first report of a true fungus forming a biofilm on a field crop host was published by Harding et al. [40]. In this report, morphological evidence of fungal biofilms was presented in scanning electron micrographs showing *Fusarium* sp. on potato and *Botrytis cinerea* on tomato. The morphology of the *in planta* biofilms was similar to those cultured *in vitro* on wood surfaces in a static biofilm reactor, and helped confirm distinct phases of filamentous fungal biofilm formation for these species. Subsequently, Peiqian et al. [41] confirmed that the fungal plant pathogen *Fusarium oxysporum* f.sp. *cucumerinum*, the causal agent of Fusarium wilt on cucumbers, was capable of forming biofilms on polystyrene strips. The biofilms grow in response to available carbon sources and were more tolerant of physical and chemical challenges when compared with planktonic cultures. Biofilms of *F. oxysporum* have also been discovered on soft contact lenses and are quite resistant to therapeutic treatments, indicating that *F. oxysporum* is quite capable of forming biofilms on a variety of surfaces [42].

It is somewhat surprising that so few reports of fungal biofilms on plants exist, considering that the vast majority of plant diseases are caused by filamentous fungi. What is not surprising is that common characteristics exist in all cases of filamentous fungi forming biofilms on plants. Regardless of the taxa, the fungi all progress through the phases of spore contact, adhesion, germination, filamentation, production of extracellular matrix, and mature biofilm formation. Mature biofilms are frequently described as three-dimensional colonies with hyphal aggregation, bundling, and/or networking encased in abundant extracellular matrix. These characteristic phases are illustrated in Figure 8.1 and discussed in detail later in this chapter.

8.4 Fungal Biofilms: What Have We Learned from the Budding Yeasts?

Most what we know about fungal biofilms comes from yeast research models, such as *Candida* spp. [11, 43–46] and *Saccharomyces* spp. [14, 47–49], or yeast-like molds such as *Aureobasidium* [50] and *Cryptococcus* [10, 51]. One reason to focus on these organisms is that they cause important clinical infections in human hosts [10, 13], and a biofilm approach is necessary for successful intervention and therapy.

While yeast biofilms have been described on plant tissues such as mung bean sprouts [52] and in mixed biofilms on leafy vegetables [53, 54], filamentous fungi are by far the predominant form of microorganism causing diseases on plants. Filamentous fungi that lack a yeast phase will have a distinct cycle and perhaps unique features when compared with yeast species, as the dynamics of hyphal growth are rather dissimilar from budding in yeasts [17], but many of the underlying genetic mechanisms and life-cycle strategies may be common in these closely related organisms.

Additionally, yeast pathogens often enter into a hyphal mode of growth when colonizing surfaces or host tissues [46] and reveal certain details about filamentous biofilms. For example, we now know a great deal about the genetic determinants of fungal biofilm formation [55], gene expression profiles [56–58], the components of the extracellular matrix [11, 19], quorum sensing in fungi [59–61], the role of the extracellular matrix in cohesion [62], adhesion [14], protection against environmental challenges, chemicals, and drugs [19, 22, 63], and nutrition and metabolism [3] from work done on yeast models. This information gives significant insight into filamentous fungal biofilms. In fact, common gene expression of major functional categories such as transcription regulators, synthesis of proteins, amino acids and cell wall components, efflux pumps and adhesins have been demonstrated for biofilms of both *Candida albicans* and *Aspergillus fumigatus* [10], confirming that yeast models can tell us much about filamentous fungi forming biofilms.

8.5 What Does a Filamentous Fungal Biofilm Look Like?

The extracellular enzyme digestion, absorptive nutrition, and exploratory apical growth of filamentous fungi are clear adaptations to a surface-associated lifestyle. Most fungal life cycles also include the production of aerial hyphae and fruiting structures, projecting from mycelial mats attached to a surface. This arrangement allows attachment, colonization, and exploration, and feeding from the surface or substratum, but also for aerial spore dispersal aiding in colonization of new host tissues or environmental niches, all of which are well-described features of biofilms.

Based on the evidence presented from both *in planta* and *in vitro* studies, and comparative analysis with bacterial and yeast models, Harding et al. [17] proposed a six-stage model for filamentous fungal biofilm formation. A representation of this model is shown diagrammatically and with SEM examples in Figure 8.1:

- In stage one, solitary or nonattached propagules, or clumps or aggregates of propagules, such as spores, or multicellular structures such as sclerotia or hyphal fragments, make contact with a suitable substrate or host. This stage is described as deposition or adsorption and the surface attachment is reversible.
- In stage two, the attachment becomes irreversible by means of secretion of adhesive substances from spores and germlings.
- In stage three, the formation of a microcolony occurs. These initial stages of microcolony growth and surface colonization involve apical elongation and branching of hyphae. Hyphae explore the substratum and ramify across and throughout surfaces. This stage involves continued production of extracellular matrix to aid in cohesion and adhesion of the microcolony, and it provides other benefits associated with an extracellular polymeric matrix.

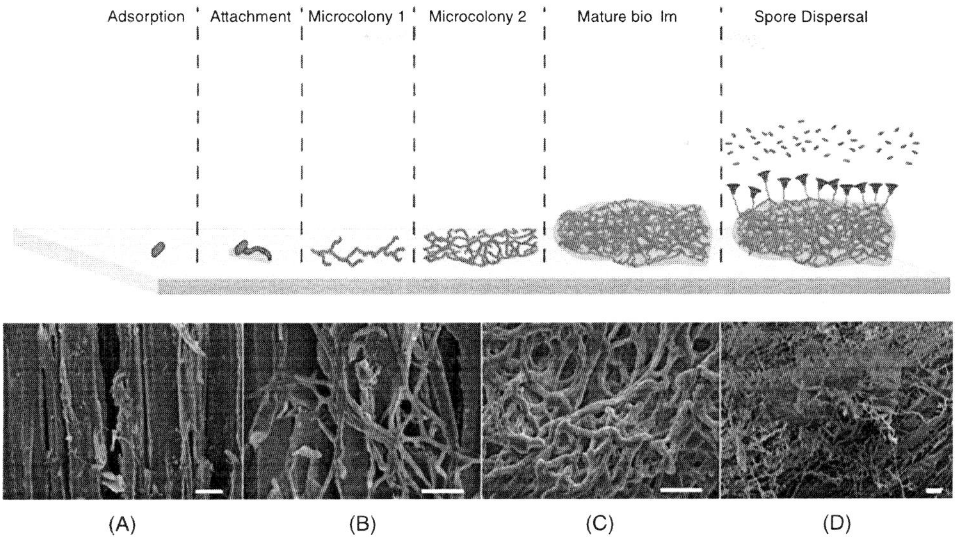

Figure 8.1 Stages of filamentous fungal biofilm formation. The upper panel shows a diagrammatic representation of six stages of fungal biofilm development. From left to right, the **Adsorption stage** involves propagules making contact with a suitable substrate or host. The deposition is reversible until the **Attachment stage** where secretion of adhesive substances from spores and germlings makes the attachment irreversible. This is followed by the **Microcolony I stage,** where initial formation of a microcolony occurs via apical elongation and branching of hyphae. Extracellular matrix is produced. **The Microcolony II** stage includes maturation of the colony where compacted hyphal networks and pervasive hyphal–hyphal adhesion resulting in hyphal bundles and cords. In submerged biofilms, fluid channels appear during this stage, via hydrophobic repulsion between hyphae. The **Mature biofilm stage** often has aerial growth and three-dimensional colony layering with near-complete encasement within the extracellular matrix. The final stage involves **Spore dispersal** via formation of fruiting bodies or sporogenous cells often formed apically on aerial hyphae. These structures produce spores or propagules for survival or dispersal and would include spores, sporangia, infective hyphal fragments or other structures. These detached cells can be translocated to new substrates or hosts and reinitiate the cycle. Scanning electron micrographs in the lower panel show *Didymella bryoniae* biofilms representing Microcolony I (A), Microcolony II (B), Mature biofilm (C) and Spore Dispersal (D). Images A-C were collected from *D. bryoniae* biofilms produced on wood surfaces in a static multiwall plate biofilm reactor (BEST™ Assay). The aerial hyphae and sporegenous cells shown in (D) were observed as part of a *D. bryoniae* biofilm on symptomatic cucumber fruit. Scale bars = 20 μm. (*See color plate section for the color representation of this figure.*)

- Stage four includes maturation of the colony and is characterized by the formation of compacted hyphal networks and pervasive hyphal–hyphal adhesion, resulting in hyphal bundles and cords. For submerged colonies, fluid channels appear during this stage as well, likely by hydrophobic repulsion between hyphae.
- Stage five is where the mature biofilm is observed. The colony will be multilayered with intensely intertwined networks of hyphae. Aerial growth may occur to give robust three-dimensional colony development, and the majority of the biofilm will display complete encasement within an abundant extracellular matrix. Additionally, the formation of fruiting bodies and sporogenous cells may be observed, often apically on aerial hyphae.
- The final stage is dispersal or release of propagules (spores, sporangia, hyphal fragments or other structures). These detached cells can be translocated to new substrates

or hosts, or remain dormant until suitable conditions reappear, after which they can reinitiate the cycle.

Other models for filamentous fungal biofilms have been presented [15, 16, 41] that, in large part, mirror that described by Harding et al. [17]. It is important to note that many of these models were based on biofilms cultures *in vitro* using bioreactors where fungal biofilms were submerged. Contrary to this condition, most filamentous fungal biofilms associated with plants or soil would be produced in environments exposed to high moisture with a major air interface, that is, unsaturated environments [64]. The nutritional, surface, and saturation parameters affect a number of biofilm characteristics and should be noted any time results from *in vitro* biofilm studies are used to predict or model those in a natural environment or on/in host tissues as there is an abundance of specific cases where naturally occurring conditions will lead to unique biofilm structure and function. Subaerial biofilms occurring naturally on rock and soil are one example. Subaerial biofilms, as defined by Gorbushina and Broughton [65], are "microbial communities that occupy significant portions of exposed terrestrial surfaces." These surfaces would include rock, mineral, and monument/building surfaces, representing environments clearly different from those used to produce the model described in Figure 8.1. In case of these subaerial biofilms, the microbial inhabitants must adapt to desiccation, solar radiation, and other harsh environmental conditions, which often leads fungi to develop unique microcolonial architectures, biofilm morphologies, and symbiotic lifestyles that might not be well-described by those biofilms formed while submerged in nutrient-rich broth in laboratory reactors.

8.6 Examples of Filamentous Fungal Biofilms

8.6.1 Ascomycete Biofilms

As previously mentioned, *Aspergillus* species biofilms are reported in clinical and industrial examples [16]. This genus can cause diseases on a number of plant species in fields and post-harvest storages. While *Aspergillus* biofilms on plants have not yet been described, *Aspergillus* biofilms are well characterized from work done in clinical and industrial microbiology. For example, the morphology of *Aspergillus* biofilms has been described using many forms of microscopy, both *in vivo* [66, 67] and *in vitro* [18, 31]. Currently, it is known that the extracellular matrix is composed of galactomannans, galactosaminogalactans, alphaglucans, hydrophobins, and melanin [67]. Initial spore adhesion is a key step in biofilm formation [68]. Cellulose production is high in biofilms [69] and extracellular DNA is present in the matrix; its release acts an antifungal resistance mechanism [70].

Biofilms formed by *Fusarium* species have also been reported. The morphologies of the biofilms were photographed at various time points and showed that the six characteristic stages of biofilm formation were evident [17, 41]. The mature biofilms displayed hyphal networks with hyphae often bundled together to form cables and glued together with extracelluar polymeric substances. Peiqian et al. [41] went on to demonstrate that the *Fusarium* biofilms were much better able to tolerate physical challenges such as high temperatures or irradiation than counterpart planktonic cultures. Harding et al. [40] also showed morphological evidence for biofilms formed by a species of *Vertillium*, and that like *Fusarium*, the mature biofilms *in planta* were morphologically similar to those grown on wood surfaces *in vitro*.

Another example of this is provided in Figure 8.2, where biofilms of *Didymella bryoniae*, the causal agent of gummy stem blight on cucumber, were formed *in vitro* on wood surfaces and then compared to biofilms observed on cucumber host tissues. The key difference seen in these studies was that the biofilms on plant surfaces had extensive aerial hyphae, often extending well above what remained of the extracellular matrix after fixation and dehydration preparations for scanning electron microscopy. This is in contrast to those biofilms formed on submerged wood surfaces, which were quite flattened to the substratum, displaying openings and channels in the biofilm, and in many cases more heavily encased in matrix. These differences are likely due to the fact that the *in planta* biofilms were formed at the host-air interface with no fluid shear while the *in vitro* biofilms were submerged in liquid media and formed under significant hydraulic shear.

Ascomycetes are also commonly found on rocks and soils. For example, melanized Ascomycetes represent a major group of fungi-forming subaerial biofilms on rocks. These communities grow on surfaces of rocks and minerals. They are particularly stress tolerant and can easily penetrate the rock substrate and interact symbiotically with phototrophic subaerial biofilm groups [65]. Some Ascomycete fungi in the soil form

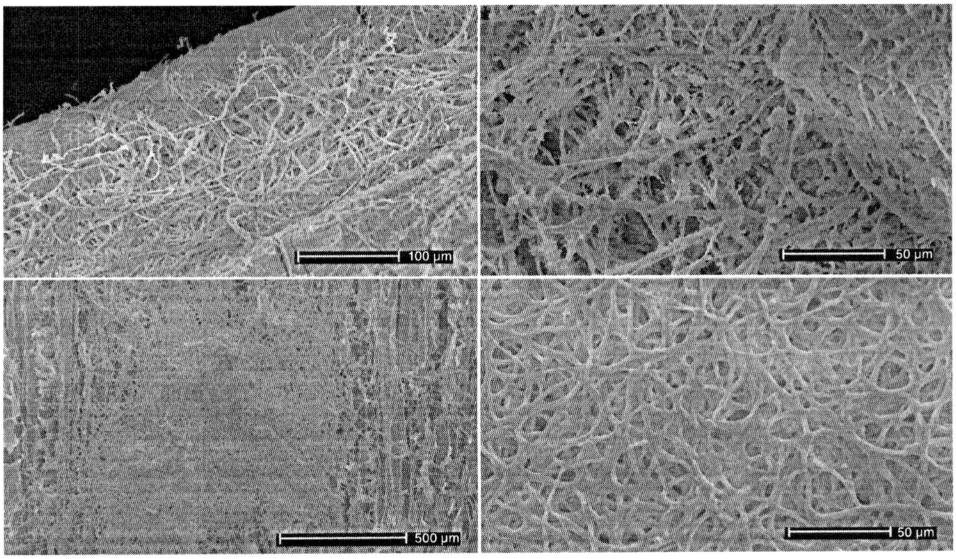

Figure 8.2 *Didymella bryoniae in vivo* biofilms on cucumber fruit (upper panels) and *in vitro* biofilms on wood surfaces (lower panels). The morphology of the biofilms cultivated on wood surfaces using the BEST™ Assay are quite similar to those on the host tissues. The main exception is that the biofilms on cucumber (upper panels) have a more aerial habit and less matrix material. This likely resulted due to their formation on cucumber fruit at the air interface with no fluid shear. Conversely, the biofilms formed using the BEST™ Assay (lower panels) were submerged in growth media and had a significant fluid shear applied throughout development. These biofilms have a more flattened morphology, lacking aerial hyphae and ample extracellular matrix with clear evidence of fluid channels for fluid movement through the biofilm.

ectomycorrhizal associations with tree roots, which include complex structures and specialized developmental stages. Additionally, mycorrhizal fungi are known to have interactions with bacterial biofilms in the soil [71, 72]. Mycorrhizal associations will be discussed further in a separate section of this chapter.

8.6.2 Zygomycete Biofilms

Fungi belonging to the class Zygomycetes are often rapidly colonizing saprophytes, but they can be opportunistic pathogens of humans, animals, and post-harvest produce. In a study aimed at elucidating the biofilm-forming potential of pathogenic zygomycetes evidence was presented for a QS mechanism, tenacious adhesion of spores, microcolony (monolayer) formation of hyphae, and extensive hyphal networking and extracellular polymeric matrix [73]. The matrices of the species investigated were positive for glucosamine and N-acetylglucosamine, but not glucose and mannose indicating that polysaccharides common in the extracellular matrices of Zygomycete fungi may be unique when compared to those of other fungal taxa.

8.6.3 Basidiomycete Biofilms

Biofilms of the wood-rotting fungus *Pleurotus ostreatus* have been well characterized [74–77]. These basidiomycete biofilms also had adherent intensely networked hyphal colonies with interstitial voids and channels, and were encased in a slimy matrix composed mainly of sugars and proteins. These characteristics were distinct from coevaluated planktonic colonies. Furthermore, the biofilms exhibited 2.4- to 6.5-fold higher tolerance to Cadmium than the planktonic cultures.

Basidiomycete fungi that form ectomycorrhizal associations with tree roots have also been studied from a biofilm perspective and are also discussed in a subsequent section of this chapter.

8.6.4 Oomycete Biofilms

While not technically fungi, the Oomycete class of organisms is composed of filamentous eukaryotic microorganisms with a hyphal growth habit very similar to true fungi. Biofilms formed by filamentous oomycetes are reported for both plant pathogens and animal pathogens. As previously mentioned, the first report of a filamentous biofilm *in planta* for a field crop was *Phythopthora parasitica* on tobacco leaves [38, 39]. The authors demonstrated that the motile zoospores of the pathogen were recruited to wound sites on the host in a directed chemotaxic manner and once a high cell density was achieved a large, adherent microcolony developed. The experimental results when inoculating with varying zoospore densities provided indirect evidence for a QS mechanism that regulated the transition from motile to sessile activity as part of biofilm formation. The microcolonies became embedded in an amorphous polymeric mucilaginous matrix that was composed of mucins, as well as polysaccharides mannose and glucose. The mature biofilm was densely packed at the core with ungerminated cysts and germ tubes. Outwardly, it contained metabolically active hyphae with irregularly distributed openings or channels that allowed for fluid to and zoospores to pass through the biofilm.

A second example of a filamentous biofilm from the Oomycete class is *Saprolegnia* species that are pathogenic on fish and fish eggs [78]. In this example, the authors showed evidence of biofilms attached to the surfaces of fish tanks and cultured biofilms on glass surfaces. The biofilms were characterized morphologically using various forms of microscopy, and evaluated for their response to malachite green and bronopol, two chemicals used to treat saprolegniosis. The results of the study showed that the *Saprolegnia* species had morphological and phenotypic characteristics of biofilms such as attachment, microcolony formation, and mature, three-dimensional colonies encased in slime that were more tolerant of chemical treatment than planktonics.

These examples indicate that the oomycetes, along with all the classes of true fungi, are fully capable of forming biofilms in association with host infection, leading one to wonder how many additional uncharacterized examples of filamentous biofilms are waiting to be described.

8.7 Examples of Fungal Biofilms in Soils and the Rhizosphere

8.7.1 Mycorrhizae

Mycorrhizae are symbiotic associations between certain soil-dwelling fungal taxa and plant roots that are classically defined as mutualisms in which the fungal symbiont gains carbon and the plant symbiont benefits from increased mineral nutrient uptake. The term *mycorrhiza* comes from the Greek *myco,* meaning mushroom/fungus, and *rhiza,* meaning root. The first mycorrhizae are thought to have originated approximately 400 to 450 million years ago, and may have facilitated the colonization of land by plants [79, 80]. While estimates vary, it is believed that greater than 80 percent of plant species form mycorrhizal associations [81], including a number of species/taxa of economic importance. As such, mycorrhizal applications for agriculture, silviculture, and reclamation/restoration purposes have received increasing focus from interested stakeholders.

The two main categories of mycorrhizae are endomycorrhizae and ectomycorrhizae (EcM). In endomycorrhizal symbioses, also known as arbuscular mycorrhizae (AM), hyphae penetrate the root cells in a non–host-specific way, forming intracellular symbioses [82, 83]. In EcM symbioses, the hyphae remain extracellular, attaching to root cells and encapsulating the root through formation of a hyphal sheath known as the mantle, and a transverse intercellular hyphal growth into the root called the Hartig net [84, 85]. EcM are formed by approximately 250 fungal genera, mostly from the Basidiomycota and Ascomycota [86]. While only a small percentage of plant species (mostly tree and shrub species) form ectomycorrhizal associations, they are of particular interest for forestry and the management of forested systems, especially the boreal ecosystem.

8.7.2 Ectomycorrhizae as a Biofilm

Clear correlations exist between EcM biofilm formation and the stages of filamentous fungal biofilm as described in Harding et al. [17]. The initial stages involve contact/adsorption, adhesion, and microcolony formation. EcM microcolonies grow through apical elongation, branching, and initial layering of hyphae, forming the mantle that colonizes root surfaces and envelops it in a sheath [84, 87, 88]. Hyphae also grow across the root surfaces by penetrating intercellular spaces between plant cells in the cortical area of the root, forming the Hartig net [81, 89, 90]. Extracellular matrix has been described in the mantle portion of the microcolony in microscopy studies [87], perhaps serving as a support element of the hyphae and forming an extracellular sheath surrounding the mycelium [91, 92]. Descriptions of *in vitro* extracellular production by EcM fungi containing water-soluble polysaccharides were reported for a number of EcM species [93].

In subsequent stages of colony maturation, the mantle develops into a more compact multilayered stack of hyphae. Extraradical hyphae start to branch from the mantle to the soil, increasing the surface area of the colonized root and spreading individually or as aggregate bundles called rhizomorph. The extraradical hyphae may form extensive underground mycelia networks and/or serve as a source of EcM infection to other nearby plants. This can lead to the formation of common mycorrhizal networks (CMNs) that facilitate the transportation of nutrients, water, and signals from plant to plant in a given area [84, 88, 90, 94–99]. This phenomenon has been called the *Wood Wide Web.*

The final stages of EcM biofilm formation are similar to other fungal models in the sense that they form fruiting bodies and spores, which can, in turn, reinitiate the process. Though mechanisms are not well known in a biofilm context, it has been demonstrated that fruiting bodies are only formed if the mutualistic EcM interaction is functional. The detachment phase will occur via fungal fruiting structures, but may be more complex based on the fact that most of the interaction and biofilm exist underground.

Four interesting questions arise regarding EcM biofilm development in relation to other fungal biofilm models:

1) What is involved in the maturation phase of the EcM biofilm?
2) Is the formation of extensive mycelial networks part of the biofilm maturation process, and a third part of the microcolony structure, along with the mantle and the Hartig net?
3) Could the formation of extensive mycelial networks also be part of the "detachment" phase, as it can lead to the colonization of other plants?
4) Does the standard definition of biofilms need to be expanded in order to include EcMs and AMs?

Additional research for EcMs within the biofilm context is needed to address these questions.

8.7.3 A Brief Look at Endomycorrhiza as a Biofilm

The sequence of events in plant colonization by AM fungi shares similarities with other fungal biofilm models, and some parallels with EcM development, but contains unique developmental features on its own. The contact/adsorption and adhesion phases are somewhat similar with initial spore germination and hyphal growth in AMs being independent of plant hosts. These fungi develop a vegetative mycelium, which, upon contact with host plant roots form appressoria on the root epidermis and directly penetrate the host root [83, 100]. Once in the root, two patterns of colonization may occur. In the Arum type, the classic depiction of AMs, extensive intercellular growth of the fungus develops as hyphae penetrate the root cortex. Once in the cortex, hyphae branch and penetrate the cortical cell walls where they differentiate into intracellular hyphal entanglements called *arbuscules* that form the interface of the symbiosis with the plant cell [101]. In the Paris type of colonization, growth is slow and primarily intracellular, with formation of coils inside each cell with rare or minimally structured arbuscules [102]. These structures remain connected to the extraradical mycelium, similar to the Hartig net and mantle in EcM. Following colonization of the root cortex, hyphae develop outward into extensive extraradical mycelia and underground networks within the soil, which is critical for acquisition of mineral nutrients from the soil and their subsequent translocation to the plant. These networks also facilitate colonization of additional roots, and, in many cases, the production of spores.

Similar to EcMs, AMs also have different types of hyphae as part of their structure: intracellular, fine runner type hyphae, and extraradical mycelium, which correlates to the intercellular Hartig net, the mantle hyphae, and the extraradical mycelium. Taken together, these facts make it unclear what precisely constitutes a mature AM biofilm, and the same questions raised for EcM biofilm development are applicable to AM biofilms.

8.8 The Mycorhizosphere

The classic definition of mycorrhizae has been rendered simplistic in light of the sophisticated interactions between fungi, plants, and bacteria unveiled in the past few years [84,87,89,97,103–105]. The term *mycorhizosphere* has been used to encompass not only the plant-fungi interaction but all other interactions occurring in this ecosystem [104,106–108]. There are a number of microbes that grow in association with AM mycelia and spores [109]. Bacterial biofilms have been detected growing on the fungal surfaces, including on the fungal biofilm's hyphae, mycelium, and spores [110]. Association of rhizobia and other nitrogen-fixing bacteria growing on AM biofilms, as well as intracellularly, is widely reported in the literature. Since some of these bacterial groups are also effective in performing certain ecosystem functions generally attributed to AMs, there is a possibility that some of these functions are actually performed by the bacterial biofilms growing on the fungal biofilms, not by the mycorrhiza directly [103, 111]. There appears to be communication between all of the members of these communities [103]. Bacterial associations with AMs are suspected to have influence on fungal development and may act in a tripartite symbiotic relationship with fungi and plants.

Bacterial biofilms have also been found in association with EcM, and there have been reports of endo/ecto underground mycelial networks connecting seemingly unrelated partners, adding another layer of complication to the picture. Until these processes are better understood in the context of biofilm formation, the best definition for these types of biofilms may be: "spatially structured communities of microbes whose function is dependent on a complex web of symbiotic interactions" (Hansen et al. 2007).

8.9 A Biofilm Approach to Plant Disease Management

The concept of microbial biofilms is rapidly reaching into all areas of microbiology. The application of the biofilm concept to crop disease management and soil health could have a profound impact, as it had in the medical and industrial fields. Early-stage efficacy testing in the agrichemical industry is mostly done following conventional antimicrobial testing against planktonic or solitary microbes in enriched media. These tests often fail to predict efficacy of fungicides and disinfectants against biofilms, which predominate in nature and are much more resistant to killing than solitary microbes. For this reason, the recommended rates on manufacturers' labels for agricultural disinfectants and fungicides may, in some instances, overestimate the efficacy of a formulation in controlling biofilms associated with plant diseases in the field or other production situations. This may explain why there are no "rescue treatments" capable of curing a fungal infection in plants, why there are no efficient treatments for vascular diseases (the "archetypal" biofilm disease in plants), and why diseases or surface contamination tend to recur after chemical treatments have been discontinued. For example, a recently published study showed that a number of chemical compounds could not disinfect biofilms formed by the bacterial ring rot pathogen of potato (*Clavibacter michiganensis* subsp. *sepidonicus*) on some surfaces [112].

Mycorrhizae biofilms primarily benefit the host plant by providing access to three things:

1) Large mycelial networks that effectively act as extended root systems, increasing the effective plant-accessible absorptive area within the soil and ability to scavenge scarce nutrients
2) Soil pores that might otherwise be too small to exploit
3) Enzymatic mobilization of mineral and organic nutrients, primarily nitrogen and phosphorus [81]

Additional benefits may include reduced water stress, pathogen resistance, barriers to uptake of heavy metals, and the enabling of symbiotic interactions with other microorganisms such as plant growth–promoting bacteria. The combination of these benefits often leads to increased seedling growth and establishment, and is essential for the establishment of some species (notably Pinus spp.), though not all EcM plant species benefit equally from colonization, with some species experiencing neutral interactions or even reduced growth [113]. Many fungi in soil and plants are vital for ecosystem and agroecosystem health in various ways.

There is a great need to revisit fungal infections and interactions *in planta* and the way we think about fungal diseases and their treatments, as well as mycorrhizae and their potential uses. The fact that fungi form biofilms in soil, plants, and on production surfaces indicates the need for new testing methods and products that will be effective against biofilms, or in encouraging the formation of mutually beneficial biofilm symbioses. Our understanding of how pathogenic microorganisms attach, infect, grow, disperse, and survive is foundational for devising disease management programs and tactics. Understanding each phase of fungal biofilm development, with its common features and species-specific elements, could provide the new tools necessary for the development of novel methods/products for control of fungal and bacterial diseases in plants. The first example of an anti-biofilm approach that improves plant disease management was published by Li and Wang [114]. In this case, the authors demonstrated increased efficacy of copper-based bactericides when co-applied with biofilm-inhibiting compounds. Perhaps future breakthroughs in fungal plant disease management will follow this example of biofilm approach to plant disease management. Additionally, a better understanding of sustainable agricultural practices involving mycorrhizae in order to promote plant and soil health and rehabilitation could enhance agricultural practices by minimizing disease through biological suppression of pests or enhancement of plant defense.

References

1 E.C. Oerke, Crop losses to pests. *The Journal of Agricultural Science*, 144(01):31–43 (2006).
2 E.C. Oerke, Estimated crop losses due to pathogens, animal pests and weeds. *Crop Production and Crop Protection*. Elsevier Science Publishing, New York, 1994.
3 M.E. Davey and G.A. O'Toole, Microbial biofilms: From ecology to molecular genetics. *Microbiology and Molecular Biology Reviews*, 64(4):847–867 (2000).
4 P. Stoodley, K. Sauer, D.G. Davies, and J.W. Costerton, Biofilms as complex differentiated communities. *Annual Reviews in Microbiology*, 56(1):187–209 (2002).
5 L. Hall-Stoodley, J.W. Costerton and P. Stoodley, Bacterial biofilms: from the natural environment to infectious diseases. *Nature Reviews Microbiology*, 2(2):95–108 (2004).
6 J. Harrison, R. Turner, L. Marques and H. Ceri, A new understanding of these microbial communities is driving a revolution that may transform the science of microbiology. *American Scientist*, 93(6):508–515 (2005).
7 N.A. Fujishige, N.N. Kapadia, and A.M. Hirsch, A feeling for the micro-organism: structure on a small scale. Biofilms on plant roots. *Botanical Journal of the Linnean Society*, 150(1):79–88 (2006).
8 T. Danhorn and C. Fuqua, Biofilm formation by plant-associated bacteria. *Annual Review of Microbiology*, 61, 401–422 (2007).
9 B.E. Ramey, M. Koutsoudis, S.B. von Bodman and C. Fuqua, Biofilm formation in plant–microbe associations. *Current Opinion in Microbiology*, 7(6):602–609 (2004).
10 S. Fanning and A.P. Mitchell, Fungal biofilms. *PLoS Pathology*, 8(4):e1002585 (2012).
11 J.L. Lopez-Ribot, Large-scale biochemical profiling of the *Candida albicans* biofilm matrix: new compositional, structural, and functional insights. *MBio*, 5(5):e01781–14 (2014).
12 G. Ramage, E. Mowat, B. Jones, C. Williams and J. Lopez-Ribot, Our current understanding of fungal biofilms. *Critical Reviews in Microbiology*, 35(4):340–355 (2009).
13 C. Williams and G. Ramage, Fungal biofilms in human disease, *in Biofilm-Based Healthcare-Associated Infections, Volume* 2, G. Donelli (Ed) Springer International Publishing, Switzerland, 2015.
14 S. Brückner and H.U. Mösch, Choosing the right lifestyle: adhesion and development in *Saccharomyces cerevisiae. FEMS Microbiology Reviews*, 36(1):25–58 (2012).
15 V.M. de Siqueira, Fungal biofilms: An overview, *in Fungal Biomolecules: Sources, Applications and Recent Developments*, V. Gupta, R. Mach, S. Sreeivasaprasad (Eds):Wiley-Blackwell, UK, 2015.
16 G. Ramage, R. Rajendran, M. Gutierrez-Correa, B. Jones and C. Williams, Aspergillus biofilms: clinical and industrial significance. *FEMS Microbiology Letters*, 324(2):89–97 (2011).
17 M.W. Harding, L.L. Marques, R.J. Howard and M.E. Olson, Can filamentous fungi form biofilms? *Trends in Microbiology*, 17(11):475–480 (2009).

18 M. Gutiérrez-Correa, Y. Ludeña, G. Ramage and G.K. Villena, Recent advances on filamentous fungal biofilms for industrial uses. *Applied Biochemistry and Biotechnology*, 167(5):1235–1253 (2012).
19 H.C. Flemming and J. Wingender, The biofilm matrix. *Nature Reviews Microbiology*, 8(9):623–633 (2010).
20 E.G. Nisbet and N.H. Sleep, The habitat and nature of early life. *Nature*, 409(6823):1083–1091 (2001).
21 D. Davies, Understanding biofilm resistance to antibacterial agents. *Nature Reviews Drug Discovery*, 2(2):114–122 (2003).
22 G. Ramage, R. Rajendran, L. Sherry and C. Williams, Fungal biofilm resistance. *International Journal of Microbiology*, 2012, 528521 (2012).
23 T.A. Clark and R.A. Hajjeh, Recent trends in the epidemiology of invasive mycoses. *Current Opinion in Infectious Diseases*, 15(6):569–574 (2002).
24 J. Perlroth, B. Choi and B. Spellberg, Nosocomial fungal infections: epidemiology, diagnosis, and treatment. *Medical Mycology*, 45(4):321–346 (2007).
25 E.J. Anaissie, S.L. Stratton, M.C. Dignani, C.K. Lee, R.C. Summerbell, J.H. Rex, T.P. Monson and T.J. Walsh, Pathogenic molds (including *Aspergillus* species) in hospital water distribution systems: a 3-year prospective study and clinical implications for patients with hematologic malignancies. *Blood*, 101(7):2542–2546 (2003).
26 V.M. Siqueira, H.M.B. Oliveira, C. Santos, R.R.M. Paterson, N.B. Gusmão and N. Lima, Biofilms from a Brazilian water distribution system include filamentous fungi. *Canadian Journal of Microbiology*, 59(3):183–188 (2012).
27 V. Siqueira and N. Lima, Surface hydrophobicity of culture and water biofilm of Penicillium spp. *Current Microbiology*, 64(2):93–99 (2012).
28 R.A. Ali, R.J. Murphy and D.J. Dickinson, Investigation of the extracellular mucilaginous materials produced by some wood decay fungi. *Mycological Research*, 103(11):1453–1461 (1999).
29 E.L. Nurmiaho-Lassila, S. Timonen, K. Haahtela and R. Sen, Bacterial colonization patterns of intact *Pinus sylvestris* mycorrhizospheres in dry pine forest soil: an electron microscopy study. *Canadian Journal of Microbiology*, 43(11):1017–1035 (1997).
30 M.B. Branzanti, E. Rocca and A. Pisi, Effect of ectomycorrhizal fungi on chestnut ink disease. *Mycorrhiza*, 9(2):103–109 (1999).
31 G.K. Villena, T. Fujikawa, S. Tsuyumu and M. Gutiérrez-Correa, Structural analysis of biofilms and pellets of *Aspergillus niger* by confocal laser scanning microscopy and cryo scanning electron microscopy. *Bioresource Technology*, 101(6):1920–1926 (2010).
32 T. Akao, K. Gomi, K. Goto, N. Okazaki and O. Akita, Subtractive cloning of cDNA from *Aspergillus oryzae* differentially regulated between solid-state culture and liquid (submerged) culture. *Current Genetics*, 41(4):275–281 (2002).
33 R. te Biesebeke, G. Ruijter, Y.S. Rahardjo, M.J. Hoogschagen, M. Heerikhuisen, A. Levin, K.G. van Driel, M.A. Schutyser, J. Dijksterhuis, Y. Zhu and F.J. Weber, *Aspergillus oryzae* in solid-state and submerged fermentations. *FEMS Yeast Research*, 2(2):245–248 (2002).
34 M. Khalesi, Q. Zune, S. Telek, D. Riveros-Galan, H. Verachtert, D. Toye, K. Gebruers, G. Derdelinckx and F. Delvigne, Fungal biofilm reactor improves the productivity of hydrophobin HFBII. *Biochemical Engineering Journal*, 88, 171–178 (2014).
35 S.R. Barratt, A.R. Ennos, M. Greenhalgh, G.D. Robson and P.S. Handley, Fungi are the predominant micro-organisms responsible for degradation of soil-buried polyester

polyurethane over a range of soil water holding capacities. *Journal of Applied Microbiology*, 95(1):78–85 (2003).

36 C.C. Gaylarde, B.O. Ortega-Morales, and P. Bartolo-Pérez, Biogenic black crusts on buildings in unpolluted environments. *Current Microbiology*, 54(2):162–166 (2007).

37 A.A. Gorbushina, Life on the rocks. *Environmental Microbiology*, 9(7):1613–1631 (2007).

38 E. Galiana, S. Fourré, and G. Engler, *Phytophthora parasitica* biofilm formation: installation and organization of microcolonies on the surface of a host plant. *Environmental Microbiology*, 10(8):2164–2171 (2008).

39 N. Theodorakopoulos, B. Govetto, L. Massi, M. Gaysinski, et al., Biology and Ecology of Biofilms formed by a plant pathogen *Phytophthora parasitica*: from biochemical Ecology to Ecological Engineering. *Procedia Environmental Sciences*, 9:178–182 (2011).

40 M.W. Harding, L.L. Marques, R.J. Howard and M.E. Olson, Biofilm morphologies of plant pathogenic fungi. *Americas Journal of Plant Science and Biotechnology*, 4:43–47 (2010).

41 L. Peiqian, P. Xiaoming, S. Huifang, Z. Jingxin, H. Ning and L. Birun, Biofilm formation by *Fusarium oxysporum* f. sp. *cucumerinum* and susceptibility to environmental stress. *FEMS Microbiology Letters*, 350(2):138–145 (2014).

42 M.C. Manganyi, T. Regnier and E.I. Olivier, Antimicrobial activities of selected essential oils against *Fusarium oxysporum* isolates and their biofilms. *South African Journal of Botany*, 99:115–121 (2015).

43 J. Chandra, P.K. Mukherjee and M.A. Ghannoum, In vitro growth and analysis of Candida biofilms. *Nature Protocols*, 3(12):1909–1924 (2008).

44 P.K. Mukherjee and J. Chandra, Candida biofilm resistance. *Drug Resistance Updates*, 7(4):301–309 (2004).

45 C.J. Seneviratne, L. Jin and L.P. Samaranayake, Biofilm lifestyle of Candida: a mini review. *Oral Diseases*, 14(7):582–590 (2008).

46 L.J. Douglas, Candida biofilms and their role in infection. *Trends in Microbiology*, 11(1):30–36 (2003).

47 K.J. Verstrepen and F.M. Klis, Flocculation, adhesion and biofilm formation in yeasts. *Molecular Microbiology*, 60(1):5–15 (2006).

48 R.K. Bojsen, K.S. Andersen and B. Regenberg, *Saccharomyces cerevisiae*—a model to uncover molecular mechanisms for yeast biofilm biology. *FEMS Immunology & Medical Microbiology*, 65(2):169–182 (2012).

49 J.A. Granek, D. Murray, Ö. Kayrkçi and P.M. Magwene, The genetic architecture of biofilm formation in a clinical isolate of *Saccharomyces cerevisiae. Genetics*, 193(2):587–600 (2013).

50 K. Czaczyk and K. Myszka, Biosynthesis of extracellular polymeric substances (EPS) and its role in microbial biofilm formation. *Polish Journal of Environmental Studies*, 16(6):799 (2007).

51 L.R. Martinez and A. Casadevall, Susceptibility of *Cryptococcus neoformans* biofilms to antifungal agents in vitro. *Antimicrobial Agents and Chemotherapy*, 50(3):1021–1033 (2006).

52 W.F. Fett, Naturally occurring biofilms on alfalfa and other types of sprouts. *Journal of Food Protection*, 63(5):625–632 (2000).

53 C.E. Morris, J. Monier and M. Jacques, Methods for observing microbial biofilms directly on leaf surfaces and recovering them for isolation of culturable microorganisms. *Applied and Environmental Microbiology*, 63(4):1570–1576 (1997).

54 C.E. Morris, J.M. Monier and M.A. Jacques, A technique to quantify the population size and composition of the biofilm component in communities of bacteria in the phyllosphere. *Applied and Environmental Microbiology*, 64(12):4789–4795 (1998).
55 J.S. Finkel and A.P. Mitchell, Genetic control of *Candida albicans* biofilm development. *Nature Reviews Microbiology*, 9(2):109–118 (2011).
56 S. García-Sánchez, S. Aubert, I. Iraqui, G. Janbon, J.M. Ghigo and C. d'Enfert, *Candida albicans* biofilms: a developmental state associated with specific and stable gene expression patterns. *Eukaryotic Cell*, 3(2):536–545 (2004).
57 L.A. Murillo, G. Newport, C.Y. Lan, S. Habelitz, J. Dungan and N.M. Agabian, Genome-wide transcription profiling of the early phase of biofilm formation by *Candida albicans*. *Eukaryotic Cell*, 4(9):1562–1573 (2005).
58 J.E. Nett, A.J. Lepak, K. Marchillo and D.R. Andes, Time course global gene expression analysis of an in vivo Candida biofilm. *Journal of Infectious Diseases*, 200(2):307–313 (2009).
59 G. Ramage, S.P. Saville, B.L. Wickes and J.L. López-Ribot, Inhibition of *Candida albicans* biofilm formation by farnesol, a quorum-sensing molecule. *Applied and Environmental Microbiology*, 68(11):5459–5463 (2002).
60 D.A. Hogan, Talking to themselves: autoregulation and quorum sensing in fungi. *Eukaryotic cell*, 5(4):613–619 (2006).
61 J.R. Blankenship and A.P. Mitchell, How to build a biofilm: a fungal perspective. *Current Opinion in Microbiology*, 9(6):588–594 (2006).
62 M.A. Al-Fattani and L.J. Douglas, Biofilm matrix of *Candida albicans* and *Candida tropicalis*: chemical composition and role in drug resistance. *Journal of Medical Microbiology*, 55(8):999–1008 (2006).
63 H.T. Taff, K.F. Mitchell, J.A. Edward and D.R. Andes, Mechanisms of Candida biofilm drug resistance. *Future Microbiology*, 8(10):1325–1337 (2013).
64 P.A. Holden, Biofilms in unsaturated environments. *Methods in Enzymology*, 337:125 (2001).
65 A.A. Gorbushina and W.J. Broughton, Microbiology of the atmosphere-rock interface: how biological interactions and physical stresses modulate a sophisticated microbial ecosystem. *Microbiology*, 63:431–450 (2009).
66 C. Loussert, C. Schmitt, M.C. Prevost, V. Balloy, E. Fadel, B. Philippe, C. Kauffmann-Lacroix, J.P. Latgé and A. Beauvais, In vivo biofilm composition of *Aspergillus fumigatus*. *Cellular Microbiology*, 12(3):405–410 (2010).
67 A. Beauvais, T. Fontaine, V. Aimanianda and J.P. Latgé, Aspergillus cell wall and biofilm. *Mycopathologia*, 178(5–6):371–377 (2014).
68 B.E. Priegnitz, A. Wargenau, U. Brandt, M. Rohde, S. Dietrich, A. Kwade, R. Krull and A. Fleißner, The role of initial spore adhesion in pellet and biofilm formation in *Aspergillus niger*. *Fungal Genetics and Biology*, 49(1):30–38 (2012).
69 N.N. Gamarra, G.K. Villena and M. Gutiérrez-Correa, Cellulase production by *Aspergillus niger* in biofilm, solid-state, and submerged fermentations. *Applied Microbiology and Biotechnology*, 87(2):545–551 (2010).
70 R. Rajendran, C. Williams, D.F. Lappin, O. Millington, M. Martins and G. Ramage, Extracellular DNA release acts as an antifungal resistance mechanism in mature *Aspergillus fumigatus* biofilms. *Eukaryotic Cell*, 12(3):420–429 (2013).
71 V. Bianciotto, S. Andreotti, P. Balestrini, P. Bonfante and S. Perotto, Mucoid mutants of the biocontrol strain *Pseudomonas fluorescens* CHA0 show increased ability in biofilm

formation on mycorrhizal and nonmycorrhizal carrot roots. *Molecular Plant-Microbe Interactions*, 14(2):255–260 (2001).

72 V. Bianciotto and P. Bonfante, Arbuscular mycorrhizal fungi: a specialised niche for rhizospheric and endocellular bacteria. *Antonie van Leeuwenhoek*, 81(1–4):365–371 (2002).

73 R. Singh, M.R. Shivaprakash and A. Chakrabarti, Biofilm formation by zygomycetes: quantification, structure and matrix composition. *Microbiology*, 157(9):2611–2618 (2011).

74 J. Wu, Y.Z. Xiao and H.Q. Yu, Degradation of lignin in pulp mill wastewaters by white-rot fungi on biofilm. *Bioresource Technology*, 96(12):1357–1363 (2005).

75 H.S. Jayasinghearachchi and G. Seneviratne, Fungal solubilization of rock phosphate is enhanced by forming fungal–rhizobial biofilms. *Soil Biology and Biochemistry*, 38(2):405–408 (2006).

76 L. Pesciaroli, M. Petruccioli, F. Federici and A. D'Annibale, *Pleurotus ostreatus* biofilm-forming ability and ultrastructure are significantly influenced by growth medium and support type. *Journal of Applied Microbiology*, 114(6):1750–1762 (2013).

77 L. Pesciaroli, M. Petruccioli, S. Fedi, A. Firrincieli, F. Federici and A. D'Annibale, Characterization of *Pleurotus ostreatus* Biofilms by Using the Calgary Biofilm Device. *Applied and Environmental Microbiology*, 79(19):6083–6092 (2013).

78 S.E. Ali, E. Thoen, T. Vrålstad, R. Kristensen, Ø. Evensen and I. Skaar, Development and reproduction of Saprolegnia species in biofilms. *Veterinary Microbiology*, 163(1):133–141 (2013).

79 K.A. Pirozynski and D.W. Malloch, The origin of land plants: a matter of mycotrophism. *Biosystems*, 6(3):153–164 (1975).

80 D. Redecker, R. Kodner and L.E. Graham, Glomalean fungi from the Ordovician. *Science*, 289(5486):1920–1921 (2000).

81 S. Smith and D. Read, *Mycorrhizal Symbiosis. Academic Press*, London, 2008.

82 P. Bonfante and S. Perotto, Tansley Review No. 82 Strategies of arbuscular mycorrhizal fungi when infecting host plants. *New Phytologist*, 130(1):3–21 (1995).

83 M.J. Harrison, Biotrophic interfaces and nutrient transport in plant/fungal symbioses. *Journal of Experimental Botany*, 50(Special Issue):1013–1022 (1999).

84 R.D. Finlay, Ecological aspects of mycorrhizal symbiosis: with special emphasis on the functional diversity of interactions involving the extraradical mycelium. *Journal of Experimental Botany*, 59(5):1115–1126 (2008).

85 P. Bonfante and I.A. Anca, Plants, mycorrhizal fungi, and bacteria: a network of interactions. *Annual Review of Microbiology*, 63:363–383 (2009).

86 L. Tedersoo and M.E. Smith, Lineages of ectomycorrhizal fungi revisited: foraging strategies and novel lineages revealed by sequences from belowground. *Fungal Biology Reviews*, 27(3):83–99 (2013).

87 E.L. Nurmiaho-Lassila, S. Timonen, K. Haahtela and R. Sen, Bacterial colonization patterns of intact *Pinus sylvestris* mycorrhizospheres in dry pine forest soil: an electron microscopy study. *Canadian Journal of Microbiology*, 43(11):1017–1035 (1997).

88 M.G. van der Heijden, F.M. Martin, M.A. Selosse and I.R. Sanders, Mycorrhizal ecology and evolution: the past, the present, and the future. *New Phytologist*, 205(4):1406–1423 (2015).

89 P. Frey-Klett, J.A. Garbaye and M. Tarkka, The mycorrhiza helper bacteria revisited. *New Phytologist*, 176(1):22–36 (2007).

90 R. Agerer, Exploration types of ectomycorrhizae. *Mycorrhiza*, 11(2):107–114 (2001).
91 C.A. Osaku, G.L. Sassaki, G.T. Zancan and M. Iacomini, Studies on neutral exopolysaccharides produced by the ectomycorrhiza *Thelephora terrestris*. *FEMS Microbiology Letters*, 216(2):145–149 (2002).
92 A. Gutierrez, A. Prieto and A.T. Martinez, Structural characterization of extracellular polysaccharides produced by fungi from the genus Pleurotus. *Carbohydrate Research*, 281:143–145 (1996).
93 J.P. Chaumeton, C. Chauveau and L. Chavant, Water-soluble polysaccharides excreted by mycelium of higher fungi: relationship with taxonomy and physiology. *Biochemical Systematics and Ecology*, 21(2):227–239 (1993).
94 V. Wiemken and T. Boller, Ectomycorrhiza: gene expression, metabolism and the wood-wide web. *Current Opinion in Plant Biology*, 5(4):355–361 (2002).
95 H.P. Bais, S.W. Park, T.L. Weir, R.M. Callaway and J.M. Vivanco, How plants communicate using the underground information superhighway. *Trends in Plant Science*, 9(1):26–32 (2004).
96 M. Giovannetti, L. Avio, P. Fortuna, E. Pellegrino, C. Sbrana and P. Strani, At the root of the wood wide web: self-recognition and nonself incompatibility in mycorrhizal networks. *Plant Signaling & Behavior*, 1(1):1–5 (2006).
97 E.K. Barto, J.D. Weidenhamer, D. Cipollini and M.C. Rillig, Fungal superhighways: do common mycorrhizal networks enhance below ground communication? *Trends in Plant Science*, 17(11):633–637 (2012).
98 J. Leake, D. Johnson, D. Donnelly, G. Muckle, L. Boddy and D. Read, Networks of power and influence: the role of mycorrhizal mycelium in controlling plant communities and agroecosystem functioning. *Canadian Journal of Botany*, 82(8):1016–1045 (2004).
99 Z. Babikova, L. Gilbert, T.J. Bruce, M. Birkett, J.C. Caulfield, C. Woodcock, J.A. Pickett and D. Johnson, Underground signals carried through common mycelial networks warn neighbouring plants of aphid attack. *Ecology Letters*, 16(7):835–843 (2013).
100 S.J. Barker, D. Tagu and G. Delp, Regulation of root and fungal morphogenesis in mycorrhizal symbioses. *Plant Physiology*, 116(4):1201–1207 (1998).
101 M.A. Abiala, O.O. Popoola, O.J. Olawuyi, J.O. Oyelude, A.O. Akanmu, A.S. Killani, O. Osonubi and A.C. Odebode, Harnessing the potentials of vesicular arbuscular mycorrhizal (VAM) fungi to plant growth–a review. *International Journal of Pure Applied Science and Technology* 14(2):61–79 (2013).
102 F.A. Smith and S.E. Smith, Structural diversity in (vesicular)–arbuscular mycorrhizal symbioses. *New Phytologist*, 137(3):373–388 (1997).
103 W.J. Wargo and D.A. Hogan, Fungal—bacterial interactions: a mixed bag of mingling microbes. *Current Opinion in Microbiology*, 9(4):359–364 (2006).
104 J.M. Barea, R. Azcón and C. Azcón-Aguilar, Mycorrhizosphere interactions to improve plant fitness and soil quality. *Antonie van Leeuwenhoek*, 81(1–4):343–351 (2002).
105 D.A. Hogan, W.J. Wargo and N. Beck, Bacterial Biofilms on Fungal Surfaces. *Lab-on-a-Chip*, 10(238):(2009).
106 R.G. Linderman, Mycorrhizal interactions with the rhizosphere microflora: the mycorrhizosphere effect. *Phytopathology*, 78(3):366–371 (1988).
107 J.F. Johansson, L.R. Paul and R.D. Finlay, Microbial interactions in the mycorrhizosphere and their significance for sustainable agriculture.*FEMS Microbiology Ecology*, 48(1):1–13 (2004).

108 G.D. Bending, T.J. Aspray and J.M. Whipps, Significance of microbial interactions in the mycorrhizosphere. *Advances in Applied Microbiology*, 60:97–133 (2006).

109 K. Mansfeld-Giese, J. Larsen and L. Bødker, Bacterial populations associated with mycelium of the arbuscular mycorrhizal fungus *Glomus intraradices*. *FEMS Microbiology Ecology*, 41(2):133–140 (2002).

110 S. Perotto, P. Angelini, V. Bianciotto, P. Bonfante, M. Girlanda, T. Kull, A. Mello, L. Pecoraro, C. Perini, A.M. Persiani and A. Saitta, Interactions of fungi with other organisms. *Plant Biosystems—An International Journal Dealing with all Aspects of Plant Biology*, 147(1):208–218 (2013).

111 M.C. Rillig, E.R. Lutgen, P.W. Ramsey, J.N. Klironomos and J.E. Gannon, Microbiota accompanying different arbuscular mycorrhizal fungal isolates influence soil aggregation. *Pedobiologia*, 49(3):251–259 (2005).

112 R.J. Howard, M.W. Harding, G.C. Daniels, S.L. Mobbs, S.L.I. Lisowski and S.H. De Boer, Efficacy of agricultural disinfectants on biofilms of the bacterial ring rot pathogen, *Clavibacter michiganensis* subsp. *sepedonicus*. *Canadian Journal of Plant Pathology*, 37(3):273–284 (2015).

113 J. Karst, L. Marczak, M.D. Jones and R. Turkington, The mutualism-parasitism continuum in ectomycorrhizas: a quantitative assessment using meta-analysis. *Ecology*, 89(4):1032–1042 (2008).

114 J. Li and N. Wang, Foliar application of biofilm formation-inhibiting compounds enhances control of citrus canker caused by *Xanthomonas citri* subsp. *citri*. *Phytopathology*, 104(2):134–142 (2014).

9

Chemical Nature of Biofilm Matrix and Its Significance

Mohd Sajjad Ahmad Khan[1], Mohd Musheer Altaf[2] and Iqbal Ahmad[2]

[1] *Department of Biology, College of Medicine, Imam Abdulrahman Bin Faisal University, Dammam, Saudi Arabia*
[2] *Department of Agricultural Microbiology, Aligarh Muslim University, Aligarh, India*

9.1 Introduction

It is now broadly acknowledged that in natural settings, microbial cells are most often found in close association with surfaces and interfaces, in the form of multicellular aggregates commonly referred to as biofilms. These are defined as a well-organized three-dimensional structured community of planktonic cells that are enclosed in a self-produced extracellular polymeric substances (EPS) or matrix. This tendency of unicellular bacterial cell to thrive for multicellularity assists bacterial cells' existence in several harsh environmental conditions [1]. Microbial biofilm development is observed on virtually all submerged surfaces in natural and industrial environments. Biofilms are also observed at interfaces as pellicles, or in the bulk of aquatic environments as flocs or granules [1, 2]. The naturally occurring biofilms are as multifarious as their integral microbes; for example, the streaming biofilms formed on submerged rocks in acid mine drainage [2] are amazingly different from the plaque formed on air-exposed surfaces of teeth [3]. But these all have one common aspect, which is the matrix that embeds cells. Formation of microbial biofilms is also observed with plant surfaces and under soil environments. There is mounting appreciation that the intensity, extent, and consequences of plant microbe interactions are significantly influenced by the conformation of adherent microbial populations [4]. It has been found that EPS produced by microorganisms in the rhizopheric region improves soil health by increasing soil's heterogeneity and aggregating ability. Aggregated soils withstand the intermediate water contents that are most favorable to plants and soil microbes. Also, EPS act as both a reservoir and a conductor of water to plant roots when bulk soil water is threatened [5].

The matrix is one of the most distinguishing features of a microbial biofilm, and the success of biofilm mode of life has been made possible because the matrix is made up of EPS. As revealed by molecular genetic studies, this matrix forms the scaffold for the three-dimensional architecture of the macroscopic biofilm and is responsible for adhesion to both inert as well as living surfaces. Therefore, the matrix becomes responsible for cohesion in the biofilm. These EPS or matrices are gel-like structures consisting of an assembly of different types of highly hydrated biopolymers that create a locally charged environment that surrounds and immobilize microbial aggregates [6]. The capability of microorganisms to make an extracellular matrix is a common feature of its multicellular communities, but a significant amount of variation is observed in the ways by which these matrices are constructed [7].

Microbial biofilms play a fundamental role in a variety of disciplines, including biotechnology, immunology, agriculture, biofouling, and biodeterioration [1, 2, 8]. Industrial pipelines, water distribution systems, nuclear power stations, space stations, and hospitals, are all prone to colonization by microorganisms growing in the form of biofilms.

Plants influence their rhizospheric environment by its exudates consisting of various organic and inorganic substances, which are rich in nitrogen and phosphorous. These nutrients attracts a range of microbial population to survive on and within their tissues, including aerial portions, the vascular network, and root tissues below ground. Therefore, plant-associated microbes establish commensal, mutualistic, and pathogenic interactions with plants in the form of biofilm growth.

Overall, the progressions of autoaggregation and biofilm development are relevant to both bacterial survival and host plant colonization [9]. These association of microbial biofilms helps in plant growth–promoting activity and also defend themselves from certain environmental stresses such as protection from protozoan predation and increased opportunity for horizontal gene transfer [4, 9]. A variety of environmental, genetic, and structural factors affect bacterial adhesion, cell–cell interactions, and plant colonization, and eventually plant–bacterial interactions [4, 9, 10]. Therefore, studying biofilm formation and structural integrity in varying microbial population and associated functions is highly acknowledged by the scientists.

It has been evident from literature that cell contact with surfaces stimulates transcription of the EPS genes [11]. Therefore monitoring the production and regulation of EPS in adherent populations enables a better understanding on the basis of biofilm phenotype and its varying characteristics [12]. These EPS are also frequently referred to as *slime,* and matrix-fenced microcolonies are termed as *stacks* or *towers.* These stacks are separated by water channels, which provide a mechanism for nutrient circulation within the biofilm matrix [13]. The matrix structure constitutes the elastic part of the biofilm. Interstitial voids and channels separating the microcolonies contain a liquid phase, mainly constituted by water, are the viscous part of the biofilm, as shown in Figure 9.1. These viscoelastic properties provide the biofilm with mechanical stability within the EPS matrix [14, 15]. Figure 9.1 clearly illustrates the water channels and heterogeneity characteristic of a mature biofilm. Liquid flow occurs in these water channels, allowing diffusion of nutrients, oxygen, and even antimicrobial agents. The constantly changing external and internal developments of the biofilm impart the heterogeneity with respect to space and time. This concept of heterogeneous biofilm architecture is expressive for both the mixed-culture biofilms (as observed in environmental biofilms) and also for pure culture biofilms common mainly on medical devices and those associated with infectious diseases. Nielsen and Molin [16] noted that every microbial biofilm community is unique, although some structural traits can usually be considered universal.

Microorganisms account for less than 10 percent of the dry weight of the biofilms, whereas the matrix can account for more than 90 percent [17]. This indicates the importance and impact of matrix on biofilm behavior and type. The studies carried out on various types of microbial biofilms have shown that the composition of the matrix differs according to the nature of the microbial entity present in it. Matrix polymers of

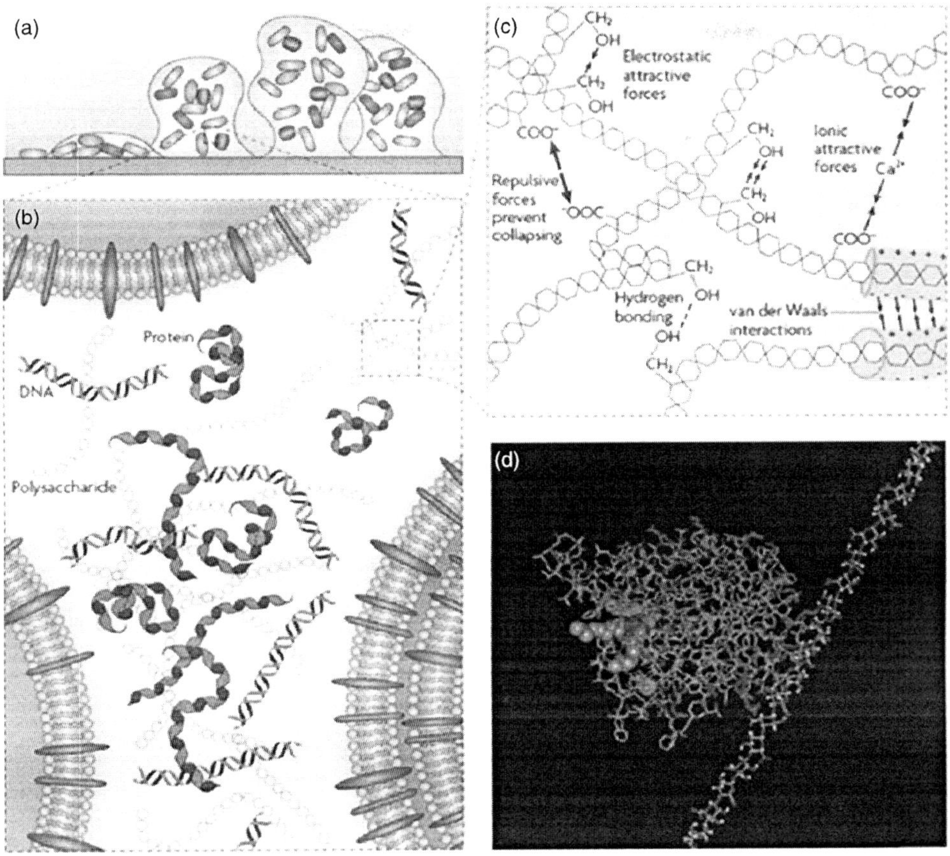

Figure 9.1 The extracellular polymeric substances matrix at different levels of resolution. (a) A model of a bacterial biofilm attached to a solid surface. Biofilm formation starts with the attachment of a cell to a surface. A microcolony forms through division of the bacterium, and production of the biofilm matrix is initiated. Other bacteria can then be recruited as the biofilm expands owing to cell division and the further production of matrix components. (b) The major matrix components—polysaccharides, proteins, and DNA—are distributed between the cells in a nonhomogeneous pattern, setting up differences between regions of the matrix. (c) The classes of weak physicochemical interactions and the entanglement of biopolymers that dominate the stability of the EPS matrix. (d) Amolecular modeling simulation of the interaction between the exo-polysaccharide alginate (right) and the extracellular enzyme lipase (left) of *Pseudomonas aeruginosa*in aqueous solution. The starting structure for the simulation of the lipase proteinwasobtained from the Protein Data Bank. The coloured spheres represent 1,2-dioctylcarbamoylglycero-3-O-octylphosphonate in the lipase active site (which was present as a part of the crystal structure), except for the green sphere, which represents a Ca^{2+} ion. The aggregate is stabilized by the interaction of the positively charged amino acids arginine and histidine (indicated in blue) with the polyanionic alginate. Water molecules are not shown. Image courtesy of H. C. Flemming, CAM-D Technologies, Essen, Germany. (With copyright permission, from [Colloids and Surfaces B: Biointerfaces, Volume 86, Issue 2, 2011, 251–259]).
(See color plate section for the color representation of this figure.)

bacterial biofilms are primarily exopolysaccharides, and many are negatively charged due to the presence of carboxyl, sulphate, or phosphate groups. Proteins, nucleic acids, and lipids are also present in lesser amounts. Although extracellular polysaccharides are considered as the major structural components of the biofilm matrix, extracellular DNA plays an important role in the establishment of biofilm structure [18–20]. Moreover, nucleases can also be the regulators of biofilm formation [21]. The polysaccharide–protein interactions in the matrix affect structural and functional assets. Actually, some of these proteins are enzymes that constitute an external digestive system for the biofilm [22]. Also, dead cells have been witnessed in some biofilms, suggesting that cell debris can be considered as part of the matrix [7, 23, 24]. Studies of the roles played by DNA and dead cells in biofilms signify exciting new directions for future studies of biofilm matrices.

Regardless of the universal presence of extracellular matrices in biofilms, it is clear that there is vast diversity in their composition and in the timing of their synthesis. This diversity is observed at several levels. For example, biofilms formed by different microbial species are usually easily notable from each other, and even by different strains of a single species. Moreover, vivid differences in biofilm architecture can result from even small changes in environmental conditions. Such differences in biofilm structure appear to reflect differences in the composition of the extracellular matrix. Characterization of EPS components is mandatory in understanding biofilm structure and function. In this chapter, however, we limit our discussion to the extracellular matrix components that have been most extensively studied: carbohydrate-rich polymers and proteins. For more detailed analysis, consult excellent review articles in this regard [4, 10, 17, 25–31].

9.2 Structural Composition of EPS

The EPS matrix is mostly 0.2 to 1.0 μm thick. In some microbial species, the thickness of the EPS layer does not exceed values from 10 to 30 nm [32]. The chemical configuration of polymeric substances secreted by the cells into the environment is varied. The actual composition of EPS can vary significantly, depending on the types of microorganisms present, environmental conditions such as nutrients, shear forces, or temperature. EPS compounds belong to various different classes of macromolecules like polysaccharides, proteins, nucleic acids, glycoproteins, and phospholipids [6, 7]. These constitutional differences are often used for identification and classification of microbial cells. Also, the antigenic properties of these extracellular molecules enable the serological characterization of the cells. The best-investigated components of the EPS layer are polysaccharides and proteins. Matrix composition has been most comprehensively studied for the biofilms produced by Gram-negative bacteria (especially pseudomonads) and by some Gram-positive bacteria. The presence of polypeptides in the matrix is the feature of a very few Gram-positive bacteria cells [6, 7, 33].

9.2.1 Exopolysaccharides of the Biofilm Matrix

Exopolysaccharides are the most important components of the biofilms matrix [22]. Mostly, the matrix exopolysaccharides are very long molecules, linear or branched, and with a molecular weight of 500 to 2000 kDa. Exopolysaccharides are varying even between strains of a single species. Vaningelgem et al. [34] observed that various *Streptococcus thermophilus* strains are producing heteropolysaccharides of different monomer compositions and ratios and different molecular masses. In general, exopolysaccharides can be either homopolymers (cellulose, curdlan, or dextran), or heteropolymers (alginate, emulsan, gellan, or xanthan). Most microbial exogenous layers contain neutral carbohydrates (mainly-hexose, seldom-pentose) and uronic acids. Various studies have reported that the structures of polysaccharides synthesized by microbial cells is varying, depending on their kinds of linkages and noncarbohydrate substituents such as acetate, pyruvate, succinate, and phosphate [8, 17, 35, 36].

9.2.1.1 Carbohydrate Content of Exopolysaccharides

Composition as well as conformation of sugar monomers may alter the properties of the exopolysaccharides and thus of the biofilm matrix. Homopolysaccharides are composed of only one monosaccharide type: D-glucose or L-fructose. Heteropolysaccharides are made from repeating units of monosaccharides such as: D-glucose, D-galactose, L-fructose, L-rhamnose, D-glucuronic acid, L-guluronic acid, and D-mannuronic acid. The type of both the linkages between monosaccharide units and the branching of the chain determines physical properties of microbial heteropolysaccharides. Most heteropolysaccharides also own substituents of pyruvates, succinates, and formats [37–39].

9.2.1.2 Polysaccharides of Gram-Negative Bacteria

P. aeruginosa, one of the best-studied models for biofilm formation, yields at least three distinct exopolysaccharides (i.e., alginate, Pel, and Psl) that add to biofilm development and architecture [40]. The other minor components of polysaccharide nature, such as cyclic β-glucans and levan, have also been reported in Gram-negative bacteria:

- *Alginate.* This is a linear polyanionic acidic heteropolysaccharide of a high molecular mass 104–106 g/mL with irregular structure. The residues of L-glucuronic and D-mannuronic acids are connected with β-1, 4 bonds, and generally acetylated in positions 2 or 3 of the mannuronic acid residue. The 1, 4 bond makes the polymer more rigid than dextrans, where the 1, 2 bond is present [22, 41]. Alginate is usually produced in the biofilms of clinical mucoid *P. aeruginosa* isolates. However, it has also been secreted by the nonpathogenic variants of pseudomonads and *Azotobacter vinelandii* [41]. Exhaustive research carried out using a variety of physicochemical techniques discovered that alginate increases EPS hydratation, decreases its flexibility, and screens the surface structures and adhesins, thus regulating the interaction between the cells and the substrate surface [42–44]. Alginates can form a gel in the presence of chelating divalent cations. The mechanical properties of alginate gels can vary depending on the amounts of guluronic acid present in the polymer [28, 45].
- *Levan.* Also known as β-polyfructan, levan is another capsular polysaccharide of pseudomonads. Levan is mainly associated with storage, rather than structural benefit to biofilm polysaccharide [28, 46]. Its possible role in resistance of *P. brassicacearum* biofilms has been hypothesized to be because of osmotic and oxidative stress caused by Cd^{2+} ions [47, 48].
- *Psl polysaccharide.* A mannose rich polymer termed Psl (polysaccharide synthesis locus) has been identified as making significant contribution to biofilm architecture of *P. aeruginosa*. Psl consists of a repeated pentasaccharide, including three D-mannose

residues and one residue each of both D-glucose and L-rhamnose bounded by 1, 3 bonds. At least two forms of this polysaccharide exist: a high molecular mass component associated with biofilms and a soluble component of relatively low molecular mass, which is present in the culture liquid [28, 30]. Psl is involved in the adherence of biofilms to abiotic and biotic surfaces and in the maintenance of biofilm architecture. During attachment, Psl is anchored to the cell surface in a helical pattern, possibly promoting cell–cell interactions [49]. During maturation of biofilms, Psl gathers in the periphery of microcolonies and forms Psl-free cavities in the micro colony centre [44, 50]. These cavities harbor swimming cells together with dead cells and eDNA which are responsible for the propagation of bacteria during dispersal of the biofilm. Chemical removal of Psl from the surface of bacterial cells causes breakdown of the matrix [49, 50].

- *Pel polysaccharide.* This is a glucose-rich polysaccharide, term derived from pellicle, and forms the structural framework of the matrix. Colvin et al. [51] using *P. aeruginosa* PA14 mutant incapable of Psl synthesis, have shown that Pel offers for the interactions of microbial cells especially at the early stages of biofilm formation. Ghafoor et al. [52] observed that phenotypically, enhanced Pel synthesis resulted in formation of wrinkled colonies. Similarly, Zhurina et al. [44] found Pel to be essential for the formation of biofilms at air–liquid interfaces and biofilms that are attached to a surface. Pel was shown to provide for microbial attachment to glass surface as well as reported by Cooley et al. [53]. Although in their *in vitro* experiments the effect was less pronounced than in the case of Psl, the synergism of Pel and Psl was witnessed in extracellular matrices of *P. aeruginosa* when both of them were present. The *P. aeruginosa* PA14 mutant unable to synthesize Pel exhibited decreased binding of the core oligosaccharide of the outer-membrane lipopolysaccharide to the cell [44, 53].
- *Cellulose.* Cellulose comprises a β-1-4 linked linear glucose and is the major polysaccharide component of enteric bacteria. The role of cellulose and colanic acid for biofilm formation was studied in great detail in these bacteria. In *E. coli*, colanic acid, a sugar polymer composed of galactose, fructose, and glucose, is commonly found in the biofilm matrix [54]. Some findings have recently demonstrated the role of cellulose as an EPS component in pseudomonads. The role of cellulose in biofilm formation was especially well studied for *P. fluorescens* and *P. putida*. Spiers and Rainey [55] observed cellulose-dependent biofilm formation by *P. fluorescens*, especially at the air–liquid interface. Nielsen et al. [56] reported that *P. putida* has no *psl* locus, and bacterial cellulose and polysaccharide A play an important role in maintaining biofilm stability. *Salmonella* and *E. coli* produce cellulose as a crucial component of the extracellular matrix [57]. The formation of cellulose fibers is provided by hydrogen bonds between the chains of glucose. These formed sheets are very stable and their number varies depending on the nature of the environment. Cellulose can form a gel at suitable temperatures. The gel organization of cellulose may explain the mechanical properties of biofilms formed by bacterial species producing this polymer [9, 22].

Cellulose also plays a key role in adherence of microbes to plant tissues, leading to biofilm formation and the support of matrix architecture. Some studies have reported its role in plant-associated bacterium *Agrobacterium tumefaciens* [58], and *Rhizobium* species [59]. As reported by Laus et al. [60], cellulose is an exopolymer with agglutinating activity in *R. leguminosarum*. After coming in contact with the host plant, this rhizobacterium aggregates on the root surface using cellulose microfibrils. In *A. tumefaciens*, a plant pathogen that persists as surface-associated populations on plants or soil particles, cellulose overproduction resulted in increased biofilm formation on roots [61].

- *Rhamnolipids.* These compounds form a class of amphiphilic surface-active glycolipids containing rhamnose and fatty acids. Rhamnolipids are mainly characterized for bacteria of the genera *Pseudomonas* and *Burkholderia.* Over 60 compounds belonging to rhamnolipids and playing role in biofilm formation have been identified [30, 44].

9.2.1.3 Polysaccharides and Related Compounds in Gram-Positive Bacteria

Among Gram-positive bacteria, biofilm formation has been extensively studied in *Bacillus* spp. (as a model organism), followed by *Staphylococcus* spp. and *Streptococcus* spp. The latter genus is of special importance due to its role in formation of multispecies biofilms (dental plaques), and in nosocomial and medical implants-associated infections [26, 62–64].

Levan is the best-studied exopolysaccharide produced by *Bacillus subtilis:* Type I consists of D-fructose residues bound with β-2, 6 bonds and type II is a fructose polymer with glucose bound to its terminal residues by α-glucoside bonds [44]. Some *B. subtilis* strains produce exopolysaccharides of different structure as reported by Marvasi et al. [65]. They observed glucose, galactose, fructose, glucuronic acid, and *O*-acetyl groups in approximate molar ratios of 2:2:1:1:1.5.

Polysaccharides from *Arabidopsis* roots and maize roots were recently found to serve as both signals for biofilm formation and a source of sugars for the synthesis of matrix EPS in the beneficial Gram-positive bacterium *B. subtilis* and *B. amyloliquefaciens,* respectively [66, 67]. *S. epidermidis* and *S. aureus* produce polysaccharide intercellular adhesion (PIA) or the related poly-N-acetyl glucosamine (PNAG) polymer, both of which depend on the ica locus for their synthesis. They serve as adhesins, and are required for biofilm formation [68]. Now it appears that PIA-like polymers are made by several Gram-negative bacterial species. For example, a PIA-like polymer was recently shown to play a role in biofilm formation by *E. coli* strain MG1655 [69]. PNAG is a positively charged linear homoglycan composed of β-1,6-Nacetylglucosamine residues with approximately 20 percent deacetylated residues [70]. PNAG forms a protective matrix around bacterial cells that is also involved in cell-to-cell interactions [71]. PNAG can also interact with eDNA, reinforcing the biofilm matrix structure [72].

In the case of dental plaques (a kind of biofilm growth) formed by *S. mutans*, polysaccharides forming the insoluble sheath of dental deposits are the main components of the matrix. This sheath prevents removal of microorganisms, rather than acid production, and is considered as the major factor accountable for the role of *S. mutans* in damage to tooth enamel [62]. Some other polymers are also found to be present in the matrix of the biofilm of *S. epidermidis* such as teichoic acid, as reported by Sadovskaya et al. [73]. There are two types of teichoic acid in *S. epidermidis*: teichoic acid associated with the bacterial membrane (CWTA) and extracellular teichoic acid (ECTA). The ECTA is responsible for the increased viscosity of the colony. The ECTA is a (1-3)-linked poly (glycerol phosphate), substituted at the 2-position with α-glucose, α-N-acetylglucose, D-alanine, and α-glucose-6-alanine [22].

9.2.2 Proteins

The protein composition of EPS has been studied to a considerably lesser degree than its polysaccharide composition. However, the protein content in some biofilms, such as activated sludge flakes and sewers, may exceed their polysaccharide content on the mass basis as reported by McSwain et al. [74]. The two organisms that are most widely studied in this regard are *P. aeruginosa* and *B. Subtilis* [17, 75]. These extracellularly secreted proteins ranges in molecular masses between 10kDa and 200kDa and comprise from 40 to 60 percent of hydrophobic amino acids. In general, proteins constituents lack sulfuric amino acids, especially in the cells of *Geobacillus stearothermophilus*. Lory [76] found that extracellular proteins synthesized by *Sulfolobus acidocalcidarius* are composed mostly of amino acids with hydroxyl groups. However, the *B. subtilis* extracellular protein layer is a composition of L- and D-glutaminosyl residues [8].

A cell surface-associated and extracellular carbohydrate-binding proteins called lectins, are involved in the formation and stabilization of the polysaccharide matrix network and constitute a link between the bacterial surface and extracellular EPS. Examples include glucan-binding proteins in biofilms of the dental pathogen *S. mutans* [77], lectin like proteins in the matrices of activated sludge flocs [78], outer-membrane lectins of *Azospirillum* brasiliense [79] and the galactose-specific lectin lecA. The secreted protein CdrA was shown to bind directly to Psl in *P. aeruginosa* biofilms [75], leading to the suggestion that extracellular CdrA cross-links Psl molecules and thereby strengthens the matrix, whereas cell-associated CdrA anchors the cells to Psl in the matrix [17].

Another group of extracellular proteins are biofilm-associated surface protein (bap) from *S. aureus* and the bap-like proteins. These are high-molecular-mass proteins on the bacterial cell surface that encourage biofilm formation in several bacterial species [80]. They contain a core domain of tandem repeats that is required for the formation of a biofilm and plays a part in bacterial infectious processes. Other ubiquitous proteinaceous components of the matrix are amyloids. These compounds consist of an orderly repeats of protein molecules arranged as fibers of indefinite length in a cross-β structure, in which the β-strands are perpendicular to the fiber axis. Amyloids are involved in adhesion to nonliving surfaces and host cells, with subsequent invasion of the host cells, and they also function as cytotoxins for both plant cells and bacteria [81].

Recent studies exposed that many proteins identified in pseudomonad biofilms were associated with membranous vesicles originating from the outer membrane of bacterial cells (OMV) [82]. OMV are spherical bubbles with an average diameter of 20 to 200 nm, surrounded by a bilayer phospholipid membrane. They usually contain proteins, lipopolysaccharides, and DNA [83]. They are believed to play important functions in bacterial populations by facilitating intercellular communication, nutrient utilization, and horizontal gene transfer, as well as participation in pathogenesis [84]. While their precise functions in biofilms have not been established explicitly, their broad existence indicates their important biological role [85].

Toyofuku et al. [86] identified dozens of proteins in the models of biofilm colonies grown on membrane filters placed on the surface of solid media. These proteins belong to secreted proteins (such as aminopeptidase), to proteins of cytoplasmic (e.g., ornithine-carbamoyl transferase and arginine deaminase), periplasmic (e.g.,TolB), and membrane origin (e.g., porin proteins). Some of them were found associated with OMV.

In addition to exopolysaccharides, the *B. subtilis* biofilm matrix consists of a secreted protein TasA and also an anionic polymer poly-D-glutamate. Although the ratio of these polymers varies in different strains, both are required for biofilm formation and structural integrity of the matrix. Especially TasA is responsible for the development of biofilm architecture in the form of fruiting body-like structures [64, 67, 75]. Furthermore, some proteinaceous appendages such as pili, fimbriae, and flagella can also act as structural elements by interacting with other EPS components of the biofilm matrix. Studies have also reported that amyloid proteins of coiled fimbria (curli) play an important role in the biofilm matrix [87, 88]. Type IV pili of *P. aeruginosa* bind DNA and so possibly act as cross-linking structures [82]. Mutants with impaired synthesis of these proteins are unable to form biofilms, in spite of their retained capacity for exopolysaccharide synthesis. Other proteins in EPS are biopolymer hydrolases, which are probably involved in metabolism and in biofilm degradation, and play no structural role in EPS [26].

9.2.3 eDNA

The matrix is also a reservoir of substantial amount of extracellular DNA (eDNA). It has been reported in biofilms of various Gram-negative microorganisms, such as *P. aeruginosa, S. intermedius, S. mutans, Enterococcus faecalis*, and many species of staphylococci. Autolysis of microbial cells is the main mechanism of its formation but active excretion of DNA cannot be ignored such as reported in *S. epidermidis* [44, 88]. The presence of eDNA in *S. aureus* biofilms has shown stabilizing effect at the early stages of biofilm formation [90]. eDNA facilitates adhesion at phase interfaces, and offers the opportunity for the horizontal transfer of genetic information. Moreover, intercellular communication is equally facilitated and exaggerated [91]. The quantity, localization, and origin of eDNA vary depending on the bacterial species. Some studies found that eDNA is arranged in certain patterns [92] and that its release is based on the lysis of certain types of bacteria [93], suggesting the occurrence of programmed cell death as reported by Wang and Bayels [94] in plant-associated microbial biofilms. In Gram-positive bacteria also, eDNA is involved in adhesion to hydrophobic surfaces and in bacterial autoaggregation. Some studies have revealed reduction in the initial adhesion and aggregation of bacteria to surfaces, if eDNA is removed [95]. Others have made known that eDNA is a chief matrix component in some species biofilm [96], including *P. aeruginosa* biofilm [97], where eDNA appeared to induce antibiotic resistance [98].

9.2.4 Surfactants and Lipids

Extracellular polysaccharides, proteins, and DNA are highly hydrated hydrophilic molecules, but EPS also exhibit hydrophobic properties. The hydrophobic character of the EPS was credited to substituents such as polysaccharide-linked methyl and acetyl groups [99]. Lipids are also found in the matrix [100]. Lipopolysaccharides are crucial for the adherence of *Thiobacillus ferrooxidans* to pyrite surfaces [101], and *Serratia marcescens* produces extracellular lipids with surface-active properties (known as *serrawettins*) [102]. Interestingly, rhamnolipids, which can act as surfactants, have also been found in the EPS matrix of *P. aeruginosa* [103]. They display surface activity and have been proposed to act in initial microcolony formation, facilitating surface-associated bacterial migration and the formation of mushroom-shaped structures, preventing colonization of channels, and playing a part in biofilm dispersion [104]. Biosurfactants possess antibacterial and antifungal properties and are important for bacterial attachment and detachment from oil droplets [105]. Biosurfactants generated by microorganisms at the air–water interface of surface waters play important role by influencing surface tension and, thus, the gas exchange between oceans and the atmosphere [106]. Other surface-active EPS include surfactin, viscosin, and emulsan, which can disperse hydrophobic substances and make them bioavailable. They may be beneficial for microbially enhanced oil recovery and for bioremediation of oil spills.

9.2.5 Water

Water is the largest component of the matrix and is responsible for a highly hydrated environment. This water dries more slowly compared to its surroundings and therefore buffers the biofilm cells against fluctuations in water potential. Many EPS are hygroscopic and seem to retain water entropically rather than through specific water-binding mechanisms. Bacteria actively respond to desiccation by producing EPS [107]. Desiccation seems to be one of the environmental conditions under which EPS delivers benefits to both EPS producers and other members of the biofilm community [108].

The stability of the biofilm structure is thus dependent on the physicochemical and biological properties of EPS and on their interactions with ions, low-molecular-weight solutes, and other macromolecules such as proteins and eDNA [6]. EPS are generally required not for preliminary adhesion but for later architectural development of the biofilm matrix [9, 17].

9.3 Properties of Matrices

However, in general microbial cell aggregates, polysaccharides, adhesin proteins, extracellular DNA, and amphiphilic compounds are involved in the processes of primary colonization of substrate surfaces. Transition from reversible to irreversible adhesion of microbial cells is accompanied by their aggregation due to the processes of mutual recognition [91, 109]. Overall, the matrix acts as an eventual recycling yard in which all constituents of lysed cells can be exploited by the residual microbial population. The nature and concentration of EPS determines the close microenvironment of biofilm cells, and, thus, their actual conditions of life [26, 110]. The most interesting fact is how microorganisms can amend the matrix once the EPS components have been excreted.

EPS are biodegradable and therefore are not accumulated over time—otherwise, biofilm organisms would be irreversibly imprisoned in the matrix if they could not use enzymes to break free and become planktonic [27]. Many extracellular enzymes have been discovered in biofilms that can potentially degrade other EPS components [49], such as water-soluble polymers (many polysaccharides, proteins, and nucleic acids) and water-insoluble compounds (cellulose, chitin, and lipids). Various enzymes target EPS made by the same bacterium, or made by other species during starvation. For instance, dextran, inulin, and levan formed by oral streptococci, or levan present in matrix voids of biofilms produced by *Pseudomonas syringae* [111, 112]. The exopolysaccharides are primarily degraded by hydrolases and lyases, and degradation is not as rapid as required for cleaning purposes [9, 27]. The presence of EPS-degrading extracellular enzymes makes the matrix a multipurpose external digestive system. This benefits the microbial communities by sequestering and breaking down the dissolved and particulate nutrients from the water phase and allowing them to be reutilized as nutrient and energy sources.

Extracellular enzymes can be efficiently retained in the biofilm matrix by their interaction with polysaccharides. For example, the association of extracellular lactonizing lipase (lipA; also known as lip) with alginate produced by *P. aeruginosa* is based on weak binding forces [113]; molecular modeling supports this hypothesis. Such interactions result in a matrix of exopolysaccharides that are biochemically activated by the attached enzymes. This arrangement retains the enzymatic activity close to the cell and keeps the diffusion distances of enzymatic products short, thereby optimizing their uptake by bacteria [82]. Moreover, the interactions between enzymes and structural exopolysaccharides enhance the thermo-stability of the enzymes and their resistance to proteolysis [114].

The term 'slime' has been used for biofilms to indicate that they are not rigid structures. Their mechanical stability is an utmost important feature that is mainly provided by the exopolysaccharides in the matrix. In general, biofilms display viscoelastic properties and undergo both reversible elastic responses and irreversible deformation, depending on the forces acting on the EPS matrix [17]. Compression experiments with *P. aeruginosa* biofilms carried out by Korstgens et al. [115] revealed that in response to pressure, the biofilms go through a phase of elastic behavior until a break point is reached, after which the biofilm behaves like a viscous fluid.

It is possible that a biofilm can increase the strength of its structural matrix in response to mechanical stresses by increasing EPS production [116]. Furthermore, a very important stage in biofilm development is the dispersion of sessile cells from the biofilms, which allows new biofilms to be formed. This dispersion occurs in response to environmental changes and can be induced by nutrient starvation [117] or sudden nutrient availability and requires modification of the matrix by enzymes secreted from the bacteria [118]. An example of an enzyme that degrades exopolysaccharides allowing for detachment and dispersal of biofilm cells is β-hexosaminidase (encoded by dspB), which is produced by the periodontal pathogen *Actinobacillus actinocetemcomitans* [77]. As observed, the dspB mutant strain was able to form biofilms but lacked the ability to release cells [77].

As shown in Figure 9.1, the porous architecture allows for convectional flow through the depth of the biofilm, while within the EPS matrix, only diffusional transport is possible. Organisms at the bottom of the biofilm, thus, can access nutrients without competing with those at the interface to the bulk water phase. Strong gradient scan occur in biofilms (e.g., caused by actively respiring aerobic heterotrophic organisms), which consume oxygen faster than it can diffuse through the matrix. This generates anaerobic habitats just below highly active aerobic colonies in distances of less than 50 μm. Other gradients, such as pH-value, redox potential, and ionic strength, are known within biofilms [109]. Biofilm organisms have to live with limitations in mass transport and have adapted to that by formation of pores and channels, to which even parts of the population may contribute by "programmed cell death," as suggested by Webb et al. [119]. Considerable proportions of the population may be dormant due to starvation, leading to "dead zones" within the matrix. Cells actually die off, but it is also possible that many of them slow down their metabolism to a point of dormancy, becoming "viable but noncultivable" [120]. The dormant organisms may be considered as a standby population because they are considerably resistant to biocides and antibiotics and may revive after antimicrobial treatment, when the more viable and susceptible parts of the community have been eliminated. Exceptionally resistant subpopulations are known as *persisters*, which can be left behind after detachment and influence further adhesion kinetics [121].

Biofilms are full of hot spots of antagonism, including chemical fighting with antibiotics and biocins, many types of competition like parasitism and predation [27]. Ecologically, competition, and cooperation in the confined space of the EPS matrix lead to a constant adaptation of population fitness [17]. EPSs immobilize biofilm cells and keep them in close proximity, thus allowing for intense interactions, including cell–cell communication, and the formation of synergistic microconsortia [6,7]. These physical and chemical properties of EPS such as stabilizing, thickening, coagulating, gelling, suspending, film forming, and water-retention ability. They result in an environment with complex interactions and a functionally structured system and are very useful in many industries like detergents, textile, paper, paints, adhesive, beverages, and food [122]. Therefore, formation of the EPS matrix can be considered as an upcoming property of microorganisms allowing them to develop a wide range of cooperative aspects.

9.4 Functions of the Extracellular Polymer Matrix: The Role of Matrix in Biofilm Biology

Bacteria attain several advantages from living in biofilm mode of life, such as defense from a variety of environmental stresses, such as UV radiation, pH shifts, osmotic shock, desiccation, predation, and exposure to antimicrobial agents, and also get benefit of enriched procurement of nutrients released in the plant environment [1,9]. Moreover, the microbial resistance to biocides and the agents of the immune system of macroorganisms is also increased in biofilm cells compared with more susceptible free-floating cells. These benefits are attributed to the presence of a matrix that forms a hydrated barrier between cells and their external environment and thereby protects against desiccation, facilitating accumulation of organic and inorganic compounds, and forms a diffusive barrier, which prevents both the loss of protective compounds and penetration of toxic materials into the biofilm. EPS is accountable for formation and maintenance of the three-dimensional biofilm structure that stabilizes the cells, providing for their contact, protecting them from the hydrodynamic shift forces at the phase interface, and offering greater opportunity for exchange of genetic information and export of cell components [22, 110].

Therefore the important roles exhibited by EPS are (i) protective, (ii) surface attachment, (iii) biofilm formation, (iv) microbial aggregation, (v) plant–microbe interaction, and (vi) bioremediation. Almost all EPS components, especially polysaccharides and proteins, are involved in these activities [29, 30]. Although, the detailed and molecular interactions of the numerous biofilm matrix polymers have not been defined, and the impacts of these components to matrix integrity are also poorly understood at a molecular level, several functions of EPS have been determined. An overview of the functions attributed to various components of EPS is described in Table 9.1.

Table 9.1 Functions Attributed to EPS in Bacterial Biofilms [modified from 27]

Functions of EPS Components	Relevance for Biofilm Organism	EPS Components Involved
Adhesion	Initial steps in colonization of abiotic and biotic surfaces by planktonic cells, long-term attachment of whole biofilms to surfaces	Polysaccharides, proteins, amphiphilic molecules, DNA
Aggregation of bacterial cells	Bridging between cells, (temporary) immobilization of bacterial populations, basis for development of high cell densities, cell–cell recognition	Polysaccharides, proteins, DNA
Cohesion of biofilms	Structural elements forming a hydrated polymer network (biofilm matrix), mediation of mechanical stability of biofilms (frequently in conjunction with multivalent cations), determination of EPS structure (capsule, slime, sheath) and biofilm architecture, generation of matrix allowing cell–cell communication	Neutral and charged polysaccharides, proteins (e.g., amyloids, lectins), DNA
Retention of water	Maintenance of highly hydrated microenvironment around biofilm organisms, desiccation tolerance in water-deficient environments	Hydrophilic polysaccharides and some proteins
Protective barrier	Resistance to nonspecific and specific host defenses during infection, tolerance to various antimicrobial agents (e.g., disinfectants, antibiotics), protection of cyanobacterial nitrogenase from harmful effects of oxygen; protection against some grazers	Polysaccharides, proteins
Sorption of organic compounds	Accumulation of nutrients from the environment, sorption of xenobiotics (detoxification)	Charged or hydrophobic polysaccharides and proteins
Sorption of inorganic ions	Promotion of polysaccharide gel formation, ion exchange, mineral formation, accumulation of toxic metal ions (detoxification)	Charged polysaccharides and proteins, including inorganic substituents such as phosphate and sulphate
Enzymatic activity	Digestion of exogenous macromolecules for nutrient acquisition, degradation of structural EPS allowing release of cells from biofilms	Proteins
Nutrient source	Source of C, N, and P compounds for utilization by biofilm community	Potentially all EPS components
Exchange of genetic information	Horizontal gene transfer between biofilm cells	DNA
Electron donor or acceptor	Redox activity in biofilm matrix, electron transport mediation to surfaces	Proteins (e.g., pili, nanowires?), humic substances?
Export of cell components	Release of cellular material as a result of metabolic turnover	Membrane vesicles (nucleic acids, enzyme proteins, lipopolysaccharides, phospholipids)

(Continued)

Table 9.1 (Continued)

Functions of EPS Components	Relevance for Biofilm Organism	EPS Components Involved
Sink for excess energy	Sink for excess carbon under unbalanced C/N metabolic conditions	Polysaccharides
Binding of enzymes	Accumulation, retention, and stabilization of enzymes through their interaction with polysaccharides	Polysaccharides, enzyme proteins

Biofilms offer survival sites for both beneficial and opportunistic pathogenic bacteria, by providing protection as mentioned above and increasing the potential of the bacteria to survive and evolve in various environments including plant soil. In this regard, biofilms results in enhancement of (i) the suitability of individual bacteria, (ii) plant health, and (iii) overall efficiency as a result of the cumulative selective advantage of the individual bacteria. Our knowledge of the identification and functions of extracellular proteins, eDNA, and lipids in the biofilm matrix of plant-associated bacteria has been remaining limited [9]. Further studies along this line will greatly enhance our understanding of the process of biofilm formation.

9.4.1 Role of EPS in Biofilm Architecture

The architecture of biofilms is influenced by many factors, including hydrodynamic conditions, concentration of nutrients, bacterial motility, and intercellular communication, as well as exopolysaccharides and proteins, as demonstrated by the altered morphology of biofilms produced by mutants lacking components of EPS [1, 17]. EPS may account for 50 to 90 percent of the total organic carbon of biofilms and can be considered the primary matrix material of the biofilm. The formation and maintenance of structured multicellular microbial communities critically depend on the production and magnitude of EPS [123].

Biofilm architecture can also be strongly influenced by the interaction of anionic EPS, containing carboxylic groups, with multivalent cations. The concentration, cohesion, charge, sorption capacity, specificity, and nature of the individual components of EPS, as well as the three-dimensional architecture of the matrix, decide the nature of life in a given biofilm. The resulting biofilm morphology can be smooth, flat, rough, fluffy, or filamentous, and the biofilm can also fluctuate in its degree of porosity, having mushroom-like macrocolonies fenced by water-filled voids [124, 125]. These highly permeable water channels interspersed throughout the biofilm in the areas surrounding the microcolonies are compared to a primitive circulatory system. This allows bacteria to exchange water, nutrients, enzymes, and signals, dispose of potentially toxic metabolites, and display enhanced metabolic cooperativity [9, 124, 126].

Sutherland [6, 37] noted two important properties of EPS that may have a distinct effect on the biofilm. First, the composition and structure of the polysaccharides determine their primary conformation. For instance, many bacterial EPS possess backbone structures that contain 1, 3- or 1,4-β-linked hexose residues and tend to be more rigid, less deformable, and in certain cases poorly soluble or insoluble. Other EPS molecules may be readily soluble in water. Second, the EPS of biofilms is not generally uniform but may vary spatially and temporally. Leriche et al. [127] used the binding specificity of lectins to simple sugars to evaluate bacterial biofilm development by different organisms. Acetyl groups are common substituents of exopolysaccharides, and they increase the adhesive and cohesive properties of EPS and change biofilm architecture. The modification of alginate with acetyl groups strongly influences the aggregation of bacteria into microcolonies and determines the structurally heterogeneous architecture of mature biofilms [128]. Some of these polysaccharides are neutral or polyanionic, as is the case for the EPS of Gram-negative bacteria. The presence of uronic acids (such as D-glucuronic, D-galacturonic, and mannuronic acids) or ketal-linked pryruvates confers the anionic property [37]. This property is important because it allows matrix to associate with metal ions and divalent cations such as calcium and magnesium. This cross-link with the polymer strands provides greater binding force in a established biofilm. For instance, Ca^{2+} can form a bridge between polyanionic alginate molecules, stimulating the development of thick and compact biofilms with increased mechanical stability [129]. EPS may also associate with other macromolecules such as proteins, DNA, lipids, and even humic substances [125].

In the case of some Gram-positive bacteria, such as the staphylococci, the chemical composition of EPS may be quite different and may be primarily cationic [125]. Hussain et al. [130] found that the slime of coagulase-negative bacteria consists of a teichoic acid mixed with small quantities of proteins. EPS is also highly hydrated because it can incorporate large amounts of water into its structure by hydrogen bonding. EPS may be either hydrophobic, hydrophilic, or both, and also vary in its solubility. EPS production is known to be affected by nutrient status of the growth medium; excess available carbon and limitation of nitrogen, potassium, or phosphate promote EPS synthesis. Slow bacterial growth will also enhance EPS production [6]. These researchers' results showed that different organisms produce differing amounts of EPS, and that the amount of EPS increases with age of the biofilm.

9.4.2 Role of EPS in Mechanisms of Antimicrobial Resistance/Tolerance to Other Toxic Substances

Bacteria experience a certain degree of shelter and homeostasis when residing within a biofilm, and one of the key components of this microniche is the surrounding extra polymeric substance matrix.The bacteria enclosed within the biofilm are extremely resistant to antibiotic treatments. Biofilm bacteria can be up to 1,000-fold more resistant to antibiotic treatment than the same organism grown planktonically [1]. In the perspectives of plant–microbe interactions, EPS are the active constituents of soil organic matter [131] and their production by bacteria in saline soil is very helpful against osmotic stress. Biofilms are established on various surfaces like roots and soil particles, respectively resulting in cementing of soil particles. This can improve crop productivity and physiochemical properties of soil. Formations of these aggregates have water-retaining capacity and sustain physiochemical properties of soil [132]. Some EPS-producing bacteria like *Pseudomona* ssp. have the ability to survive even under drought stress due to the production of their EPS [133]. EPS of bacteria are hydrated compounds with 97 percent of water in polymer matrix, which impart protection against desiccation under drought stress by enhancing the water retention and by regulation of organic carbon source's diffusion [134]. Due to enzymatic activities of EPS, they help in heavy metal transformation and degradation of organic recalcitrant [135]. Plants treated with EPS-producing bacteria *Azospirillum* showed resistance to water stress [136]. Alami et al. [137] found that inoculation of sunflowers with EPS-producing bacterial strain YAAF34 under drought stress exhibited increased growth in root tissue. Naseem and Bano [138] have demonstrated the role of plant growth–promoting rhizobacteria and their exopolysaccharide in drought tolerance of maize.

EPS has also been held responsible for bioremediation of many metals. Biosorption of metals by the organisms at the surface by the EPS secreted to form the biofilms enables organisms to tolerate metals [139]. Haloarchaea synthesize EPS as a protective mechanism for survival under adverse conditions such as nutrient starvation, temperature fluctuation, and presence of toxic compounds. Similarly, the hyperthermophilic archaeon, *Archaeoglobus fulgidus*, was found to form a biofilm in response to toxic concentrations of metals, where the toxic metal was proposed to be trapped within the EPS matrix [35]. In a study by Kawakami et al. [140], it has been found that *Halobacterium salinarum* CCM 2090 has a Ca(II)-dependent aggregation system, where the Ca(II) binds to certain aggregation factors present on the cell surface and induces ionic cross bridging between the EPS, resulting in aggregation of the haloarchaeal cells.

Such resistance/tolerance mechanisms can be explained by hypotheses, not necessarily limited to the following five:

1) *The EPS matrix physically prevents access of certain antimicrobial agents (antibiotics/antibodies) into the biofilm.* This most likely occurs by binding directly to these agents and thereby restricting mass transport of these compounds from the surrounding milieu into the biofilm [110, 141, 142]. Moreover, EPS is negatively charged and functions as an ion-exchange resin, which is capable of binding a large number of the antibiotic molecules that are attempting to reach the embedded biofilm cells. The effect appears to be most pronounced with antibiotics that are hydrophilic and positively charged, such as the aminoglycosides. However, a number of studies have demonstrated that reductions in the diffusion coefficients of antibiotics within biofilms are insufficient to account solely for the observed changes in susceptibility [143]. This mechanism might be more relevant for reactive (bleach or superoxides), charged (metals), or large (immunoglobulin) antimicrobial agents that are neutralized or bound by the EPS and are effectively diluted to sublethal concentrations before they can reach all of the individual bacterial cells within the biofilm. Drug access is also assisted by the presence of water channels in the biofilm structure. Nevertheless, matrix components could retard access to such an extent that cells lying deep within a microcolony escape exposure. This would occur via drug adsorption or neutralization, and would depend on the thickness of the biofilm and on the chemical nature of both the antimicrobial agent and the matrix material. It is known, for example, that fluoroquinolones penetrate *P. aeruginosa* biofilms readily, whereas penetration by positively charged aminoglycosides is retarded [144]. Similarly, fluconazole permeates single-species *Candida* biofilms more rapidly than flucytosine [145]. Rates of drug diffusion through biofilms of *Candida glabrata* or *Candida krusei* are faster than those through biofilms of *Candida parapsilosis* or *Candida tropicalis*, while drug diffusion through mixed-species biofilms of *C. albicans* and *S. epidermidis* is very slow.

2) *The physiological state of biofilm organisms could provide protective mechanisms.* Studies with environmental and *in vitro* biofilms have revealed that the oxygen concentration is high at the surface but low in the center of the biofilm where anaerobic conditions are present [124]. Similarly, growth, protein synthesis, and metabolic activity are radiantly distributed in biofilms—that is, a high level of activity at the surface and a low level and dormant or no growth in the center [146]. These cells embedded deep in the biofilm matrix are generally not actively involved in cell division, and are also smaller in size and therefore less pervious to antibiotics. As a principle, only actively growing cells are susceptible to antimicrobials because the mechanism of action of most antibiotics involves disruption of a microbial process. Therefore, these subpopulations exhibiting resistant phenotypes, referred as *persisters,* explain another mechanism of protection [121]. It remains unclear whether these organisms do indeed represent a distinct phenotype or are simply the most resistant cells within a population distribution. The formation of starved, stationary phase dormant zones in biofilms seems to be a significant factor in the resistance of biofilm populations to antimicrobials [147, 148]. This is true particularly against antibiotics such as β–lactams, which are effective against rapidly dividing Gram-positive bacteria by interruption of cell-wall synthesis. Moreover, transition from exponential to slow or no growth is generally accompanied by expression of antibiotic-resistant factors [149]. Slow growth triggers the RelA-dependent synthesis of ppGpp, which inhibits anabolic processes in bacterial cells. ppGpp curbed the activity of a major *E. coli* autolysin, SLT, which would make the cells more resistant to autolysis. This could also explain the mechanism of tolerance to antibiotics in slowly growing cells. ppGpp also obstructs peptidoglycan synthesis, which could explain the decreased levels of activity of cell-wall synthesis inhibitors under starvation conditions [150, 151].
3) *The mutation frequency in the microbial cells growing under biofilm stage is significantly increased compared with planktonically growing isogenic bacteria* [152]. In addition, biofilms also provide an ideal niche for the exchange of extrachromosomal DNA responsible for antibiotic resistance, virulence factors, and environmental survival capabilities at accelerated rates, making it a prefect milieu for emergence of drug-resistant pathogens [153]. These physiological alterations may also explain why biofilm-growing bacteria easily become multidrug resistant, mainly against β-lactam antibiotics, aminoglycosides, and fluoroquinolones. Plasmid-carrying strains have also been shown to transfer plasmids to recipient organisms to promote biofilm formation. Without plasmids, the same organisms produce only microcolonies without any further development. Higher levels of conjugation and therefore increased rate of horizontal gene transfer is observed in the biofilms compared to planktonic cells [154]. The probable reason for enhanced conjugation is that the biofilm environment provides minimal shear and closer cell-to-cell contact. Since plasmids carry markers resistance to multiple antimicrobial agents, biofilm association provides a mechanism for selecting and promoting the spread of bacterial resistance to antimicrobial agents.
4) *The antimicrobial agent is deactivated in the outer layers of the biofilm faster than it diffuses.* This is true for reactive oxidants such as hypochlorite and H_2O_2 [149]. These antimicrobial oxidants are products of the oxidative burst of phagocytic cells, and poor penetration of these may partially account for the inability of phagocytic cells to destroy biofilm microorganisms. Furthermore, cells in biofilms produce

enzymes that degrade antibiotics and also display low affinity antibiotic targets. Antibiotic-degrading enzymes such as β-lactamase may also be arrested in the EPS matrix, so that the incoming antibiotic molecules can be inactivated effectively. For instance, extracellular β-lactamase enzymatic activity against *P. aeruginosa* occurs within the matrix [155]. It is reported by Bagge et al. [156] that biofilm cells of the *P. aeruginosa* produce 32-fold more β-lactamase than cells of the same strain grown planktonically.

5) The overexpression of efflux pumps in biofilm cells compared to planktonic cells, that have a broad range of substrates. It has been found that up to 40 percent of the cell-wall protein composition of bacteria in biofilms is altered from that of its planktonic brethren [157]. The membranes of microbial cells from biofilms are better fortified to pump out antibiotics before they can cause damage, or even antibiotics targets may disappear.

Although the relative contribution of each of these mechanisms varies with the type of biofilm and the nature of the environmental stress, overall, the net result is to provide defense to biofilm forming microbial cells. In general, the EPS network confers mechanical stability, allows for temporary immobilization of cells, and plays a crucial role in most matrix functions, including water retention, protection from environmental stresses, adsorption of compounds, and nutrient availability [17].

9.5 Conclusion

There is no biofilm without an EPS or matrix, which is essential for biofilm formation and makes possible a lifestyle that is entirely different from the planktonic state. Microbes produce diverse nature of extracellular matrix, contributing significantly to the organization of the community. Microbial exopolysaccharides play important roles in biofilm formation, architecture, and cellular persistence, and have several applications in the food, agriculture, medical, and other industries. The recalcitrance of biofilms to antimicrobial agents is often attributed to the failure of these agents to penetrate the biofilm matrix. EPS provides favorable conditions for exchange of genetic information and horizontal gene transfer. The rate of these processes in biofilms is quite higher than in planktonic microbial cells. Cell-to-cell communication mechanisms, such as quorum sensing, and the intracellular level of the second messenger cyclic di-GMP are involved in regulating biofilm formation and the production of matrix components such as certain polysaccharides and proteins, DNA, and rhamnolipids. An enhanced understanding of the regulation of EPS production in mono or mixed-species biofilms, as well as a three-dimensional and temporal examination of the phases in EPS production, will disclose important aspects of the oldest, most successful and pervasive form of life on Earth.

Acknowledgments

We acknowledge research support received from department of Scientific Research from Imam Abdulrahman Bin Faisal University for this work.

References

1 M.E. Davey and G.A. O'Toole, Microbial biofilms: from ecology to molecular genetics, *Microbiol. Mol. Biol. Rev.*, 64, 847–867 (2000).

2 K.J. Edwards, P.L. Bond, T.M. Gihring, J.F. Banfield, An archaeal iron-oxidizing extreme acidophile important in acid mine drainage, *Science*, 287, 1796–1799 (2000).

3 P.E. Kolenbrander and J. London, Adhere today, here tomorrow: oral bacterial adherence, *J. Bacteriol.*, 175, 3247–3252 (1993).

4 T. Danhorn and C. Fuqua, Biofilm formation by Plant-associated bacteria, *Annu. Rev. Microbiol.*, 61, 401–422 (2007).

5 J. Deng, E.P. Orner, J.F. Chau, E.M. Anderson, A.L. Kadilak, R.L. Rubinstein, G. M. Bouchillon, R.A. Goodwin, D.J. Gage, L.M. Shor, Synergistic effects of soil microstructure and bacterial EPS on drying rate in emulated soil micromodels, *Soil Biol. Biochem.*, 83, 116–124 (2015).

6 I.W. Sutherland, The biofilm matrix—an immobilized but dynamic microbial environment, *Trends Microbiol.*, 9, 222–227 (2001a).

7 S.S. Branda, A.S. Vik, L. Friedman, R. Kolter, Biofilms: The matrix revisited, *Trends Microbiol.*, 13, 20–26 (2005).

8 K. Czaczyk and K. Myszka, Biosynthesis of extracellular polymeric substances (EPS) and its role in microbial biofilm formation, *Polish J. of Environ.*, 16, 799–806 (2007).

9 P.C. Bogino, M.M. Oliva, F.G. Sorroche, W. Giordano, The role of bacterial biofilms and surface components in plant-bacterial associations, *Int. J. Mol. Sci.*, 14, 15838–15859 (2013).

10 T. Rudrappa, M.L. Biedrzycki, H.P. Bais, Causes and consequences of plant-associated biofilms, *FEMS Microbiol. Ecol.*, 64, 153–166 (2008).

11 W. M. Dunne, Bacterial adhesion: seen any good biofilms lately? *Clin. Microbiol. Rev.*, 15, 155–166 (2002).

12 F.C. Petersen, L. Tao, A.A. Scheie, DNA binding uptake system: a link between cell-to-cell communication and biofilm formation, *J. Bacteriol.*, 187, 4392–4400 (2005).

13 R.M. Donlan and J.W. Costerton, Biofilms: survival mechanisms of clinically relevant microorganisms, *Clin. Microbiol. Rev.*, 15, 167–193 (2002).

14 T. Shaw, M. Winston, C.J. Rupp, I. Klapper, P. Stoodley, Commonality of elastic relaxion times in biofilms, *Phys. Rev. Lett.*, 93, 1–4 (2004).

15 B.W. Peterson, Y. He, Y. Ren, A. Zerdoum, M.R. Libera, P.K. Sharma, A.J. Winkelhoff, D. Neut, P. Stoodley, H.C. van der Mei, H.J. Busscher, Viscoelasticity of biofilms and their recalcitrance to mechanical and chemical challenges, *FEMS Microbiol. Rev.*, 39, 234–245 (2015).

16 T.T. Nielsen and S. Molin, Spatial organization of microbial biofilm communities, *Microb. Ecol.*, 40, 75–84 (2000).

17 H.C. Flemming and J. Wingender, The biofilm matrix, *Nat. Rev. Microbiol.*, 8, 623–633 (2010).

18 C.B. Whitchurch, T.T. Nielsen, P.C. Ragas, J.S. Mattick, Extracellular DNA required for bacterial biofilm formation, *Science*, 295, 1487 (2002).

19 M. Harmsen, M. Lappann, S. Knochel, S. Molin, Role of Extracellular DNA during biofilm formation by *Listeria monocytogenes*, *Appl. Environ. Microbiol.*, 76, 2271–2279 (2010).

20 I. Gilan and A. Sivan, Extracellular DNA plays an important structural role in the biofilm of the plastic degrading actinomycete Rhodococcus ruber, *Advan. Microbiol.*, 3, 543–551 (2013).

21 M.R. Kiedrowski, J.S. Kavanaugh, C.L. Malone, J.M. Mootz, J.M. Voyich, M.S. Smeltzer, K.W. Bayles, A.R. Horswill, Nuclease modulates biofilm formation in community associated methicillin-resistant *Staphylococcus aureus*, *PLoS One*, 6, e26714 (2011).

22 P. Lembre, C. Lorentz, P. Di Martino, Exopolysaccharides of the biofilm matrix: A complex biophysical world, *Intech Open Scie.*, 371–392 (2012).

23 R.C. Hunter and T.J. Beveridge, High-resolution visualization of *Pseudomonas aeruginosa* PAO1 biofilms by freeze-substitution transmission electron microscopy, *J. Bacteriol.*, 187, 7619–7630 (2005).

24 N. Billings, M.R. Millan, M. Caldara, R. Rusconi, Y. Tarasova, R. Stocker, K. Ribbeck, The extracellular matrix component Psl provides fast-acting antibiotic defense in *Pseudomonasaeruginosa* biofilms, *PLoS Pathog.*, 9, e1003526 (2013).

25 L. Hall-Stoodley, J.W. Costerton, P. Stoodley, Bacterial biofilms: from the natural environment to infectious diseases, *Nature*, 2, 95–108 (2004).

26 D. Lopez, H. Vlamakis, R. Kolter, Biofilms, *Cold Spring Harb. Perspect. Biol.*, 2, a000398(2010).

27 H.C. Flemming, The perfect slime, *Colloid. Surf. B: Biointer.*, 86, 251–259 (2011).

28 E.E. Mann and D.J. Wozniak, Pseudomonas biofilm matrix composition and niche biology, *FEMS Microbiol. Rev.*, 36, 893–916 (2012).

29 M.R. Parsek, Controlling the connections of cells to the biofilm matrix, *J. Bacteriol.*, 98, 12–14 (2015).

30 L. Hobley, C. Harkins, C.E. MacPhee, N.R. Stanley-Wall, Giving structure to the biofilm matrix: an overview of individual strategies and emerging common themes, *FEMS Microbiol. Rev.*, 39, 649–669 (2015).

31 J.K. Teschler, D.Z. Sanchez, A.S. Utada, C.J.A. Warner, G.C.L. Wong, R.G. Linington, F.H. Yildiz, Living in the matrix: assembly and control of *Vibrio cholera* biofilms, *Nat. Rev. Microbiol.*, 13, 255–268 (2015).

32 U.B. Sleytr, Basic and applied S-layer research: An overview, *FEMS Microbiol. Rev.*, 20, 5–12 (1997).

33 U.B. Sleytr, B. Schuster, E.M. Egelseer, D. Pum, S-layers: Principles and applications, *FEMS Microbiol. Rev.*, 38, 823–864 (2014).

34 F. Vaningelgem, M. Zamfir, F. Mozzi, T. Adriany, M. Vancanneyt, J. Swings, L. De Vuyst, Biodiversity of exopolysaccharides produced by *Streptococcus thermophilus* strains is reflected in their production and their molecular and functional characteristics, *Appl. Environ. Microbiol.*, 70, 900–912 (2004).

35 A. Poli, P. Di Donato, G.R. Abbamondi, B. Nicolaus, Synthesis, production, and biotechnological applications of exopolysaccharides and polyhydroxyalkanoates by archaea, *Archaea*, 2011: 693253 (2011).

36 C. Delbarre-Ladrat, C. Sinquin, L. Lebellenger, A. Zykwinska, S. Colliec-Jouault, Exopolysaccharides produced by marine bacteria and their applications as glycosaminoglycan-like molecules, *Front. Chem.*, 2, 85 (2014).

37 I.W. Sutherland, Microbial polysaccharides from Gram-negative bacteria, *Int. Dairy J.*, 11, 663–674 (2001b).

38 P. Monsan, S. Bozonnet, C. Albenne, G. Joucla, R.M. Willemot, M.R. Simeon, Homopolysaccharides from lactic acid bacteria, *Int. Dairy J.*, 11, 675–685 (2001).

39 U.U. Nwodo, E. Green, A.I. Okoh, Bacterial exopolysaccharides: functionality and prospects, *Int. J. Mol. Sci.*, 13, 14002–14015 (2012).

40 C. Ryder, M. Byrd, D.J. Wozniak, Role of exopolysaccharides in *Pseudomonas aeruginosa* biofilm development, *Curr. Opin. Microbiol.*, 10, 644–648 (2007).

41 I.D. Hay, Z. Rehman, M.F. Moradali, Y. Wang, B.H.A. Rehm, Microbial alginate production, modification and its applications, *Microb. Biotechnol.*, 6, 637–650 (2013).

42 G.G. d'Ayala, M. Malinconico, P. Laurienzo, Marine derived polysaccharides for biomedical applications: chemical modification approaches, *Molecules*, 13, 2069–2106 (2008).

43 K.Y. Leeand D.J. Mooney, Alginate: properties and biomedical applications, *Prog. Polym. Sci.*, 37,106–126 (2012).

44 M.V. Zhurina, A.V. Gannesen, E.L. Zdorovenko, V.K. Plakunov, Composition and functions of the extracellular polymer matrix of bacterial biofilms, *Microbiology*, 83, 713–722 (2014).

45 E. Hermanssona, E. Schuster, L. Lindgren, A. Altskar, A. Strom, Impact of solvent quality on the network strength and structure of alginate gels, *Carbohyd. Polym.*, 144, 289–296 (2016).

46 H. Laue, A. Schenk, H. Li, L. Lambertsen, T.R. Neu, S. Molin, M.S. Ullrich, Contribution of alginate and levan production to biofilm formation by *Pseudomonas syringae, Microbiology,* 152, 2909–2918 (2006).

47 D. Pages, L. Sanchez, S. Conrod, X. Gidrol, A. Fekete, P.S. Kopplin, T. Heulin, W. Achouak, Exploration of intraclonal adaptation mechanisms of *Pseudomonas brassicacearum* facing cadmium toxicity, *Environ. Microbiol.*, 9, 2820–2835 (2007).

48 O. Orgad, Y. Oren, S.L. Walker, M. Herzberg, The role of alginate in *Pseudomonas aeruginosa* EPS adherence, viscoelastic properties and cell attachment, *Biofouling*, 27, 787–798 (2011).

49 L. Ma, M. Conover, H. Lu, M.R. Parsek, K. Bayles, D.J. Wozniak, Assembly and development of the *Pseudomonas aeruginosa* biofilm matrix, *PloS Pathog.*, 5, e1000354 (2009).

50 S. Periasamy, H.A.S. Nair, K.W.K. Lee, J. Ong, J.Q.J. Goh, S. Kjelleberg, S.A. Rice, *Pseudomonas aeruginosa* PAO1 exopolysaccharides are important for mixed species biofilm community development and stress tolerance, *Front. Microbiol.*, 6, 851 (2015).

51 K.M. Colvin, V.D. Gordon, K. Murakami, B.R. Borlee, D.J. Wozniak, G.C.L. Wong, M.R. Parsek, The Pel polysaccharide can serve a structural and protective role in the biofilm matrix of *Pseudomonas aeruginosa*, *PLoSPathog.*, 7, e1001264 (2011).

52 A. Ghafoor, I.D. Hay, B.H.A. Rehm, Role of exopolysaccharides in *Pseudomonas aeruginosa* biofilm formation and architecture, *Appl. Environ. Microbiol.*, 77, 5238–5246 (2011).

53 B.J. Cooley, T.W. Thatcher, S.M. Hashmi, G. L'Her, H.H. Le, D.A. Hurwitz, D. Provenzano, A. Touhami, V.D. Gordon, The extracellular polysaccharide Pel makes the attachment of *P. aeruginosa* to surfaces symmetric and short ranged, *Soft Matter*, 9, 3871–3876 (2013).

54 D.O. Serra, A.M. Richter, G. Klauck, F. Mika, R. Hengge, Microanatomy at cellular resolution and spatial order of physiological differentiation in a bacterial biofilm, *mBio.*,4(2), e00103–13 (2013).

55 A.J. Spiers and P.B. Rainey, The *Pseudomonas fluorescens* SBW25 wrinkley spreader biofilm requires attachment factor, cellulose fibre and LPS interactions to maintain strength and integrity, *Microbiology.*, 151, 2829–2839 (2005).

56 L. Nielsen, X. Li, L.J. Halverson, Cell–cell and cell–surface interactions mediated by cellulose and a novel exopolysaccharide contribute to *Pseudomonas putida* biofilm formation and fitness under water limiting conditions, *Environ. Microbiol.*, 13, 1342–1356 (2011).

57 X. Zogaj, M. Nimtz, M. Rohde, W. Bokranz, U. Romling, The multicellular morphotypes of *Salmonella typhimurium* and *Escherichia coli* produce cellulose as the second component of the extracellular matrix, *Mol. Microbiol.*, 39, 1452–1463 (2001).

58 A.G. Matthysse, S. White, R. Lightfoot, Genes required for cellulose synthesis in *Agrobacterium tumefaciens*, *J. Bacteriol.*, 177, 1069–1075 (1995).

59 M. Robledo, L. Rivera, J.I. Jimenez-Zurdo, R. Rivas, F. Dazzo, E. Velazquez, E. Martinez-Molina, A.M. Hirsch, P.F. Mateos, Role of *Rhizobium* endoglucanase CelC2 in cellulose biosynthesis and biofilm formation on plant roots and abiotic surfaces, *Microb. Cell Fact.*, 11, 125 (2012).

60 M.C. Laus, A.A. van Brussel, J.W. Kijne, Role of cellulose fibrils and exopolysaccharides of *Rhizobium leguminosarum* in attachment to and infection of *Vicia sativa* root hairs, *Mol. Plant Microbe. Interact.*, 18, 533–538 (2005).

61 A.G. Matthysse, M. Marry, L. Krall, M. Kaye, B.E. Ramey, C. Fuqua, A.R. White, The effect of cellulose overproduction on binding and biofilm formation on roots by *Agrobacterium tumefaciens*, *Mol. Plant Microbe. Interact.*, 18, 1002–1010 (2005).

62 J.A. Lemos, R.G. Quivey, H. Koo, J. Abranches, *Streptococcus mutans*: a new Gram-positive paradigm? *Microbiology*,159, 436–445 (2013).

63 H. Vlamakis, Y. Chai, P. Beauregard, R. Losick, R. Kolter, Sticking together: building a biofilm the *Bacillus subtilis* way, *Nat. Rev. Microbiol.*, 11, 157–168 (2013).

64 P. Sanchez-Vizuete, D. Le Coq, A. Bridier, J.M. Herry, S. Aymerich, R. Briandet, Identification of ypqP as a new *Bacillus subtilis* biofilm determinant that mediates the protection of *Staphylococcus aureus* against antimicrobial agents in mixed-species communities, *Appl. Environ.Microbiol.*, 81, 109–118 (2015).

65 M. Marvasi, P.T. Visscher, L.C. Martinez, Exopolymeric substances (EPS) from *Bacillus subtilis*: polymers and genes encoding their synthesis, *FEMS Microbiol. Lett.*, 313, 1–9 (2010).

66 P.B. Beauregard, Y. Chai, H. Vlamakis, R. Losick, R. Kolter, *Bacillus subtilis* biofilm induction by plant polysaccharides, *Proc. Natl. Acad. Sci. USA.*,110, 1621–1630 (2013).

67 N. Zhang, D. Yang, D. Wang, Y. Miao, J. Shao, X. Zhou, Z. Xu, Q. Li, H. Feng, S. Li, Q. Shen, R. Zhang, Whole transcriptomic analysis of the plant-beneficial rhizobacterium *Bacillus amyloliquefaciens* SQR9 during enhanced biofilm formation regulated by maize root exudates, *BMC Genom.*,16, 685 (2015).

68 J.P. O'Gara, Ica and beyond: Biofilm mechanisms and regulation in *Staphylococcus epidermidis* and *Staphylococcus aureus*, *FEMS Microbiol. Lett.*, 270, 179–188 (2007).

69 X. Wang, J.F. Preston, T. Romeo, The pgaABCD locus of *Escherichia coli* promotes the synthesis of a polysaccharide adhesin required for biofilm formation, *J. Bacteriol.*,186, 2724–2734 (2004).

70 F. Gotz, *Staphylococcus* and biofilm, *Mol. Microbiol.*, 43, 1367–1378 (2002).

71 N. Cerca, T. Maira-Litran, K.K. Jefferson, M. Grout, D.A. Goldmann, G.B. Pier, Protection against *Escherichia coli* infection by antibody to the *Staphylococcus aureus* poly-N-acetylglucosamine surface polysaccharide, *PNAS*, 104, 7528–7533 (2007).

72 E.A. Izano, M.A. Amarante, W.B. Kher, J.B. Kaplan, Differential roles of poly-Nacetylglucosamine surface polysaccharide and extracellular DNA in *Staphylococcus aureus* and *Staphylococcus epidermidis* biofilms, *Appl. Environ. Microbiol.*, 74, 470–476 (2008).

73 I. Sadovskaya, E. Vinogradov, J. Li, S. Jabbouri, Structural elucidation of the extracellular and cell-wall teichoic acids of *Staphylococcus epidermidis* RP62A, a reference biofilm-positive strain, *Carbohyd. Res.*, 339, 1467–1473 (2004).

74 B.S. McSwain, R.L. Irvine, M. Hausner, P.A. Wilderer, Composition and distribution of extracellular polymeric substances in aerobic flocs and granular sludge, *Appl. Environ. Microbiol.*, 71, 1051–1057 (2005).

75 S.S. Branda, F. Chu, D.B. Kearns, R. Losick, R.A. Kolter, Major protein component of the *Bacillus subtilis* biofilm matrix, *Mol. Microbiol.*, 59, 1229–1238 (2006).

76 S. Lory, Determinants of extracellular protein secretion in Gram-negative bacteria, *J. Bacteriol.*, 174, 3423–3428(1992).

77 D.J. Lynch, T.L. Fountain, J.E. Mazurkiewicz, J.A. Banas, Glucan-binding proteins are essential for shaping *Streptococcus mutans* biofilm architecture, *FEMS Microbiol. Lett.*, 268, 158–165 (2007).

78 M.J. Higgins and J.T. Novak, Characterization of exocellular protein and its role in bioflocculation, *J. Environ. Eng.*, 123, 479–485 (1997).

79 P. Mora, F. Rosconi, L.F. Fraguas, S. Castro-Sowinski, *Azospirillum brasilense* Sp7 produces an outer-membrane lectin that specifically binds to surface-exposed extracellular polysaccharide produced by the bacterium, *Arch. Microbiol.*, 189, 519–524 (2008).

80 I. Lasa and J.R. Penades, Bap: a family of surface proteins involved in biofilm formation, *Res. Microbiol.*, 157, 99–107 (2006).

81 D. Otzen and P.H. Nielsen, We find them here, we find them there: functional bacterial amyloid, *Cell. Mol. Life Sci.*, 65, 910–927 (2007).

82 J.N.C. Fong and F.H. Yildiz, Biofilm matrix proteins, *Microbiol. Spect.*, 3 (2015). MB–0004–2014.

83 A. Kulp and M.J. Kuehn, Biological functions and biogenesis of secreted bacterial outer membrane vesicles, *Annu. Rev. Microbiol.*, 64, 163–184 (2010).

84 Y. Tashiro, H. Uchiyama, N. Nomura, Multi-functional membrane vesicles in *Pseudomonas aeruginosa*, *Environ. Microbiol.*, 14, 1349–1362 (2012).

85 S.R. Schooling and T.J. Beveridge, Membrane vesicles: an overlooked component of the matrices of biofilms, *J. Bacteriol.*, 188, 5945–5957 (2006).

86 M. Toyofuku, B. Roschitzki, K. Riedel, L. Eberl, Identification of proteins associated with the *Pseudomonas aeruginosa* biofilm extracellular matrix, *J. Proteom. Res.*, 11, 4906–4915 (2012).

87 D. Romero, C. Aguilar, R. Losick, R. Kolter, Amyloid fibers provide structural integrity to *Bacillus subtilis* biofilms, *PNAS*, 107, 2230–2234 (2010).

88 D. Romero and R. Kolter, Functional amyloids in bacteria, *Int. Microbiol.*,17, 65–73 (2014).

89 Z. Qin, Y. Ou, L. Yang, Y. Zhu, T. Tolker-Nielsen, S. Molin, D. Qu, Role of autolysin mediated DNA release in biofilm formation of *Staphylococcus epidermidis*, *Microbiology*, 153, 2083–2092 (2007).

90 T. Das, S. Sehar, M. Manefield, The roles of extracellular DNA in the structural integrity of extracellular polymeric substance and bacterial biofilm development, *Environ. Microbiol. Rep.*, 5, 778–786 (2013).

91 M. Okshevsky and R.L. Meyer, The role of extracellular DNA in the establishment, maintenance and perpetuation of bacterial biofilms, *Crit. Rev. Microbiol.*, 41, 341–352 (2013).

92 U. Bockelmann, A. Janke, R. Kuhn, T.R. Neu, J. Wecke, J.R. Lawrence, U. Szewzyk, Bacterial extracellular DNA forming a defined network-like structure, *FEMS Microbiol. Lett.*, 262, 31–38 (2006).

93 M. Allesen-Holm, K.B. Barken, L. Yang, M. Klausen, J.S. Webb, S. Kjelleberg, S. Molin, M. Givskov, T.A. Tolker-Nielsen, A characterization of DNA release in *Pseudomonas aeruginosa* cultures and biofilms, *Mol. Microbiol.*, 59, 1114–1128 (2006).

94 J. Wang, K.W. Bayles, Programmed cell death in plants: lessons from bacteria? *Trends Plant Sci.*, 18, 133–139 (2013).

95 T. Das, S.K. Prashant, H.J. Busscher, H.C. van der Mei, B.P. Krom, Role of extracellular DNA in initial bacterial adhesion and surface aggregation, *Appl. Environ. Microbiol.*, 76, 3405–3408 (2010).

96 H.C. Flemming, T.R. Neu, D.J. Wozniak, The EPS matrix: the "house of biofilm cells", *J. Bacteriol.*, 189, 7945–7947 (2007).

97 L. Yang, K.B. Barken, M.E. Skindersoe, A.B. Christensen, M. Givskov, T. Tolker-Nielsen, Effects of iron on DNA release and biofilm development by *Pseudomonas aeruginosa*, *Microbiology*,153, 1318–1328 (2007).

98 H. Mulcahy, L. Charron-Mazenod, S. Lewenza, Extracellular DNA chelates cations and induces antibiotic resistance in *Pseudomonas aeruginosa* biofilms, *PLoSPathog.*,4, e1000213 (2008).

99 T.R. Neu, T. Dengler, B. Jann, K. Poralla, Structural studies of an emulsion-stabilizing exopolysaccharide produced by an adhesive, hydrophobic *Rhodococcus* strain, *J. Gen. Microbiol.*,138, 2531–2537 (1992).

100 A. Conrad, M.K. Suutari, M.M. Keinanen, A. Cadoret, P. Faure, L. Mansuy-Huault, J.C. Block, Fatty acid lipid fractions in extracellular polymeric substances of activated sludge flocs, *Lipids*, 38, 1093–1105 (2003).

101 W. Sand and T. Gehrke, Extracellular polymeric substances mediate bioleaching/ biocorrosion via interfacial processes involving iron (III) ions and acidophilic bacteria, *Res. Microbiol.*, 157, 49–56 (2006).

102 T. Matsuyama and Y. Nakagawa, Surface-active exolipids: analysis of absolute chemical structures and biological functions, *J. Microbiol. Methods*, 25, 165–175 (1996).

103 M.E. Davey, N.C. Cajazza, G.A. O′Toole, Rhamnolipid surfactant production affects biofilm architecture in *Pseudomonas aeruginosa* PAO1, *J. Bacteriol.*, 185, 1027–1036 (2003).

104 B.R. Boles, M. Thoendel, P.K. Singh, Self-generated diversity produces "insurance effects" in biofilms communities, *Proc. Natl Acad. Sci. USA.*,101,16630–16635 (2004).

105 E.Z. Ron and E. Rosenberg, Natural role of biosurfactants, *Environ. Microbiol.*, 3, 229–236 (2001).

106 C. Leck and E.K. Bigg, Biogenic particles in the surface microlayer and overlaying atmosphere in the central Arctic Ocean during summer, *Tellus B*, 57, 305–316 (2005).

107 E.B. Roberson and M.K. Firestone, Relationship between desiccation and exopolysaccharide production in a soil *Pseudomonas* sp, *Appl. Environ. Microbiol.*, 58, 1284–1291 (1992).

108 M. Potts, Desiccation tolerance of prokaryotes, *Microbiol. Rev.*, 58, 755–805 (1994).

109 P. Stewart and M. Franklin, Physiological heterogeneity in biofilms, *Nat. Rev. Microbiol.*, 6, 199–210 (2008).

110 O. Ates, Systems biology of microbial exopolysaccharides production, *Front. Bioeng. Biotechnol.*, 3, 200 (2015).

111 H. Laue, A. Schenk, L. Hongqiao, L. Lambertsen, T.R. Neu, S. Molin, M. Ullrich, Contribution of alginate and levan production to biofilm formation by *Pseudomonas syringae, Microbiology,*152, 2909–2918 (2006).

112 A. Ogawa, S. Furukawa, S. Fujita, J. Mitobe, T. Kawarai, N. Narisawa, T. Sekizuka, M. Kuroda, K. Ochiai, H. Ogihara, S. Kosono, S. Yoneda, H. Watanabe, Y. Morinaga, H. Uematsu, H. Senpuku, Inhibition of *Streptococcus mutans* biofilm formation by *Streptococcus salivarius* FruA, *Appl. Environ. Microbiol.*, 77, 1572–1580 (2011).

113 C. Mayer, R. Moritz, C. Kirschner, W. Borchard, R. Maibaum, J. Wingender, H.C. Flemming, The role of intermolecular interactions studies on model systems for bacterial biofilms, *Int. J. Biol. Macromol.*, 26, 3–16 (1999).

114 L. Skillman, I.W. Sutherland, M.V. Jonse, The role of exopolysaccharides in dual species biofilm development, *J. Appl. Microbiol.*, 85, S13–S18 (1999).

115 V. Korstgens, H.C. Flemming, J. Wingender, W. Borchard, Influence of calcium ions on the mechanical properties of a model biofilm of mucoid *Pseudomonas aeruginosa, Water Sci. Technol.*,43, 49–57 (2001).

116 T. Shaw, M. Winston, C.J. Rupp, I. Klapper, P. Stoodley, Commonality of elastic relaxation times in biofilms, *Phys. Rev. Let.*, 93, 098102 (2004).

117 J.B. Kaplan, K. Velliyagounder, C. Ragunath, H. Rohde, D. Mack, K.-M. Knobloch, K.M.N. Ramasubbu, Genes involved in the synthesis and degradation of matrix polysaccharide in *Actinobacillus actinomycetemcomitans* and *Actinobacillus pleuropneumoniae* biofilms, *J. Bacteriol.*, 186, 8213–8220 (2004).

118 M. Gjermansen, P. Ragas, C. Sternberg, S. Molin, T. Tolker-Nielsen, Characterization of starvation-induced dispersion in *Pseudomonas putida* biofilms, *Environ. Microbiol.*, 7, 894–906 (2005).

119 J. Webb, L.S. Thompson, S. James, T. Charlton, T. Tolker-Nielsen, B. Koch, M. Givskov, S. Kjelleberg, Cell death in *Pseudomonas aeruginosa* biofilm development, *J. Bacteriol.*,185, 4585–4592 (2003).

120 J.D. Oliver, Recent findings on the viable but nonculturable state in pathogenic bacteria, *FEMS Microbiol. Rev.*, 34, 415–425 (2010).

121 T.K. Wood, S.J. Knabel, B.W. Kwan, Bacterial persister cell formation and dormancy, *Appl. Environ. Microbiol.*, 79, 7116–7121 (2013).

122 I.W. Sutherland, Extracellular polysaccharides, *Biotechnology,* 200, 615–657 (1996).

123 D.H. Limoli, C.J. Jones, D.J. Wozniak, Bacterial extracellular polysaccharides in biofilm formation and function, *Microbiol.Spectr.*,3, MB-0011-2014 (2015).

124 J.W. Costerton, Z. Lewandowski, D.E. Caldwell, D.R. Korber, H.M. Lappin-Scott, Microbial biofilms, *Annu. Rev. Microbiol.*, 49, 711–745 (1995).

125 L.D. Renner and D.B. Weibel, Physicochemical regulation of biofilm formation, *MRS Bull.*, 36, 347–355 (2011).

126 N.R. Stanley, B.A. Lazazzera, Environmental signals and regulatory pathways that influence biofilm formation, *Mol. Microbiol.*, 52, 917–924 (2004).

127 V. Leriche, P. Sibille, B. Carpentier, Use of an enzyme-linked lectin sorbent assay to monitor the shift in polysaccharide composition in bacterial biofilms, *Appl. Environ. Microbiol.*, 66, 1851–1856 (2000).

128 P. Tielen, M. Strathmann, K.E. Jaeger, H.C. Flemming, J. Wingender, Alginate acetylation influences initial surface colonization by mucoid *Pseudomonas aeruginosa*, *Microbiol. Res.*, 160, 165–176 (2005).

129 V. Korstgens, H.C. Flemming, J. Wingender, W. Borchard, Influence of calcium ions on the mechanical properties of a model biofilm of mucoid *Pseudomonas aeruginosa*, *Water Sci. Technol.*,43, 49–57 (2001).

130 M. Hussain, M.H. Wilcox, P.J. White, The slime of coagulase-negative staphylococci: biochemistry and relation to adherence, *FEMS Microbiol. Rev.*, 104, 191–208 (1993).

131 L. Gouzou, G. Burtin, R. Philippy, F. Bartoli, T. Heulin, Effect of inoculation with *Bacillus polymyxa* on soil aggregation in the wheat rhizosphere: preliminary examination, *Geoderma.*, 56, 479–490 (1993).

132 R. Batool and S. Hasnain, Growth stimulatory effects of *Enterobacter* and *Serratia* located from biofilms on plant growth and soil aggregation, *Biotechnol.*, 4, 347–353 (2005).

133 V. Sandhya, S.K.Z. Ali, M. Grover, G. Reddy, B. Venkateswarlu, Alleviation of drought stress effects in sunflower seedlings by exopolysaccharides producing *Pseudomonas putida* strain GAP-P45, *Biol. Fertil. Soil*, 46, 17–26 (2009).

134 C. Chenu and E.B. Roberson, Diffusion of glucose in microbial extracellular polysaccharide as affected by water potential, *Soil Biol. Biochem.*, 28, 877–884 (1996).

135 A. Pal and A.K. Paul, Microbial extracellular polymeric substances: central elements in heavy metal bioremediation, *Ind. J.Microbiol.*, 48, 49–64 (2008).

136 S. Bensalim, J. Nowak, S.K. Asiedu, A plant growth promoting rhizobacterium and temperature effects on performance of 18 clones of potato, *Am. J. Potato. Res.*, 75, 145–152 (1998).

137 Y. Alami, L. Champolivier, A. Merrien, T. Heulin, The role of *Rhizobium* sp. rhizobacterium that produces exopolysaccharide in the aggregation of the rhizospherical soil of the sunflower: Effects on plant growth and resistance to hydric constraint, *OCL- Oleag. Corps Gras Lipid.*, 6, 524–528 (2000).

138 H. Naseem and A. Bano, Role of plant growth–promoting rhizobacteria and their exopolysaccharide in drought tolerance of maize, *J. Plant Inter.*, 9, 689–701 (2014).

139 J.J. Harrison, R.J. Turner, H. Ceri, Metal tolerance in bacterial biofilms, *Rec. Res. Dev. Microbiol.*, 9, 33–55 (2009).

140 Y. Kawakami, N. Hayashi, M. Ema, M. Nakayama, Effects of divalent cations on *Halobacterium salinarum* cell aggregation, *J. Bioscie. Bioengin.*, 104, 42–46 (2007).

141 P. Gilbert, J. Das, I. Foley, Biofilms susceptibility to antimicrobials, *Adv. Dent. Res.*, 11, 160–167 (1997).

142 R.M. Donlan, Role of biofilms in antimicrobial resistance, *ASAIO J.*, 46, S47–S52 (2000).

143 P. Gilbert, T. Maira-Litran, A.J. McBain, A.H. Rickard, F.W. Whyte, The physiology and collective recalcitrance of microbial biofilm communities, *Adv. Microb. Physiol.*, 46, 203–256 (2002).

144 E. Drenkard, Antimicrobial resistance of *Pseudomonas aeruginosa* biofilms, *Microbes Infect.*, 5, 1213–1219 (2003).

145 M.A. Al-Fattani and L.J. Douglas, Penetration of *Candida* biofilms by antifungal agents, *Antimicrob. Agents Chemother.*, 48, 3291–3297 (2004).

146 L. Hall-Stoodley, J.W. Costerton, P. Stoodley, Bacterial biofilms: from the Natural environment to infectious diseases, *Nat. Rev. Microbiol.*, 2, 95–108 (2004).

147 J.N. Anderl, J. Zahller, F. Roe, P.S. Stewart, Role of nutrient limitation and stationary-phase existence in *Klebsiella pneumoniae* biofilm resistance to ampicillin and ciprofloxacin, *Antimicrob. Agents Chemother.*, 47, 1251–1256 (2003).

148 M.C. Walters, F. Roe, A. Bugnicourt, M.J. Franklin, P.S. Stewart, Contributions of antibiotic penetration, oxygen limitation, and low metabolic activity to tolerance of *Pseudomonas aeruginosa* biofilms to ciprofloxacin and tobramycin, *Antimicrob. Agents Chemother.*, 47, 317–323(2003).

149 F.C.M. Thien and G.A. O'Toole, Mechanisms of biofilm resistance to antimicrobial agents, *Trends Microbiol.*, 9, 34–39 (2001).

150 A.S. Betzner, L.C. Ferreira, J.V. Holtje, W. Keck, Control of the activity of the soluble lytic transglycosylase by the stringent response in *Escherichia coli*, *FEMS Microbiol. Lett.*, 55, 161–164 (1990).

151 K. Lewis, Riddle of Biofilm Resistance, *Antimicrob. Agents Chemother.*, 45, 999–1007 (2001).

152 K. Driffield, K. Miller, J.M. Bostock, A.J. O'Neill, I. Chopra, Increased mutability of *Pseudomonas aeruginosa* in biofilms, *J. Antimicrob. Chemother.*, 61, 1053–1056 (2008).

153 R.M. Donlan, Biofilms: microbial life on surfaces, *Emerg. Infect. Dis.*, 8, 881–890 (2002).

154 S. Molin and T. Tolker-Nielsen, Gene transfer occurs with enhanced efficiency in biofilms and induces enhanced stabilisation of the biofilm structure, *Curr.Opin. Biotechnol.*,14, 255–261 (2003).

155 G.H. Dibdin, S.J. Assinder, W.W. Nichols, P.A. Lambert, Mathematical model of β-lactam penetration into a biofilm of *Pseudomonas aeruginosa* while undergoing simultaneousinactivation by released β-lactamases, *J. Antimicrob. Chemother.*, 38, 757–769 (1996).

156 N. Bagge, M. Schuster, M. Hentzer, O. Ciofu, M. Givskov, E. P. Greenberg, N. Hoiby, *Pseudomonas aeruginosa* biofilms exposed to imipenem exhibit changes in global gene expression and β-Lactamase and alginate production, *Antimicrob. Agents Chemother.*, 48, 1175–1187 (2004).

157 G. A. O'Toole, H.B. Kaplan, R. Kolter, Biofilm formation as microbial development, *Annu. Rev. Microbiol.*, 54, 49–79 (2000).

10

Root Exudates: Composition and Impact on Plant–Microbe Interaction

Shamsul Hayat, Ahmad Faraz and Mohammad Faizan

Department of Botany, Aligarh Muslim University, Aligarh, India

10.1 Introduction

The roots are the most important underground part of plants, as they help in absorption of water and mineral from soil and supports the plants. Apart from these primary roles (water/nutrient uptake), roots also perform other function such as storage organ in beets and carrots, respiratory organ in the mangrove's plant roots, and anchoring in potatoes [1]. Plants synthesize many chemical compounds by their metabolic reaction which is helpful in plants development. Release of organic compound from the roots of living plant to their surrounding soil is termed as root exudates and it is a ubiquitous phenomenon [2].

Root exudates may be the result of root pressure, which causes oozing of these substances into the rhizosphere zone and is responsible for attracting microbes. Roots exudation rates may vary among plants species with environmental conditions [3]. Structures of roots are very complex morphologically and physiologically, and their metabolites are repeatedly released in huge quantities into the soil rhizosphere zone from these living roots, which may be hairy or fibrous. Root exudates contain root-specific metabolites and have significant ecological impacts on soil macro and micro-biota, as well as on the entire plant.

These root exudates serve many purposes such as affect the soil microbial community in their immediate locality; manipulate the confrontation to pests; hold up important symbioses; modify the chemical and physical properties of the soil; and slow the growth of competing plant species [4]. The process of root exudation in the rhizosphere zone is directly related to physiology and metabolism of plants. It has been reported that, on average, 20 percent of carbon (C) assimilated by higher plants through photosynthesis is released by roots as exudates in the form of sugars and polysaccharides, organic and amino acids, peptides, and proteins [5].

10.2 Chemical Composition of Root Exudates and Their Significance

In recent years, emphasis has been placed on understanding the chemical composition of root exudates and its influence on soil microbe and microbe plant interaction. In this chapter, we have addressed of some of these issues and attempted to provide a comprehensive literature survey. Soil organic matter mainly contains root exudates which are deposited near the rhizhosphere zone which are ultimately secreted by the plants roots [6, 7]. The quantity and superiority of root exudates are depends on plant species, the age of an individual plant and external environmental factors like biotic (e.g., bacteria, nematode, insect, etc.) and abiotic stresses (e.g., temperature, drought, heavy metal, etc.).

The main composition of root exudates are organic compounds that may be divided into two classes: organic compounds that have low molecular weight and organic compound that have higher molecular weight. Low-molecular compound includes monomer of proteins (amino acids), sugars (monosaccharide disaccharides), phenolics, and some secondary metabolite. Organic compound like polypeptides (proteins), polysaccharides, and mucilage come under the high-molecular weight classes. Organic components in root exudates are malic acid, citric acid, and so on, and one of the important factors for microbe attraction [8].

Different organic compound secreted by plants as root exudates are listed in Table 10.1, they cover both low-molecular and high-molecular weight compound.

10.3 Root Exudates in Mediating Plant–Microbe Interaction in Rhizosphere (Negative and Positive Interactions)

Root exudates secreted by plants are useful in the microbial interaction (Figure 10.1). Plant roots constantly exude compounds into the rhizosphere; this includes exudation or discharge of ions, free oxygen and water, enzymes, mucilage, and a miscellaneous range of carbon containing primary and secondary metabolites [11].

Root-secreted chemicals mediate the multipartite interactions in the rhizosphere, where plant roots continually act in response to and alter their immediate environment. Increasing data suggest that root exudates instigate and modulate channel of communication between roots and soil microbes. For example, root exudates provide signals that initiate symbiosis with rhizobia and mycorrhizal fungi. In addition, root exudates continue and support a specific diversity of microbes in the rhizosphere of a given particular plant species, which suggests a close evolutionary link [10], as shown in Figure 10.1.

Plant root exudates are essential factors that make up the rhizosphere bacterial community [12] and perform important functions such as providing both resistance against pathogenic organisms [20] and a basis for chemotaxis to attract and repel (particular) microbial species and populations [14]. Recently, Bais et al. (2002) identified (±)-catechin as the root-secreted phytotoxin responsible for the offensive behavior of knapweed in the rhizosphere. Interestingly, (−)-catechin was shown to account for the allele chemical activity, whereas (+)-catechin was inhibitory to soil borne bacteria [15].

Table 10.1 Chemical composition of root exudates.

Chemical nature of compounds	example of organic compound in individual form
Carbohydrates:	Arabinose(5C*), glucose(6C), galactose(6C), fructose(6C), sucrose(12C), pentose(5C), rhamnose(6C), raffinose(18), ribose(5C), xylose(5C) and mannitol(6C) *nNumber of carbon atoms present in that compound*
Sterols:	Campesterol, cholesterol, sitosterol, stigmasterol
Amino acids:	α and β-Alanine, γ-aminobutyric, α-aminoadipic, arginine, asparagine, aspartic, citrulline, cystathionine, cysteine, cystine, deoxymugineic, 3-epihydroxymugineic, glutamine, glutamic, glycine, histidine, homoserine, isoleucine, leucine, lysine, methionine, mugineic, ornithine, phenylalanine, praline, proline, serine, threonine, tryptophan, tyrosine, valine
Fatty acid:	Linoleic, linolenic, oleic, palmitic, stearic
Enzymes:	Amylase, invertase, peroxidase, phenolase, acid/alkaline phosphatase, polygalacturonase, protease
Flavonones and nucleotides	Adenine, flavonone, guanine, uridine/cytidine
Lignins	Catechol, benzoic acid, nicotinic acid, phloroglucinol, cinnamic acid, gallic acid, ferulic acid, syringic acid, sinapoyl aldehyde, chlorogenic acid, coumaric acid, vanillin, sinapyl alcohol, quinic acid, pyroglutamic acid
Indole compounds	Indole-3-acetic acid, brassitin, sinalexin, brassilexin, methyl indole carboxylate, camalexin glucoside
Anthocyanins	Cyanidin, delphinidin, pelargonidin and their substitutes with sugar molecules
Glucosinolates	Cyclobrassinone,desuphoguconapin, desulphoprogoitrin, desulphonapoleiferin, desulphoglucoalyssin
Organic acids:	Acetic, aconitic, ascorbic, aldonic, benzoic, butyric, caffeic, citric, pcoumaric, erythronic, ferulic, formic, fumaric, glutaric, glycolic, lactic, glyoxilic, malic, malonic, oxalacetic, oxalic, p-hydroxybenzoic, piscidic, propionic, pyruvic, succinic, syringic, tartaric, tetronic, valeric, vanillic

Source: Adapted from Dennis et al. (2010) [9], Badri and Vivanco (2009) [10].

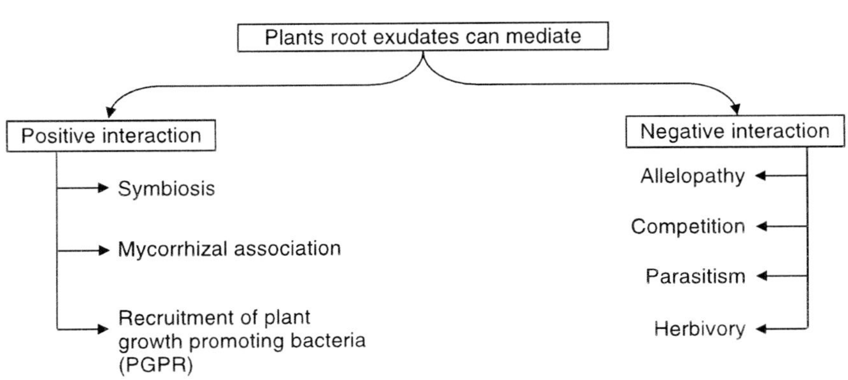

Figure 10.1 Flow chart shows type of plants root interaction with microbe due the secretion of root exudates.

Interactions between plants mediated by root exudates showed a positive role. Beside this, some root exudates encourage defense responses in neighboring plants. In some cases, the plant defenses induced by root exudates simply reduce susceptibility to pathogen infection, whereas in other cases these defenses initiate production and release of leafy volatiles that attract predators of plant enemies. One negative interaction between plants is *allelopathy*, a chemical mediated response between plants. Plants release phytotoxins in decomposing leaf and root tissue, in leachates from live tissue, in green leafy volatiles, and in root exudates [4, 16]. Phototoxic root exudates can mediate negative plant–plant interactions only if present at sufficient amounts to affect plant growth and survival. Plant–microbe interaction can positively influence plant growth by various mechanisms (see Figure 10.2).

10.4 Direct and Indirect Effect of Root Exudates on PGPR, Root Colonization, and in Stress Tolerance

Root colonization is a process in which soil-growing bacteria aggregated toward rhizosphere zone and create suitable environment for plants as well as their growth and survival. Root exudates directly or indirectly play an important role in mediating root colonization by plant growth–promoting rhizobacteria (PGPR) and increase the stress tolerance. Direct and indirect effects of root exudates on plant-growth activity are achieved by modifying root structures to absorb minerals and water from soil.

Interactions between roots and PGPR are mediated by root exudates, which are considered as the first line of communication between roots and PGPR in the rhizosphere [17].

Dobbelaere et al. (2003) classify the plant-associated bacteria into valuable, vicious, and impartial groups on the basis of their effects on plant growth. PGPR colonize plant roots and promote plant growth. Only a fraction of PGPR(1 to 2 percent) promotes

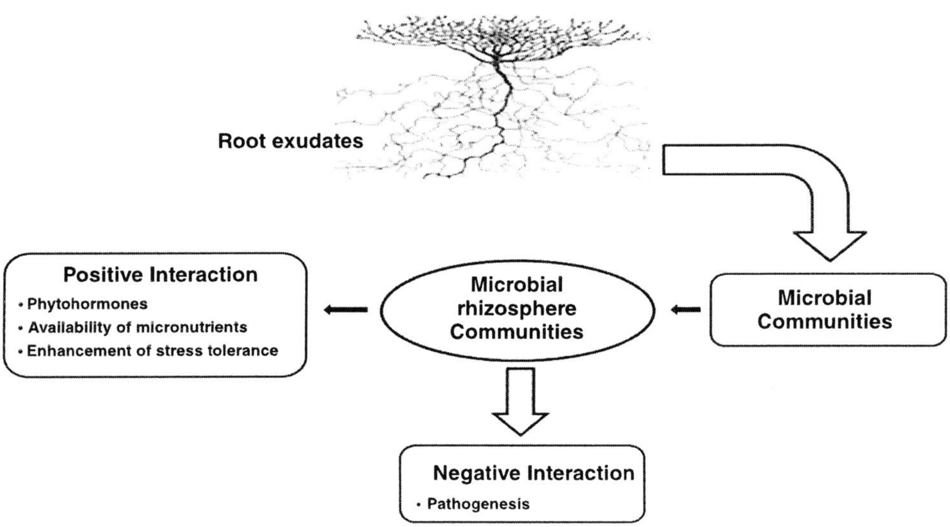

Figure 10.2 Possible interactions of microorganism due to the secretion of root exudates.

plant growth in rhizosphere. Microbes colonize the roots of plants due to the presence of root exudates, which contain organic chemical, amino acid, ions, and microbes attracted to rhizosphere zone by chemotactic movement [18].

PGPR shows different types of relationship with plants, which involve biotic and abiotic factors of the rhizosphere region other than the two partners. Success of PGPR is dependent on their endurance and firm with the root/rhizosphere [19]. Unfavorable conditions create stress, hindering plant growth and reproduction. *Stress tolerance* is the adaptation of plants to this stress environment.

Stress in plants may be due to the biotic factor such as insects or microbes, or abiotic factors such as air quality, temperature, water availability, mineral nutrient, and light intensity. These stress factors have a profound effect on plants growth and reproduction. Abiotic factors change with geographic region; some have a negative effect on plants life, but plants that adapt for all these changes will survive in extreme conditions [20].

Root exudates are very useful in overcoming the stress environments and increasing plant growth and development. Allelochemicals and other metabolites that are secreted by plant roots have important role in rhizosphere signaling, plant defense, and in response to abiotic stresses. Plants have adopted a variety of sequestration and transport mechanisms to move and export bioactive products securely into the rhizosphere [21]. Root exudates are helpful to prevent the herbivores and microbes, and by stimulating symbiotic interaction, which may alter soil properties and inhibit the growth of competing species [22].

Most rhizobacteria belong to the genera *Pseudomonas* and *Bacillus* and are well known for their antagonistic effects and their ability to trigger induction of systemic resistance (ISR). Resistance-inducing and antagonistic rhizobacteria might be useful in formulating new inoculants with combinations of different mechanisms of action, leading to a more efficient use for biocontrol strategies to improve cropping systems [23].

PGPR can also colonize roots of monocots and dicots, and by doing this they may enhance the plant growth directly and indirectly. In colonization root system of plants, PGPR modify the root architecture and production of phytohormones and other signals, leading to the improved lateral root branching and improvement of root hairs. As we know that roots are the principle organs for plants development, which absorbs the water and nutrition, PGPR also modify root functioning, which ultimately improve plant nutrition and influence the physiology of the whole plant [24]. The most commonly investigated PGPR are pseudomonas and bacillus genera, which come from dominating bacterial group in rhizosphere zone [25].

Plant secondary metabolism may be altered by rhizobacteria, which results in changed plant–insect relationships as reported in root colonization of cucumber by four different PGPR. The research shows a reduced level of cucurbitacin (secondary metabolite), which acts as a feeding stimulant to cucumber beetles [26]. Commercially, PGPRs are used in plant disease management as biocontrol agent, as biofertilizer, and also as phytostimulation or in phytoremediation [27].

10.4.1 Root Colonization

Root colonization involved the interactions of bacterium with its physical, chemical, and biological environment, and it is a complex process. Movement of rhizobacteria toward roots by chemotaxis method assumed to be the first step of bacterial colonization [28]. To understand the colonization process, it is required to screen or develop strains possessing good growth-promoting and biocontrol activity in the field [11]. Colonization is very important for successful plants–microbe interactions [29]. There are several steps involved in the colonization process, such as attraction of microbes toward root, adherence, and colonization and growth, for endophyte invasion also involved in colonization [30]. PGPR reach the roots by chemotactic response, and this motility is provided by flagella cells [31]. Root colonization by *Azospirillum brasilense* also showed that active bacterial motility toward the root hair zone is an important factor for initiating colonization at these sites [32]. Recently, root exudates of banana were analyzed with the help of high-pressure liquid chromatography (HPLC), which shows the chemical composition of root exudates, and it is found that exudates contained several organic acids (OAs) including oxalic, malic, and fumaric acid [33].

10.4.2 Root Exudates and Stress Tolerance

Plants face many stress conditions during growth and development. Stresses may arise by natural processes or by anthropogenic activity. Stresses may be biotic, such as plant pathogens and pests (viruses, bacteria, fungi, insects, nematodes, etc.), or abiotic. Abiotic stresses include salinity (excess salt), drought (less water) and flooding, heavy metals (Cd Ni Pb etc), temperature, some gases and nutrient deficiency or excess. Generally, there are three environmental stimulus that may enhanced root exudation of organic anions from plants roots [34–36]:

1) Nutrient deficiency (predominantly associated with phosphorus)
2) Exposure of toxic cations
3) Waterlogging, which causes anoxia (absence of oxygen)

Biotic and abiotic stresses are two most important constrains to agricultural production. Plant growth is affected by stress conditions, which include a large number of factors such as hormonal and nutritional imbalance, ion toxicity, physiological disorders, and susceptibility to diseases. PGPR and mycorrhizal fungi may boost the plant growth under stress conditions by their inoculation into the soil. These microbes regulate nutritional and hormonal balance–produced plant-growth regulators, solubilizing the nutrients and inducing resistance against plant pathogens that can promote plant growth. Other than plants, these microbes also show synergistic as well as antagonistic interactions with other microbes in the soil environment, and these interactions may be vital for sustainable agriculture as they depend on biological processes rather than on agrochemicals to maintain plant growth and development as well as proper soil health under stress conditions. Research shows the function of rhizobacteria and mycorrhizae alone and/or in combination in enhancing plant growth under stress conditions [37].

PGPR proliferate into the rhizosphere zone due to physical, chemical, and biological processes of root, which make an ideal place for these microbes [38]. Root exudates are the main reason for these microorganisms, which survive more or less in the vicinity of the roots, and root exudates are used by microbe for their growth as a source of nutrients [39]. Survival of various microorganisms depends on plant root exudates [40]. In a number of reviews, it was found that many of the rhizobacterial (PGPR) strains can also improve plant life against salinity, drought, flooding stress, and heavy-metal toxicity stress and increase the plant tolerance, thus enabling plants to endure under adverse environmental conditions [41–45]. PGPR may also enhance plant growth and development by the asset of their key enzymes (ACC-deaminase, chitinase) and also by the producing the substances such as exopolysaccharides and rhizobitoxines to facilitate the plants to withstand stressful conditions [44, 46].

Heavy metals are released into the enviorment from a variety of sources, which ultimately reaches to earth accumulate in soil. When plants interact with these heavy metals, their growth performance is retarded due to the toxic effect of heavy metals, and the toxic effect may be due the increased free radicals or may be due to cell damage. Many studies report that the heavy metals aluminium (Al) and Cadmium (Cd) have a toxic effect on plants growth. Understanding the nature of these tolerance mechanisms has been the focus of ongoing research in the area of stress physiology. Aluminium cation (Al^{3+}) is toxic to many plants at micromolar (μM) concentrations. However, some plant species have evolved mechanisms that facilitate their growth on acidic soils with toxic concentrations of Al^{3+}. An important factor is root exudates (organic acids), which participate in these aluminium tolerance mechanisms. Some plants detoxify aluminium in the rhizosphere by releasing organic acids that chelate aluminium [47].

Nutrients are a very important factor for plants growth and development. Deficiency of mineral nutrients causes several diseases in plants. So availability of nutrient elements become a major constraint to plant growth in many environments of the world, especially the tropics, where soils are extremely low in nutrients. Root exudates play an important role in nutrient accusation. Plants take up most mineral nutrients through the rhizosphere, where microorganisms interact with plant products in root exudates [48].

Salt stress is now common due to excess irrigation. Most of the agricultural soil becomes saline, which has consequences for plant growth and development. Strains of PGPR also have the ability to protect the plants from the harmful effects of high Na^+ concentration in the saline soil environment by their ability to produce exopolysaccharides. Production of exopolysaccharides reduces Na^+ uptake in the plant by binding it and also by biofilm formation [49]. Elevated levels of ethylene production have been reported under salinity, drought, and waterlogging stress [50, 51], and excess ethylene level may have negative effects on plants and also inhibit the root growth of plants [52, 53].

There is more than one mechanism of some strains that acquire and can endure not only the normal but also stressful environment. Interaction of PGPR with host plant and soil environment evaluates the effectiveness of these microbes for promoting plant growth. It also depends on their inherent capabilities. The performances and survival of various PGPR and fungi may become dependent on root exudates and therefore directly or indirectly influence rhizobacteria.

10.5 Role of Root Exudates in Biofilm Formation by PGPR

The diverse role of root exudates also includes biofilm formation by rhizobacteria, as root exudates are thought of as adherence molecules. Microbial cells that adhere to a surface and are surrounded by an extracellular polymeric matrix and that form structurally complex communities are called biofilms. Biofilms formation is a complex process whose composition depends on the system. Microorganisms that form biofilms synthesize a wide spectrum of multifunctional polysaccharides, including intracellular polysaccharides, structural polysaccharides, and extracellular polysaccharides or exopolysaccharides

(EPS). Production of exopolysaccharide takes part in biofilm formation, and equally can affect the interaction of microbes with roots and root appendages [54].

Formation of biofilm plays important roles in attachment and colonization of plant surfaces by both beneficial bacteria (e.g., PGPR) and phytopathogenic (negative interaction) bacteria. During the process of biofilm development and maturation, surface-attached cells undergo aggregation to form microcolonies. Biosurfactants are produced by many plant-associated bacterial species and play essential roles in bacterial motility, signaling, biofilm formation, and control of plant–bacteria interactions [55].

Biofilms helps to protect the bacteria from deleterious conditions and the biofilms formation also appears to be an essential factor in the disease cycle of bacterial pathogens in both animals and plants [56]. Biofilm formation on root surface have been reported in the case of *Agrobacterium tumefaciens* and rhizobia, which can form dense, structurally complex biofilms on root surfaces, extensively coating the epidermis and root hairs, and these bacteria can form elaborate biofilms on abiotic surfaces also [57]. Attachment is required for biofilm formation, and for this bacteria interact with plant tissues through adhesions protein, which include polysaccharides and surface proteins. Active motility starts the initial contact. Specificity in biofilm formation is created due to the recognition between lectins and their cognate carbohydrates. Biofilm improvement and the resulting intimate interactions with plants often require cell–cell communication between colonizing bacteria [58].

Various advantages are provided by bacterial biofilm. Increased resistance to certain environmental stresses as well as antimicrobial tolerance, protection from protozoan predation, consortial metabolism, or the opportunity for horizontal gene transfer (HGT) are also provided by biofilm. Single cells cannot accomplish these functions efficiently, but they are easily done by high population density. Certain processes, such as the production of excreted metabolites or exoenzymes, are only effective above certain threshold concentration.

Dense biofilm formed by these pseudomonas on the biotic and abiotic surfaces [59]. Several signal and genes are responsible for the biofilm formation, and they may be present in root exudates or in rhizobacteria. It was reported that in *A. Thaliana* some of the known gram-positive biocontrol PGPBs (such as B. subtilis) form protective biofilm and assist plants in evading a Gram-negative plant pathogen, *Pseudomonas syringae* pv. Tomato DC3000, and these PGPB also produces an antimicrobial cyclic lipopeptide surfactin [60].

10.6 Role of Root Exudates in Protecting Plants Pathogenic Biofilm, Quorum Sensing Inhibition

It has been reported that plant root exudates attract many Gram-negative bacteria that produce acylhomoserine lactones (AHLs), which initiate quorum sensing (QS) actions and biofilm formation and reduce the pathogenicity [61]. A number of studies have shown that certain virulence factors are required for bacterial pathogenesis, and these factors are also found in both mammalian and plant systems [62, 63]. Initiation of the production and secretion of these virulence factors in large population of bacteria is controlled by quorum sensing. In 1981, Eberhardt et al. reported QS phenomenon in the aquatic bacteria *Vibrio fischeri* and described QS as a density-dependent regulatory mechanism and their signal mediates the induction of the *lux* gene, which is responsible for bioluminescence in these bacteria [64]. Activation of QS mechanism in bacteria is mediated by small autoinducer (AI) molecules, which initiate the cell–cell communication, and it also coordinated the action of many other bacteria, including plant-associated bacteria.

The most commonly reported type of autoinducer signals are N-acyl homoserine lactones (AHLs) [65]. Several different AHL signals commonly synthesize by a given bacterial species that will differ from one another in the length of the N-acyl side chain (4 to 14 carbons). They may also differ by the presence or absence of double bonds, or side chain substituents (i.e., keto or hydroxyl group) [66]. Compound furanones are inhibitors of induction of AHL-stimulated behaviors because furanones are structurally similar to AHLs, which binds competitively to the AHL receptor protein [67]. Like furanones, plants root exudates contain other compounds that mimic the AHLs system of bacteria and inhibit the ptahogenicity of these Gram-negative bacteria. Exudates from pea (*Pisum sativum*) seedlings and other higher plants showed that they contain several separable activities that mimicked AHL signals in well-characterized bacterial reporter strains, stimulating AHL-regulated behaviors in some strains while inhibiting such behaviors in others. These AHL signal-mimic compounds may perhaps prove to be vital in the outcome of interactions between higher plants and a diversity of pathogenic, symbiotic, and saprophytic bacteria [68]. Undoubtedly, a sensible body of indication suggests that cross talk among plants and bacteria may occur through QS signal mimics. QS mediates several plant–microbe interactions, both pathogenic and beneficial. *Rhizobium etli,* the Gram-negative nitrogen-fixing soil bacterium is the bacterial symbiotic partner of the common bean plant. *R. etli* CNPAF512 produces at least seven different QS signal molecules, produced by the cinIR and raiIR QS system. cinI and raiI code for the AHL synthases, and cinR and raiR code for the transcriptional regulators that bind the AHLs [69,70]. Fatima et al. (2010) have reported the QS behavior of bacteria from extracts of seven leguminous plants (*Pisum sativum, Vigna radiata, Vigna mungo, Cajanus cajan, Lentil culinaris, Cicer arietinum,* and *Trigonella foenum graecum*). Germinating seedling and seedling extracts of only *P. sativum* (pea) showed inhibition of violacein production. The most important thing noted in the study is that the *T. foenum graecum* (fenugreek) seed extracts enhance the pigment production. Quorum sensing regulated swarming motility in *Pseudomonas aerugionsa* PAO1 was reduced by pea seedling extract while enhanced by the fenugreek seed extracts. These findings suggest that plant metabolites of some legumes interact actively with bacterial quorum sensing and could modulate its associated functions [71].

10.7 Isolation of Root Exudates

Isolation and characterization of root exudates are very important. Analysis of root exudates can be exploited to expand understanding of the nature of compounds in root exudates and their possible function in rhizosphere and also in the microbial community. Chemicals secreted by plant root have an allelopathic effect on their surrounding environment in the soil—that is, chemicals secreted by roots may inhibit or promote the growth of neighboring plants [72]. Plant root exudates are complex compound and contain large numbers of organic chemicals. These chemicals are generally the result of secondary metabolic products, so their characterization is necessary to find out the chemical composition of these root exudates. Their composition depends on the type of the species of plants, environmental factors, and many other processes that also affect the nature of root exudation.

Different methods are applied to isolate these chemicals from plants, depending on the plant species. Chemicals secreted by root deposits in the rhizosphere zone are called rhizodeposition. Qualitative composition of water-soluble root exudates and exudation rate can be obtained by creating an artificial rhizosphere in which different plants are grown. Consequent filtration or centrifugation are needed for root exudates collection. This process usually requires removing solids, root detritus, and microbial cell debris. This is followed by consequent concentration using an evaporator, lyophilizator, or ultrafiltration. Analysis of total groups of compounds (total proteins and total carbohydrates) and total organic carbon are finished by simple methods. By contrast, for analysis for individual low-molecular-weight organic molecules (such as sugars, organic acids, and amino acids), HPLC or GS/MS are commonly used with separation into different columns [73]. Under these conditions, rhizodeposition can be measured chemically by growing plants in solution culture.

Chemical sampling reflects the target of the experiment and can range from thorough analyses of a single class of chemicals to much more broad observations. For example, analysis of dicarboxylic acids for wheat and flax in sterile solution culture were quantified in one study of exudates [74], and isoflavonoids in exudates of white lupin were measured in another [75]. Changes in gross classes of compounds, such as low- and high-molecular-weight materials and particulate matter, have been examined in exudates derived from maize that is grown axenically in solution culture [76]. In 1982, Tang and Young developed a collection of root exudates from allelopathy substance using XAD-4 resin to isolates the root exudates.

10.8 Conclusion

This chapter describes the exudation of root exudates from plants in soil and their possible role in rhizosphere soil. Plant roots are important organs, as they help in the mineral uptake from the soil and absorb other inorganic ions to improve plant growth and development. Roots also function as anchors for the plant. In addition, roots secrete a large number of organic compounds into the soil, which deposit in the rhizosphere zone. These secreted compounds are called root exudates and have diverse role in plants as well as in soil.

Plants synthesize many compounds by the process of photosynthesis, and about 30 to 40 percent of these are secreted in the form of root exudates. PGPR are attracted to these exudates and improve plant growth and development. Root exudates mediate both positive and negative interaction by microorganisms to plants. Positive interactions are helpful in plant development and reproduction processes, as they increase the mineral uptake from soil. Negative interaction plays a harmful role for plants, as they disturb plant growth and development.

Root exudates directly or indirectly play an important role in mediating root colonization by PGPR and increase the stress tolerance. The direct and indirect effects of root exudates on plants growth activity are achieved by modifying root structure to absorb minerals and water from soil. The diverse role of root exudates also include biofilm formation by rhizobacteria, as root exudates are considered adherence molecules. Microbial cells that adhere to a surface and are surrounded by an extracellular polymeric matrix and formed structurally complex communities are called biofilms.

Quorum sensing inhibition by root exudates is also an important function, as they inhibit the growth of many Gram-negative bacteria and protect the plant from their harmful effects. In addition, root exudates play an important role in the isolation and characterization of plants, which enables us to identify the specific compounds in plant root secretion. More information is still being uncovered about root exudates secretion. More study is required at the molecular level to discover the genes responsible for their secretion from plants with varying levels. Effects of root exudates can become an advanced area of research.

References

1 H.E. Flores, J.M. Vivanco, V.M. Loyola-Vargas, "Rad icle" biochemistry: the biology of root-specific metabolism, *Trends Plant Sci.*, 4, 220–226 (1999).
2 D.L. Jones, P.R. Darrah, Influx and efflux of organic acids across the soil-root interface of *Zea mays* L. and its implications in rhizosphere C flow, *Plant Soil*, 173, 103–109 (1995).
3 L.V. Kochian, M.A. Pineros, O.A. Hoekenga, The physiology, genetics and molecular biology of plant aluminum resistance and toxicity, *Plant Soil*, 274, 175–195 (2005).
4 C. Bertin, X.H. Yang, L.A. Weston, The role of root exudates and allelochemicals in the rhizosphere, *Plant and Soil*, 256, 67–83 (2003).
5 P. Hinsinger, G.R. Gobran, P.J. Greogry, W.W. Wenzel, Rhizosphere geometry and heterogeneity arising from root-mediated physical and chemical processes, *New Phytologist* 168, 293–303 (2005).
6 B.W. Hutsch, J. Augustin, W. Merbach, Plant rhizodeposition an important source for carbon turnover in soils, *J. Plant Nut. Soil Sci.*, 165, 397–407 (2000).
7 C. Nguyen, Rhizodeposition of organic C by plants: mechanisms and controls, *Agron*, 23, 375–396 (2003).
8 N. Ling, W. Raza, J.H. Ma, Q.W. Huang, Q.R. Shen, Identification and role of organic acids in watermelon root exudates for recruiting *Paenibacillus polymyxa* SQR-21 in the rhizosphere, *Eur J Soil Biol.*, 47, 374–379 (2011).
9 P.G. Dennis, A.J. Miller, P.R. Hirsch, Are root exudates more important than other sources of rhizodeposits in structuring rhizosphere bacterial communities, *FEMS Microbiol Ecol.*, 72, 313–327 (2010).
10 D.V. Badri, J.M. Vivanco, Regulation and function of root exudates, *Plant Cell Environ.*, 32, 666–681 (2009).
11 S. Nardi, G. Concheri, D. Pizzeghello, A. Sturaro, R. Rella, G. Parvoli, Soil organic matter mobilization by root exudates, *Chemosphere*, 5, 653–658 (2000).
12 T.S. Walker, H.P. Bais, E. Grotewold, J.M. Vivanco, Root exudation and rhizosphere biology, *Plant Physiol.*, 2003, 132: 44–51 (2003).
13 L. Abbott, D. Murphy, Soil biology fertility: A key to sustainable land use in agriculture, *Kluwer Academic Publishers*, 187–203 (2003).
14 R. Kumar, R. Bhatia, K. Kukreja, R.K. Behl, S.S. Dudeja, N. Narula, Establishment of Azotobacter on plant roots: Chemotactic response, development and analysis of root

exudates of cotton (*G. hirusitum* L.) and wheat (*T. aestivum* L.), *J. Basic Microbiol.*, 47, 436–439 (2007).
15 H.P. Bais, T.S. Walker, F.R. Stermitz, R.A. Hufbauer, J.M. Vivanco, Enantiomeric dependent phytotoxic and antimicrobial activity of (±)-catechin; a rhizosecreted racemic mixture from Centaurea maculosa (spotted knapweed), *Plant Physiol.*, 128, 1173–1179 (2002).
16 T.L. Weir, H.P. Bais, J.M. Vivanco, Intraspecific and interspecific interactions mediated by a phytotoxin, (±)-catechin, secreted by the roots of *Centaurea maculosa* (spotted knapweed), *J Chem Ecol.*, 29, 2397–2412 (2003).
17 H.P. Bais, V.V.M. Loyola, H.E. Flores, J.M. Vivanco, Root specific metabolism: the biology and biochemistry of underground organs, *In vitro Cell Dev Biol Plant*, 37, 730–741 (2001).
18 S. Dobbelaere, J. Vanderleyden, Y. Okon, Plant growth-promoting effects of diazotrophs in the rhizosphere, *Crit. Rev. Plant Sci.*, 22, 107–149 (2003).
19 A.R. Podile, R.V.N.R. Vukanti, A. Sravani, S. Kalam, S. Dutta, P. Durgeshwar, V. P. Rao, *Root Colonization and Quorum Sensing are the Driving Forces of Plant Growth*, 2013.
20 L. Taiz, E. Zeiger, *Plant Physiology, Fourth Edition*, Sinauer Associates, Sunderland, MA, 2006.
21 L.A. Weston, P.R. Ryan, M. Watt, Mechanisms for cellular transport and release of allelochemicals from plant roots into the rhizosphere, *J. Exp. Bot.* (2012).
22 U. Mathesius, M. Watt, Rhizosphere signals for plant–microbe interactions: implications for field-grown plants. In: Lüttge U, Beyschlag W, Büdel B, Francis D, eds. *Progress in Botany 72*. Berlin, Springer-Verlag, 125–161 (2011).
23 A, Beneduzi, A. Ambrosini, L.M. Passaglia, Plant growth-promoting rhizobacteria (PGPR): their potential as antagonists and biocontrol agents, *Genet Mol Biol.*, 35, 1044–1051 (2012).
24 J. Vacheron, G. Desbrosses, M.L. Bouffaud, B. Touraine, Y. Moënne-Loccoz, D. Muller, L. Legendre, F. Wisniewski-Dyé, C. Prigent-Combaret, Plant growth promoting rhizobacteria and root system functioning, *Frontiers in Plant Science*, 4 (2013).
25 J.A.W. Morgan, J.M. Whipps, Methodological approaches to the study of rhizosphere carbon flow and microbial population dynamics. In: *The rhizosphere, Biochemistry and organic substances at the soil-plant interface*, A. Pinton, Z. Varanini, P. Nannipieri eds. New York, USA, Marcel Dekker, 373–410 (2001).
26 G. Zehnder, J. Kloepper, C. Yao, G. Wei, Induction of systemic resistance against cucumber beetles (Coleoptera: Chrysomelidae) by plant growth-promoting rhizobacteria, *Journal of Economic Entomology*, 90, 391–396 (1997).
27 B. Lugtenberg, F. Kamilova, Plant-growth-promoting rhizobacteria, *Annu. Rev. Microbiol.*, 63, 541–556 (2009).
28 S. G. Sood, Chemotactic response of plant-growthpromoting bacteria towards roots of vesicular–arbuscular mycorrhizal tomato plants, *FEMS Microbiology Ecology*, 45, 219–227 (2003).
29 F. Kamilova, S. Validov, T. Azarova, I. Mulders, B. Lugtenberg, Enrichment for enhanced competitive root tip colonizers selects for a new class of biocontrol bacteria, *Environ Microbiol*, 7, 1809–1817 (2005).
30 V. Nihorimbere, M. Ongena, H. Cawoy, Y. Brostaux, P. Kakana, E. Jourdan, P. Thonart, Beneficial effects of Bacillus subtilis on field-grown tomato in Burundi: Reduction of local Fusarium disease and growth promotion, *African Journal of Microbiology Research*, 4 [11], 1135–1142 (2010).

31 R. Pinton, Z. Veranini, P. Nannipieri, *The rhizosphere. Biochemistry and organic substances at the soil-plant interface.* Taylor & Francis Group, LLC New York, USA 2007.

32 A.V.D. Broek, M. Lambrecht, J. Vanderleyden, Bacterial chemotactic motility is important for the initiation of wheat root colonization by Azospirillum brasilense, *Microbiology,* 144, 2599–2606 (1998).

33 J. Yuan, N. Zhang, Q. Huang, W. Raza, R. Li, J.M. Vivanco, Q. Shen, Organic acids from root exudates of banana help root colonization of PGPR strain *Bacillus amyloliquefaciens* NJN-6, *Scientific Reports* 5, 13438 (2015).

34 B.B. Yosef, Root excretions and their environmental effects. In: *Plant Roots The Hidden Half,* Y. Waisel Eds, A. Eshel, U. Kafkafi. Marcel Dekker Inc., New York 1991.

35 H. Marschner, *Mineral Nutrition of Higher Plants 2nd Edn.* Academic Press, London, 1995.

36 G. Neumann, V. Römheld, The release of root exudates as affected by the plant physiological status. In: *The Rhizosphere: Biochemistry and Organic Substances at the Soil-Plant Interface, 2nd Edn.,* R. Pinton eds, Z. Varanini, P. Nannipieri, editors, 23–72 (2007).

37 S.M. Nadeem, M. Ahmad, Z.A. Zahir, A. Javaid, M. Ashraf, The role of mycorrhizae and plant growth promoting rhizobacteria (PGPR) in improving crop productivity under stressful environments, *Biotechnol.,* 32,429–448 (2014).

38 J. Sorensen, The rhizosphere as a habitat for soil microorganisms. In: *Modern soil microbiology,* J.D. Elsas van, J.T. Trevors, E.M.H. Wellington eds. Marcel Dekker, New York, 1997.

39 R.F. Doornbos, L.C. van Loon, P.A.H.M. Bakker, Impact of root exudates and plant defense signaling on bacterial communities in the rhizosphere, *A Rev. Agron. Sustain. Dev.* 32, 227–243 (2012).

40 B.R. Glick, D.M. Penrose, J.P. Li, A model for the lowering of plant ethylene concentrations by plant growth-promoting bacteria, *J Theor Biol.* 190, 63–68 (1998).

41 A.A. Belimov, V.I. Safronova, T.A. Sergeyeva, T.N. Egorova, V.A. Matveyeva, V.E. Tsyganov, A.Y. Borisov, I.A. Tikhonovich, C. Kluge, A. Preisfeld, K.-J. Dietz, V.V. Stepanok, Characterisation of plant growth-promoting rhizobacteria isolated from polluted soils and containing 1-aminocyclopropane-1-carboxylate deaminase, *Canadian Journal of Microbiology,* 47, 642–652 (2001).

42 S. Mayak, T. Tirosh, B.R. Glick, Plant growth-promoting bacteria confer resistance in tomato plants to salt stress, *Plant Physiol. Biochem.,* 42, 565–572 (2004).

43 Z.A. Zahir, A. Munir, H.N. Asghar, B. Shaharoona, M. Arshad, Effectiveness of rhizobacteria containing ACC-deaminase for growth promotion of pea (*Pisum sativum*) under drought conditions, *J. Microbiol. Biotechnol.,* 18, 958–963 (2008).

44 V. Sandhya, S.K.Z. Ali, M. Grover, G. Reddy, B. Venkateswarlu, Alleviation of drought stress effects in sunflower seedlings by the exopolysaccharides producing Pseudomonas putida strain GAP-P45, *Biol. Fertil. Soils,* 46, 17–26 (2009).

45 B.R. Glick, *Plant Growth-Promoting Bacteria: Mechanisms and Applications,* Hindawi Publishing Corporation, Scientifica, 2012.

46 M. Ashraf, S.H. Berge, O.T. Mahmood, Inoculating wheat seedling with exopolysaccharide-producing bacteria restricts sodium uptake and stimulates plant growth under salt stress, *Biol Fertil Soils,* 40, 157–162 (2004).

47 J.F. Ma, P.R. Ryan, E. Delhaize, Aluminum tolerance in plants and the complexing role of organic acids, *Trends Plant Sci.,* 6, 273–278 (2001).

48 F.D. Dakora, D.A. Phillips, Root exudates as mediators of mineral acquisition in low-nutrient environments, *Plant Soil.*, 245, 35–47 (2002).

49 A.W. Qurashi, A.S. Sabri, Bacterial exopolysaccharide and biofilm formation stimulate chickpea growth and soil aggregation under salt stress, *Brazilian Journal of Microbiology,* 43 [3], 1183–1191 (2012).

50 B.R. Glick, B. Todorovic, J. Czarny, Z. Cheng, J. Duan, B. McConkey, Promotion of Plant Growth by Bacterial ACC Deaminase, *Critical Rev Plant Sci.,* 26, 227–242 (2007).

51 P.J. Zapata, M. Serrano, M.T. Pretel, A. Amoros, M.A. Botella, Changes in ethylene evolution and polyamine profiles of seedlings of nine cultivars of *Lactuca sativa* L. in response to salt stress during germination, *Plant Soil.,* 164, 557–563 (2003).

52 A.A. Belimov, V.I. Safranova, T. Mimura, Response of spring rape (*Brassica napus*) to inoculation with PGPR containing ACC-deaminase depends on nutrient status of plant, *Can J Microbiol.,* 48, 189–199 (2002).

53 P. Sun, Q. Tian, M. Zao, X. Dai, J. Huang, L. Li, Aluminum-induced ethylene production is associated with inhibition of root elongation in *Lotus japonicus* L., *Plant Cell Physiol,* 48, 1229–1265 (2007).

54 V. Bianciotto, A. Genre, P. Jargeat, E. Lumini, G. Becard, P. Bonfante, Vertical transmission of endobacteria in the arbuscular mycorrhizal fungus Gigaspora margarita through generation of vegetative spores, *Appl Environ Microbiol.,* 70, 3600–3608 (2004).

55 E.D. Primo, F. Ruiz, O. Masciarelli, W. Giordano, Biofilm Formation and Biosurfactant Activity in Plant-Associated Bacteria, *Bacterial Metabolites in Sustainable Agroecosystem,* 12, 337–349 (2015).

56 R. Dey, K.K. Pal, D.M. Bhatt, S.M. Chauhan, Growth promotion and yield enhancement of peanut (Aracis hypoggaea L.) by application of plant growth promoting rhizobacteria, *Microbiological Research,* 159, 371–394 (2004).

57 B.E. Ramey, M. Koutsoudis, S.B. von Bodman, C. Fuqua, Biofilm formation in plant–microbe associations, *Curr Opin Microbiol.,* 7, 602–609 (2004).

58 T. Danhorn, C. Fuqua, Biofilm formation by plant associated bacteria, *Annu Rev Microbiol.,* 61, 401–422 (2007).

59 M.R. Parsek, C. Fuqua, Biofilms 2003: emerging themes and challenges in studies of surface-associated microbial life, *Journal of Bacteriology,* 186, 4427–4440 (2004).

60 H.P. Bais, S.W. Park, T.L. Weir, R.M. Callaway, J.M. Vivanco, How plants communicate using the underground information superhighway, *Trends Plant Sci.,* 9, 26–32 (2004).

61 N.A. Whitehead, A.M. Barnard, H. Slater, N.J. Simpson, G.P. Salmond, Quorum-sensing in Gram-negative bacteria, *FEMS Microbiol Rev.,* 25, 365–404 (2001).

62 L.G. Rahme, E.J. Stevens, S.F. Wolfort, J. Shao, R.G. Tompkins, F.M. Ausubel, Common virulence factors for bacterial pathogenicity in plants and animals, *Science,* 286, 1899–1902 (1995).

63 A.K. Jha, H.P. Bais, J.M. Vivanco, *Enterococcus faecalis* uses mammalian virulencerelated factors to exhibit potent pathogenicity in the *Arabidopsis thaliana* plant, *Model. Infect. Immun.,* 73, 464–475 (2005).

64 A, Eberhardt, A.L. Burlingame, C. Eberhardt, G.L. Kenyon, K.H. Nealson, N.J. Oppenheimer, Structural identification of autoinducer of Phytobacterium fischeri luciferase, *Biochemistry,* 20, 2444–2449 (1981).

65 S.B. Von Bodman, W.D. Bauer, D.L. Coplin, Quorum sensing in plant-pathogenic bacteria, *Annu Rev Phytopathol.,* 41, 455–482 (2003).

66 S. Swift, M.J. Lynch, L. Fish, D.F. Kirke, J.M. Tomás, G.S.A.B. Stewart, Quorum sensing-dependent regulation and blockade of exoprotease production in *Aeromonas hydrophila*, Infect, *Immun.*, 67, 5192–5199 (1999).
67 M. Manefield, R. de Nys, N. Kumar, R. Read, M. Givskov, Evidence that halogenated furanones from *Delisea pulchra* inhibit acylated homoserine lactone (AHL)-mediated gene expression by displacing the AHL signal from its receptor protein, *Microbiology*, 145, 283–291 (1999).
68 M. Teplitski, J.B. Robinson, W.D. Bauer, Plants secrete substances that mimic bacterial N-acyl homoserine lactone signal activities and affect population density dependent behaviors in associated bacteria, *Mol. Plant–Microbe Interact.*, 13, 637–648 (2000).
69 V. Rosemeyeer, J. Michiels, C. Verreth, J. Vanderlryden, luxI and luxR- homologous genes of Rhizobium etli CNPAF512 contribute to synthesis of autoinducer molecules and nodulation of Phaseolus vulgaris, *J. Bacteriol.*, 180, 815–821 (1998).
70 R. Daniels, D.E. De Vos, J. Desair, G. Raedschelders, E. Luyten, V. Rosemeyer, C. Verreth, E. Schoeters, J. Vanderleyden, J. Michiels, The cin quorum sensing locus of Rhizobium etli CN-PAF512 affects growth and symbiotic nitrogen fixation, *J. Biol. Chem.*, 277, 462–468 (2002).
71 Q. Fatima, M. Zahin, M. S. A. Khan, I. Ahmad, Modulation of quorum sensing controlled behaviour of bacteria by growing seedling, seed and seedling extracts of leguminous plants, *Indian J. Microbiol.*, 50, 238–242 (2010).
72 C. Murrell, E. Gerber, C. Krebs, M. Parepa, U. Schaffner, O. Bossdorf, Invasive knotweed affects native plants through allelopathy, *American Journal of Botany*, 98, 38–43 (2011).
73 P. Dundek, L. Holik, T. Rohlik, V. Vranova, K. Rejsek, P. Formanek, Methods of plant root exudates analysis: a review, *Acta univ. agric. et silvic. Mendel. Brun.*, 3, 241–246 (2011).
74 A.M. Szmigielska, K.C.J. Van Rees, G. Cieslinski, P.M. Huang, D.R. Knott, Determination of low molecular weight dicarboxylic acids in root exudates by gas chromatography, *J. Agric. Food Chem.*, 44, 956–959 (1995).
75 H. Gagnon, S. Tahara, R. Ibrahim, Biosynthesis, accumulation and secretion of isoflavonoids during germination and development of white lupin (*Lupinus albus* L.), *Journal of Experimental Botany*, 46, 609–613 (1995).
76 A, Chaboud, M. Rougier, Effect of Root Density in Incubation Medium on Root Exudate Composition of Axenic Maize Seedlings, *J. Plant Physiol.*, 137, 602–606 (1991).

11

Biochemical and Molecular Mechanisms in Biofilm Formation of Plant-Associated Bacteria

Alwar Ramanujam Padmavathi[1,2], *Dhamodharan Bakkiyaraj*[1,3] *and Shunmugiah Karutha Pandian*[1]

[1] *Department of Biotechnology, Alagappa University, Karaikudi, India*
[2] *Nanotec-PSU Center of Excellence on Drug Delivery System and Department of Pharmaceutical Technology, Faculty of Pharmaceutical Sciences, Prince of Songkla University, Hat Yai, Songkhla, Thailand*
[3] *Department of Microbiology, Faculty of Science, Prince of Songkla University, Hat Yai, Songkhla, Thailand*

11.1 Introduction

Biofilm is an intricate mass of microbial cells enveloped by the extracellular polymeric substances (EPSs). This structure provides resistance to microbial cells towards unfavorable conditions. Biofilms play a crucial role in aiding the pathogenesis of microbes. Biofilm life cycle encompasses various developmental stages that include microbial attachment, irreversible adsorption to the substratum, microcolony formation, maturation, and dispersion. Biofilms are instigated by single species and transform into multispecies that strengthen the biofilm ambience and facilitate interspecies communication and transfer of genetic materials. Several biotic and abiotic factors influence each stages of biofilm formation. Among them, the pathogen's genetic mechanism plays imperative role.

Biofilms have been shown to be formed by various bacterial and fungal pathogens in clinical settings. The impact of biofilms in the field of agriculture is vast but an underrated thing. Recent studies have started to unveil the biofilm-forming potentials of numerous phytopathogens. Though the ability of phytopathogens to forms biofilms is clear, the mechanisms by which the plant pathogens institute biofilm mode of growth and subsequent infections are less studied.

Quorum sensing (QS), the cell–cell communication, is one such mechanism that plays pivotal role in biofilm formation of plant bacteria. QS, a density-dependent communication system, is regulated by small signaling molecules known as autoinducers. *Burkholderia* spp. produces three signaling molecules—namely, N-octanoyl HSL (C8-HSL), N-3-hydroxy-decanoyl HSL (3OHC10-HSL), and N-3-hydroxy-octanoyl HSL (3OHC8-HSL) [1]. Phytopathogens secrete several extracellular enzymes to degrade the plant cell wall and invade hosts, which are also controlled by QS mechanism.

Basic understanding of cellular processes involved in biofilm formation will enable readers to identify an efficient control mechanism to prevent or control pathogenic biofilms in agriculture to overcome the loss of plant productivity. This chapter discusses the molecular and biochemical mechanisms of biofilm formation employed by key plant pathogens.

11.2 Plant-Associated Bacteria

Plants are the fundamental source of life on the Earth, providing energy to human and animals. It occupies a pivotal role in food chain and maintenance of a functional ecosystem. Apart from macroconsumers, they provide ample energy for microconsumers, such as bacteria, fungi, and viruses. The interaction between plants and microbes can be either beneficial or detrimental to plant systems. Overall plant productivity is controlled by various biotic factors that include bacteria, fungi, viruses, oomycetes, phytoplasmas, protozoans, and nematodes. Among them, bacteria play an important role in controlling plant productivity. They actively take part in nutrient cycles and bring back the energy to the environment, which is the major base for plants. They follow symbiotic relationships with plants and promote plant growth [2]. They help in several key functions such as nitrogen fixation and can be utilized as probiotics or biocontrol agents and biofertilizers to control pathogenic infections and growth promoters [3, 4].

Beside these benefits, they cause various diseases in plants that result in wilt, rot, leaf curl, leaf spot, canker, blights, and crown galls [5]. Even though fungal genus such as *Plasmopara, Colletotrichum, Plasmodiophora, Fusarium,* and *Taphrina* also contribute to numerous plant diseases, bacterial infections garner much attention due to the huge economic loss incurred each year [6]. Bacteria from the genus *Agrobacterium, Xanthomonas, Xylella, Erwinia, Pseudomonas, Pectobacterium, Burkholderia, Ralstonia,* and *Dickeya* cause devastating plant diseases that curb economic growth. Plant bacterial pathogens follow various virulence mechanisms to invade plant parts and establish severe infections. They secrete several degrading enzymes that digest the plant cell wall and other structures and enable the pathogen to infect the plant. Biofilm is a pivotal virulence mechanism of plant pathogens that aids in initial colonization and later results in prolonged survival in the host. Most of the bacterial pathogens adapt biofilm mode of life cycle to establish a serious infection.

11.3 Biofilms and Plant Pathogens

Plant-associated bacteria forms biofilm for several reasons including attachment to the host plant, to prevent washing off, to protect from the harsh environments and chemicals, and so on. Plants possess vast area that provides ample opportunities for bacterial colonization. Bacteria form biofilms on various plant parts such as vasculature, rhizosphere, xylem vessels, leaves, and meristem [5, 7]. Bacterial biofilms are even found in seeds and sprouts [8, 9]. *Pseudomonas* spp. causing halo blight, brown spot, holcus spot, root rot, brown rot and bacterial speck form biofilms on tomato, clove, bean, and corn [10–17]. Fire blight and bacterial rot causing *Erwinia* spp. form biofilms on potato and members of *Rosaceae* family [18–20]. Biofilm is responsible for virulence and pathogenicity of *Spiroplasma* which causes corn stunt [21]. *Acidovorax* is an important pathogen causing bacterial blotch in which the biofilm formation was identified as an important virulence factor required for pathogenicity [22]. *Burkholderia* spp. causing sour skin of onion was also found to be associated with biofilms [23]. *Ralstonia solanacearum* forms biofilm that causes lethal wilt on tomato [24]. *Pectobacterium* form biofilms to cause soft rot of fruits and vegetables [25]. *Pantoea* spp. form biofilms that cause white spot disease, blight, and wilt on maize [26, 27]. *Xanthomonas* spp. form biofilms to cause citrus canker [28–30]. Biofilm formation and pathogenicity of plant bacterial pathogens are inseparable traits. Understanding the molecular mechanisms involved in biofilm formation will give a better idea about the pathogenesis of bacterial infections.

11.4 Molecular and Biochemical Mechanisms Involved in Biofilm Formation

Biofilm formation is a process orchestrated by several genetic and biochemical factors. Even single mutation in a gene affects the whole processes and hinders biofilm formation and impairs the virulence of pathogens.

11.4.1 *Pseudomonas*

Pseudomonas causes bacterial apical necrosis. *P. syringae* produces 3-oxohexanoyl-DL-homoserine lactone (3O-HSL), which controls biofilm formation and colony morphology [31]. *P. syringae* pv. *tabaci* possesses a homolog of 3-oxoacyl-(acyl carrier protein) synthase III which is involved in AHL synthesis and fatty acid synthesis. Mutant strains exhibited reduced EPS production and were devoid of AHL production, proving its indispensable role in biofilm formation [32]. *FabAB* operon consists of *fabA* and *fabB*, which encodes beta-hydroxyacyl-acyl carrier protein dehydratase and beta-ketoacyl-acyl carrier protein synthase I, respectively. *FabAB* operon regulates the synthesis of unsaturated fatty acid that promote biofilm formation in *P. aeruginosa* [33, 34].

Cyclic-di-GMP (bis-(3′–5′)-cyclic dimeric guanosine monophosphate or c-di-GMP) is an ubiquitous secondary messenger in bacteria that regulates biofilm formation and virulence. It plays a key role in deciding motile and biofilm life cycle of bacteria. Elevated levels of c-di-GMP enhanced biofilm formation in *P. savastanoi* pv. *savastanoi* that resulted in larger knots on olive plants with reduced necrosis [35]. Transposon mutant of c-di-GMP was identified from *P. savastanoi* pv. *savastanoi* and found to have similarity with *bifA* that encodes BifA, active phosphodiesterase. Elevated intracellular c-di-GMP levels have correlation with the virulence, and the connection between phosphodiesterase and diseases in olives was well established. Loss of *bifA* resulted in reduced motility, which reciprocally regulated biofilm formation [36]. Diguanylate cyclases (DGC) and phosphodiesterases (PDE) regulate the c-di-GMP levels. DGC increases EPS production in *P. syringae* and decreases major pathogen-associated molecular pattern (PAMP) flagellin [37]. *Chp8* present in *hrp* regulon, regulated by Hrp (hypersensitive response and pathogenicity) protein, exhibited c-di-GMP synthase activity in *P. syringae* and is upregulated by plant signals. Chp8 downregulates flagellin production and increases EPS production, besides determining the lifestyle [37]. Wsp operon consists of genes responsible for biofilm formation viz. N-terminal phosphor receiver, which receives signals and a C-terminal DGC that controls the levels of c-di-GMP [38]. *galU* that encodes for a uridine diphosphate-glucose pyrophosphorylase plays important role in carbohydrate synthesis. Even though mutation in *galU* resulted in a complete loss of functions of type 3 secretion system (T3SS), loss of motility and impaired biofilm formation, and failed to produced necrosis of the host leafs, thus explaining the role of GalU in biofilm formation and pathogenesis of *P. syringae* [39]. GacA/GacS is a master two-component regulatory system of *Pseudomonas* sp. It regulates the expression of *rsmX, rsmY,* and *rsmZ,* which competitively antagonize RsmA of this two-component system (TCS). *rsmA* negatively regulates biofilm formation and alginate production. Overexpression of rsmA resulted in diminished biofilm formation and alginate production in *P. fluorescens* [40]. Flagellin (*fliC*) and peptidoglycan hydrolase (*flgJ*) were found to be important for biofilm formation of *P. fluorescens.* Glycosyl transferase (*fgtA1*), cephalosporin hydroxylase (*flhH*) and chemotaxis protein (*cheW*) mutants were also found to be deficient for biofilms [41]. Nudixpyrophosphatase (*nudC*) hydrolyses NADH, maintains intracellular redox balance and is important for vital cellular functions. Lack of *nudC* in *P. syringae* leads to decreased cellular fitness and affected growth, motility and biofilm formation, whereas deletion of *nudC* did not affect the biofilm formation of *P. aeruginosa* explaining the variation among different species of same genus [42]. ATP binding cassette (ABC) transporter and associated cell surface protein are required for biofilm formation by *P. fluorescens.* Mutation in *lapB, lapC,* and *lapE* that encode outer membrane components LapB, LapC, and LapE, respectively, affected irreversible attachment and thickness of biofilm [43].

PelF is a putative glycosyltransferase and PelG is a component of polysaccharide transporter (PST) family. *Pel* cluster consists of seven genes namely *pelA, pelB, pelC, pelD, pelE, pelF,* and *pelG* which are involved in pellicle formation in *P. aeruginosa. pelA, pelC, pelD, pelE* and *pelF* are involved in carbohydrate processing and produce glucose rich matrix. *Pel* mutation produced nonadherent cells *in vitro,* and affected the initial attachment and biofilm formation. All *pel* genes individually contribute to biofilm formation and deletion of any of them produced non-biofilm formers [44]. Alginate is an important factor of bacterial biofilms, which is composed of β (1–4) linked, unbranched chains. Alginate (*alg*) stabilizes biofilm structure and hydrates the biofilm in water-limiting conditions in addition to providing stress tolerance to *P. putida* biofilms [45]. Alginate lyase (*algL*) cleaves alginate and regulates later stages of biofilm formation in *P. syringae* pv. *phaseolicola.* It cleaves the mucoid biofilms and releases the cells leading to dispersion and pave way for nutrient in flow as well [46]. *hrpM* (regulate synthesis of periplasmic glucans) and *algD* are involved in biofilm formation of *P. syringae* [47]. Low temperature (18°C) exerts an oxidative stress in *P. syringae* that represses alginate production, which contributes to biofilm formation [48]. Cellulose acts as a major matrix component of *P. asplenii, P. corrugate, P. fluorescens, P. marginalis, P. putida, P. savastanoi,* and *P. syringae* [49]. Cellulose production is a key determinant for biofilm mode of life. Overproduction of cellulose by *P. syringae* leads to adaptation of biofilm life cycle, whereas deficient cellulose producers tend to follow planktonic state [50]. Levan along with alginate stabilize the EPS and contribute to biofilm formation of *P. syringae.* Levansucrase (*lsc*) is an enzyme that catalyses sucrose and forms levan. Concavalin A was unable to bind to the EPS of levan mutant and the strain grown in the absence of sucrose [51].

Pyoverdine is a siderophore found in *Pseudomonas* spp., which is involved in iron uptake. Three pyoverdine synthesis genes—*pvdJ, pvdL* and *fpvA*—were identified in *P. syringae* pv. *tabaci,* which encode peptide synthetase III, chemosphere synthetase and ferripyoverdine receptor, respectively. These pyoverdine mutants produced scanty EPS and the biofilms were also easily disrupted [52].

11.4.2 *Xanthomonas*

Xanthomonas spp. are Gram-negative bacteria that commonly cause diseases in plants, including blight disease, leaf streak, black rot, and citrus cankers, thus resulting in heavy economic loss. *Xanthomonas oryzae* pv. *oryzae*, is the causative agent of rice blight disease, resulting in huge economic loss each year. It enters through wounds and water pores of rice leaf and invades xylem. It produces cis-11-methyl-2-dodecenoic acid, a diffusible signal factor (DSF) required by the QS system to control virulence. DSF regulates the attachment and biofilm formation in *X. Oryzae* and negatively regulates its motility [53]. DSF-regulated extracellular protein also controls the biofilm formation in *X. oryzae* pv. *oryzicola*. Mutation in *XOC_0319* that encodes Ax21-like protein, a small-protein-type QS signal, resulted in reduced biofilm formation explaining the roles of DSF and DSF-regulated proteins in the biofilm formation of *X. oryzae* [54]. In several pathogenic bacteria, the biofilm life cycle was positively regulated by high levels of c-di-GMP. *filP* encodes for FilP, a c-di-GMP regulator, which binds with c-di-GMP and regulates bacterial virulence. Deletion of *filP* affects the virulence and T3SS, whereas EPS and biofilm were not affected [55]. *X. oryzae* pv. *oryzicola* causes leaf streak on rice. *rpfG* encodes a HD-GYP domain protein RpfG, that exhibits PDE activity and is important for the virulence of *X. oryzae* pv. *oryzicola*. RpfG regulator acts with RpfC sensor kinase in DSF signal transduction. RpfG is essential for the full virulence of *X. oryzae* pv. *oryzicola*, but it negatively regulates biofilm formation. Deletion of *rpfG* yielded Δ*rpfG* with doubled biofilm formation, whereas EPS production was reduced [56]. ThiG, a component of thiamine biosynthesis pathway, regulates biofilm formation. Mutation of *thiG* downregulates *rpfG* and *rpfC* and induces biofilm [57]. Flagellin is important for the formation of a functional flagellum, a prerequisite for motility in pathogens. Acetyltransferase, which is involved in flagellin glycosylation of *X. oryzae* pv. *oryzae*, regulate the biofilm formation. Mutation in a putative acetyltransferase (PXO_00987) of flagellin regulon afflicted biofilm formation [58]. Adhesins (XadM) play an important role in early attachment of pathogens on a solid substratum to enable them to form a confluent biofilm in the longer run. Deletion of *xadM* impaired attachment and biofilm formation of *X. oryzae* pv. *oryzae* [59].

X. campestris pv. *campestris* causes black rot of cruciferous plant. Cell–cell signaling and DSF govern biofilm formation in *X. campestris* and overproduction of DSF prompt biofilm dispersion. *rpfF* (DSF synthase) regulates DSF synthesis, whereas *rpfG* and *rpfC* are involved in DSF perception [60]. DSF system governs the dispersion of *X. campestris* biofilm [61], and the deletion of *rpf* gene cluster impairs DSF production and hence *rpfF*, *rpfG*, *rpfC*, and *rpfGHC* mutants form aggregates encased with xanthan matrix, which are devoid of dispersion signals [62]. Synthesis of xanthan, an extracellular polysaccharide, enables *X. campestris* pv. *campestris* to form aggregates and DSF-controlled endo-β-1,4-mannanase stimulates the dispersion of aggregates, to switch back to planktonic state from biofilm lifestyle [62, 63].

Transcriptional analysis of Δ*rpfF*Δ*manA* unveiled a gene locus *xagABC* encodes for glycosyl transferase that promotes exopolysaccharide production and biofilm formation. DSF negatively regulates *xagABC*, and ManA disperses the biofilms formed by XagABC [64]. Post-transcriptional regulator, RsmA also negatively regulates *X. campestris* biofilms by the expression of DGCs that synthesize c-di-GMP. Δ*rsmA* upregulated *xag* and produced increased levels of c-di-GMP and reduced ManA due to the downregulation of *manA* [65]. Clp, a putative nucleotide receptor protein, encoded by *clp* regulates the expression of DSF and deletion of *clp* produced opposite effects of Δ*rsmA* [65, 66].

11.4.3 *Erwinia*

Erwinia amylovora causes fire blight disease in members of *Rosaceae* family and creates huge economic hurdle. *E. amylovora* chiefly forms biofilms to colonize xylem vessels and abolish water transport that leads to systemic infections. Similar to *Xanthomonas* and *Dickeya* species, the biofilm formation in *E. amylovora* is regulated by c-di-GMP level. *Edc* (Erwinia diguanylate cyclase) genes, *edcA, edcC* and *edcE* encode the active DGCs and positively regulate c-di-GMP. c-di-GMP directly influences amylovoran (*ams*) exopolysaccharide production in *E. amylovora* besides downregulating flagellar motility [67]. Amylovoran and levan (*lsc*) production increase and strengthen the structured biofilm formation and promote occlusion of plant vessels [68, 69]. Besides promoting biofilm formation and colonization, c-di-GMP represses *hrpA* that forms a part of T3SS and reduces the virulence. Among the three DGCs that synthesize c-di-GMP, *edcC* and *edcE* regulate biofilm formation, whereas deletion of *edcA* does not alter the biofilms formed by *E. amylovora in vitro* [67]. Bacterial cell structures aid in attachment and play an indispensable role in biofilm formation. Type I fimbriae (*fim* and *fimD*) is involved in initial attachment, and type IV pilus (*hof* and *hofC*) is pivotal in later stages of attachment and maturation of biofilms. Curli (encoded by *clr* gene) plays an important role in cell adhesion, invasion, and host responses. Mutation in *clr* drastically reduced the biofilms formed by *E. amylovora* [70].

Hfq, a global small RNA chaperone, binds to sRNAs and post-transcriptionally regulates the expression of various traits of *E. amylovora*. Δ*hfq* resulted in altered biofilm formation with fewer cell aggregates and inhibited amylovoran production, explaining its pivotal role in maturation as well as limiting initial attachment [71, 72]. *HrpL* is an alternative sigma factor that regulates T3SS via *hrp* promoter elements. *HrpL* is regulated by *HrpX/HrpY* TCS as well as *RsmA*/*rsmB* [73, 74]. H*rpL* negatively regulates *nlpI*, which encodes a lipoprotein with tetra-tricopeptide repeats, which is homologous to *Escherichia coli nlpI*, a negative regulator of extracellular DNA (eDNA). Δ*nlpI* exhibits enhanced biofilm formation, which proves the role of eDNA in biofilm formation [75]. QS system plays a crucial role in biofilm formation of *E. carotovora*, a causative agent of soft rot disease in various plants. 3-oxo-HSL and C6-HSL are the autoinducers synthesized by *ExpI/R* QS system that control biofilm formation in *E. carotovora* [76].

11.4.4 *Ralstonia*

Ralstonia solanacearum is the causative agent of bacterial wilt disease of economically important crops, which grossly affects more than 200 plant species. Several virulence factors such as plant cell wall–degrading enzymes, exopolysaccharides, T3SS, and biofilm formation help in pathogenesis and disease progression of the pathogen. The genes *lecM* and *lecX* encode for carbohydrate-binding lectins, deletion of which affected the attachment to tomato roots. *LecM* is regulated by the SolI/R QS system and contributes to host colonization and xylem invasion [77]. Colonization of *R. solanacearum* at intercellular spaces is regulated by a lectin encoded by *lecM* [78].

Type IV pili have also been found essential for the attachment, biofilm formation, and full pathogenicity of *R. solanacearum*. Deletion of six genes responsible for the biogenesis of Tfp resulted in reduced virulence on potato [79]. Disruption of O-oligosaccharyl transferases that O-glycosylate proteins like type IV pilin, afflicts biofilm formation and results in reduced infection of tomato plants [80]. Deletion of *pilA* and *pilQ* that encode type IV pili also abrogated biofilm formation [24, 81]. Flagella, regulated by FlhDC, play an important role in early attachment and invasion of *R. solanacearum*. MotN negatively regulated flagellar motility, and the *MotN* mutant was hyper-flagellated and produced weaker biofilms, which substantiate its role in early attachment, an initial step in a confluent biofilm formation [82]. Aerotaxis genes *aer1* and *aer2* are required for biofilm formation on tomato plants by sensing high-energy regions through electron transport chain [24].

Lipopolysaccharide (LPS) is also a chief determinant of virulence and biofilm formation in this pathogen [83]. T6SS regulates Hcp and biofilm formation in *R. solanacearum*. Mutation in *tssB*, which encodes TssB of T6SS, revokes biofilm formation and displays reduced virulence on tomato plants [84].

11.4.5 *Pectobacterium carotovorum*

P. carotovorum subsp. *carotovorum* causes soft rot in fresh vegetables and creates considerable economic burden. It secretes several extracellular enzymes such as protease, pectate lyase, xylanase, cellulase, and polygalacturonase, which degrade the plant cell wall structures and causing soft rot. QS system (ExpI/R) governs the virulence and pathogenesis in this pathogen. Biofilm formation is an important virulence factor, which is governed by *expI*, *expR*, and *qseC*. Deletion of *expI*, *expR*, and *qseC* drastically reduced biofilm formation and yielded as little as 20 percent of biofilms compared to that of the wild type [25]. *P. carotovorum* subsp. *brasiliense* formed biofilm-like aggregates inside xylem vessels of potato [85]. Mutation of *expI* affected AHL and plant cell wall–degrading enzymes production. The mutant-formed cell aggregates at intracellular spaces and could not invade xylem and occlude the vessels [86].

PleD mutant independent of phosphorylation produced elevated c-di-GMP levels, which induced biofilm formation in *P. atrosepticum*, which causes blackleg of potato [87]. Similar to other pathogens, elevated c-di-GMP levels positively regulated biofilm formation in *P. atrosepticum*. C-di-GMP induced the poly-β-1,6-N-acetyl-glucosamine dependent flocculation while downregulating the motility and virulence [88]. It releases rhamnogalacturonan I into the xylem, which acts as a matrix for bacterial *emboli*—a multicellular structure of bacteria [89].

11.4.6 *Xylella fastidiosa*

Xylella fastidiosa, belonging to *Xanthomonadaceae* family, is an important Gram-negative phytopathogen colonizing the xylem vessels of economically important crops, resulting in huge economic losses every year. It causes varied plant diseases such as Pierce's disease, leaf scorch, alfalfa dwarf, and citrus variegated chlorosis [90]. It is transmitted by sharpshooter insects, which feed on xylem, thus introducing the pathogens into xylem vessels, in turn facilitating the bacterium to colonize and form biofilms, which lead to xylem blocking and nutritional deprivation in the infected plants. *X. fastidiosa* biofilms block the water transport of xylem vessels. Pili (*pil*A/*fim*A and *pil*C), *slp* and *psp*A adhesins play crucial role in initial attachment on solid substratum and biofilm initiation [91]. This was further confirmed by *fimA/pilO* double mutation that completely obliterated biofilm formation [92]. *gum* operon (*gum*BCDEFHJKM) was found in *X. fastidiosa*, consisting of nine genes that synthesize polysaccharide with tetrasaccharide repeating units similar to xanthan gum in *Xanthomonas campestris* [93]. Deletion of *gum*B and *gum*F did not affect EPS production, instead affect biofilm formation [94]. EPS knock out mutants Δ*gumD* and Δ*gumH* resulted in lack of EPS production, biofilm formation, and virulence in grapevines, and were also found deficient in vector transmission, explaining the crucial role of EPS in biofilm formation, pathogenesis, and disease transmission. *X. fastidiosa* EPS is made of acetyl-D-mannose and D-glucuronic acid residues. The *gumH* mutant produced unpolymerized EPS lacking these residues [95].

The biofilm stage possesses high virulence and provides resistance to toxic substances. Totally 456 proteins were reported to be expressed during biofilm state [96]. *rpfF* gene encodes diffusible signal molecule in *X. fastidiosa* which is responsible for cell–cell signaling and interaction with insects and plants [97]. *popP* is required for pathogenicity of *X. fastidiosa* and deletion of *popP* resulted in hindered cell aggregation and cell-matrix adherence [98]. GacA regulates several regulating factors such as *xadA*, *hsf*, *cvaC* and *gumC* which control attachment and biofilm formation of *X. fastidiosa*. *gacA* mutation hindered biofilm formation and *gacA* mutants were unable to infect grapevines [99]. *algG* mutant exhibited lack of cell–cell aggregation and biofilm formation in addition to reduced virulence in grapevines similar to *gacA* mutants [100]. PilA2 and PilC, which form type IV pili, were also found to be important for biofilm formation and were upregulated during initial attachment and dispersion. XadA1 is essential in all the stages of biofilm, whereas XadA2 is needed in biofilm maturation [101]. Mutation of *xhpT*, a response regulator, impaired the surface attachment, cell–cell aggregation, biofilm formation, and virulence in grapevine despite of increased EPS production [102]. BigR (Biofilm growth associated repressor) is a winged-helix homodimeric repressor of ArsR family metal sensor. It regulates the *bigR* operon involved in biofilm formation, which detoxify hydrogen sulfide and promote growth under oxygen limiting conditions [103]. *pglA* codes for a functional polygalacturonase which is crucial for degrading the host membranes and successful colonization [93].

In vitro biofilm studies have revealed that the resilient biofilms were formed in the presence of metals such as zinc, manganese, calcium, potassium and copper [104, 105]. *X. fastidiosa* utilizes the ionome of infected plants and accumulates them into biofilm [106]. Calcium promoted biofilm formation of *X. fastidiosa*, while the removal of intracellular and extracellular calcium reduced the biofilm formation. But, the planktonic cells remained unaffected and their growth was independent of calcium level. *fimA pilO* double mutant was also unaffected by calcium and unable to form biofilm even in the presence of calcium [107].

11.4.7 *Agrobacterium tumefaciens*

Agrobacterium spp are Gram-negative bacteria that belong to *Rhizobiaceae* family. *A. tumefaciens* (AT) is the most predominant pathogen from this genus, causing crown galls. It possesses Ti plasmid that transfers tumor-inducing genes to the plants and leads to tumor formation, disease progression, and loss of plant productivity. AT initially attached to infected sites of plants and forms biofilms. Surface interactions regulated by SinR are important for biofilm maturation [108]. Various molecular mechanisms govern the biofilm formation in this tumor-inducing bacterium [109]. It possesses a CtrA master regulator, which controls cell cycle and division and contributes to multicellularity. CtrA is regulated by PleC/DivJ-DivK phosphorelays. AT also possesses sensor kinases encoded by *pdhS1* and *pdhS2* that positively regulate biofilm formation [110]. The AT genome contains a cluster of genes *ctpABCDEFGHI* that encode pili, which play a pivotal role in initial attachment and biofilm formation [111]. ExoR is a periplasmic regulator regulated through ChvG-ChvI acid-sensing TCS. ExoR promotes biofilm formation and represses succinoglycan exopolysaccharide production [112, 113].

Iron limitation upregulates *exoQ, exoF, exoY, exoP, exoN, exoM, exoL, exoK, exoH, exoT, exoW,* and *exoU* genes required for succinoglycan synthesis. Iron acts as a cofactor for several enzymes and regulates essential cellular processes. In addition to the physiological processes, it influences the attachment and biofilm formation of *A. tumefaciens*. Iron limitation severely impairs the biofilm-forming ability of this debilitating tumor-forming pathogen [114]. Cellulose is one of the polysaccharides that forms EPS of AT. *CelA* and *celB* genes that encode cellulose and glucans positively regulate biofilm formation [115]. AT possesses phosphatidylcholine (PC), which negatively regulates biofilm formation [116, 117]. Phosphorous limitation activates the PhoR-PhoB two-component regulatory system and induces biofilm formation [118]. QS also accelerates biofilm formation mediated by AHL signals [119].

11.4.8 *Dickeya*

Dickeya belongs to the *Enterobacteriaceae* family and consists of pathogens causing wilt and rots afflicting herbaceous plants. *Dickeya dadantii* (formerly *Erwinia chrysanthemi)* secretes pectin-degrading exoenzymes that can degrade the plant cell wall. *D. dadantii* utilizes T3SS to suppress the host immune system. GacS/GacA two-component signal transduction system plays important role in *D. dadantii* pathogenesis and virulence. Deletion of *gacA* delayed biofilm-pellicle formation *in vitro*. *gacA* deletion afflicted pectate lyase (*pelD* and *pelL*), cellulose, and protease production [120]. Deletion of *slyA* transcription factor from SlyA/MarR family in *D. zeae*, causative agent of root rot, resulted in reduced biofilm formation and loss of pathogenicity in rice [121].

D. dadantii forms surface–air–liquid interface biofilm (SAL biofilm) and a biofilm at the air–liquid interface (pellicle). These biofilms consist of cellulose nanofibers. Cellulose is required for pellicle-biofilm formation in *D. dadantii* and provides resistance to chlorine treatments. *bcs* operon (*bcsABCD*) encodes cellulose biosynthesis genes [122]. *bcsA, bcsC* mutants produced fragile biofilms that are resistant to cellulase. On the other hand, *ardA* mutants produced biofilm similar to wild types, which was partially degraded by cellulase. This explains the composition of *D. dadantii* biofilm as partially composed of cellulose and some other material, which is a cellulase-resistant polymer. Sigma factors *fliA* and *HrpL* involved in flagella production and T3SS, respectively, are required for pellicle formation. Deletion of *fliA* and *hrpL* produced SAL biofilms, but their structure varied [123]. *fliA* mutation affected pellicle formation and *fliA* mutant did not produce pellicles [124]. Nucleoid-associated protein (Fis) master regulators repress the expression of *bcs* operon. *fis* mutants produced thicker cell aggregates in liquid media containing cellulose [122].

GGDEF/EAL domain containing PDEs have the potential to regulate c-di-GMP levels in *D. dadantii*. EcpB (GGDEF and EAL), and EcpC (EAL domain) have shown to regulate various virulence factors of this pathogen and negatively affect biofilm formation. Deletion of PDE coding genes *ecpB* and *ecpC* resulted in increased c-di-GMP levels and enhanced biofilm formation up to fivefold that of the wild type. Double mutation resulted in tenfold increase in biofilms. *EcpB* and *EcpC* were found to repress *hrp* that regulate T3SS and virulence [125]. *HrpX/HrpY* is a TCS, encoded by *hrp* cluster, in which *HrpX* and *Y* act as transmembrane sensor protein and response regulator respectively. It is encoded within T3SS gene cluster of *D. dadantii* and is required for activation of downstream genes of T3SS such as *hrpA*, *hrpN*, and *dspE* via *hrpS* and *hrpL*, which encode σ^{54}(RpoN)-enhancer binding protein and σ factor, respectively [126]. *hrpY* deletion resulted in loss of pellicle formation and enabled researchers to identify the genes controlled by arabinose-regulated HrpY. However, biofilms formed at air–liquid interface by *D. dadantii* were not affected by *hrpY* deletion. In contrast to most bacterial pathogens, AHL-mediated QS systems did not contribute to the biofilm formation in *D. dadantii* [127].

11.4.9 *Clavibacter michiganensis*

C. michiganensis is a Gram-positive actinomycetes causing bacterial wilt and canker in tomatoes and potatoes leading to huge economic loss worldwide. *In vitro* biofilm studies have revealed that *C. michiganensis* quantitatively produces more biofilm than other plant pathogens [128]. *C. michiganensis* subsp. *michiganensis* colonizes in xylem vessels of tomato, causing xylem clogging and affect water transport. Colonization pattern and biofilm formation was studied using GFP tagged strains and found that *C. michiganensis* moves acropetally to systemically colonize and reach the apical region of tomato. It firmly adheres to the spiral secondary wall, resulting in thickening of xylem vessels, which might be mediated by adhesins such as pili and fimbriae proteins. It migrates through apoplast and preferentially colonizes protoxylem but avoids metaxylem. Chromosomal *chp/tomA* pathogenicity island (PAI) and plasmid-borne virulence factors are important for colonization and formation of cellular aggregates, leading to biofilm formation [129].

11.4.10 *Bacillus subtilis*

Biofilm formation plays an indispensable role in plant-associated bacteria and is not limited to pathogens. *B. subtilis* also forms biofilms to protect the plants from other pathogenic bacterial and fungal colonizations, thus preventing diseases. It is a Gram-positive rod-shaped bacterium, found in soil and plants, which mostly plays a beneficial role in plant systems. *B. subtilis* forms biofilm and produces surfactin that protects plants from pathogenic infections [14]. *B. subtilis* biofilm formation in *Arabidopsis thaliana* was found to be induced by plant polysaccharides [130]. Plant polysaccharides act as signal to trigger the biofilm formation via kinases, which control the phosphorylation of the master regulator Spo0A. TasA, a major protein component of the extracellular matrix and EPS, is important for *in vitro* colonization and robust biofilm formation. Spo0A plays a central role in signal transduction pathways and it activates the matrix genes responsible for biofilm formation. When Spo0A-P reaches threshold level, SinI is produced that curbs the SinR repressor and activates the transcription of matrix genes. *tapA-sipW-tasA* and *epsA-O* are the two matrix gene operons that code for amyloid-like fibers and an exopolysaccharide in *B. subtilis*, which are responsible for biofilm formation and are directly controlled by SinR [131].

Plant polysaccharides act as an environmental cue and induce biofilm formation of *B. subtilis* wherein other nonplant polysaccharides failed to. Histidine kinases (KinA-KinE) sense these cues and phosphorylate Spo0A, which, in turn, facilitate biofilm formation. *kinCD* double mutant failed to form pellicle in the presence of plant polysaccharides, which reveal the importance of KinC and KinD in bacterial colonization on root. *A. thaliana* arabinogalactan is one such example that potentially induces biofilm formation by *B. subtilis* [130].

11.5 Conclusion

Most biofilm-forming plant-associated bacteria being phytopathogens have started to attract the interest of researchers. Application of antimicrobials being one way of biofilm control measure has been found effective, but in turn has had set backs such as the presence of residual antimicrobials on the food products and their ineffectiveness against bacteria that thrive inside the plants. Biofilm-forming beneficial bacterium like *Bacillus subtilis* could be a viable alternative that provides protection by acting as probiotic biocontrol agents that compete with the pathogens for space and nutrition besides producing antagonistic agents. Researchers now look to the development of more such beneficial bacteria with the ability to form biofilms that can thrive in harder environments and can protect crops and plants from pathogens. Understanding the biochemical and molecular mechanisms underlying biofilm formation in these plant-associated bacteria (both beneficial and harmful) will aid in the development of robust control measures such as target-specific biocontrol agents or antimicrobials.

References

1 B.S. Tseng, C.D Majerczyk, D. Passos da Silva, J.R. Chandler, E.P. Greenberg and M.R. Parsek, Quorum sensing influences *Burkholderia thailandensis* biofilm development and matrix production, *J. Bacteriol.*, pii, JB.00047-16 (2016).

2 S. Compant, B. Duffy, J. Nowak, C. Clement and E.A. Barka, Use of plant growth-promoting bacteria for biocontrol of plant diseases: principles, mechanisms of action, and future prospects, *Appl. Environ. Microbiol.*, 71:4951–4959 (2005).

3 G.E. Dawwam, A. Elbeltagy, H.M. Emara, I.H. Abbas and M.M. Hassan, Beneficial effect of plant growth promoting bacteria isolated from the roots of potato plant, *Ann. Agric. Sci.*, 58:195–201 (2013).

4 J.D. Flores-Felix, L.R. Silva, L.P. Rivera, M. Marcos-Garcia, P. Garcia-Fraile, E. Martinez-Molina, P.F. Mateos, E. Velazquez, P. Andrade and R. Rivas, Plants probiotics as a tool to produce highly functional fruits: the case of *Phyllobacterium* and vitamin C in strawberries, *PLoS One*, 10:e0122281 (2015).

5 A.R. Padmavathi, D. Bakkiyaraj and S.K. Pandian, Significance of biosurfactants as antibiofilm agents in eradicating phytopathogens, *Bacterial Metabolites in Sustainable Agroecosystem*, 12:319–336 (2015).

6 M. Hawes and P. Ronald, Focus: plant interactions with bacterial pathogens, *Plant Physiol.* 150:1621–1622 (2009).

7 B.E. Ramey, M. Koutsoudis, S.B. von Bodman and C. Fuqua, Biofilm formation in plant–microbe associations, *Curr. Opin. Microbiol.*, 7:602–609 (2004a).

8 W.F. Fett, Naturally occurring biofilms on alfalfa and other types of sprouts, *J. Food Prot.*, 63:625–632 (2000).

9 W.F. Fett and P.H. Cooke, Scanning electron microscopy of native biofilms on mung bean sprouts, *Can. J. Microbiol.* 49:45–50 (2003).

10 G. Surico and J.E. DeVay, Effect of syringomycin and syringotoxin produced by *Pseudomonas syringae*pv. *syringae* on structure and function of mitochondria isolated from holcus spot resistant and susceptible maize lines, *Physiol. Plant Pathol.* 21:49–53 (1982).

11 S.J. Roberts, S.J. Eden-Green, P. Jones and D.J. Ambler, *Pseudomonas syzygii*, sp. nov., the Cause of Sumatra Disease of Cloves, *Syst. Appl. Microbiol.* 13:34–43 (1990).

12 J.J. Rich and D.K. Willis, Multiple loci of *Pseudomonas syringae*pv. *syringae* are involved in pathogenicity on bean: restoration of one lesion-deficient mutant requires two tRNA genes, *J. Bacteriol.*, 179:2247–2258 (1997).

13 A.J. Gonzalez, E. Landeras and M.C. Mendoza, Pathovars of *Pseudomonas syringae* causing bacterial brown spot and halo blight in *Phaseolus vulgaris* L. are distinguishable by ribotyping, *Appl. Environ. Microbiol.*, 66:850–854 (2000).

14 H.P. Bais, R. Fall and J.M. Vivanco, Biocontrol of *Bacillus subtilis* against infection of Arabidopsis roots by *Pseudomonas syringae* is facilitated by biofilm formation and surfactin production, *Plant Physiol.*, 134:307–319 (2004).

15 T.S. Walker, H.P. Bais, E. Deziel, H.P. Schweizer, L.G. Rahme, R. Fall and J.M. Vivanco, *Pseudomonas aeruginosa*-plant root interactions. Pathogenicity, biofilm formation, and root exudation, *Plant Physiol.*, 134:320–331 (2004).

16 A. Keisa, K. Kanberga-Silina, I. Nakurte, L. Kunga and N. Rostoks, Differential disease resistance response in the barley necrotic mutant necl, *BMC Plant Biol.*, 11:66 (2011).

17 S.P. Thapa and G. Coaker, Genome sequences of two *Pseudomonas syringae*pv. *tomato* Race 1 strains, isolated from tomato fields in California, *Genome Announc.* 4, pii:e01671–15 (2016).

18 P.E. Axelrood, M. Rella and M.N.Schroth, Role of antibiosis in competition of erwinia strains in potato infection courts, *Appl. Environ. Microbiol.*, 54:1222–1229 (1988).

19 F. Rezzonico and B. Duffy, The role of luxS in the fire blight pathogen *Erwinia amylovora* is limited to metabolism and does not involve quorum sensing, *Mol. Plant Microbe. Interact.*, 20:1284–1297 (2007).

20 K. Vrancken, M. Holtappels, H. Schoofs, T. Deckers and R. Valcke, Pathogenicity and infection strategies of the fire blight pathogen *Erwinia amylovora* in Rosaceae: state of the art, *Microbiology*, 159:823–832 (2013).

21 F.O. Bastian, P.H. Elzer and X. Wu, *Spiroplasma* spp. biofilm formation is instrumental for their role in the pathogenesis of plant, insect and animal diseases, *Exp. Mol. Pathol.* 93:116–128 (2012).

22 O. Bahar, T. Goffer and S. Burdman, Type IV Pili are required for virulence, twitching motility, and biofilm formation of *Acidovoraxavenae* subsp. *Citrulli*, *Mol. Plant Microbe. Interact.*, 22:909–920 (2009).

23 B. Ryall, X. Lee, J.E. Zlosnik, S. Hoshino and H.D. Williams, Bacteria of the *Burkholderia cepacia* complex are cyanogenic under biofilm and colonial growth conditions. *BMC Microbiol.*, 8:108 (2008).

24 J. Yao and C. Allen, The plant pathogen *Ralstonia solanacearum* needs aerotaxis for normal biofilm formation and interactions with its tomato host, *J. Bacteriol.*, 189:6415–6424(2007).

25 D.H. Lee, J.A. Lim, J. Lee, E. Roh, K. Jung, M. Choi, C. Oh, S. Ryu, J. Yun and S. Heu, Characterization of genes required for the pathogenicity of *Pectobacterium carotovorum* subsp. *carotovorum* Pcc21 in Chinese cabbage, *Microbiology*, 159:1487–1496 (2013).
26 C.M. Herrera, M.D. Koutsoudis, X. Wang and S.B. von Bodman, *Pantoeastewartii* subsp. *stewartii* exhibits surface motility, which is a critical aspect of Stewart's wilt disease development on maize, *Mol. Plant Microbe. Interact.*, 21:1359–1370 (2008).
27 J.A. Asselin, J. Lin, A.L. Perez-Quintero, I. Gentzel, D. Majerczak, S.O. Opiyo, W. Zhao, S.M. Paek, M.G. Kim, D.L. Coplin, J.J. Blakeslee and D. Mackey, Perturbation of maize phenylpropanoid metabolism by an AvrE family type III effector from *Pantoeastewartii*, *Plant Physiol.*, 167:1117–1135 (2015).
28 A. Obradovic, A. Mavridis, K. Rudolph, J.D. Janse, M. Arsenijevic, J.B. Jones, G.V. Minsavage and J. Wang, Characterization and PCR-based Typing of *Xanthomonas campestris* pv.*vesicatoria* from Peppers and Tomatoes in Serbia, *Euro. J. Plant Pathol.*, 110:285–292 (2004).
29 L.A. Rigano, F. Siciliano, R. Enrique, L. Sendin, P. Filippone, P.S. Torres, J. Questa, J.M. Dow, A.P. Castagnaro, A.A. Vojnov and M.R. Marano, Biofilm formation, epiphytic fitness, and canker development in *Xanthomonas axonopodis* pv.*citri*, *Mol. Plant Microbe. Interact.*, 20:1222–1230 (2007).
30 Y. Zhang, N. Jalan, X. Zhou, E. Goss, J.B. Jones, J.C. Setubal, X. Deng and N. Wang, Positive selection is the main driving force for evolution of citrus canker-causing *Xanthomonas*, *ISME J.*, 9:2128–2138 (2015).
31 R.W. Shepherd and S.E. Lindow, Two Dissimilar N-Acyl-Homoserine Lactone Acylases of *Pseudomonas syringae* Influence Colony and Biofilm Morphology, *Appl. Environ. Microbiol.*, 75:45–53 (2009).
32 F. Taguchi, Y. Ogawa, K. Takeuchi, T. Suzuki, K. Toyoda, T. Shiraishi and Y. Ichinose, A Homologue of the 3-Oxoacyl-(Acyl Carrier Protein) Synthase III Gene Located in the Glycosylation Island of *Pseudomonas syringae*pv. *tabaci* Regulates Virulence Factors via N-Acyl Homoserine Lactone and Fatty Acid Synthesis, *J. Bacteriol.*, 188:8376–8384 (2006).
33 T.T. Hoang and H.P. Schweizer, Fatty acid biosynthesis in *Pseudomonas aeruginosa*: cloning and characterization of the*fabAB* operon encoding beta-hydroxyacyl-acyl carrier protein dehydratase (FabA) and beta-ketoacyl-acyl carrier protein synthase I (FabB), *J. Bacteriol.*, 179:5326–5332 (1997).
34 H.P. Schweizer and K. Choi, Characterization of Molecular Mechanisms Controlling fabAB Transcription in *Pseudomonas aeruginosa*, *PLoS One*, 7:e45646 (2012).
35 D. Perez-Mendoza, I.M. Aragon, H.A. Prada-Ramirez, L. Romero-Jimenez, C. Ramos, M.T. Gallegos and J. Sanjuan, Responses to elevated c-di-GMP levels in mutualistic and pathogenic plant-interacting bacteria, *PLoS One*, 9:e91645 (2014).
36 I.M. Aragon, D. Perez-Mendoza, M.T. Gallegos and C. Ramos, The c-di-GMP phosphodiesterase BifA is involved in the virulence of bacteria from the *Pseudomonas syringae* complex, *Mol. Plant Pathol.*, 16:604–615 (2015).
37 C. Engl, C.J. Waite, J.F. McKenna, M.H. Bennett, T. Hamann and M. Bucka, Chp8, a Diguanylate Cyclase from *Pseudomonas syringae*pv. *Tomato* DC3000, Suppresses the Pathogen-Associated Molecular Pattern Flagellin, Increases Extracellular Polysaccharides, and Promotes Plant Immune Evasion, *mBio.*, 5:e01168-01114 (2014).
38 N. De, M.V.A.S. Navarro, R.V. Raghavan and H. Sondermann, Determinants for the Activation and Autoinhibition of the Diguanylate Cyclase Response Regulator WspR, *J. Mol. Biol.*, 393:619–633 (2009).

39 W.L. Deng, Y.C. Lin, R.H. Lin, C.F. Wei, Y.C. Huang, H.L. Peng and H.C. Huang, Effects of *galU* Mutation on *Pseudomonas syringae*–Plant Interactions, *Mol. Plant Microbe. Interact.*, 23:1184–1196 (2010).

40 H.S. Kong, D.P. Roberts, C.D. Patterson, S.A. Kuehne, S. Heeb, D.K. Lakshman and J. Lydon, Effect of Overexpressing *rsmA* from *Pseudomonas aeruginosa* on Virulence of Select Phytotoxin-Producing Strains of *P. syringae*, *Phytopathol.*, 102:575–587 (2012).

41 M.D. Mastropaolo, M.W. Silby, J.S. Nicoll and S.B. Levya, Novel Genes Involved in *Pseudomonas fluorescens* Pf0-1 Motility and Biofilm Formation, *Appl. Environ. Microbiol.*, 78:4318–4329 (2012).

42 M. Modzelan, M. Kujawa, K. Głabski, G. Jagura-Burdzy and E. Kraszewska, NudC Nudix hydrolase from *Pseudomonas syringae*, but not its counterpart from *Pseudomonas aeruginosa*, is a novel regulator of intracellular redox balance required for growth, motility and biofilm formation, *Mol. Microbiol.*, 93:867–882 (2014).

43 S.M. Hinsa, M. Espinosa-Urgel, J.L. Ramos and G.A. O'Toole, Transition from reversible to irreversible attachment during biofilm formation by *Pseudomonas fluorescens* WCS365 requires an ABC transporter and a large secreted protein, *Mol. Microbiol.*, 49:905–918 (2003).

44 P. Vasseur, I. Vallet-Gely, C. Soscia, S. Genin and A. Filloux, The *pel* genes of the *Pseudomonas aeruginosa* PAK strain are involved at early and late stages of biofilm formation. *Microbiology*, 151:985–997 (2005).

45 W, Chang, M, van de Mortel, L, Nielsen, G.N. de Guzman, X. Li and L.J. Halverson, Alginate Production by *Pseudomonas putida* Creates a Hydrated Microenvironment and Contributes to Biofilm Architecture and Stress Tolerance under Water-Limiting Conditions, *J. Bacteriol.*, 189:8290–8299 (2007).

46 C.M. Ott, D.F. Day, D.W. Koenig and D.L. Pierson, The release of alginate lyase from growing *Pseudomonas syringae* pathovar*phaseolicola*, *Curr. Microbiol.*, 42:78–81(2001).

47 A. Penaloza-Vazquez, A, Sreedharan, C.L. Bender, Transcriptional studies of the hrpM/opgH gene in *Pseudomonas syringae* during biofilm formation and in response to different environmental challenges, *Environ. Microbiol.* 12:1452–1467 (2010).

48 J.L. Arvizu-Gomez, A. Hernandez-Morales, J.R. Aguilar and A. Alvarez-Morales, Transcriptional profile of *P. syringae* pv. *phaseolicola* NPS3121 at low temperature: physiology of phytopathogenic bacteria, *BMC Microbiol.*, 13:81(2013).

49 S. Ude, D.L. Arnold, C.D. Moon, T. Timms-Wilson and A.J. Spiers, Biofilm formation and cellulose expression among diverse environmental *Pseudomonas* isolates, *Environ. Microbiol.*, 8:1997–2011 (2006).

50 E. Arrebola, V.J. Carrion, J.A. Gutiérrez-Barranquero, A. Pérez-García, P. Rodríguez-Palenzuela, F.M. Cazorla, FM and A. de Vicente, Cellulose production in *Pseudomonas syringae*pv. *syringae*: a compromise between epiphytic and pathogenic lifestyles, *FEMS Microbiol. Ecol.*, 91, pii:fiv071 (2015).

51 K. Laue, A. Schenk, H. Li, L. Lambertsen, T.R. Neu, S. Molin and M.S. Ullrich, Contribution of alginate and levan production to biofilm formation by *Pseudomonas syringae*, *Microbiology*, 152:2909–2918 (2006).

52 F. Taguchi, T. Suzuki, Y. Inagaki, K. Toyoda, T. Shiraishi and Y. Ichinose, The siderophore pyoverdine of *Pseudomonas syringae* pv. *tabaci* 6605 is an intrinsic virulence factor in host tobacco infection, *J. Bacteriol.*, 192:117–126 (2010).

53 R. Rai, M. Ranjan, B.B. Pradhan and S. Chatterjee, A typical regulation of virulence-associated functions by a diffusible signal factor in *Xanthomonas oryzae*pv. *Oryzae, Mol. Plant Microbe. Interact.*, 25:789–801 (2012).

54 G. Qian, Y. Zhou, Y. Zhao, Z. Song, S. Wang, J. Fan, B. Hu, V. Venturi and F. Liu, Proteomic analysis reveals novel extracellular virulence-associated proteins and functions regulated by the diffusible signal factor (DSF) in *Xanthomonas oryzae* pv. *Oryzicola, J. Proteome Res.*, 12:3327–3341 (2013).

55 F. Yang, F. Tian, X. Li, S. Fan, H. Chen, M. Wu, C.H. Yang and C. He, The degenerate EAL-GGDEF domain protein Filp functions as a cyclic di-GMP receptor and specifically interacts with the PilZ-domain protein PXO_02715 to regulate virulence in *Xanthomonas oryzae*pv. *Oryzae, Mol. Plant Microbe. Interact.*, 27:578–589(2014).

56 Y. Zhang, C. Wei, W. Jiang, L. Wang, C. Li, Y. Wang, J.M. Dow and W. Sun, The HD-GYP domain protein RpfG of *Xanthomonas oryzae*pv. *oryzicola* regulates synthesis of extracellular polysaccharides that contribute to biofilm formation and virulence on rice, *PLoS One*, 8:e59428 (2013).

57 X. Yu, X. Liang, K. Liu, W. Dong, J. Wang and M.G. Zhou, The *thiG* Gene Is Required for Full Virulence of *Xanthomonas oryzae*pv. *oryzae* by Preventing Cell Aggregation, *PLoS One*, 10:e0134237 (2015).

58 H. Li, C. Yu, H. Chen, F. Tian and C. He, PXO_00987, a putative acetyltransferase, is required for flagellin glycosylation, and regulates flagellar motility, exopolysaccharide production, and biofilm formation in *Xanthomonas oryzae* pv.*oryzae, Microb. Pathog.*, 85:50–57 (2015).

59 B.B. Pradhan, M. Ranjan and S. Chatterjee, XadM, a novel adhesin of *Xanthomonas oryzae*pv. *oryzae*, exhibits similarity to Rhs family proteins and is required for optimum attachment, biofilm formation, and virulence, *Mol. Plant Microbe. Interact.*, 25:1157–1170 (2012).

60 P.S. Torres, F. Malamud, L.A. Rigano, D.M. Russo, M.R. Marano, A.P. Castagnaro, A. Zorreguieta, K. Bouarab, J.M. Dow and A.A. Vojnov, Controlled synthesis of the DSF cell–cell signal is required for biofilm formation and virulence in *Xanthomonas campestris, Environ. Microbiol.*, 9:2101–2109 (2007).

61 L. Crossman and J.M. Dow, Biofilm formation and dispersal in *Xanthomonas campestris, Microbes Infect.*, 6:623–629(2004).

62 J.M. Dow, L. Crossman, K. Findlay, Y.Q. He, J.X. Feng and J.L. Tang, Biofilm dispersal in *Xanthomonas campestris* is controlled by cell–cell signaling and is required for full virulence to plants, *Proc. Natl. Acad. Sci. USA.*, 100:10995–11000 (2003).

63 Y.M. Hsiao, Y.F. Liu, M.C. Fang and Y.H. Tseng, Transcriptional regulation and molecular characterization of the *manA* gene encoding the biofilm dispersing enzyme mannan endo-1,4-beta-mannosidase in *Xanthomonas campestris, J. Agric. Food Chem.*, 1653–1663 (2010).

64 F. Tao, S. Swarup and L.H. Zhang, Quorum sensing modulation of a putative glycosyltransferase gene cluster essential for *Xanthomonas campestris* biofilm formation, *Environ, Microbiol.*, 12:3159–3170 (2010).

65 X.H. Lu, S.Q. An, D.J. Tang, Y. McCarthy, J.L. Tang, J.M. Dow and R.P. Ryan, RsmA regulates biofilm formation in *Xanthomonas campestris* through a regulatory network involving cyclic di-GMP and the Clp transcription factor, *PLoS One*, 7:e52646(2012).

66 Y.W. He, A.Y. Ng, M. Xu, K. Lin, L.H. Wang, Y.H. Dong and L.H. Zhang, *Xanthomonas campestris* cell–cell communication involves a putative nucleotide receptor protein Clp and a hierarchical signalling network, *Mol. Microbiol.*, 64:281–292(2007).
67 A.C. Edmunds, L.F. Castiblanco, G.W. Sundin and C.M. Waters, Cyclic Di-GMP modulates the disease progression of *Erwinia amylovora*, *J. Bacteriol.*, 195:2155–2165(2013).
68 M. Gross, G. Geier, K. Rudolph and K. Geider, Levan and levansucrase synthesized by the fireblight pathogen *Erwinia amylovora*, *Physiol. Mol. Plant Pathol.*, 40:371–381(1992).
69 J.M. Koczan, M.J. McGrath, Y. Zhao and G.W. Sundin, Contribution of *Erwinia amylovora* exopolysaccharides amylovoran and levan to biofilm formation: implications in pathogenicity, *Phytopathol.*, 99:1237–1244 (2009).
70 J.M. Koczan, B.R. Lenneman, M.J. McGrath and G.W. Sundin, Cell surface attachment structures contribute to biofilm formation and xylem colonization by *Erwinia amylovora*, *Appl. Environ. Microbiol.*, 77:7031–7039 (2011).
71 Q. Zeng, R.R. McNally and G.W. Sundin, Global small RNA chaperone Hfq and regulatory small RNAs are important virulence regulators in *Erwinia amylovora*, *J. Bacteriol.*, 195:1706–1717 (2013).
72 Q. Zeng and G.W. Sundin, Genome-wide identification of Hfq-regulated small RNAs in the fire blight pathogen *Erwinia amylovora* discovered small RNAs with virulence regulatory function, *BMC Genomics*, 15:414 (2014).
73 Z. Wei, J.F. Kim, S.V. Beer, Regulation of hrp genes and type III protein secretion in *Erwinia amylovora* by HrpX/HrpY, a novel two-component system, and HrpS, *Mol. Plant Microbe Interact.*, 13:1251–1262 (2000).
74 A. Chatterjee, Y. Cui, A.K. Chatterjee, Regulation of *Erwinia carotovorahrpL*(Ecc) (sigma-L(Ecc)), which encodes an extracytoplasmic function subfamily of sigma factor required for expression of the HRP regulon, *Mol. Plant Microbe. Interact.*, 15:971–980 (2002).
75 R.R. McNally, I.K. Toth, P.J. Cock, L. Pritchard, P.E. Hedley, J.A. Morris, Y. Zhao and G.W. Sundin, Genetic characterization of the HrpL regulon of the fire blight pathogen *Erwinia amylovora* reveals novel virulence factors, *Mol. Plant Pathol.*, 13:160–173(2012).
76 H. Hasegawa, A. Chatterjee, Y. Cui and A.K. Chatterjee, Elevated temperature enhances virulence of *Erwinia carotovora* subsp. carotovora strain EC153 to plants and stimulates production of the quorum sensing signal, N-acyl homoserine lactone, and extracellular proteins, *Appl Environ. Microbiol.*, 71:4655–4663 (2005).
77 F. Meng, L. Babujee, J.M. Jacobs and C. Allen, Comparative Transcriptome Analysis Reveals Cool Virulence Factors of *Ralstonia solanacearum* Race 3 Biovar 2, *PLoS One.*, 10:e0139090 (2016).
78 Y. Mori, K. Inoue, K. Ikeda, H. Nakayashiki, C. Higashimoto, K. Ohnishi, A. Kiba and Y. Hikichi, The vascular plant pathogenic bacterium *Ralstonia solanacearum* produces biofilms required for its virulence on the surfaces of tomato cells adjacent to intercellular spaces, *Mol. Plant Pathol.*, DOI: 10.1111/mpp.12335 (2015).
79 M.I. Siri, A. Sanabria, C. Boucher and M.J. Pianzzola, New type IV pili-related genes involved in early stages of *Ralstonia solanacearum* potato infection, *Mol. Plant Microbe. Interact.*, 27:712–724 (2014).
80 W. Elhenawy, N.E. Scott, M.L. Tondo, E.G. Orellano, L.J. Foster and M.F. Feldman, Protein O-linked glycosylation in the plant pathogen *Ralstonia solanacearum*, *Glycobiology*, 26:301–311 (2016).

81 Y. Kang, H. Liu, S. Genin, M.A. Schell and T.P. Denny, *Ralstonia solanacearum* requires type 4 pili to adhere to multiple surfaces and for natural transformation and virulence, *Mol. Microbiol.,* 46:427–437 (2002).

82 F. Meng, J. Yao and C. Allen, A MotN mutant of *Ralstonia solanacearum* is hypermotile and has reduced virulence, *J. Bacteriol.,* 193, 2477–2486 (2011).

83 W.C. Yang, Y.M. Lin, Y.S. Cheng and C.P. Cheng, *Ralstonia solanacearum* RSc0411 (lptC) is a determinant for full virulence and has a strain-specific novel function in the T3SS activity, *Microbiology,* 159:1136–1148 (2013).

84 L. Zhang, J. Xu, H. Zhang, L. He and J. Feng, TssB is essential for virulence and required for type VI secretion system in *Ralstonia solanacearum, Microb. Pathog.,* 74:1–7 (2014).

85 G.C. Kubheka, T.A. Coutinho, N. Moleleki and L.N. Moleleki, Colonization patterns of an mCherry-tagged *Pectobacterium carotovorum* subsp. *brasiliense* strain in potato plants, *Phytopathology,* 103:1268–1279 (2013).

86 L.N. Moleleki, R.G. Pretorius, C.K. Tanui, G. Mosina and J. Theron, A quorum sensing-defective mutant of *Pectobacterium carotovorum* subsp. *brasiliense* 1692 is attenuated in virulence and unable to occlude xylem tissue of susceptible potato plant stems, *Mol. Plant Pathol.,* doi: 10.1111/mpp.12372 (2016).

87 D. Perez-Mendoza, S.J. Coulthurst, J. Sanjuan and G.P. Salmond, N-Acetylglucosamine-dependent biofilm formation in *Pectobacterium atrosepticum* is cryptic and activated by elevated c-di-GMP levels, *Microbiology,* 157:3340–3348 (2011).

88 H. Tan, J.A. West, J.P. Ramsay, R.E. Monson, J.L. Griffin, I.K. Toth and G.P. Salmond, Comprehensive overexpression analysis of cyclic-di-GMP signalling proteins in the phytopathogen *Pectobacterium atrosepticum* reveals diverse effects on motility and virulence phenotypes, *Microbiology,* 160:1427–1439 (2014).

89 V.Y. Gorshkov, A.G. Daminova, P.V. Mikshina, O.E. Petrova, M.V. Ageeva, V.V. Salnikov, T.A. Gorshkova and Y.V. Gogole, Pathogen-induced conditioning of the primary xylem vessels—A prerequisite for the formation of bacterial emboli by *Pectobacterium atrosepticum, Plant Biol (Stuttg).,* doi: 10.1111/plb.12448 (2016).

90 M. Hendson, A.H. Purcell, D. Chen, C. Smart, M. Guilhabert and B Kirkpatrick, Genetic diversity of Pierce's disease strains and other pathotypes of *Xylella fastidiosa, Appl Environ Microbiol.,* 67:895–903 (2001).

91 A.A. de Souza, M.A. Takita, H.D. Coletta-Filho, C. Caldana, G.M. Yanai, N.H. Muto, R.C. de Oliveira, L.R. Nunes and M.A. Machado, Gene expression profile of the plant pathogen *Xylella fastidiosa* during biofilm formation *in vitro, FEMS Microbiol Lett.,* 237:341–353 (2004).

92 Y. Li, G. Hao, C.D. Galvani, Y. Meng, L. De La Fuente, H.C. Hoch and T.J. Burr, Type I and type IV pili of *Xylella fastidiosa* affect twitching motility, biofilm formation and cell–cell aggregation, *Microbiology,* 153:719–726 (2007).

93 M.C. Roper, L.C. Greve, J.G. Warren, J.M. Labavitch and B.C. Kirkpatrick, *Xylella fastidiosa* requires polygalacturonase for colonization and pathogenicity in *Vitis vinifera*grapevines, *Mol. Plant Microbe. Interact.,* 20:411–419 (2007).

94 L.C. Souza, N.A. Wulff, P. Gaurivaud, A.G. Mariano, A.C. Virgilio, J.L. Azevedo and P.B. Monteiro, Disruption of *Xylella fastidiosa* CVC gumB and gumF genes affects biofilm formation without a detectable influence on exopolysaccharide production, *FEMS Microbiol. Lett.,* 257:236–242 (2006).

95 N. Killiny, R.H. Martinez, C.K. Dumenyo, D.A. Cooksey and R.P. Almeida, The exopolysaccharide of *Xylella fastidiosa* is essential for biofilm formation, plant virulence, and vector transmission, *Mol. Plant Microbe. Interact.,* 26:1044–1053 (2013).

96 M.S. Silva, A.A. De Souza, M.A. Takita, C.A. Labate and M.A. Machado, Analysis of the biofilm proteome of *Xylella fastidiosa*, *Proteome Sci.*, 9:58 (2011).

97 K.L. Newman, R.P. Almeida, A.H. Purcell and S.E. Lindow, Cell–cell signaling controls *Xylella fastidiosa* interactions with both insects and plants, *Proc. Natl. Acad. Sci. USA*, 101:1737–1742 (2004).

98 X. Shi and H Lin, Characterization of *Xylella fastidiosapopP* Gene Required for Pathogenicity, *J. Plant Pathol. Microb.*, 6:1–7 (2015).

99 X.Y. Shi, C.K. Dumenyo, R. Hernandez-Martinez, H. Azad and D.A. Cooksey, Characterization of regulatory pathways in *Xylella fastidiosa*: genes and phenotypes controlled by *gacA*, *Appl. Environ. Microbiol.*, 75:2275–2283 (2009).

100 X.Y. Shi, C.K. Dumenyo, R. Hernandez-Martinez, H. Azad and D.A. Cooksey, Characterization of regulatory pathways in *Xylella fastidiosa*: genes and phenotypes controlled by *algU*, *Appl. Environ. Microbiol.*, 73:6748–6756 (2007).

101 R. Caserta, M.A. Takita, M.L. Targon, L.K. Rosselli-Murai, A.P. de Souza, L. Peroni, D.R. Stach-Machado, A. Andrade, C.A. Labate, E.W. Kitajima, M.A. Machado and A.A. de Souza, Expression of *Xylella fastidiosa* fimbrial and afimbrial proteins during biofilm formation, *Appl. Environ. Microbiol.*, 76:4250–4259 (2010).

102 T.M. Voegel, H. Doddapaneni, D.W. Cheng, H. Lin, D.C. Stenger, B.C. Kirkpatrick and M.C. Roper, Identification of a response regulator involved in surface attachment, cell–cell aggregation, exopolysaccharide production and virulence in the plant pathogen *Xylella fastidiosa*, *Mol. Plant. Pathol.*, 14:256–264 (2012).

103 B.G. Guimaraes, R.L. Barbosa, A.S. Soprano, B.M. Campos, T.A. de Souza, C.C. Tonoli, A.F. Leme, M.T. Murakami and C.E. Benedetti, Plant pathogenic bacteria utilize biofilm growth-associated repressor (BigR), a novel winged-helix redox switch, to control hydrogen sulfide detoxification under hypoxia, *J. Biol. Chem.*, 286:26148–26157 (2011).

104 P.A. Cobine, L.F. Cruz, F. Navarrete, D. Duncan, M. Tygart and L. De La Fuente, *Xylella fastidiosa* differentially accumulates mineral elements in biofilm and planktonic cells, *PLoS One.*, 8:e54936 (2013).

105 F. Navarrete and L. De La Fuente, Response of *Xylella fastidiosa* to zinc: decreased culturability, increased exopolysaccharide production, and formation of resilient biofilms under flow conditions, *Appl. Environ. Microbiol.*, 80:1097–1107 (2014).

106 L. De La Fuente, J.K. Parker, J.E. Oliver, S. Granger, P.M. Brannen, E. van Santen and P.A. Cobine, The bacterial pathogen *Xylella fastidiosa* affects the leaf ionome of plant hosts during infection, *PLoS One.*, 8:e62945 (2013).

107 L.F. Cruz, P.A. Cobine and L. De La Fuente, Calcium increases *Xylella fastidiosa* surface attachment, biofilm formation, and twitching motility, *Appl. Environ. Microbiol.*, 78:1321–1331 (2011).

108 B.E. Ramey, M. Koutsoudis, S.B. von Bodman and C Fuqua, Biofilm formation in plant–microbe associations, *Curr. Opin. Microbiol.*, 7:602–609 (2004a).

109 J.E. Heindl, Y. Wang, B.C. Heckel, B. Mohari, N. Feirer and C. Fuqua, Mechanisms and regulation of surface interactions and biofilm formation in *Agrobacterium*, *Front Plant Sci.*, 5:176 (2014).

110 J. Kim, J.E. Heindl and C. Fuqua, Coordination of division and development influences complex multicellular behavior in *Agrobacterium tumefaciens*, *PLoS One.*, 8:e56682 (2013).

111 Y. Wang, C.H. Haitjema and C. Fuqua, The Ctp type IVb pilus locus of *Agrobacterium tumefaciens* directs formation of the common pili and contributes to reversible surface attachment, *J. Bacteriol.*, 196:2979–2988 (2014).

112 A.D. Tomlinson, B. Ramey-Hartung, T.W. Day, P.M. Merritt, C. Fuqua, *Agrobacterium tumefaciens* ExoR represses succinoglycan biosynthesis and is required for biofilm formation and motility, *Microbiology*, 156:2670–2681 (2010).

113 B.C. Heckel, A.D. Tomlinson, E.R. Morton, J.H. Choi and C. Fuqua, *Agrobacterium tumefaciensexoR* controls acid response genes and impacts exopolysaccharide synthesis, horizontal gene transfer, and virulence gene expression, *J. Bacteriol.*, 196:3221–3233 (2014).

114 J.E. Heindl, M.E. Hibbing, J. Xu, R. Natarajan, A.M. Buechlein and C. Fuqua, Discrete Responses to Limitation for Iron and Manganese in *Agrobacterium tumefaciens*: Influence on Attachment and Biofilm Formation, *J. Bacteriol.*, 198:816–829 (2015).

115 A.G. Matthysse, M. Marry, L. Krall, M. Kaye, B.E. Ramey, C. Fuqua and A.R. White, The effect of cellulose overproduction on binding and biofilm formation on roots by *Agrobacterium tumefaciens*, *Mol. Plant Microbe. Interact.*, 18:1002–1010 (2005).

116 S. Klusener, M. Aktas, K.M. Thormann, M. Wessel and F. Narberhaus, Expression and physiological relevance of *Agrobacterium tumefaciens* phosphatidylcholine biosynthesis genes, *J. Bacteriol.*, 191:365–374 (2009).

117 M. Aktas, L. Danne, P. Moller and F. Narberhaus, Membrane lipids in *Agrobacterium tumefaciens*: biosynthetic pathways and importance for pathogenesis, *Front Plant Sci.*, 5:109 (2014).

118 T. Danhorn, M. Hentzer, M. Givskov, M.R. Parsek and C Fuqua, Phosphorus limitation enhances biofilm formation of the plant pathogen *Agrobacterium tumefaciens* through the PhoR-PhoB regulatory system, *J. Bacteriol.*, 186:4492–4501(2004).

119 D. An, T. Danhorn, C. Fuqua and M.R. Parsek, Quorum sensing and motility mediate interactions between *Pseudomonas aeruginosa* and *Agrobacterium tumefaciens* in biofilm cocultures, *Proc. Natl. Acad. Sci. USA*, 103:3828–3833 (2006).

120 S. Yang, Q. Peng, Q. Zhang, X. Yi, C.J. Choi, R.M. Reedy, A.O. Charkowski and C.H. Yang, Dynamic regulation of GacA in type III secretion, pectinase gene expression, pellicle formation, and pathogenicity of *Dickeya dadantii* (*Erwinia chrysanthemi* 3937), *Mol. Plant Microbe. Interact.*, 21:133–142 (2008).

121 J.N. Zhou, H.B. Zhang, M.F. Lv, Y.F. Chen, L.S. Liao, Y.Y. Cheng, S.Y. Liu, S.H. Chen, F. He, Z.N. Cui, Z.D. Jiang, C. Chang and L.H. Zhang, SlyA regulates phytotoxin production and virulence in *Dickeya zeae* EC1, *Mol. Plant Pathol.*, doi: 10.1111/mpp.12376 (2016).

122 C. Prigent-Combaret, O. Zghidi-Abouzid, G. Effantin, P. Lejeune, S. Reverchon and W. Nasser, The nucleoid-associated protein F is directly modulates the synthesis of cellulose, an essential component of pellicle-biofilms in the phytopathogenic bacterium *Dickeya dadantii*, *Mol. Microbiol.*, 86:172–186 (2012).

123 C.E. Jahn, D.A. Selimi, J.D. Barak and A.O. Charkowski, The *Dickeya dadantii* biofilm matrix consists of cellulose nanofibres, and is an emergent property dependent upon the type III secretion system and the cellulose synthesis operon, *Microbiology*, 157:2733–2744 (2011).

124 C.E. Jahn, D.K. Willis and A.O. Charkowski, The flagellar sigma factor *fliA* is required for *Dickeya dadantii* virulence, *Mol. Plant Microbe. Interact.*, 21:1431–1442 (2008).

125 X. Yi, A. Yamazaki, E. Biddle, Q. Zeng and C.H. Yang, Genetic analysis of two phosphodiesterases reveals cyclic diguanylate regulation of virulence factors in *Dickeya dadantii*, *Mol. Microbiol.*, 77:787–800 (2010).

126 M.N. Yap, C.H. Yang, J.D. Barak, C.E. Jahn and A.O. Charkowski, The *Erwinia chrysanthemi* type III secretion system is required for multicellular behaviour, *J. Bacteriol.*, 187:639–648 (2005).

127 M.N. Yap and CH Yang and A.O. Charkowski, The Response regulator HrpY of *Dickeya dadantii* 3937 regulates virulence genes not linked to the *hrp* cluster, *Mol. Plant. Microbe. Interact.*, 21:304–314 (2008).

128 Y.M. Bae, L. Zheng, J.E. Hyun, K.S. Jung, S. Heu and S.Y. Lee, Growth characteristics and biofilm formation of various spoilage bacteria isolated from fresh produce, *J. Food Sci.*, 79:M2072–2080 (2014).

129 L. Chalupowicz, E.M. Zellermann, M. Fluegel, O. Dror, R. Eichenlaub, K.H. Gartemann, A. Savidor, G. Sessa, N. Iraki, I. Barash and S. Manulis-Sasson, Colonization and movement of GFP-labeled *Clavibacter michiganensis* subsp. *michiganensis* during tomato infection, *Phytopathology*, 102:23–31 (2012).

130 P.B. Beauregard, Y. Chai, H. Vlamakis, R. Losick and R. Kolter, *Bacillus subtilis* biofilm induction by plant polysaccharides, *Proc. Natl. Acad. Sci. USA*, 110:E1621–1630 (2013).

131 Y. Chen, F. Yan, Y. Chai, H. Liu, R. Kolter, R. Losick and J.H. Guo, Biocontrol of tomato wilt disease by *Bacillus subtilis* isolates from natural environments depends on conserved genes mediating biofilm formation, *Environ. Microbiol.*, 15:848–864 (2013).

12

Techniques in Studying Biofilms and Their Characterization: Microscopy to Advanced Imaging System *in vitro* and *in situ*

Elisabeth Grohmann and Ankita Vaishampayan

Department of Life Sciences and Technology, Beuth University of Applied Sciences Berlin, Berlin, Germany

12.1 Introduction

It is now well accepted that the default mode of growth of microorganisms in nature is that of association with interphases of any kind [1, 2]. Microorganisms also occur as mobile bioaggregates, flocs, granules, as sediment or soil biofilms, or symbiotic plant root biofilms [2]. Due to their ubiquity and importance in health care as well as in nature and technical processes (e.g., wastewater treatment plants, membrane reactors), it is of primordial interest to understand their development and life cycle, to identify and quantify their microbial "inhabitants," and to understand the regulatory processes governing their formation, maturation, and dispersal. The study of microbial biofilms is an extremely interdisciplinary field. As physical, chemical, and diverse biological processes are involved in the development of a biofilm, they affect its thickness, its heterogeneity, the metabolic activity of the different biofilm layers, and its resistance to biotic and abiotic stressors. As complex as the formation and as diverse as the composition of biofilms are, the research questions in relation to biofilms will determine the examination method.

The first breakthrough in the study of microbial biofilms was the development of fluorescence *in situ* hybridization (FISH) in the 1980s [3–5]. In classical FISH, rRNA-targeted fluorescent probes are applied to biofilm samples to visualize the microorganisms. The probes can be designed in a way that virtually all bacteria present can be detected (EUB338 probe, 3) or they can be highly specific and only detect bacteria of a single species.

FISH can be combined with standard epifluorescence microscopy. However, to achieve structural insights in hydrated complex microbiological samples, such as biofilms, a more sophisticated microscopic technique is required, such as confocal laser scanning microscopy (CLSM), which has revolutionized the structural investigation of biofilms since the early 1990s.

Due to their complexity, natural microbial communities are challenging objects of investigation. Furthermore, biofilms are often located at places that are difficult to access, making direct and continuous examinations difficult [6]. To reduce complexity and facilitate investigations in the laboratory under controlled and reproducible conditions, various biofilm model systems have been established. The most widely used system is that of biofilm flow cells [6–8]. The gold standard in biofilm research is an approach that involves biofilm flow-cell technology in combination with CLSM. This methodology allows researchers to gain insight into details of developmental processes, spatial organization, and function of laboratory-grown biofilms in real time under continuous and noninvasive conditions down to single-cell level [6, 7, 9, 10].

The resolution of light microscopy is determined by the law of Ernst Abbe, meaning the resolution limit is about half the wavelength of light used for imaging. Thus, objects smaller than 200 to 300 nm cannot be resolved. The microscopes employed in modern biofilm studies depend on the scale one is interested in. For microscale studies, laser scanning microscopes (LSM), for example, in combination with structural sensors, such as nucleic acid specific fluorochromes, are the method of choice even though their maximum resolution is limited to objects larger than 200 to 300 nm. If one wants to go beyond this resolution, nanoscale techniques, such as structured illumination microscopy (SIM) and stimulated emission depletion (STED) microscopy, are appropriate techniques. The term *nanoscopy* was created for these techniques and was suggested as the "Method of the Year 2008" [2, 11].

Microbial biofilms can be microscopically resolved at the cell and microcolony dimension. However, due to biofilm heterogeneity and statistical issues, the biofilm structure often must be examined over a larger area [2, 12]. This can be achieved by mesoscale techniques that allow imaging across several mm. For example, optical coherence tomography (OCT), a technique derived from medical applications, has been successfully used for imaging biofilms [2, 13].

For LSM imaging, specific advanced digital image analysis is required, as the result of biofilm LSM imaging can be a series of hundreds of images in up to five separate channels that need to be optimized for data extraction [2]. Digital image analysis may include three different aspects: visualization, quantification, and deconvolution. For all of them, specific software tools have been developed (overview in 2).

The chapter intends to give a brief overview over the most important techniques employed in biofilm analysis, highlighting their advantages and restraints. It concludes with emerging technologies, which are often combinations of known techniques that will bring us closer to the ideal technique, providing us with compositional, structural, and metabolic data of the biofilm in one combined approach.

12.2 Classical Techniques to Study Biofilms

Quantification of biofilm formation can be performed in a simple microtiter-plate test using different stains, such as crystal violet, the most widely used, and subsequent measurement of the absorbance of the biofilm using a microplate reader [14, 15]. This staining method is primarily used for monitoring biofilm formation of microorganisms *in vitro*. Other frequently used stains target microbial nucleic acids. The most common method to identify microbes in a biofilm is a technique called FISH.

12.2.1 Nucleic Acid Stains and FISH (in Combination with Epifluorescence Microscopy)

Direct staining of adherent microorganisms is possible by using DAPI (4,'6-diamidino-2-phenylindole) binding to the AT-rich regions of double-stranded DNA of vital and dead cells [15, 16]. However, no differentiation of bacterial species is possible with this simple technique. For this purpose, a specific hybridization and fluorescence-based approach, FISH [3] is required. FISH is based on oligonucleotide probes labeled with fluorescent dyes binding specifically to rRNA. The oligonucleotides can be either selected so that they bind 16S or 23S rRNA from virtually all bacteria (EUB338 probe), or they can be species-specific so that only one single species is detected. A large number of intact ribosomes representing the biological activity of the tested cell is required for the FISH technique—thus, apparently, only vital bacteria are stained [3–5, 15]. The FISH technique is often used in combination with epifluorescence microscopy.

12.2.2 FISH and Confocal Laser Scanning Microscopy (CLSM)

The invention of laser scanning microscopy (LSM) in the 1980s caused a revolution in light microscopy [17]. Since the first publication in the field of microbiology in 1991, LSM has developed into an important and indispensable technique for three-dimensional *in situ* imaging of microbial communities [17, 18].

The LSM technique can be performed in various ways. The two most common are the following setups. First, the system can be equipped with lasers using one-photon excitation. This is done by using continuous UV and visible lasers or, more recently, by using white lasers also known as super-continuum light sources [17]. This method is usually called confocal laser scanning microscopy (CLSM). Alternatively, the system can be set up with a laser using two-photon excitation. This method is known as two-photon or multiphoton laser scanning microscopy (2PLSM).

Advantages and major applications of LSM in microbiology will be summarized in the following section [modified from 17]. LSM can be used to analyze living, fully hydrated samples, it is suited for multichannel (up to 5) analysis; it can be employed for noninvasive optical sectioning, and it is used in reflection or fluorescence mode. It is suitable for intensity and lifetime imaging. The technique is very useful for three-dimensional analysis of macromolecules, cells, and microbial communities. In addition, LSM gives information on structure, processes, and microhabitats, and it can be combined with a large range of associated analytical tools [17].

FISH is the most frequently used fluorescent technique to study the composition as well as the metabolic activities of soil microbiota. The percentage of metabolically active bacteria varies over a wide range in different ecosystems, as the count of active cells is at least tenfold more variable than the total number of bacteria [19, 20]. The fluctuations observed in the percentage of active bacteria may suggest that changes occurred in the ecosystems under study. An important factor limiting microbial activity is stress caused by water availability or lack of nutrients [20 and references therein]. A thorough description of complex metabolic activities and dependencies occurring in natural environments, if only based on culture data, can be extremely troublesome [20], as it is known that only 0.3 percent of the soil microorganisms and less than 0.1 percent of the microorganisms in water are culturable on standard culture media [4, 21, 22]. Molecular ecology methods such as FISH enable researchers to identify the so-called unculturable fraction in a natural community. As intact structures of ribosomes are the target site of the probes used in FISH, this method can be an indicator of the physiological state of a cell [20, 23]. Only viable and highly active cells have a sufficient number of ribosomes that make performance of FISH with a specific probe possible [24–26].

Specific hybridization protocols have been developed for diverse habitats, for the identification of the soil microflora [24, 27–29].

The probes used in FISH are usually 15 to 30 nucleotides long and are labeled at the 5′ end with a fluorochrome (e.g., fluorescein, Texas red, TAMRA, Alexa448, TRITC, CY3, CY5, or CY7). They can be used to identify large phylogenetic groups or individual species [20, 30–32]. There is a constantly extending online database, Probe Base (http://www.microbial-ecology.net/probebase), containing characteristics of 2788 probes that have been applied in a variety of environments (database accessed on March, 22, 2016).

One limitation of the FISH method is its low sensitivity in the case of a small number of ribosomes in a cell, low metabolic activity of the microorganism, or small cells [33]. Therefore, new solutions have been implemented to amplify the signal obtained in the standard FISH technique. One of the most successful signal-amplifying techniques is the so-called CARD-FISH (catalyzed reporter deposition) method [34], which applies oligonucleotide probes with HRP enzyme (horseradish peroxidase). The method is based on the deposition of a large amount of fluorescein-labeled tyramide due to the catalytic activity of HRP [35, 36]. Another method, employed for the recognition of individual cells, is RING-FISH [35]. In this technique, polynucleotide probes (100 or several hundred base pairs long) of a high concentration are used. After hybridization, the fluorescent signal is visible in the form of a ring.

With FISH, the composition of microbial communities in different habitats can be elucidated. However, metabolically active bacteria present (e.g., in soil) are directly identified using a combination of FISH with microradiography, MAR-FISH [20, 22, 37]. MAR-FISH enables *in situ* identification of microbes and determination of pathways of substrate uptake in individual cells, as the substrate is applied in radiolabeled form. Thus, MAR-FISH combines two independent measurements: the content and quality of ribosomes in a cell and uptake of substrates [20, 38].

12.3 The Gold Standard: Flow-Cell Technology and Confocal Laser Scanning Microscopy (CLSM)

In lab studies, a number of biofilm model systems have been employed to study biofilms under controlled and reproducible conditions. Studies using CLSM of biofilms formed in lab flow-chamber systems by genetically color-coded bacteria have resulted in detailed knowledge about biofilm development, cell differentiations, spatial organization, and function of laboratory-grown biofilms—in some cases, down to the single-cell level [6]. This combination of flow-cell technology with CLSM, which allows getting insight into details of developmental processes and functions of biofilms in real-time under continuous and noninvasive conditions—at its best down to the single-cell level—is the "gold standard" in biofilm research [6, 7, 9, 10]. In the following sections, the technique and its versatile applications will be briefly described.

12.4 The Biofilm Flow Cell

A central component of the biofilm flow-cell system is the flow cell, which provides chambers for biofilm cultivation. One widely used type of flow cells is illustrated in

Figure 12.1. It is a modified version of the flow cell originally developed by [39, summarized in 40]. The flow cell is designed so that it can be mounted on almost any optical microscope. It consists of two parts, a flow-cell base and a conventional microscopy glass coverslip. The flow-cell base is made of polycarbonate in which parallel channels with individual dimensions of 40 × 4 ×1 mm have been drilled. In the flow cell shown in Figure 12.1, three individual channels have been drilled, meaning that in one flow cell three individual biofilm experiments can be carried out [6]. To both ends of each channel, ports of 1 mm diameter have been drilled, which are the medium inlet and effluent outlet, respectively [6, 8]. A microscopy glass coverslip is put on top of the flow-cell base, thereby forming closed channels. The coverslip has three functions: It serves as one wall of the channels thereby building a closed flow channel, it serves as substratum/carrier for biofilm formation, and it is optically suited for microscopic examinations [6–9, 41].

For biofilm growth, the flow-cell system is filled with medium using a peristaltic pump. The microorganisms to be studied are introduced at desired initial optical density into the flow chambers via a syringe. Flow rate in the biofilm systems is set to around 0.2 mm/s. While fresh medium is continuously transported into the flow cell, effluent is transported out and collected in a reservoir [6].

The biofilm setup described is compatible with many visualization and quantification techniques. Cell–cell interactions in a biofilm can be analyzed between (i) cells of a single strain, (ii) cells of different strains of a single species, (iii) cells from different bacterial species, as well as (iv) cells belonging to different domains (e.g., between bacterial and eukaryotic cells)[summarized in 6]. Activities in natural multicellular microbial communities can also be examined in flow cells after collecting samples from the environment and transferring the cells into flow chambers where they are further analyzed [6, 39, 42].

Confocal laser scanning microscopes have helped overcoming the shortcomings of the conventional light microscope by introducing point illumination and a pinhole, which permits optical sectioning of the sample [6]. The individual optical sections are subsequently assembled by the help of advanced computer software to a virtual three-dimensional image (for more details, see section 12.4). Typically, a biofilm with a thickness of more than approximately 150 μm cannot be rendered with a reasonable detail with standard CLSM due to physical limitations. Tremendous technological improvements have been implemented to overcome these problems. These new technologies will be briefly summarized in section 12.5.

Confocal microscopy and derived techniques require fluorescent specimens. Thus, the biofilm must either be autofluorescent by means of indigenous fluorescent molecules or the cells in the biofilm must express a fluorescent protein (e.g., the green fluorescent protein, GFP [43] or individual cells or other components of the multicellular structures must be stained [6]). Stains such as the Syto stains (Invitrogen, Carlsbad, USA) can efficiently stain cells in virtually any color. In combination with propidium iodide (PI), it is possible to specifically stain and discriminate between live and dead cells. Syto 9 stains all cells green regardless if they are dead or alive, while the red PI dye stains only cells with a damaged membrane, indicating dead cells. However, for environmental bacteria, the usefulness of the PI stain should be carefully checked prior to the experiment [6]. Dyes binding to the extracellular biofilm matrix such as lectins [44, 45] or calcofluor white [46, 47] can be also used to visualize the surrounding of the

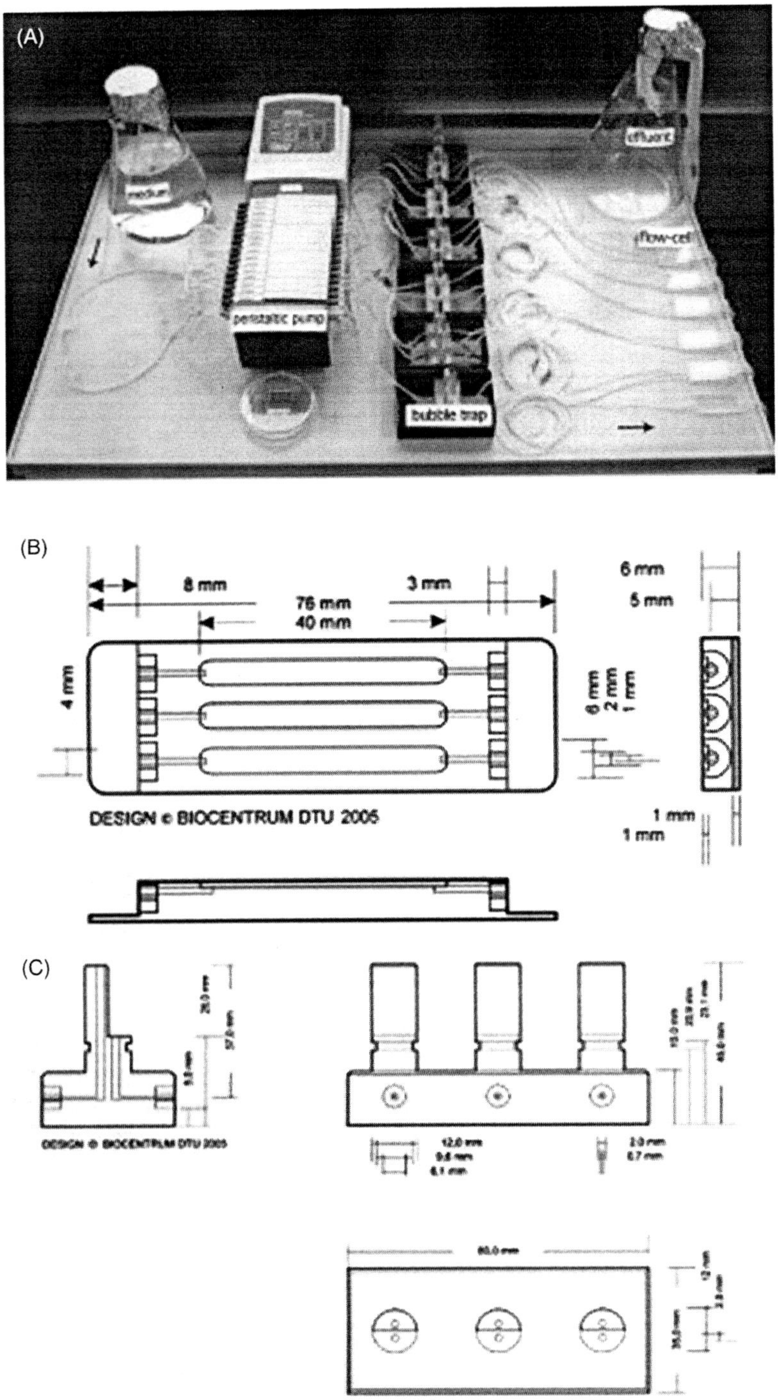
(A)
medium
peristaltic pump
bubble trap
effluent
flow-cell
(B)
8 mm
3 mm
76 mm
40 mm
6 mm
5 mm
4 mm
6 mm
2 mm
1 mm
DESIGN © BIOCENTRUM DTU 2005
1 mm
1 mm
(C)

biofilm cells. Extracellular DNA in the matrix can be visualized by using different DNA-binding fluorophores [6, 48].

If the biofilm cells can be genetically manipulated, chromosomal tagging of cells with a gene cassette encoding GFP can be performed [49]. As an alternative, plasmids encoding GFP can be transformed in the cells prior to the biofilm studies. Fluorescence tagging of *E. coli* and *Pseudomonas* spp. strains has been performed to monitor physiological activity in biofilms by introducing constructs encoding GFP derivatives with a short half-life, under transcriptional control of a ribosomal promoter [6, 50]. Cells with high physiological activity showed high expression and emitted a high fluorescence signal, whereas cells with low physiological activity exhibited low or no expression of the fluorescent protein [6, 40, 50]. Another way of fluorescently labeling cells in a biofilm is by using FISH (see section 12.2).

12.5 Advanced Digital Analysis of Confocal Microscopy Images

This section will provide a brief overview on modern digital image analysis methods, which are continuously under further improvement. The recorded confocal microscopic images can be used immediately or processed further for presentation or quantitative analysis. The images are usually grayscale bitmap images, one image from each focal plane and one image for each detection channel (color) [6]. Standard images from a confocal microscopy analysis are generally relatively large, resulting in files approximately of 45 MB. Special software is necessary to handle these files and to generate the impressive biofilm representations (Figure 12.2). Many analysis software packages are available, with a few dominating the market, such as Imaris (Bitplane, Bern, Switzerland), Amira (Visage Imaging, Carlsbad, CA), and Volocity (Improvision, Coventry, UK) [for a review, see 6, 51].

Due to the small size of bacteria, an average Gram-negative cell has a dimension of 1 μm by 2–3 μm. Three-dimensional reconstruction of a single cell from confocal images is difficult, as the optical resolution of a confocal image recorded with one-photon excitation is at best 0.48 μm [6, 52]. However, the biofilm as a whole is much larger and can thus be shown in 3D, although it may not be possible to locate the individual bacterial cells. The main features of the image analysis software packages are the capability to visualize the spatial organization of the recorded data. They can provide perspective 3D or 2D images in all three axes. Sequential recordings over time can be also rendered, providing a four-dimensional data set, *x-y-z-t* [6].

Figure 12.1 The biofilm flow-cell system allows the cultivation and analysis of biofilm cells under continuous hydrodynamic conditions. The system consists of five major components: a medium reservoir, a multichannel peristaltic pump, bubble traps, flow-cells, and an effluent reservoir. All parts are consecutively connected via silicone tubes, splitters and connectors (A). Working drawing of the flow-cell base (design copyright DTU-Biosys). Reproduced from *Curr. Protoc. Microbiol.* 1B.2.1—1B.2.15 with permission from John Wiley & Sons, Inc. (B). Working drawing of the advanced bubble trap base (design copyright DTU-Biosys). Reproduced from *Curr. Protoc. Microbiol.* 1B.2.1—1B.2.15 with permission from John Wiley & Sons, Inc. (C). The whole figure is reproduced from Pamp et al., *Cytometry Part A*, 75A, 90–103 (2009) with permission from John Wiley & Sons, Inc.

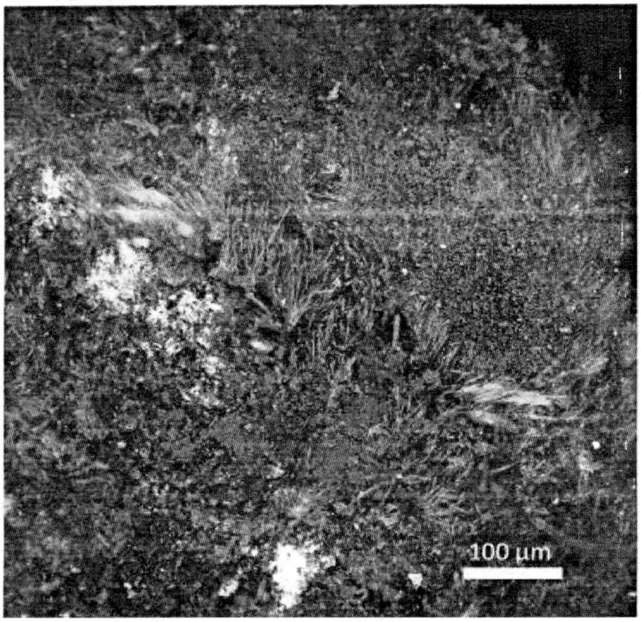

Figure 12.2 Maximum intensity projection of a multichannel dataset showing a complex river biofilm landscape. The sample was stained for nucleic acids, lectin-specific glycoconjugates, and β-D-glucans (with a 1→3 or 1→4 linkage). For multichannel imaging of reflection, autofluorescence and fluorescence staining, the data were recorded in five separate detector channels in combination with colocalization. Color allocation: green, nucleic acids; red, glycoconjugates; blue, algae; pink, cyanobacteria; yellow, β-D-glucans; white, reflection. Reproduced from Neu and Lawrence, *Trends in Microbiol.*, 23, 233–242 (2015) with permission from Elsevier. (*See color plate section for the color representation of this figure.*)

Quantitative analysis of three-dimensional images is challenging; various special software packages have been developed for this purpose. In all of them, the algorithms start by measuring the extent of the biomass by thresholding each focal plane image [6]. Yang et al. (2000) [53] and Heydorn et al. (2000) [54] developed the first robust quantification software, the programs ISA and COMSTAT, respectively. Both programs extract several parameters, which can be used to characterize the biofilm (e.g., biomass, biofilm height, height distribution, roughness coefficient, and diffusion distances). Both programs are developed in MATLAB (MathWorks, Natik, MA). Also, web-based quantification software is available, such as PHILIP [55]. PHILIP has a higher level of automation than the ISA and COMSTAT packages [6]. Further improvements and development of new software packages are underway.

12.6 Biofilm Studies at Different Scales

Depending on the research question, biofilm studies are performed at different scales. Standard analyzes are LSM studies at the microscale to investigate single cells in biofilms. If intracellular and extracellular domains of biofilms shall be resolved, high-resolution microscopy techniques (nanoscale techniques) applied under *in situ* conditions are required [2]. In addition, observing and understanding the structure and performance of a biofilm at larger scale can be of great interest. For this purpose, laser-based techniques at the mesoscale can be employed [2].

12.6.1 Microscale: LSM and Structural Fluorescent Sensors

LSM in combination with structural fluorescent sensors can be applied to investigate intracellular and extracellular domains within microbial biofilms. The various intracellular and extracellular targets must be addressed with different probes and different staining approaches to be able to separate potential interference of fluorochromes [2, 56]. The intracellular space, which represents the bacteria, is usually stained with nucleic acid specific fluorochromes (e.g., Syto series) or fluorochromes sensing intracellular enzyme activity (e.g., CellTrace or CellTracker) [2]. These fluorochromes are used to determine cell morphology and distribution, as well as biofilm structure. Other dyes (e.g., FM dyes) are employed to specifically target bacterial membranes. Sypro dyes specifically target proteins [2]. Fluorescent dyes used in FISH, MAR-FISH, or CARD-FISH technology target specific bacterial groups, as described in section 12.2.2.

The external space, which represents the biofilm matrix, is examined by a variety of probes targeting specific components of the biofilm matrix [56]. The so-called fluorescence lectin-binding analysis (FLBA) is one of the most common techniques used to detect the glycoconjugate distribution in relation to bacterial cells and the overall biofilm structure. An elegant technique, which is applied to single bacterial biofilms, is genetically labeling the bacteria of interest with a fluorescent dye, such as GFP or mCherry (see also section 12.3.1.). This technique allows monitoring gene expression or metabolic activity of the cells [2, 57]. Quantum dots (Q-dots) are semiconductor nanocrystals whose surface is functionalized to be applicable for cellular experiments [58]. In microbial studies, Q-dots were used as markers for imaging their binding [59] and as labels for phages [60, 61], lectins, and antibodies [2, 62].

As probes that allow three-dimensional sensing of the overall structure and fabric of the biofilm, fluorescent spheres or particles are used as well as fluor-labeled dextrans or ficols [2]. These probes can detect different features such as porosity and diffusion coefficients throughout the biofilm matrix.

For deep biofilm imaging (two-photon excitation), LSM setups with a pulsed infrared laser can be used [2].

12.6.2 Nanoscale: Structured Illumination Microscopy (SIM) and Stimulated Emission Depletion (STED) Microscopy

The resolution of light microscopy is limited to about half the wavelength of light. Thus, objects smaller than 200 to 300 nm cannot be resolved. During the last 20 years, new concepts in light and laser imaging have been developed that revolutionized biological imaging for a second time [2]. On the basis of these concepts, new, high-resolution techniques have been developed and are opening a new field of biological imaging, nanoscopy. Structured illumination microscopy (SIM), stimulated emission depletion (STED) microscopy, and localization, stochastic, or blink microscopy are nanoscopy methods [2, 63–66]. More than 20 acronyms refer to blink microscopy; however, it is mostly known as PALM or STORM [2].

In SIM, the sample is illuminated by a pattern of light at three different angles. Thereby, high-frequency information can be extracted that results in double the conventional resolution of about 120 nm [2, 67, 68]. Although SIM only doubles the resolution of light and laser microscopy, it is a very useful technique because it is compatible with all common fluorochromes. STED microscopy is based on two lasers, the first for excitation generating a confocal spot, the second for depletion of the signal in the outer region, thereby shaping and focusing the point-spread function [2, 63]. Commercial STED microscopes achieve a resolution of 90 to 70 nm, with the newest instruments reaching a resolution of 25 nm. Localization, stochastic, or blink microscopy is based on the excitation of individual fluorochromes, which are switched on and off in various ways [2, 69–73]. The resulting blinking pattern is recorded over time. The final resolution under optimum conditions is in the range of 50 to 30 nm [2]. SIM has been employed to resolve structural details in bacterial cells, particularly, *Bacillus subtilis* ultrastructure [74], *Pseudomonas fluorescens* adhesins [75], and *B. subtilis* flagella assembly were studied [2, 76, 77]. In contrast, very few publications are available where STED microscopy—in modified form— was used to visualize bacterial cells [78] whereas, the high-resolution potential of blink microscopy has been employed in several microbiological studies, including the analysis of *Vibrio cholerae* biofilms [79]. In spite of their high resolution, nanoscopy methods show some limitations: Compared to LSM, their limitations lie in imaging of thick samples and the small number of fluorochromes that can be applied [2, 80].

12.6.3 Mesoscale: Optical Coherence Tomography (OCT) and Scanning Laser Optical Tomography (SLOTy)

Biofilms can be microscopically resolved at the cell and microcolony dimension. However, due to heterogeneity and statistical issues, the biofilm structure often has to be analyzed over a larger area [2, 12]. On one hand, this can be accomplished with LSM using mosaic or tile scan options resulting in rather large data sets, on the other hand, mesoscale techniques allow imaging across several mm [2]. One of these techniques, optical coherence tomography (OCT), a technique originating in the medical field, has already been employed for biofilm imaging [2, 13]. OCT is performed with infrared laser sources; no labels are required and the reflection signal can be recorded over several mm^2 [2]. Large biofilm areas have also been examined by using ultrasound [81]. Another mesoscale technique, derived from optical projection tomography, is scanning laser optical tomography (SLOTy), which was demonstrated to be applicable in biofilm imaging [2, 82].

12.7 Conclusions and Perspectives

In summary, the tools to examine biofilms have advanced tremendously in the last decades. They reach from confocal microscopy in combination with particular labeling techniques such as FISH to identify the inhabitants of the biofilm to nanoscopy techniques, such as SIM and STED, which allow visualization of ultrastructures of microbial cells in biofilms. Still, a very useful tool to examine bacterial biofilms *in situ* is the microbial flow cell, where the biofilm grows on a microscopic coverslip. At selected time intervals during biofilm growth, features of the biofilm or the spatial distribution of microbial cells in the biofilm can be studied by confocal microscopy (e.g., in combination with FISH). To study larger areas of a biofilm, mesoscale techniques such as OCT or SLOTy can be applied, which allow imaging of the biofilm across several mm^2.

Although we have been experiencing the development of exciting new microscopic techniques in the last years, the ideal (combined) technique, which enables us to identify the microbes and to analyze their physiological activity and their surface structures as well as the extracellular environment, the biofilm matrix, has not been invented so far. We think that the future will teach us that combined approaches between microscopic techniques, such as CLSM or nanoscale/mesoscale methods with new generation sequencing methods, will bring us quite close to the full picture of complex microbial communities including cell–cell interactions between microbes and microbes and plants, which are of uttermost importance in soil ecosystems.

Acknowledgments

The authors sincerely thank Dr. Karsten Arends for critical reading of the manuscript. We regret that not all valuable contributions of colleagues in the field could have been included in this chapter due to space limitation.

References

1 D. Mc Dougald, S. A. Rice, N. Barraud, P. D. Steinberg, S. Kjelleberg, Should we stay or should we go: mechanisms and ecological consequences for biofilm dispersal, *Nat. Rev. Microbiol.*, 10, 39–50 (2012).
2 T.R. Neu, and J.R. Lawrence, Innovative techniques, sensors, and approaches for imaging biofilms at different scales, *Trends in Microbiol.*, 23, 233–242 (2015).
3 R.I. Amann, B.J. Binder, R.J. Olson, S.W. Chisholm, R. Devereux, D.A. Stahl, Combination of 16S rRNA-targeted oligonucleotide probes with flow cytometry for analyzing mixed microbial populations, *Appl. Environ. Microbiol.*, 56, 1919–1925 (1990).
4 R.I. Amann, W. Ludwig, K.H. Schleifer, Phylogenetic identification and *in situ* detection of individual microbial cells without cultivation, *Microbiol. Rev.*, 59, 143–169 (1995).
5 R.I. Amann, *In situ* identification of microorganisms by whole cell hybridization with rRNA-targeted nucleic acid probes, *In Molecular Microbial Ecology Manual*, A.D.L. Akkermans, J.D. van Elsas and F.J. de Bruijin (Eds), Dordrecht: Kluwer press, 1995.
6 S.J. Pamp, C. Sternberg, T. Tolker-Nielsen, Insight into the microbial multicellular lifestyle via flow-cell technology and confocal microscopy, *Cytometry Part A.*, 75A, 90–103 (2009).
7 B.B. Christensen, C. Sternberg, J.B. Andersen, R.J. Palmer, A.T. Nielsen, M. Givskov, M. Soren, Molecular tools for study of biofilm physiology. *in Methods in Enzymology*, R.J. Doyle (Ed), Biofilms 310, Academic Press, San Diego,1999a.
8 C. Sternberg, and T. Tolker-Nielsen, Growing and analyzing biofilms in flow cells, *in Curr Protoc Microbiol*, R. Coico, T. Kowalik, J. Quarles, B. Stevenson, R. Taylor (Eds), John Wiley, New Jersey, 2005.

9 R.J. Palmer, Microscopy flow cells: Perfusion chambers for real-time study of biofilms, *in Methods in Enzymology*, R. J. Doyle (Ed), Biofilms 310, Academic Press, San Diego, 1999.

10 R.J. Palmer, J.A. Haagensen, T.R. Neu, C. Sternberg, Confocal microscopy of biofilms—spatiotemporal approaches. *in Handbook of Biological Confocal Microscopy*, J.B. Pawley (Ed), Springer Science+Business Media, New York, 2006.

11 Editorial Method of the Year, *Nat. Methods*, 6, 1 (2009).

12 E. Morgenroth, and K. Milferstedt, Biofilm engineering: Linking biofilm development at different length and time scales, *Rev. Environ. Sci. Biotechnol.*, 8, 203–208 (2009).

13 M. Wagner, D. Taherzadeh, C. Haisch, H. Horn, Investigation of the mesoscale structure and volumetric features of biofilms using optical coherence tomography, *Biotechnol. Bioeng.*, 107, 844–853 (2010).

14 A. Al-Ahmad, M. Wiedmann-Al-Ahmad, T. M. Auschill, M. Follo, G. Braun, E. Hellwig, N.B. Arweiler, Effects of commonly used food preservatives on biofilm formation of *Streptococcus mutans, Arch. Oral Biol.*, 53, 765–772 (2008).

15 C. Hannig, M. Follo, E. Hellwig, A. Al-Ahmad. Visualization of adherent micro-organisms using different techniques, *J. Med. Microbiol.*,59, 1–7 (2010).

16 T. Schwartz, S. Hoffmann, U. Obst, Formation of natural biofilms during chlorine dioxide and u.v. disinfection in a public drinking water distribution system, *J. Appl. Microbiol.*, 95, 591–601 (2003).

17 T.R. Neu, B. Manz, F. Volke, J.D. Dynes, A.P. Hitchcock, J.R. Lawrence, Advanced imaging techniques for assessment of structure, composition and function in biofilm systems, *FEMS Microbiol. Ecol.*, 72, 1–21 (2010).

18 J.R. Lawrence, D.R. Korber, B.D. Hoyle, J.W. Costeron, D.E. Caldwell, Optical sectioning of microbial biofilms, *J. Bacteriol.*, 173, 6558–6567 (1991).

19 P.A. Del Giorgio, CY.T. Prairie, D.F. Bird, Coupling between rates of bacterial production and the abundance of metabolically active bacteria in lakes, enumerated using CTC reduction and flow cytometry, *Microb. Ecol.*, 34, 144–154 (1997).

20 S. Lew, T. Mieszczynski, J. Szarek, Selected fluorescent techniques for identification of the physiological state of individual water and soil bacterial cells—review, *Folia Microbiol.*, 55 (2), 107–118 (2010).

21 P.H. Janssen. Identifying the dominant soil bacterial taxa in libraries of 16S rRNA and 16SrRNA genes, *Appl. Environ. Microbiol.*, 72, 1719–1728 (2006).

22 S.W. Rogers, T.B. Moorman, S.K. Ong, Fluorescent *in situ* hybridization and micro-autoradiography applied to ecophysiology in soil, *Soil Sci.*, 71, 620–631 (2007).

23 T. Bouvier, and P.A. Del Giorgio, Factors influencing the detection of bacterial cells using fluorescence *in situ* hybridization (FISH): a quantitative review of published reports, *FEMS Microbiol. Ecol.*, 44, 3–15 (2003).

24 H. Christensen, M. Hansen, J. Sorensen, Counting and size classification of active soil bacteria by fluorescence *in situ* hybridization with an rRNA oligonucleotide probe, *Appl. Environ. Microbiol.*, 65, 1753–1761 (1999(b)).

25 C.L. Gruden, S. Fevig, M. Abu-Dalo, M. Hernandez, 5-Cyano-2, 3-ditolyl tetrazolium chloride (CTC) reduction in a mesophilic anaerobic digester: measuring redox behavior, differentiating abiotic reduction, and comparing FISH response as an activity indicator, *J. Microbiol. Methods.*, 52, 59–68 (2003).

26 S. Lew, and A. Swiatecki, Using fluorescence techniques in identification of metabolically active bacteria, *Limnolog. Papers.*, 2, 9–19 (2007).

27 A. Barra Caracciolo, P. Grenni, R. Ciccoli, G. Di Landa, C. Cremisini, Simazine biodegradation in soil: analysis of bacterial community structure by *in situ* hybridization, *Pest. Manag. Sci.*, 61, 863–869 (2005(a)).

28 A. Barra Caracciolo, P. Grenni, C. Cupo, S. Rossetti, In situ analysis of native microbial communities in complex samples with high particulate loads, *FEMS Microbiol. Lett.*,253, 55–58 (2005(b)).

29 J. Bertaux, U. Gloger, M. Schmid, A. Hartmann, S. Scheu, Routine fluorescence *in situ* hybridization in soil. *J. Microbiol. Methods.*,69, 451–460 (2007).

30 B. Bottari, D. Ercolini, M. Gatti, N. Neviani, Application of FISH technology for microbiological analysis: current state and prospects, *Appl. Microbiol. Biotechnol.*, 73, 485–494 (2006).

31 M. Gatti, V. Bernini, E. Neviani, Fluorescence microscopy for studying the viability of microorganisms *in natura* whey starters, *Lett. Appl. Microbiol.*, 42, 338–343 (2006).

32 A. Skowronska, Distribution of microbial-selected populations in lake north Mamry by fluorescent *in situ* hybridization. *Polish. J. Environ. Stud.*, 16, 123–128 (2007).

33 K. Zwirglmaier, Fluorescence *in situ* hybridization (FISH) – the next generation, *FEMS Microbiol. Lett.*, 246, 151–158 (2005).

34 A. Pernthaler, J. Pernthaler, R. Amann, Fluorescence *in situ* hybridization and catalyzed reporter deposition for the identification of marine bacteria, *Appl. Environ. Microbiol.*, 68, 3094–3101 (2002).

35 K. Zwirglmaier, W. Ludwig, K.H. Schleifer, Recognition of individual genes in a single bacterial cell by fluorescence *in situ* hybridization –RING-FISH, *Mol. Microbiol.*, 51, 89–96 (2004).

36 A. Skowronska, I. Zmyslowska, Current methods of identification of bacteria in aquatic microbial ecology—fluorescent *in situ* hybridization (FISH). (In Polish) *Post. Mirobiol.*, 45, 183–193 (2006).

37 M. Wagner, M. Horn, H. Daims, Fluorescence *in situ* hybridization for the identification and characterization of prokaryotes, *Curr. Opin. Microbiol*, 6, 302–309 (2003).

38 N. Lee, P. H. Nielsen, K.H. Andreasen, S. Juretschko, J. L. Nielsen, K.H. Schleifer, M. Wagner, Combination of fluorescent *in situ* hybridization and microautoradiography—a new tool for structure-function analyses in microbial ecology, *Appl. Environ. Microbiol.*, 65(3), 1289–1297 (1999).

39 G.M. Wolfaardt, J.R. Lawrence, R.D. Robarts, S.J. Caldwell, D.E. Caldwell, Multicellular organization in a degradative biofilm community, *Appl. Environ. Microbiol.*, 60 (2), 434–446 (1994(a)).

40 S.J. Pamp, M. Gjermansen, H.K. Johansen, T. Tolker-Nielsen, Tolerance to the antimicrobial peptide colistin in *Pseudomonas aeruginosa* biofilms is linked to metabolically active cells, and depends on the *pmr* and *mexAB-oprM* genes, *Mol. Microbiol.*, 68, 223–240 (2008).

41 Center for Systems Microbiology. Biofilm flow-cell setup. Lyngby, Denmark: Center for Systems Microbiology, Technical University of Denmark; 2005. http://www.dtu.dk/centre/CSM/English/Instrument%20Center/Resources/Biofilm%20Setup.aspx.

42 G.M. Wolfaardt, J.R. Lawrence, R.D. Robarts, D.E. Caldwell, The role of interactions, sessile growth and nutrient amendments on the degradative efficiency of a microbial consortium, *Can. J. Microbiol.*, 40 (5), 331–340 (1994(b)).

43 M. Chalfie, Y. Tu, G. Euskirchen, W.W. Ward, Green fluorescent protein as a marker for gene expression, *Science*, 263 (5148), 802–805 (1994).

44 J.R. Lawrence, T.R. Neu, G.D.W. Swerhone, Application of multiple parameter imaging for the quantification of algal, bacterial and exopolymer components of microbial biofilms, *J. Microbiol. Methods.*, 32, 253–261 (1998).

45 M. Strathmann, J. Wingender, H.C. Flemming, Application of fluorescently labelled lectins for the visualization and biochemical characterization of polysaccharides in biofilms of *Pseudomonas aeruginosa*, *J. Microbiol. Methods.*, 50, 237–248 (2002).

46 P.S. Stewart, R. Murga, R. Srinivasan, D. de Beer, Biofilm structural heterogeneity visualized by three microscopic methods, *Water Res.*, 8, 2006–2009 (1995).

47 L. Yang, J.A.J. Haagensen, L. Jelsbak, H.K. Johansen, C. Sternberg, N. Hoiby, S. Molin, *In situ* growth rates and biofilm development of *Pseudomonas aeruginosa* populations in chronic lung infections, *J. Bacteriol.*,190, 2767–2776 (2008).

48 M. Allesen-Holm, K.B. Barken, L. Lang, M. Klausen, J.S. Webb, S. Molin, M. Givskov, T. Tolker-Nielsen, A characterization of DNA release in *Pseudomonas aeruginosa* cultures and biofilms, *Mol. Microbiol.*, 59, 1114–1128 (2006).

49 L. Lambertsen, C. Sternberg, S. Molin, Mini-Tn*7* transposons for site-specific tagging of bacteria with fluorescent proteins, *Environ. Microbiol.*, 6, 726–732 (2004).

50 C. Sternberg, B.B. Christensen, T. Johansen, T. Na, J.B. Andersen, M. Givskov, S. Molin, Distribution of bacterial growth activity in flow-chamber biofilms, *Appl. Environ. Microbiol.*, 65 (9), 4108–4117 (1999).

51 T.R. Neu, and J.R. Lawrence, One-photon versus two-photon laser scanning microscopy and digital image analysis of microbial biofilms, *Methods Microbiol.*, 34, 89–136 (2005).

52 S.W. Hell, G. Reiner, C. Cremer, E.H.K. Stelzer, Aberrations in confocal fluorescence microscopy induced by mistakes in refractive index, *J. Microsc.*,169, 391–405 (1993).

53 X. Yang, H. Beyenal, G. Harkin, Z. Lewandowski, Quantifying biofilm structure using image analysis, *J. Microbiol. Methods.*, 39, 109–119 (2000).

54 A. Heydorn, B.K. Ersboll, M. Hentzer, M.R. Parsek, M. Givskov, S. Molin, Experimental reproducibility in flow-chamber biofilms, *Microbiology.*, 146, 2409–2415 (2000).

55 J.B. Xavier, D.C. White, J.S. Almeida, Automated biofilm morphology quantification from confocal laser scanning microscopy imaging, *Water Sci. Technol.*, 47, 31–37 (2003).

56 T.R. Neu, and J.R. Lawrence, Investigation of microbial biofilm structure by laser scanning microscopy. *Adv. Biochem. Eng. Biotechnol.*, 146, 1–51 (2014).

57 S. Schamfuss, T. R. Neu, J. R. van der Meer, R. Tecon, H. Harms, L. Y. Wick, Impact of mycelia on the accessibility of fluorene to PAH-degrading bacteria, *Environ. Sci. Technol.*, 47, 6908–6915 (2013).

58 U. Resch-Genger, M. Grabolle, S. Cavaliere-Jaricot, R. Nitschke, T. Nann, Quantum dots versus organic dyes as fluorescent labels, *Nat. Methods.*, 5, 763–775 (2008).

59 Y. Yang, H. Zu, V.L. Colvin, P.J. Alvarez, Cellular and transcriptional response of *Pseudomonas stutzeri* to quantum dots under aerobic and denitrifying conditions. *Environ. Sci. Technol.*, 45, 4988–4994 (2011).

60 R. Edgar, M. McKinstry, J. Hwang, A. B. Oppenheim, R. A. Fekete, G. Giulian, C. Merril, K. Nagashima, S. Adhya, High-sensitivity bacterial detection using biotin-tagged phage and quantum-dot nanocomplexes, *Proc. Natl. Acad. Sci. U.S.A.*,103, 4841–4845 (2007).

61 P.B. Yim, M. L. Clarke, M. McKinstry, S. H. De Paoli Lacerda, L. F. Pease III, M. A. Dobrovolskaia, H. Kang, T. D. Read, S. Sozamannan, J. Hwang, Quantitative characterisation of quantum dot-labelled lambda phage for *Escherichia coli* detection, *Biotechnol. Bioeng.*, 104, 1059–1067 (2009).

62 N.I. Chalmers, R. J. Palmer Jr, L. Du-Thumm, R. Sullivan, W. Shi, P. E. Kolenbrander, Use of quantum dot luminescent probes to achieve single-cell resolution of human oral bacteria in biofilms, *Appl. Environ. Microbiol.*, 73, 630–636 (2007).

63 S.W. Hell, Microscopy and its focal switch, *Nat. Methods.*, 6, 24–32 (2009).

64 B. Huang, M. Bates, X. Zhuang, Super-resolution fluorescence microscopy, *Annu. Rev. Biochem.*, 78, 993–1016 (2009).

65 M. Heilemann, Fluorescence microscopy beyond the diffraction limit, *J. Biotechnol.*, 149, 243–251 (2010).

66 G. Patterson, M. Davidson, S. Manley, J. Lippincott-Schwartz, Superresolution imaging using single-molecule localization, *Annu. Rev. Phys. Chem.*, 61, 345–367 (2010).

67 M.G.L. Gustafsson. L. Shao, P. M. Carlton, C. J. Rahcel Wang, I. N. Golubovskaya, W. Zacheus Cande, D. A. Agard, J. W. Sedat, Three-dimensional resolution doubling in wide-field fluorescence microscopy by structured illumination, *Biophys. J.*, 94, 4957–4970 (2008).

68 L. Schermelleh, P. M. Carlton, S. Hasse, L. Shao, L. Winoto, P. Kner, B. Burke, M. C. Cardoso, D. A. Agard, M. G. L. Gustafsson, H. Leonhardt, J. W. Sedat, Subdiffraction multicolor imaging of the nuclear periphery with 3D structured illumination microscopy, *Science*, 320, 1332–1336 (2008).

69 E. Betzig, G. H. Patterson, R. Sougrat, O. Wolf Lindwasser, S. Olenych, J. S. Bonifacino, M. W. Davidson, J. Lippincott-Schwartz, H. F. Hess, Imaging intracellular fluorescent proteins at nanometer resolution, *Science*, 313, 1642–1645 (2006).

70 S.T. Hess, T. P. K. Girirajan, M. D. Mason, Ultra-high resolution imaging by fluorescence photoactivation localization microscopy, *Biophys. J.*, 91, 4258–4272 (2006).

71 M.J. Rust, M. Bates, X. Zhuang, Sub-diffraction-limit imaging by stochastic optical reconstruction microscopy (STORM), *Nat. Methods.*, 3, 793–796 (2006).

72 J. Fölling, M. Bossi, H. Bock, R. Medda, C. A. Wurm, B. Hein, S. Jakobs, C. Eggeling, S. W. Hell, Fluorescence nanoscopy by ground-state depletion and single-molecule return, *Nat. Methods.*,5, 943–945 (2008).

73 M. Heilemann, S. Van de Linde, M. Schüttpelz, R. Kasper, B. Seefeldt, A. Mukherjee, P. Tinnefeld, M. Sauer, Subdiffraction-resolution fluorescence imaging with conventional fluorescent probes. *Angew. Chem. Int. Ed. Engl.*, 47, 6172–6176 (2008).

74 P. Eswaramoorthy, M. L. Erb, J. A. Gregory, J. Silverman, K. Pogliano, K. S. Ramamurthi, Cellular architecture mediates DivIVA ultrastructure and regulates Min activity in *Bacillus subtilis, MBio* 2, e00257-e311 (2011).

75 I.E. Ivanov, C. D. Boyd, P. D. Newell, M. E. Shwartz, L. Turnbull, M. S. Johnson, C. B. Whitchurch, G. A. O'Toole, T. A. Camesano, Atomic force and super-resolution microscopy support a role for LapA as a cell-surface biofilm adhesin of *Pseudomonas fluorescens, Res. Microbiol.*, 163, 685–691 (2012).

76 S.B. Guttenplan, S. Shaw, D. B. Kearns, The cell biology of peritrichous flagella in *Bacillus subtilis, Mol. Microbiol.*,87, 211–229 (2013).

77 S. Mukherjee, P. Babitzke, DB. Kearns, FliW and FliS function independently to control cytoplasmic flagellin levels in *Bacillus subtilis, J. Bacteriol.*, 195, 297–306 (2013).

78 P. Delcanale, F. Pennacchietti, G. Maestrini, B. Rodríguez-Amigo, P. Bianchini, A. Diaspro, A. Iagatti, B. Patrizi, P. Foggi, M. Agut, S. Nonell, S. Abbruzzetti, C. Viappiani,

Subdiffraction localization of a nanostructured photosensitizer in bacterial cells, *Sci. Rep.*, 5, 15564 (2015).

79 V. Berk, J. C. Fong, G. T. Dempsey, O. N. Develioglu, X. Zhuang, J. Liphardt, F. H. Yildiz, S. Chu, Molecular architecture and assembly principles of *Vibrio cholerae* biofilms, *Science*, 337, 236–239 (2012).

80 L. Schermelleh, R. Heintzmann, H. Leonhardt, A guide to super-resolution fluorescence microscopy. *J. Cell Biol.*, 190, 165–175 (2010).

81 H. Shemesh, R. Heintzmann, H. Leonhardt, High frequency ultrasound imaging of a single-species biofilm, *J. Dent.*, 35, 673–678 (2007).

82 M. Heidrich, M. P. Kühnel, M. Kellner, R-A. Lorbeer, T. Lange, A. Winkel, M. Stiesch, H. Meyer, A. Heisterkamp, 3D imaging of biofilms on implants by detection of scattered light with a scanning laser optical tomograph, *Biomed. Opt. Express.*, 2, 2982–2994 (2011).

13

Gene Expression and Enhanced Antimicrobial Resistance in Biofilms

Daniel Padilla-Chacón[1], Israel Castillo-Juárez[1*], Naybi Muñoz-Cazares[2] and Rodolfo García-Contreras[3]*

[1] *Colegio de Postgraduados, Campus Montecillo, Posgrado de Botánica, Montecillo, Mexico City, Mexico*
[2] *Colegio de Postgraduados, Campus Montecillo, Posgrado de Botánica, Montecillo, Mexico City, Mexico*
[3] *Universidad Nacional Autónoma de México, Department of Microbiology and Parasitology, Faculty of Medicine, Mexico City, Mexico*

13.1 Introduction

Biofilms constitute the preferred mode of life of bacteria in nature, having broad importance in the fields of ecology; industry; and human, animal, and plant health. They mediate both mutualism and pathological relationships between bacteria and their hosts. According to the International Union of Pure and Applied Chemistry (IUPAC), a biofilm is an aggregate of microorganisms in which cells that are frequently embedded within a self-produced matrix of extracellular polymeric substance (EPS) adhere to each other and/or to a surface. Most of the known bacterial species are able to form biofilms.

Biofilms provide multiple advantages to bacterial populations such as conferring protection from several forms of environmental stress, including the actions of toxic organic compounds [1], harmful metals [2], antimicrobials [3], and antibiotics [4]. Furthermore, biofilms protect bacteria against important biotic stressors such as predation from protozoa and lyses by bacteriophages [5]. In addition, biofilm formation facilitates conjugation, horizontal gene transfer [6], and consortial metabolism, since the high cell density of biofilms allow the efficient production of excreted metabolites and exoenzymes. In addition to being widespread, biofilms are probably also one of the more ancient complex biological communities, as evidenced by fossil stromatolites. Despite their importance (e.g., for microbial physiology, ecology, human, animal, and plant health), biofilms were not extensively studied until the early 1980s, when the Canadian researcher William Costerton coined the word *biofilm* [7]; three years later, Costerton's group was the first that associated biofilm formation with human health, describing their formation in intravenous and intraarterial catheters [8]. Now, scientific papers related to biofilm formation are published by the thousands each year.

Biofilm formation is a complex process that comprises several steps, from the initial reversible attachment mediated by electrostatic weak interactions such as Van der

Waals forces and hydrophobic interactions, to a phase of irreversible attachment mediated by specific adhesion structures such as fimbria and curli, the formation of microcolonies and the maturation of biofilms, that involve the synthesis of a complex extracellular matrix generally composed by exopolysacharides, specific proteins, glycolipids, and extracellular DNA.

Such maturation process often requires the expression of quorum sensing (QS), a mechanism that coordinate the reprogramming of gene expression at high cell densities [9], hence restricting the production of metabolically costly products to situations in which there are sufficient bacterial loads and the diffusion of those metabolites is low [10]. Notably, the production of important biofilm matrix exopolysaccharides is regulated by QS in several bacterial species, including plant pathogens like *Pseudomonas syringae* [11], *Pseudomonas aeruginosa* [12], *Xanthomonas oryzae* [13], and *Pantoea stewartii*, as well as in plant symbionts like *Sinorhizobium meliloti* [14]. Remarkably, QS also triggers the release of extracellular DNA that is a major component in some bacterial biofilms [15]. Such DNA release is mediated by cell lyses induced by respiratory poisoning, triggering a process that resembles apoptosis [16]. Other factors that greatly stimulate bacterial biofilm formation include the presence of conjugative plasmids [17], after maturation biofilms could remain stable for ample periods of time or otherwise experience the phenomenon of dispersal which consist in the detachment of a few or many cells from the biofilm, than can then colonize new niches restarting the biofilm formation cycle, biofilm dispersal requires the enzymatic degradation of the biofilm matrix and is triggered by sudden changes in nutrient availability, and influenced by QS, nitric oxide, and other environmental cues [18].

13.2 Biofilms in the Plant–Microbe Relationship

Today, it is widely recognized that most bacteria found in natural, clinical, and industrial settings persist in association with surfaces by forming biofilms [9]. Biofilms are the natural multicellular aggregates of bacteria that adhere to surfaces producing multicellular assemblies. Biofilms vary in size as a function of the nutrient availability [20]. Several studies had demonstrated that bacterial cells adhere to surfaces through a complex matrix comprising a variety of extracellular polymeric substances (EPS) including exopolysaccharides, protein, and extracellular DNA [21]. In this complex, plants offer nutrients-rich oases, hence allowing bacterial growth. Bacterial colonize and form biofilms on different tissues, including leaves, stems, vasculature, seeds, and roots [22].

It is important to note that biofilm formation in plants depend of many key enzymes regulated by mechanisms associated with the formation of multimeric protein complexes and global changes in gene expression [23]; and the composition of biofilms varies depending on the system. Here, we focus in the mayor bacterial plant pathogens that form biofilms in the xylem and roots and describe their biofilm formation mechanisms in plants.

13.2.1 Biofilm Formation in the Vascular System (Xylem)

Biofilm development contributes to the virulence of phytopathogenic bacteria through various mechanisms, including blockage of xylem vessels, increased resistance to plant antimicrobial compounds, and/or enhanced colonization of specific habitats [22].

Xyllela fastidiosa is a major plant pathogen that forms biofilms inside xylem vessels, is deposited into the xylem of plants by sap-feeding leafhoppers, and can induce Pierce's disease of grapevine and citrus variegated chlorosis [24]. Hence, biofilm formation has been shown to be a major contributor to disease development caused by this bacterium within xylem vessels of many economically important crops such sweet orange, olive, peach, and others [25]. For the formation of biofilms, *X. fastidiosa* express the fimbrial adhesins XadA1 and HxfA, which play a critical role in the initial attachment and formation of cell aggregates on abiotic surfaces, XadA1 also was detected in all biofilm stages, corroborating its role in attachment and biofilm development [26]. Additionally, previous reports demonstrated that the concentration of Calcium (Ca^{2+}) enhanced the virulence [27], due the differential expression of several genes in response to Ca^{2+} concentration, including those related to attachment, motility, exopolysaccharide synthesis, biofilm formation, peptidoglycan synthesis, regulatory functions, iron homeostasis, and phages. Furthermore, the proteins XfTolB and XfPal that exhibit putative adhesion properties are involved in membrane integrity and demonstrate dynamic, coordinated expression during the *X. fastidiosa* biofilm development process [28].

Similarly, other bacterial species form biofilms in xylem—for example, *E. amylovora* produces biofilm through the expression of genes that encode proteins related to amylovoran production [29]. Recent studies suggest that genes that encode type I fimbriae, flagella, type IV pili, and curli contribute to biofilm formation and virulence in plants [30, 31].

Xanthomonas campestris pathovar *campestris* (hereafter *Xcc*) is the causative agent of black rot disease of cruciferous plants, which is a significant disease globally [32]. A number of factors and several regulatory pathways have been implicated in the formation or dispersal of biofilms by *Xcc* [33]. Biofilm formation requires the synthesis of the extracellular polysaccharide xanthan and of an uncharacterized polysaccharide whose synthesis is directed by the products of the *xag* gene cluster. Conversely, the extracellular enzyme beta (1,4)-mannanase has been implicated in biofilm dispersal [34].

In 2012, Lu and coworkers demonstrated that the post-transcriptional regulator RsmA exerts a negative regulatory influence on biofilm formation in *Xcc* [32]. RsmA binds to the transcripts of three genes encoding GGDEF domain diguanylate cyclases to influence their expression. Accordingly, deletion of *rsmA* leads to an increase in cellular levels of cyclic di-GMP. This effect is associated with a downregulation of transcription of *manA*, but an upregulation of *xag* gene transcription. Mutation of *clp*, which encodes a cyclic di-GMP-responsive transcriptional regulator of the CRP-FNR family, has similar divergent effects on the expression of *manA* and *xag*. Nevertheless, Clp binding to *manA* and *xag* promoters is inhibited by cyclic di-GMP. On the other hand, recently it was demonstrated that in *Pectobacterium carotovorum* subsp. *brasiliense* 1692 (Pcb1692), an emerging pathogen of potatoes causing blackleg in the field and soft rot during post-harvest storage, flagella are part of the quorum sensing regulon, while fimbriae and pili appear to be negatively regulated by quorum sensing [35]. In *Pectobacterium* spp., the LuxI homolog (ExpI) synthesizes acyl homoserine lactones (AHLs) when AHLs reach threshold levels bind to ExpR, and the resulting ExpR-AHL complex is unable to repress expression of these virulence factors.

Ralstonia solanacearum is a soil-borne pathogen that causes wilt of diverse plant species [21–23]. It colonizes intercellular spaces of the root cortex and vascular parenchyma, and eventually enters in xylem vessels and spreads into the stems and leaves through the xylem [36, 37]. Recently, its mechanism of colonization was investigated, analyzing the behavior of the soil-borne and vascular plant-pathogenic strain OE1-1. The cells formed microcolonies on the surfaces of tomato cells adjacent to intercellular spaces, and then aggregated surrounded by an extracellular matrix, forming mature biofilm structures [38]. Mutation of *lecM* encoding a lectin, RS-IIL, which exhibits affinity for these sugars, led to a significant decrease in biofilm formation. Colonization in intercellular spaces was significantly decreased in the *lecM* mutant, leading to a loss of virulence on tomato plants. Complementation of the *lecM* mutant with native *lecM* resulted in the recovery of mushroom-type biofilms and virulence on tomato plants [38].

Other studies provide insights into the pathogenesis of *Ralstonia solanacearum*, specifically in strains belonging to phylotype IIB, sequevar 1 (IIB-1) interacting with potato, its natural host. In 2014, Siri and coworkers made a comparative genomic analysis among IIB-1 *R. solanacearum* strains with different levels of virulence in order to identify candidate virulence genes. With this approach, they identified a 33.7-kb deletion in a strain showing reduced virulence on potato. This region contains a cluster of six genes putatively involved in type IV pili (Tfp) biogenesis. Functional analysis suggested that these proteins contribute to several Tfp-related functions such as twitching motility and biofilm formation. In addition, this genetic cluster was found to contribute to early bacterial wilt pathogenesis and colonization of potato roots [39]. The specific signals that attract this pathogen were unknown; nevertheless, previous findings indicate that one candidate is aerotaxis, or energy taxis, which guides bacteria toward optimal intracellular energy levels. The *R. solanacearum* genome encodes two putative aerotaxis transducers Aer1 and Aer2; accordingly, cloned *aer1* and *aer2* genes restored aerotaxis in an *Escherichia coli aer* mutant, demonstrating that both genes encode functional aerotaxis transducers. In addition, mutants lacking *aer1*, *aer2*, or the double mutant were significantly less able to move up in an oxygen gradient than the wild-type parent strain; in fact, the aerotaxis of the *aer* mutants was indistinguishable from that of a completely nonmotile strain [37].

13.2.2 Biofilm Formation in Rizosphere (Roots)

The rhizosphere is the soil niche influenced by plant roots [40], since soil microorganisms depend on organic materials derived from plant roots [23]. Soil bacteria need various specific microenvironments, and rhizosphere is the main example of bacteria living as biofilms adhered to various soil surfaces, including those rich in nutrients derived from root exudates and bulk soil (deficient in nitrogen, phosphorus, water, and other nutrients) [41]. The mechanism of colonization of intercellular spaces by the root-associated pseudomonads has been studied extensively, and many of these promote the growth of host plants [20]. Species of *Pseudomonas* form dense biofilms on both abiotic and biotic surfaces, and are a primary model in biofilm research [42]. For example, biofilms of *Pseudomonas putida* were shown to undergo changes in architecture and exopolysaccharide (EPS) composition in order to create a more hydrated microenvironment in response to water-limiting conditions [43, 44].

Similar to other pseudomonads, *P. syringae* pv. *actinidiae* can form biofilms as a physiological response to environmental stresses mediated by the regulation of specific sets of genes, including genes encoding virulence factors. Biofilm formation protects cells against host defense mechanisms as well as various chemical agents. Previously, it was shown that the biofilm growth mode of *P. syringae* is involved in plant pathogenicity [45]. *P. syringae* pv. *actinidiae* bacterial communities can be established outside and inside host tissues and are composed of bacterial cells embedded in a dense matrix of extracellular polymeric substances (EPSs) often composed of exopolysaccharides, proteins, and extracellular DNA [44].

Species of *Agrobacterium* and genera of symbiotic rhizobia not only cause neoplasia and symbiotic nodules on roots but are also effective root colonizers. In *Agrobacterium tumefaciens*, a plant pathogen that persists as surface-associated populations on plants or soil particles, cellulose overproduction resulted in increased biofilm formation on roots [46]. *Azospirillum brasilense* and related species are motile, heterotrophic proteobacteria that interact with roots of a variety of cereals such as wheat and maize, and often promote the growth of their host plants [22].

The fluorescent *Pseudomonas* group includes several species of rhizobacteria that have been used as model strains for rhizosphere colonization experiments [44].

Interestingly, in *P. fluorescens* mutants affected in important regulatory genes that modulate biofilm development, namely *gacS* (G), *sadB* (S) and *wspR* (W), as well as the hypermotile derivatives from the *P. fluorescens* F113 strain and the hypermotile variant (V35) isolated from the rhizosphere are severely impaired in biofilm formation on abiotic surfaces, but are as able to colonize alfalfa roots as the wild-type strain. These findings indicate that in *P. fluorescens*, biofilm formation on abiotic surfaces follows different regulatory pathways than rhizosphere colonization [47]. Table 13.1 summarizes the important biofilm determinants in plant pathogenic bacteria.

Table 13.1 Biofilm Molecular Determinants in Important Plant-Associated Bacteria.

Bacterial Specie	Relationship to Plant	Niche of Biofilm Formation	Biofilm Mechanisms	References
Xillela fastidiosa	Xylem	*Vitis vinifera*	Vessel-to-vessel movement is a key colonization strategy.	[103]
			FimA and FimF adhesins are important for attachment but not essential for virulence.	[104]
Pseudomonas putida	Roots	*Triticum aestivum*	Motility contribute to survival in soil and the initial phase of colonization.	[105]
	Seeds	*Zea mays*	Putative surface and membrane proteins, including a calcium-binding protein, a hemolysin, peptide transporter, and a potential multidrug efflux pump.	[106]
Pseudomonas fluorescens	Roots	*Daucus carota*	Mucoid strains formed a dense and patchy bacterial layer on the roots.	[107]
Pseudomonas syringae	Leaf	*Nicotiana benthamiana, Solanum lycopersicum*	T3SS is important for epiphytic survival and growth on leafs.	[108]

(Continued)

Table 13.1 (Continued)

Bacterial Specie	Relationship to Plant	Niche of Biofilm Formation	Biofilm Mechanisms	References
Pseudomonas aeruginosa	Roots	*Arabidopsis thaliana*	QS controlled virulence factors contribute to biofilm formation and virulence.	[109]
Actinobacteria Streptomyces spp.	Roots	*Triticum aestivum*	No specific factors had been described yet.	[110]
Ralstonia solanacearum	Roots	*Solanum lycopersicum*	Flagella is required for host initial invasion and colonization.	[111]
	Roots	*Solanum tuberosum*	Functional analysis suggests type IV pili contribute to twitching motility and biofilm formation.	[39]
Pectobacterium carotovorum	Stem	*Solanum tuberosum*	Forms aggregates within xylem tissue of potato stems. Gene expression analyses confirmed that flagella are part of the quorum sensing regulon.	[35]
Erwinia amylovora	Xylem	*Malus domestica*	Boinformatic approach and the recently sequenced genome of *E. amylovora* had identified genes encoding putative cell surface attachment structures.	[29]
Xhantomonas campestris	Leaf	*Raphanus sativus* var. *niger*	β,4-mannanase encoded by the Xcc genome is part of the biofilm matrix.	[34]

13.3 Stress Induces Biofilm Formation

Since bacteria living in biofilms are naturally much more resistant to stress than planktonic cells, it is not surprising that an enhancement of its formation as a response to stress could be a common feature in prokaryotic organisms. In fact, biofilm induction by stress was observed earlier in the thermophilic anaerobic archaeon *Archaeoglobus fulgidus*, that do not form an appreciable biofilm in the laboratory optimal conditions but otherwise form robust biofilms if the cells were grown under nonoptimal alkaline pH (higher than 7.5) low or high temperature (55° to 60°C, or $\geq 90^{\circ}$), by the addition of high concentrations of heavy metals such as chromium and copper, and by addition of antibiotics like rifampicin, xenobiotics like 5-fluorouracil, or oxygen.

As expected, biofilm formation enhanced the survival of *A. fulgidus* under the tested stress conditions. To date, the induction of biofilm formation by stress has been also observed in other archaea species like the methanogenic *Methanosarcina acetivorans* that forms a robust biofilm upon cadmium exposure [48]. Similarly, biofilm formation is induced by stress in eubacteria, *like Escherichia coli*, which form much more robust biofilms in the presence of H_2O_2, cadmium, low pH, and low temperature [49], and in the presence of sublethal concentrations of aminoglycoside antibiotics, which also induce biofilm formation of *Pseudomonas aeruginosa* [50]. Moreover, osmotic stress enhances biofilm formation of *Staphylococcus aureus* [51] and stress also increases biofilm formation of *Staphylococcus epidermidis* [52].

13.4 Relevance for Bacterial-Associated Plants

The ability of plant-associated bacteria to enhance biofilm formation upon stress is relevant for both symbiotic relationships and pathogenic ones. For instance, recently it was shown that two plant-associated isolates *Halomonas variabilis* (HT1) and *Planococcus rifietoensis* (RT4) that are able to tolerate saline stress increase the synthesis of extracellular exopolysaccharide and increase biofilm formation during cultivations with high salt concentrations. Moreover, the *in vitro* ability is also presented *in vivo* when the bacteria were added to the chickpea plant cultures (*Cicer arietinum* Var. CM-98). In such conditions, biofilm formation in the plant roots significantly increased. When the plant-bacteria co-cultures were challenged with up to 100 mM of NaCl, these increment in biofilm formation enhanced plant growth and allowed its development in the otherwise hostile environment [53].

13.5 Enhanced Antimicrobial Resistance in Biofilms Is Mediated by Biofilm Physicochemical Characteristics and Specific Changes in Gene Expression

Bacterial cells are often 10 to 1,000 more resistant to antibiotics and antimicrobials than their planktonic counterparts; this great enhancement of antimicrobial tolerance in biofilms is partially a consequence of the biofilm intrinsic architecture and their physicochemical characteristics. Biofilms act as a shield for bacteria that are protected by the biofilm matrix, which decreases antimicrobial permeability [54]. In addition, specific biofilm matrix components can bind and inactivate some antimicrobials, in this regard, it is known than extracellular DNA binds positively charged antimicrobials such as aminoglycosides and that is able to sequester them, providing protection to the biofilm-associated bacteria [55]. Moreover, in *Salmonella enterica,* Mg^{2+} chelation by eDNA activates the PhoPQ two-component system that, in turn, promotes the expression of the polymyxin resistance *(pmr)* operon that mediates the aminoarabinose, modification of which protects the outer membrane from the action of antimicrobial peptides, hence increasing bacterial survival [56]. Similarly, exopolysaccharides can protect *Pseudomonas aeruginosa* against the action of positive charged antibiotics like aminoglycosides [57], and *Acinetobacter baumannii* against peptidic antibiotics [58].

Another important effect of biofilms for enhancing antimicrobial resistance is that as nutrients and oxygen gradients are established through the biofilm depth, a substantial proportion of the cells living in biofilms have either low metabolism or are metabolically inactive, and so far the majority of the available antimicrobials such as the antibiotics had metabolic active processes such as cell wall, protein and nucleic acid synthesis as targets; hence, most of the cells inhabiting biofilms are intrinsically more tolerant to antimicrobials [59].

The fraction of cells that display almost no metabolic activity, also known as *persisters*, are of particular importance, since they can lie dormant and thus survive the action of antimicrobials and resume growth after the challenge has passed. Those cells are responsible for perpetuating recalcitrant chronic infections [60]. The generation of bacterial persister cells could be stochastic or otherwise triggered by stress and depend on the actions of toxins from toxin-antitoxin (TA) modules [61].

As discussed previously, an important factor that contributes to the enhanced antimicrobial resistance of bacterial biofilms is the physicochemical characteristics of biofilm. However, there are also several specific changes in gene expression in biofilms that also confer high antimicrobial resistance. For instance, in 2003, Mah and coworkers screened a transposon mutant library in a *P. aeruginosa* to identify genes in which disruption abolished the specific higher tolerance to tobramycin without decreasing tobramycin tolerance of planktonic cells. Mah et al. found that interrupting the gene *ndvB*, which codify for a glucosyl transferase required for the synthesis of cyclic-b-(1,3)-glucans, abolished the enhancement of tobramycin resistance in biofilms. Later on, they demonstrated that cyclic glucans can actually bind and sequester aminoglycosides, hence decreasing their toxic effects [62]. Another important role of *ndvB* in providing antimicrobial resistance to biofilms is that this gene is required for a biofilm-specific induction of ethanol oxidation genes such as the gene encoding a quinoprotein ethanol dehydrogenase (*exaA*) and genes involved in the synthesis of its cofactor pyrroloquinoline-quinone (PQQ) (*pqqBCE*), importantly deletion of *exaA*, *exaC*, *pqqC*, and *eraR* genes decrease *P. aeruginosa* biofilm tolerance against tobramycin [63].

Other important processes that confer antimicrobial tolerance to bacterial biofilms is the activation of multidrug efflux pumps. Such pumps are found in almost all known bacterial species and genes and are either codified in plasmids or in the chromosomes. So far, at least five efflux families are known: MFS (major facilitator superfamily), SMR (small multidrug resistance), RND (resistance-nodulation-division), the ABC (ATP binding-cassette), and the MATE family (multidrug and toxic compound extrusion) [64]. Recently, it was discovered that some efflux pumps are primarily active in bacterial biofilms—for example, the MexAB-OprM pump that is able to pump out several classes of antibiotics like quinolones, β-lactams, tetracycline, chloramphenicol and novobiocin, macrolides, and biocides [65]. Even quorum sensing inhibitors like brominated furanone C-30 [66] confer *P. aeruginosa* biofilm cells resistance against the antibacterial peptide colistin [67]. More recently, it was discovered that in *P. aeruginosa* biofilms, the MerR-Like transcriptional regulator BrlR binds the promoter regions of the efflux pumps operons *mexAB-oprM* and *mexEF-oprN*, activating their expression and hence conferring resistance to multiple antibiotics, including tobramycin, trimethoprim, and chloramphenicol [68].

Remarkably, although the specific changes in gene expression linked to a higher antimicrobial tolerance of plant-associated bacteria are scarce, in 2012 Muranaka et al. [69] studied the global gene expression profile of biofilms of the plant pathogen *Xylella fastidiosa* exposed to the commonly used antimicrobial agent copper and against the antibiotic tetracycline. Exposure to subinhibitory cooper concentration (3 mM) increased the expression of 223 genes while repressing 150. The most induced genes included those belong to phages, virulence, and adaptation such as hemagglutinins, hemolysin secretion/activation protein, copper homeostasis proteins, and efflux pumps, while the main repressed genes included those related to RNA and protein synthesis and type IV pilus. Remarkably, increasing the cooper concentration to 7 mM was linked to a differential expression of more genes (461 induced and 407 repressed), belonging to the same categories but including general metabolic function and movement in the repressed ones. While for tetracycline (100 and 800 µg/mL), clearly most of the differential expressed genes upon tetracycline treatments were specific, showing only little overlap with those affected by cooper, nevertheless strikingly, among the common repressed genes were those encoding ribosomal proteins and general metabolic proteins, indicating that *Xylella fastidiosa* decreases its metabolic functions, translation, and cell movement as a response against antimicrobials. Interestingly, the differentially expressed genes included 12 TA modules, and the authors demonstrated that *X. fastidiosa* is able to form persister cells as a response of both tetracycline and copper treatment (0.05 percent of the total population). Moreover, exposing the cells to a subinhibitory concentration of copper increased 26-fold the number of persisters [69], and although not proved, these persisters were likely produced by the activation of the TA modules. Muranaka et al. demonstrated for the first time that the expression of TA systems may be associated with persister formation as a main biofilm survival strategy for a plant pathogenic bacterium; however to date, there are no further studies to evaluate the generation of persisters as a response against antimicrobials in more bacterial species associated with plants.

It is important to note that the use of antibiotics in agriculture has been a common practice. Streptomycin is used in several countries to combat *Erwinia amylovora*, the causative agent of fire blight, the most important bacterial disease of apple, pear, and related ornamental plants, while gentamicin is used in Mexico to control fire blight of apple and pear, as well other bacterial plant pathogens such as *Pectobacterium*, *Pseudomonas*, *Ralstonia*, and *Xanthomonas*, while oxytetracycline is used in the United States primarily on pear and apple for fire blight management and oxolinic acid is used in Israel for the same reason and in Japan to control *Burkholderia glumae* infections in rice. Unfortunately, plant pathogens have already evolved resistance mechanisms against these antibiotics [70]. Since *Erwinia amylovora* is able to form biofilm [71], it is expected that some of all the specific biofilm antimicrobial mechanisms known to date are important in this pathogen to adapt against antimicrobial treatment; nevertheless, there is a lack of information about such mechanisms in this and many other important plants pathogenic bacteria. Table 13.2 summarizes the genes related to the specific enhancement of antimicrobial resistance in biofilms of plant pathogenic bacteria.

13.6 Potential for Implementing Antibiofilm Strategies to Protect Crops

Providing food for the human population has become a challenge due to the reduction in plantation areas, as well as the presence of pests and diseases affecting crops [72]. In this regard, efforts are underway to increase cultivation areas and in the development of methods to increase agricultural productivity [73]. However, it is expected that the current situation will become more complicated due to global climate change, since it is predicted that droughts and severe rains will intensify, hence affecting crops [74, 75]. These changes will also contribute to the establishment of plant pathogens, particularly bacterial diseases that are difficult to control and often result in severe financial losses [76]. Within management practices and eradication of bacterial diseases in agricultural

Table 13.2 Genes Involved in Biofilm Resistance against Antimicrobials.

Bacterial Specie	Gene (s)	Antimicrobial Compounds	Resistance Mechanisms	References
Pseudomonas aeruginosa	*ndvB*	Aminoglycosides such as tobramycin	Production of cyclic glycans that bind and inactivate aminoglycosides Enhances ethanol oxidation that by an unknown mechanism increase antibiotic tolerance	[62]
Salmonella enterica serovar Typhimurium	Pmr operon	Antibacterial peptides such as polimyxin	Modificate the aminoarabinose in the outer membrane to protect of action of the antimicrobial peptides	[56]
Acinetobacter baumanii	K locus	Antibacterial peptides	Production of complex exopolysaccharides, including the capsule for protect against antimicrobials	[58]
Pseudomonas aeruginosa	MexAB-OprM MexEF-oprN	Antibiotics like tobramycin, trimethoprim, chloramphenicol Antibacterial peptide colistin Quorum sensing inhibitors like furanone C-30	Activate multidrug efflux pump to confer resistance	[66, 67]
Xylella fastidiosa	pspA	Copper Antibiotics like tetracycline	Activate greater expression of hemagglutinin for multilayer biofilm	[69]
Xylella fastidiosa	pilP and pilT	Copper Antibiotics like tetracycline	Reduced movement and metabolism for entering into a resistant physiological state	[69]
Xylella fastidiosa	XF1695, XFa0045, XF2080, XF1589, XF1597, XF1709 XFa0046, XF2081, XF1596, XF710, XF2491, XFXF2067	Copper Antibiotics like tetracycline	Toxin-antitoxin systems for formation of persister cells that are highly tolerant to antibiotics	[69]

crops, the use of antibiotics is limited and restricted, because the resistance generated can be transmitted to bacteria that interact with humans who eat the antibiotic-treated foods [77, 78].

Although the utilization of antimicrobial agents in the clinic to treat infections, and in the fields of agriculture, meat, and poultry production has been very valuable, it is an inevitable feature that eventually bacteria will develop resistance against them. Thus, other strategies that include the utilization of QS inhibitors and biofilm disruption chemicals have been also proposed as alternatives to the current antimicrobials. Interestingly, QS inhibition (quorum quenching or QQ) has been shown to inhibit virulence factor production (including biofilm formation and robustness) without affecting bacterial growth *in vitro*, hence it was speculated that QQ treatments do not impose a harsh selective pressure for the selection of bacterial resistance [79]. Nevertheless, it was demonstrated that resistance against single QQ like the brominated furanone C-30 arises easily by the activation of the antibiotic efflux pump MexAB-OmpR [66, 80] and by decreasing its permeability [80]. In addition, it was found that QS enhances the stress response its disruption by C-30 select for QQ resistance, since those resistant bacteria are also more tolerant to a wide range of stressors including oxidative, hyperosmotic, and heavy metal exposure [81]. Nevertheless, in nature, marine algae like *Delisea pulchra* protect their blades against bacterial biofilm formation by the simultaneous production of several different halogenated furanones [82], suggesting a key factor for avoiding the selection of bacterial resistance may be the utilization of multiple QQ compounds instead of a single one. In a pioneer study, des Essarts et al. screened a library of QS inhibitors to identify those useful against *Pectobacterium atrosepticum*, the causative agent of the blackleg and soft-rot diseases, identifying four different N,N′-bisalkylated imidazolium (Figure 13.1A–D) useful compounds. Remarkably, two of them decreased the severity of the symptoms provoked by *P. atrosepticum* in potato tuber assays [83], notably there is a need for more similar studies to identify suitable QS inhibitors to combat biofilm formation of bacterial plant pathogens.

Another potential advantage of disrupting QS and or biofilm formation is that it will severely decrease the pathogen tolerance to classical antimicrobials [81, 84, 85]; hence, combination schemes could be a more potent alternative to protect crops against biofilm-forming bacterial pathogens. Remarkably, although resistance against specific antibiofilm compounds have not been extensively studied to date. In 2013, Travier and coworkers demonstrated that for *Eschericha coli*, the development of resistance against the biofilm inhibitor group 2 capsule polysaccharide (G2cps), although possible, is rare, since it is mediated by the combination of several different mutations that modify the surface physicochemical properties of the cells, hence counteracting the changes in ionic charge and Lewis-base properties mediated by G2cps [86]. Since the simultaneous appearance of those mutations is unlikely, it is appealing to think that biofilm disruptors may be a robust alternative for protecting commercially important crops from bacterial pathogens.

Although biofilms are regulated by QS, inhibition by small molecules targeting other processes can have antibiofilm activity as well, such is the case of D-leucine (Figure 13.1E) and 3-indolylacetonitrile (Figure 13.1F), which inhibit biofilm formation of *Xanthomonas citri* subsp. *citri*, at not bactericide concentrations; furthermore, the *in vivo* applications of these compounds decrease lesions and significantly reduce the bacterial population when applied directly to leaves. In addition, both molecules increase bacteria

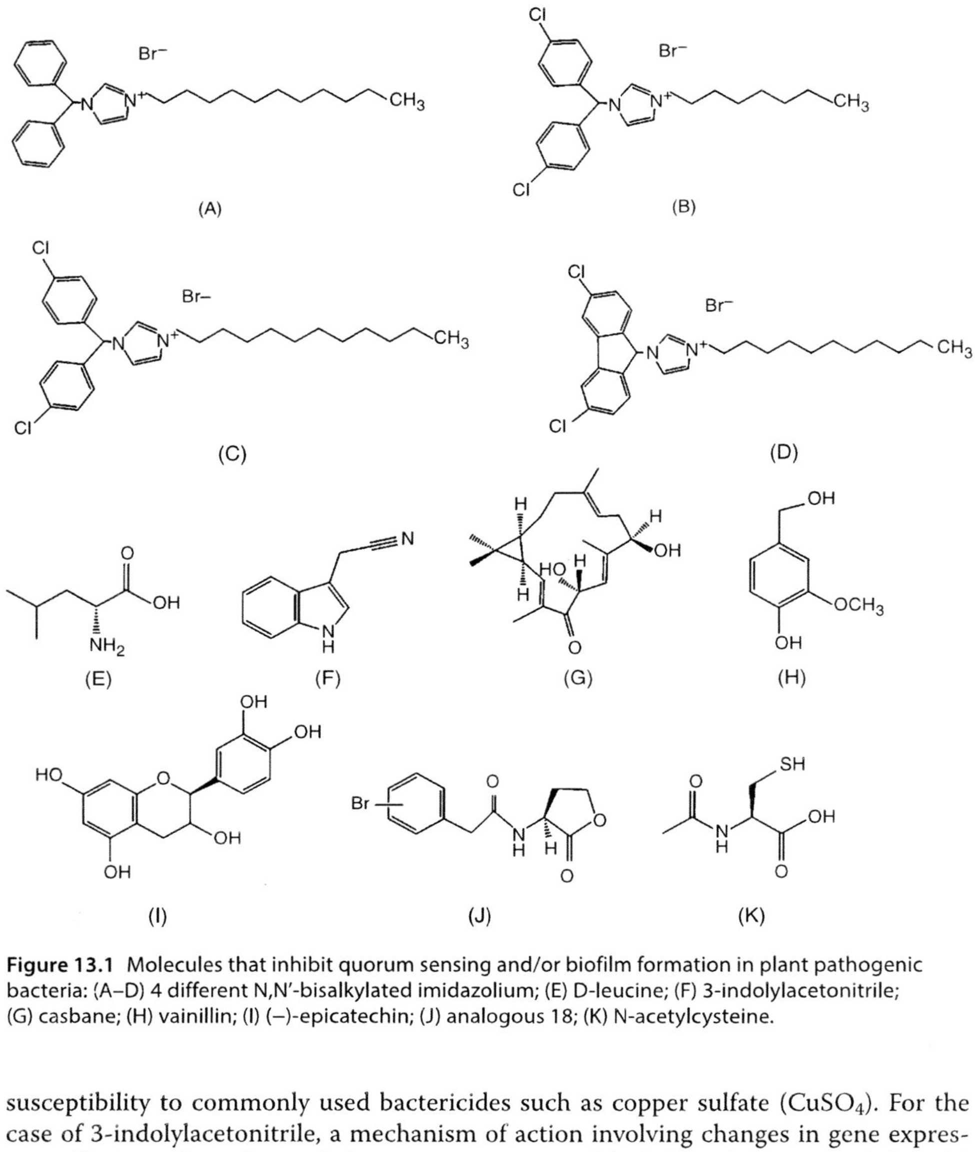

Figure 13.1 Molecules that inhibit quorum sensing and/or biofilm formation in plant pathogenic bacteria: (A–D) 4 different N,N′-bisalkylated imidazolium; (E) D-leucine; (F) 3-indolylacetonitrile; (G) casbane; (H) vainillin; (I) (−)-epicatechin; (J) analogous 18; (K) N-acetylcysteine.

susceptibility to commonly used bactericides such as copper sulfate ($CuSO_4$). For the case of 3-indolylacetonitrile, a mechanism of action involving changes in gene expression affecting of motility and chemotaxis is suggested [87]. The ability to modulate QS (either inhibiting or stimulating), using small molecules, represents a viable strategy for controlling virulence and biofilm formation not only for phytopathogenic bacteria, but also for symbiotic species [88]. However, although these kinds of strategies are promising, more studies are needed before achieving its implementation.

Although plants lack an immune system for combating bacterial infections, thousands of years of evolution have allowed them to develop several alternative metabolic pathways, producing an arsenal of diverse metabolites to protect themselves. In the last decade, several studies have identified phytochemicals and others natural products that interfere with the process of QS and biofilm formation of pathogenic bacteria [84, 89, 90].

The exact mechanism of action is not completely understood; the effect on bacteria that interact with plants requires further investigation. Some of these compounds are: 3-indolylacetonitrile (Figure 13.1F), which inhibit biofilm formation of *X. citri* [87], and casbane diterpene (Figure 13.1G) in *Pseudomonas fluorescens* [91]. Moreover, phytochemicals that enhance the formation of biofilms have also been identified. Examples of them are vainillin (Figure 13.1E) and (−)-epicatechin (Figure 13.1F) in *Agrobacterium tumefaciens* [92]. In addition, the use of small molecules that induce the formation of biofilms prematurely sometimes promotes abnormal biofilm maturation, which renders them more sensitive to antimicrobials. Furthermore, induction of QS is proposed to control the expression of certain bacterial phenotypes, which are beneficial to associated crop [93]. This is the case of *Pseudomonas aureofaciens,* a symbiotic bacterium that uses QS for expression of antifungal phenazines. Similarly, this strategy can also be applied to nitrogen-fixing bacteria to enhance nitrogen fixation, thus reducing the use of fertilizers [93].

To date, several synthetic compounds have been evaluated for their QS and biofilm inhibitory properties over different pathogenic bacterial [94], among them Palmer and coworkers determined the effect of 23 AHL synthetic analogs over plant pathogenic bacteria, showing differences *in vitro* among the effects of the compounds on the QS regulation (stimulation/inhibition) of *Pectobacterium carotovora* and *Pseudomonas syringae* B2784. Interestingly, the activity patterns also correlated with *in vivo* models used *Solanum tuberosum* (potato) and *Phaseolus vulgaris* (green bean). This study stresses the potent QQ effect shown by the analogous 18 (Figure 13.1 J), suggesting that the application of this class of compounds could be able to counteract the bacterial infections and the associated damage in a wide range of organisms and environments [95].

Another approach for the control of biofilms and its application in agriculture is the evaluation of molecules commonly used in other areas, by instance a study by Muranaka and colleagues explored the utilization of N-acetylcysteine (NAC) (Figure 13.1K), which is a molecule with antibiofilm activity widely used in the clinic. NAC is an analog of cysteine that decreases the viscosity of bronchial secretions due its property of breaking disulfide bonds [96]. These researchers evaluated NAC in *Xylella fastidiosa,* a bacterium that infects citrus (citrus variegated chlorosis) and whose main pathogenic mechanism is the formation of biofilms in the xylem of plants. Importantly, *in vitro* studies show that NAC reduced adhesion, exopolysaccharides production, and biofilm formation. Furthermore, using an *in vivo* model with *Citrus sinensis* (orange), they showed remission of symptoms and decreased presence of the bacteria. These results are encouraging, since currently there are no approved methods to reduce *X. fastidiosa* presence in crops [97].

Other strategies that have been proposed for controlling plant pathogens in agriculture are derived from studies of the interaction between QS bacterial systems and plants. In this regard, it has been shown that some bacterial autoinducers are able to stimulate plant growth, providing crop protection [98, 99]. Current studies show that the implementation of exogenous AHL can regulate growth and resistance against bacterial infections in *A. thaliana*. Interestingly, for each of these phenomena, the type AHL and the size of its acyl chain is important, since AHL with short acyl chains (C6 and oxo C8) are involved primary in root elongation, while AHL with longer chains (C10 oxo) in root hair induction and (oxo C14) increase resistance [100].

Schenk et al. reported that *A. thaliana* that are treated for resistance against *P. syringae* pv. *tomato* DC3000 reduce their C.F.U. numbers [101]. Recently, it has been shown that this defense mechanism depends on the production of oxylipins (jasmonic acid) and salicylic acid, the latter being a key element in the systemic plant defense [102].

13.6 Conclusions

The relevant findings in this chapter encourage further studies to test whether the same effects could be achieved in plants of agronomic interest. Although various strategies related to QS and biofilm manipulations to protect crops are under development, it is advisable to concentrate our efforts on the most promising lines of research to ensure their application in a shorter time. As recent studies demonstrate that the bacteria-plant communication systems are a very complex, the application of QQ and antibiofilms molecules to counteract bacterial infections in plants is still challenging. Hence, we propose that more effort must be made to understand the mechanism of action of these molecules, and the possible side effects over the symbiotic bacterial communities present in soils and in human microbiota. One possible solution would be the identification of highly specific molecules with short half-lives to minimize their interactions with other bacterial species and side effects. In this sense, is necessary to further extend the evaluation of QS-antibiofilm activities using *in vivo* relevant models, preferably with wild bacterial strains interacting with their native eukaryotic hosts.

In conclusion, despite the progress that has been made so far in developing the strategies discussed in this chapter, their application in agriculture is still at an early stage. However, current evidence shows an enormous potential; hence, significant advances in these fields are expected in the future.

Acknowledgments

R-GC research is funded by SEP-CONACYT 152794 and by PAPIIT-UNAM IA201116, I-CJ research is supported by Fideicomiso-COLPOS 167304 and Cátedras-CONACyT program, D-PC research is supported by Cátedras-CONACyT program, and N-MC research is supported by the CONACYT PhD grant 376049.

References

1 X.Z. Li, J.S. Webb, S. Kjelleberg and B. Rosche, Enhanced benzaldehyde tolerance in *Zymomonas mobilis* biofilms and the potential of biofilm applications in fine-chemical production, *Appl Env. Microbiol.*, 72, 1639–1644 (2006).
2 S.C. Booth, M.L. Workentine, J. Wen, R. Shaykhudinov, H.J. Vogel, H. Ceri, R. J. Turner and A. M. Weljie, Differences in metabolism between the biofilm and planktonic response to metal stress, *J Proteome Res.*, 10, 3190–3199 (2011).
3 P.S. Stewart, Antimicrobial tolerance in biofilms, *Microbiol Spectr.*, 3 (2015).

4 N. Hoiby, T. Bjarnsholt, M. Givskov, S. Molin and O. Ciofu, Antibiotic resistance of bacterial biofilms, *Int J Antimicrob Agents.*, 35, 322–332 (2010).

5 J. Zhang, A.M. Ormala-Odegrip, J. Mappes and J. Laakso, Top-down effects of a lytic bacteriophage and protozoa on bacteria in aqueous and biofilm phases, *Ecol Evol.*, 4, 4444–4453 (2014).

6 J. S. Madsen, M. Burmolle, L. H. Hansen, and S. J. Sorensen, The interconnection between biofilm formation and horizontal gene transfer, *FEMS Immunol Med Microbiol.*, 65, 183–195 (2012).

7 W. F. McCoy, J. D. Bryers, J. Robbins and J. W. Costerton, Observations of fouling biofilm formation, *Can J Microbiol.*, 27, 910–917 (1981).

8 T. J. Marrie and J. W. Costerton, Scanning and transmission electron microscopy of *in situ* bacterial colonization of intravenous and intraarterial catheters, *J Clin Microbiol.*, 19, 687–693 (1984).

9 I. Castillo-Juarez, T. Maeda, E. A. Mandujano-Tinoco, M. Tomás, B. Pérez-Eretza, S. J. García-Contreras, T.K. Wood and R. García-Contreras, Role of quorum sensing in bacterial infections, *World J Clin Cases.*, 3, 575–598 (2015).

10 B. A. Hense, C. Kuttler, J. Müller, M. Rothballer, A. Hartmann and J.U. Kreft, Does efficiency sensing unify diffusion and quorum sensing? *Nat Rev Microbiol.*, 5, 230–239 (2007).

11 B. Quiñones, G. Dulla and S. E. Lindow, Quorum sensing regulates exopolysaccharide production, motility, and virulence in *Pseudomonas syringae*, *Mol. Plant. Microbe. Interact.*, 18, 682–693 (2005).

12 D. G. Davies, M. R. Parsek, J. P. Pearson, B. H. Iglewski, J. W. Costerton and E. P. Greenberg, The involvement of cell-to-cell signals in the development of a bacterial biofilm, *Science*, 280, 295–298 (1998).

13 Y. Zhao, G. Quian, F. Yin, J. Fan, Z. Zhai, C. Liu, B. Hu and F. Liu, Proteomic analysis of the regulatory function of DSF-dependent quorum sensing in *Xanthomonas oryzae* pv. *oryzicola*, *Microb Pathog.*, 50, 48–55 (2011).

14 M. M. Marketon, S. A. Glenn, A. Eberhard and J. E. Gonzalez, Quorum sensing controls exopolysaccharide production in *Sinorhizobium meliloti*, *J Bacteriol.*, 185, 325–331 (2003).

15 L. Montanaro, A. Poggi, L. Visai, S. Ravaioli, D. Campoccia, P. Speziale and C.R. Arciola, Extracellular DNA in biofilms, *Int J Artif Organs.*, 34, 824–831 (2011).

16 R. Hazan, Y. A. Que, D. Maura, B. Strobel, P. A. Majcherczyk, L.R. Hopper, D.J. Wilbur, T.N. Hreha, B. Barquera and L.G. Rahme, Auto poisoning of the respiratory chain by a quorum-sensing-regulated molecule favors biofilm formation and antibiotic tolerance, *Curr Biol.*, 26, 195–206 (2016).

17 J. M. Ghigo, Natural conjugative plasmids induce bacterial biofilm development, *Nature.*, 412, 442–445 (2001).

18 J. B. Kaplan, Biofilm dispersal: mechanisms, clinical implications, and potential therapeutic uses, *J Dent Res.*, 89, 205–218 (2010).

19 M. E. Davey and G. A. O'toole, Microbial Biofilms: from ecology to molecular genetics, *Microbiol Mol Biol Rev.*, 64, 847–867 (2000).

20 T. Danhorn and C. Fuqua, Biofilm formation by plant-associated bacteria, *Annu Rev Microbiol.*, 61, 401–422 (2007).

21 P. Bogino, M. Oliva, F. Sorroche and W. Giordano, The role of bacterial biofilms and surface components in plant-bacterial associations, *Int. J. Mol. Sci.*, 14, 15838–15859 (2013).

22 B. E. Ramey, M. Koutsoudis, S. B. von Bodman and C. Fuqua, Biofilm formation in plant–microbe associations, *Curr. Opin. Microbiol.*, 7, 602–609 (2004).
23 L. F. Castiblanco and G. W. Sundin, New insights on molecular regulation of biofilm formation in plant-associated bacteria, *J. Integr. Plant Biol.*, 58, 362–372 (2015).
24 C. M. Wallis, A. K. Wallingford and J. Chen, Grapevine rootstock effects on scion sap phenolic levels, resistance to *Xylella fastidiosa* infection, and progression of Pierce's disease, *Front. Plant Sci.*, 4, 502 (2013).
25 R. P. Purcino, C. L. Medina, D. M. de Souza, F. V. Winck, E. C. Machado, J. C. Novello, M. A. Machado and P. Mazzafera, *Xylella fastidiosa* disturbs nitrogen metabolism and causes a stress response in sweet orange *Citrus sinensis* cv. Pera, *J. Exp. Bot.*, 58, 2733–2744 (2007).
26 R. Caserta, M.A. Takita, M.L. Tragon, L.K. Roselli-Murai, A.P. de Souza, L. Peroni, D.R. Stach-Machado, A. Andrade, C.A. Labate, E.W. Kitajima, M.A. Machado and A. A. de Souza, Expression of *Xylella fastidiosa* fimbrial and afimbrial proteins during biofilm formation, *Appl. Environ. Microbiol.*, 76, 4250–4259 (2010).
27 J. K. Parker, H. Chen, S. E. McCarty, L. Y. Liu and L. de La Fuente, Calcium transcriptionally regulates the biofilm machinery of *Xylella fastidiosa* to promote continued biofilm development in batch cultures, *Environ. Microbiol.*, 18, 1620–1634 (2016).
28 C. M. Rojas, J. H. Ham, W. L. Deng, J. J. Doyle and A. Collmer, HecA, a member of a class of adhesins produced by diverse pathogenic bacteria, contributes to the attachment, aggregation, epidermal cell killing, and virulence phenotypes of *Erwinia chrysanthemi* EC16 on *Nicotiana clevelandii* seedlings, *Proc. Natl. Acad. Sci. U. S. A.*, 99, 13142–13147 (2002).
29 J. M. Koczan, B. R. Lenneman, M. J. McGrath and G. W. Sundin, Cell surface attachment structures contribute to biofilm formation and xylem colonization by *Erwinia amylovora*, *Appl. Environ. Microbiol.*, 77, 7031–7039 (2011).
30 J. Bogs, I. Bruchmüller, C. Erbar and K. Geider, Colonization of host plants by the fire blight pathogen *Erwinia amylovora* marked with genes for bioluminescence and fluorescence, *Phytopathology.*, 88, 416–421 (1998).
31 N. Piqué, D. Miñana-Galbis, S. Merino and J. Tomás, Virulence factors of *Erwinia amylovora*: A Review, *Int. J. Mol. Sci.*, 16, 12836–12854 (2015).
32 X. H. Lu, S.Q. An, D.J. Tang, Y. McCarthy, J.L. Tang, J.M. Dow and R.P. Ryan, RsmA regulates biofilm formation in *Xanthomonas campestris* through a regulatory network involving cyclic di-GMP and the Clp transcription factor, *PLoS One.*, 7, e52646 (2012).
33 L. C. Martínez and V. Vadyvaloo, Mechanisms of post-transcriptional gene regulation in bacterial biofilms, *Front. Cell. Infect. Microbiol.*, 4, 1–15 (2014).
34 J. M. Dow, L. Crossman, K. Findlay, Y.Q. He, J.X. Feng and J.L. Tang, Biofilm dispersal in *Xanthomonas campestris* is controlled by cell-cell signaling and is required for full virulence to plants, *Proc. Natl. Acad. Sci. U. S. A.*, 100, 10995–11000 (2003).
35 L. Moleleki, R. Pretorius, C. Tanui, G. Mosina and J. A. Theron, A quorum sensing-defective mutant of *Pectobacterium carotovorum* subsp. *brasiliense* 1692 is attenuated in virulence and unable to occlude xylem tissue of susceptible potato plant stems, *Mol. Plant Pathol.*, doi:10.1111/mpp.12372 (2016).
36 J. Vasse, Microscopic studies of intercellular infection and protoxylem invasion of tomato roots by *Pseudomonas solanacearum*, *Mol. Plant-Microbe Interact.*, 8, 241 (1995).

37 J. Yao and C. Allen, The plant pathogen *Ralstonia solanacearum* needs aerotaxis for normal biofilm formation and interactions with its tomato host, *J. Bacteriol.*, 189, 6415–24 (2007).

38 Y. Mori, The vascular plant pathogenic bacterium *Ralstonia solanacearum* produces biofilms required for its virulence on the surfaces of tomato cells adjacent to intercellular spaces, *Mol. Plant Pathol.*, 17, 890-902 (2015).

39 M. I. Siri, A. Sanabria, C. Boucher and M. J. Pianzzola, New type IV pili-related genes involved in early stages of *Ralstonia solanacearum* potato infection, *Mol. Plant. Microbe. Interact.*, 27, 712–724 (2014).

40 M. Frederix, A. Edwards, A. Swiderska, A. Stanger, R. Karunakaran, A. Williams, P. Abbruscato, M. Sanchez-Contreras, P.S. Poole and J.A. Downie, Mutation of *praR* in *Rhizobium leguminosarum* enhances root biofilms, improving nodulation competitiveness by increased expression of attachment proteins, *Mol. Microbiol,*. 93, 464–478 (2014).

41 D. N. Rodríguez-Navarro, M. S. Dardanelli and J. E. Ruíz-Saínz, Attachment of bacteria to the roots of higher plants, *FEMS Microbiol. Lett.*, 272, 127–136 (2007).

42 I. Z. Zerrouk, M. Benchabane, L. Khelifi, K. Yokawa, J. Ludwing-Müller and F. Baluska, A *Pseudomonas* strain isolated from date-palm rhizospheres improves root growth and promotes root formation in maize exposed to salt and aluminum stress, *J. Plant Physiol.*, 191, 111–119 (2016).

43 C. Fuqua, Passing the baton between laps: adhesion and cohesion in *Pseudomonas putida* biofilms, *Mol. Microbiol.*, 77, 533–536 (2010).

44 A. Ueda and H. Saneoka, Characterization of the ability to form biofilms by plant-associated *Pseudomonas* species, *Curr. Microbiol.*, 70, 506–513 (2015).

45 S. Ghods, I. M. Sims, M. F. Moradali and B. H. A Rehm, Bactericidal compounds controlling growth of the plant pathogen *Pseudomonas syringae* pv. *actinidiae*, which forms biofilms composed of a novel exopolysaccharide, *Appl. Environ. Microbiol.*, 81, 4026–4036 (2015).

46 P. M. Merritt, T. Danhorn and C. Fuqua, Motility and chemotaxis in *Agrobacterium tumefaciens* surface attachment and biofilm formation, *J. Bacteriol.*, 189, 8005–8014 (2007).

47 E. Barahona, A. Navazo, F. Yousef-Coronado, D. Aguirre de Cárcer, F. Martínez-Granero, M. Espinosa-Urgel, M. Martín and R. Rivilla, Efficient rhizosphere colonization by *Pseudomonas fluorescens* f113 mutants unable to form biofilms on abiotic surfaces. *Environ. Microbiol.* 12, 3185–95 (2010).

48 E. Lira-Silva, M. G. Santiago-Martínez, R. García-Contreras, A. Marín-Hernández,R. Moreno-Sánchez, R. Jasso-Chávez, Cd2+ resistance mechanisms in *Methanosarcina acetivorans* involve the increase in the coenzyme M content and induction of biofilm synthesis, *Env. Microbiol Rep.*, 5, 799–808 (2013).

49 X. S. Zhang, R. Garcia-Contreras and T. K. Wood, YcfR (BhsA) influences *Escherichia coli* biofilm formation through stress response and surface hydrophobicity, *J Bacteriol.*, 189, 3051–3062 (2007).

50 L. R. Hoffman, D. A. D'Argenio, M. J, MacCoss, Z. Zhang, R. A, Jones and S. A. Miller, Aminoglycoside antibiotics induce bacterial biofilm formation, *Nature.*, 436, 1171–1175 (2005).

51 S. Rachid, K. Ohlsen, U. Wallner, J. Hacker, M. Hecker and W. Ziebuhr, Alternative transcription factor sigma(B) is involved in regulation of biofilm expression in a *Staphylococcus aureus* mucosal isolate, *J Bacteriol.*, 182, 6824–6826 (2000).

52 S. Rachid, S. Cho, K. Ohlsen, J. Hacker and W. Ziebuhr, Induction of *Staphylococcus epidermidis* biofilm formation by environmental factors: the possible involvement of the alternative transcription factor sigB, *Adv Exp Med Biol.*, 485, 159–166 (2000).
53 A. W. Qurashi and A. N. Sabri, Bacterial exopolysaccharide and biofilm formation stimulate chickpea growth and soil aggregation under salt stress, *Braz J Microbiol.*, 43, 1183–1191 (2012).
54 P. A. Suci, M. W. Mittelman, F. P. Yu and G. G. Geesey, Investigation of ciprofloxacin penetration into *Pseudomonas aeruginosa* biofilms, *Antimicrob Agents Chemother.*, 38, 2125–2133 (1994).
55 W. C. Chiang, M. Nilsson, P. Ø Jensen, N. HØiby, T. E. Nielsen, M. Givskov and T. Tolker-Nielsen, Extracellular DNA shields against aminoglycosides in *Pseudomonas aeruginosa* biofilms, *Antimicrob Agents Chemother.*, 57, 2352–2361 (2013).
56 L. Johnson, S. R. Horsman, L. Charron-Mazenod, A. L. Turnbull, H. Mulcahy, M. G. Surrete and S. Lewenza, Extracellular DNA-induced antimicrobial peptide resistance in *Salmonella enterica* serovar Typhimurium, *BMC Microbiol.*, 13, 115 (2013).
57 K. Lewis, Riddle of biofilm resistance, *Antimicrob Agents Chemother.*, 45, 999–1007 (2001).
58 E. Geisinger and R. R. Isberg, Antibiotic modulation of capsular exopolysaccharide and virulence in *Acinetobacter baumannii*, *PLoS Pathog.*, 11, e1004691 (2015).
59 T. Bjarnsholt, O. Ciofu, S. Molin, M. Givskov and N. Hoiby, Applying insights from biofilm biology to drug development - can a new approach be developed? *Nat Rev Drug Discov.*, 12, 791–808 (2013).
60 K. Lewis, Persister cells, *Annu Rev Microbiol.*, 64, 357–372 (2010).
61 T. K. Wood, Combatting bacterial persister cells, *Biotechnol Bioeng.*, 113, 476–483 (2015).
62 T. F. Mah, B. Pitts, B. Pellock, G. C. Walker and P. S. Stewart, A genetic basis for *Pseudomonas aeruginosa* biofilm antibiotic resistance, *Nature.*, 426, 306–310 (2003).
63 T. Beaudoin, L. Zhang, A. J. Hinz, C. J. Parr and T. F. Mah, The biofilm-specific antibiotic resistance gene ndvB is important for expression of ethanol oxidation genes in *Pseudomonas aeruginosa* biofilms, *J Bacteriol.*, 194, 3128–3136 (2012).
64 J. Sun, Z. Deng and A. Yan, Bacterial multidrug efflux pumps: mechanisms, physiology and pharmacological exploitations, *Biochem Biophys Res Commun.*, 453, 254–267 (2014).
65 H. Chen, C. Yi, J. Zhang, W. Zhang, Z. Ge, C. G. Yang and C. He, Structural insight into the oxidation-sensing mechanism of the antibiotic resistance of regulator MexR, *EMBO Rep.*,11, 685–690 (2010).
66 T. Maeda, R. García-Contreras, M. Pu, L. Sheng, L. R. García, M. Tomás and T. K. Wood, Quorum quenching quandary: resistance to antivirulence compounds, *ISME J.*, 6, 493–501 (2012).
67 S. J. Pamp, M. Gjermansen, H. K. Johansen and T. T. Tolker-Nielsen, Tolerance to the antimicrobial peptide colistin in *Pseudomonas aeruginosa* biofilms is linked to metabolically active cells, and depends on the pmr and mexAB-oprM genes, *Mol Microbiol.*, 68, 223–240 (2008).
68 J. Liao, M. J. Schurr and K. Sauer, The MerR-like regulator BrlR confers biofilm tolerance by activating multidrug efflux pumps in *Pseudomonas aeruginosa* biofilms, *J Bacteriol.*, 195, 3352–3363 (2013).
69 L. S. Muranaka, M. A. Takita, J. C. Olivato, L. T. Kishi and A. A. de Souza, Global expression profile of biofilm resistance to antimicrobial compounds in the

plant-pathogenic bacterium *Xylella fastidiosa* reveals evidence of persister cells, *J Bacteriol.*, 194, 4561–4569 (2012).
70 V. O. Stockwell and B. Duffy, Use of antibiotics in plant agriculture, *Rev Sci Tech.*, 31, 199–210 (2012).
71 J. M. Koczan, M. J. McGrath, Y. Zhao and G. W. Sundin, Contribution of *Erwinia amylovora* exopolysaccharides amylovoran and levan to biofilm formation: implications in pathogenicity. *Phytopathology.*, 99, 1237–1244 (2009).
72 D. Tilman, C. Balzer, J. Hill and B. L. Befort, Global food demand and the sustainable intensification of agriculture, *Proc. Natl. Acad. Sci.*, 108, 20260–20264 (2011).
73 M.D. Edgerton, Increasing crop productivity to meet global needs for feed, food, and fuel, *Plant Physiol.*,149, 7–13 (2009).
74 A. J. McMichael, Impact of climatic and other environmental changes on food production and population health in the coming decades, *Proc. Nutr. Soc.*, 60, 195–201 (2001).
75 B. Sánchez, A. Rasmussen and J. R. Porter, Temperatures and the growth and development of maize and rice: a review, *Glob. Chang. Biol.*, 20, 408–417 (2014).
76 D. P. Bebber, Range-expanding pests and pathogens in a warming world, *Annu. Rev. Phytopathol.*, 53, 335–356 (2015).
77 J. Davies and D. Davies, Origins and evolution of antibiotic resistance. *Microbiol. Mol. Biol. Rev.*, 74, 417–433 (2010).
78 M. P. Francino, Antibiotics and the human gut microbiome: dysbioses and accumulation of resistances, *Front. Microbiol.*, 6, 1543 (2016).
79 D. A. Rasko and V. Sperandio, Anti-virulence strategies to combat bacteria-mediated disease, *Nat Rev Drug Discov.*, 9, 117–128 (2010).
80 R. GarcÍa-Contreras, B. Pérez-Eretza, R. Jasso-Chávez, E. Lira-Silva, J. A. Roldán-Sánchez, A. González-Valdez, G. Soberón-Chávez, R. Coria-Jiménez, M. Martínez-Vázquez, L.D. Alcaraz, T. Maeda and T. K. Wood, High variability in quorum quenching and growth inhibition by furanone C-30 in *Pseudomonas aeruginosa* clinical isolates from cystic fibrosis patients, *Pathog Dis.*, 73, ftv040 (2015).
81 R. García-Contreras, L. Nuñez-López, R. Jasso-Chávez, B. K. Kwuan, J. A. Belmont, A. Rangel-Vega, T. Maeda and T.K. Wood, Quorum sensing enhancement of the stress response promotes resistance to quorum quenching and prevents social cheating, *ISME J.*, 9, 115–125 (2014).
82 M. Hentzer, H. Wu, J. B. Andersen, K. Riedel, T. B. Rasmussen, N. Bagge, N. Kumar, M. A. Schembri, Z. Song, P. Kristoffersen, M. Manefield, J. W. Costerton, S. Molin, L. Eberl, P. Steinberg, S. Kjelleberg, N. HØiby and M. Givskov, Attenuation of *Pseudomonas aeruginosa* virulence by quorum sensing inhibitors, *EMBO J.*, 22, 3803–3815 (2003).
83 Y. R. des Essarts, M. Sabbah, A. Comte, L. Soulére, Y. Queneau, Y. Dessaux, V. Hélias and D. Faure, N,N′-alkylated Imidazolium-derivatives act as quorum-sensing inhibitors targeting the *Pectobacterium atrosepticum*-induced symptoms on potato tubers, *Int J Mol Sci.*, 14, 19976–19986 (2013).
84 G. Brackman, K. Breyne, R. de Rycke, A. Vermote, F. van Nieuwerburgh, E. Meyer, S. van Calenbergh and T. Coenye, The quorum sensing inhibitor hamamelitannin increases antibiotic susceptibility of *Staphylococcus aureus* biofilms by affecting peptidoglycan biosynthesis and eDNA release, *Sci Repk.*, 6, 20321 (2016).

85 S. Kiran, P. Sharma, K. Harjai and N. Capalash, Enzymatic quorum quenching increases antibiotic susceptibility of multidrug resistant *Pseudomonas aeruginosa, Iran J Microbiol.*, 3, 1–12 (2012).

86 L. Travier, O. Rendueles, L. Ferrieres, J. M. Herry and J. M. Ghigo, *Escherichia coli* resistance to nonbiocidal antibiofilm polysaccharides is rare and mediated by multiple mutations leading to surface physicochemical modifications, *Antimicrob Agents Chemother.*, 57, 3960–3968 (2013).

87 J. Li and N. Wang, Foliar application of biofilm formation-inhibiting compounds enhances control of citrus canker caused by *Xanthomonas citri* subsp. *citri, Phytopathology.*, 104, 134–142 (2014).

88 E. Buommino, M. Scognamiglio, G. Donnarumma, A. Fiorentino and B. D'Abrosca, Recent advances in natural product-based anti-biofilm approaches to control infections, *Mini Rev. Med. Chem.*, 14, 1169–1182 (2014).

89 W. R. Galloway, J. T. Hodgkinson, S. D. Bowden, M. Welch and D. R. Spring, Quorum sensing in gram-negative bacteria: small-molecule modulation of AHL and AI-2 quorum sensing pathways, *Chem. Rev.*, 111, 28–67 (2011).

90 C. A. Ta and J. T. Arnason, Mini review of phytochemicals and plant taxa with activity as microbial biofilm and quorum sensing inhibitors, *Molecules.*, 21, E29 (2015).

91 V. A. Carneiro, H. S. Santos, F. V. Arruda, P. N. Bandeira, M. R. Albuquerque, M. O. Pereira, M. Henriques, B. S. Cavada, E. H. Teixeira, Casbane diterpene as a promising natural antimicrobial agent against biofilm-associated infections, *Molecules.*, 16, 190–201 (2011).

92 V. Plyuta, J. Zaitseva, E. Lobakova, N. Zagoskina, A. Kuznetsov and I. Khmel, Effect of plant phenolic compounds on biofilm formation by *Pseudomonas aeruginosa, APMIS.*, 121, 1073–81 (2013).

93 W. R. Galloway, J. T. Hodgkinson, S. Bowden, M. Welch and D. R. Spring, Applications of small molecule activators and inhibitors of quorum sensing in Gram-negative bacteria, *Trends Microbiol.*, 20, 449–458 (2012).

94 N. Rabin, Y. Zheng, C. Opoku-Temeng, Y. Du, E. Bonsu and H. O. Sintim, Agents that inhibit bacterial biofilm formation, *Future Med. Chem.*, 7, 647–71 (2015).

95 A. G. Palmer, E. Streng and H. E. Blackwell, Attenuation of virulence in pathogenic bacteria using synthetic quorum-qensing modulators under native conditions on plant hosts, *ACS Chem. Biol.*, 6, 1348–1356 (2011).

96 J. P. Pintucci, S. Corno and M. Garotta, Biofilms and infections of the upper respiratory tract, *Eur. Rev. Med. Pharmacol. Sci.*, 14, 683–690 (2010).

97 L. S. Muranaka, T. E. Giorgiano, M. A. Takita, M. R. Forim, L. F. C. Silva, H. D. Coletta-Filho, M. A. Machado and A. A. de Souza, N-Acetylcysteine in agriculture, a novel use for an old molecule: focus on controlling the plant–pathogen *Xylella fastidiosa, PLoS One.*, 8, e72937 (2013).

98 U. Mathesius, S. Mulders, M. Gao, M. Teplitski, G. Caetano-Anollés, B. G. Rolfe and W. D. Bauer, Extensive and specific responses of a eukaryote to bacterial quorum-sensing signals, *Proc. Natl. Acad. Sci.*, 100, 1444–1449 (2003).

99 R. Ortíz-Castro, M. Martínez-Trujillo and J. López-Bucio, N-acyl-L-homoserine lactones: a class of bacterial quorum-sensing signals alter post-embryonic root development in *Arabidopsis thaliana, Plant. Cell Environ.*, 31, 1497–509 (2008).

100 A. Schikora, S. T. Schenk, E. Stein, A. Molitor, A. Zuccaro and K. H. Kogel, N-acyl-homoserine lactone confers resistance toward biotrophic and hemibiotrophic pathogens via altered activation of AtMPK6, *Plant Physiol.*, 157, 1407–18 (2011).

101 S. T. Schenk, E. Stein, K. H. Kogel and A. Schikora, *Arabidopsis* growth and defense are modulated by bacterial quorum sensing molecules, *Plant Signal. Behav.*, 7, 178–81 (2012).

102 S. T. Schenk and A. Schikora, AHL-priming functions via oxylipin and salicylic acid, *Front. Plant Sci.*, 5, 784 (2014).

103 K. L. Newman, R. P. Almeida, A. H. Purcell and S. E. Lindow, Use of a green fluorescent strain for analysis of *Xylella fastidiosa* colonization of *Vitis vinifera*, *Appl. Environ. Microbiol.*, 69, 7319–7327 (2003).

104 H. Feil, W. S. Feil and A. H. Purcell, Effects of date of inoculation on the within-plant movement of *Xylella fastidiosa* and persistence of pierce's disease within field grapevines, *Phytopathology.*, 93, 244–251 (2003).

105 G. A. Turnbull, J. A. Morgan, J. M. Whipps and J. R. Saunders, The role of bacterial motility in the survival and spread of *Pseudomonas fluorescens* in soil and in the attachment and colonisation of wheat roots, *FEMS Microbiol. Ecol.*, 36, 21–31 (2001).

106 M. Espinosa-Urgel, A. Salido and J. L. Ramos, Genetic analysis of functions involved in adhesion of *Pseudomonas putida* to seeds, *J. Bacteriol.*, 182, 2363–2369 (2000).

107 V. Bianciotto, S. Andreotti, R. Balestrini, P. Bonfante and S. Perotto, Mucoid mutants of the biocontrol strain *Pseudomonas fluorescens* CHA0 show increased ability in biofilm formation on mycorrhizal and nonmycorrhizal carrot roots, *Mol. Plant. Microbe. Interact.*, 14, 255–260 (2001).

108 J. Lee, G. M. Teitzel and J. T. Greenberg, SGT1b is required for HopZ3-mediated suppression of the epiphytic growth of *Pseudomonas syringae* on *N. benthamiana*, *Plant Signal. Behav.*, 7, 1129–1131 (2012).

109 B. Prithiviraj, H. P. Baiss, T. Weir, B. Suresh, E. H. Najarro, B. V. Dayakar, H. P. Schweizer and J. M. Vivanco, Down regulation of virulence factors of *Pseudomonas aeruginosa* by salicylic acid attenuates its virulence on *Arabidopsis thaliana* and *Caenorhabditis elegans*, *Infect. Immun.*, 73, 5319–28 (2005).

110 J. T. Coombs and C. M. Franco, Isolation and identification of actinobacteria from surface-sterilized wheat roots, *Appl. Environ. Microbiol.*, 69, 5603–8 (2003).

111 J. Tans-Kersten, H. Huang and C. Allen, *Ralstonia solanacearum* needs motility for invasive virulence on tomato, *J. Bacteriol.*,183, 3597–605 (2001).

14

In Vitro Assessment of Biofilm Formation by Soil- and Plant-Associated Microorganisms

Michael W. Harding and G.C. Daniels

Alberta Agriculture and Forestry, Crop Diversification Centre South, Brooks, Canada

14.1 Introduction

Brown and Gilbert [1] described biofilms as "functional consortia of microbial cells within extracellular polymer matrices and associated with surfaces." The ability to form biofilms is an extremely common trait selected very early in the evolutionary history of microorganisms [2, 3]. Biofilms are commonly found in the natural environment, in clinical and dental infections of humans and animals, in biofouling of pipelines, structures and industrial equipment, and on plants [4–9]. The fact that biofilms are the predominant form of microbial populations in nature, industry and disease indicates that their formation of is a very important trait. In fact, biofilm populations often have increased capacities for adhesion, virulence, and tolerance to chemical treatments and environmental stresses. For these reasons, one would predict that microbiologists focus experimental protocols on culturing and studying biofilms. Contrary to this, routine culturing of microorganisms is most often done in nutrient-rich broths encouraging free-floating, solitary planktonic cells, or loosely attached colonies on semi-solid agar gels. Broths and agar gels rarely encourage the formation of biofilms and as a result, a great deal of microbiological research has been done on planktonic populations, not biofilms. One of the early observations associated with biofilm research was that biofilms are very difficult to disinfect, treat, or eradicate [10]. Furthermore, laboratory studies often overestimated the antimicrobial doses necessary to treat or eradicate biofilms, because the planktonic cells tested in the laboratory were much easier to kill. Observations such as these have initiated a new branch of microbiology involving culturing, describing and evaluating microbial biofilms. Many laboratory studies using a wide array of biofilm culturing methods have been reviewed [11–14], and while all these results have added to the body of knowledge, the diversity in culturing formats and equipment, and lack of standardized methods, means that not all results are comparable.

This chapter reviews examples of laboratory methods successfully used for cultivating microbial biofilms in vitro. For each in vitro biofilm reactor, the system's strengths and weaknesses will be discussed and examples of protocols, diagrams, or schematics provided. The chapter will conclude with a summary of the methods that have been used to describe plant-associated microbial biofilms.

14.2 How to Make a Biofilm

Technically, all that is needed to produce a biofilm is living microbial inoculum, nutrients for microbial growth, and a substrate. In the simplest terms, a biofilm is microbial growth on a surface. However, there are other characteristics that are frequently ascribed to biofilms. Perhaps the most consistently applied requirement is the formation of, and encasement within, a self-produced extracellular polymeric matrix—in other words, a slimy coating. The extracellular matrix has been shown repeatedly to be a key feature in a biofilm's ability to attach, as well as withstand challenging conditions or chemical treatments [15, 16]. Another concept that is often discussed in the context of biofilm formation, structure, and function is hydrodynamics or fluid shear. A number of studies have shown that biofilm formation and morphology are responsive to fluid shear [17–21]. There may be additional host or environmental cues that induce or affect biofilm structure and function, but in order to accommodate a broad discussion of in vitro biofilm formation, it is sufficient to say that there are three requirements for biofilm formation:

1) Living microbial inoculum
2) Nutrition sufficient for microbial growth
3) A solid or semi-solid substrate

Fluid shear will play a role in biofilm formation in many cases, but not all.

14.3 What Is the Best Way to Make a Biofilm in Vitro?

Biofilm culturing methods are generally aimed at a common goal—to produce a microbial biofilm in controlled, laboratory culture conditions. Despite the common ground, there are literally dozens of methods, or variations of methods, used to form biofilms in vitro. The diversity of methods arises from the variability in individual experimental goals. Some experiments require time-course sampling, others biological or clinical relevance. In other cases, biochemical measurements, genetic screens, or microscopic analyses will be the focus. Some experiments use a microorganism in pure culture, while others use a defined microcosm, or an unpurified sampling. Once the experimental goals and objectives are determined, a second consideration in selecting an in vitro biofilm method is the cost, ease of use, and consistency or reproducibility of each system. With these two considerations in mind, a biofilm culturing method is frequently tailored or modified specifically to fulfill the experimental purpose(s) and to align with project or lab budget(s). Despite the wide array of potential in vitro biofilm reactors, all of them can be grouped into two broad categories: open, flow systems and closed, static reactors.

14.4 Flow Systems

Flow systems are open systems that provide growth medium continuously or semi-continuously. They are also called flow displacement systems to indicate that the flow of fresh medium displaces spent media and waste products. The constant flow acts to remove planktonic cells from the system so that only sessile cells remain and multiply (i.e., those that attach to a surface). For this reason, flow systems create ideal conditions to encourage and evaluate biofilms.

The constant flow of fresh medium also means that flow systems are capable of maintaining a steady state, or at least a great degree of homeostasis, so that variables such as surface type, hydrostatic forces, temperature, and so forth can be studied without any complications due to changes in nutrition or flow. It may be of interest for some to note that reactors with rods or coupons suspended into the reaction vessel are not chemostats in the strict sense based on the definition of Herbert et al. [22]. Regardless of mechanistic details and definitions, flow systems can be categorized into two types—continuous flow through a stirred tank and plug flow.

14.4.1 Continuous Plug Flow Reactors

In plug flow systems, the medium moves in one direction through the system and mixing occurs only by radial diffusion. As such, the conditions may not be identical throughout the reactor, but incremental changes can occur along the reactor [23]. One of the first flow systems developed was the Robbins device [24] and various modifications of it [25, 26]. The device contains a linear array of openings along a rectangular channel. The channel has ports at each end for fluid entry and exit. Each opening in the array can receive a plug that holds a disk in the fluid chamber and biofilms form on the disks. A schematic of the device is given in McCoy et al. [24] and photos of the components in Coeyne and Nelis [12].

Modified Robbins devices have been used in many types of biofilm experiments, but perhaps with the greatest degree of utility in experiments that evaluate modified surfaces, or surface coatings under flow conditions, as well as in artificial throat experiments [12]. Additional adaptations have allowed examination of biofilms on biologically relevant surfaces [27,28]. The main limitation of the modified Robbins devices are the difficulties in scaling up to experimental designs that require many replications or high-throughput capacities.

14.4.1.1 Flow Cells

Flow-cell systems are composed of a biofilm chamber through which fresh liquid medium is continuously pumped (Figure 14.1). A coverslip or coated glass surface on the biofilm chamber serves as the attachment site for the biofilm, while planktonic cells are washed through the system. Flow-cell systems can be purchased commercially from a number of suppliers [12]. A basic protocol is provided by Peterson et al. [13]. Originally, these devices were used mainly in environmental biofilm studies to evaluate biodegradation potential [29], but have also been used to study bacteria causing human infections [12].

The greatest advantage of flow-cell reactors is the opportunity for microscopic evaluations of the cells attached to the cover slip, including live imaging. Flow-cell systems have been used by biofilm researchers to visually evaluate the effects of nutrition and

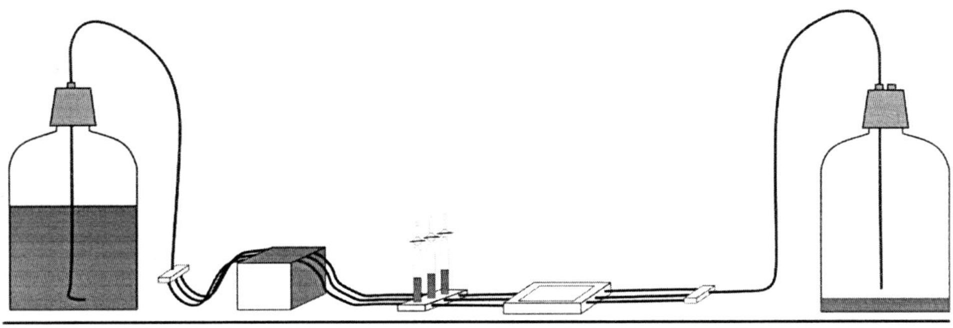

Figure 14.1 The flow-cell reactor is an example of a continuous plug flow reactor. In this system, a flow-cell biofilm reactor is continuously fed broth culture. Biofilms form within the flow cell and/or on the surface of the glass or plastic covering slip. This system is often used to visualize biofilms formed under continuous flow.

hydrodynamics on biofilm structure and function. The use of optical enhancers, dyes, and fluorescent molecules aided in the ability to visualize and describe biofilms formed in flow cells. Furthermore, microorganisms expressing fluorescent reporter genes have been used in flow cells to study spacio-temporal gene expression in biofilms [30].

However, Peterson et al. [13] point out three limitations of these devices. First, they are not readily comparable with in vivo pathosystems. As a result, it is difficult to draw conclusions from these in vitro biofilms with any certainty of clinical relevance. Second, while these systems make it easy to visualize the biofilm, they do not allow for the ability to sample significant biomass for additional analyses. Therefore, they are not well-suited for studies of gene expression, antimicrobial challenge, or microbial metabolism where biomass sampling or analyses are required. Finally, the apparatus is cumbersome when many experimental replications, treatments, or microbial strains are evaluated, or when high-throughput capacities are required.

A specialized subset of flow cells are used to study biofilm formation in porous media. In general terms, these reactors are porous columns composed of media such as sand, rocks, glass beads, and so forth. Examples include capillary flow cells, artificial fracture cells, and column and two flow-cell systems. These reactors range in scale from laboratory-size reactors of millimeters to centimeters long, up to meters-long pilot-scale systems. One of the major problems with these reactors is the opaque nature of the media used makes visualization and imaging of the biofilms near impossible. A very thorough review of porous substrate biofilm reactors, with schematics, detailed descriptions, and strengths/weaknesses is provided by Gerlach and Cunningham [31].

14.4.1.2 Tube Biofilms

Tube biofilm reactors function using the same basic system platform as flow cells and are additional examples of continuous plug flow reactors. However, in tube reactors, the biofilm reactor chamber is replaced with tubing (not shown). The system has a continuous flow, but biofilms form on the interior surfaces of tubing, rather than on a glass coverslip. A basic protocol is given in Peterson et al. [13]. This system primarily utilizes silicon tubing commonly used in medical and clinical settings. Tube reactors are very useful for studies of antimicrobial coatings used to prevent clinical infections associated with medical indwelling devices such as catheters and stents. Tube flow systems have also been used to characterize gene expression and quorum sensing in *Pseudomonas aeruginosa* biofilms [32, 33]. As with other continuous flow systems, tube reactors are not useful when many experimental replications or high-throughput capacities are required. The other limitation is that destructive sampling of the biofilm chamber is necessary when one desires to directly assay or analyze the biofilm because the biomass accumulates on the interior surfaces of the tube. The experiment must be terminated and tubes cut open in order to access the biofilm.

14.4.1.3 Drip-Flow Reactor

Drip-flow biofilm reactors are also continuous plug flow systems, with the main difference being that they create a much lower fluid shear. These reactors pump growth medium onto a biofilm chamber in a drop-wise fashion. The reactor is placed at an angle so that gravity pulls the medium across the chamber and collects it below in a waste container (see Figure 14.2).

A protocol for drip-flow bioreactors is given by Schwartz et al. [34]. This type of chamber has reduced hydrodynamic forces and dispersive mixing so that a slight gradient of conditions and concentrations may exist across the reactor. However, like other flow systems, the drip-flow reactor creates an excellent environment for culturing and evaluating biofilms and is consistent and reproducible to the degree that it is a standard method for quantification of *Pseudomonas aeruginosa* biofilms (ASTM Standard Method E2647-08). Drip-flow reactors are useful when biofilms in low fluid shear environments are to be modeled, and also when a significant liquid-air interface is desired [12, 35]. Drip-flow systems are also well suited when high biomass production is

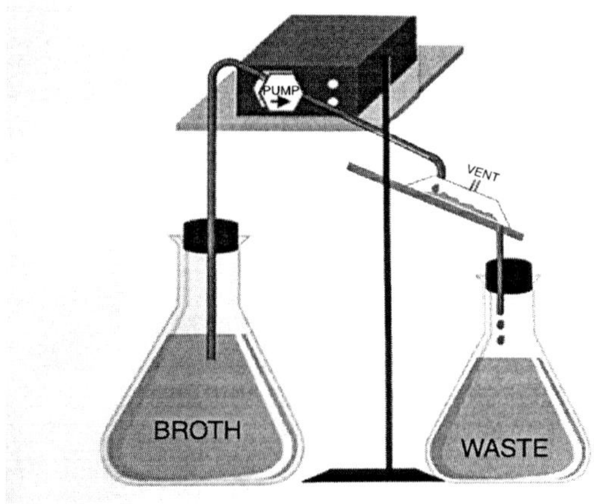

Figure 14.2 The drip-flow reactor is an example of a continuous plug flow reactor. In this system, a biofilm reactor is continuously fed broth culture drop wise onto the upper end of an inclined reaction chamber. Gravitational forces pull the culture down the reaction chamber surface and a biofilm forms along the chamber surface. This reactor can be used to produce biofilms in a low fluid shear environment.

required, or when cryosectioning biofilms, or when evaluating medical materials and indwelling devices [34].

An interesting miniaturized system similar to a drip-fed reactor was described by Merritt et al. [36]. It uses a six-well microtiter plate with each well converted to a low-flow reactor. Fresh media is pumped into each well via a needle inserted through the lid, and waste is removed via a second needle inserted into the well. Although media is exchanged, it is a very low-flow system, more comparable to a drip-flow device, and intermediate between continuous flow and static reactors. The six-well format is highly adaptable and capable of producing enough biomass for DNA microarray and proteomics studies. The disadvantages include the need for frequent monitoring due to the chances of overflowing or drying out of wells if systems are unattended for extended periods, or if a blockage occurs.

14.4.1.4 Perfused Biofilm Fermenters

Perfused biofilm fermenters also contain a chamber through which an input and output of nutrients is continuously pumped. However, the input is forced through a permeable membrane composed of either cellulose acetate, nylon or Sorbarod filters [11]. Four variations of the membrane-based systems utilizing nylon or cellulose membranes have been described, namely the baby machine [37], the perfused biofilm fermenter [38], the swinnex biofilm fermenter [39], and the Sartorius filter apparatus [40]. All four are reviewed, described, and diagrammed in McBain [11]. These reactors have been used to control the growth rates within biofilms, produce synchronous cultures, and study biofilm recalcitrance [41] and the bacterial cell cycle [40].

The Sorbarod biofilm fermenters are basically the same system with the exception that Sorbarod filters are used rather than cellulose acetate and nylon. The Sorbarod filter device is simple and inexpensive to construct, and can be used in nondestructive sampling studies of population dynamics and antibiotic efficacy testing [11]. The multiple Sorbarod filter device was developed to allow destructive sampling for real-time determination of biocide efficacy [42] and has also been used to grow oral biofilms [11]. One of the potential disadvantages of the Sorbarod devices is that they tend to produce heterogeneous biofilms. However, this may be an advantage if a heterogenous biofilm is more clinically or environmentally relevant. The Sorbarod devices are described and shown schematically in McBain [11].

14.4.2 Continuous Flow Stirred Tank Reactors

Some continuous flow reactors also incorporate stirring or mixing of the contents of the biofilm reactor vessel in addition to continuous flow of medium through it. The biofilm reactor system is fundamentally the same as those previously described for plug flow reactors, but in contrast to the continuous plug flow systems these reactors add constant mixing to the reaction vessel via a magnetic stir plate and bar. The effect is thorough mixing and equal distribution of the contents throughout the entire reaction vessel. Additionally, the stirring creates a constant and uniform fluid shear across the entire reactor vessel.

14.4.2.1 CDC Biofilm Reactor

The Centers for Disease Control (CDC) biofilm reactor is an example of a continuous flow, stirred tank reactor. This reactor is composed of a cylindrical reaction vessel with

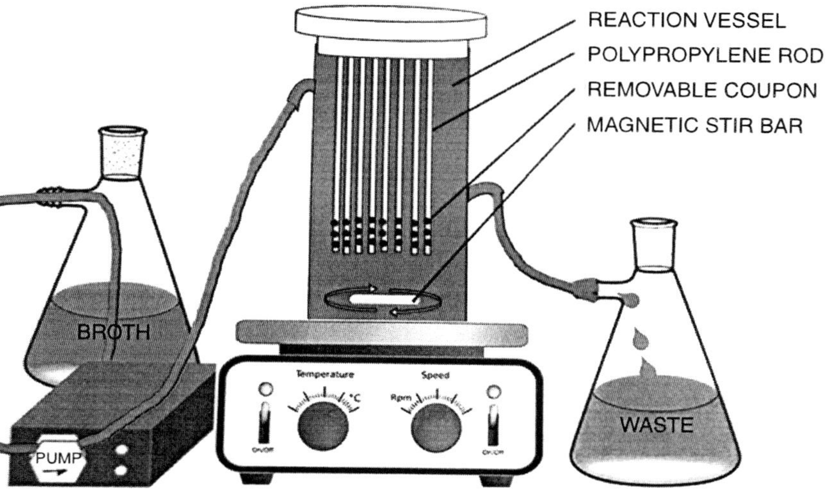

Figure 14.3 CDC biofilm reactor is an example of a continuous plug flow reactor with constant stirring. In this system, a biofilm reactor is continuously fed broth culture. Polypropylene rods each containing three removable coupons are suspended in the reaction chamber. The contents of the reaction vessel are constantly stirred using a magnetic stir bar. Biofilms form on the surface of the removable coupons.

input and output ports and a plastic top that suspends eight polypropylene rods into the vessel. The rods each hold three removable coupons on which biofilms form. A diagrammatic representation of the CDC device is shown in Figure 14.3. The vessel has a continuous plug flow of media through it, but also a magnetic stir bar at the bottom. Stirring of the reaction vessel provides a constant, steady state within the reactor for both fluid shear around each coupon, and for diffusion of nutrients and waste. As such, identical conditions will exist around each of the coupons and the result is 24 identical biofilms that can be evaluated or tested. This makes the CDC biofilm reactor a very reliable system for standardized testing [43]; in fact, it is an ASTM standard method (E5262-07). It has been used extensively to evaluate temporal stages of biofilm formation [44, 45], disinfectants, biocides, sanitizers, antimicrobial coatings [46–48] and the biofilm matrix [49].

14.4.2.2 Rotating Disk, Concentric Cylinder, and Annular Reactors

Rotating disk, concentric cylinder, and annular biofilm reactors are all examples of continuous flow, stirred tank systems. The rotating disk reactor utilizes a large rotating Teflon disk that fits inside a cylindrical reaction vessel. The reaction vessel is often a 1000 mL glass container with a side arm. The disk has openings (usually 18) that accept removable coupons such that they are surface-flush. The base of the disk has a bar magnet secured to it that allows the disk to rotate in controlled fashion inside the reactor using a magnetic stir plate. A diagrammatic representation of a rotating disk reactor is shown in Figure 14.4 and basic protocol can be found in Peterson et al. [13].

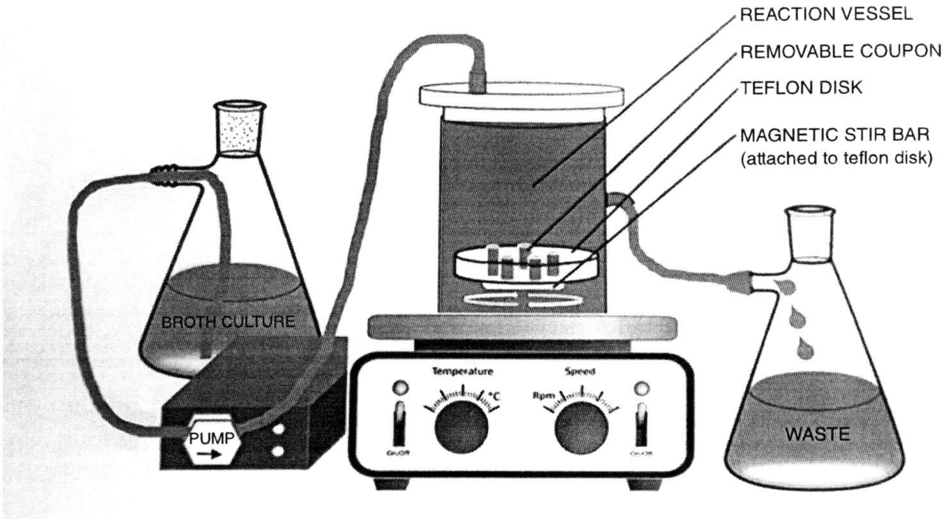

Figure 14.4 The rotating disk reactor is an example of a continuous plug flow reactor with constant stirring. In this system, a biofilm reactor is continuously fed broth culture. Inside the reactor, coupons are flush-mounted within a Teflon disk and spun via a magnetic stir bar attached to the bottom of the disk. Biofilms form on the surfaces of the flush-mounted coupon.

The rotation of the disk creates fluid shear across the coupons and encourages formation of biofilms. The principle behind the rotating disk reactor is the same as the CDC biofilm reactor except that instead of using magnetic stirring to move the growth medium around the coupons, it is used to move the coupons through the medium. In both cases, fluid shear is created on the coupons to create consistent, reproducible conditions ideal for biofilm formation. The rotating disk reactors have been used to evaluate the efficacies of biocides and anti-fouling compounds, and in biofilm removal studies [13, 34]. Advantages of this system include the ability to establish a steady-state environment for the evaluation of biofilm structure and function. It creates a consistent and reproducible environment for biofilm formation and evaluation in vitro. In 2005 it was suggested that this method be adopted as a standard for biofilm antimicrobial susceptibility studies due to its reliable, reproducible results [50]. Subsequently, it was adopted as an ASTM standard method (E2196-07) for evaluation of *P. aeruginosa* biofilms. Additionally, because the biofilm surface coupons can be removed and replaced, a wide array of surface materials can be evaluated, as well as surfaces with coatings or treatments.

The concentric cylinder reactor was first described by Willcock et al. [51], and a protocol, a schematic, and photographs of the apparatus can be found within their report. The device is quite similar in principle to the rotating disk reactor in that a plug flow is established from a nutrient nurse vessel through the reactor and collecting in a waste container. However, the cylinder reactor has no removable coupons, the biofilms form directly on the stainless steel cylinders. Additionally, the radial arrangement and movement of the rotating cylinders within the collection chambers makes it possible to very accurately calculate and control the liquid shear force applied within the reactor. As a result, detailed evaluations of the effects of fluid dynamic force on biofilm biology are possible [51, 52]. The limitations of rotating disk and concentric ring device are similar to those describe for the CDC biofilm reactor, mainly being the lack of high-throughput analysis or evaluations of large numbers of strains or mutants simultaneously. Additionally, the concentric ring reactor cannot employ coupons of various types, but can only evaluate biofilms on stainless steel surfaces.

The final device in this category is the annular biofilm reactor. This device is also a plug flow system with a spinning cylinder. It comprises an inner cylinder that spins mechanically with slides or coupons mounted within (not shown). The spinning cylinder creates the appropriate conditions for biofilm formation is it produces fluid shear around the slides. This apparatus has been used predominantly in studies of water treatment [53] or quality in distribution systems [54].

14.5 Static Reactors

Unlike flow systems, static reactors are closed devices with finite nutrients that do not provide for replenishment or continuous flow through the device. As a result, the chambers create an environment that evolves from abundant resource availability initially, through to resource exhaustion and accumulation of wastes, unless the experiment is paused to recharge the reactor cell(s). Therefore, the static reactor will create an environment where microorganisms undergo dynamic growth similar growth curves demonstrated by planktonic cells in finite broth cultures. However, a static reactor's lack in steady-state homeostasis, or chemostat status, is compensated by its simplicity, capacity, and cost.

14.5.1 Microtiter Plate Assay

Microtiter plate-based biofilm reactors are one of the earliest, and the most commonly chosen classes of biofilm reactors around the world [12, 13, 36, 55, 56]. They are quite simple in design, basically utilizing a standard 96-well microtiter plate as a biofilm reactor. Each well is given growth medium and inoculated. The biofilms form on the bottoms and walls of the plate wells. If desired, fluid shear can be created using an orbital shaker/incubater. A basic protocol for microtiter plate biofilm assay is given in Peterson et al. [13] and Merritt et al. [36], and a photograph is shown in Figure 14.5. There are many published reports on biofilm biology, antimicrobial efficacy (biocides, anti-quorum sensing chemicals, antibiotics, coatings), mutant screening, genetic mechanisms of biofilm form and function, and the effects of environmental parameters and nutrition on biofilms, using the microtiter plate assay [reviewed in reference 11–13]. There are three main reasons why the microtiter plate assay is so widely used:

1) The reactor is small, inexpensive, and easy to obtain. Additionally, it requires small quantities of medium and experimental chemicals, and it is disposable.
2) Reaction setup is fast and easy, as it does not require multiple feed, waste, and reactor vessels, pumps or tubes.
3) One reactor can produce 96 biofilms, so it can accommodate high-throughput screens or experimental designs with large numbers of treatments or replications.

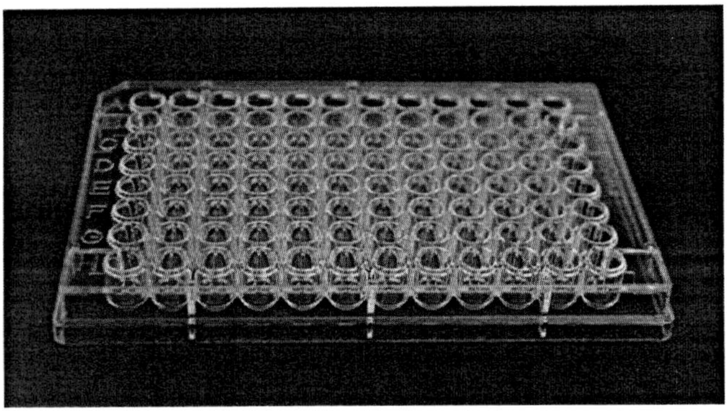

Figure 14.5 A 96-well microtiter plate with flat-bottom wells can be used as a closed, static biofilm reactor.

The standard 96-well microtiter plate is easily accessible around the world from most of the major laboratory supply retailers, and is relatively inexpensive. Therefore, most microbiology labs in the world can obtain it and budget for it. The experimental setup is minimal, compared to those of continuous flow reactors, as the plates generally arrive in a sterile ready-to-use condition. There are no additional containers, tubes or pumps required so they can be immediately used as biofilm reactors. Additionally, the reactor is disposable when the experiment is complete such that no washing of glassware or sterilizing of vessels or tubes is required. Despite the fact that cheap and easy rarely coincides with high quality, the microtiter plate assay has proven to be a consistent reactor capable of routinely delivering quality results.

A major step forward in standardizing biofilm biomass quantification was achieved with this system when biofilms were stained with crystal violet, which binds electrostatically to charged components of microbial surfaces and some extracellular matrix components. The biomass of the biofilms can be directly represented by the amount of dye taken up. The amount of dye in a stained biofilm can be quantified spectrophotometrically via absorbance at 595 nm [55]. While staining and absorbance provided a quantification method, biofilms in a microtiter plate are not easily removed or visualized from within the wells. Additionally, crystal violet staining cannot discriminate between live and dead cells, and obscures the ability to see cells within the matrix, since both may stain equally. As a result of these limitations, other staining techniques based on cell metabolism have been employed to distinguish live versus dead cells within the biofilm [57].

More recently, additional metabolic stains as well as genetic methods (PCR and FISH) for population estimates have been employed [14]. It is important to note that, like other biofilm reactors, results from microtiter plate assays are not always clinically relevant. One example of this fact showed a mutant of *Staphlococcus aureus* that was unable to form biofilms in a microtiter plate reactor was still infectious in a mouse model and showed no loss of virulence [58].

A variation of the microtiter plate assay is described by Caiazza and O'Toole [59], which allows for evaluation of early-stage biofilms (about 4 to 48 hr, or attachment and early microcolony stage) formed at the air-liquid interface. In this case, a 12-well format microtiter plate is placed at a 30° to 50° angle relative to horizontal, and inoculated growth media is slowly applied to the wells (not shown). The bacteria grow at the center of the well bottoms, after which they can be viewed by phase-contrast microscopy. Alternatively, the plate can be inoculated without being placed at an angle and a plastic or glass coverslip placed on the surface of the inoculated medium in each well. In this case, the biofilm on the cover slip can be rinsed, stained and viewed under any microscope. Protocols for these methods are provided in Merritt et al. [36]. These methods are very fast, easy, and inexpensive. Disadvantages include the inability to evaluate later stage biofilm characteristics and the small amount of biomass produced is not sufficient for metabolic, genomic, or proteomic evaluations.

14.5.2 MBEC™ Assay

The MBEC™ Assay began as the "Calgary Biofilm Device" [60]. It is a microtiter plate format with a unique and innovative twist. The reactor consists of a 96-well plate bottom, but the lid contains 96 pegs that extend downward into each corresponding well. The reactor is primed and inoculated in the same manner as a standard microtiter plate assay, but instead of forming biofilms on the walls and bottoms of the well, the biofilms form on the pegs. This is a significant advantage, because the 96 biofilms can be serially treated by simply removing and replacing the lid to subsequent plate bottoms. For example, antibiotic susceptibility testing, phenotypic screening, or biocide efficacy testing experiments can be serially rinsed, treated, neutralized, and recovered by simply removing and replacing the peg lid onto a series of pre-prepared plate bottoms [12, 61–64]. The experimental protocol, and a flow diagram, for the MBEC™ Assay are presented in Peterson et al. [13] (see Figure 14.6).

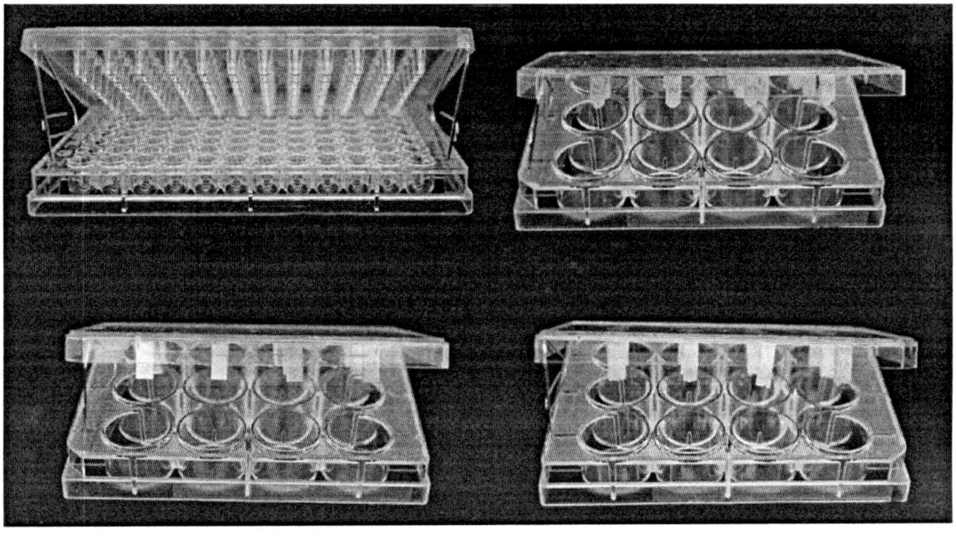

Figure 14.6 Examples of static biofilm reactors with peg lids. The MBEC™ Assay with polystyrene pegs (top left) and the BEST™ Assay with concrete pegs (top right), stainless steel pegs (bottom left) and wood pegs (bottom right).

The ability to easily remove and re-expose the biofilms means that a wide variety of experimental design options are opened to the user. It adds a second dimension to the high-throughput capacity of the microtiter plate system. Pegs can be removed as part of time-course experiments or for sampling of biofilms. Another major advantage of this system is the ability to evaluate both planktonic and sessile cells in the same reactor. The sessile cells end up on the pegs, while planktonics accumulate in the wells, therefore susceptibility testing on both types of cells, grown within an identical reaction chamber, can be done simultaneously [65–67]. The system is highly reproducible and rugged [68] and was the first approved standard method for disinfectant efficacy testing of biofilms in a static reactor (ASTM Standard Method 2799). Finally, unlike biofilms formed within plate wells, biofilms on pegs can be evaluated microscopically because the pegs are easily removed from the lid [67]. This feature allows experiments to include microscopic evaluations of general effects on biofilm morphology or structure in conjunction with the treatment effects or experimental variables tested using light, fluorescence, or electron microscopy. The limitations of the assay include the need to optimize growth and experimental conditions for each strain, and antimicrobial chemical, respectively, when multiple strains, species, or treatments are employed in a single reactor. Also, a limited amount of biomass accumulates on the pegs, preventing rigorous biochemical or physiological analyses. Producing sufficient biomass for biochemical and genetic studies requires large batches such as those described by Crouzet et al. [69], where 1 g of glass wool was used as a solid substrate in 100 to 500 mL of growth medium. Here we see that the limited reactor volumes of the microtiter plate-based assays are perhaps the most significant drawback.

14.5.3 Colony Biofilm Assay

Some authors have reported that colonies formed on the surfaces of solid agar growth media can encourage characteristics associated with biofilms [11]. While this may be true in some instances, there are clearly others where agar plate methods do not encourage phenotypes associated with biofilms [70]. For the purposes of this chapter, it is worth considering the agar plate as a simple, useful method for culturing and evaluating colonies—with, however, the strong caution that some reports have indicated that results may or may not reflect those of clinically or environmentally relevant biofilms. We will consider one method of this type, the colony biofilm assay. Additional agar plate methods are reviewed by McBain [11].

Colony biofilms are formed in the absence of fluid and liquid shear. The protocol involves inoculating a semi-permeable polycarbonate filter that is placed on a solid growth medium. This method allows one to study biofilm structure and function when a major surface-air interface exists, and without fluid shear. A description of the method is given in Peterson et al. [13] and Merritt et al. [36], and a diagrammatic representation of a colony biofilm in Figure 14.7. The carbon source or chemical treatment can be changed quickly and easily so this method is frequently used for testing survival after chemical or drug treatments or antimicrobial susceptibility. It is also useful when one wants to separate the role of tenacious attachment from antimicrobial efficacy. It is the only reactor that does not have attachment as a prerequisite for successful biofilm formation. Further to this, changes in recovery will be attributed exclusively to cell death, whereas other reactors will potentially have changes in recovery due to detachment.

Figure 14.7 Colony biofilm reactor utilizes a standard Petri dish with semi-solid agar medium as a platform. A polycarbonate filter is placed on the agar surface and inoculated. A second filter is placed on the bacterial biofilm and capped with a paper disc impregnated with a test material. This type of reactor can be used to produce biofilms in the absence of fluid shear, and can evaluate antimicrobial susceptibility without the requirement of tenacious surface attachment.

Finally, the colony biofilms are exclusively clonal/isogenic because there is no recruitment or deposition of other cells. Drawbacks include difficulty in manipulating membranes when colony biomass becomes large and that some species adept in surface motility do remain confined to the membranes [13].

14.6 Special Considerations for Filamentous Fungal Biofilms

There are unique challenges associated with culturing fungal biofilms in vitro. The first challenge is that many fungi do not adhere to, or grow well attached to, the hydrophobic surfaces commonly used in biofilm reactors. So while glass and plastics are suitable for bacterial biofilms, these surfaces do not promote attachment and biofilm formation by many fungi. For example, Pesciaroli et al. [71] grew *Pleurotus ostreatus* on a number of hydrophobic, or mildly hydrophobic, surfaces (Teflon, polyurethane, polycarbonate, and nylon) as well as hydrophilic surfaces (glass and hydroxyapatite). The fungal biofilm was best on the hydrophilic followed by mildly hydrophobic and worst on hydrophobic surfaces. This was further confirmed in a subsequent study of *P. ostreatus* in the MBEC™ Assay static reactor where a coating of hydroxyapatite on the MBEC™ reactor pegs was necessary to enable biofilm formation [72]. Another example of surface modification affecting fungal biofilm formation was reported by Wesenberg et al. [73] where *Candida albicans* attachment to polystyrene was significantly reduced by coating with Pluronic F127. In addition to coating plastic surfaces with hydrophyllic coatings such as hydroxyapatite, others have used wood pegs in static multiwell plate reactors to encourage attachment and biofilm formation of fungi [74–78].

The second challenge faced by researchers growing fungal biofilms is that fungal colonies are composed of complex networks of hyphal filaments that are very difficult to quantify. Unlike bacterial colonies that can be separated into individual cells and quantified by colony-forming units in standard plate counts, fungal filaments or mycelia cannot be separated into individual cells or quantified by plant counts. This challenge was partially overcome by using the protocol developed for bacterial biofilm quantification (i.e., staining with crystal violet and quantification by absorbance spectrophotometry) [55], but further refined for fungal biofilms by Pierce et al. [79] employing metabolic reduction of tetrazolium salt and colorimetric reading of absorbance at

490 nm. The Pierce method was suggested as a standard method for antifungal biofilm susceptibility testing.

For industrial fermentation using fungi such as *Aspergillus, Trichoderma,* and *Penicillium,* bioreactor design and methodology has been impacted by the concept of microbial biofilm biology. Solid-state, submerged, and biofilm reactors are beginning to move from basic laboratory benchtop studies to adaptive pilot and commercial scale evaluations and show potential to revolutionize traditional fermentation methods with respect to cultivation and productivity of systems producing desirable fungal metabolites [80].

14.7 Biofilm Reactors Used to Characterize Plant-Associated Biofilms

When one considers the diversity of biofilm reactors that have been engineered and utilized, it is interesting that relatively few have been used to characterize plant-associated biofilms. The first plant-pathogenic biofilm characterized in vitro was that of *Xylella fastidiosa*, the causal agent of vascular diseases in a number of plant hosts [81]. In this example, the biofilms were produced using inoculated broth cultures incubated on an orbital shaker in glass tubes, or in glass tubes containing pine wood sticks. The biofilms attached to glass and wood surfaces were characterized. Since this first report of plant-pathogenic bacteria-forming biofilms in vitro, subsequent studies have used flow cells [82,83] or coverslips floating on media in a 12-well plate format [83]. Additionally, plant-pathogenic biofilms formed by fungi and bacteria have been cultured on coupons made of wood, metal, plastic, and concrete in a variation of the MBEC™ Assay called the BEST™ Assay [75–78]. The results of these all these experiments with plant-pathogenic bacteria and fungi have confirmed that plant-associated microorganisms do form biofilms, that biofilm populations are much more tolerant to treatments with biocides and disinfectants than their planktonic counterparts, that quorum sensing mechanisms exist in these species, and that biofilm formation is often a key component of attachment, invasion, virulence, or survival of harsh conditions.

While much progress has been made, the body of knowledge for plant-associated microbial biofilms lags far behind those of medical, environmental, and industrial biofilms, especially for filamentous fungal biofilms in association with plants. A significant amount of work is yet to be done characterizing the roles of filamentous fungal biofilms on mechanisms of attachment, infection, virulence, production of host toxins or effectors, survival structures, dispersal, and resource utilization.

14.8 Value-Added Products from Biofilm Reactors

Many industrial processes for manufacturing desirable metabolites utilize bacterial or fungal species in large-scale fermentation systems. Industrial fermentation has been used for hundreds of years to produce value-added products. More recently, advances in fermentation such as solid-state or semi-solid, submerged and biofilm reactors have been developed and evaluated for their abilities to improve productivity [80]. Biofilm reactors had been primarily used in methods of water purification, wastewater treatment [84, 85], environmental bioremediation [86], and detoxification [87, 88]. However, more recently, biofilm formation and manipulation in specialized reactors has opened up possibilities of continuous fermentation in large-scale production of value-added products. For example, renewable bioenergy products, organic acids, antibiotics, enzymes, and polysaccharides can all be produced via biofilm fermentation [89]. Biofilm reactors have improved fermentation productivities of such products because they can increase the biomass of productive microbial mass within the batch. Five to 10 times higher biomass means increased production and reduces washing out, dilution, elimination, and need for reinoculation. As a result, improved operation stability is frequently associated with biofilm reactors. Additional scale-up work is still needed for certain parameters critical to each production process [80], but it is clear that biofilm reactors in laboratory and pilot-scale investigation have created a new paradigm for the production of value-added microbial-derived, value-added products.

References

1 M.R. Brown and P. Gilbert (eds.), *Microbiological quality assurance: a guide towards relevance and reproducibility of inocula*. CRC Press, New York, 1995.

2 E.G. Nisbet and N.H. Sleep, The habitat and nature of early life. *Nature*, 409 (6823), 1083–1091 (2001).

3 D. Davies, Understanding biofilm resistance to antibacterial agents. *Nature reviews Drug discovery*, 2(2), 114–122 (2003).

4 P. Jorge, A. Lourenço and M.O. Pereira, New trends in peptide-based anti-biofilm strategies: a review of recent achievements and bioinformatic approaches. *Biofouling*, 28(10), 1033–1061 (2012).

5 L. Hall-Stoodley, J.W. Costerton and P. Stoodley, Bacterial biofilms: from the natural environment to infectious diseases. *Nature Reviews Microbiology*, 2(2), 95–108 (2004).

6 V.M. Siqueira, H.M.B. Oliveira, C. Santos, R.R.M. Paterson, N.B. Gusmão and N. Lima, Biofilms from a Brazilian water distribution system include filamentous fungi. *Canadian Journal of Microbiology*, 59(3), 183–188 (2012).

7 K.M. Salli and A.C. Ouwehand, The use of in vitro model systems to study dental biofilms associated with caries: a short review. *Journal of Oral Microbiology*, 7 (2015).

8 T. Danhorn and C. Fuqua, Biofilm formation by plant-associated bacteria. *Annual Review of Microbiology*, 61, 401–422 (2007).

9 J. Harrison, R. Turner, L. Marques and H. Ceri, A new understanding of these microbial communities is driving a revolution that may transform the science of microbiology. *American Scientist* 93(6), 508–515 (2005).

10 H. Anwar, M.K. Dasgupta and J.W. Costerton, Testing the susceptibility of bacteria in biofilms to antibacterial agents. *Antimicrobial Agents and Chemotherapy*, 34(11), 2043 (1990).

11 A.J. McBain, In vitro biofilm models: an overview. *Advances in Applied Microbiology*, 69, 99–132 (2009).

12 T. Coenye and H.J. Nelis, In vitro and in vivo model systems to study microbial biofilm formation. *Journal of Microbiological Methods*, 83(2), 89–105 (2010).

13 S.B. Peterson, Y. Irie, B.R. Borlee, K. Murakami, J.J. Harrison, K.M. Colvin and M.R. Parsek, Different methods for culturing biofilms in vitro. In: *Biofilm Infections*, Springer, New York, 2011.

14 F. Pantanella, P. Valenti, T. Natalizi, D. Passeri and F. Berlutti, Analytical techniques to study microbial biofilm on abiotic surfaces: pros and cons of the main techniques currently in use. *Ann Ig*, 25(1), 31–42 (2013).

15 J.L. Lopez-Ribot, Large-scale biochemical profiling of the *Candida albicans* biofilm matrix: new compositional, structural, and functional insights. *MBio*, 5(5), e01781–14 (2014).

16 L. Hobley, C. Harkins, C.E. MacPhee and N.R. Stanley-Wall, Giving structure to the biofilm matrix: an overview of individual strategies and emerging common themes. *FEMS Microbiology Reviews*, 39(5), 649–669 (2015).

17 B.F. Picologlou, W.G. Characklis and N. Zelver, Biofilm growth and hydraulic performance. *Journal of the Hydraulics Division*, 106(5), 733–746 (1980).

18 J. Bryers and W. Characklis, Early fouling biofilm formation in a turbulent flow system: overall kinetics. *Water Research*, 15(4), 483–491 (1981).

19 P. Stoodley, J.D. Boyle and H.M. Lappin-Scott, Biofilm structure and behaviour: influence of hydrodynamics and nutrients. In: *Dental Plaque Revisited: Oral Biofilms in Health and Disease*. H.H. Newman and M. Wilson (eds.), Bioline, Cardiff, GB, 1999.

20 B.C. Dunsmore, A. Jacobsen, L. Hall-Stoodley, C.J. Bass, H.M. Lappin-Scott and P. Stoodley, The influence of fluid shear on the structure and material properties of sulphate-reducing bacterial biofilms. *Journal of Industrial Microbiology and Biotechnology*, 29(6), 347–353 (2002).

21 R. Fink, M. Oder, D. Rangus, P. Raspor and K. Bohinc, Microbial adhesion capacity. Influence of shear and temperature stress. *International Journal of Environmental Health Research*, 25(6), 656–669 (2015).

22 D. Herbert, R. Elsworth and R.C. Telling, The continuous culture of bacteria; a theoretical and experimental study. *Microbiology*, 14(3), 601–622 (1956).

23 J. Heersink and D. Goeres, Reactor design considerations, *in The Biofilm Laboratory: Step-by-step Protocols for Experimental Design, Analysis, and Data Interpretation*, M. Hamilton, J. Heersink, K. Buckingham-Meyer and D. Goeres (eds.), Cytergy Publishing, Bozeman, 2003.

24 W.F. McCoy, J.D. Bryers, J. Robbins and J.W. Costerton, Observations of fouling biofilm formation. *Canadian Journal of Microbiology*, 27(9), 910–917 (1981).

25 J.C. Nickel, J.B. Wright, I. Ruseska, T.J. Marrie, C. Whitfield and J.W. Costerton, Antibiotic resistance of *Pseudomonas aeruginosa* colonizing a urinary catheter in vitro. *European Journal of Clinical Microbiology*, 4(2), 213–218 (1985).

26 G. Ramage, B.L. Wickes and J.L. López-Ribot, A seed and feed model for the formation of *Candida albicans* biofilms under flow conditions using an improved modified Robbins device. *Revista Iberoamericana de Micologia*, 25(1), 37 (2008).

27 R.M. Landry, D. An, J.T. Hupp, P.K. Singh and M.R. Parsek, Mucin–*Pseudomonas aeruginosa* interactions promote biofilm formation and antibiotic resistance. *Molecular Microbiology*, 59(1), 142–151 (2006).

28 S. Moreau-Marquis, J.M. Bomberger, G.G. Anderson, A. Swiatecka-Urban, S. Ye, G.A. O'Toole and B.A. Stanton, The ΔF508-CFTR mutation results in increased biofilm formation by *Pseudomonas aeruginosa* by increasing iron availability. *American Journal of Physiology-Lung Cellular and Molecular Physiology*, 295(1), L25–L37 (2008).

29 S. Moller, D.R. Korber, G.M. Wolfaardt, S. Molin and D.E. Caldwell, Impact of nutrient composition on a degradative biofilm community. *Applied and Environmental Microbiology*, 63(6), 2432–2438 (1997).

30 P.S. Stewart and M.J. Franklin, Physiological heterogeneity in biofilms. *Nature Reviews Microbiology*, 6(3), 199–210 (2008).
31 R. Gerlach and A.B. Cunningham, *Influence of biofilms on porous media hydrodynamics*, CFC Press: New York, 2010.
32 M. Whiteley, M.G. Bangera, R.E. Bumgarner, M.R. Parsek, G.M. Teitzel, S. Lory and E.P. Greenberg, Gene expression in *Pseudomonas aeruginosa* biofilms. *Nature*, 413(6858), 860–864 (2001).
33 M.J. Kirisits, J.J. Margolis, B.L. Purevdorj-Gage, B. Vaughan, D.L. Chopp, P. Stoodley and M.R. Parsek, Influence of the hydrodynamic environment on quorum sensing in *Pseudomonas aeruginosa* biofilms.*Journal of Bacteriology*, 189(22), 8357–8360 (2007).
34 K. Schwartz, R. Stephenson, M. Hernandez, N. Jambang and B.R. Boles, The use of drip flow and rotating disk reactors for *Staphylococcus aureus* biofilm analysis. *Journal of Visualized Experiments*, 46, e2470 (2010).
35 D.M. Goeres, M.A. Hamilton, N.A. Beck, K. Buckingham-Meyer, J.D. Hilyard, L.R. Loetterle, L.A. Lorenz, D.K. Walker and P.S. Stewart, A method for growing a biofilm under low shear at the air-liquid interface using the drip flow biofilm reactor. *Nature Protocols*, 4(5), 783–788 (2009).
36 J.H. Merritt, D.E. Kadouri and G.A. O'Toole, Growing and analyzing static biofilms. *Current Protocols in Microbiology*, 1B-1 (2005).
37 C.E. Helmstetter and D.J. Cummings, Bacterial synchronization by selection of cells at division. *Proceedings of the National Academy of Sciences*, 50(4), 767–774 (1963).
38 P. Gilbert, D.G. Allison, D.J. Evans, P.S. Handley and M.R. Brown, Growth rate control of adherent bacterial populations. *Applied and Environmental Microbiology*, 55(5), 1308–1311 (1989).
39 S. Gander and P. Gilbert, The development of a small-scale biofilm model suitable for studying the effects of antibiotics on biofilms of gram-negative bacteria. *Journal of Antimicrobial Chemotherapy*, 40(3), 329–334 (1997).
40 A. Omar and P. Gilbert, Towards a molecular analyses of whole genome expression in *E. coli, in Biofilms: Persistence and Ubiquity*, A.J. McBain, D.G. Allison, J. Pratten, D. Spratt, M. Upton, and J. Verran (Eds), Vol. 7, Biofilm Club, Manchester, 2005.
41 P. Gilbert, D.G. Allison, D.J. Evans, P.S. Handley and M.R. Brown, Growth rate control of adherent bacterial populations. *Applied and Environmental Microbiology*, 55(5), 1308–1311 (1989).
42 J. Coulburn, *Characterisation and in Vitro Modelling of Biofilm Formation and Fouling in Paper Manufacturing*. University of Manchester, 2002.
43 D.M. Goeres, L.R. Loetterle, M.A. Hamilton, R. Murga, D.W. Kirby and R.M. Donlan, Statistical assessment of a laboratory method for growing biofilms. *Microbiology*, 151(3), 757–762 (2005).
44 K. Honraet, E. Goetghebeur and H.J. Nelis, Comparison of three assays for the quantification of *Candida biomass* in suspension and CDC reactor-grown biofilms. *Journal of Microbiological Methods*, 63(3), 287–295 (2005).
45 H. Nailis, R. Vandenbroucke, K. Tilleman, D. Deforce, H. Nelis and T. Coenye, Monitoring ALS1 and ALS3 gene expression during in vitro *Candida albicans* biofilm formation under continuous flow conditions. *Mycopathologia*, 167(1), 9–17 (2009).
46 K. Buckingham-Meyer, D.M. Goeres and M.A. Hamilton, Comparative evaluation of biofilm disinfectant efficacy tests. *Journal of Microbiological Methods*, 70(2), 236–244 (2007).

47 A. Agostinho, G. James, O. Wazni, M. Citron and B.D. Wilkoff, Inhibition of *Staphylococcus aureus* biofilms by a novel antibacterial envelope for use with implantable cardiac devices. *Clinical and Translational Science*, 2(3), 193–198 (2009).

48 R. Hadi, K. Vickery, A. Deva and T. Charlton, Biofilm removal by medical device cleaners: comparison of two bioreactor detection assays. *Journal of Hospital Infection*, 74(2), 160–167 (2010).

49 D.L. Williams and R.D. Bloebaum, Observing the biofilm matrix of *Staphylococcus epidermidis* ATCC 35984 grown using the CDC biofilm reactor. *Microscopy and Microanalysis*, 16(02), 143–152 (2010).

50 B.E. Ramey and M.R. Parsek, Growing and analyzing biofilms in fermenters. *Current Protocols in Microbiology*, 1B–3 (2005).

51 L. Willcock, P. Gilbert, J. Holah, G. Wirtanen and D.G. Allison, A new technique for the performance evaluation of clean-in-place disinfection of biofilms. *Journal of Industrial Microbiology and Biotechnology*, 25(5), 235–241 (2000).

52 A.H. Rickard, A.J. McBain, A.T. Stead and P. Gilbert, Shear rate moderates community diversity in freshwater biofilms. *Applied and Environmental Microbiology*, 70(12), 7426–7435 (2004).

53 M.M. Williams, J.W. Santo Domingo and M.C. Meckes, Population diversity in model potable water biofilms receiving chlorine or chloramine residual. *Biofouling*, 21(5–6), 279–288 (2005).

54 P. Declerck, J. Behets, A. Margineanu, V. van Hoef, B. De Keersmaecker and F. Ollevier, Replication of *Legionella pneumophila* in biofilms of water distribution pipes. *Microbiological Research*, 164(6), 593–603 (2009).

55 B.E. Christensen, J. Kjosbakken and O. Smidsrød, Partial chemical and physical characterization of two extracellular polysaccharides produced by marine, periphytic Pseudomonas sp. strain NCMB 2021. *Applied and Environmental Microbiology*, 50(4), 837–845 (1985).

56 G.A. O'Toole and R. Kolter, Initiation of biofilm formation in *Pseudomonas fluorescens* WCS365 proceeds via multiple, convergent signalling pathways: a genetic analysis. *Molecular Microbiology*, 28(3), 449–461 (1998).

57 E. Peeters, H.J. Nelis and T. Coenye, Comparison of multiple methods for quantification of microbial biofilms grown in microtiter plates. *Journal of Microbiological Methods*, 72(2), 157–165 (2008).

58 S.A. Kristian, T. Golda, F. Ferracin, S.E. Cramton, B. Neumeister, A. Peschel, F. Götz and R. Landmann, The ability of biofilm formation does not influence virulence of *Staphylococcus aureus* and host response in a mouse tissue cage infection model. *Microbial Pathogenesis*, 36(5), 237–245 (2004).

59 N.C. Caiazza and G.A. O'Toole, SadB is required for the transition from reversible to irreversible attachment during biofilm formation by *Pseudomonas aeruginosa* PA14. *Journal of Bacteriology*, 186(14), 4476–4485 (2004).

60 H. Ceri, M.E. Olson, C. Stremick, R.R. Read, D. Morck and A. Buret, The Calgary Biofilm Device: new technology for rapid determination of antibiotic susceptibilities of bacterial biofilms. *Journal of Clinical Microbiology*, 37(6), 1771–1776 (1999).

61 J.J. Harrison, R.J. Turner and H. Ceri, Persister cells, the biofilm matrix and tolerance to metal cations in biofilm and planktonic *Pseudomonas aeruginosa*. *Environmental Microbiology*, 7(7), 981–994 (2005).

62 J.J. Harrison, R.J. Turner, D.A. Joo, M.A. Stan, C.S. Chan, N.D. Allan, H.A. Vrionis, M.E. Olson and H. Ceri, Copper and quaternary ammonium cations exert synergistic bactericidal and antibiofilm activity against *Pseudomonas aeruginosa. Antimicrobial Agents and Chemotherapy*, 52(8), 2870–2881 (2008).
63 J.J. Harrison, W.D. Wade, S. Akierman, C. Vacchi-Suzzi, C.A. Stremick, R.J. Turner and H. Ceri, The chromosomal toxin gene yafQ is a determinant of multidrug tolerance for *Escherichia coli* growing in a biofilm. *Antimicrobial Agents and Chemotherapy*, 53(6), 2253–2258 (2009).
64 N.D. Allan, A. Omar, M.W. Harding and M.E. Olson, A rapid, high-throughput method for culturing, characterizing and biocide efficacy testing of both planktonic cells and biofilms, *in Science against Microbial Pathogens: Communicating Current Research and Technological Advances*, A. Méndez-Vilas (Ed), Formatex Research Centre, Badajoz, Spain, 2011.
65 M.E. Olson, H. Ceri, D.W. Morck, A.G. Buret and R.R. Read, Biofilm bacteria: formation and comparative susceptibility to antibiotics. *Canadian Journal of Veterinary Research*, 66(2), 86 (2002).
66 J.J. Harrison, H. Ceri, C.A. Stremick and R.J. Turner, Biofilm susceptibility to metal toxicity. *Environmental Microbiology*, 6(12), 1220–1227 (2004).
67 M.W. Harding, M.E. Olson, L.L.R. Marques and R.J. Howard, Biology and management of microbial biofilms on plant surfaces, *in ISHS Acta Horticulturae 905: International Symposium on Biological Control of Postharvest Diseases: Challenges and Opportunities*, M. Wisniewski and S. Droby (Eds), ISHS, Leuven, Belgium, 2011.
68 A.E. Parker, D.K. Walker, D.M. Goeres, N. Allan, M.E. Olson and A. Omar, Ruggedness and reproducibility of the MBEC biofilm disinfectant efficacy test. *Journal of Microbiological Methods*, 102, 55–64 (2014).
69 M. Crouzet, C. Le Senechal, V.S. Brözel, P. Costaglioli, C. Barthe, M. Bonneu, B. Garbay and S. Vilain, Exploring early steps in biofilm formation: set-up of an experimental system for molecular studies. *BMC microbiology*, 14(1), 253 (2014).
70 R. Bayston, W. Ashraf, R. Barker-Davies, E. Tucker, R. Clement, J. Clayton, B.J. Freeman and B. Nuradeen, Biofilm formation by *Propionibacterium acnes* on biomaterials in vitro and in vivo: impact on diagnosis and treatment. *Journal of Biomedical Materials Research Part A*, 81(3), 705–709 (2007).
71 L. Pesciaroli, M. Petruccioli, F. Federici and A. D'Annibale, *Pleurotus ostreatus* biofilm-forming ability and ultrastructure are significantly influenced by growth medium and support type. *Journal of Applied Microbiology*, 114(6), 1750–1762 (2013).
72 L. Pesciaroli, M. Petruccioli, S. Fedi, A. Firrincieli, F. Federici and A. D'Annibale, Characterization of *Pleurotus ostreatus* Biofilms by Using the Calgary Biofilm Device. *Applied and Environmental Microbiology*, 79(19), 6083–6092 (2013).
73 K.E. Wesenberg-Ward, B.J. Tyler and J.T. Sears, Adhesion and biofilm formation of *Candida albicans* on native and Pluronic-treated polystyrene. *Biofilms*, 2(01), 63–71 (2005).
74 M.W. Harding, R.J. Howard, G.D. Daniels, S.L. Mobbs, S.L.I. Lisowski, N.D. Allan, A. Omar and M.E. Olson, A multi-well plate method for rapid growth, characterization and biocide sensitivity testing of microbial biofilms on various surface materials, *in Science against microbial pathogens: communicating current research and technological advances*, A. Méndez-Vilas (Ed), Formatex Research Centre, Badajoz, Spain, 2011.

75 M.W. Harding, G.D. Daniels, R.J. Howard, L.L.R. Marques, N.D. Allan, A. Omar and M.E. Olson, *In vitro* evaluations of microbial biofilms and their responses to chemical disinfectants, *in*: *ISHS Acta Horticulturae 1053: II International Symposium on Discovery and Development of Innovative Strategies for Postharvest Disease Management*, M. Wisniewski, S. Droby and P. Kinay-Teksur (eds). ISHS, Leuven, Belgium, 2014.

76 M.W. Harding, L.L.R. Marques, R.J. Howard and M.E. Olson, Biofilm morphologies of plant pathogenic fungi, *in Plant science and biotechnology in North America: Focus on Canada I*, K Al-Mughrabi (ed.). *The Americas Journal of Plant Science and Biotechnology* 4 (Special Issue 2), 43–47 (2010).

77 M.W. Harding, R.J. Howard, G.D. Daniels, S.L. Mobbs, S.L.I. Lisowski, N.D. Allan, A. Omar and M.E. Olson, A multi-well plate method for rapid growth, characterization and biocide sensitivity testing of microbial biofilms on various surface materials, *in Science against microbial pathogens: communicating current research and technological advances*, A. Méndez-Vilas (ed.), Formatex Research Centre, Badajoz, Spain, 2011.

78 R.J. Howard, M.W. Harding, G.C. Daniels, S.L. Mobbs, S.L.I. Lisowski and S.H. De Boer, Efficacy of agricultural disinfectants on biofilms of the bacterial ring rot pathogen, *Clavibacter michiganensis* subsp. *sepedonicus*. *Canadian Journal of Plant Pathology*, 37(3), 273–284 (2015).

79 C.G. Pierce, P. Uppuluri, A.R. Tristan, F.L. Wormley, E. Mowat, G. Ramage and J.L. Lopez-Ribot, A simple and reproducible 96-well plate-based method for the formation of fungal biofilms and its application to antifungal susceptibility testing. *Nature protocols*, 3(9), 1494–1500 (2008).

80 M. Musoni, J. Destain, P. Thonart, J.B. Bahama and F. Delvigne, Bioreactor design and implementation strategies for the cultivation of filamentous fungi and the production of fungal metabolites: from traditional methods to engineered systems/Conception de bioréacteurs et mise en oeuvre de stratégies pour la culture de champignons filamenteux et la production de métabolites d'origine fongique: des méthodes traditionnelles aux technologies actuelles. *Biotechnologie, Agronomie, Société et Environnement*, 19(4), 430 (2015).

81 L.L.R. Marques, H. Ceri, G.P. Manfio, D.M. Reid and M.E. Olson, Characterization of biofilm formation by *Xylella fastidiosa* in vitro. *Plant Disease*, 86(6), 633–638 (2002).

82 D. An, T. Danhorn, C. Fuqua and M.R. Parsek, Quorum sensing and motility mediate interactions between *Pseudomonas aeruginosa* and *Agrobacterium tumefaciens* in biofilm cocultures. *Proceedings of the National Academy of Sciences of the United States of America*, 103(10), 3828–3833 (2006).

83 A.D. Tomlinson, B. Ramey-Hartung, T.W. Day, P.M. Merritt and C. Fuqua, *Agrobacterium tumefaciens* ExoR represses succinoglycan biosynthesis and is required for biofilm formation and motility. *Microbiology*, 156(9), 2670–2681 (2010).

84 V. Lazarova and J. Manem, Innovative biofilm treatment technologies for water and wastewater treatment. *ChemInform*, 31(32) (2000).

85 V. O'Flaherty and P.N.L. Lens, Biofilms, *in Biofilms in Medicine, Industry and Environmental Biotechnology: characteristics, analysis and control*, P. Lens (Ed), IWA Publishing, 2003.

86 H. Von Canstein, S. Kelly, Y. Li and I. Wagner-Döbler, Species diversity improves the efficiency of mercury-reducing biofilms under changing environmental conditions. *Applied and Environmental Microbiology*, 68(6), 2829–2837 (2002).

87 T. Ebihara and P.L. Bishop, Effect of acetate on biofilms utilized in PAH bioremediation. *Environmental Engineering Science*, 19(5), 305–319 (2002).

88 G. Jin, A.J. Englande Jr and Y.L. Qiu, An integrated treatability protocol for biotreatment/bioremediation of toxic pollutants generated by chemical industries. *Journal of Environmental Science and Health, Part A*, 38(4), 597–607 (2003).

89 K.C. Cheng, A. Demirci and J.M. Catchmark, Advances in biofilm reactors for production of value-added products. *Applied Microbiology and Biotechnology*, 87(2), 445–456 (2010).

15

Factors Affecting Biofilm Formation in *in vitro* and in the Rhizosphere

Firoz Ahmad Ansari[1], Huma Jafri[1], Iqbal Ahmad[1] and Hussein H Abulreesh[2]

[1] Department of Agricultural Microbiology, Aligarh Muslim University, Aligarh, India
[2] Department of Biology, Faculty of Applied Science, Umm Al-Qura, Makkah, Saudi Arabia

15.1 Introduction

Biofilms are bacterial communities enclosed within self-produced extracellular polymeric substances (EPS). In nature, biofilms constitute a protected growth modality allowing bacteria to survive in hostile environments [1]. This cellular colonization confirms the nutrient utilization, expression of surface molecules, and virulence factors, and furnishes bacteria with an arsenal of properties that facilitate their survival in unfavorable conditions. It has been now well understood that bacterial ability to grow adhered to almost every surface-forming biofilm, although the mechanisms involved in the process of biofilm formation differ depending on characteristics of bacterial strain and environmental conditions [2]. Various organisms are studied extensively in this regards, such as *Pseudomonas aeruginosa*, *Bacillus subtilis*, *Staphylococcus aeureus*, *Escherichia coli*, and other environmental bacteria as reviewed by Lopez et al. [3]. Similarly, biofilm formation under natural setting such as plant and associated soil are now the subject of intense investigation.

One of the major hotspot for biofilm interaction is at the interface between plant roots and soil, rhizosphere and rhizoplane. Rhizosphere microorganisms are known to have significant impact on plant health through various type of interactions with plants, as well as within other soil microflora. Considering the complex and fluctuating conditions in soil and presence of an overwhelming number of micro and microorganisms, it is much more challenging to study biofilm in rhizosphere [4]. Rhizosphere is the soil area around the plant roots where convoluted biological and ecological processes take place, and it forms an environment suited for biofilm formation, including sufficient moisture and inventory of nutrients, which are mainly contributed by the plant. Biofilm formation on plant surface can take place in a response of associative, symbiotic, or pathogenic association (negative interaction).

It is still a mystery how plants regulate the microbial association. The main function of biofilm is to provide a resistant structure against stress factors such as antibiosis, UV radiation, desiccation, and predation [5].

In the rhizosphere, roots liberate large amount of metabolites from root hairs or fibrous root systems in the form of root exudates. These metabolites act as chemical signals to provide motility to the bacteria toward the root surfaces but also the main nutrient sources to support the growth and endurance in the rhizosphere. Some microbes that occupy plant rhizosphere are bacteria that are adequate to colonize very efficiently on the root surface or the rhizospheric soil [6]. These bacteria are attributed as plant growth–promoting rhizobacteria (PGPR). PGPR fulfill dominant responsibility for plant growth promotion and plant health protection by disparate manners. Direct plant growth promotion may results either from upgraded nutrient acquisition and/or from hormonal stimulation. Disparate mechanisms are involved in the defeat of plant pathogens, which is often incidentally linked with plant growth promotion [7].

In nature, interaction among different types of microorganism may result in multispecies biofilm. Mixed biofilm is the interaction between the different types of microorganisms that critically influence the development and shape of the community. Generally, interspecies interactions involve communication, commonly through quorum sensing, and metabolic cooperation or competition. Some interactions such as antagonistic, synergistic, or competition for the nutrients and growth inhibition take place among different bacterial species within the biofilm. These comprise the encouragement of biofilm formation through co-aggregation and metabolic cooperation where one species utilizes a metabolite produced by a neighboring species. This beneficial cooperation in mixed biofilm has important environmental, clinical, and industrial implications [8]. The bacterial colony in biofilm shows comparatively more resistance against various kinds of stress than the planktonic cells. The biofilm-associated protection is explained by several factors, often operating in concert, including structural changes and reductions in the diffusion rates of compounds in the biofilm matrix, changes in gene expression patterns, and low rates of growth of the biofilm cells [9].

In this chapter, we emphasize the process of biofilm formation briefly and provide an overview of how different factors—both biotic and abiotic—might influence the biofilm and its functions under rhizospheric conditions.

15.2 Process of Biofilm Formation

Biofilm formation process is multistep process involving attachment, maturation of biofilm, and detachment and return to the planktonic growth [10]. These processes of biofilm formation on the surfaces are briefly summarized here.

15.2.1 Attachment

Microbial cell attachment to both abiotic and biotic surfaces is the first interaction that may be reversible or irreversible. It is turning point from planktonic life to the biofilm mode. Initial reversible attachment is mediated by cell structure such as flagella, pili, and fimbriae.

During the attachment stage, the aggregation of rhizobacteria undergoes physiological changes that lead to EPS production and fix the cells to root surface. The cells then divide and form microcolonies by clonal propagation. Bacterial cells following maturation of biofilm consist of reversible and irreversible processes and involve diverse conserved and/or species-specific factors. Initially, bacterial cells are introduced to a surface, driven by Brownian movement and gravitational forces and simultaneously influenced by surrounding hydrodynamic forces [11]. Within a niche, bacteria encounter attractive or repelling forces that vary, depending on nutrient levels, pH, ionic strength, and temperature. Medium properties, along with bacterial cell-surface composition, affect velocity and direction toward or away from the contact surface [12]. Motile bacteria have a competitive advantage, utilizing flagella to overcome hydrodynamic and repulsive forces. The flagellar motility are important for initial attachment, as has been chronicled for many bacteria [12]. Chemotaxis also plays a role in directing attachment in response to nutrient composition in some bacterial species [13]. Attachment of microbial cells to biotic surfaces involve more cooperative / complex interaction with plant root [14]. It has been demonstrated that exopolysaccharides production is a key factor in determining optimal cell-to-cell and cell-to-surface interaction and biofilm formation by *Pseudomonas putida*. Similarly, rhizobial adhesion protein RapA1 was found to play a specific role in colonization of biotic surfaces [15]. Similarly, the role of root exudates in stimulation of attachment of rhizobia through production of acid EPS and arabinogalactans proteins from both legumes and non-legumes has been evident [16]. The role of secondary messenger is well described, and the concentrations of cAMP and cyclic diGMP are controlled by various environmental factors, such as carbon and oxygen, and thus regulate surface attachment [17].

15.2.2 Maturation of the Biofilm

Biofilm maturation requires two factors: (i) QS signal and (ii) EPS accumulation through continued cell division. Differential gene expression between the two bacterial growth states that is planktonic / sessile is related to adhesive needs of the population during surface colonization. For example, production of surface appendages is inhibited in sessile forms because motility is no longer necessary. Expression of genes involved in production of cell surface proteins and excretion products increases concomitantly. Transport of extracellular products in the cell is facilitated by surface proteins (porins) such as *Opr*C and *Opr*E, whereas transport of excretion products out of the cell is facilitated by certain polysaccharides [18]. Extracellular matrix composition has been more extensively investigated in *P. aeruginosa*, and has been shown to vary, depending on the environmental conditions [19]. As the biofilm matures, eDNA amounts increase through lysis of a bacterial subpopulation in response to the *P. aeruginosa* quinolone signal (Pqs) quorum sensing system [19]. Harmensen et al. [19] demonstrated that eDNA is organized in distinct patterns and localizes in the stalk portion of the mushroom-shaped biofilms. This localization may act as a scaffold for the formation of the mushroom structure, as type IV pili show high eDNA binding affinity, inducing the accumulation of migrating bacteria toward the areas of high eDNA concentration [20].

15.2.3 Detachment and Return to the Planktonic Growth Mode

Within the mature biofilm there is a bustling community that actively exchanges and shares products that play a pivotal role in maintaining biofilm architecture and providing a favorable living environment for the resident bacteria. However, as biofilm matures, dispersal becomes an option. Besides passive dispersal, brought about by shear stresses, bacteria have evolved ways to perceive environmental changes and gauge whether it is still beneficial to reside within the biofilm or whether it is time to resume a planktonic lifestyle. Biofilm dispersal can be the result of several cues, such as alterations in nutrient availability, oxygen fluctuations, and increase in the toxic products or other stress-inducing conditions [20]. Several sensory systems monitor the levels of small molecules, as a proxy to environmental changes, and alter gene expression accordingly, promoting dispersal [21]. Among other signals, the universal c-di-GMP has been extensively implicated in the shift between sessility and motility in bacteria; typically, an increase in c-di-GMP favors sessility, whereas reduced c-di-GMP leads to upregulation of motility [22]. EPS-degrading enzymes, such as alginate lyase in *P. aeruginosa*, also contribute to bacterial detachment from the matrix [23], although a controlled rhamnolipid production take place that contributes to channel formation within mature *P. aeruginosa* biofilm. An increase in the level of rhamnolipid aids bacterial dispersal [24].

15.3 Factor Influencing Biofilm Formation

Various factors in *in vitro* and in soil are known to influence the growth, survival, root colonization, and activities of microorganisms. Such factors are contributed by microbial cell structure and physiology, synthesis of exopolysaccharides, quorum sensing interference, interaction with other microorganism, influence of plant root and root exudates, as well as physico-chemical characteristics of soil and organic matter in soil [25]. In the rhizosphere, plant roots cope with both pathogenic and beneficial bacterial interactions. The exometabolite production in certain bacterial species regulates root growth and other root–microbe interactions in the rhizosphere [26].

Various microbial products can also influence the process of biofilm formation. The role of cyanide production in pseudomonad virulence affecting plant root growth and other rhizospheric processes has been demonstrated by Rudrappa et al. [27]. They used model plant *Arabidopsis thaliana* Col-0 seedlings and treated with both direct (with KCN) and indirect forms of cyanide from different pseudomonad strains. The treatment causes significant inhibition of primary root growth due to suppression of an auxin responsive gene, specifically at the root tip region by *pseudomonad cyanogenesis*. Additionally, *Pseudomonad cyanogenesis* also affected other beneficial rhizospheric processes such as *Bacillus subtilis* colonization by biofilm formation on *A. thaliana* Col-0 roots. The effect of cyanogenesis on *B. subtilis* biofilm formation was further established by the downregulation of important *B. subtilis* biofilm operons epsA and yqxM. The authors demonstrated the functional significance of pseudomonad cyanogenesis in regulating the multitrophic rhizospheric interactions [27].

An important advantage of the biofilm lifestyle for soil bacteria (rhizobacteria) is the protection against water deprivation (desiccation or osmotic effect) [28]. The composition and functions of bacterial biofilms in soil microniches are poorly understood. In one study, multibacterial communities established as biofilm-like structures in the rhizosphere of *Medicago sativa* (alfalfa) exposed to triple experimental conditions of water limitation. It was observed that the whole biofilm-forming ability (WBFA) for rhizospheric communities exposed to desiccation is higher than that of communities exposed to saline or nonstressful conditions [29].

The effect of various factors on biofilm formation/disruption is briefly summarized in the following subsections and also presented in Table 15.1 and Figure 15.1.

15.3.1 Surfaces

Under a rhizospheric environment, various interacting factors contributed by soil environment, microbial characteristics, and plant surfaces are influencing some of the attachment leading to biofilm formation. For the sake of convenience, we briefly describe the individual factors known to influence microbial adherence on abiotic or biotic surfaces as follows.

Biofilm formation is dependent on the surrounding environmental conditions and substratum parameters. Cell adhesion to a surface is a prerequisite for colonization. Physicochemical parameters are known to affect initial attachment of cell [29]. Once the cells attach, the surface chemistry will influence cell adhesion, while topographic features allow maximum cell-surface binding, enhancing strength of attachment and thus retention. In an aqueous environment (liquid–solid), bacterial attachment to a surface such as material surface, plant surface including root and shoot, animal tissues, and soil occurs rapidly, over a few seconds to a few minutes. Moreover, the binding of microorganisms to a surface can confer advantages to cell survival—for example, the attachment of cells to solid surfaces has been reported to immediately upregulate alginate synthesis in a strain of *Pseudomonas aeruginosa* [30]. The metal surfaces are susceptible for the microbial attack and hence for biofilm formation. The attachment on the metal surfaces crucially depends on growth medium, characteristics of cell surface, and substratum [31]. The microbial cell attachment and thus biofilm formation can occur on metal surfaces also, including aluminium [32], stainless steel [33], and copper as well. However, some metals, such as aluminium or copper, are considered toxic to bacteria [34]. It has been suggested that microbial resistance to some metals (e.g., lead acetate) can be attributed to the high lead content of disinfectants and antiseptics, while resistance to copper sulphate may be due to its use as an algicide [35].

15.3.2 Temperature and Moisture Content

Terrestrial bacterial communities are exposed to various environmental stress factors, of which limited water availability is typically the most critical factor to exhibit the greatest effect on survival and activity of these communities [36]. The availability of water in soils (water potential, ψ) depends on dissolved solutes (osmotic potential) and characteristics of the matrix environment (matric potential; water retention force on the ground). These two potentials represent different types of water deprivation that may affect bacterial physiology in different ways.

Understanding the role of temperature and water stress in protocooperation between the plants and beneficial rhizobacteria may enhance the efficacy of biocontrol agents in reducing plant diseases. The influence of low or high temperature, combined with a normal and reduced water regime on the interaction between *Bacillus amyloliquefaciens* strain S499 and plants, results in the induction of systemic resistance (ISR). A reduction in ISR level was observed when plants were subjected to stress before bacterization; however, root treatment with S499 prior to stress exposure attenuated this negative effect. Further investigation revealed that relative production of surfactin by

Table 15.1 Factors Influencing Rhizobacterial Biofim Formation and Root Colonization in the Rhizosphere.

Name of Bacteria	Phenotype	Relevant Characteristics	Factors Affecting Biofilm	Reference
Pseudomonas chlororaphis	Root colonization of wheat	PGPB, bioremediation	Microbial products (Phenazine)	[72]
Bacillus subtilis	Root colonization of *Arabidopsis thaliana*	Biocontrol	Microbial products (Enzymes)	[73]
Bacillus spp.	Root colonization of wheat	PGPR, biocontrol	Microbial products (EPS)	[74]
Paenibacillus polymyxa	Colonization and biofilm formation on pea nut	PGPR, biocontrol	Microbial products (EPS)	[75]
Bacillus cereus and Bacillus Pumilus	Root colonization of wild barley found in the Evolution Canyon, Israel	Salt, heat, and desiccation tolerance	Salinity, temperature, and desiccation	[76]
Bacillus sp.	Root colonization of wheat	PGPB, salt tolerance	Salinity (>100M)	[82]
Azospirillum brasilense, Sp7	Biofilm formation	PGPB, desiccation and osmotic pressure	Starvation, salinity, and osmotic pressure	[83]
Bacillus amyloliquefaciens SQR9	Colonization on cucumber	PGPR	Root exudates (organic acids)	[84]
Stenotrophomonas rhizophila	Enhanced mixed-species biofilm formation with *Rhizobium, Azotobacter*	Various PGPB activities	Root exudates (organic acids)	[85]
Pseudomonas putida	Root colonization of maize and *Arabidopsis thaliana*	Bioremediation and desiccation tolerance	High temperature (≤45°C)	[80, 13]
Pantoea agglomerans	Root colonization of chickpea and wheat	PGPR and moisture control	High salinity (≥150 M)	[79, 80]
Bacillus amyloliquefaciens strain S499	Biocontrol	Temperature tolerance	High temperature (≤45°C)	[36]
Agrobacterium sp.	Biofilm formation on *alfalfa*	Drought tolerance	High temperature (≥40°C)	[28]
Rhizobium alamii	Root colonization of *Arabidopsis thaliana* and rapeseed	Heavy metal tolerance	Microbial products (exopolysaccharides)	[78]
Staphylococcus saprophyticus	Biofilm on *Lens esculenta*	Salt tolerance	Salinity (≥200 M)	[40]
Bacillus amyloliquefaciens NJN-6	Colonization on banana root	PGPR, biocontrol	Root exudates (organic acids)	[77]

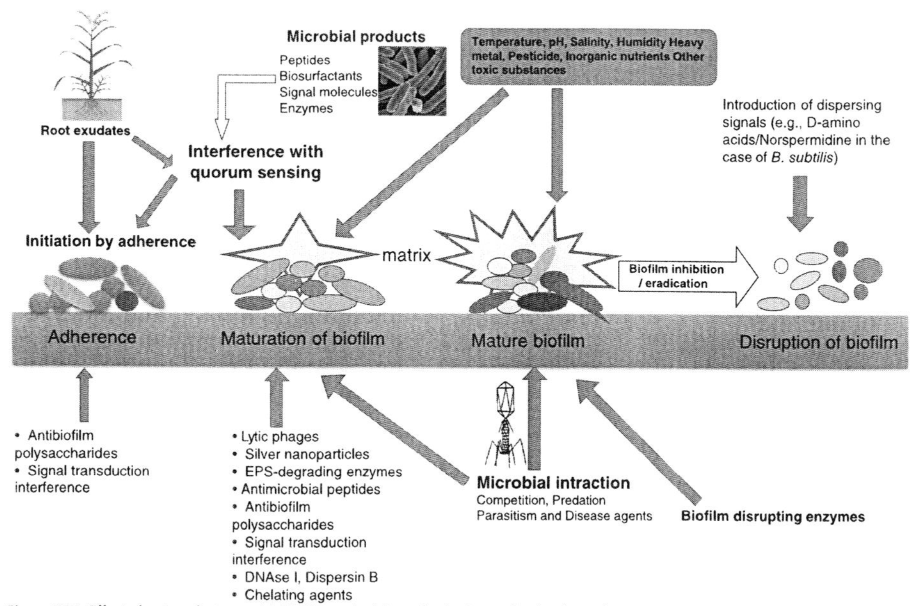

Figure 15.1 Effect of various factors on biofilm formation/disruption *in vitro* and in the rhizosphere.

S499 was clearly enhanced at low temperatures, making it possible to counter-balance the negative effect on the traits associated with rhizosphere fitness (colonization, motility, and biofilm formation) observed *in vitro* in cold conditions [37]. In anaerobic bacteria like *clostridium perfringens*, biofilm formation is drastically affected by temperature. The morphology, thickness, and cell density reflect the temperature-dependent regulation of EPS production [38].

15.3.3 Salinity

One-third of the world's arable land resources are affected by salinity [39]. Salt tolerance in plants depends mainly on the capability of roots for (i) restricted or controlled uptake of Na^+ and Cl^-, and (ii) continued uptake of essential elements, particularly K+ and NO3-. Consequently, the preferential uptake of K^+ over Na^+ has generally been considered as an important trait contributing to salt tolerance in various halophytes and nonhalophytes. Considering the potential of bacterial exopolysaccharides (EPSs) to bind cations including Na^+ [40], it may be envisaged that increasing the population density of EPS-producing bacteria in the root zone would decrease the content of Na^+ available for plant uptake, and thus help alleviating salt stress in plants growing in saline environments.

Establishment of biofilm, production of exopolysacharides (EPS), and accumulation of endogenous osmolytes under varying stress conditions are significant strategies adopted by bacterial strains for their successful survival in plant rhizosphere [41]. A study was conducted on determining the osmoadaptation strategies used by two native salt-tolerant strains *Oceanobacillus profundus* (Pmt2) and *Staphylococcus saprophyticus* (ST1) and their plant growth–promoting abilities. The ability of these strains to be used as inoculants for *Lens esculenta Var.* masoor 93 under salt stress was tested in the laboratory. Unlike the bacterial growth, biofilm formation, exopolysaccharide production, and endogenous osmolytes (proline, glycine, and betaine) accumulation increased at higher salt stress. Biofilm formation and endogenous osmolytes accumulation increased with increasing salt concentrations. The maximum increase in EPS accumulation was observed at maximum NaCl stress for ST1. Bacterial inoculation improved growth parameters and endogenous osmolytes accumulation of plants under salt stress compared to noninoculated control plants. The ST1 strain efficiently produced biofilm and exopolysacharide and accumulated osmolytes in response to NaCl stress. It is suggested that these strategies reverse the detrimental effects of high osmolarity in soil and are helpful for improving crop under salt stress [42].

15.3.4 Nutrient Availability

Nutrient availability is one of the major factor influencing growth and activities of microorganisms in soil and other habitats. Under rhizosphere conditions, nutrient availability is more compared to bulk soil due to nutrients release in root exudates by plants. However, there is tough competition between various microbial communities of soil. The effect of various nutrients and their influence on biofilm formation under natural conditions has been poorly exposed [43]. Biofilm bacteria acquire nutrients by concentrating trace organics on surfaces by the extracellular polymers, using the waste products from their neighbors and secondary colonizers, and by using different enzymes to break down food supplies. Biofilm matrix is often negatively charged; many nutrients (particulary cations) are attracted to the biofilm surface. Besides, nutrients with negative charge can exchange with ions on the surface. This provides bacterial cells within the biofilm with plenty of food compared to the surrounding [44]. Various factors, including carbon source, amount of nitrate, phosphate, calcium, and magnesium as well as the effects of osmolarity and pH, have been investigated on biofilm on *Sinorhizobium meliloti in vitro.* Nutrients such as sucrose, phosphate, and calcium enhance biofilm formation as their concentrations increase, whereas extreme temperatures and pH negatively affect biofilm formation of rhizobia. Similarly, in case of *B. subtilis,* growth was not limited in any of the conditions that did not result in pellicle formation. Similarly, galactose, arabinose, fucose, xylose, and glucuronic acid added at a 0.5 percent concentration and did not induce pellicle formation, which further demonstrated that the effect of plant polysaccharides as an environmental cue inducing biofilm formation is not due to increased growth attributed to the presence of additional sugars [45].

15.3.5 Microbial Products

Various exometabolites of microorganisms are known to influence biofilm formation. Some of the well-known examples are briefly mentioned here.

15.3.5.1 QS Signal Molecules in Biofilm Formation

Quorum sensing plays an important role in the communication between neighboring bacterial cells via signal molecules. It is a social behavior that enables interactions within the mono and mixed bacterial communities. Quorum sensing depends on the release and production of signal molecules, called *autoinducers,* and it increases in concentration of cell density while physiological conditions may also play significant role [46]. The quorum sensing system assists bacteria to express specific genes in a hormonize fashion [47]. Quorum sensing plays an important role in the development of biofilms in terms of induction vs. repression of biofilm formation, which varies depending on the bacterial species and environmental conditions [48]. In Gram-positive bacteria, the autoinducers are often peptides.

Production of several extracellular proteases involved in dispersal of biofilm is regulated by Agr QS system in *S. aureus* [49]. Similarly, in *B. subtilis* the production and secretion of QS molecule and surfactin is important for biofilm formation [50]. Specifically, the role of QS signals that is acylhomoserinelactone (AHL) produced by Gram-negative bacteria and their role in the induction of biofilm have been demonstrated through regulation of eDNA and production of PEL polysaccharides [51]. In QS the AI-2 system is studied globally and can mediate interspecies communication [52]. This system was identified in several Gram-negative and Gram-positive bacterial species. Many bacterial species are found in intimate association with one another in natural environmental condition [53].

In addition to QS signal molecules, other microbial products act as non-QS signal molecules that influence biofilm formation. Microbial products of this group include secondary metabolites like antibiotics, pigments, and siderophores.

Antibiotics are naturally produced by soil microorganisms and have a regulatory role in soil bacterial population. Major producers of such compounds include actinomycetes, fungi, and bacteria such as *Bacillus, Pseudomonas* [54]. *In vitro* experiment biofilm modulation as well QS interference have been documented concerning various antibiotics at sub-MICs of Imipenem, aminoglycoside, tobramycin induced biofilm in *Pseudomonas aeuroginosa,* and *E. coli.* Many QS inhibitory molecules such as furanoses may favor biofilm formation in *S. aeureus.* However, antibiotics such as Doxycyclin and Azithromycin at sub-MIC inhibit QS as well as biofilm formation at sub-MIC [55].

Within the biofilms, the phenazine pyocyanin functions in extracellular electron transfer to generate energy for growth. Small amounts of diffusible molecules shuttling electrons in a biofilm where the diffusion solubility may be limited is beneficial for the community [56]. Phenazines in *P. aeruginosa* also function as signaling molecules in biofilm formation, as a mutant unable to produce phenazines produced dramatically more wrinkled colony morphology than a wild-type strain [57].

15.3.5.2 Antimicrobial Peptides

Soil rhizosphere contains several habitats with functional microbial communities, where some microbial communities defend themselves from others by producing antimicrobial metabolites. These antimicrobial compounds produced by bacteria are found in all major bacterial lineages [58] and are produced by both Gram-negative and Gram-positive bacteria [59]. The antimicrobial peptides are the cystine-rich low-molecular-weight compounds, also called *host defense peptides.* The lytic peptides are assessed for their effects on biofilm formation. Lytic peptides bind the LPS (lipopolysaccharide) moieties of the bacterial cell membrane and disrupting membrane stability. Studies on *Staphylococcus aureus* have shown that the lytic peptide PTP-7 prevented *in vitro* biofilm formation and was also capable of diffusing into the deep layer of preformed biofilm that results killing of 99.9 percent of biofilm-forming bacteria. This peptide retained activity under highly acidic environments and in the presence of excess of metals, conditions that mimic the *S. aureus* biofilm environment [60].

15.3.5.3 Exopolysaccarides

The synthesis of exopolysaccharides by bacteria is well known and researchers have documented the relationship between exopolysaccharides production and biofilm formation [61]. Exopolysaccharides mediate cell-to-surface and cell-to-cell interactions that are critical for biofilm formation and stabilization. Mutants that are typically deficient in adherence and biofilm formation could not synthesize or export such polysaccharides and thus are highly sensitive to killing by antibiotics and host immune defenses [62]. However, recent evidences indicate that some bacterial exopolysaccharides inhibit or destabilize biofilm formation by other species [63]. Polysaccharides with nonbiocidal antibiofilm properties have also been isolated from cell-free biofilm extracts of several species. Their antibiofilm properties are believed to depend on their ability to (i) alter the physical characteristics of bacterial cells or abiotic surfaces; (ii) act as signaling molecules that impact the gene expression patterns of susceptible bacteria; or (iii) competitively inhibit multivalent carbohydrate–protein interactions, thereby interfering with adhesion [64]. Polymers are basically high molecular mass compounds formed by joining together a large number of repeating units of simple molecules, called monomers. On the basis of their origin, they may be natural polymers or synthetic polymers. Synthetic polymers may contain many additive chemicals, such as antioxidants, light stabilizers, lubricants, pigments, and plasticizers, added to improve the desired physical and chemical properties of the material [65]. However, these additives may leach into the surrounding environment and provide nutrients for microorganisms present. Phosphorus has been shown to increase the formation of biofilms on polyvinyl chloride in phosphorus-limited water [66]. Several studies have shown that plastic materials can support the growth of biofilms, but it has been suggested that growth in plastic pipes is usually comparable with that on iron, steel, or cement [67]. However, Bachmann et al. [68] used *Aquabacterium commune* cells under continuous cultivation with stainless steel and medium density polyethylene (MDPE) surfaces and found that biofilm cell density on MDPE slides was four times greater than on stainless steel.

15.3.6 Soil Enzymes

In the soil ecosystem, there are several extracellular enzymes mainly contributed by microorganism as well as plant roots. These enzymes are also abundant in aquatic and terrestrial ecosystems and may be present in significant amounts in soil and water biofilm. Complexes between enzymes and humic matter from soil have been reported to be extremely resistant to thermal denaturation, dehydration, and proteolysis [69]. N-acetyl-D-glucosamine-1-phosphate acetyltransferase (GlmU), which is involved in the biosynthesis of activated UDP-GlcNAc, an essential peptidoglycan and lipopolysaccharide (LPS) precursor in Gram-positive and Gram-negative bacteria, respectively, is among the enzymes targeted for matrix disruption. The effects of GlmU inhibitors, including N-ethyl maleimide (NEM), and its analogs showed antibiofilm activity against *E. coli, P. aeruginosa, K. pneumoniae,* and *E. faecalis* [70]. The enzymes DNase-I and Dispersin-B have also recently gained attention as potential antibiofilm agents, particularly against Gram-positive bacteria. The effects of DNase-I are linked with its ability to digest the eDNA found within the biofilm structure [71].

15.4 Conclusions and Future Direction

Biofilm formation by bacteria on various surfaces, both living and nonliving, is a common phenomenon, provided that conditions are suitable. The majority of bacteria are known to form biofilm under natural conditions / habitats. The ability of bacteria to form biofilm largely depends on the characteristics of bacterial strains and environmental factors. Many bacteria of soil and environments such as *Bacillus subtilis* and *Pseudomonas spp.* have been widely studied as model organisms for biofilm studies. Typical characteristics of bacteria responsible for biofilm formation includes presence of specific adhesins, ability to express various gene products such as EPS, quorum sensing, and regulation and switching ability from planktonic to sessile and vice versa. However other major factors include environmental factors such as temperature, moisture, host surface chemistry, nutrient availability, as well as interfering agents, both of biotic and abiotic nature. Under soil–plant systems, these factors are more complex, and various interacting factors affect biofilm. Mixed biofilm formed under natural conditions is least understood. Further new insight on the molecular basis of gene expression under suitable biofilm model is needed to study the impact of fluctuating environmental conditions on biofilm. Further contribution of plant genotype and its role in recruiting root microbes specifically in biofilm state have to be explored to understand the mechanism recruitment of microorganisms by plant.

References

1 L.V. Rinaudi and G. Walter, An integrated view of biofilm formation in *rhizobia., FEMS microbiolo lett.*, 304, 1–11 (2010).
2 D. Lopez and R. Kolter, Biofilms: Published by Cold Spring Harbor Laboratory Press. *Prespective in Biology,* 2 (2010).
3 A.A. Annette and A.M. Hirsch, Biofilm formation and attachment to roots, *Molecular Microbial Ecology of the Rhizosphere, Volume 2,* Section 8 (2013).
4 L.V. Rinaudi and J.E. Gonźalez, The low-molecular-weight fraction of the exopolysaccharide II from Sinorhizobium meliloti is a crucial determinant of biofilm formation, *J Bacteriol*, 191, 7216–7224 (2009).
5 T. Rudrappa, M.L. Biedrzycki, and H.P. Bais, Causes and consequences of plant-associated biofilms, *FEMS. Microbiol. Ecol.*, 64, 153–166. (2008).
6 V. Nihorimbere, P. Fickers, P. Thonart, and M. Ongena, Ecological fitness of *Bacillus subtilis* BGS3 regarding production of the surfactin lipopeptides in the rhizosphere, *Environ Microbiol Rep* 1, 124–130 (2009).
7 E. Sivan and B. Ehud, Multi-species biofilms: living with friendly neighbors, *FEMS Microbiol Riv.*, 990–1004 (2012).
8 T.F. Mah, and G.A. O'Toole, Mechanisms of biofilm resistance to antimicrobial agents, *Trends in Microbiology.*, 1913–1912 (2001).
9 H.C. Flemming and J. Wingender, The biofilm matrix, *Nat. Rev. Microbiol.*, 8, 623–633 (2010).
10 M.R. Donlan, Biofilms: Microbial Life on Surfaces, *Emerging Infectious Disease.*,volume 8, 1080–6059 (2002).
11 C.M. Toutain, N.C. Caizza, M.E. Zegans, and G.A. O'Toole, Roles for flagellar stators in biofilm formation by *Pseudomonas aeruginosa. Res Microbiol* 158, 471–477 (2007).
12 J. Schmidt, M. Musken, T. Becker, Z. Magnowska, et al. The *Pseudomonas aeruginosa* chemotaxis methyltransferase CheR1 impacts on bacterial surface sampling. *PLoS ONE* 6, 18184 (2011).
13 L. Nielsen, X. Li, and L.J. Halverson, Cell–cell and cell–surface interactions mediated by cellulose and a novel exopolysaccharide contribute to *Pseudomonas putida* biofilm formation and fitness under water limiting conditions, *Environ Microbiol.*, 13, 1342–1356 (2011).
14 E.J. Mongiardini, N. Ausmees, J. Perez-Gimenez, M. Julia Althabegoiti, J. Ignacio Quelas, S.L. Lopez-Garcia, The rhizobial adhesion protein RapA1 is involved in adsorption of rhizobia to plant roots but not in nodulation. *FEMS Microbiol Ecol.*, 65, 279–288 (2008).
15 H.P. Bais, T.L. Weir, L.G. Perry, S. Gilroy, and J.M. Vivanco, The Role of Root Exudates in Rhizosphere Interactions with Plants and Other Organisms, *The Annual Review of Plant Biology.* 57, 032905–105159 (2006).
16 D. Davies, Understanding biofilm resistance to antibacterial agents. *Nat. Rev. Drug Discov.* 2, 114–122 (2003).
17 K.A. McDonough, and A. Rodriguez, The myriad roles of cyclic AMP in microbial pathogens: from signal to sword, *Nat. Rev. Microbiol.* 10, 27–38 (2012).
18 M. Allesen-Holm, K.B. Barken, L. Yang, M. Klausen, et al., A characterization of DNA release in *Pseudomonas aeruginosa* cultures and biofilms. *Mol Microbiol.*, 59, 1114–1128 (2006).

19 K.B. Barken, S.J. Pamp, L. Yang, M. Gjermansen, et al., Roles of type IV pili, flagellum-mediated motility and extracellular DNA in the formation of mature multicellular structures in *Pseudomonas aeruginosa* biofilms. *Environ Microbiol.*, 10, 2331–2343 (2008).
20 S.H. Hong, J. Lee, and T.K. Wood, Engineering global regulator Hha of *Escherichia coli* to control biofilm dispersal. *Microb Biotechnol.*, 3, 717–728 (2010).
21 J.B. Kaplan, Biofilm dispersal: Mechanisms, clinical implications, and potential therapeutic uses. *J Dent Res* 89, 205–218 (2010).
22 T.K. Wood, S.H. Hong, and Q. Ma, Engineering biofilm formation and dispersal. *Trends Biotechnol.*, 29, 87–94 (2010).
23 A.M. Chakrabarty, and A. Boyd, Role of alginate lyase in cell detachment of *Pseudomonas aeruginosa. Appl Environ Microbiol.*, 60, 2355–2359 (1994).
24 M. Harmsen, L. Yang, S.J. Pamp, and T. Tolker-Nielsen An update on *Pseudomonas aeruginosa* biofilm formation, tolerance, and dispersal, *FEMS Immunol Med Microbiol* 59, 253–268 (2010).
25 S. Haldar and S. Sengupta, Plant-microbe Cross-talk in the Rhizosphere: Insight and Biotechnological Potential, *The Open Microbiology Journal*, 9, 1–7 (2015).
26 T. Rudrappa, M.L. Biedrzycki, H.P. Bais, Causes and consequences of plant-associated biofilms. *FEMS Microbiol Ecol*, 64, 153–166. (2008).
27 T. Rudrappa, Cyanogenic *Pseudomonads* Influence Multitrophic Interactions in the Rhizosphere, *PLoS ONE*, 4, 2073 (2008).
28 P. Bogino, A. Abod, F. Nievas, and W. Giordano, Water-Limiting Conditions Alter the Structure and Biofilm-Forming Ability of Bacterial Multispecies Communities in the *Alfalfa* Rhizosphere. *PLoS ONE*, 11, 79614 (2013).
29 K.A. Whitehead and J. Verran, The Effect of Substratum Properties on the Survival of Attached Microorganisms on Inert Surfaces, *Springer Series on Biofilms*, 10, 7123–7142 (2008).
30 D. Davies, A. Chakrabarty, G. Geesey, Exopolysaccharide production in biofilms: substratum activation of alginate gene expression by *Pseudomonas aeruginosa*, *Appl Environ Microbiol* 59, 1181–1186 (1993).
31 M.D. Rodney, Biofilms: Microbial Life on Surfaces. *J. Emerg Infect Dis*, 8, 881–890 (2002).
32 J.S. Nickels, R.J. Bobbie, D.F. Lott, R.F. Martz, P.H. Benson, and D.C. White, Effect of manual brush cleaning on biomass and community structure of micro-fouling film formed on aluminum and titanium surfaces exposed to rapidly flowing seawater. *Appl Environ Microbiol.*, 41, 1442–1453 (1981).
33 M.W. Mittelman, D.E. Nivens, C. Low, D.C. White Differential adhesion, activity, and carbohydrate–protein ratios of *Pseudomonas-atlantica* monocultures attaching to stainless-steel in a linear shear gradient, *Microb Ecol.*, 19, 269–278 (1990).
34 S.V. Avery, N.G. Howlett, and S. Radice, Copper toxicity towards *Saccharomyces cerevisiae*: dependence on plasma membrane fatty acid composition. *Appl Environ Microbiol.*, 62, 3960–3966 (1996).
35 K. Hiramatsu, H. Hanaki, T. Ino, K. Yabuta, T. Oguri, and F. Tenover, Methicillin resistant *Staphylococcus aureus* clinical strain with reduced vancomycin susceptibility, *J Antimicrob Chemo*, 40, 135–136 (1997).
36 M. Potts, Desiccation tolerance of prokaryotes, *Microbiol Rev* 58, 755–805 (1994).
37 Ilaria Pertot, Gerardo Puopolo, Taha Hosni, Lorenzo Pedrotti, Emmanuel Jourdan & Marc Ongena, Limited impact of abiotic stress on surfactin production in planta and

on disease resistance induced by *Bacillus amyloliquefaciens* S499 in tomato and bean, *FEMS Microbiol Ecol* 86, 505–519 (2013).
38 N. Obana, K. Nakamura, and N. Nomura, A sporulation factor is involved in the morphological change of *Clostridium perfringens* biofilms in response to temperature, *J. Bacteriol.* 196, 1540–1550 (2014).
39 M. Qadir, A. Ghafoor, and G. Murtaza, Amelioration strategies for saline soils: a review. *Land.Degrad. Dev.*, 11, 501–521 (2000).
40 J.L. Geddie and I.W. Sutherland, Uptake of metals by bacterial polysaccharides, *J Appl Bacteriol.*, 74, 467–472 (1993).
41 C. Fernandez-Aunión, T. Ben-Hamouda, F. Iglesias-Guerra, M. Argandona, R.M. Nieto, J.J. Bueno, M.E. Aouani, C. Vargas, Biosynthesis of compatible solutes in rhizobial strains isolated from *Phaseolus vulgaris* nodules in Tunisian fields, *BMC Microbiol.*, 19, 2180–10 (2010).
42 A.M. Qurashi and A.N. Sabri, Osmoadaptation and plant growth promotion by salt tolerant bacteria under salt stress, *African Journal of Microbiology Research* 21, 3546–3554 (2011).
43 J. López-Bucio, A. Cruz-Ramírez, and L. Herrera-Estrella, The role of nutrient availability in regulating root architecture, *Current Opinion in Plant Biology* 6, 280–287 (2003).
44 B. Prakash, B.M. Veeregowda, and G. Krishnappa, *Biofilms: A survival strategy of bacteria Curr Sci.*, 9, 1299–1307 (2003).
45 Pascale, B. Beauregarda, Yunrong Chaib, Hera Vlamakisa, Richard Losickb, and Roberto Koltera, *Bacillus subtilis* biofilm induction by plant polysaccharides, *PNAS.*, 17, 1621–1630 (2013).
46 W.L. Ng, B.L. Bassler, Bacterial quorum-sensing network architectures, *Annu Rev Genet*, 43, 197–222 (2009).
47 M.B. Miller, B.L. Bassler, Quorum sensing in bacteria, *Annu Rev Microbiol*, 55, 165–199 (2001).
48 Y. Sakuragi and R. Kolter, Quorum-sensing regulation of the biofilm matrix genes (pel) of *Pseudomonas aeruginosa*, *J. Bacteriol.*, 189, 5383–5386 (2007).
49 J.P. O'Gara, *Ica* and beyond: Biofilm mechanisms and regulation in *Staphylococcus epidermidis* and *Staphylococcus aureus*, *FEMS Microbiol Lett.*, 270, 179–188 (2007).
50 D. Lopez, M.A. Fischbach, F. Chu, R. Losick, and R. Kolter, Structurally diverse natural products that cause potassium leakage trigger multicellularity in *Bacillus subtilis*, *Proc Natl Acad Sci* 106, 280–285 (2009a).
51 M. Allesen-Holm, K.B. Barken, L. Yang, M. Klausen, et al., A characterization of DNA release in *Pseudomonas aeruginosa* cultures and biofilms. *Mol Microbiol.*, 59, 1114–1128 (2006).
52 M.J. Federle, Autoinducer-2-based chemical communication in bacteria: complexities of interspecies signalling, *Contrib Microbiol* 16, 18–32 (2009).
53 D. McDougald, S. Srinivasan, S.A. Rice, and S. Kjelleberg, Signal-mediated cross-talk regulates stress adaptation in *Vibrio* species, *Microbiology*, 149, 1923–1933 (2003).
54 M. Alexander, Most probable number method for microbial population, *Amer. Soc. Agron.* 9, 815–820 (1982).
55 F.M. Husain, I. Ahmad, M. Asif, and Q. Tahseen, Influence of clove oil on certain quorum sensing regulated functions and biofilm of *Pseudomonas aeruginosa* and *Aeromonas hydrophila*,. *J. Biosci.*, 38, 1–10 (2013).

56 M.E. Hernandez and D.K. Newman, Extracellular electron transfer, *Cell Mol Life Sci* 58, 1562–1571. (2001).

57 L.E. Dietrich, T.K. Teal, A. Price-Whelan, and D.K. Newman, Redox-active antibiotics control gene expression and community behaviour in divergent bacteria. *Science* 321, 1203–1206 (2008).

58 M.A. Riley and J.E. Wertz, Bacteriocins: Evolution, Ecology, and Application, *Annu. Rev. Microbiol*, 56, 117–137 (2002).

59 A. Savadogo, C.A.T. Ouattara, I.H.N. Basssole, S.A. Traoer, Bacteriocins and lactic acid bacteria- a minireview. *Afr J Biotechnol* 5, 678–683 (2006).

60 R. Kharidia, J.F. Liang, The activity of a small lytic peptide PTP-7 on *Staphylococcus aureus* biofilms, *J. Microbiol.*, 49, 663–668 (2011).

61 R. Oliveira, Polysaccharide production and biofilm formation by *Pseudomonas fluorescens*: effects of pH and surface material. *J. Colloids and Surfaces B, Biointerfaces*, 2, 41–46 (1994).

62 O. Rendueles, J.B. Kaplan, and J.M. Ghigo, Antibiofilm polysaccharides, *Environ. Microbiol.*, 10, 1462–2920 (2012).

63 Z. Qin, L. Yang, Qu. D. S. Molin, and T. Tolker-Nielsen, Pseudomonas aeruginosa extracellular products inhibit staphylococcal growth, and disrupt established biofilms produced by *Staphylococcus epidermidis*. *Microbiology* 155, 2148–2156 (2009).

64 O. Rendueles, J.B. Kaplan, and J.M. Ghigo, Antibiofilm polysaccharides, *Environ. Microbiol.*, 10, 1462–2920 (2012).

65 D. Brocca, E. Arvin, and H. Mosbæk, Identification of organic compounds migrating from polyethylene pipelines into drinking water, *Water Res* 15, 3675–3680 (2002).

66 M.J. Lehtola, I.T. Miettinen, and P.J. Martikainen, Biofilm formation in drinking water affected by low concentrations of phosphorus, *Can J Microbiol* 48, 494–499 (2002).

67 P. Niquette, P. Servais, and R. Savoir, Impacts of pipe materials on densities of fixed bacterial biomass in a drinking water distribution system, *Water Res* 6, 1952–1956 (2000).

68 R.T. Bachmann and R.G.J. Edyvean, AFM study of the colonisation of stainless steel by *Aquabacterium commune.*, *Int Biodeter Biodeg* 58, 112–118 (2006).

69 R.G. Burns and S. Alstrom, Cyanide production by rhizobacteria as a possible mechanism of plant growth inhibition. *Biology and Fertility of Soils*, 3, 232–238 (1989).

70 E. Burton, P.V. Gawande, N. Yakandawala, K. LoVetri, et al., Antibiofilm activity of GlmU enzyme inhibitors against catheter-associated uropathogens, *Antimicrob. Agents. Chemother.*, 50, 1835–1840 (2006).

71 P.S. Guiton, C.S. Hung, K.A. Kline, R. Roth, et al., Contribution of autolysin and Sortase A during *Enterococcus faecalis* DNA-dependent biofilm development, *Infect. Immun.*, 77, 3626–3638. (2009).

72 A.W.T.F. Chin, G.V. Bloemberg, I.H. Mulders, L.C. Dekkers, and B.J. Lugtenberg, Root colonization by phenazine-1-carboxamide-producing bacterium *Pseudomonas chlororaphis* PCL1391 is essential for biocontrol of tomato foot and root rot, *Mol Plant-Microbe Interact.*, 13, 1340–1345 (2000).

73 H.P. Bais, R. Fall, and J.M. Vivanco, Biocontrol of *Bacillus subtilis* against infection of *Arabidopsis* roots by *Pseudomonas syringae* is facilitated by biofilm formation and surfactin production, *Plant Physiol.*, 134, 307–319 (2004).

74 G. Gang, W. Bizun, M. Weihong, L. Xiaofen, Y. Xiaolin, Z. Chaohua, M. Jianhong and Z. Huicai, Biocontrol of Fusarium wilt of banana: Key influence factors and strategies., 41, 4835–4843 (2013).

75 W.M. Haggag and S. Timmusk, Colonization of peanut roots by biofilm forming *Paenibacillus polymyxa* initiates biocontrol against crown rot disease, *J Appl Microbiol.*, 104, 961–969 (2008).

76 S. Timmusk, V. Paalme, T. Pavlicek, J. Bergquist, J. Vangala, and T. Danilas, Bacterial distribution in the rhizosphere of wild barley under contrasting microclimates., *PLoS One* 6, 17968 (2011).

77 J. Yuan, N. Zhang, Q. Huang, W. Raza, R. Li, J.M. Vivanco, and Q. Shen, Organic acids from root exudates of banana help root colonization of PGPR strain *Bacillus amyloliquefaciens NJN-6, Scientific Reports* (2015).

78 M. Schue, A. Fekete, P. Ortet, C, Brutesco, T. Heulin, P. Schmitt-Kopplin, Modulation of metabolism and switching to biofilm prevail over exopolysaccharides production in the response of Rhizobium alamii to cadmium, *PLoS One.*, 6, 26771. (2011).

79 P.S. Chauhan and C.S. Nautiyal, The purB gene controls rhizosphere colonization by *Pantoea agglomerans, Lett Appl Microbiol.*, 50, 205–210 (2010).

80 M.A. Matilla et al. Complete genome of the plant growth promoting rhizobacterium *Pseudomonas putida BIRD-1, J. Bacteriol.*, 193, 1290 (2011).

81 M. Ashraf, S. Hasnain, O. Berge, and T. Mahmood, Inoculating wheat seedlings with exopolysaccharide-producing bacteria restricts sodium uptake and stimulates plant growth under salt stress, *Biol Fertil Soils.*, 40, 157–162 (2004).

82 A. Lerner, S. Castro-Sowinski, H. Lerner, Y. Okon, and S. Burdman, Glycogen phosphorylase is involved in stress endurance and bio¢lm formation in *Azospirillum brasilense Sp7, FEMS Microbiol Lett.*, 300, 75–82 (2009).

83 N. Zhang, D. Wang, Y. Liu, S. Li, Q. Shen, and R. Zhang, Effects of different plant root exudates and their organic acid components on chemotaxis, biofilm formation and colonization by beneficial rhizosphere-associated bacterial strains, *Plant Soil.*, 374, 689–700 (2014).

84 P. Alavi, M.R. Starcher, C. Zachow, H. Müller and G. Berg, Root microbe systems: the effect and mode of interaction of Stress Protecting Agent (SPA) *Stenotrohomonas rhizophila* DSM14405T, 4 (141) (2013).

16

Ecological Significance of Soil-Associated Plant Growth–Promoting Biofilm-Forming Microbes for Stress Management

Arpita Singh and Puneet Singh Chauhan

Division of Plant Microbe Interactions, CSIR-National Botanical Research Institute, Lucknow, India

16.1 Introduction

Life on the earth is affected by changes in climatic conditions, and these conditions directly or indirectly not only affect human and animal life but also significantly impact plants and microbes. Due to climatic changes, extreme conditions such as prolonged drought, flooding, and temperature extremes will have a significant impact on plant and soil microorganisms. These varying climatic conditions and other abiotic and biotic stress factors cause an intense loss in agricultural crop productivity on the basis of reduced growth and development, including reduced seed germination, photosynthesis inhibition, loss of water potential and nutrient imbalance, male and female sterility, altered gene expression, and protein content. The microbial world includes both beneficial and harmful microorganisms, and these beneficial microorganisms help in ameliorating stresses in plants, thus enhancing the plant's immunity and ability to cope up with abiotic and biotic stresses. These beneficial microbes are associated with plant roots and are called plant growth–promoting rhizobacteria (PGPR) and help in promoting plant growth and development through several mechanisms such as production of exopolysaccharides and biofilm formation. During crop production, microorganisms serve many purposes such as (i) biological activity monitoring in soil (microbial number, enzymatic activity and biodiversity); (ii) as indicators of soil health/quality; (iii) abiotic and biotic stress amelioration in plants; and (iv) as inoculants of beneficial microorganisms.

Bacterial biofilms involve assemblage of bacterial cells to abiotic and biotic surface required for cooperation among themselves. A biofilm consists of surface-attached microbial cells (algal, fungal, bacterial and/or other microbial) enclosed within a self-produced extracellular polymeric substance (EPS), which provides shelter, morphology, and protection to the bacterial community [1, 2]. Bacterial cells residing in the biofilm matrix are protected from various environmental stress factors, such as extreme pH, UV radiation, antimicrobial substances, dehydration, and osmotic shock, predators [3–5]. Naturally, biofilms occur in animals, plants and the environment and can be beneficial or harmful/pathogenic [6]. Beneficial biofilm formaton on plant roots of some crops may help plants in ameliorating abiotic as well as biotic stress, enhancing nutrient uptake and improving crop productivity. Quorum sensing plays a vital role in biofilm formation [7]. The aim of this chapter is to provide an insight on microbial biofilm formation and its ecological significance in abiotic and biotic stress amelioration for sustainable development in agriculture by enhancing crop productivity.

16.2 Rhizosphere Hub of Plant–Microbe Interactions

The term *rhizosphere* was introduced by Hiltner [8], derived from the Greek word *rhiza*, meaning root, and *sphere*, meaning field of influence; the rhizosphere includes plant roots surrounded by soil and microfauna interacting with each other. Biochemical interactions and the exchange of signal molecules take place between them as previously reviewed [9]. Microorganisms living in close association with plant roots compete for water, nutrients, and space [10], and their activities assist plants in nutrient uptake and endowing protection against pathogen attacks [11]. In the rhizosphere hub, interactions not only occur between soil and microfauna but also occur microbe–microbe, plant–microbe, and plant–plant (Figure 16.1). In fact, many microbiological studies are limited since the majority of soil microorganisms are non-culturable by using standard lab practices [12–14]. A more complete understanding of the rhizosphere microbiome has been aided through the advances in metagenomics [15–17], and microbiome microbial activities have been revealed through transcriptomic studies [18–20].

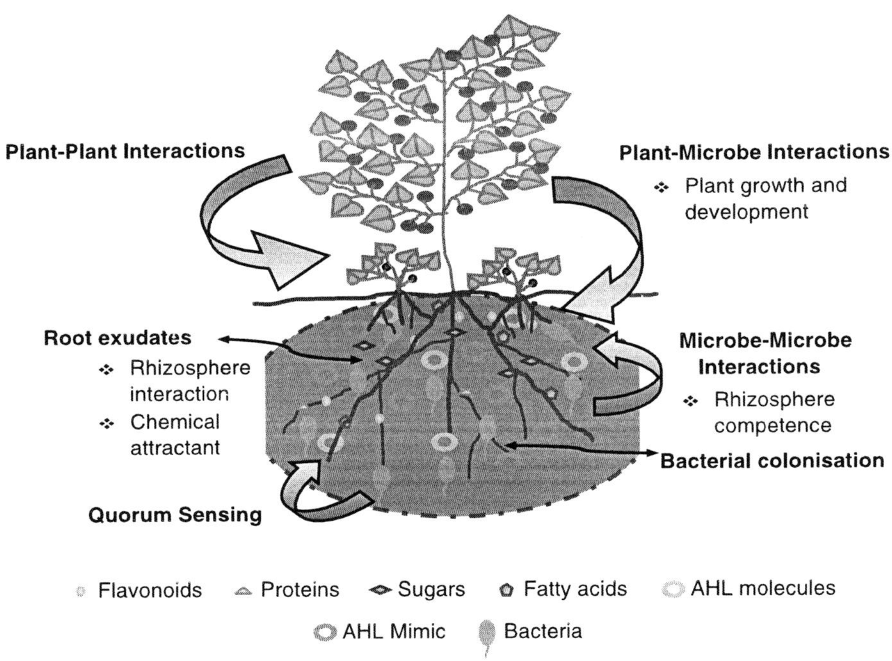

Figure 16.1 Rhizosphere hub of plant-microbe interactions.

16.3 Commencement of Rhizosphere Effect and Bacterial Colonization by Root Exudates

16.3.1 Rhizosphere Effect

Rhizodeposits are the substances or compounds released by plant roots commonly called root exudates and include water-soluble ions, carbohydrates, proteins, and lipids (high-molecular-mass compounds), monosaccharides, amino acids, and organic acids (low-molecular-mass compounds), mucilage, secondary metabolites like antimicrobial compounds, flavonoids, and enzymes. The term *rhizosphere effect* is defined as the processes occurring in a zone of few mm of thickness in soil surrounding plant roots, including root exudation, residing microbial activities, genetic exchange, and transformation of nutrients. All nutrients are unlimited, whereas the carbon source is limited in other than rhizosphere soil. In the rhizosphere, priming effect organic carbon released by plant roots is converted to carbon dioxide (CO_2); out of the total carbon, one-third to half is assigned to below-ground and 15 to 25 percent is exuded from plant roots in rhizosphere for its conversion. The release of huge amounts of carbon from plant roots increases microbial numbers and their activities [10, 21, 22], and proliferation of soil microorganism's takes place after chemotactically attraction toward root exudates. Through the lipid bilayer of the plasma membrane, direct or passive diffusion of root exudates is determined by membrane permeability, depending on the physiological state of the root cell and on the polarity of the compounds [23]. Root exudation involves two processes. First, it includes root excretion, in which some waste materials with unknown functions are exuded out from the roots. Second, it includes secretion of compounds responsible for lubrication, attractactant, and defense purposes.

Various factors, including pH, soil type, oxygen status, light intensity, soil temperature, nutrient availability, and the presence of microorganisms have a greater impact on root exudation than differences due to the plant species. In the rhizosphere, hypoxia occurs due to high moisture content of the soil, which increases the rate of anaerobic respiration, ethanol, lactic acid, and alanine accumulation. In the rhizosphere, the composition of root exudates is also affected by minerals and toxic metals. For example, aluminium is detoxified by the secretion of citric, malic, and oxalic acids whereas in phosphorous (P) and nitrogen deficient soil, secretion of phenolic compounds is increased in the former, whereas signaling molecules such as flavonones are increased in the latter. Mineral deficiency increases elicitor production that exhibits a detrimental impact on the process of root exudation such that defense responses are mediated by increased jasmonic acid facilitated by mineral deficiency. Light and temperature also exhibits a detrimental effect on root exudates—for example, under light conditions, elevated levels of flavonoids could be observed, whereas at lower temperatures, secretion of root exudates is diminished. Specifically, in *Vicia faba*, secretion of phenolic and tannins was found to be reduced at 4°C. In the rhizhosphere, the presence of microbes and plant species also affects the root exudation. For example, the existence of other plant species regulates the secretion of glucosinolates from *Arabidopsis*.

16.3.2 Rhizosphere Competence

To initiate an interaction with an appropriate plant host root, the microorganism in question must survive, participate, and somehow arrive at the root surface as a part of the rhizosphere population in the rhizosphere until the first interaction is consolidated. That the transition to the rhizosphere is not made by all soil microorganisms is evidenced by the wealth of data that document the qualitative differences between the population of the rhizosphere and that of soil at large. The microorganism's adaptation toward the rhizosphere begins as the seed germination takes place. Due to the inhibitory effects of diffusates from the seed coat, results in delayed establishment of some soil bacteria in the vicinity of the germinating legume seed, but the significance or duration of this effect under highly artificial conditions has not been established. As the root grows in contact with the soil, microorganisms residing in the vicinity immediately respond to it by enhancing growth for commencing the rhizosphere effect. The controlling factors in root zone colonization depend on the nature of the root exudates, numerous soil properties, and whatever complex cellular attributes contribute to the competitiveness of a microorganism in nature. But nothing is known of growth rates or substrate concentration effects for specific microorganisms in any complex natural environment. As compared to the non-rhizosphere, the rhizosphere is the zone of soil region having intense microbial activity where microbial interactions must be a dominant factor for a specific microorganism in its colonization of, or disappearance from, a rhizosphere [24–30].

16.3.3 Involvement of Genes and Traits in Rhizosphere Colonization

It has been widely accepted that, in addition to the ability to produce biologically active compounds, successful PGPR also must establish and maintain themselves in the presence of the large, metabolically active resident microbial population supported by the exudates and other organic metabolites provided by the root. It can be concluded from studies over the past 30 years that in rhizosphere colonization, many attributes are exhibited by bacterial strains; the significance of each and every attribute varies, depending on the individual bacterium characteristics, host plant and age, environmental conditions, soil type, and assay type used. Root colonization is defined as an active process involving the multiplication and survival of the introduced bacteria for several weeks on or around the roots in the presence of the indigenous bacteria leading to biofilm formation. This process includes colonization of the rhizosphere, inside of the root, the root surface, and the rhizosphere soil. Root colonization is a complex process. It includes interactions among the resident rhizosphere microflora in a constantly changing environment. Root colonization by introduced bacteria generally is considered essential to successfully establish, proliferate, and persist in the rhizosphere. Bacteria colonizing in or near the roots are ideally positioned to limit the establishment and/or spread of harmful microorganisms. Microbial colonization of the root has been an area under passionate discussion as well as heavily researched during the past two decades because colonization by numerous bacteria (variable colonization) remains one of the major hinderances to the extensive use of rhizobacteria for commercial agriculture purposes. In order to perform in the field efficiently, a PGPR strain needs to successfully establish, proliferate, and persist in the rhizosphere. Intensive studies over the last decade have revealed several traits critical for the root colonization:

- Attachment to roots
- Motility and chemotaxis
- Polysaccharide production
- Catabolism of nutritional sources
- A two-component signaling system
- Site-specific recombinase
- NADH dehydrogenase I

In the rhizosphere hub of plant–microbe interactions, after colonization at different locations of plant parts, bacterial biofilm formation takes place (Figure 16.2).

16.4 Quorum Sensing as a Way of Interaction between Bacteria and Host Plant

Quorum sensing is a mechanism through which bacteria are able to communicate with each other in a population density–dependent manner to behave in a coordinate mode that enables bacteria to express genes related to some traits necessary for biofilm formation [31,32]. Quorum sensing systems have been found to be present in both

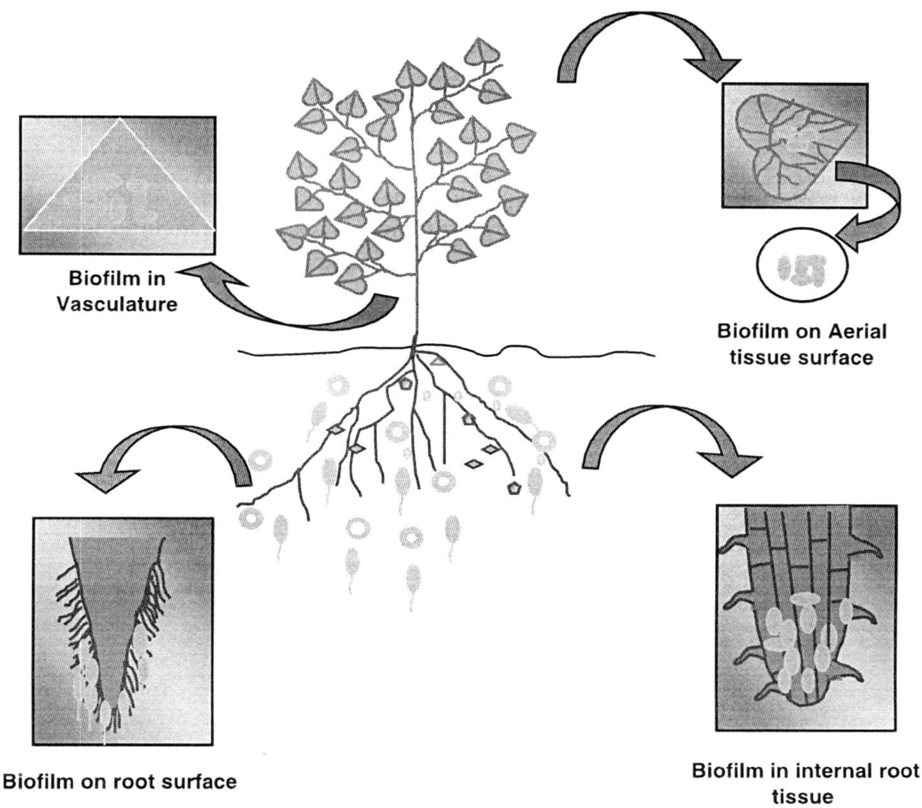

Figure 16.2 Rhizosphere colonization and biofilm formation.

Gram-positive and Gram-negative bacteria, but the only difference is that short peptide pheromones are produced by Gram-positive bacteria, where as acyl homoserine lactones are produced by Gram-negative bacteria. These AHL-signaling molecules were found to regulate various traits, including bioluminescence, conjugation, antibiotic production, toxin production, and extracellular hydrolytic enzymes, motility, production of virulence factors, and biosurfactant production.

In *Pseudomonas Xuorescens* 2P24, the *pcoI* gene responsible for the C6-HSL and 3-oxo-C8-HSL production was deleted and the resulting mutant was unable to colonize the wheat rhizosphere, defective in biofilm formation and biocontrol ability against wheat take-all disease [33]. These AHL molecules are mostly produced by plant-associated bacteria, and some of them belongs to the genera *Agrobacterium*, *Erwinia*, *Pantoea*, *Rhizobium*, and *Pseudomonas*. About 12 percent of the 300 bacterial isolates from the tomato rhizosphere were found to be AHL producers [34], 700 from wheat using different AHL biosensors [35]. So it can be inferred that in plant-associated communities, AHL-mediated QS is a common mechanism. Several studies revealed that in different ways eukaryotes respond to quorum sensing signals. Recently, it has been predicted that these AHL molecules alter expression of some genes in roots and shoots of plants, thereby amending defense and growth responses. Some of the PGPR produces acetoin and 2, 3-butanediol (bacterial volatiles) for their communication with plants and triggering plant growth promotion (Figure 16.3).

16.5 Biofilms

Biofilms are surface-attached bacterial communities in which planktonic cells are embedded in a self-produced extracellular polymeric matrix [36,37]. Genomics and

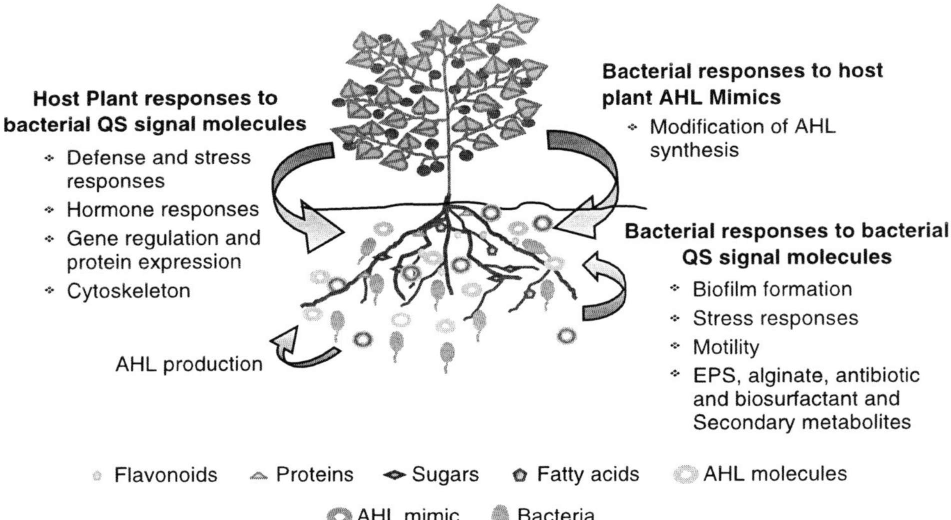

Figure 16.3 Bacterial quorum sensing signaling mediated interaction between host plants and bacteria.

proteomics studies reveal that the physiology of bacterial biofilms is different from those of planktonic cells due to their different expression patterns of genes and proteins.

16.5.1 Why Microorganisms Form Biofilms

Bacterial biofilms mainly consists of water and the bacterial cells, followed by the matrix made of exopolysaccharide. As compared to freely swimming planktonic stage cells, microbial cells residing in biofilms get many advantages, and that's the reason for them to prefer the biofilm mode of living [36, 38–40]. Some of these potential advantages are:

- Microorganisms existing in biofilms are protected from environmental stresses such as extreme pH, oxygen, osmotic shock, heat, freezing, UV radiation, predators, and antibiotics.
- Bacterial exopolysaccharides (EPS) secrete extracellular polymeric matrices, which increases the binding of water. It helps in protection against a common stress condition of planktonic cell due to decrease in dehydration (desiccation) of the bacterial cells.
- Biofilms are resistant to phagocytosis.
- The adherent nature of microbial cells in biofilms allows rapid exchange of nutrients, metabolites, and genetic material. As compared with planktonic cells, process of conjugation has been demonstrated in bacteria existing in biofilms.

16.5.2 Composition of Biofilms

The composition of biofilms plays a significant role as they are responsible for regulating and enhancing various metabolic and physiological activities. Water is the major component of biofilms and constitutes up to 97 percent, whereas 3 percent of the dry weight is constituted by bacterial cells. In addition to minerals, some compounds such as polysaccharides, lipopolysaccharides (LPS), DNA, proteins, and lipids together form the biofilm matrix [36] (Figure 16.4).

16.5.2.1 Extrapolymeric Substance

The stability of the biofilm structure is dependent on the properties of extrapolymeric substances interacting with macromolecules such as DNA and proteins, solutes, and ions. The self-produced extra polymeric matrix generates positive environmental conditions for the endurance of bacterial existence by providing essential nutrients and architectural integrity, enabling genetic transfer and intracellular communication.

16.5.2.2 Water

Water is the major component of biofilms, which allows the easy flow of nutrients, oxygen, and microorganisms from one site to another, accessing the sheltered microorganisms. This hydrated environment evades the harmful effects of desiccation from natural condition. Hydrogen bonding is favored by the presence of water between the embedded microbial cells enclosed in extra polymeric matrices. Microcolonies of bacterial cells in biofilms are separated by water channels.

16.5.2.3 Biomolecules

Biomolecules includes proteins, lipids, and nucleic acids. Proteins in the biofilm matrix have both a physiological and structural purpose. Some of the matrix proteins function

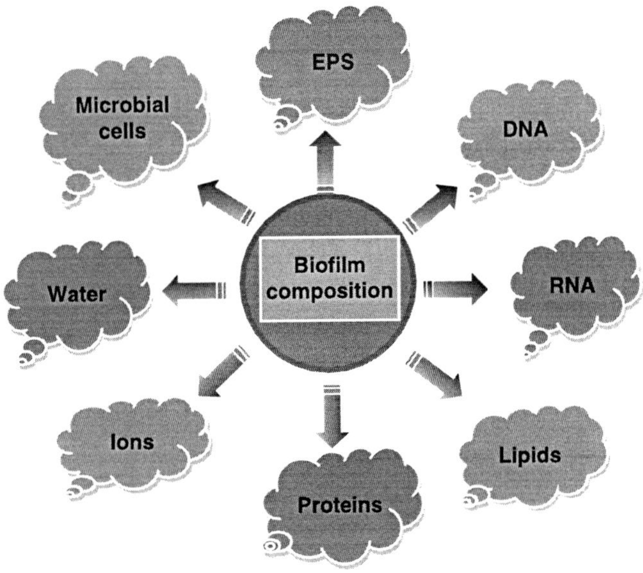

Figure 16.4 Composition of biofilm.

as enzymes, including lipases, hydrolases, lyases, and glucanases [41–43] and are responsible for the degradation and recycling of biopolymers for use as nutrients and modifying exopolymers for shaping or dispersion of cells from the biofilm structure. Pathogenic bacteria release specific enzymes that may act as virulence factors [44, 45].

16.5.3 Mechanism of Biofilm Formation

Biofilm formation is a multistep process (Figure 16.5).

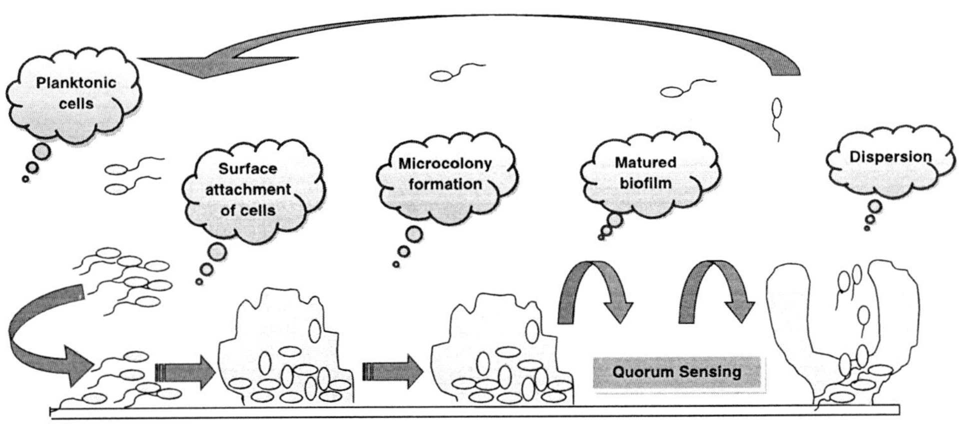

Figure 16.5 Stages involved in biofilm development.

16.5.3.1 Surface Attachment of Bacteria

This initial step of biofilm formation involves reversible attachment of planktonic bacterial cells to a surface by using adhesins. For example, polysaccharide adhesin (PS/A) of *S. epidermidis* helps in adhesion on naked or coated polymer surfaces (expression is controlled by the intercellular adhesion operon (Ica), whereas in case of *Streptococcus pyogenes* proteins or lipoteichoic acid, the various cell surface molecules helps in adhesion on cultured human cells but with the help of lateral flagella, adhesion of *Vibrios* could be possible. The adhesion process is affected by the physiological state of the organism, as some organisms have high attachment during log phase while others do in their stationary phase. At this stage, bacteria are still susceptible to antibiotics.

16.5.3.2 Microcolony Formation

The next step in biofilm formation involves irreversible binding of bacterial cells to a surface, followed by the multiplication of the cells, which results in microcolony formation. Then a polymer matrix around the microcolony is produced resulting in a matured biofilm.

16.5.3.3 Matured Biofilm and Dispersion

In matured biofilms, microbial cells are organized in mushroom-like or tower-like structures, varying as opposed to flat homogenous layer of cells. Biofilm development involves dispersal of bacterial cells (erosion) so that they can further initiate biofilm formation to new surface.

Quorum sensing (QS) system controls the maturation stage of biofilm formation in both Gram-positive and Gram-negative bacteria. QS systems, such as N-acyl-homoserine lactone (AHL) and 4-quinolone systems (in Gram negatives), AgrD peptide systems (in Gram positives), AI2/LuxS system (in both gram negatives and Gram-positives), and the farnesol systems (in fungi) control the maturation stage [46, 47]. The biofilm dispersion process may be prompted by bacteriophage activity. Gene regulation and expression differs between biofilm and planktonic growth for members of the same species. Davies and Geesey [48) showed that in *P. aeruginosa*, the gene algC controlling phosphomannomutase, involved in alginate (exopolysaccharide) synthesis, is upregulated within minutes of adhesion to a solid surface. Recently, it has been shown that algD, algU, rpoS, and the genes controlling polyphosphokinase synthesis are all upregulated in biofilm formation, and the expression of 45 genes differs between sessile cells and their planktonic counterparts [49]. Ram-age et al. [50] reported that in *C. albicans*, expression of CDR1, CDR2, and MDR1 genes was increased during biofilm formation.

16.5.4 Dynamics of Biofilms

Historical studies of biofilms indicate a variety of factors such as gene expression and environmental conditions that influence biofilm's surface properties—shear stress, exopolysaccharide production, quorum sensing signaling molecules, and aqueous medium (Figure 16.6). Some of the factors influencing biofilm development are still under debate among researchers.

16.5.4.1 Nutritional Conditions

For biofilm development, O_2 and CO_2 are two important constituents; out of these two, O_2 has a detrimental impact on initial adherence of planktonic cells and EPS production.

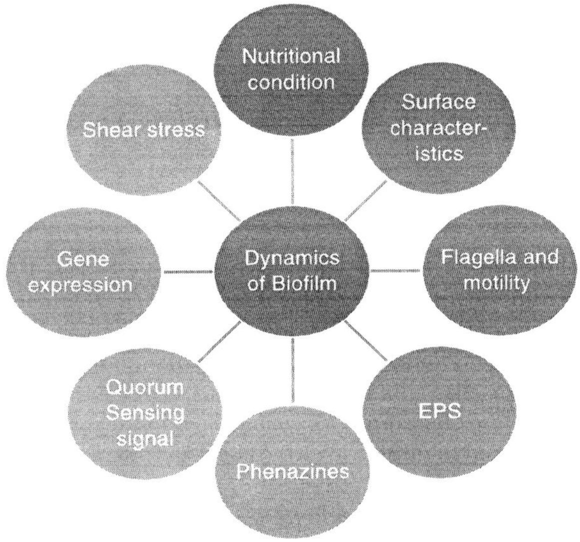

Figure 16.6 Dynamics of biofilm.

Undernourished conditions create void areas in centers of the cells residing in biofilm, leading to its dispersion. Hunt et al. propose a hypothesis modeling the mechanism behind this [51].

16.5.4.2 Surface Characteristics

All surface characteristics like structure, morphology, and composition have a detrimental impact on the development of the biofilm at all stages, from initial attachment of cells to the final stage of dispersion. Rough surfaces promote colonization, leading to biofilm formation as a result of pockets with diminished shear stress that provide a protective environment. Additionally, cells are more capable of adhearing to nonpolar and hydrophobic surfaces, as compared to hydrophilic surfaces. Further, porous, permeable materials lead to the formation of biofilm to a greater extent compared to dense materials. Calcium and magnesium act as lipopolysaccharide cross-linkers and positively contribute to biofilm development and cell membrane integrity [51].

16.5.4.3 Exopolysaccharides

Bacterial cells residing in a self-produced extracellular matrix is essentially composed of EPS and its biosynthesis is a QS-dependent process. For example, exopolysaccharides are involved in the *Azospirillum brasilense* and *Rhizobium leguminosarum* attachment to arbuscular mycorrhizal structures, as EPS production was found to be partially impaired in the *Azospirillum brasilense* mutants, AB7001 and AB7002, and completely lacks in 7030::Tn5-101 and 7030::Tn5-23. Only a few cells of AB7001 and AB7002 mutants were found to be adhered to fungal structures, whereas very rare *R. Leguminosarum* EPS mutant cells could be seen on *Gi. Margarita* hyphae [52]. Beauregard [53] demonstrated that *B. subtilis* is able to colonize the roots of *A. thaliana* and form biofilms, triggered by plant polysaccharides serving as a carbon source for extracellular polysaccharide matrix synthesis and signal for biofilm formation regulated by SpoA. SpoA activity depends on phosphorylation, which is controlled by five histidine kinases, KinA to KinE. It responds to different environmental cues. Phosphorylated Spo0A produces SinI, which is a SinR antagonist that shuts off the matrix genes when environmental conditions are not favorable for biofilm formation and development.

16.5.4.4 Flagella and Motility

Bacterial flagella have a detrimental impact on biofilm formation. For example: In *P. putida* IsoF strain, a gene responsible for flagellum production was mutated and named as GC25. More compactness of the cells could be observed in microcolony formation of the mutant strain, as compared to the wild type suggesting that within aggregates, flagella plays a significant impact on positioning of cells [54].

Swarming motility is regulated by quorum sensing and influences the biofilm formation; slow motility results in bacterial aggregate formation with resulting structured biofilms, whereas cells are found to spread on surfaces with high motility, giving rise to flat biofilms.

16.5.4.5 Quorum Sensing Signals

Quorum sensing is a process that enables the bacteria to produce signaling molecules, known as autoinducers, to regulate their gene expression in a population density dependent manner. For example, in *Pseudomonas putida* PCL1445, the quorum sensing system composed of *ppuI-rsaL-ppuR*, is involved in regulating biofilm formation. An experimental study shows that *ppuI* and *ppuR* mutants produce thick biofilm, whereas the *rsaL* mutant shows a delay in biofim formation compared to the wild-type strain. This correlates with the putisolvin I and II biosynthesis that controls biofilm formation. The reason behind this is that both *ppuI* and *ppuR* are responsible for AHL production that regulates putisolvin I and II biosynthesis. In polyvinylchloride (PVC) titer plate, biofilm formation has been reduced in case of *ppu I* and *ppuR* mutants, and this same phenotype could be observed in the putisolvin biosynthetic mutant by forming a thicker biofilm. Results have shown that in strain PCL1445, biofilm formation is regulated by putisolvin production in a cell-density-dependent manner. Upon addition of synthetic AHL molecules to the medium, the *ppuI* mutant started producing biosurfactant and lost the ability to form thick biofilms, as compared to the putisolvin biosynthetic mutant [55].

16.5.4.6 Gene Expression

Day by day, due to the development of the technologies, it is now easier to get the scientific information deeply. So with the help of molecular approaches, like transposon mutagenesis and gene knockout based studies, some of the genes responsible for the formation of biofilms have been revealed.

16.5.4.7 Shear Stress

Whether shear stress improves or inhibits biofilm formation is still under debate. Many experimental works have demonstrated that shear forces negatively affect biofilm formation and development, as these stresses delay maturation and tend to form a thin-layer biofilm, leading to a diminished diversity of bacterial cells residing in extracellular polymeric matrix. Positive impacts of low and high shear stress could be revealed, as the former tends to form multilayered, dense, thick biofilms and the latter one increases the susceptibility of cells to antibiotics like gentamicin. It also increases the residential period of cells and motility and non uniform stress prevents biofilm formation [51].

16.5.4.8 Phenazines

Both bacterial and synthetic phenazines serve as an electron shuttle acting as a signal regulating gene expression pattern leading to biofilm formation for their survival. It has significant impact on eukaryotic hosts and host tissues by modifying their cellular responses for example, in plants it induces systemic resistance.

16.6 Effects of Stress on Plants

Plant stress can be defined as the response of plants towards changing environmental conditions which affects their rate of photosynthesis ultimately reducing the plants ability to convert energy into biomass. Within a few minutes some environmental factors become stressful like temperature, air etc and some take days to weeks like water status of the soil and mineral deficiencies. There are mainly two types of stresses, abiotic stress and biotic stress (Figure 16.7).

16.6.1 Abiotic Stress

Abiotic stress occurs due to drought, salinity, high temperature, and cold.

16.6.1.1 Drought Stress in Plants

Drought is the most serious threat for crop productivity in meeting future needs of the world population due to the reduced rainfall and water content of the soil. The major consequences of drought on plants can be depicted in Figure 16.8.

Drought stress greatly reduces seed germination, which was demonstrated in a pea study. When the drought effect on seed germination was checked using five cultivars, then reduction in seed germination, hypocotyl length, and fresh and dry weight of root and shoot could be observed, while in others increase in root length was observed. However, in rice during vegetative growth, drought stress reduces its growth and development [56].

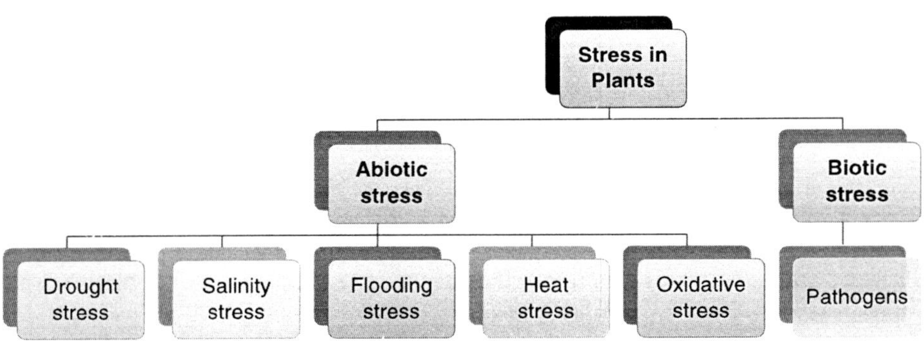

Figure 16.7 Stress in plants.

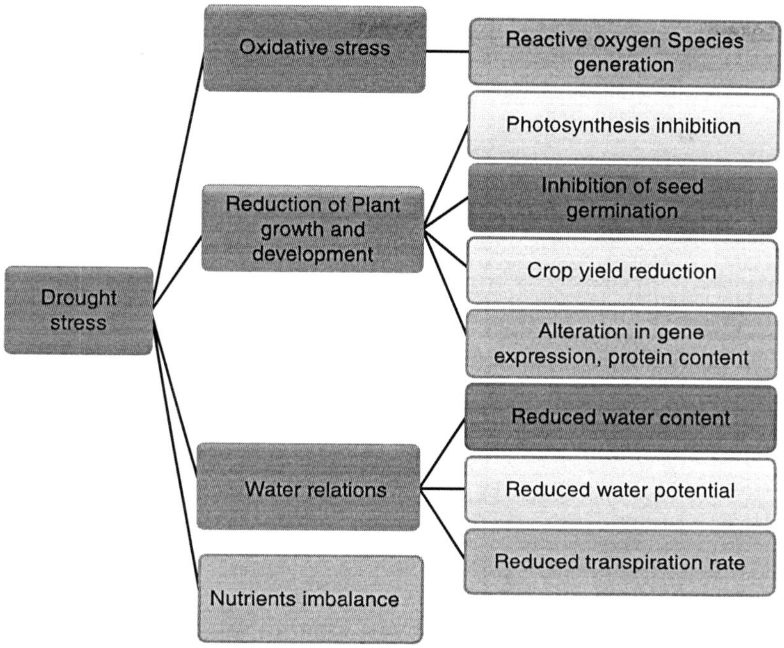

Figure 16.8 Major consequences of drought stress.

In plants, cell elongation can be inhibited in reduced water conditions due to disruption of water flow in elongating cells from xylem. It was demonstrated that mitosis, cell elongation and expansion was impaired under drought stress conditions, which results in inhibition of plant height, leaf area and crop growth. At the flowering stage, this stress causes barrenness. In whole, it can be stated that flowering and grain filling in drought conditions was reduced due to reduced assimilation, enzymatic activities, and growth and development of plants [56].

Some key factors influencing the plant–water relationships include relative water content, leaf water potential and temperature, stomatal resistance, transpiration rate, and canopy temperature. Initially in the development of wheat leaves, relative water content was higher and subsequently decreased as the leaves matured. In wheat, the water use efficiency was found to be greater and was correlated with the stomatal closure for reducing the transpiration rate. However, in a water-deficit condition study conducted on clover (*Trifolium alexandrinum*), increased water-use efficiency could be observed due to decreased water loss because of the reduced rate of transpiration and leaf area. In short, it can be inferred that leaf area, stomatal opening and closure, transpiration rate, and some other factors are responsible for controlling and maintaining water content under drought stress [56].

Drought conditions restrict the nutrient uptake and root to shoot translocation, thereby inhibiting nutrient concentration in crop plants. Increase in nitrogen and decreases in phosphorous were induced by moisture stress, but no effect was observe with K levels. In cotton, uptake of nitrogen and potassium was diminished in plant tissues; lowering PO^{3-4} mobility diminishes P and PO^{3-4}. In potassium-applied plants, lower water potential lowers the diffusive resistance in its leaves, as compared to non-K-applied controls. In whole, it can be summarized that acquisition, uptake, translocation, metabolism of nutrients, and transpiration rate are all reduced under drought conditions, and a reduced transpiration rate reduces the absorption of nutrients [56].

One of the foremost effects of drought is photosynthesis reduction, which comes up by reduced leaf area, impaired photosynthetic machinery, early leaf senescence, and decline in RuBisCO enzymatic activity. Drought stress induces stomatal closure so that carbon dioxide (CO_2) uptake is restricted by leaves' increased sensitivity to photo-damage. Crop yields get reduced by inhibited enzymatic activities of calvin cycle, altered photosynthetic pigments and components, reactive oxygen species generated oxidative stress in membrane lipids, proteins in other cellular components [56].

Through the process of respiration, the amount of carbohydrate loss determines the plant's overall metabolic efficiency. Mostly, carbon is consumed by roots, utilized for growth, stored for plant maintenance, and used in the production of dry matter [56].

16.6.1.2 Salinity Stress in Plants

In agriculture salinity, stress can be defined as the presence of soluble salts in high concentrations in the rhizospheric soil affecting plant growth by inhibiting water uptake and absorption of nutrients. Salt stress also affects seed germination, which is an important phase in the plant growth and developmental cycle. However, salinity strongly influences the seed germination process in a variety of plants, including *Posidonia*, *Oryza sativa*, *Triticum aestivum*, *Zea mays*, and *Brassica* spp. Many alterations in enzymatic activities responsible for metabolism of nucleic acids, proteins, and disturbance in hormonal balance are caused by altered water absorption due to lower osmotic potential of germination media. Recently, studies conducted on *Z. mays* by Khodarahmpour et al. [57] found that at 240 mM concentration, NaCl reduces seed germination by 32 percent, radicle length by 80 percent, plumule by 78 percent, and seedling length by 78 percent. In agriculture, effects like lowering photosynthesis rate, seed germination, dehydration of cells, altered cytoplasmic enzymatic activities, salt toxicity, senescence, and total reduced biomass reduce the crop yield (Figure 16.9).

Salt stress also affects plant growth and reduces crop yield, in two ways. First, it influences the plant's ability to uptake water, causing water-deficit conditions leading to many physiological disorders. Second, the ion excess or salt-specific effect occurs when sodium and chloride ions enter the plants via transpiration, causing injury to leaves. The two effects give rise to a two-phase growth response to salinity [58]:

- In *Phase 1*, plant growth is affected from the outer side; that means the soil solution containing salts decreases the root and leaf growth. The elongating cells can accumulate salt that can reach xylem-containing growing vacuoles.
- In *Phase 2*, plant growth is affected when salt is taken up by the plants, accumulating in the leaves so that the salt concentration eventually arises and the leaves die. They might be accumulating in the cell wall, causing their dehydration and inhibiting enzymatic activities in the cytoplasm. Cl^- is more harmful as compared to Na^+, as the former causes chlorotic toxicity by interfering with chlorophyll production whereas Na^+ ion interferes with potassium (K^+) ion uptake, hindering regulation of stomata, causing loss of water.

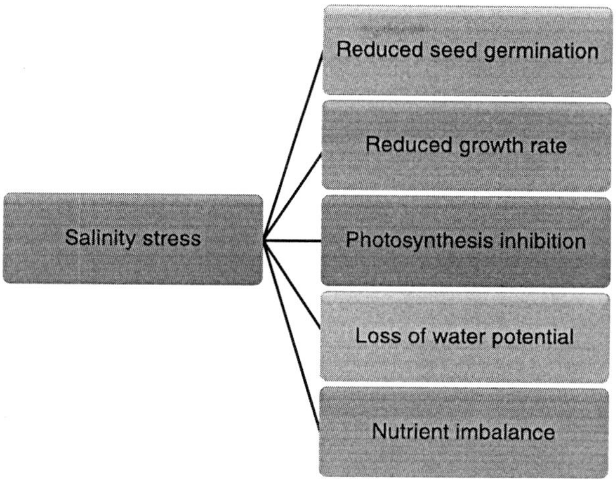

Figure 16.9 Major consequences of salinity stress.

In salinity stress, it has been reported that increased sodium chloride concentration increases sodium (Na^+) and chloride (Cl^-) ions, which causes nutrient deficiency due to their competition with nutrients like Ca^{2+}, K^+, and NO_3^- leading to decreased levels of N, P, K, Ca, and Mg level in fennel [59]; *Trachyspermum ammi* [60]; peppermint and *lemonverbena* [61]; *Matricariarecutita* [62]; and *Achillea fragratissima* [63].

Photosynthesis is inhibited by the reduction in water potential and chlorophyll content on the accumulation of both or one of the Na^+ and Cl^- ions in chloroplast. Chlorophyll a, b content were found to be reduced by 33 percent and 41 percent, irrespectively, in *O. sativa* leaves at 200 mM NaCl, 14 days [64], whereas at 100 mMNaCl chlorophyll a and b, and carotenoids were reduced by 30, 45, and 36 percent, irrespectively [65]. Efficiency of Photosystem II (PS II), a sensitive component to salt stress, electron transport chain (ETC), and CO_2 assimilation rate has been reduced. Under salt stress, the photosynthetic rate, stomatal conductance, and PSII efficiency was reduced in citrus [66]. Due to salt stress the altered chlorophyll fluorescence (PS II) affects the growth of barley [67]; (PS II), whereas electron transport rates, and D1 protein were also affected which influences the *Brassica juncea* growth [68]. In the whole, it can be inferred that under salt stress growth yield can be reduced as a result of lowered photosynthesis rate, cell dehydration, altered cytoplasmic enzymatic activities, salt toxicity, and senescence.

16.6.1.3 Flooding Stress in Plants

Excessive soil moisture causes wilting, epinasty, and senescence there by inhibiting overall plant growth and development. Excess water leads to an inadequate oxygen supply in the submerged tissues and the diffusion of water is 10^4 fold slower through water compared to air. Oxygen deficiency influences the ethylene levels and accumulation of Mn^{2+}, Fe^{2+}, S^{2-}, H_2S and carboxylic acid products.

16.6.1.4 Heat Stress in Plants

In plants, high temperature or heat stress adversely affects its physiological and metabolic processes resulting in altered plant growth and development and reduced crop productivity (Figure 16.10). ROS is produced as a major consequence that causes oxidative damage in plants. This stress alters the expression of some genes responsible for osmoprotection, regulatory proteins, and detoxifying enzymes involved in the cellular protection of the plants and maintenance of cytoskeleton structure, protein stability, and membrane proteins [69–72].

Heat stress causes many changes in plant morphology like discoloration and sunburns of plant parts (leaves and twigs, branches, and stems), senescence, and abscission of leaves, and inhibition of root and shoots growth. For example, in sugarcane, curling, drying, necrosis, and damage to leaves could be observed while in some plants at 28/29°C, temperature hyponasty and inhibition of total biomass could be observed. In wheat at 30/25°C day/night, reduced leaf area and tiller numbers could be observed [73].

During heat stress, firstly, among all the growth stages of the plant seed germination is affected, leading to abnormal seedlings, reduced radical, and plumule growth and seed germination also reduced by induction of abscissic acid. For example: In wheat at 45°C, germination rate was inhibited causing cell death, whereas in the case of rice, total biomass was reduced. Cell growth is diminished due to its reduced water content. In maize, millet, and sugarcane, reduced assimilation rate reduces relative growth rate. At extreme levels, stress due to denaturation or protein aggregation, programmed cell death in specific cells or tissues could be observed, and at moderate stress levels plant injuries including shedding of leaves, abortion of flower and fruit, or even death of the entire plant [73].

Heat stress influences more photosynthesis in C3 plants as compared to C4 plants. Metabolic activities like carbon metabolism and photochemical reactions occurring in the chloroplast and thylakoid lamellae are greatly influenced or impaired. It also reduces photosystem II (PSII), photosynthetic pigments, carbon dioxide assimilation and

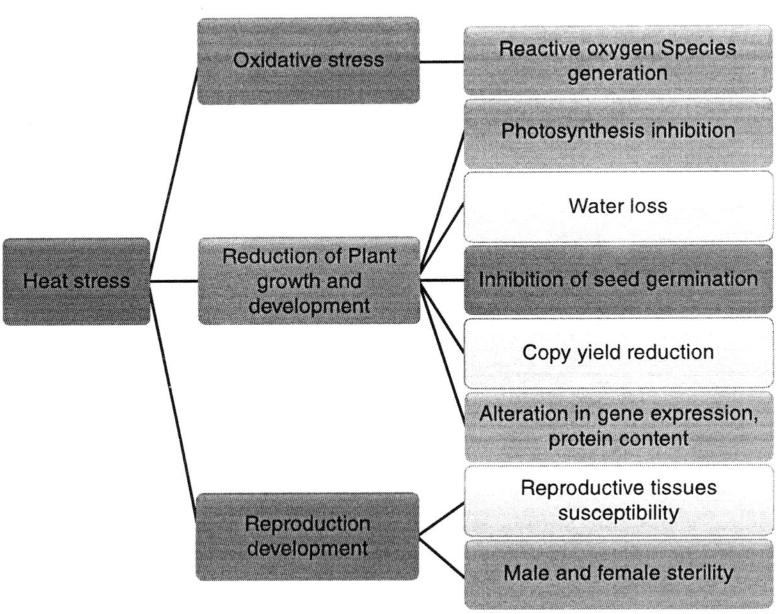

Figure 16.10 Major consequences of heat stress

concentration, water status of the leaf, and stomatal conductance. In soybeans, decreases in total chlorophyll content (18 percent), chlorophyll *a* content (7 percent), chlorophyll *a*/*b* ratio (3 percent), Fv/Fm ratio (5 percent), Pn (20 percent) and *gs* (16 percent) result in decreases in sucrose content by 9 percent and increases in reducing sugar content by 47 percent and in leaves by 36 percent.

In Shuanggui and T219 varieties of rice plant, the photosynthetic rate decreased by 16 percent in the former and 15 percent in latter at (33°C, 5 days) [74]. When temperature increases from 25°C to 45°C, the rate of photosynthesis decreases by 60 percent in leaves of *Vitis vinifera,* which is attributed to stomatal closure by 15 to 30 percent. Some other factors diminishing photosynthesis include the reduction of soluble proteins, RuBisCO binding proteins (RBP), and large and small subunits of RuBisCO in darkness [75], whereas starch and sucrose synthesis are influenced by the reduced activity of sucrose phosphate synthase, ADP-glucose pyrophosphorylase, and invertase [76, 77].

Growth and developmental stages of plants are adversely affected by heat stress, as reproductive tissues are highly susceptible; heat stress lends to delayed flowering, fruiting, and seed production, resulting in loss of crop productivity. There are several reasons for causing sterility in male and female flowering plants, including impaired meiosis, impaired pollen germination and pollen tube growth, reduced ovule viability, abnormality in stigmatic and style positions, reduced number of pollen grains, disturbed fertilization processes, hinderance in growth of the endosperm, and unfertilized embryos. Elevated levels of ethylene cause male sterility in rice, and it is hypothesized that ethylene inhibits the key enzymes in sugar–starch metabolism, which weakens sink strength and restricts grain filling and ultimately produces sterile grain [73].

Heat stress greatly reduces crop yield, including cereals (e.g., rice, wheat, barley, sorghum, maize), pulse (e.g., chickpea, cowpea), oil-yielding crops (mustard, canola), and so on. It was demonstrated that at 35–40°C in rice cultivar Shuanggui the 1000-grain weight was reduced by 7.0 to 7.9 percent as compared to heat-tolerant cultivar Huanghuazhan (21.7 to 24.5 percent). In sorghum, seed weight and seed size were reduced by 53 percent and 51 percent, respectively, while in canola (*Brassica* spp.), seed yield reduced by 89 percent, but the overall yield loss was about 52 percent at 30°C. Heat stress reduces the quality of grains. In barley, for example, carbohydrates, starch, fructose, raffinose, lipids, and aluminum content were reduced; proteinogenic amino acids concentrations and maltose content were increased [73].

16.6.1.5 Oxidative Stress in Plants

In plants, oxidative damage occurs due to the generation of reactive oxygen species by the environmental stress conditions. Reactive oxygen species include superoxide anion radicals (O^{-2}), hydroxyl radicals (OH), hydrogen peroxide (H_2O_2), alkoxy radicals (RO), and singlet oxygen (O^{12}) [78] (Figure 16.11). These reactive oxygen species are produced in many cellular compartments but chloroplasts can be an important source, as in the thylakoid membrane, pigments react with oxygen to form oxidants. In the mitochondria, ROS is generated due to the interaction of oxygen (O_2) with reduced components of the electron transport chain. Hydrogen peroxide (H_2O_2) is produced in peroxisomes during photorespiration when oxidation of glycolate into glyoxylic acid occurs, which impairs the normal functioning of cells by reacting with proteins, lipids, and DNA, causing oxidative damage. These ROS cause lipid peroxidation, membrane injuries, protein degradation, enzyme inactivation, and protein oxidation.

16.6.2 Biotic Stress in Plants

Biotic stress includes infection and development of disease in plants caused by fungi, bacteria, viruses, phytoplasmas, nematodes, and parasitic plants. Their infection causes heavy loss in crop productivity, which is one of the major challenge for feeding the world population in the future.

Fungi are organisms that cannot make their food due to the lack of chlorophyll. Fungi feed on dead and decaying matter (saprophyte) and are also capable of deriving their nutrition from their living host (parasitic). In favorable conditions, fungal spores come in contact with the host plant for extracting nutrients after the development of hyphae. They are spread by soil, wind, and water and enter into the plant cells via stomata, lenticels (natural openings), or through wounds from mechanical damage. Their infection causes a variety of symptoms, such as "damping off" seedling, leaf spots, leaf curling, cankers, and root rot.

The other causal agent of biotic stress is single-celled bacteria capable of infecting all parts of the plant. Like fungi, bacteria are able to enter the plant cell through natural openings and are able to move from one plant to another via soil or water. Their infections cause stem rot, cankers, leaf blight, and leaf spots.

Viruses are microscopic obligate parasites, having either ribonucleic acid or deoxyribonucleic acid and attaching themselves to a live host. With the help of vectors, viruses are transmitted from one plant to another plant, causing heavy loss to crop productivity. Viruses cause mottling, spots, mosaic-like patterns, crinkling, and other malformations on leaves and fruits, and may stunt plants.

Nematodes are microscopic roundworms that live in the soil, plant material, and water. They cause stunted growth, yellowing and wilting of plant tissues.

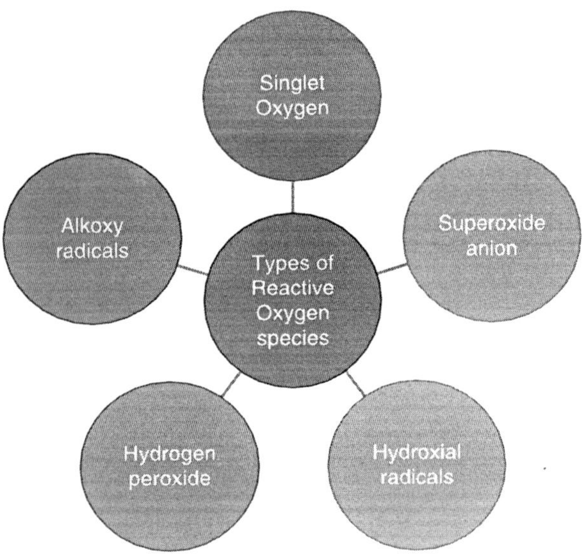

Figure 16.11 Reactive oxygen species.

16.7 Stress Tolerance in Plants

Plants are able to tolerate various types of stresses either by self-adaptation mechanisms or with the aid of plant growth–promoting rhizobacteria.

16.7.1 Adaptation Mechanisms of Plants Toward Abiotic Stress

In general, plants use avoidance, tolerance, and acclimation strategies for adaptation against abiotic stress based on the nature of the particular stress (Figure 16.12).

16.7.2 Management of Abiotic and Biotic Stresses in Plants

For rescuing plant growth and enhancing crop productivity under abiotic and biotic stresses, PGPR helps in mitigating the impact of both types of stresses on plants, which include both physiological and biochemical changes. The mechanisms in making plants tolerant against stresses are not fully understood but research has shown that some mechanisms directly or indirectly stimulate plant growth and recovery from environmental stress conditions:

- *Direct mechanisms*: These mechanisms include nitrogen fixation, production of growth regulators, solubilization of minerals phosphorous, zinc, iron, siderophore production, enzymes, and induced systemic resistance (Figure 16.13).
- *Indirect mechanisms*: These mechanisms include antibiotic production and extracellular enzyme production responsible for the hydrolysis of the fungal cell wall, which are attributes related to biocontrol (Figure 16.13).

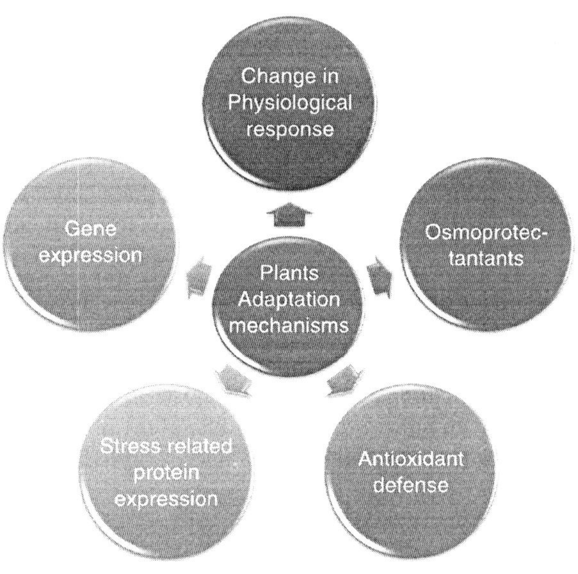

Figure 16.12 Adaptation mechanisms of plants toward abiotic stress.

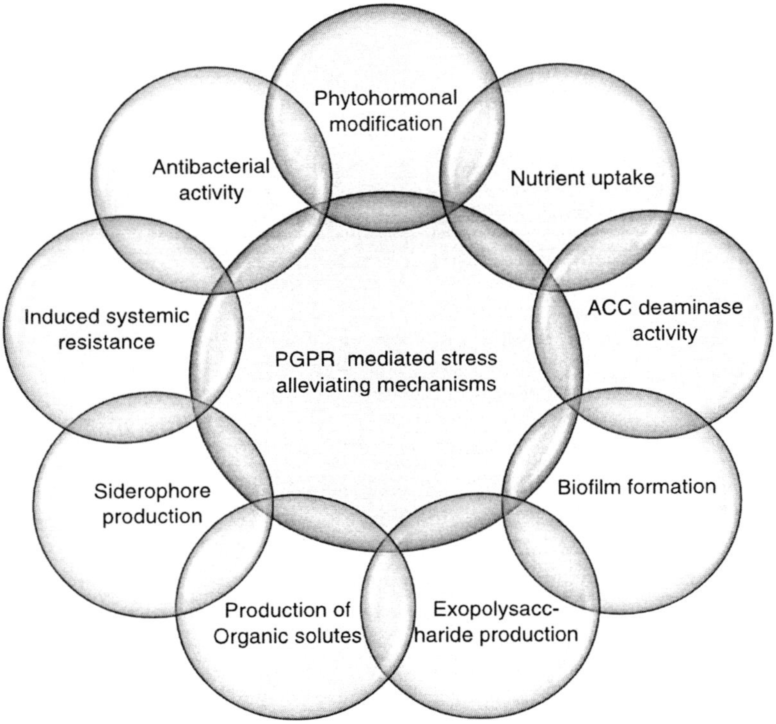

Figure 16.13 Plant growth promoting rhizobacteria-mediated abiotic and biotic stress amelioration.

16.7.2.1 Phytohormone Production

PGPR are able to produce phytohormones such as auxins, cytokinin, ethylene, abscisic acid (ABA) and gibberellins, which are beneficial for plants in ameliorating abiotic stress. For bacteria, it has been reported that indole-3-acetic acid (IAA) is an effective phytohormone in providing tolerance toward osmotic stress. PGPR are able to produce IAA, which modifies root architecture by enhancing the root surface area and number of lateral roots. It increases the uptake of water and nutrients from the surrounding medium, which helps to cope with drought conditions and nutrient deficiency. It has been revealed that volatile organic compounds produced by *Bacillus subtilis* strain (GB03) enhance *Arabidopsis* growth by upregulation of auxin responsive genes. It has also been reported that upon inoculation of wheat seedlings, that *Azospirillum* modifies the xylem architecture and is attributed to upregulation of the indole-3-pyruvate decarboxylase gene and enhanced IAA synthesis which helps in ameliorating osmotic stress [79]. Upon inoculation of PGPR on tomato plants, increases in the biomass of roots and numbers of root hair by IAA-mediated ethylene production were observed. Cytokinin involved in the initiation of root cell division and enlargement resulted in adventitious and lateral root formation, leading to the increased surface area of the root.

16.7.2.2 Maintenance of Nutrient Content

Plant growth and yield are reduced by nutrient deficiency. So, inoculation of PGPR influences the nutritional content of the plants by enhancing the uptake of macro as well as micro nutrients, including nitrogen, phosphorous, potassium and iron, zinc, manganese, calcium, and copper. Many reports are available showing that the co-inoculation of *Microbacterium* and *Bacillus* enhances the mineral uptake by crop plants [80].

16.7.2.3 Nitrogen Fixation

Like phosphorous, nitrogen is also a major nutrient for the plants. Its deficiency, which can result from heavy rainfall resulting in mineral leaching, causes great loss in crop productivity. PGPR like *Rhizobium spp., Pantoea agglomerans, Klebsiella pneumoniae, Beijerinckia spp.*, and *Azoarcus spp.*, are able to fix atmospheric nitrogen in the soil and make it available to plants. *Fluorescent pseudomonads* are able to induce nodulation in the chickpea and later on were shown to play a role in nitrogen (N_2) fixation. PGPR like *Bacillus* spp., *Azotobacter* spp., *Beijerinckia* spp., and others are able to fix atmospheric nitrogen symbiotically, whereas free-living diazotrophs, belonging to genera like *Pseudomonas, Azoarcus, Gluconacetobacter* and *Azospirillum,* are able to fix asymbiotically [81].

Many microbes are able to convert organic amino nitrogen to ammonia (ammonification). During nitrification, the ammonium ions are oxidized to nitrite ions by *Nitrosomonas, Nitrosospira, Nitrosococcus, Nitrosolobus,* and then to nitrate ions by *Nitrobacter, Nitrospira,* and *Nitrococcus.* Denitrification is an undesirable process in which fixed forms of nitrogen are converted to molecular nitrogen by *Pseudomonas, Thobacillus, Bacillus,* and *Moraxella* (Figure 16.14).

16.7.2.4 Phosphorous Solubilization

In plants, phosphorous is a major nutrient. It is available in very small amounts to the plants, as it is present in insoluble form, while only soluble forms such as di-hydrogen phosphate ($H_2PO_4^-$) and hydrogen phosphate (HPO_4^{2-}) are absorbed by the plants. Numerous phosphate solubilizing microorganisms are known to convert the insoluble form of phosphorous to the soluble form via acidification, exchange reactions, chelation, and secreated organic acids, which then makes it available to the plants. Most significant

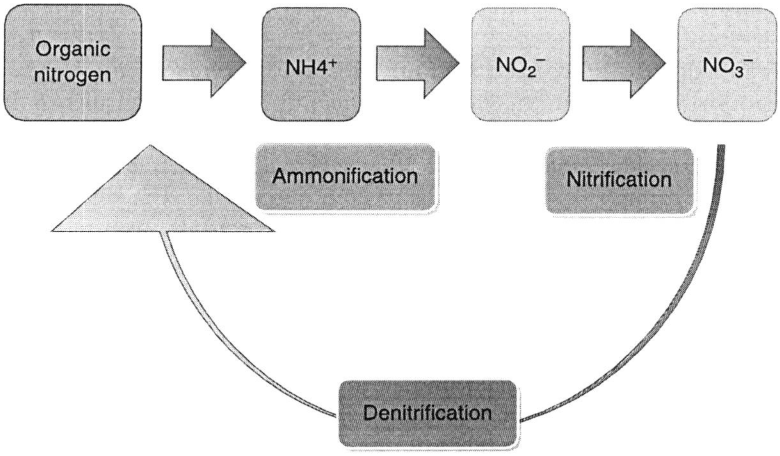

Figure 16.14 Nitrogen fixation and conversion.

phosphate solubilizing bacteria belong to the generas *Pseudomonas, Azospirillum, Bacillus, Azotobacter, Rhizobium, Beijerinckia, Burkholderia, Enterobacter, Erwinia, Flavobacterium, Microbacterium,* and *Serratia*. Organic ligands release from plant roots, as root exudates are capable of amending the phosphorous concentration in soil solution [81].

16.7.2.5 Siderophore Production

Siderophores are low-molecular-weight compounds produced by PGPR ranging from 400 to 1,500 Dalton. Siderophores chelate iron from their surrounding medium and make it available to the plants, thereby increasing the iron content in plants. Iron, which is essential for bacterial metabolism and growth, is taken up by the bacteria and transported inside the cell. In soil, iron is present as Fe^{+3} and binds the siderophore. Thus, it limits the availability of iron to plant pathogens, resulting in growth inhibition, sporulation inhibition, and changes in cell morphology [80].

16.7.2.6 Exopolysaccharide (EPS) Production

Numerous PGPR secreate exopolysaccharides, which are a carbohydrate polymer that play a vital role in biofilm formation, bacterial cell protection from dessication, antibacterial activity, degradation of pollutants, and normal functioning of cells and bioremediation activity.

16.7.2.7 ACC Deaminase Activity

Ethylene phytohormone is necessary for inducing plant growth and normal development. Above its normal concentration, it causes defoliation and alters cellular processes leading to a loss in crop productivity. With the help of their 1-amino cyclo-propane-1-carboxylic acid (ACC) deaminase enzymatic activity, PGPR are able to suppress the ethylene levels in plants by acting on ACC (the precursor of ethylene) and converting it into α-ketobutyrate and ammonia which reestablishes the plant's normal growth and function. Most significant ACC-producing bacteria belong to the genera *Achromobacter, Azospirillum, Bacillus, Enterobacter, Pseudomonas,* and *Rhizobium* [81].

16.7.2.8 Volatile Organic Compounds (VOCs)

Volatile organic compounds are produced by PGPR for promoting plant growth and development. Ryu et al. [82] documented some of the PGPR strains like *Bacillus subtilis* GB03, *B. amyloliquefaciens* IN937a and *Enterobactercloacae* JM22. They produce 2, 3-butanediol and acetoin volatile organic compound, able to enhance *A. thaliana* growth. Now it has also been ascertained that plant-microbe interactions can be mediated by these VOCs acting as a signaling molecule, triggering positive responses in plants [81].

16.7.2.9 PGPR as Biotic Elicitors

Elicitors are biological or chemical factors that are able to prompt physiological and morphological responses as well as accumulation of phytoalexins in plants. They can be abiotic, including metal ions or inorganic compounds. Biotic elicitors are produced by bacteria, viruses, fungi, cell-wall components of plants, and chemicals that are released by plants when exposed to pathogenic or herbivore attack. Elicitors help in the accumulation of phytoalexins (bioactive molecules) and induce gene expression for antimicrobial compounds synthesis. The best biotic elicitors are microbes residing in the rhizosphere and capable of inducing plants' secondary products. Elicitors produced by PGPR include serpentine, hyoscyamine, and ajmalicine. They induce physiological and morphological responses in crops.

16.7.2.10 Induction of Systemic Disease Resistance

Plant immunization is a process in which resistance is increased by infection stimulated by pathogens. Necrotizing pathogens are able to trigger induced resistance and chemical responses—for example, salicylic acid triggers systemic acquired resistance. In plants, systemic resistance is induced on application of mixtures of rhizobacterial strains to the seeds. In 1994, Maurhofer et al. demonstrated that *Pseudomonas fluorescens* strain CHA0 induces ISR by accumulation of pathogenesis-related (PR) protein accumulation. Later reports suggested that an SA-independent pathway, jasmonic acid, bacterial lipopolysaccharides, siderophores, and ethylene all are involved in inducing systemic resistance [81]. *Pseudomonas fluorescens* strain WCS417r is able to induce systemic resistance in plants via SA-independent, JA-ethylene dependent signaling, ISR-related gene expression, and NPR 1-dependent signaling pathways [81].

16.7.3 Management of Abiotic and Biotic Stress in Plants via Biofilm-Forming Rhizobacteria

Abiotic and biotic stress can be managed via eco-friendly, stress-tolerant, biofilm-forming rhizobacteria by ameliorating different stresses in plants, thereby promoting plant health and crop productivity.

16.7.3.1 Salt Stress Amelioration

Microorganisms use different types of mechanisms to cope with salinity stress in crop plants. Some rhizobacterial strains (PGPR) affect the growth and development of tomatoes, peppers, beans, and lettuce grown in high-salinity environments [83, 84]. Inoculation of wheat seedlings with exopolysacharides (EPS)-producing bacteria restricts the uptake of sodium and stimulates plant growth under high salt stress condition [83, 85]. Corn, beans, and clover inoculated with *Arbuscular Mycorrhiza* fungi improved their osmoregulation and increased proline accumulation, which results in salinity resistance.

Strategies adapted by bacteria for successful survival in the plant rhizosphere include compatible solutes, which are used for the osmotic adjustment of bacterial cells, and EPS production, which protects the cells against desiccation, high temperature, oxygen radicals, salt stress, and UV radiations. Under salt stress, inoculation of *Lens esculenta* Var. masoor with biofilm-forming bacterial strains *Oceanobacillus profundus* (Pmt2) and *Staphylococcus saprophyticus* (ST1) improved plant growth parameters and endogenous osmolytes accumulation in plants [86].

16.7.3.2 Drought Stress Amelioration

Crops inoculated with bacteria- (e.g., *Bacillus amyloliquifaciens*) producing exopolysaccharides (EPS) improves the soil structure by facilitating the formation of macro aggregates, which results in increased plant resistance against stress arises due to water scarcity. Over a longer period of time, soil fertility is reduced as a result of the presence of these small aggregates, which are responsible for causing poor aeration and evacuation of water from soil pores. Macro aggregates are protectors of soil fertility, because they maintain a balance between aerobic and anaerobic conditions and ensure a gradual uptake of nutrients from soil reserves. Water availability in the soil also has an effect on the soil structure. Under water stress conditions, plants treated with EPS-producing bacteria like *Azospirillum* showed resistance related to the improvement in the soil structure and soil aggregation because of its water-holding capacity. Due to EPS production by bacteria such as *Pseudomonas*, plants are able to survive under drought stress. Bacterial EPS are hydrated compounds with 97 percent water in the polymer matrix, which provides protection against desiccation by enhancing the water retention and by source regulation nutrients such as of organic carbon diffusion. During drought stress, inoculation of *Zea mays L.* with EPS-producing and biofilm-forming bacterial strains *Proteus penneri* (Pp1), *Pseudomonas aeruginosa* (Pa2), and *Alcaligenes faecalis* (AF3) showed an increase in relative water, protein, and sugar content though the proline content and the activities of antioxidant enzymes declined. The Pa2 strain was found to be the most potent PGPR isolated from the semiarid region under drought stress. Consortia of inocula and their respective EPS production showed superior potential to drought tolerance as compared to PGPR inocula used alone [87]. In wheat, drought tolerance can be enhanced from bacterial biofilm formation by the inactivation of Sfp-type PPTase [88].

In recent studies it was demonstrated that *P. polymyxa* A26 mutant Sfp-type PPTase gene was inactivated and the resulting A261 *sfp* mutant strain is incapable of producing the enzyme 4′-phosphopantetheinyl transferase. Biofilm production by this mutant strain was enhanced and promoted the plant growth by increasing the dry weight of plants under drought stress conditions.

Under harsh abiotic stresses especially drought, *Rhizobia* are able to tolerate and survive as they are able to produce exopolysaccharides, catalase (antioxidant enzyme), siderophores, osmolytes [89,90], sugars, phosphate solubilization [91], phytohormones, trehalose, and ACC-deaminase [92]. *Rhizobia* that produce exopolysaccharides and catalase help in amelioration of drought stress in wheat [93]. Under stress conditions, wheat growth was rescued due to soil aggregation by the exopolysaccharides production capability of *Rhizobia* (KYGT207) [94]; whereas, under drought conditions, the dry biomass of sorghum was improved by auxin and cytokinin producing *Rhizobium* and *Bradyrhizobium* [95]. Similarly, sunflower, inoculated with exopolysaccharides producing rhizobial strain YAS34, showed improved dry biomass and nitrogen, nutrition, and water uptake under drought conditions [96].

16.7.3.3 Temperature

High temperature promotes plant growth and development while low temperature is an important limiting factor to the productivity and geographic distribution of agricultural crops.

Some bacterial species and strains help in increasing plant tolerance to high temperatures [83]. *Pseudomonas* species strain NBRI0987 stimulates thermo tolerance in sorghum seedlings by synthesizing high-molecular-weight proteins in leaves, resulting in increased plant biomass. The bacterium *Burkholderia phytofirmans* PSJN colonizes grapevine and protects the plant against heat and frost by increasing the starch, proline, and phenol levels. At low temperatures, inoculation of wheat seeds with *Serratia marscescens* strain SRM and *Pantoea dispesa* strain 1A increases the seedling biomass and nutrient uptake.

16.7.3.4 Metal Transformation

The enzymatic activities of bacterial exopolysaccharides help in heavy metal transformation and degradation of organic recalcitrant.

16.7.3.5 Biocontrol Activity

Biological control activity (BCA) involves the restraint of plant diseases with the help of plant growth–promoting rhizobacteria by their successful colonization, biofilm formation, and production of anti-pathogenic compounds.

Some strains of *B. subtilis* are well skilled to form biofilms and suppresses the *Ralstonia* wilt disease in numerous host plant [97–99], thus, acting as a biocontrol agent. Antibacterial agents including lipopeptides like surfactin via *B. subitillis* act as antibacterial agents against many fungal pathogens, thereby enhances crop productivity. In *Bacillus cereus*, gene ptsI is responsible for phosphotransferase activity and is involved in colonization, biofilm formation, and acts as a biocontrol agent against wheat (*Triticum aestivum* L.) sharp eyespot [100]; this is a serious plant disease in China caused by *Rhizoctonia cerealis* [101, 102]. Bacterial phosphotransferase (PTS) is responsible for the transport of sugar and phosphorylation, for carbon catabolism, and also for their colonization and biofilm formation [103–106]. Shuqing Li et al. [107] showed that biofilm-forming *Bacillus subtilis* HJ5 controls the Verticillium wilt disease of cotton. On roots of *Arabidopsis, B. subtilis* strain (ATCC 6051) forms biofilms and protects the host plant from *Pseudomonas syringae* infections [108]. Numerous strains of *Bacillus* spp. have been reported and considered as biocontrol agents against *Aspergillus flavus* such that *B. subtilis* NK-330 produces aflatoxin and inhibits the growth of *A. flavus*, strains AU195 and B-FS06 produces some compounds as bacillomycin (antifungal activity) against *A. flavus* and from Iranian pistachio fruits, strain UTB1 isolated that also acts against *A. flavus* [109]. Mucoid mutants of *Pseudomonas fluorescens* strain CHA0 showed enhanced colonizing capacity and biofilm formation, and also acts as a potent biocontrol agent as compared to wild-type strain [110]. Jayasinghearachchi and Seneviratne [111] demonstrated that endophytic colonization has been increased by 1,000 percent in case of *Pleurotusostreatus–Pseudomonas fluorescens* biofilm as compared to solitary inoculation of *P. fluorescens*. In biofilms, it was observed that between indole acetic acid and pH there was always a negative relationship. So that biofilm having high acidity will have higher production of IAA, and this high acidity plays a vital role in pathogen containment and rock phosphate solubilization. Haggag and Timmusk [112] showed that *Paenibacillus polymyxa* acts as a biocontrol against crown rot disease of peanuts. In agriculture, for crop productivity, *Anabaena* biofilms were also found to show evidence of plant growth promoting and biocontrol traits in various crops [113–115] and Trichoderma (saprophytic fungus) also acts as a biological control agent [116]. When inoculation of biofilmed culture was compared with monoculture inoculation, it was found that biofilmed fertilizer enhances rice growth and dry weight by 25 percent [117]. *Anabena* and *Trichoderma* were found to be potent against *Macrophomina phaseolina* of cotton crops [118].

16.7.4 Stress Management via Quorum Sensing Signals Producing PGPR

Gram-negative bacteria utilizes acyl homoserine lactones (AHL) for regulating diverse rhizospheric activities and can suffer interferences by the plant metabolites. There are different types of AHL molecules found functioning in plant defenses and stress responses, metabolic activities, hormone responses, transcriptional regulation, protein processing, and cytoskeletal activities [119]. Salicylic acid in the leaves of tomato plants and systemic resistance against *Alternaria alternate*, a fungal leaf pathogen, is increased by the presence of AHL-producing bacteria in its rhizosphere [120], and macroarray analyses illustrates that the salicylic acid and ethylene-dependent defense genes were induced by the use of synthetic AHLs. The application of homoserine lactones to bean roots augments the stomatal conductance and transpiration, benefiting the plant and bacteria through increased uptake of nutrients [121].

Exogenous application of AHLs affects the root developmental processes in *Arabidopsis* [122]. It was found that low-micro-molar concentrations of C4-HSL and C6-HSL increase root growth while C10-HSL decreases root growth and rosettes [122]. No systemic resistance response was induced in *A. thaliana* when roots were stimulated with short side-chain AHLs [123]. It was found that C10-homoserine lactone induces some developmental changes in the root system of *Arabidopsis* plants by altering the expression of cell division and cell differentiation–related genes. When plants were transformed to produce, which on diffusion in the rhizosphere affects bacterial AHL regulated processes.

Tobacco plants have been genetically modified to produce short- and long-chain AHL compounds that could be detected in substantial amounts on leaf and roots surfaces as well as in soil. Tobacco plants were transformed to express of QS genes, leading to altered systemic resistance elicited by the rhizobacterium *Serratia marcescens* [124]. In mung bean, oxo C10-homoserine lactone activates auxin-induced adventitious root formation via H_2O_2 and NO-dependent cyclic GMP signaling.

On the other hand, C12 and C14 AHLs induce systemic resistance against biotrophic fungus *Golovinomyces orontii* in *A. thaliana* and to *Blumeria graminis* sp. *Hordei* in barley (*Hordeum vulgare*) [125]. This response is arbitrated through altered activation of *AtMPK6*. The mitogen-activated protein kinases AtMPK3 and AtMPK6 are activated by the model elicitor flg22 in the presence of C12 or C14 AHL compounds, which results in a higher expression of the defense-related transcription factors WRKY26 and WRKY29 as well as the PR1 gene [125]. Thus, short and medium AHL molecules induce developmental effects on root architecture; while in *A. thaliana*, long side-chain AHLs induce systemic resistance [126].

16.8 Conclusion

This chapter aims to provide knowledge about the biofilm-forming, plant growth–promoting rhizobacteria and its ecological significance in abiotic and biotic stress amelioration. This chapter also provides additional information regarding the rhizosphere hub, the effects of abiotic and biotic stress on plants, and its amelioration through stress-tolerant and plant growth–promoting rhizobacteria. These beneficial eco-friendly rhizobacteria colonize or adhere to the plant roots surface and form biofilm by secreting an extracellular polymeric matrix and may increase crop productivity through some of the mechanisms that improves nutrient uptake, phytohormone production, and disease suppression. In PGPR, quorum sensing regulated exopolysaccharide traits help in drought stress alleviation, whereas biosurfactant production in *Bacillus* species helps in biocontrol activity. PGPR are able to produce EPS and form biofilms, which helps in amelioration of drought stress in plants, as it provides protection against desiccation by enhancing the water retention and by source regulation nutrients such as of organic carbon diffusion. Consortia of inocula and their respective EPS production showed superior potential to drought tolerance as compared to PGPR inocula used alone. EPS-producing beneficial rhizobacteria restricts the uptake of sodium and stimulates plant growth under high salt stress condition. Some of the commercial strains are available in market which acts as a potent biocontrol agent against fungal pathogens. For example, Serenade biofungicide contains *B. subitilis,* which is effective against variety of pathogens including *Pseudomonas, Xanthomonas,* and *Erwina* strains. The mechanisms behind this antibacterial agent are in vague, even though it is well known that the production of lipopeptides such as surfactin, having antibacterial properties, imparts protection against fungal pathogens. Generally, PGPR with their ability for biofilm formation and quorum sensing provides a sustainable development in agriculture by enhancing plant growth and crop productivity after abiotic and biotic stress amelioration. In summary, it can be inferred that this enhanced crop productivity via biofilm-forming rhizobacteria, will be able to meet one of the world's future challenges by fulfilling food-feed requirements of the increasing population while minimizing the use of pesticides.

16.9 Future Perspectives

Further investigations to decipher the specific interactions occurring in the rhizosphere with native or introduced microbes/microflora will benefit the fundamental understanding of plant biology and provide stability for food production. For PGPR, one or more of these attributes could be achieved: enhancing proliferation, persistence, and organization of PGPR toward roots, and genetic manipulation in host plants for secreting root-associated chemical compounds. Rhizobacterial biofilm-mediated abiotic stress amelioration is still a topic for exploration.

Acknowledgments

Thanks are due to the Director, CSIR-NBRI, Lucknow for facilities to accomplish this work. A. Singh is thankful to Department of Science and Technology, New Delhi, India, for her INSPIRE fellowship. The study was supported by the network project Plant Microbe and Soil Interactions (PMSI; BSC-0117) funded by Council of Scientific and Industrial Research, New Delhi, India.

List of Abbreviations

LPS	Lipopolysaccharide
UV	Ultraviolet
O_2	Oxygen
CO_2	Carbon dioxide
EPS	Exopolysaccharide

PO_4^{3-}	Phosphate
P	Phosphorous
Z. Mays	*Zeamays*
NaCl	Sodium chloride
Na^+	Sodium ion
Cl^-	Chloride ion
N	Nitrogen
P	Phosphorous
K	Potassium
Ca	Calcium
Mg	Magnesium
Ca^{2+}	Calcium ion
K^+	Potassium ion
NO_3^-	Nitrate
Mn^{2+}	Manganesium ion
Fe^{2+}	Ferrous
S^{2-}	Sulphudryl
H_2S	Hydrogen sulphide
ROS	Reactive oxygen species
PGPR	Plant growth promoting rhizobacteria
$H_2PO_4^-$	Dihydrogen phosphate
HPO_4^{2-}	Hydrogen phosphate
Fe^{+3}	Ferricion
VOCs	Volatile organic compound
ISR	Induced systemic resistance
HSL	Homoserine lactones
AHL	Acyl homoserine lactones
QS	Quorum sensing
RuBisCO	Ribulose-1, 5-bisphosphate carboxylase/oxygenase

References

1 P. Vandevivere and D.L. Kirchman, Attachment stimulates exopolysaccharide synthesis by a bacterium. *Appl. Environ. Microbiol.*, 59, 3280–3286 (1993).

2 G. Seneviratne, Development of eco-friendly, beneficial microbial biofilms. *Curr. Sci.*, 85, 1395–1396 (2003).

3 J.W. Costerton, G.G. Geesy and K.J. Cheng, Bacterial biofilms in nature and disease. *Annu. Rev. Microbiol.*, 41, 435–464 (1987).

4 P. Stewart and J.W. Costerton, Antibiotic resistance of bacteria in biofilms. *Lancet.*, 358, 135–138 (2001).

5 Y.M. Romanova, T.A. Smirnova, A.L. Andreev, T.S. Il'ina, L.V. Didenko and A.L. Gintsburg, Formation of biofilms as an example of the social behaviour of bacteria. *Microbiology.*, 75, 481–448 (2006).

6 M. Morikawa, Beneficial biofilm formation by industrial bacteria *Bacillus subtilis* and related species. *Journal of Biosciences and Bioengineering.*, 101, 1–8 (2006).

7 K.F. Kong, C. Vuong and M. Otto, Staphylococcus quorum sensing in biofilm formation and infection. *Int. J. Med. Microbiol.*, 296, 133–139 (2006).
8 L. Hiltner, Überneuere Erfahrungen und Probleme auf demGebiete der Bodenbakteriologieunterbesonderer Berücksichtigung der Gründüngung und Brache. *Arb. Dtsch. Landwirtsch. Gesellschaft*, 98 59–78 (1904).
9 R. Pinton, Z. Veranini and P. Nannipieri, The rhizosphere. Biochemistry and organic substances at the soil-plant interface. New York, USA: Taylor & Francis Group, LLC (2007).
10 A. Hartmann, M. Schmid, D. van Tuinen and G. Berg, Plant-driven selection of microbes. *Plant Soil.*, 321, 235–257 (2009).
11 R.L. Berendsen, C.M.J. Pieterse and P.A.H.M. Bakker, The rhizosphere microbiome and plant health. *Trends in Plant Sci.*, 17, 478–486 (2012).
12 R.I. Amann, W. Ludwig and K.H. Schleifer, Phylogenetic identification and in situ detection of individual microbial cells without cultivation. *Microbiol. Rev.*, 59, 143–169 (1995).
13 A.D. Kent and E.W. Triplett, Microbial communities and their interactions in soil and rhizosphere ecosystems. *Annu. Rev. Microbiol.*, 56, 211–236 (2002).
14 R.F. Doornbos, L.C. Van Loon and P.A.H.M. Bakker, Impact of root exudates and plant defense signaling on bacterial communities in the rhizosphere. *Agron. Sustain. Dev.*, 32, 227–243 (2012).
15 J.H.J. Leveau, The magic and menace of metagenomics: prospects for the study of plant growth promoting rhizobacteria. *Eur. J. Plant Pathol.*, 119, 279–300 (2007).
16 J. Sorensen, M. Hauberg Nicolaisen, E. Ron and P. Simonet, Molecular tools in rhizosphere microbiology–from single-cell to whole-community analysis. *Plant Soil.*, 321, 483–512 (2009).
17 P.R. Hirsch and T.H. Mauchline, Who's who in the plant root microbiome? *Nat. Biotechnol.*, 30, 961–962 (2012).
18 T. Urich, A. Lanzen, J. Qi, D.H. Huson, C. Schleper and S.C. Schuster, Simultaneous assessment of soil microbial community structure and function through analysis of the meta-transcriptome. *PLOS ONE*, 3 (2008).
19 M.J. Gosalbes, J.J. Abellan, A. Durban, A.E. Perez-Cobas, A. Latorre, and A. Moya, Metagenomics of human microbiome: beyond16s rDNA. *Clin. Microbiol. Infect.*, 18, 47–49 (2012).
20 J.K. Jansson, J.D. Neufeld, M.A. Moran and J.A. Gilbert, Omics for understanding microbial functional dynamics. *Environmental Microbiology.*, 14, 1–3 (2012).
21 T.S. Walker, H.P. Bais, E. Grote-wold and J.M. Vivanco, Root exudation and rhizosphere biology. *Plant Physiol.*, 132, 44–51 (2003).
22 H.P. Bais, T.L. Weir, L.G. Perry, S. Gilroy and J.M. Vivanco, The role of root exudates in rhizosphere interactions with plants and other organisms. *Annu. Rev. Plant Biol.*, 57, 233–266 (2006).
23 T. Rudrappan, W.J. Quinn, N.R. Stanley-Wall and H.P. Bais, A degradation product of the salicylic acid pathway triggers oxidative stress resulting in down-regulation of *Bacillus subtilis* biofilm formation on *Arabidopsis thaliana* roots. *Planta.*, 226, 283–297 (2007).
24 E.L. Schmidt, Initiation of plant root microbe interactions. *Annu. Rev. Microbiol.*, 33, 355–376 (1979).
25 C.S. Nautiyal, A method for selection and characterization of rhizosphere- competent bacteria of chickpea. *Curr. Microbiol.*, 34, 12–17 (1997a).

26 C.S. Nautiyal, Rhizosphere competence of *Pseudomonas* sp. NBRI9926 and *Rhizobium* sp. NBRI9513 involved in the suppression of chickpea (*Cicer arietinum* L) pathogenic fungi. *Fems Microbiol. Ecol.*, 23, 145 (1997b).
27 C.S. Nautiyal, J.K. Johri and H.B. Singh, Survival of the rhizosphere- competent biocontrol strain *Pseudomonas fluorescens* NBRI2650 in the soil and phytosphere. *Can. J. Microbiol.*, 48, 588–601 (2002).
28 C.S. Nautiyal, Biological control of plant diseases by natural and genetically engineered fluorescent *Pseudomonas* spp. In: Ray RC, Owen PW (eds.) Microbial Biotech. in Horticult., Science Publishers, Enfield (NH) USA, pp 125–162 (2006).
29 C.S. Nautiyal, S. Srivastava, P.S. Chauhan, Rhizosphere colonization: molecular determinants from plant-microbe coexistence perspective. In: Nautiyal CS and Dion P (eds.), Molecular Mechanisms of Plant and Microbe Coexistence., pp. 99–124. Soil Biology Series, Vol. 15 (2008).
30 A.J. Bennett and J.M. Whipps, Beneficial microorganism survival on seed, roots and in rhizosphere soil following application to seed during drum priming. *Biol. Control.*, 44, 349–361 (2008).
31 D.G. Davies, M.R. Parsek, J.P. Pearson, B.H. Iglewski, J.W. Costerton and E.P. Greenberg, The involvement of cell-to-cell signals in the development of a bacterial biofilm. *Science*, 280, 295–298 (1998).
32 B. Huber, K. Riedel, M. Hentzer, A. Heydorn, A. Gotschlich, M. Givskov, S. Molin and L. Eberl, The *cep* quorum-sensing system of *Burkholderia cepacia* H111 controls biofilm formation and swarming motility. *Microbiology.*, 147, 2517–2528 (2001).
33 H.L. Wei and L.Q. Zhang, Quorum-sensing system influences root colonization and biological control ability in *Pseudomonas Xuorescens* 2P24. *Antonie Van Leeuwenhoek.*, 89, 267–280 (2006).
34 A. Steidle, K. Sigl, R. Schuhegger, A. Ihring, M. Schmid, S. Gantner, M. Stoffels, K. Riedel, M. Givskov and A. Hartmann, *et al.*, Visualization of *N*-acyl homoserine lactone-mediated cell–cell communication between bacteria colonizing the tomato rhizosphere. *Appl. Environ. Microbiol.*, 67, 5761–5770 (2001).
35 E.A. Pierson, D.W. Wood, J.A. Cannon, F.M. Blachere and L.S. Pierson. Interpopulation signalling via N-acyl-homoserine lactones among bacteria in the wheat rhizosphere. *Mol. Plant Microbe Interact.*, 11, 1078–1084 (1998).
36 J.W. Costerton, P.S. Stewart and E.P. Greenberg, Bacterial biofilms: a common cause of persistent infections. *Science*, 284, 1318–1322 (1999).
37 S.S. Branda, S. Vik, L. Friedman and R. Kolter, Biofilms: The matrix revisited. *Trends Microbiol.*, 13, 20–26 (2005).
38 R.M. Donlan, Biofilms: microbial life on surfaces, *Emerg. Infect. Dis.*, 8, 881–890 (2002).
39 B. Annous, P. Fratamico and J.L. Smith, Quorum sensing in biofilms: why bacteria behave the way they do. *J. Food Sci.*, 74, 24–37 (2009).
40 B. Vu, M. Chen, R.J. Crawford and E.P. Ivanova, Bacterial extracellular polysaccharides involved in Biofilm formation. *Molecules.*, 14, 2535–2554 (2009).
41 X. Zhang and P. Bishop, Biodegradability of biofilm extracellular polymeric substances. *Chemosphere.*, 50, 63–69 (2003).
42 D.M. Russo, A. Williams, A. Edwards, D.M. Posadas, C. Finnie, M. Dankert, J.A. Downie, A. Zorreguieta, Proteins exported via the PrsD-PrsE type I secretion system and the acidic exopolysaccharide are involved in biofilm formation by *Rhizobium leguminosarum*. *J. Bacteriol.*, 188, 4474–4486 (2006).

43 H. Laue, A. Schenk, H. Li, L. Lambertsen, T.R. Neu, S. Molin and M.S. Ullrich, Contribution of alginate and levan production to biofilm formation by *Pseudomonas syringae. Microbiology.*, 152, 2909–2918 (2006).
44 J. Hu, W. Qian and C. He, The *Xanthomonas oryzae* pv. *Oryzaee gl Xo Bendoglucanase gene is required for virulence to rice. FEMS Microbiol. Lett.*, 269, 273–279 (2007).
45 T.A. Dafna, T. Rosenberg, N. Navon and S. Burdman, A secreted lipolytic enzyme from *Xanthomonas campestris* pv. *Vesicatoria is expressed in plant and contributes to its virulence. Mol. Plant. Pathol.*, 13, 556–567 (2012).
46 N. Høiby, O. Ciofu, H.K. Johansen, Z.J. Song, C. Moser, P. Jensen and T. Bjarnsholt, The clinical impact of bacterial biofilms. *IJOS.*, 3, 55–65 (2011).
47 L. Chen and Y. Wen, The role of bacterial biofilm in persistent infections and control strategies, *IJOS.*, 3, 66–73 (2011).
48 D.G. Davies and G.G. Geesey, Regulation of the alginate biosynthesis gene *alg C* in *Pseudomonas aeruginosa* during biofilm development in continuous culture. *Appl. Environ. Microbiol.*, 61, 860–867 (1995).
49 R.M. Donlan and J.W. Costerton, Biofilms: survival mechanisms of clinically relevant microorganisms, *Clin. Microbiol. Rev.*, 15, 167–193 (2002).
50 G. Ramage, S. Bachmann, T.F. Patterson, B.L. Wickes and J.L. Lopez- Ribot, Investigation of multidrug efflux pumps in relation to fluconazole resistance in *Candida albicans* biofilms. *J. Antimicrob. Chemother.*, 49, 973–980 (2002).
51 J. Kim, H.D. Park and S. Chung, Microfluidic approaches to bacterial biofilm formation, *Molecules.*, 17, 9818–9834 (2012).
52 V. Bianciotto, S. Andreotti, R. Balestrini, P. Bonfante and S. Perotto, Extracellular polysaccharides are involved in the attachment of *Azospirillum brasilense* and *Rhizobium leguminosarum* to arbuscular mycorrhizal structures. *Eur. J. Histochem.*, 45, 39–49 (2001).
53 P.B. Beauregarda, Y. Chai, H. Vlamakisa, R. Losick and R. Koltera, *Bacillus subtilis* biofilm induction by plant polysaccharides. *PNAS.*, 110 (2013).
54 G. Cárcamo-Oyarcel, P. Lumjiaktase, R. Kümmerli and L. Eberl, Quorum sensing triggers the stochastic escape of individual cells from *Pseudomonas putida* biofilms. *Nat. Commun.*, 6, 5945 (2015).
55 J.F. Dubern, B.J.J. Lugtenberg and G.V. Bloemberg, The *ppuI-rsaL-ppuR* Quorum-Sensing System Regulates Biofilm Formation of *Pseudomonas putida* PCL1445 by Controlling Biosynthesis of the Cyclic Lipopeptides Putisolvins I and II. *J. Bacteriol.*, 188, 2898–2906 (2006).
56 M. Farooq, A. Wahid, N. Kobayashi, D. Fujita and S.M.A. Basra, Plant drought stress: effects, mechanisms and management. *Agron. Sustain. Dev.*, 29, 185–212 (2009).
57 Z. Khodarahmpour, M. Ifar and M. Motamedi, Effects of NaCl salinity on maize (*Zea mays* L.) at germination and early seedling stage. *Afr. J. Biotechnol.*, 11, 298–304 (2012).
58 P. Parihar, S. Singh, R. Singh, V.P. Singh and S.M. Prasad, Effect of salinity stress on plants and its tolerance strategies: a review *Environ. Sci. Pollut. Res.*, 22, 4056–4075 (2015).
59 M.A. Abd El-Wahab, The efficiency of using saline and fresh water irrigation as alternating methods of irrigation on the productivity of Foeniculum vulgare Mill subsp. vulgare var. vulgare under North Sinai conditions. *Res. j. agric. biol. sci.*, 2, 571–577 (2006).
60 M. Ashraf and A. Orooj, Salt stress effects on growth, ion accumulation and seed oil concentration in an arid zone traditional medicinal plant ajwain (*Trachyspermum ammi* L. Sprague). *J. Arid Environ.*, 64, 209–220 (2006).

61 S.J. Tabatabaie and J. Nazari, Influence of nutrient concentration and NaCl salinity on growth, photosynthesis and essential oil content of peppermint and lemon verbena. *Turk. J. Agric.*, 31, 245–253 (2007).
62 K. Baghalian, A. Haghiry, M.R. Naghavi and A. Mohammadi, Effect of saline irrigation water on agronomical and phytochemical characters of chamomile (*Matricaria recutita* L.). *Sci. Hort.*, 116, 437–441 (2008).
63 W.M. Abd EL-Azim and S. Ahmed, The Effect of salinity and cutting date on growth and chemical constituents of Achillea fragratissima Forssk, under Ras Sudr conditions. *Res. j. agric. biol. sci.*, 5, 1121–1129 (2009).
64 M.R. Amirjani, Effect of salinity stress on growth, sugar content, pigments and enzyme activity of rice. *Int. J. Bot.*, 7, 73–81 (2011).
65 S. Chutipaijit, S. Cha-um and K. Sompornpailin, High contents of proline and anthocyanin increase protective response to salinity in *Oryza sativa* L. spp. *indica*. *Aust. J. Crop Sci.*, 5, 1191–1198 (2011).
66 M.F. Lòpez-Climent, V. Arbona, R.M. Pérez-Clemente and A. Gòmez-Cadenas, Relationship between salt tolerance and photosynthetic machinery performance in citrus. *Environ. Exper. Bot.*, 62, 176–184 (2008).
67 H.M. Kalaji, B.K. Govindjee, J. Koscielniakd and K. Zük-Gołaszewska, Effects of salt stress on photosystem II efficiency and CO2 assimilation of two Syrian barley landraces. *Environ. Exper. Bot.*, 73, 64–72 (2011).
68 S. Mittal, N. Kumari and V. Sharma, Differential response of salt stress on *Brassica juncea*: photosynthetic performance, pigment, proline, D1 and antioxidant enzymes. *Plant Physiol. Biochem.*, 54, 17–26 (2012).
69 P. Pagamas and E. Nawata, Sensitive stages of fruit and seed development of chili pepper (*Capsicum annuum* L. var. Shishito) exposed to high-temperature stress. *Sci. Hort.*, 117, 21–25 (2008).
70 E. Ruelland and A. Zachowski, How plants sense temperature. *Environ. Exper. Bot.*, 69, 225–232 (2010).
71 N. Suzuki, G. Miller, J. Morales, V. Shulaev, M.A. Torres and R. Mittler, Respiratory burst oxidases: The engines of ROS signaling. *Curr. Opin. Plant Biol.*, 14, 691–699 (2011).
72 N. Suzuki, S. Koussevitzky, R. Mittler and G. Miller, ROS and redox signalling in the response of plants to abiotic stress. *Plant Cell Environ.*, 35, 259–270 (2012).
73 M. Hasanuzzaman, M.A. Hossain and M. Fujita, Physiological and biochemical mechanisms of nitric oxide induced abiotic stress tolerance in plants. *Am. J. Plant Physiol.*, 5, 295–324 (2010).
74 W.J. Hurkman, W.H. Vensel, C.K. Tanaka, L. Whitehand and S.B. Altenbach, Effect of high temperature on albumin and globulin accumulation in the endosperm proteome of the developing wheat grain. *J. Cereal Sci.*, 49, 12–23 (2009).
75 K.V. Sumesh, P. Sharma-Natu and M.C. Ghildiyal, Starch synthase activity and heat shock protein in relation to thermal tolerance of developing wheat grains. *Biol. Plantarum.*, 52, 749–753 (2008).
76 M. Djanaguiraman, J.A. Sheeba, D.D. Devi and U. Bangarusamy, Cotton leaf senescence can be delayed by nitrophenolate spray through enhanced antioxidant defense system. *J. Agron. Crop Sci.*, 195, 213–224 (2009).
77 M. Rodríguez, E. Canales and O. Borrás-Hidalgo, Molecular aspects of abiotic stress in plants. *Biotechnol. Appl.*, 22, 1–10 (2005).

78 S. Munné-Bosch and J. Penuelas, Photo and antioxidative protection, and a role for salicylic acid during drought and recovery in field grown *Phillyrea angustifolia* plants. *Planta.*, 217, 758–766 (2003).

79 M. Kaushal and S.P. Wani, Plant-growth-promoting rhizobacteria: drought stress alleviators to ameliorate crop production in drylands, *Ann. Microbiol.*, (2015).

80 C.K. Jha and M. Saraf, Plant growth promoting Rhizobacteria (PGPR): a review E3 *J. Agric. Res. Dev.*, 5, 0108-0119 (2015).

81 P.N. Bhattacharyya and D.K. Jha, Plant growth-promoting rhizobacteria (PGPR): emergence in agriculture World. *J. Microbiol Biotechnol.*, 28, 1327–1350 (2012).

82 C.M. Ryu, M.A. Farag, C.H. Hu, M.S. Reddy, H.X. Wei, P.W. Pare and J.W. Kloepper, Bacterial volatiles promote growth in *Arabidopsis. PNAS USA.*, 100, 4927–4932 (2003).

83 M. Grover, S.Z. Ali, V. Sandhya, A. Rasul and B. Venkateswarlu, Role of microorganisms in adaptation of agriculture crops to abiotic stress. *World J. Microbiol Biotechnol.*, 27, 1231–1240 (2010).

84 E. Yildirium and A.G. Taylor, Effect of biological treatment on growth of bean plants under salt stress. *Annu. Rep. Bean Improv. Coop.*, 48, 176–177 (2005).

85 M. Ashraf and M. Hafeez, Thermo tolerance of pearl millet and maize at early growth stages: Growth and nutrient relations. *Biologia. Plantarum.*, 48, 81–86 (2004).

86 A.W. Qurashi and A.N. Sabri, Osmo adaptation and plant growth promotion by salt tolerant bacteria under salt stress. *Afr. J. Microbiol. Res.*, 5, 3546–3554 (2011).

87 H. Naseem and A. Bano, Role of plant growth-promoting rhizobacteria and their exopolysaccharide in drought tolerance of maize. *Int. J. Plant Sci.*, 9, 689–701 (2014).

88 S. Timmusk, S.B. Kim, E. Nevo, AbdEl Daim, B. Ek, J. Bergquist and L. Behers . Sfp type PPTase inactivation promotes bacterial biofilm formation and ability to enhance wheat drought tolerance. *Front. Microbiol.*, 21;6:387 (2015).

89 E.M. Vanderlinde, J.J. Harrison, A. Muszynski, R.W. Carlson, R.J. Turner and C.K. Yost, Identification of a novel ABC-transporter required for desiccation tolerance, and biofilm formation in *Rhizobium leguminosarum* bv. *Viciae* 3841. *FEMS Microbiol. Ecol.*, 71, 327–340 (2010).

90 V. Goyal, S. Chetal and H.S. Nainawatee, Alteration in *Rhizobium trifolii* catalase under water stress. *Folia Microbiol.*, 31, 164–166 (1986).

91 A. Zaidi, M.S. Khan, M. Ahemad and M. Oves, Plant growth promotion by phosphate solubilizing bacteria. *Acta. Microbiol. Immunol. Hung.*, 56, 283–284 (2009).

92 Z.A. Zahir, U. Ghani, M. Naveed, S.M. Nadeem and H.N. Asghar, Comparative effectiveness of *Pseudomonas* and *Serratia* sp. containing ACC-deaminase for improving growth and yield of wheat (*Triticum aestivum L.*) under salt-stressed conditions. *Arch. Microbiol.*, 191, 415–424 (2009).

93 M.B. Hussain, Z.A. Zahir, H.N. Asghar and M. Asgher, Can Catalase and Exopolysaccharides Producing Rhizobia Ameliorate Drought Stress in Wheat? *IJAB.*, 1560–8530 (2014).

94 Y. Kaci, A. Heyraud, M. Barakat and T. Heulin, Isolation and identification of an EPS-producing *Rhizobium* strain from arid soil (Algeria): characterization of its EPS and the effect of inoculation on wheat rhizosphere soil structure. *Res. Microbiol.*, 156, 522–531 (2005).

95 M. Rashad, A. Ragab and S. Salem, The influence of some *Bradyrhizobium* and *Rhizobium* strains as plant growth promoting rhizobacteria on the growth and yield of sorghum (*Sorghum bicolour* L.) plants under drought stress. *J. Plant Nutr.*, 92, 664–665 (2002).

96 Y. Alami, W.A. Achouak, C. Marol and T. Heulin, Rhizosphere soil aggregation and plant growth promotion of sunflower by an exopolysaccharides producing *Rhizobium* sp. strain isolated from sunflower roots. *Appl. Environ. Microbiol.*, 66, 3393–3398 (2000).

97 F. Lemessa and W. Zeller, Screening rhizobacteria for biological control of *Ralstonia solanacearum* in Ethiopia. *Biol. Control.*, 42, 336–344 (2007).

98 N. Aliye, C. Fininsa and Y. Hiskias, Evaluation of rhizosphere bacterial antagonists for their potential to bio- protect potato (*Solanum tuberosum*) against bacterial wilt (*Ralstonia solanacearum*). *Biol. Control.*, 47, 282–288 (2008).

99 X.L. Ji, G.B. Lu, Y.P. Gai, C.C. Zheng and Z.M. Mu, Biological control against bacterial wilt and colonization of mulberry by an endophytic *Bacillus subtilis* strain. *FEMS Microbiol Ecol.*, 65, 565–573 (2008).

100 Y.B. Xu, M. Chen, Y. Zhang, M. Wang, Y. Wang, Qb. Huang, X. Wang and G. Wang, The phosphotransferase system gene ptsI in the endophytic bacterium *Bacillus cereus* is required for biofilm formation, colonization, and biocontrol against wheat sharp eyespot. *Fed. Eur. Microbiol.Soc.*, 354, 142–152 (2014).

101 L. Chen, Z.Y. Zhang, H.X. Liang, H.X. Liu, L.P. Du, H.J. Xu and Z.Y. Xin, Overexpression of TiERF1 enhances resistance to sharp eyespot in transgenic wheat. *J. Exp. Bot.*, 59, 4195–4204 (2008).

102 B.Y. Liu, Y. Lu, Z.Y. Xin and Z.Y. Zhang, Identification and antifungal assay of a wheat b-1, 3-glucanase. *Biotechnol. Lett.*, 31, 1005–1010 (2009a).

103 L. Houot, S. Chang, C. Absalon and P.I. Watnick, *Vibrio cholerae* phosphoenol pyruvate phosphotransferase system control of carbohydrate transport, biofilm formation, and colonization of the germ free mouse intestine. *Infect. Immun.*, 78, 1482–1494 (2010a).

104 L. Houot, S. Chang, B.S. Pickering, C. Absalon and P.I. Watnick, The phosphoenolpyruvate phosphotransferase system regulates *Vibrio cholerae* biofilm formation through multiple independent pathways. *J. Bacteriol.*, 192, 3055–3067 (2010b).

105 Y. Fujita, Carbon catabolite control of the metabolic network in *Bacillus subtilis*. *Biosci. Biotechnol. Biochem.*, 73, 245–259 (2009).

106 I. Comas, F. Gonzalez-Candelas and M. Zuniga, Unraveling the evolutionary history of the phosphoryl-transfer chain of the phosphoenol pyruvate: phosphotransferase system through phylogenetic analyses and genome context. *Bio. Med. Central.*, 147 (2008).

107 S. Li, N. Zhang, Z. Zhang, J. Luo, B. Shen, R. Zhang and Q. Shen, Antagonist *Bacillus subtilis* HJ5 controls Verticillium wilt of cotton by root colonization and biofilm formation. *Biol. Fert. Soils.*, 49, 295–303 (2013).

108 H.P. Bais, R. Fall and J.M. Vivanco, Biocontrol of *Bacillus subtilis* against infection of *Arabidopsis* roots by *Pseudomonas syringae* is facilitated by biofilm formation and surfactin production. *Plant Physiol.*, 134, 307–319 (2004).

109 M. Farzaneh, Z.Q. Shi, A. Ghassempour, N. Sedaghat, M. Ahmadzadeh, M. Mirabolfathy and M. Javan- Nikkhah, Aflatoxin B1 degradation by *Bacillus subtilis* UTBSP1 isolated from pistachio nuts of Iran. *Food Control.*, 23, 100–106 (2012).

110 V. Bianciotto, S. Andreotti, R. Balestrini, P. Bonfante and S. Perotto, Mucoid mutants of the biocontrol strain *Pseudomonas fluorescens* CHA0 show increased ability in biofilm formation on mycorrhizal and nonmycorrhizal carrot roots. *Mol. Plant Microbe Interact.*, 14, 255–260 (2001).

111 H.S. Jayasinghearachchi and G. Seneviratne (2006b) A mushroom-fungus helps improve endophytic colonization of tomato by *Pseudomonas fluorescence* through biofilm formation. *Res J Microbiol.*, 1, 83–89.

112 W.M. Haggag and S. Timmusk (2007) Colonization of peanut roots by biofilm-forming *Paenibacillus polymyxa* initiates biocontrol against crown rot disease. *J. Appl. Microbiol.*, ISSN 1364-5072 doi:10.1111/j.1365-2672.2007.03611.x

113 R. Prasanna, S. Pattnaik, T. K. Sugitha, L. Nain and A. Saxena, Development of cyanobacterium-based biofilms and their in vitro evaluation for agriculturally useful traits. *Folia Microbiol.*, 56, 49–58 (2011).

114 R. Prasanna, A. Kumar, S. Babu, G. Chawla, V. Chaudhary, S. Singh and A. K Saxena, Deciphering the biochemical spectrum of novel cyanobacterium-based biofilms for use as inoculants. *Biol. Agric. Hortic.*, 29, 145–158 (2013).

115 R. Prasanna, S. Triveni, N. Bidyarani, S. Babu, K. Yadav and A. K. Saxena, Evaluating the efficacy of cyanobacterial formulations and biofilmed inoculants for leguminous crops. *Arch. Agron. Soil Sci.*, 60, 349–366 (2014).

116 F. Scala, A. Raio, A. Zoina, M. Lorito, S. Chincholkar and K. Mukerji, Biological control of fruit and vegetable diseases with fungal and bacterial antagonists: *Trichoderma* and *Agrobacterium*. In S. B. Chincholkar and K. G. Mukerji (eds.), Biological control of plant diseases. pp. 150–190. Binghamton: Howorth Press (2007).

117 G. Seneviratne, R. Thilakaratne, A. Jayasekara, K. Seneviratne, Padmathilake and K.R.E., De Silva, M. Developing beneficial microbial biofilms on roots of Non legumes: A novel biofertilizing technique. In M. S. Khan, A. Zaidi, and J. Musarrat (Eds.), Microbial strategies for crop improvement. pp. 51–62. Berlin, Heidelberg: Springer (2009).

118 S. Triveni, R. Prasannaa, A. Kumar, N. Bidyarania, R. Singh and A.K. Saxena. Evaluating the promise of *Trichoderma* and *Anabaena* based biofilms as multifunctional agents in *Macrophomina phaseolina-infected cotton crop*. 25, 6, 656–670 (2015).

119 U. Mathesius, S. Mulders, M.S. Gao, M. Teplitski, G. Caetano-Anolles, B.G Rolfe, and W.D. Bauer, Extensive and specific response of a eukaryote to bacterial quorum-sensing signals. *PNAS USA.*, 100, 1444-9 (2003).

120 R. Schuhegger, A. Ihring, S. Gantner, G. Bahnweg, G. KnappeVogg, P. Hutzler, M. Schmid, M. Van Breusegem, L. Eberl, A. Hartmann and C. Langebartels, Induction of systemic plant resistance by *N*-acyl homoserine lactone-producing rhizosphere bacteria. *Plant Cell Environ.*, 29, 90–918 (2006).

121 C.M. Joseph and D.A. Phillips. Metabolites from soil bacteria affect plant water relations. *Plant Physiol. Biochem.*, 41, 189–192 (2003).

122 U. Von Rad, I. Klein, P.I. Dobrev, J. Kottova, E. Zazimalova, A. Fekete, A. Hartmann, P. Schmitt-Kopplin and J. Durner, Response of *Arabidopsis thaliana* to *N*-hexanoyl-DL-homoserine lactone, a bacterial quorum sensing molecule produced in the rhizosphere. *Planta.*, 229, 73–85 (2008).

123 R. Ortíz-Castro, M. Martínez-Trujillo and J. López-Bucio, *N*-acyl-L homoserine lactones: a class of bacterial quorum-sensing signals alter post-embryonic root development in *Arabidopsis thaliana*. *Plant Cell Environ.*, 31, 1497–1509 (2008).

124 C.M. Ryu, H.K. Choi, C.H. Lee, J.F. Murphy, J.K. Lee and J.W. Kloepper, Modulation of quorum sensing in acyl–homoserine lactone- producing or—degrading tobacco plants leads to alteration of induced systemic resistance elicited by the rhizobacterium *Serratia marcescens* 90–166. *Plant Pathol.*, 29, 182–192 (2013).

125 A. Schikora, S.T. Schenk, E. Stein, A. Molitor, A. Zuccaro and K. H. Kogel, *N*-acyl-homoserine lactone confers resistance toward biotrophic and hemi biotrophic pathogens via altered activation of AtMPK6. *Plant Physiol.*, 157, 1407–1418 (2011).

126 S.T. Schenk, E. Stein, K.H. Kogel and A. Schikora, *Arabidopsis* growth and defense are modulated by bacterial quorum sensing molecules. *Plant Signal Behav.*, 7, 178–181 (2012).

17

Developed Biofilm-Based Microbial Ameliorators for Remediating Degraded Agroecosystems and the Environment

G. Seneviratne, P.C. Wijepala and K.P.N.K. Chandrasiri

Microbial Biotechnology Unit, National Institute of Fundamental Studies, Kandy, Sri Lanka

17.1 Introduction

The environment, with all its ecosystem services, comprises the entire basis for all plants and animals. With the recognition that environmental change resulting from both natural and anthropogenic disturbances is causing a rapid decline in biodiversity, much attention has been paid to understanding how changes in biodiversity may alter levels of ecosystem functioning. Many species have become extinct in areas dominated by human influences. Extinction is a natural process, but it is occurring at an unnaturally rapid rate as a consequence of human activities [1]. Studies show that preservation of biodiversity is essential for the maintenance of stable productivity in ecosystems [2]. Most of the studies have aimed to evaluate different land-use and land-use change strategies, since the greatest threat to biodiversity is the conversion of natural habitats to agricultural lands [3].

Similarly, many toxic substances have been introduced into the environment through human activities, particularly in agriculture. Many of these substances either immediately or ultimately come in contact with or are sequestered by soil, plants, and water. This results in soil quality degradation, crop yield reduction, and poor quality of agricultural products, while making significant damage to human, animal, and ecosystem health. Many strategies, such as land sharing, land sparing, and crop rotation have been introduced to reduce these mentioned issues. However, there are limits to those practices [3], and at times they have ended up with some issues in the amount of crop harvest etc. [4]. Therefore, we have to look for alternative methods that reinstate the degraded ecosystems with improved productivity.

Beyond agriculture, the use of toxic substances in industries damages the land resources, thus rapidly losing their overall quality and converting them to contaminated sites due to carelessness in using the substances. It is well-known that a contaminated site is a potential threat to human health. Scientists are now seeking novel remedies to replace conventional techniques used for cleaning sites, such as landfills, chemical application, and high-temperature incineration [5].

Bioremediation is a promising alternative, as the conventional techniques consist of some drawbacks. The term *bioremediation* simply can be defined as the degradation of hazardous or contaminated pollutants to nonhazardous substances by using biological agents [6]. The process of bioremediation can be executed either in the contaminated site, that is, *in situ*, or by taking out the substances from the contaminated site, *ex situ* conditions, by using fauna and flora in an ecosystem [7]. Some plants have the ability to absorb toxic materials or heavy metals, the process of which is known as phytoremediation. However, microbial processes are more efficient than phytoremediation, because energy produced during the formation of nonhazardous substances by microbes is used as an impetus for their metabolic activities, and also elemental ions mediated in the energy production are used as nutrients [8]. These naturally occurring beneficial bacteria and fungi in the soil system, which are involved in the bioremediation process, are called effective microorganisms (EM). During conversion of the contaminants to nonhazardous substances, EMs should attack the pollutants enzymetically [5]. However, these microbes can survive within a limited range of chemical contaminants, because utilizations variably depend on the contaminants [9]. Further, there should be favorable environmental conditions for them to enhance growth and activity, ultimately resulting in an effective biodegradation [10]. As such, there are certain constraints for their proper function in the environment.

Microorganisms play a major role in balancing the ecosystem functions and structure. For example, soil microbial communities have been reported to determine plant species diversity in natural ecosystems [11]. Thus, changes in the composition and activity of microbial communities may have immediate or lasting effects on ecosystem structure and functioning. Further, microbial communities have a great impact on biosynthesis and biodegradation in nature. They can play an extremely important role in rehabilitation of degraded agricultural ecosystems and environs. Conventionally, monocultures of microbes or their mixtures have been used to revive degraded systems. This chapter describes the potential of using developed microbial communities to remediate degraded agroecosystems and also the environment.

17.2 Developed Microbial Communities as a Potential Tool to Regenerate Degraded Agroecosystems

Soil microbes play an immense role in the ecosystem, though they are not seen by our naked eyes. In nature, they do not live as single cells. Patel et al. [12] explains that microorganisms prefer to exist in the form of communities, due to various advantages they can gain with that. They tend to form film-like structures by sticking together. These structures are called microbial biofilms. Microbes from a wide range of genera are capable of forming biofilms using their effective communication. In bacteria and fungi, three types of biofilms can be seen in the soil system: bacterial biofilms, fungal biofilms, and fungal-bacterial biofilms (FBBs), [13]. Fungal surface-attached bacteria, particularly nitrogen fixers encased from extracellular polymeric substances (EPS), are generally named FBBs (Figure 17.1). Development of FBBs *in vitro* was first reported by Seneviratne and Jayasinghearachchi [14] for fungal mycelial colonization by bradyrhizobial and azorhizobial strains.

The EPS coating of the FBBs, being less permeable to gases, creates a micro-aerobic condition around N_2 fixing bacteria (Figure 17.2). That helps them to biologically fix N_2 that is slightly dissolved in the EPS. Resultant NH_4^+, amino acids, and perhaps proteins

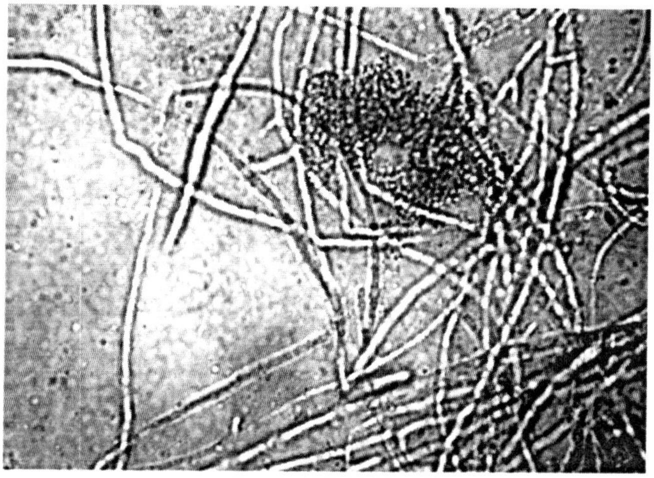

Figure 17.1 *Colletotrichum* sp. mycelium colonized by *Azotobacter* sp. forming fungal–bacterial biofilms (FBBs), when developed under *in vitro* conditions and stained with lactophenol cotton blue. Magnification, x 400. (*See color plate section for the color representation of this figure.*)

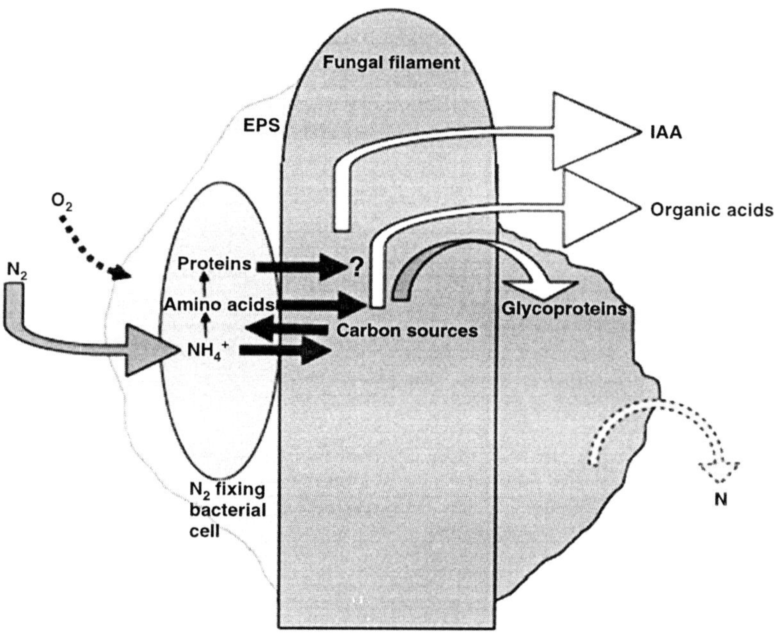

Figure 17.2 A conceptual model of fungal-bacterial biofilm (FBB) and its processes.

in the N_2 fixers are transferred to associated fungal mycelium for its nutrition. In return, the fungal mycelium provides carbon sources to the N_2 fixers as the energy source. In this manner, fungi and bacteria in the FBBs establish a metabolic cooperation in the symbiosis. In addition, it is also reported that the FBBs secrete IAA, other organic acids and glycolproteins to the vicinity [15]. When glycoproteins decompose, their nutrients are gradually released to the soil, which can, in turn, be taken up by plants and microbes. Thus, the FBBs play an important role in plant growth and nutrition. The FBBs are also capable of attaching to plant root systems, creating root–biofilm associations, which benefit both plant and biofilm microbes [16]. This association supplies more nitrogen to the soil–plant system and enhances nutrient absorption from the surrounding soil, thus resulting in prominent crop growth. It was recently found that exudates of the FBBs help to break dormancy of soil microbial seed bank that was developed under stress conditions such as conventional agricultural practices [17]. This leads to reinstating the lost microbial diversity in agroecosystems, and hence ecosystem functioning and sustainability. It has also been shown that the inoculation of FBBs helps to maintain a higher cell density of rhizobia on the root system of legumes than the inoculation of rhizobial monocultures [18]. On the roots of nonlegumes, nodule-like structures are formed by FBBs, which act as *pseudonodules*, fixing N_2 biologically. Moreover, these microbial communities on the root system protect the plant from adverse environmental conditions and pathogenic infections [19].

17.3 Biochemistry of Fungal-Bacterial Biofilms

Gene expression of biofilm is different from their non–biofilm forming stages [20]. During growth and development, biofilms engage in novel gene expressions compared with their planktonic stage [21]. As a result, biochemical diversity of biomolecules in exudates is higher in FBBs in comparison to that of fungal or bacterial monocultures, as evidenced from novel functional groups emerged with the FBBs [22]. It has also been observed that the biochemical expression of the microbial biofilms is substrate specific. Biotic interaction of the fungal surface-attached bacterial biofilms has led to increased biochemical expressions compared with the counterparts' monocultures and also association of the microbes with abiotic surfaces [23]. Further, the diverse biochemicals expressed in the FBBs have been reported to support culturing as yet unculturable bacteria and fungi by providing them with multiple growth requirements that are absent in non–biofilm forming conditions [24].

17.4 Endophytic Microbial Colonization with the Application of Fungal-Bacterial Biofilms

When hydroponically grown rice was applied with FBBs, increased endophytic colonization by diverse bacteria and their *in planta* biofilm induction were observed, possibly due to dormancy breaking of endophytic microbial seed bank by the biofilm-specific biomolecules [25]. Soil application of FBBs in the form of biofilm biofertilizers (BFBFs) to a maize crop has also been observed to increase endophytic colonization of bacteria due to reinstating their diversity from soil microbial seed dormancy breaking process [26], in addition to the similar endophytic process explained above. Generally, the endophytic microbial colonization is known to be important in environment stress tolerance of plants and their increased growth and yield. For example, application of BFBFs to rice plants under soil moisture stress showed improved seedling growth and root growth, and also reduction of leaf drying and rolling [27]. In a multi-location experiment of tea cultivations, BFBFs application showed significantly reduced leaf transpiration during dry periods [28]. These results could be attributed to increased endophytic colonization of microbes with the BFBFs applications. Improving drought resistance of crops through selective breeding as well as molecular methods is in practice and also in continued research. However, biological methods such as the use of biofertilizers, which warrant the extensive popularization, have gained little attention.

In addition to water stress, seasonal differences in light intensity affect productivity of some crops (e.g., strawberries). Under low light conditions, there is a natural reduction of plant growth with retarded photosynthesis, which causes low productivity of crops [29]. However, application of BFBFs has been shown to reinstate the reduced plant growth with increased chlorophyll production under limited light regimes, especially rainy and cloudy weather conditions [30]. In this study, the BFBFs application reduced soil pH under low light compared with a relatively high-light regime. Lowering of soil pH has been observed to help dominate fungal endophytes [31]. In general, fungal endophytes produce diverse organic acids [32], which could facilitate an H^+ gradient around the thylakoid membrane, once entered into plant cells, and enhance Mg^{2+} and Ca^{2+} exchange through the membrane, leading to an improved photosynthesis rate [33]. All in all, these evidences prove that application of BFBFs can defeat plant environmental stresses while contributing to healthy plant growth.

17.5 Biofilm Biofertilizers for Restoration of Conventional Agroecosystems

BFBFs are important in addressing many issues that affect the sustainability of agroecosystems. They render numerous biochemical and physiological benefits to plant growth, particularly under stress conditions, and improve soil quality, thus leading to a reduction of chemical fertilizer NPK use by 50 percent of recommended doses in various crops [34]. The other important feature of this fertilizer reduction is that it does not lead to reduced crop yields, but instead marginally increases the yields of many crops. Also, this potential of BFBFs in reducing the use of fertilizers contributes to numerous health, economic, and environmental benefits in agroecosystems, particularly in reinstating the lost biodiversity. Thus, the use BFBFs is a restoration strategy of degraded croplands rather than just a biofertilizing method, and hence the application of BFBFs is extremely important, despite the use of various nutrient sources, chemical or organic, if we are to sustain productivity of croplands.

17.6 Developed Microbial Biofilms for Environmental Bioremediation

In conventional bioremediation of the environment, there are certain constraints of microbial tools for proper function in the environment. Thus, considerable attention has been focused recently on microbial biofilms and their potential on bioremediation in contaminated environments, because they have better opportunities for survival during stress periods, as they are safe within EPS, and hence that results in more effective bioremediation [35]. Thus, the use of developed biofilms has been proposed and is being tested in environmental applications [36]. As such, this field of research is still in the early stages.

17.6.1 Fungal-Bacterial Biofilms for Heavy Metal Bioremediation in Soil–Plant Environment

Use of developed FBBs in heavy metal bioremediation is a novel field of research, introduced recently. Role of FBBs on heavy metal uptake by plants was studied using *Zea mays* as the test plant [37]. Here, serpentine soil was planted with *Zea mays*, and it was applied with FBBs in a pot experiment. Results showed that the plant roots took up Ni in higher concentrations than other heavy metals. Ni and Cr concentrations in shoots were significantly higher than that of Co. When translocation of heavy metals in the soil–plant system was considered, heavy metal bioavailability in the soil was increased, whereas translocation of Ni, Mn, and Cr in the plant was decreased in the presence of FBBs. FBBs application showed the maximum plant biomass. Catalase and polyphenol oxidase activities were significantly high with the FBBs. As such, FBBs played an important role in plant growth promotion as well as soil quality enhancement in the presence of heavy metals.

Heavy metal bioremediation by using biochar (BC) is an upcoming field [38]. BC is known as a carbon-rich substance made as a result of pyrolysis of organic materials [39], and is considered as a diligent soil amendment for immobilizing heavy metals [40]. With the addition of BC, Cr, Ni, and Mn in serpentine soil were immobilized by reducing heavy metal toxicity for tomato plants [38]. However, polyphenol oxidase, dehydrogenase, and catalase activities were reduced gradually with the increase of the rate of BC application. When BC was coupled with FBBs, soil dehydrogenase activity was increased and plant growth was improved, indicating that the FBBs can overcome adverse effects of BC in heavy metal bioremediation and plant growth in contaminated soils.

17.6.2 Fungal-Bacterial Biofilms for Heavy Metal Bioremediation in Wastewater

Use of developed FBBs for Ni removal in wastewater was first reported by Seneviratne et al. [41]. Initially, Ni resistant microorganisms were isolated from Ni-rich serpentine soils. Five FBBs were developed from the isolates. They were challenged with a series of Ni concentrations. Results showed that FBBs were more capable than the monoculture isolates in Ni biosorption. Further, it was found that amino groups in proteins of EPS in FBBs were involved in Ni biosorption.

Bioremediation of hexavalent chromium [Cr(VI)] in wastewater by using FBBs is also a novel technique. A study was conducted to evaluate tolerability and removal of Cr (VI) by using developed FBBs, glass-wool-attached bacterial biofilms (BBs), and their monocultures [42]. Bacteria and fungi were isolated from sediments taken from a wastewater channel. This study showed that FBBs were able to tolerate significantly higher level (500 $\mu g\ mL^{-1}$) of Cr(VI) than BBs and the monocultures. Also, the FBBs removed up to 90 percent of Cr(VI) after 10 days.

17.7 Conclusion

Developed microbial biofilms, particularly FBBs, result in more effective bioremediation than the conventional methods due to reinstating lost microbes by breaking dormancy of soil seed bank in degraded agroecosystems and contaminated environments. However, there are numerous processes other than that, which contribute to effective functioning of the FBBs. Thus, it is important to open the black box of this concept in order to understand the processes involved. For that, advance modern technologies at the molecular level should be employed under real field conditions. Therefore, it is concluded that future studies on this line should be directed to explore simultaneous mechanisms operating at the ecosystem level with the application of FBBs, and also their complex interactions that lead to effective bioremediation, simulating natural environmental processes.

References

1 S.L. Pimm, G.J. Russell, J.L. Gittleman and T.M. Brooks, The future of biodiversity, *Science*, 269, 347–350 (2015).
2 D. Tilman and J.A. Downing, Biodiversity and stability in grasslands, *Nature*, 367, 363–365 (1994).
3 B. Phalan, M. Onial, A. Balmford and R. Green, Reconciling food production and biodiversity conservation: land sparing and land sharing compared, *Science*, 333, 1289–1291 (2011).
4 Y. Clough, J. Barkmann, J. Juhrbandt, M. Kessler and T.C. Wanger, Combining high biodiversity with high yields in tropical agroforests. *Proc. Natl. Acad. Sci. USA*, 108, 8311–8316 (2011).
5 M. Vidali, Bioremediation: an overview. *Pure Appl. Chem.*, 73, 1163–1172 (2001).
6 S.C. Sasikumar and T. Papinazath, Environmental management—bioremediation of polluted environment, *in Proceedings of the Third International Conference on Environment and Health*, J. Martin, V. Bunch, M. Suresh and T.V. Kumaran (eds.), Department of Geography, University of Madras, Chennai and Faculty of Environmental Studies, York University, New York, 465–469 (2003).
7 S. Sharma, Bioremediation: features, strategies and applications. *Asian J. Phil. and Life Sci.*, 2, 202–213 (2012).
8 C.Y. Tang, Q. S. Criddle, C.S. Fu and J.O. Leckie, Effect of flux (transmembrane pressure) and membranes properties on fouling and rejection of reverse osmosis and nanofiltration membranes treating perfluoro octanesulfonate containing waste water, *J. Environ. Sci. Tech.*, 4, 2008–2014 (2007).
9 L.M. Prescott, J.P. Harley and D.A. Klein, *Microbiology*, 5th Edition, McGraw Hill, New York, 2002.
10 A. Kumar, B.S. Bisht, V.D. Joshi and T. Dhewa, Review on bioremediation of polluted environment: a management tool, *Int. J. Environ. Sci.*, 1, 1079–1083 (2011).
11 S.A. Mangan, S.A. Schnitzer, E.A. Herre, K.M.L. Mack, M.C. Valencia, E.I. Sanchez and J.D. Bever, Negative plant–soil feedback predicts tree-species relative abundance in a tropical forest, *Nature*, 466, 752–755 (2010).
12 I. Patel, V. Patel, A. Thakkar and V. Kothari, Microbial biofilms: microbes in social mode, *Int. J. Agric. Food Res.*, 3, 34–49 (2014).

13 G. Seneviratne, J.S. Zavahir, W.M.M.S. Bandara and M.L.M.A.W. Weerasekara, Fungal-bacterial biofilms: their development for novel biotechnological applications, *World J. Microbiol. Biotechnol.*, 24, 739–743 (2008).
14 G. Seneviratne and H.S. Jayasinghearachchi, Mycelial colonization by Bradyrhizobia and Azorhizobia, *J. Biosci.*, 28, 243–247 (2003).
15 W.M.M.S. Bandara, G. Seneviratne and S.A. Kulasooriya, Interactions among endophytic bacteria and fungi: effects and potentials, *J. Biosci.*, 31, 645–650 (2006).
16 G. Seneviratne, R.M.M.S. Thilakaratne, A.P.D.A. Jayasekara, K.A.C.N. Seneviratne, K.R.E. Padmathilake and M.S.D.L. De Silva, Developing beneficial microbial biofilms on roots of non-legumes: a novel biofertilizer technique, *in Microbial Strategies for Crop Improvement*, M.S. Khan, A. Zaidi and J. Mussarat (Eds), Springer Verlag, Berlin, Heidelberg, 2009.
17 G. Seneviratne G and S.A. Kulasooriya, Reinstating soil microbial diversity in agroecosystems: the need of the hour for sustainability and health, *Agric. Ecosyst. Environ.*, 164, 181–182 (2013).
18 G. Seneviratne and H.S. Jayasinghearachchi, A rhizobial biofilm with nitrogenase activity alters nutrient availability in a soil, *Soil Biol. Biochem.*, 37, 1975–1978 (2005).
19 G. Seneviratne, M.L.M.A.W. Weerasekara, K.A.C.N. Seneviratne, J.S. Zavahir, M.L. Kecskés and I.R. Kennedy, Importance of biofilm formation in plant growth promoting Rhizobacterial action, *in Plant Growth and Health Promoting Bacteria*, D.K. Maheshwari (Ed), Springer Verlag, Berlin, Heidelberg, 2010.
20 S. Vilain and V.S. Brözel, Multivariate approach to comparing whole-cell proteomes of *Bacillus cereus* indicates a biofilm specific proteome, *J. Proteome Res.*, 5, 1924–1930 (2006).
21 D.G. Davies, A.M. Chakrabarty and G.G. Geesey, Exopolysaccharide production in biofilms: substratum activation of alginate gene expression by *Pseudomonas aeruginosa*, *Appl. Environ. Microbiol.*, 59, 1181–1186 (1993).
22 H.M.L.I. Herath, D.M.N. Senanayeke, G. Seneviratne and D.C. Bandara, Variation of biochemical expressions of developed fungal-bacterial biofilms over their monocultures and its effect on plant growth, *Trop. Agric. Res.*, 24, 186–192 (2013).
23 N.A.D. Lakmali, G. Seneviratne, I.D. Singhalage and S. Ediriweera, Effect of substrate on biochemical expression of bacterial biofilms, *in Proceedings of Uva Wellassa University Annual Research Symposium*, Uva Wellassa University, Badulla, 2015.
24 G. Seneviratne, H.M.L.I. Herath and A.S.F. Rifana, Developing a method to culture yet unculturable soil bacteria and fungi using exudates of developed fungal-bacterial biofilms, *in Proceedings of NIFS Annual Review*, National Institute of Fundamental Studies, Kandy, 2012.
25 I.S. Manawasinghe, G. Seneviratne, M.C.M. Zakeel and I.D. Singhalage, Evolution of an introduced biofilmed biofertilizer in a microbial environment, *in* 6th *Annual Research Symposium Proceedings of Rajarata University of Sri Lanka*, Rajarata University, Mihintale, 2014.
26 U.V.A. Buddhika, G. Seneviratne, Biofilmed biofertilizers: a recent trend in biofertilizer application, *in Fertilizer Technology II: Biofertilizers*, S. Sinha, K.K. Pant, S. Bajpai and J.N. Govil (Eds), Studium Press LLC, Texas, 2015.
27 N. Weerarathne and G. Seneviratne, Biofilmed biofertilizers induce drought tolerance of rice, *in Interdrought –IV Conference*, Crown Perth, Perth, 2013.
28 M.S.D.L. De Silva, A.P.D.A. Jayasekera, G. Seneviratne, U.P. Abeysekera, E.W.T.P. Premathunga and S.N. Wijesekera, Soil fertility improvement through biofilmed

biofertilizers: potential for field applications in tea cultivations, *Sri Lanka J. Tea Sci.*, 79, 46–61 (2014).

29 A. Polthanee, K. Promsaena and A. Laoken, Influence of low light intensity on growth and yield of four soybean cultivars during wet and dry seasons of Northeast Thailand, *Agric. Sci.*, 2, 61–67 (2011).

30 P.C. Wijepala, G. Seneviratne, G. Chandrasena, I.D. Singhalage and M. Nugaliyadda, Plant response to biofertilizers under low light regime, *in Proceedings of Uva Wellassa University Annual Research Symposium*, Uva Wellassa University, Badulla, 2013.

31 T.O. Adejumo and O. Orole, Effect of pH and moisture content on endophytic colonization of maize roots, *Sci. Res. Essays*, 5, 1655–1661 (2010).

32 M. Waqas, A.L. Khan, M. Kamran, M. Hamayun, S.M. Kang, Y.H. Kim and I.J. Lee, Endophytic fungi produce Gibberellins and Indoleacetic acid and promotes host plant growth during stress, *Molecules*, 17, 10754–10773 (2012).

33 W.F. Ettinger, A.M. Clear, K.J. Fanning and M.L. Peck, Identification of a Ca^{2+}/H^{+} antiport in the plant chloroplast thylakoid membrane, *Plant Physiol.*, 119, 1379–1385 (1999).

34 U.V.A. Buddhika, G. Seneviratne, Ekanayake EMHGS, D.M.N. Senanayake, A.D. Igalavithane, N. Weeraratne, A.P.D.A. Jayasekara, W.L. Weerakoon, A. Indrajith, H.M.A.C. Gunaratne, R.K.G.K. Kumara, M.S.D.L. De Silva and I.R. Kennedy, Biofilmed biofertilizers: application in agroecosystems, *in The Handbook of Microbial Bioresources*, V.K. Gupta, G.D. Sharma, M.G. Tuohy and R. Gaur (eds.), CAB International, Wallingford, 2016.

35 A.W. Decho, Microbial biofilms in intertidal systems: an overview, *Continental Shelf Res.*, 20, 1257–1273 (2000).

36 G. Seneviratne, J.S. Zavahir, W.M.M.S. Bandara and M.L.M.A.W. Weerasekara, Fungal-bacterial biofilms: their development for novel biotechnological applications, *World J. Microbiol. Biotechnol.*, 24, 739–743 (2008).

37 M. Seneviratne, G. Seneviratne, H.M.S.P. Madawala, M.C.M. Iqbal, N. Rajakaruna, T. Bandara and M. Vithanage, A preliminary study of the role of bacterial–fungal co-inoculation on heavy metal phytotoxicity in serpentine soil, *Aust. J. Bot.*, 63, 261–268 (2015).

38 I. Herath, P. Kumarathilaka, A. Navaratne, N. Rajakaruna and M. Vithanage, Immobilization and phytotoxicity reduction of heavy metals in serpentine soil using biochar, *J. Soils Sediments*, 15, 126–138 (2015).

39 M. Ahmad, A. Rajapaksha, J.E.Z.M. Lim, N. Bolan, D. Mohan, M. Vithanage, S.S. Lee and Y.S. Ok, Biochar as a sorbent for contaminant management in soil and water: a review, *Chemosphere*, 99, 19–23 (2014).

40 M. Ahmad, S.S. Lee, J,E. Yang, H.M. Ro, Y.H. Lee and Y.S. Ok, Effects of soil dilution and amendments (mussel shell, cow bone, and biochar) on Pb availability and phytotoxicity in military shooting range soil, *Ecotoxicol. Environ. Safety*, 79, 225–231 (2012).

41 M. Seneviratne, M. Vithanage, H.M.S.P. Madawala and G. Seneviratne, The potential of biofilms developed from heavy metal resistant microbes to remove nickel ions in aqueous media, *in Proceedings of the Peradeniya University International Research Sessions*, University of Peradeniya, Peradeniya, 2014.

42 H.M.L.I. Herath, A.U. Rajapaksha, M. Vithanage and G. Seneviratne, Developed fungal-bacterial biofilms as a novel tool for bioremoval of hexavelant chromium from wastewater, *Chem. Ecol.*, 30, 418-427 (2014).

18

Plant Root–Associated Biofilms in Bioremediation

Sadaf Kalam, Anirban Basu and Sravani Ankati

Department of Plant Sciences, School of Life Sciences, University of Hyderabad, Hyderabad, India

18.1 Introduction

Microorganisms are ubiquitous in nature, thriving in all environmental niches and playing vital roles in maintaining ecological integrity. Soil is an important playground for microbes where several interactions exist between plants and microorganisms [1]. Bacterial biofilms are widely distributed, playing cardinal roles in soil environments ranging from nutrient-rich rhizosphere to soils deficient in nitrogen, phosphates, water, and nutrients [2]. Biofilms can rightly be referred to as *islands of microbes* or *flocks of microbes* inhabiting a particular environmental niche like the rhizosphere. Research on soil microbial biofilms is gaining impetus specially targeting biofilm-associated organisms and evaluating their practical applications in agriculture and bioremediation. Soil contaminants like heavy metals and xenobiotics pose severe detrimental threats to native microbiota, inhibiting their activity and leading to the loss of biodiversity and functions, including recycling of nutrients in biogeochemical cycles [3]. Persistence of these contaminants necessitates the use of eco-friendly means of environmental remediation.

Bioremediation is a sophisticated *in situ* technology leading to mitigation of environmental pollutants [4]. Since the inception of the concept of biofilms, scientists have been exploring strategies for improving bioremediation efficiency of microorganisms, which include genetically engineered strains with improved chemotactic ability and the use of mixed population biofilms [5]. Soil and plant root–associated biofilms represent a microenvironment of heterogeneous microbial communities with an inherent potential to degrade pollutants and aid in reclamation of contaminated soils [6]. This chapter presents an overview of plant root–associated biofilms and bioremediation.

18.2 Biofilms: Definition and Biochemical Composition

The term *biofilm* was coined and described by Costerton et al. [7] as bacterial communities surrounded by a self-produced matrix and reversibly attached to an inert (abiotic) or biotic surface. International Union of Pure and Applied Chemistry (IUPAC) defines biofilms as aggregate of microorganisms in which cells that are frequently embedded within a self-produced matrix of extracellular polymeric substances (EPS) adhere to each other and/or to a surface [8]. It is known that microorganisms can be planktonic, freely suspended, or occurring in close association with surfaces and interfaces forming multicellular aggregates glued together with the EPS slime they secrete [9]. Biofilm architecture is heterogeneous as well as dynamic, inside which cycling of various nutrients (e.g., nitrogen, sulfur, and carbon) occurs through redox reactions. Biofilms shield the native microbial community from environmental and biotic stresses [10, 11].

In nutritionally rich culture media or a moist environment incorporating good nutrient reserve, surface attachment by the microbes becomes easier. The biotic components of a biofilm can be a single bacterial species or it can be several species of bacteria, fungi, algae, and protozoa. The composition of biofilm matrix is usually 97 percent water that is bound to the microbial cells, or it can also be some solvent whose physicochemical properties can be assessed by solutes dissolved in it [12]. Biofilm matrix harbors certain diffusion processes that are dependent on the water binding capacity and mobility of the biofilm [10], while the EPS provides a physical barrier against diffusion of compounds *viz.*, antibiotics and defense substances from the host and protection against environmental factors [13].

Biochemically, the biofilm matrix comprises absorbed nutrients and metabolites, cell lysis products, and possibly particulate material and detritus from the immediate surrounding environment [14]. A biofilm environment canopies all major biopolymers (proteins, polysaccharides, DNA, and RNA) in addition to peptidoglycan lipids, phospholipids [12], carbohydrate-binding proteins [15], pili, flagella, other adhesive fibers and extracellular DNA [16]. Within the biofilms, living conditions for resident microorganisms are determined solely by exopolysaccharides as they affect the polarity, density, water content, charge, hydrophobicity and mechanical stability of biofilms [17]. Biofilms are anionic in nature due to the presence of uronic acids (*viz.*, D-glucuronic, D-galacturonic and mannuronic acids) or ketal-linked pyruvates [14]. The formation of biofilm involves the key process of microbial chemotaxis, which initiates surface colonization by the microbial cells culminating into formation of multicellular aggregates being glued to form functional biofilm [18].

18.3 Bioremediation and Its Significance

Bioremediation is a process of utilizing specific microorganisms for conversion of toxic and hazardous contaminants in soil/water to nontoxic/nonhazardous waste products [4]. Holistically, bioremediation could be defined as a process that uses biological systems, either plants or microbes for remediation, and also involves, most importantly, the plant–microbe interactions existing in root zone. Researchers have also defined bioremediation to be a technique that exploits the genetic diversity and metabolic versatility of living organisms to decontaminate or remediate polluted environments, thus leading to the elimination, attenuation, or transformation of polluting substances by the use of biological processes into lesser toxic forms [19]. Broadly, bioremediation can be classified into direct and indirect mechanisms, depending on involvement of the relevant agency (microbe/plant) in plant root-associated soil as depicted in Figure 18.1.

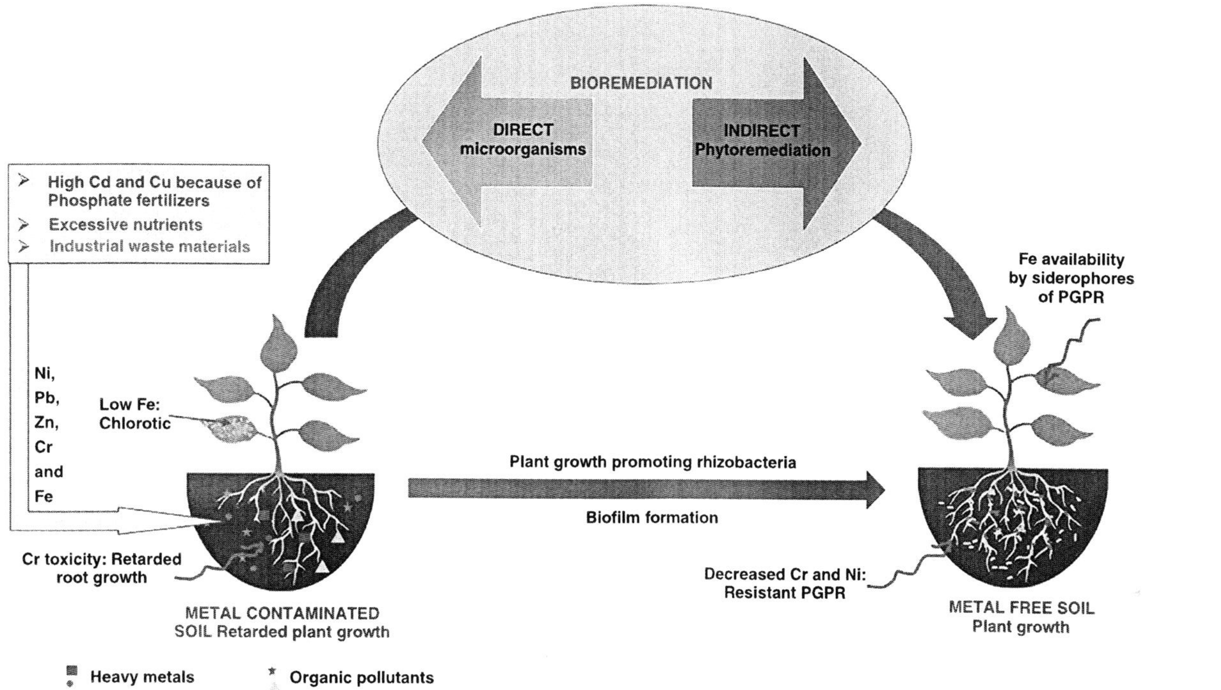

Figure 18.1 Plant and microbe mediated bioremediation in plant root associated soil.

The process of bioremediation by microbes (mainly bacteria or fungi) involves microbial enzymatic machineries that catalyze degradative reactions digesting toxins, thus degrading the targeted contaminants and mopping the environment. Achievement of a successful bioremediation attempt depends on the utilization of the appropriate microorganisms. Such microbial populations can, in theory, be consortia of naturally existing species or genetically engineered microorganisms [20].

Biological approaches of remediation include the use of microorganisms for detoxification of metal contaminants by various chemical processes in soil by inactivating metals in the rhizosphere or translocating them to the aerial parts. Another bioremediation methodology, phytoremediation, exploits the engineered use of vegetation to include, sequester, extract, accumulate, remove, degrade, and/or detoxify inorganic and organic contaminants from soils, sediments, surface waters, and groundwater [21]. More than 400 plant species have been identified to have potential for soil and water remediation [22]. Phytoremediation finds its wide application for soils contaminated with relatively immobile contaminants to shallow depths and in the case of organic pollutants decontamination [23].

18.4 Root-Associated Biofilms

18.4.1 Microbial Biofilm Associations on Plant Root Surface

The rhizosphere provides a microenvironment possessing intense biological, chemical, and physical activities, harboring diverse microorganisms involved in various positive and negative plant–microbial interactions [24]. Cell-to-cell signaling has been documented to play a key role in cell attachment and detachment from biofilms. Quorum sensing (QS) plays a vital role in root–microbe interactions, which facilitate bacteria to communicate and coordinate behavior via signaling molecules to colonize roots. QS is a regulatory process by which bacteria monitor their population density affecting population dynamics in association with host plants [25]. These signaling molecules are unique among the microbial species. Direct observations of bacteria adhering to plant surfaces have identified multicellular congregations, preferably microcolonies and cell clusters [26, 27]. During biofilm formation, the bacteria communicate chemically with each other by quorum sensing [28]. Some examples of diffusible cues include N-acyl-homoserine lactones (AHLs), 2-heptyl-3-hydroxy-4-quinoline, and autoinducer-2 that aid cell-to-cell communication within the bacterial community and facilitate synchronization of some actions [29].

18.4.2 Formation of Rhizospheric Biofilms by PGPR and Their Application

Rhizosphere supports rhizobacterial community sheltering diverse groups of bacteria. Those rhizobacteria that stimulate plant growth are termed as plant growth–promoting rhizobacteria (PGPR). They are a heterogeneous group of free-living soil bacteria that aggressively colonize plant roots and benefit plants by aiding in plant growth promotion [30, 31]. Some PGPR are capable of forming biofilms due to successful plant–microbe interaction. PGPR influence plant growth via production and secretion of various regulatory chemicals in the vicinity of rhizosphere, which enhance plant growth and improve health of crops [32]. PGPR and other root-associated microorganisms thus are able to establish a synergism with root/rhizosphere of host plants and thus aid in absorption of nutrients leading to improved plant growth.

Biofilm-forming PGPR can be effectively used to increase crop yields through a range of plant growth–promoting mechanisms and plant–microbe interactions. They can also be used as biofertilizers, plant growth regulators and biocontrol agents [33]. PGPR upon successful root colonization further multiply into microcolonies, which are often accompanied by biofilm formation [33]. Plant-associated biofilms serve various functions, including protection from external stress, mitigating microbial competition, and providing beneficial effects to the host plant, facilitating growth, yield, and crop quality [34].

Biofilm formation on root surface involves a complex cascade, starting from the initial colonization step and finally culminating into development of a mature biofilm microenvironment. Plant root surfaces are dynamic, wherein interplay of several physicochemical variations exists. These, along with other abiotic factors such as nutrient availability, temperature, and relative humidity influence root biofilm associations [35]. Albeit root-surface colonizing bacteria face these harsh challenges, but diverse bacterial species circumvent these adverse conditions and start the multifactorial cascade of colonization by forming microcolonies glued in an EPS matrix on different root areas from tip to elongation zone. These microaggregates grow into large population sizes to form functional biofilms [36, 37]. Bacterial EPSs have been known to exhibit many functions in plant signaling [38], but Beauregard et al. [39] for the first time gave previously undescribed evidence of plant polysaccharides (major components of the plant's cell wall) acting as a key signal for bacterial biofilm formation.

Based on the biochemistry, physiology, and genetics of biofilms, it is evident that an intricate mechanism is involved between bacterial interactions, growth, and microaggregate formation on the root surface [40]. Earlier studies have focused largely on biofilm formation on several biotic and abiotic surfaces, but the study of bacterial biofilm formation on root surfaces still remains less explored, as evident by available literature.

Reports document that many *Pseudomonas* spp. and *Bacillus* spp. are able to colonize plant leaves or root surfaces and are capable of biofilm formation [41]. *Pseudomonas* spp. discontinuously colonize the root surface and develop as small biofilms along epidermal fissures [42]. Species of *Pseudomonas* are considered to be a primary model in biofilm research as they form dense biofilms on both abiotic and biotic surfaces [43]. Experimental evidence suggests that *Pseudomonas putida* tends to rapidly respond to the signal of root exudates in soils, congregating at root colonization sites and finally culminating in the establishment of stable biofilms [44]. Another plant growth-promoting bacterium, *Bacillus subtilis,* has been frequently found associated with plant roots, where it protects plants from infection. Beauregard et al. [39] demonstrated that attachment of *B. subtilis* to roots of *Arabidopsis* occurs through biofilm formation.

Studies also document the importance of a 14 kDa calcium-binding protein involved in bacterial attachment to the plant roots, as evidenced by symbiosis existing between *Bradyrhizobium* sp. and peanut roots [45]. Rhizobial adhesion proteins (Rap) have been isolated from *Rhizobium leguminosarum* bv. *trifolii* [46], which include extracellular calcium-binding proteins like RapA1 that promote rhizobial autoaggregation, attachment and rhizosphere colonization [47]. Function of EPS in colonization and biofilm formation by *Rhizobium* sp. YAS34 on two nonlegume plant roots (*Arabidopsis thaliana* and *Brassica napus*) was determined, and it was revealed that EPS production is not necessarily essential for biofilm formation on biotic and abiotic surfaces [48]. Attachment of bacteria to the roots of higher plants has also been well documented by formation of biofilms [49].

18.4.3 Role of Root Exudates in Triggering Biofilm Formation

Major plant-derived factors responsible for affecting rhizosphere bacterial community and triggering root colonization are served by root exudates (REs) [50, 51]. Plants secrete around 10 to 44 percent of their photosynthates as root exudates, which tend to provide a rich nutrient source for the microorganisms encroaching toward root surfaces [50]. REs provide nutrients to the growing plant, thereby increasing plant growth, and also facilitate chemotaxis, thus enabling formation of a rhizosphere/rhizoplane community [52]. REs include low- and high-molecular-weight compounds secreted from plant roots, including amino acids, carboxylic acids, sugars, simple flavonoid type phenolics, polysaccharides, proteins, and lipids [53]. In addition to REs, certain compounds are also continuously exuded from plant roots into the rhizosphere, including free oxygen, water, enzymes, mucilage, and a wide range of carbonaceous primary and secondary metabolites. Walker et al. [54] have described the effective role of root exudates in biofilm associations in sweet basil roots, while Plotnikova et al. [55] have reported the same in *Arabidopsis*. A definite complicated correlation seems to exist between root exudates, biofilm formation, and chemotaxis, as represented in Figure 18.2.

18.4.4 Consequences of Root-Associated Biofilms on Plant Growth

In the rhizosphere, biofilms are formed on the surfaces of roots and soil particles, respectively, resulting in enhanced root colonization by the bacteria and cementing of soil particles. This can improve crop productivity and sustain physicochemical properties of soil. Biofilms can retain moisture and protect plant roots from a number of phytopathogens [56]. In case of leguminous hosts, biofilm mode of life offers a means for survival of bacteria as well as for establishment of symbiosis with the host legume [57]. Successful biofilm formation has been reported in *Mesorhizobium* species—*M. huakuii* and *M. tianshanense* [49, 58].

There has been a current upsurge targeting roles and functions of biofilms in the plant host system and in the rhizosphere region [57]. Matthysse [59] has published an extensive review targeting the attachment of *Agrobacterium* to plant surfaces. It is also documented that *A. tumefaciens* binds to the surfaces of inanimate objects, plants, and fungi, and is an excellent root-surface colonizer. They can also effectively bind to soil particles and to other abiotic surfaces [60].

Biofilm-inoculated plants showed significantly higher shoot and root nitrogen accumulation, indicating the efficacy of using nitrogen-fixing biofilms as inoculants that may promote soil nitrogen fertility and plant growth. Seneviratne and Jayasinghearachchi [61] elaborately studied the effect of application of the *Bradyrhizobium elkanii* SEMIA 5019-*Penicillium* spp. mix, which resulted in enhanced phosphate mineralization in addition to increased nitrogen availability in soil. Thus, an application of such a microbial association as a biofilm inoculum may prove to be effective for maintenance of soil fertility and survival of soil rhizobia in the absence of their hosts.

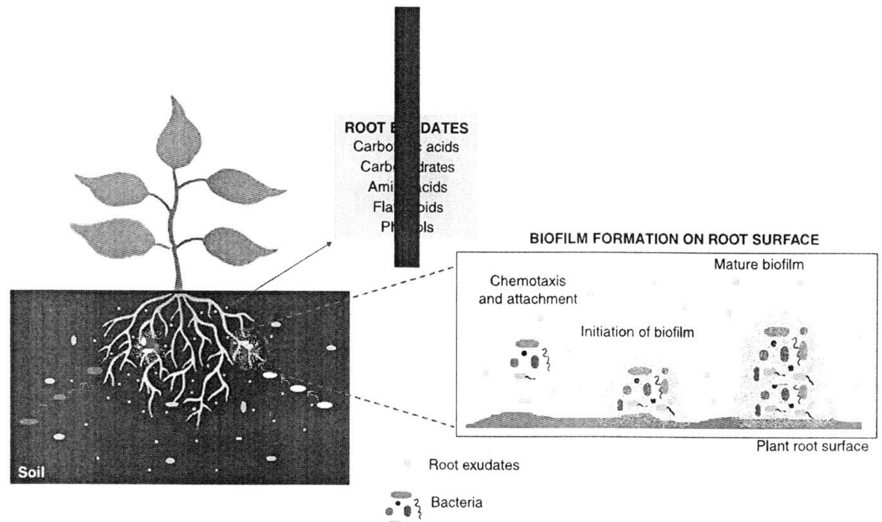

Figure 18.2 Intricate correlation between root exudates, biofilm formation, and chemotaxis. (*See color plate section for the color representation of this figure.*)

18.5 Bioremediation of Contaminants in Rhizospheric Soils

18.5.1 Rhizosphere, Rhizodeposition, and Bioremediation

Rhizosphere is *a nesting site for a rich and dynamic community of microorganisms*, a zone at the root soil interface controlled by root and rhizometabolites where an array of mutually interacting physico-chemical-biological intricacies operate [62]. During the last hundred years, attempts have been made to study this black box. Based on the proximity to the mature root, rhizosphere could be segregated in three different zones: endorhizosphere, rhizoplane, and ectorhizosphere. Out of these three zones, the endorhizosphere constitutes an inner important zone including cortex and endodermis, because microbes and cations (heavy metals) tend to occupy interstitial spaces. Hence, this zone plays a key role to study microbe-metal interactions and similar processes. Plant roots strongly influence the rhizosphere through rhizodeposition of root exudates, mucilage, and sloughed-off root cells [63]. Hiltner [62] has also presumed that if plants could attract beneficial bacteria in the rhizosphere, due to their rhizodepositions, they would also possess the ability to attract alien molecules, which would be treated similar to useful ones. This necessitates research on transformation of soil pollutants in rhizosphere [64]. Anderson et al. [65] asserted that in the rhizosphere, microbe-mediated humification processes might have an important influence on the persistence and bioavailability of toxicants in surface soils, and also further affirmed that numerous interacting factors such as moisture, pH, temperature, oxygen levels, bioavailability and bacterial nutrient requirements, microbiota, electron acceptors, microenvironment, nutrients, co-substrates, soil properties, and contaminant characteristics affect degradation of soil pollutants.

18.5.2 Bioremediation of Xenobiotics

Xenobiotic compounds comprise highly hazardous and persistent organic pollutants like hydrocarbons, polythene, phenols, and polychlorinated biphenyls. Although these are generally present in industrial effluents, nevertheless, they are capable of rapidly percolating through soil particles. In addition to these compounds, certain other xenobiotics like chemical fertilizers, herbicides, and pesticides largely contaminate agricultural soils [66]. Presence of these xenobiotic compounds in soils is a major environmental concern, as most of them are recalcitrant in nature and are carcinogenic even at very low concentrations [10]. Decontamination of xenobiotics-contaminated soils can be achieved through the promising application of bioremediation. Enhanced microbial activity in the rhizosphere could be associated with decontamination of soil polluted by polycyclic aromatic hydrocarbons (PAHs) and 2,4,5-trichlorophenoxyacetic acid [67]. *Pseudomonas* sp. and *Rhodococcus* sp. have been reported to degrade chlorinated aromatic compounds like 2,4-dichlorophenol (DCP), 2,4,6-trichlorophenol, 2,3,4,6-tetrachlorophenol, and pentachlorophenol [68, 69].

18.5.3 Bioremediation of Heavy Metal(loid)s

Elements (both metals and metalloids) with an atomic density greater than 6 g/cc excluding As, B, and Se are grouped under heavy metal(loid)s. They canopy both biologically essential (Co, Cu, Cr, Mn, and Zn) and nonessential (Cd, Pb, and Hg) elements

[70]. Essential elements are required in low concentrations and are henceforth called as *trace elements* or *micronutrients* [71]. Nonessential elements are toxic elements that are either phytotoxic and/or zootoxic. Soil harbors these nutrients required for plant growth, but it also acts as a sink for the removal of contaminants from waste materials and polluted areas. Common heavy metal contaminants in the soil include Cr, Cd, Fe, and Ni. Indiscriminate and uncontrolled discharge of industrial and urban wastes into the environmental sink has led to disturbances in soil ecosystems limiting crop production [72]. Safe disposal of such wastes generated due to agricultural and industrial activities is needed for waste management as heavy metal(loid)s are reaching the food chain. These metal(loid)s are not chemically degraded, and hence they persist for a long time, posing severe threats to soil fertility and environmental health sustainability [73].

Bioremediation of heavy metal contaminants from the polluted ecosystems with the help of microorganisms is again a natural process. Response of microbial communities to heavy metals depends on the concentration and availability of heavy metals and is an intricate process controlled by metal type, nature of medium, and microbial species [71]. Biofilm developed from consortium of *Bacillus subtilis* and *B. cereus* on coarse sand has been reported to efficiently remove Cr^{3+} ions [74].

Transformation of heavy metal(loid)s in soils involves some general mechanisms leading to either retention (mediated by sorption, precipitation and complexation reactions) or loss (via plant uptake, leaching, and volatilization) of these elements in the soil. Microorganisms in soils and sediments act as biologically active methylators, which induce bio-methylation of toxic elements into methyl derivatives that are subsequently removed by volatilization [75].

18.5.4 Rhizobacteria Facilitating Bioremediation

PGPR can improve host plant growth and development in heavy metal contaminated soils by decreasing the toxic effects of heavy metals on plants [76]. Sufficiently high concentration of heavy metals in the environment can prove to be toxic even for the metal-accumulating and metal-tolerant plants. This can be exemplified to iron deficiency in range of different plant species [77, 78] in heavy metal–contaminated soils. If plants with low Fe content are grown in the presence of high levels of heavy metals, then these plants tend to be chlorotic, as Fe deficiency inhibits both chloroplast development and chlorophyll biosynthesis [79]. If microbial Fe-siderophore complex can be taken up by plants, then they can serve as efficient Fe source for plant growth and development [80].

Henceforth, prevention of plants from becoming chlorotic in presence of high levels of heavy metals could be done by associating them with siderophore-producing bacteria [81]. PGPR can significantly enhance plant growth in the presence of heavy metals, including Ni, Pb, and Zn [82, 83]. Cr is a highly toxic nonessential transition metal for microorganisms and plants [84]. Scientific evidences confirm the presence of Cr in the soil environment and indicate its role in selection of microbial and plant variants that can tolerate high level of Cr compounds [85, 86]. Burd et al. [82] studied Ni resistant ACC deaminase containing PGPR, which could decrease Ni toxicity to canola plants. Several reports speculate that inoculation of PGPR facilitates plant growth and enhanced metal(loid)s uptake by host plant [87].

18.5.5 Metal Accumulating Rhizobacteria

Certain plants involve the use of PGPR to mitigate the deleterious effects of heavy metals [81]. Metal accumulating and hyperaccumulating plants harbor rhizobacteria, which play a crucial role in tolerance to and uptake of heavy metals. Thus, such rhizobacteria facilitate mobilization or immobilization of heavy metals [88]. In the similar context, chromium-resistant pseudomonads (isolated from paint industry effluents) were able to stimulate seed germination and growth of wheat in the presence of potassium dichromate [89], wherein the enhancement of seedling growth due to bacteria was associated with reduced chromium uptake. de Souza et al. [90] reported the involvement of rhizospheric bacteria in accumulation of potentially toxic trace elements into plant tissues.

Earlier field studies have demonstrated that elevated metal loadings in metal-contaminated soils can lead to mitigated microbial size [91]. On the contrary, recent research indicates that some elevated levels of heavy metals in soil exert significant impact on size, structure, and overall activity of the soil microbial communities, resulting in significantly higher population of rhizobacteria than that in the bulk soil [81, 92]. Several scientists have reported the existence of a high proportion of metal-resistant bacteria in the rhizosphere of the hyperaccumulators *Thalaspi caerulescens* [93], *Alyssum bertolonii* [94], and *Alyssum murale* [95] grown in soil contaminated with Zn and Ni, respectively.

18.5.6 Role of Root Exudates in Heavy Metal Decontamination and Degradation of Organic Pollutants

Plants secrete certain metabolites through their roots as root exudates and select beneficial microbial communities. Components of the REs may come from stable complexes with metal cations in the soil solution matrix and can also modify metal availability in the rhizosphere [96]. Carboxylates present in REs (malate, citrate, and oxalate) are known to play an important role in influencing metal biogeochemical cycles as they form complexes with metal(loid)s and are judiciously involved in ligand exchange [97]. During plant growth, roots secrete mucilage, which additionally aids in attracting an array of beneficial microorganisms, along with root pathogens and also sequesters metal(loid)s [98].

Several studies indicate higher solubility of heavy metal(loid)s *viz.*, As, Cu, Zn, and Pd in soil with plants than in control soil without plants [99, 100]. In soil contaminated by metal(loid)s, plant growth is stunted and root development is poor, leading to limited phytoaccumulation of metal(loid)s. In order to circumvent these obstacles in rhizosphere, microbial activity improvement and organic amendments could be a rescuing force.

Bioremediation using plants and associated rhizospheric exudates to decontaminate polluted soil is an important method that utilizes the catabolic potential of rhizosphere microbiota, which are supported by a pool of organic substrates in root exudates [101]. Plants used for bioremediation enhance degradation of organic contaminants in soil, either by increasing the number of sites in the organic matrix available for contaminant absorption or by releasing REs that not only support and stimulate the growth of microorganisms actively involved in degradation but also enhance contaminant bioavailability by promoting their desorption from soil matrix [102]. By providing excellent carbonaceous primary substrates, the REs can also change the structure and function of rhizosphere microbial communities and thereby aid in co-metabolic degradation of pollutants [102].

Chaineau et al. [103] reported that root exudates of corn grown in hydroponic culture tend to change hydrocarbon bioavailability in solution or stimulate degradative microbial populations that enhance biodegradation of petroleum hydrocarbons. Corgie et al. [104] studied the biodegradation rate of phenanthrene in the rhizosphere and observed a strong degradation gradient existing toward the roots that could be attributed to the presence of root exudates, which, along with phenanthrene, induced spatial modification of degrading bacterial communities.

Thus, knowledge regarding root exudation by different plants and their qualitative composition might aid in formulating these compounds for soil application and to mimic rhizosphere environment for decontamination of polluted soils, as a part of sustainable agriculture.

18.6 Implications of Rhizospheric Biofilm Formation on Bioremediation

Plants actively select bacteria from rhizosphere and aid in creating a habitat more favorable for the growing plant. Some of these selected bacteria get aggregated into microcolonies of clonal cells and finally into mature and robust biofilms on the root surface [105, 106]. Although plant-beneficial bacteria get established on plant roots effectively, bacterial strains exhibit a higher preference to their native host plants [107], improving plant growth. Heterogeneous bacterial colonization has also been reported in rhizosphere, which interact metabolically and might confer several selective advantages to the constituent bacteria [9].

Microbial biofilm developed in the rhizosphere and roots is known to alter parameters of the soil environment (e.g., reduction/oxidation reactions, moisture, and aeration), which favor degradation of pollutants [2]. Environmental stress tolerance, communication through quorum sensing, capability of horizontal gene transfer, metabolic diversity to utilize waste products/pollutants, redox and electron acceptor diversity enabling diverse metabolic functions involving electron acceptor reduction, porous physical structure with water channels allowing free movement of nutrients/metabolites, and production of surfactants aiding in solubilizing hydrophobic xenobiotics are a few of the major characteristics of rhizospheric biofilms that are associated with bioremediation in plant root–associated soils [66].

Under limiting conditions of carbon and energy sources, chemotactic ability of biofilm-forming bacteria guides them to move toward surface-adsorbed nutrients (or hydrophobic pollutants), supporting bacterial survival and enhancing xenobiotic pollutant degradation in contaminated sites [10]. The dense population of bacteria and packed structure of biofilms provide enhanced opportunities for the horizontal transfer of genes encoding bioremediation-related traits such as heavy metal resistance and pollutant degradation [108]. Current advances in bioremediation field are initiating footsteps to successfully study rhizospheric biofilm formation and highlight their effective role in bioremediation. Several biofilm-forming soil bacteria have been documented to be potential bioremediation agents, as summarized in Table 18.1.

Table 18.1 Biofilm-Forming Soil Bacteria with Bioremediation Capability.

Bacterium	Relevant Features
Acinetobacter calcoaceticus	PGPR, bioremediation of hydrocarbons
Bacillus cereus	PGPR, Cr remediation
Bacillus subtilis	PGPR, Cr remediation
Burkholderia vietnamiensis	Toluene bioremediation
Kluyvera ascorbata	Resistance to Ni, Cr, Pb, Zn
Paenibacillus lentimorbus	Heavy metal tolerance
Pseudomonas flourescens	PGPR, resistant to Zn, Cd
Pseudomonas putida	Bioremediation of Zn, DCP
Pseudomonas tolaasii	PGPR, resistant to Zn, Cd
Rhizobium alamii	Heavy metal tolerance
Rhodococcus ruber	Polythene bioremediation
Variovorax paradoxus	PGPR, Cd tolerant

Compiled from Angus and Hirsch [29]; Jing et al. [81]; Das et al. [109].

Rhizospheric soil biofilms may be successively employed in breaking down several classes of xenobiotics. Diverse communities containing biofilm-forming bacteria are efficient degraders as observed in case of PAHs, which are the most common environmental pollutants [110], whereas multispecies diversity in soil biofilms exhibited higher biodegradation efficiency than the monospecies biofilms.

18.7 Conclusion and Future Prospects

The hallmark of this chapter is the importance of plant root–associated biofilms in bioremediation, which offers an eco-friendly approach for the remediation of polluted soils and aids in sustainable agriculture. Biofilms are microbial assemblages that induce plant growth and protect plants through plant–microbial interactions, which play a vital role in soil ecosystems. Plant roots secrete a variety of compounds, including REs, that help in selection of biofilm-forming microorganisms, supporting them to compete and thrive in the rhizospheric niche. Bioremediation, a process for conversion of toxic and hazardous contaminants in soil/water to nontoxic/nonhazardous waste products, exploits utilization of specific microorganisms and/or plants.

Employing biofilm-forming microorganisms for bioremediation of polluted soil ecosystems promises to be a proficient, cost-effective, safer, and greener strategy for eliminating heavy metals and xenobiotic compounds from polluted soil environments. Strain improvement of biofilm-forming microorganisms could be achieved by engineering metabolic pathways and degradative enzymes or by enhancing the copy number of genes involved in degradation that would finally enhance biofilm-mediated bioremediation. Similarly, genetically engineered biofilm-forming bacterial strains produced by introduction of metal transformation activities could also be an eco-friendly approach to ameliorate heavy metal contamination from soil. In the same context, the use of specific consortia of biofilm-forming microorganisms could be a boon for bioremediation of hydrocarbons, organic pollutants, pesticides, and heavy metals. Another promising technology for reclaming polluted soils would be the introduction of biofilm-forming and contaminant-degrading bacteria at the contaminated sites.

Acknowledgments

Authors would like to earnestly thank their mentor, Prof. Appa Rao Podile, Vice Chancellor, University of Hyderabad, for his incessant support and encouragement.

References

1 G.A. Kowalchuk, E. Yergeau, J.H.J. Leveau, A. Sessitch and M. Bailey, Plant-associated microbial communities, *in* Environmental Molecular Microbiology, W.T. Liu and J.K. Jansson (Eds), Caister Academic Press, Norwich, UK, 2010.
2 D. Pearce, M.J. Bazin and J.M. Lynch, The rhizosphere as a biofilm, *in* Microbial Biofilms, H.M. Lappin-Scott and J.W. Costerton (eds.), Cambridge University Press, New York, 1995.
3 C. Su, L. Jiang and W. Zhang, A review on heavy metal contamination in the soil worldwide: Situation, impact and remediation techniques, *Environ. Skeptics Critics*, 3(2), 24–38 (2014).
4 O. Murali and S.K. Mehar, Bioremediation of heavy metals using spirulina, *Int. J. Geol. Earth Environ. Sci.*, 4(1), 244–249 (2014).
5 M. Burmolle, L.H. Hansen and J. Sorensen, Establishment and early succession of a multispecies biofilm composed of soil bacteria, *Microb. Ecol.*, 54, 352–362 (2007).
6 H.C. Flemming and J. Wingender, The biofilm matrix, *Nat. Rev. Microbiol.* 8, 623–633 (2010).
7 J.W. Costerton, G.G. Geesey and K.J. Cheng, How bacteria stick, *Sci. Am.*, 238, 86–95 (1978).
8 M. Vert, Y. Doi, K.H. Hellwich, M. Hess, P. Hodge, P. Kubisa, et al., Terminology for biorelated polymers and applications (IUPAC Recommendations 2012), *Pure Appl. Chem.*, 84(2): 377–410 (2012).
9 J. Wimpenny, W. Manz and U. Szewzyk, Heterogeneity in biofilms, *FEMS Microbiol. Rev.*, 24(5), 661–671 (2000).
10 R. Singh, D. Paul and R.K. Jain, Biofilms: implications in bioremediation, *Trends Microbiol.*, 14(9), 389–397 (2006).
11 F. Ahimou, M.J. Semmens, G. Haugstad and P.J. Novak, Effect of protein, polysaccharide, and oxygen concentration profiles on biofilm cohesiveness, *Appl. Environ. Microbiol.*, 73, 2905–2910 (2007).
12 I.W. Sutherland, Biofilm exopolysaccharides: A strong and sticky framework, *Microbiology*, 147, 3–9 (2001).
13 P. Gilbert, J. Das and I. Foley, Biofilms susceptibility to antimicrobials, *Adv. Dent. Res.*, 11, 160–167(1997).
14 H.C. Flemming, Sorption sites in biofilm, *Wat. Sci. Tech.*, 32(8), 27–33 (1995).

15 D. Tielker, S. Hacker, R. Loris, M. Strathmann, J. Wingender, S. Wilhelm, et al., *Pseudomonas aeruginosa* lectin LecB is located in the outer membrane and is involved in biofilm formation, *Microbiology*, 151, 1313–1323 (2005).
16 S. Vilain, J.M. Pretorius, J. Theron and V.S. Brozel, DNA as an adhesin: *Bacillus cereus* requires extracellular DNA to form biofilms, *Appl. Environ. Microbiol.* 75, 2861–2868 (2009).
17 H.C. Flemming and J. Wingender, Extracellular polymeric substances: structure, ecological functions, technical relevance, *in* Encyclopedia of Environmental Microbiology, Vol. 3, G. Bitton (ed.), Wiley, New York, 2002.
18 R. Morgan, S. Kohn, S.H. Hwang, D.J. Hassett and K. Sauer, BdlA, a chemotaxis regulator essential for biofilm dispersion in *Pseudomonas aeruginosa*, *J. Bacteriol.*, 188(21), 7335–7343 (2006).
19 R. Singh, P. Singh and R. Sharma, Microorganism as a tool of bioremediation technology for cleaning environment: A review, *Proc. Int. Acad. Ecol. Environ. Sci.* 4(1), 1–6 (2014).
20 J.L. Morel, G. Echevarria and N. Goncharova (eds.), Phytoremediation of Metal Contaminated Soils, IOS Press, Amsterdam, and Springer in conjunction with the NATO Public Diplomacy Division–Springer, 2002.
21 P. Ziarati and S. Alaedini, The phytoremediation technique for cleaning up contaminated soil by *Amaranthus* sp., *Environ. Anal. Toxicol.*, 4(2), 208 (2014).
22 M.I. Lone, Z.H. He, J. Stoffella and X. Yang, Phytoremediation of heavy metal polluted soils and water: progresses and perspectives, *J. Zhejiang Univ. Sci. B.* 9, 210–220 (2008).
23 H.B. Bradl, Heavy Metals in the Environment: Origin, Interaction and Remediation. Elsevier Academic Press, Amsterdam/Boston, 2005.
24 F. Kamilova, L.V. Kravchenko, A.I. Shaposhnikov, T. Azarova, N. Makarova and B. Lugtenberg, Organic acids, sugars, and L-tryptophane in exudates of vegetables growing on stonewool and their effects on activities of rhizosphere bacteria, *Mol. Plant Microbe Interact.*, 19, 250–256 (2006).
25 X. He, W. Chang, D.L. Pierce, L.O. Seib, J. Wagner and C. Fuqua, Quorum sensing in *Rhizobium* sp. strain NGR234 regulates conjugal transfer (*tra*) gene expression and influences growth rate, *J. Bacteriol.*, 185(3), 809–822 (2003).
26 C.E. Morris and J.M. Monier, The ecological significance of biofilm formation by plant-associated bacteria, *Annu. Rev. Phytopathol.*, 41, 429–453 (2003).
27 G.V. Bloemberg and B.J.J. Lugtenberg, Bacterial biofilms on plants: relevance and phenotypic aspects, *in* Microbial Biofilms, M. Ghannoum and G.A. O'Toole (Eds), ASM Press, Washington, 2004.
28 T.R. De Kievit, R. Gillis, S. Marx, C. Brown and B.H. Iglewski, Quorum-sensing genes in *Pseudomonas aeruginosa* biofilm: their role and expression patterns, *Appl. Environ. Microbiol.*, 67, 1865–1873 (2001).
29 A.A. Angus and A.M. Hirsch, Biofilm Formation in the Rhizosphere: Multispecies Interactions and Implications, *in* Molecular Microbial Ecology of the Rhizosphere, Volume 2, F.J. de Bruijn (ed.), John Wiley & Sons, Inc., 2013.
30 J.W. Kloepper and M.N. Schroth, Plant growth-promoting rhizobacteria on radishes, *in* Station de Pathologie Végétale et Phyto-Bactériologie (ed.), Proceedings of the 4th International Conference on Plant Pathogenic Bacteria, vol. II. Gilbert-Clarey, Tours, France, 1978.
31 S. Dutta and A.R. Podile, Plant Growth Promoting Rhizobacteria (PGPR): the bugs to debug the root zone, *Crit. Rev. Microbiol.*, 36(3), 232–244 (2010).

32 G.V. Bloemberg and B.J. Lugtenberg, Molecular basis of plant growth promotion and biocontrol by rhizobacteria, *Curr. Opin. Plant Biol.*, 4(4), 343–350 (2001).
33 S. Saleh-Lakha and B.R. Glick, Plant growth-promoting bacteria, *in* Modern Soil Microbiology, J.D. van Elsas, J.K. Jansson and J.T. Trevors (eds.), CRC/Thomson Publishing, Boca Raton, FL/UK, 2006.
34 B.E. Ramey, M. Koutsoudis, S.B. von Bodman and C. Fuqua, Biofilm formation in plant–microbe associations, *Curr. Opin. Microbiol.*, 7, 602–609 (2004).
35 N.R. Stanley and B.A. Lazazzera, Environmental signals and regulatory pathways that influence biofilm formation, *Mol. Microbiol.*, 52, 917–924 (2004).
36 T. Rudrappa, M.L. Biedrzycki, H.P. Bais, Causes and consequences of plant- associated biofilms, *FEMS Microbiol. Ecol.*, 64, 153–166 (2008).
37 N.R. Stanley and B.A. Lazazzera, Defining the genetic differences between wild and domestic strains of *Bacillus subtilis* that affect poly-γ-DL-glutamic acid production and biofilm formation, *Mol. Microbiol.*, 57, 1143–1158 (2005).
38 J.A. Downie, The roles of extracellular proteins, polysaccharides and signals in the interactions of rhizobia with legume roots, *FEMS Microbiol. Rev.*, 34(2), 150–170 (2010).
39 P.B. Beauregard, Y. Chai, H. Vlamakis, R. Losick and R. Kolter, *Bacillus subtilis* biofilm induction by plant polysaccharides, *Proc. Natl. Acad. Sci. USA*, 110(17), E1621–E1630 (2013).
40 T. Rudrappa, W.J. Quinn, N.R. Stanley-Wall and H.P. Bais, A degradation product of the salicylic acid pathway triggers oxidative stress resulting in down-regulation of *Bacillus subtilis* biofilm formation on *Arabidopsis thaliana* roots, *Planta*, 226, 283–297 (2007).
41 S. Ude, D.L. Arnold, C.D. Moon, T. Timms-Wilson and A.J. Spiers, Biofilm formation and cellulose expression among diverse environmental *Pseudomonas* isolates, *Environ. Microbiol.*, 8, 1997–2011 (2006).
42 G.V. Bloemberg, A.H.M. Wijfjes, G.E.M. Lamers, N. Stuurman and B.J.J. Lugtenberg, Simultaneous imaging of *Pseudomonas fluorescens* WCS365 populations expressing three different autofluorescent proteins in the rhizosphere: new perspectives for studying microbial communities, *Mol. Plant Microbe Interact.*, 13, 1170–1176 (2000).
43 M.R. Parsek and C. Fuqua, Biofilms 2003: Emerging themes and challenges in studies of surface-associated microbial life, *J. Bacteriol.* 186, 4427–4440 (2004).
44 M. Espinosa-Urgel, R. Kolter and J.L. Ramos, Root colonization by *Pseudomonas putida*: love at first sight, *Microbiology*, 148, 341–343 (2002).
45 M. Dardanelli, J. Angelini and A. Fabra, A calcium-dependent bacterial surface protein is involved in the attachment of rhizobia to peanut roots, *Can. J. Microbiol.*, 49, 576–582 (2003).
46 N. Ausmees, K. Jacobsson and M. Lindberg, A unipolarly located, cell-surface-associated agglutinin RapA belongs to a family of Rhizobium-adhering proteins (Rap) in *Rhizobium leguminosarum bv. trifolii, Microbiology*, 147, 549–559 (2001).
47 E.J. Mongiardini, N. Ausmees, J. Pérez-Giménez, M.J. Althabegoiti, J.I. Quelas, S.L. López-Garcia, et al., The rhizobial adhesion protein RapA1 is involved in adsorption of rhizobia to plant roots but not in nodulation, *FEMS Microbiol. Ecol.*, 65, 279–288 (2008).
48 C. Santaella, M. Schue, O. Berge, T. Heulin and W. Achouak, The exopolysaccharide of *Rhizobium* sp. YAS34 is not necessary for biofilm formation on *Arabidopsis thaliana* and *Brassica napus* roots but contributes to root colonization, *Environ. Microbiol.*, 10, 2150–2163 (2008).

49 P. Wang, Z. Zhong, J. Zhou, T. Cai and J. Zhu, Exopolysaccharide biosynthesis is important for *Mesorhizobium tianshanense*: Plant host interaction, *Arch. Microbiol.*, 189, 525–530 (2008).

50 B.J.J. Lugtenberg, L.V. Kravchenko and M. Simons, Tomato seed and root exudate sugars: composition, utilization by *Pseudomonas* biocontrol strains and role in rhizosphere colonization, *Environ. Microbiol.* 1, 439–446 (1999).

51 T.S. Walker, H.P. Bais, E. Grotewold and J.M. Vivanco, Root exudation and rhizosphere biology, *Plant Physiol.*, 132, 44–51 (2003).

52 R. Kumar, R. Bhatia, K. Kukreja, R.K. Behl, S.S. Dudeja and N. Narula, Establishment of *Azotobacter* on plant roots: chemotactic response, development and analysis of root exudates of cotton (*G. hirusitum* L.) and wheat (*T. aestivum* L.), *J. Basic Microbiol.* 47, 436–439 (2007).

53 N.C. Uren, Type, amounts, and possible functions of compounds released into the rhizosphere by soil-grown plants, *in* The rhizosphere: Biochemistry and organic substances at the soil-plant interface, R. Pinton, Z. Varanini and P. Nannipieri (Eds.), CRC Press, Boca Raton, FL, 2007.

54 T.S. Walker, H.P. Bais, E. Deziel, H.P. Schweitzer, L.G. Rahme, R. Fall, et al., *Pseudomonas aeruginosa*-plant root interactions: Pathogenicity, biofilm formations, and root exudation, *Plant Physiol*, 134, 3210–3331 (2004).

55 J.M. Plotnikova, L.G. Rahme, and F.M. Ausubel, Pathogenesis of the human opportunistic pathogen *Pseudomonas aeruginosa* PA14 in *Arabidopsis*, *Plant Physiol.*, 124, 1766–1774 (2000).

56 A.W. Qurashi and A.N. Sabri, Bacterial exopolysaccharide and biofilm formation stimulate chickpea growth and soil aggregation under salt stress, *Braz. J. Microbiol.*, 43(3), 1183–1191 (2012).

57 L.V. Rinaudi and W. Giordano, An integrated view of biofilm formation in rhizobia. *FEMS Microbiol. Lett.*, 304, 1–11 (2010).

58 H. Wang, Z. Zhong, T. Cai, S. Li and J. Zhu, Heterologous overexpression of quorum-sensing regulators to study cell density-dependent phenotypes in a symbiotic plant bacterium *Mesorhizobium huakuii*, *Arch. Microbiol.*, 182, 520–525 (2004).

59 A.G. Matthysse, Attachment of *Agrobacterium* to plant surfaces, *Front. Plant. Sci.*, 5, 252 (2014).

60 A.D. Tomlinson and C. Fuqua, Mechanisms and regulation of polar surface attachment in *Agrobacterium tumefaciens*, *Curr. Opin. Microbiol.*, 12, 708–714 (2009).

61 G. Seneviratne and H.S. Jayasinghearachchi, A rhizobial biofilm with nitrogenase activity alters nutrient availability in a soil. *Soil Biol. Biochem.*, 37, 1975–1978 (2005).

62 L. Hiltner, Über neuere Erfahrungen und Probleme auf dem Gebiete der Bodenbakteriologie unter bessonderer Berücksichtigung der Gründung und Brache [About our experiences and problems in the area of soil bacteriology under special regards of cultivated and uncultivated land] *Arbeiten der Deutschen Landwirtschaftlichen Gesellschaft*, 98, 59–78 (1904).

63 V.C. Coats and M.E. Rumpho, The rhizosphere microbiota of plant invaders: an overview of recent advances in the microbiomics of invasive plants, *Frontiers Microbiol.* 5, 368 (2014).

64 S. Yang, S. Xing, C. Liu, Z. Du, H. Wang and Y. Xu, Effects of root pruning on the vegetative growth and fruit quality of Zhanhuadongzao trees, *Hort. Sci.*, 1, 14–21 (2010).

65 T.A. Anderson, E.A. Guthrie and B.T. Walton, Bioremediation in the rhizosphere: plant roots and associated microbes clean contaminated soil. *Environ. Sci. Technol.*, 27(13), 2630–2636 (1993).

66 S.J. Edwards and B.V. Kjellerup, Applications of biofilms in bioremediation and biotransformation of persistent organic pollutants, pharmaceuticals/personal care products, and heavy metals, *Appl. Microbiol. Biotechnol.*, 97, 9909–9921 (2013).

67 J.J. Boyle and J.R. Shann, The influence of planting and soil characteristics on mineralization of 2,4,5-T in rhizosphere soil, *J. Environ. Qual.*, 27, 704–709 (1998).

68 J.A. Puhakka, E.S. Melin, K.T. Järvinen, P.M. Koro, J.A. Rintala, P. Hartikainen, et al., Fluidized-bed biofilms for chlorophenol mineralization, *Wat. Sci. Technol.*, 42, 227–235 (1995).

69 F. Kargi and S. Eker, Removal of 2,4-dichlorophenol and toxicity from synthetic wastewater in a rotating perforated tube biofilm reactor, *Process Biochem.*, 40, 2105–2111 (2005).

70 B. Seshadri, N.S. Bolan and R. Naidu, Rhizosphere-induced heavy metal(loid) transformation in relation to bioavailability and remediation, *J. Plant Nutr. Soil Sci.*, 15(2), 524–548 (2015).

71 G. Girma, Microbial bioremediation of some heavy metals in soils: an updated review, *Indian J. Sci. Res.*, 6(1), 147–161 (2015).

72 S.P. Kamaludeen, M. Megharaj, R. Naidu, I. Singleton, A.L. Juhasz and B.G. Hawke, Microbial activity and phospholipid fatty acid pattern in long-term tannery waste-contaminated soil, *Ecotoxicol. Environ. Saf.*, 56, 302–310 (2003).

73 D.C. Adriano, W.W. Wenzel, J. Vangronsveld and N.S. Bolan, Role of assisted natural remediation in environmental cleanup, *Geoderma*, 122, 121–142 (2004).

74 K. Sundar, A. Mukherjee, M. Sadiq and N. Chandrasekaran, Cr (III) bioremoval capacities of indigenous and adapted bacterial strains from Palar river basin, *J. Hazard. Mater.*, 187, 553–561 (2011).

75 W.T. Frankenberger Jr. and M. Arshad, Bioremediation of selenium-contaminated sediments and water, *Biofactors*, 14, 241–254 (2001).

76 A.A. Belimov, A.M. Kunakova, V.I. Safronova, V.V. Stepanok, L.Y. Yudkin, Y.V. Alekseev, et al., Employment of rhizobacteria for the inoculation of barley plants cultivated in soil contaminated with lead and cadmium, *Microbiology (Moscow)*, 73(1), 99–106 (2004).

77 A. Wallace, G.A. Wallace and J.W. Cha, Some modifications in trace elements toxicities and deficiencies in plants resulting from interactions with other elements and chelating agents—The special case of iron, *J. Plant Nutr.*, 15(2), 1589–1598 (1992).

78 J.F. Ma and K. Nomoto, Inhibition of mugineic acid-ferric complex in barley by copper, zinc and cobalt, *Physiologia Plantarum*, 89(2), 331–334 (1993).

79 J. Imsande, Iron, sulfur, and chlorophyll deficiencies: a need for an integrative approach in plant physiology, *Physiologia Plantarum*, 103(1), 139–144 (1998).

80 Y. Wang, H.N. Brown, D.E. Crowley and P.J. Szaniszlo, Evidence for direct utilization of a siderophore, ferroxamine B, in axenically grown cucumber, *Plant Cell Environ.*, 16(5), 579–585 (1993).

81 Y. Jing, Z. He and X. Yang, Role of soil rhizobacteria in phytoremediation of heavy metal contaminated soils, *J. Zhejiang Univ. Sci. B.*, 8, 192–207 (2007).

82 G.I. Burd, D.G. Dixon and B.R. Glick, A plant growth-promoting bacterium that decreases nickel toxicity in seedlings, *Appl. Environ. Microbiol.*, 64(3), 3663–3668 (1998).

83 G.I. Burd, D.G. Dixon and B.R. Glick, Plant growth-promoting bacteria that decrease heavy metal toxicity in plants, *Can. J. Microbiol.*, 46(3), 237–245 (2000).

84 H. Oliveira, Chromium as an environmental pollutant: insights on induced plant toxicity, *J. Bot.*, doi:10.1155/2012/375843 (2012).

85 C. Cervantes, J. Campos-García, S. Devars, F. Gutiérrez-Corona, H. Loza-Tavera, J.C. Torres-Guzmán, et al., Interactions of chromium with microorganisms and plants, *FEMS Microbiol. Rev.*, 25(3), 335–347 (2001).

86 R.O. Castro, M.M. Trujillo, J.L. Bucio, C. Cervantes and J. Dubrovsky, Effects of dichromate on growth and root system architecture of *Arabidopsis thaliana* seedlings, *Plant Sci.*, 172, 684–691 (2007).

87 S.K. Shukla, R.K. Mishra, M. Pandey, V. Mishra, A. Pathak, A. Pandey, et al., Land reformation using plant growth promoting rhizobacteria in context of heavy metal contamination, *in* Plant metal interaction emerging remediation techniques, P. Ahmad (Ed), Elsevier Inc., Netherlands, 2016.

88 G.M. Gadd, Metals, minerals and microbes: geomicrobiology and bioremediation, *Microbiology*, 156, 609–643 (2010).

89 S. Hasnain and A.N. Sabri, Growth stimulation of *Triticum Aestivum* seedlings under Cr-stresses by non rhizospheric pseudomonad strains, *In* Abstracts of the 7th International Symposium on Biological Nitrogen Fixation with Non-Legumes, Kluwer Academic Publishers, Netherlands, 1996.

90 M.P. de Souza, C.P. Huang, N. Chee and N. Terry, Rhizosphere bacteria enhance the accumulation of selenium and mercury in wetland plants, *Planta*, 209(2), 259–263 (1999).

91 A. Konopka, T. Zakharova, M. Bischoff, L. Oliver, C. Nakatsu and R.F. Turco, Microbial biomass and activity in lead-contaminated soil, *Appl. Environ. Microbiol.*, 65(5), 2256–2259 (1999).

92 R.A. Abou-Shanab, H. Ghozlan, K. Ghanem and H. Moawad, Behaviour of bacterial populations isolated from rhizosphere of *Diplachne fusca* dominant in industrial sites, *World J. Microbiol. Biotechnol.*, 21, 1095–1101 (2005).

93 T.A. Delorme, J.V. Gagliardi, J.S. Angle and R.L. Chaney, Influence of the zinc hyperaccumulator *Thalaspi caerulescens* J. and C. Presl and the nonmetal accumulator *Trifolium pratense* L. on soil microbial populations, *Can. J. Microbiol.*, 47(8), 773–776 (2001).

94 A. Mengoni, R. Barzanti, C. Gonnelli, R. Gabbrielli and M. Bazzicalupo, Characterization of nickel-resistant bacteria isolated from serpentine soil, *Environ. Microbiol.*, 3(11), 691–698 (2001).

95 R.A. Abou-Shanab, J.S. Angle, T.A. Delorme, R.L Chaney, P. van Berkum, H. Moawad, et al., Rhizobacterial effects on nickel extraction from soil and uptake by *Alyssum murale*, *New Phytol.*, 158(1), 219–224 (2003).

96 D.L. Jones, Organic acids in the rhizosphere—a critical review, *Plant Soil*, 205, 25–44 (1998).

97 P. Hinsinger, Bioavailability of soil inorganic P in the rhizosphere as affected by root-induced chemical changes: a review, *Plant Soil*, 237, 173–195 (2001).

98 M.C. Hawes, U. Gunawardena, S. Miyasaka and X. Zhao, The role of root border cells in plant defense, *Trends Plant Sci.*, 5, 128–133 (2000).

99 G.S. Banuelos and Z.Q. Lin, Acceleration of selenium volatilization in seleniferous agricultural drainage sediments amended with methionine and casein, *Environ. Pollut.*, 150(3), 306–312 (2007).

100 N.S. Bolan, G. Choppala, A. Kunhikrishnan, J.H. Park and R. Naidu, Microbial transformation of trace elements in soils in relation to bioavailability and remediation, *Rev. Environ. Contam. Toxicol.*, 225, 1–56 (2013).

101 P. Dundek, L. Holik, L. Hromádko, T. Rohlík, V. Vranová, K. Rejšek, et al., Action of plant root exudates in bioremediations: a review, *Acta. Univ. Agric. et Silvic. Mendel Brun.*, 59(1), 303–308 (2014).

102 K.J. Yoshitomi and J.R. Shann, Corn (*Zea mays* L.) root exudates and their impact on ^{14}C-pyrene mineralization, *Soil Biol. Biochem.*, 33, 1769–1776 (2001).

103 C.H. Chaineau, J.L. Morel and J. Oudot, Biodegradation of fuel oil hydrocarbons in the rhizosphere of maize, *J. Env. Qual.*, 29, 569–578 (2000).

104 S.C. Corgie, T. Beguiristain and C. Leyval, Spatial distribution of bacterial communities and phenanthrene degradation in the rhizosphere of *Lolium perenne* L., *Appl. Environ. Microbiol.*, 70(6), 3552–3557 (2004).

105 P. Garbeva, J.A. van Veen and J.D. van Elsas, Microbial diversity in soil: selection of microbial populations by plant and soil type and implications for disease suppressiveness, *Ann. Rev. Phytopathol.*, 42, 243–270 (2004).

106 T. Rudrappa, M.L. Biedrzycki, S.G. Kunjeti, N.M. Donofri, K.J. Czymmek, P.W. Pare, et al., The rhizobacterial elicitor action induces system resistance in *Arabidopsis thaliana*, *Integr. Biol.*, 3, 130–138 (2010).

107 M. Bacilio-Jimenez, S. Aguilar-Flores, E. Ventura-Zapata, E. Pérez-Campos, S. Bouquelet and E. Zenteno, Chemical characterization of root exudates from rice (*Oryza sativa*) and their effect on the chemotactic response of endophytic bacteria, *Plant Soil*, 249(2), 271–277 (2003).

108 C.A. Fux, J.W. Costerton, P.S. Stewart and P. Stoodley, Survival strategies of infectious biofilms, *Trends Microbiol.*, 13, 34–40 (2005).

109 N. Das, V. Lakshmi, G. Basak, J.A. Salam, E. Alice and M. Abigail, Application of biofilms on remediation of pollutants – an overview, *J. Microbiol. Biotech. Res.*, 2(5), 783–790 (2012).

110 A.R. Johnsen, L.Y. Wick and H. Harms, Principles of microbial PAH-degradation in soil, *Environ. Pollut.*, 133, 71–84 (2005).

19

Biofilms for Remediation of Xenobiotic Hydrocarbons—A Technical Review

John Pichtel

Ball State University, Natural Resources and Environmental Management, Muncie, USA

19.1 Introduction

Many terrestrial and aquatic environments in the industrialized nations have become contaminated with anthropogenic contaminants. Among these, polycyclic aromatic hydrocarbons (PAHs); chlorinated solvents including trichloroethylene (TCE) and perchloroethylene (PCE); chlorinated phenols; polychlorinated biphenyls (PCBs); and polychlorinated dibenzodioxins (PCDDs) pose significant risks to public health and the environment [1, 2]. These compounds have been released to the biosphere via discharges of industrial and municipal wastewaters, leachates from landfills, leaking underground storage tanks, and urban runoff [3].

Xenobiotic hydrocarbons comprise several of the most prioritized compounds on the US Environmental Protection Agency (EPA) and European Union (EU) lists of harmful and/or toxic contaminants [4, 5]. Many are acute or chronic toxins; some are carcinogenic, mutagenic, or teratogenic; others are endocrine disruptors [5–9]. Given their unique chemical structures, xenobiotic hydrocarbons tend to be recalcitrant in the biosphere. The adverse effects of these compounds on public health and the environment, including their slow biodegradation rates and accumulation in the food chain, have been described elsewhere [10–13].

19.1.1 Conventional Bioremediation Technologies

Conventional remediation methods for contaminated soil, sediments, and groundwater include soil excavation and removal, groundwater pump-and-treat, off-site treatment by solvent extraction, ozonation, photocatalytic oxidation, and high temperature incineration [14, 15]. Unfortunately, however, many of these techniques require highly trained personnel and use of hazardous reagents, are costly, and may themselves impart adverse effects to soil and aquatic environments [15]. Other, less-invasive methods of management include dredging, capping, and monitored natural remediation [16]. The need exists for innovative, efficient, environmentally friendly, and cost-effective strategies for remediation of xenobiotic soil and groundwater contaminants.

A number of indigenous bacterial communities have been identified that are capable of metabolizing xenobiotic hydrocarbon pollutants. In natural settings, however, reactions carried out by these species are inefficient due to their low abundance and activity combined with limited access to contaminant molecules and to available nutrients [5, 17–19]. Furthermore, microbial survival in the environment is tenuous due to the presence of high levels of toxins, shear stresses, and minimal protection from predation [5].

In recent years, the use of biofilms for soil and groundwater treatment has drawn increased interest for research and development. Based on the literature to date, biofilms offer significant potential for removal of xenobiotic compounds from soil, sediments, natural waters, and industrial wastewaters containing low levels of pollutants.

19.1.2 Composition and Properties of Biofilms

Biofilms are structured sessile microbial communities in which microbial cells attach to a surface and become fixed within an agglomeration of extracellular polymeric substances (EPS) secreted by the cells [20–23]. Surfaces may include soil, sediment, and aquatic colloids; living tissue such as root surfaces; anthropogenic equipment such as industrial or portable water piping systems and medical devices; and others.

Biofilms may consist of single or multiple species of microorganisms from one or more kingdoms such as bacteria, fungi, algae, and archaea. Members will vary in terms of environmental requirements such as nutrient types and concentrations, and electron acceptors and donors [24, 25]. Microbial numbers are substantial; populations may number between 10^9 and 10^{11} cells/ml biofilm mass [26].

The biofilm matrix is estimated to be predominantly water (85–95% wet weight), exopolysaccharides (1–2%), and polypeptide polymers [27]. Within the EPS, the biofilm network matures with time to incorporate channels that allow transport of nutrients and water and electron acceptors such as oxygen or reduced compounds. The EPS contains surfactants that help to solubilize hydrophobic substrates that would otherwise be unavailable. This phenomenon has been well documented in biofilms occurring in diverse locales such as agricultural soils, tidal flats, stream beds, and fabricated systems such as plumbing [28–30].

19.1.3 Unique Properties of Biofilms

Biofilms are dynamic systems in which varied components are assembled, synthesized, transformed, decomposed, and released to the local environment [27, 31]. Biofilms can be considered microenvironments that possess distinct characteristics compared with the bulk water phase. The microbial types that construct biofilms are typically able to grow either in the free-living (planktonic) or the attached state, depending on environmental conditions [32]. In nature, the biofilm phase is the predominant lifestyle of microbes [33].

Aggregation of microorganisms in sessile communities has the advantage of increased tolerance toward changes in environmental conditions such as fluctuations in nutrient levels, predation, exposure to toxic chemicals (e.g., antibiotics or pollutants in high concentrations), or other environmental stressors such as shifts in pH, temperature, and salinity levels [5, 34–36]. Aggregated microbial structures can better overcome external stresses such as desiccation and UV irradiation compared to their planktonic counterparts [21, 37, 38]. The ability for microbial cells to associate in structured biofilm networks therefore offers numerous advantages over free-living planktonic cells [23, 39].

Biofilms enhance the ability of cells to communicate, and may serve as sites for transfer of genetic material [40]. Furthermore, horizontal transfer of genes on conjugative plasmids has been shown to induce planktonic bacteria to form biofilm communities, which promotes further gene transfer [41]. Therefore, it is possible that conjugation, and the evolution of new genetic traits, will be concentrated in biofilm communities [42]. In addition to genetic diversification in biofilms containing a single species, biofilms harbor diverse species of both aerobic and anaerobic organisms, allowing them to survive under diverse conditions [19, 43]. This suite of characteristics offers a complex and resilient system for biodegradation of pollutants [10, 44, 45].

19.1.4 Significance of Biofilms to Environmental Remediation

Bioremediation is defined as the engineered application of microbial processes for the destruction of hazardous pollutants [15]. In this context, microbial cultures have been exploited for decomposition of PAHs [46, 47]; solvents such as TCE and PCE [48, 49]; PCBs [50]; and other toxic compounds.

In recent years, biofilms have become a focus of research in the remediation of xenobiotic hydrocarbon compounds. Research in naturally occurring biofilms—for example, in soil, sediments, and wetland vegetation—has demonstrated the potential for biofilms to treat wastewaters containing anthropogenic pollutants and stimulated research and development in large-scale reactors for treatment of hazardous chemicals. Biofilm-based reactors are now used for treatment of large volumes of industrial and municipal wastewaters.

19.1.5 Objectives

By virtue of the intimate, mutually beneficial synergistic interactions among organisms in biofilms, the decomposition and mineralization of xenobiotics is accelerated. It is for these reasons that biofilms are employed in industrial facilities to enhance immobilization and degradation of pollutants. This chapter provides an overview of the role of biofilms for remediation of xenobiotic hydrocarbon pollutants and considers the role of biofilms in aerobic and anaerobic biofilm treatment of pollutants using various reactor types.

19.2 Polycyclic Aromatic Hydrocarbons

Polycyclic aromatic hydrocarbons (PAHs) are organic compounds consisting of two or more fused aromatic rings (Figure 19.1). Although PAHs occur naturally in fossil fuels, they are also significant environmental contaminants, occurring in soils at former manufactured gas plants and creosote wood treatment facilities [29, 51]. Over 100 PAH compounds exist, and various mixtures occur in affected environments [52].

Many PAHs are known carcinogens or mutagens and are included in lists of significant pollutants by the US Environmental Protection Agency and the EU Water Framework Directives [47, 53, 54]. PAHs can be consumed by aquatic organisms and be transported through food chains, accumulating in fatty tissues with potential long-term

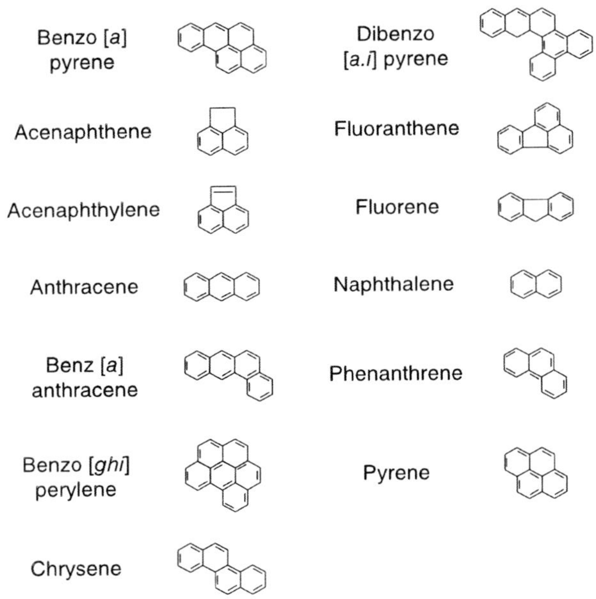

Figure 19.1 Structure of selected polycyclic aromatic hydrocarbons.

effects [5]. Although PAHs undergo sorption, volatilization, photolysis, and chemical degradation in nature, microbial degradation is considered the major degradation process [47].

19.2.1 Microbiology of PAH Degradation

PAHs are microbially transformed into less complex metabolites, and through mineralization into H_2O and CO_2 (aerobic conditions) or CH_4 (anaerobic). Several bacterial genera have been isolated from contaminated soil or sediments which are capable of degrading PAHs. Species of *Pseudomonas, Alcaligenes, Mycobacterium, Rhodococcus, Sphingomonas* and *Cycloclasticus* are known to decompose PAHs [46]. *Pseudomonas aeruginosa, P. fluoresens, Mycobacterium* spp., *Haemophilus* spp., *Rhodococcus* spp., and *Paenibacillus* spp. are some commonly studied PAH-degrading bacteria. Lignolytic fungi also are capable of PAH degradation; examples include *Phanerochaete chrysosporium, Bjerkandera adusta*, and *Pleurotus ostreatus* [47].

Microbial decomposition of PAHs progresses slowly in natural systems. PAH structure strongly influences biodegradability—those with more than three rings pose challenges to microbial degradation [33]. This is due to their low solubility in water and high affinity to organic materials. Sorbed PAHs, crystalline forms and PAHs dissolved in nonaqueous phase liquids tend to be of limited availability to PAH-degrading microbes [33, 55–60].

19.2.2 Biofilm Processes and PAH Degradation

Recent studies of biofilm properties and development have greatly augmented our knowledge of PAH transformations [33]. Some microbes are capable of growing on hydrophobic PAHs by direct contact [61], thus surmounting the limiting steps of substrate solubilization and diffusion through the aqueous phase to the cell surface. Biofilms are now recognized as optimal environs for PAH degradation, as compared with planktonic environments [62, 63].

A mechanism for increased mass transfer from crystalline PAHs to bacterial cells was demonstrated by Wick et al. [64, 65]. Cells of *Mycobacterium frederiksbergense* LB501T formed biofilms on anthracene crystals; the nearest cells were less than 1 mm from the crystal surface. According to Fick's first law of diffusion, the small distance between the PAHs and the biofilm cells, combined with a high PAH affinity, strongly favors diffusive mass transfer of PAHs to cells by steepening the aqueous concentration gradient [63].

Using confocal laser scanning microscopy, the colonization pattern of a gfp-labeled strain of *Pseudomonas putida* was monitored on fluorene and phenanthrene crystals [62]. *P. putida* was found to grow directly on phenanthrene, forming a biofilm on accessible crystalline surfaces. In contrast, however, no significant biofilm formation was observed in the presence of fluorene [62].

Ortega-Calvo and Alexander [63] found that naphthalene occurring in nonaqueous-phase liquid (NAPL) was degraded at a faster rate by bacteria growing at the water–NAPL interface compared with suspended bacteria. Johnsen and Karlson [58] showed that biofilm formation was the principal mechanism by bacteria to overcome mass transfer limitations when growing on poorly soluble PAHs. The authors screened a number of PAH degraders for formation of biofilms on PAH crystals and suggested that hydrophobic bacteria (i.e., *Mycobacterium* and *Nocardia*) may overcome low aqueous phase substrate concentrations by efficient substrate-to-cell contact mechanisms.

Biofilm formation and PAH degradation by a strain of the marine bacterium *Stenotrophomonas acidaminihila* was investigated by Mangwani et al. [60]. Dense growth was observed in BSM media supplemented with PAHs, indicating its ability to utilize PAHs as an energy source. The rate of biofilm-mediated degradation was greatest during the initial phase (two days) and decreased after three to five days. By seven days, the biofilm culture degraded 71 percent and 40 percent of phenanthrene and pyrene, respectively, as compared with 39 percent of phenanthrene and 30 percent of pyrene in planktonic culture.

19.2.3 Microbial Production of Surfactant Molecules

Bacteria occurring in biofilms may enhance PAH bioavailability by secreting biosurfactants and via production of EPS [67, 68]. Three semi-colloid *Sphingomonas* polysaccharides (gellan, welan, and rhamsan) all increased solubility of PAHs [58]. The increases in apparent solubility were most pronounced for four-ring PAHs (pyrene and fluoranthene) versus three-ring PAHs, presumably because the former compounds are more hydrophobic. Several PAH-degrading bacteria that have been isolated from soil belong to the sphingomonads [46, 69–72], which produce an outer gluco-sphingolipid cell membrane consisting of a hydrophilic mono- or tetra-saccharide and a lipophilic dihydrosphingosine residue [73, 74]. The structure of the gluco-sphingolipids resembles the structure of surfactants [58].

Biosurfactants produced by *Pseudomonas aeruginosa* grown on phenanthrene and naphthalene increased PAH solubility, suggesting that the organism was promoting the bioavailability of its substrate [75]. In contrast, however, Willumsen and Karlson [76] screened 57 PAH-degrading isolates for production of biosurfactants but found no correlation between biosurfactant production and PAH mineralization. Johnsen and Karlson [58] found that most strains of PAH-degrading bacterial cultures did not substantially reduce surface tension when grown on PAHs in liquid shaken cultures. According to these authors, therefore, so-called *pseudo-solubilization* of PAHs in biosurfactant micelles does not appear to be a general strategy to enhance PAH bioavailability [58].

19.2.4 Application of Surfactants

In order to overcome the low solubility and strong sorption of PAHs to solids, one possible enhancement of bioremediation technology may be to solubilize contaminants by introduction of surfactants to biofilm reactors [51, 77]. Added surfactants have been reported to increase the solubility and mass transfer of hydrophobic organic compounds, including PAHs [29, 80–84]. At high concentrations, particularly above the surfactant's critical micelle concentration, surfactants can alter the distribution of the contaminant between aqueous and solid phases, in which spherical micelle clusters form. The interior is a nonpolar phase and may dissolve substantial quantities of nonpolar solutes such as PAHs [52].

In the presence of Triton X-100, a modest enhancement of three- and four-ring PAH degradation occurred when they were present as sole substrates [52]. This was due to the higher solubility of the PAHs in the presence of the surfactant. Degradation of the two-ring PAH, however, was not significantly enhanced. Biofilms responded well to mixtures of phenanthrene/naphthalene and pyrene/naphthalene, with removals of 45 and 24 percent, respectively, in the presence of surfactant; however, higher biodegradation was always achieved in the presence of PAH mixtures without surfactant, indicating the importance of cometabolic mechanisms over improved solubilization of PAHs [52].

19.2.5 Degradation of PAHs in Biofilm Reactors

A packed-bed column reactor containing mixed microflora was tested for degradation of acenaphthene, phenanthrene, and pyrene [78]. The three PAHs were mostly eliminated in the inoculated reactor after 10 hours hydraulic residence time (HRT). Acenaphthene and phenanthrene were removed by > 99 percent from the reactor while pyrene was removed by 90 percent. PAH disappearance was probably caused by both sorption to the reactor as well as microbial degradation [78]. An example biofilm reactor system appears in Figure 19.2.

Cometabolism was demonstrated in biofilms in the Biozo® process, where a combination of biological treatment and ozone application was used to treat PAH contamination from landfill leachate [79]. A staged moving-bed biofilm reactor combined with ozone injection that was installed between a preanoxic zone, where the majority of the PAH removal occurred, and an aerobic zone, exhibited optimal PAH removal [79].

A laboratory-scale mulch biofilm barrier experienced 97–99 percent removal efficiency for phenanthrene and 99.9 percent for pyrene over 150 d [29]. Sorption and biodegradation of PAHs resulted in stable and consistent operation of the system. Addition of a nonionic surfactant increased the solubility of phenanthrene and pyrene

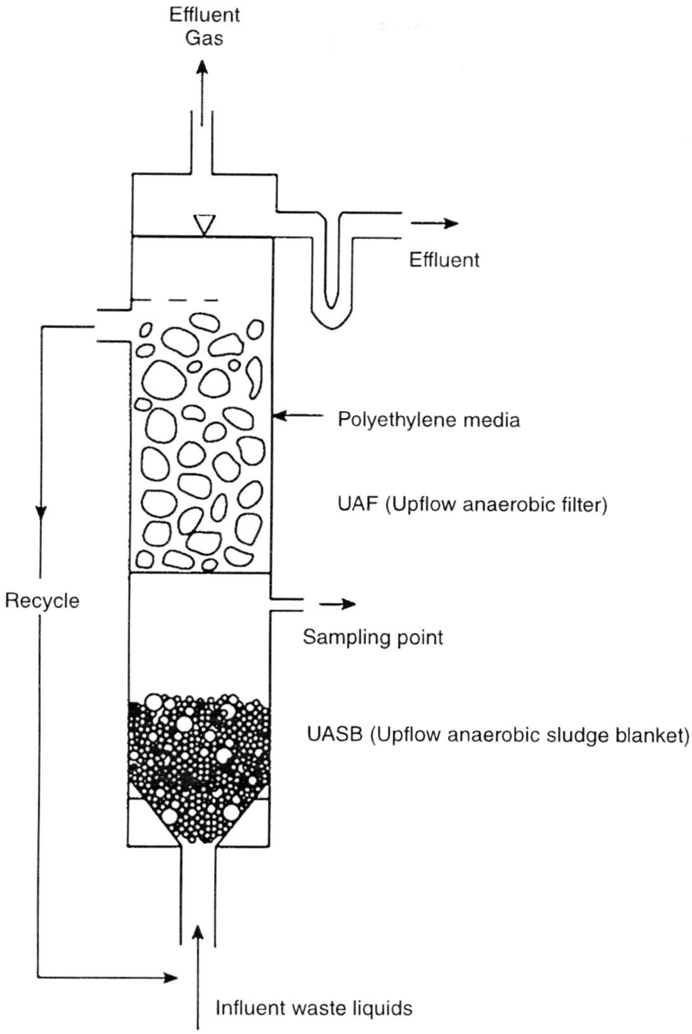

Figure 19.2 Schematic of a sample bench-top biofilter apparatus.

significantly. The presence of surfactant and the resultant increased phenanthrene or pyrene concentration did not cause toxic effects to the biofilm. However, the presence of surfactant altered biofilm structural composition [29].

Fungi have been found to produce biofilms that are effective for PAH decomposition. Formation of permeable reactive biobarriers (PRBBs) using *Trichoderma longibrachiatum* on nylon sponge was studied for PAH removal by Cobas et al. [54]. After inoculation of the fungus into the sponge, a substantial and strongly adhesive biofilm formed. A 90 percent reduction of phenanthrene concentration was observed after 14 days. In other experiments, mixtures of phenanthrene, benzo[*a*]anthracene, and pyrene at concentrations from 100 to 400 μM were treated, and total PAH removals were achieved [54].

In the natural environment, PAH-degrading microbes are now known to be relatively common [33]. If a reactor system is employed for PAH decomposition, contaminated soils and sediments should first be screened for microorganisms possessing PAH degradation potential. In addition to aerobic environments, many potential PAH-degrading bacteria have been isolated from anaerobic environments such as marine sediments [85] and municipal sewage sludges [83]. Bacterial isolates having an inherent property of PAH degradation [60], as well as genetically modified microorganisms, can be introduced and exploited in a bioremediation reactor system.

19.3 Chlorinated Ethanes, Ethenes, and Aromatics

Chlorinated hydrocarbon compounds comprise some of the most useful and economically important chemicals available to industry and agriculture; they occur in herbicides, insecticides, fungicides, heat transfer media, insulators, and lubricants [86]. Chlorinated hydrocarbons are essential for the manufacture of plastics and solvents. In addition, some are generated as wastes—for example, from pulp and paper manufacture and from chlorination of water for disinfection purposes [14, 87]. Several chlorinated compounds have been designated priority environmental pollutants by the US EPA and the EEC.

Widespread use has led to the presence of chlorinated hydrocarbons in industry and municipal effluents with eventual release into terrestrial and aquatic environments [88]. Contaminant plumes containing PCE, TCE and 1,1,1-trichloroethane (TCA) have been identified in aquifers in the United States and Canada [3, 89, 90]. Water in sediments in proximity to some industrial sites has been found to contain chlorinated hydrocarbons in excess of 1,000 mg/l [91]. Such plumes pose problems regarding containment and remediation [3].

In natural environments, biotransformation is one of the most important processes influencing the fate of chlorinated hydrocarbons. For purposes of industrial wastewater treatment and environmental remediation, adsorption and biological decomposition have become preferred approaches for treatment of these pollutants [86, 88, 92, 93].

19.3.1 Chlorinated Ethanes

1,2-Dichloroethane (1,2-DCA) and 1,1,2-trichloroethane (1,1,2-TCA) have been used extensively in industrial processes—1,2-DCA primarily as a feedstock for plastics production and 1,1,2-TCA as a degreasing agent [48]. Structures of selected chloroethanes are shown in Figure 19.3.

19.3.1.1 Microbiology of Biodegradation of Chlorinated Ethanes

Chlorinated ethanes are susceptible to biodegradation under aerobic, cometabolic, and anaerobic conditions. Therefore, bioremediation is a viable option for their removal from contaminated soil and water [48, 92].

Lower-chlorinated methanes, which include chloroethane (CA), 1,1-dichloroethane (1,1-DCA), and 1,2-DCA, are utilized as primary growth substrates by aerobic microorganisms. They are only cometabolized by anaerobes. Higher chlorinated ethanes are only known to be cometabolized. The higher chlorinated ethanes include 1,1,1-TCA,

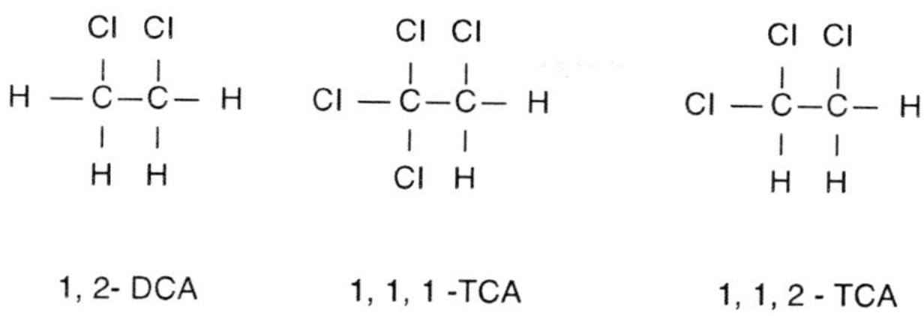

Figure 19.3 Structures of selected chlorinated ethanes.

1,1,2-TCA, various isomers of tetrachloroethane (TeCA), pentachloroethane (PCA), and hexachloroethane (HCA). The biodegradation of PCA and HCA occurs primarily only under anaerobic conditions [92].

Mixed anaerobic microbial cultures enriched from an aquifer at a former chlorinated solvent disposal facility in Louisiana, USA, were examined to identify the organisms involved in dechlorination of 1,2-DCA and 1,1,2-TCA. *Dehalobacter* and *Dehalococcoides*, two genera of anaerobic bacteria known to respire with chlorinated ethenes, were detected. *Dehalobacter* grew during the conversion of 1,1,2-TCA to vinyl chloride (VC) but not during the subsequent reductive dechlorination of VC to ethane; in contrast, *Dehalococcoides* grew only during the reductive dechlorination of VC. These results demonstrated that in mixed cultures containing multiple dechlorinating microorganisms, complementary dechlorination activities can occur, depending on substrate [48].

A Gram-positive, strictly anaerobic bacterium, tentatively identified as a proteolytic *Clostridium* sp., transformed 1,1,1-TCA, trichloromethane, and tetrachloromethane [49]. 1,1,1-TCA was completely transformed ($\geq$99.5 percent) by reductive dehalogenation to 1,1-DCA (30 to 40 percent), acetic acid (7 percent) and unidentified products [49].

19.3.1.2 Degradation of Chlorinated Ethanes in Biofilm Reactors

Suspended culture systems such as conventional activated sludge processes often fail to remove high concentrations of chlorinated hydrocarbons from wastewater due to their toxicity. Recent investigations on biodegradation of chlorinated hydrocarbons have focused on the use of biofilm reactors [94]. Biofilm reactors are more resistant to high concentrations of such compounds because of high biomass concentrations and diffusion barriers for the toxic compounds within the biofilm. Therefore, biofilm systems usually result in better reactor performance as compared to suspended growth systems and often support high removal efficiencies [94].

Engineered cometabolic degradation under aerobic conditions has been demonstrated for 1,1,1-TCA. The compound was degraded in a methanotrophic biofilm reactor with methane as the primary substrate [95]. In a closed suspended-growth reactor, a mixed methane-oxidizing culture also degraded 1,1,1-TCA [96]. Conversion of radiolabeled 1,1,1-TCA by biofilms from an anaerobic bioreactor was evaluated by Vogel et al. [97]. In addition to 1,1-DCA and CA formation, $^{14}CO_2$ was identified as a product. During incubations of up to 80 days, 17 percent of the label was recovered as $^{14}CO_2$.

A membrane-aerated biofilm reactor (MBR) with a biofilm of *Pseudomonas* sp. strain DCA1 was studied for removal of 1,2-DCA from water [98]. A hydrophobic membrane was used to create a barrier between the liquid and gas phases. A stable and active

DCA-degrading biofilm formed on the membrane. The maximum removal rate was reached at a DCA concentration of 80 μM [98]. A continuous-flow system composed of two anaerobic bioreactor (ABR) columns (based on engineered wetland systems) connected in series was tested to treat multiple volatile organic compounds (VOCs) [99]. Dechlorination of *cis*-dichloroethene (*c*DCE) inhibited the dechlorination of 1,1-DCA; furthermore, 1,1-DCA dechlorination proceeded only when *c*DCE had been degraded to ethene. The greatest extent of 1,1-DCA dechlorination was observed near the effluent of the second ABR due to conditions favoring a higher population of 1,1-DCA degrading bacteria [99].

19.3.2 Chlorinated Ethenes

In this discussion, chlorinated ethenes will be divided into categories of lower and higher chlorinated compounds due to unique behaviors regarding their biodegradability. Lower-chlorinated ethenes include monochlorinated ethene, VC, 1,1-DCE, *trans*-dichloroethene (*t*DCE) and *c*DCE. Higher-chlorinated ethenes include PCE and TCE. The lower-chlorinated ethenes tend to be more susceptible to aerobic degradation and less prone to anaerobic degradation compared to the higher-chlorinated ethenes. However, there is evidence for aerobic degradation of TCE as well [92]. Structures of selected chloroethenes are shown in Figure 19.4.

Many chlorinated ethenes can be anaerobically transformed to less-chlorinated or nonchlorinated compounds via reductive reactions carried out by mixed bacterial

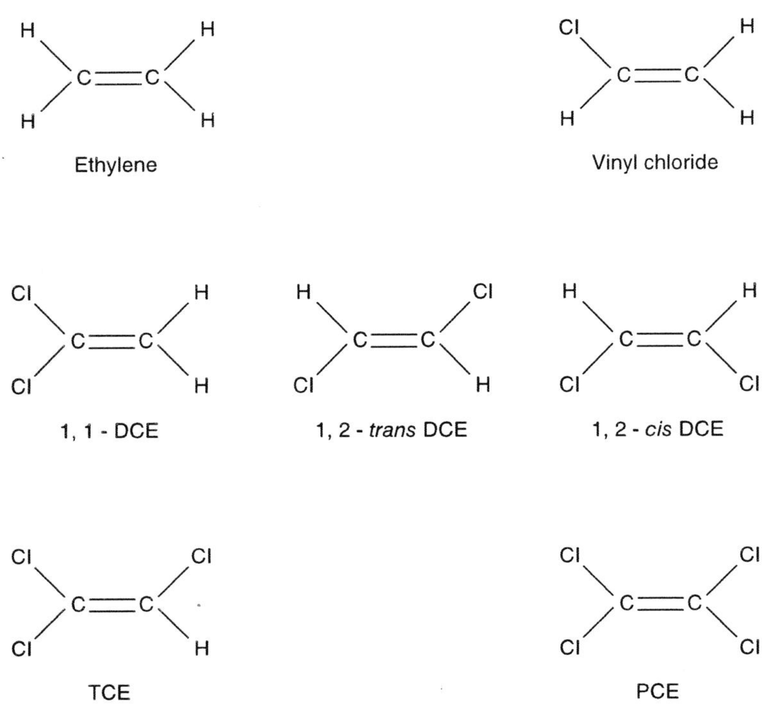

Figure 19.4 Structures of selected chlorinated ethenes.

consortia. Reductive dechlorination can occur under methanogenic and sulfate-reducing conditions or under chloro-respiring conditions, where TCE is used as an electron acceptor. In reductive dechlorination, TCE is sequentially reduced by two electrons to 1,1-DCE, 1,2-*cis*-DCE, or 1,2-*trans*-DCE isomers to VC and eventually to ethene. In some cases, ethane is subsequently mineralized to CH_4 and CO_2 [100–105].

19.3.3 Degradation of Chlorinated Ethenes in Biofilm Reactors

A membrane bioreactor was used to treat synthetic wastewater containing 1,2-DCE [106]. Biofilms growing on the surface of membrane tubes biodegraded DCE while avoiding direct contact between the DCE and the aerating gas. This effect significantly reduces air stripping as would occur in conventional aerated bioreactors. Over 99 percent removal of DCE from a wastewater containing 1,600 mg/L DCE was achieved at wastewater residence times of 0.75 hours [106].

For biological reduction of chlorinated ethenes to occur, a microbiologically available electron donor must be incorporated. The biodegradation of TCE and intermediates has been achieved using different electron donors, such as methanol, acetate, lactate, butanol, fructose, and hydrogen gas (H_2) [107]. Hydrogen gas is considered to be an ideal electron donor as it is nontoxic and relatively inexpensive [108–110].

Chung et al. [107] employed a hydrogen-based, denitrifying membrane biofilm reactor (MBfR) to reductively dechlorinate trichloroethenes. The MBfR delivers H_2 directly as electron donor by diffusion through the wall of a bubble-free gas-transfer membrane. Biofilm bacteria living on the outer wall of the membrane oxidize the H_2, and the electrons are used by the bacteria to reduce electron acceptors present in the water. When TCE was added to the MBfR, reductive dechlorination occurred immediately and increased over 18 weeks [107]. TCE was completely dechlorinated to ethene by 120 days. It was reported that *Dehalococcoides* was naturally present at the beginning and end of the study and that they were enriched by exposure to TCE. At least two *Dehalococcoides* strains were present in the enriched biofilm. Their numbers increased after TCE addition [107].

Research in recent decades suggests that H_2 is the sole electron donor that bacteria can use to reduce all chlorinated solvents [110, 112, 113]. Thus, using H_2 may allow for simultaneous reduction of mixtures of chlorinated solvents [113]. Contaminated water often contains mixtures of chlorinated solvents; therefore, a treatment technology that detoxifies them simultaneously is preferred. Chung and Rittmann [113] studied simultaneous dechlorination of TCE, chloroethane and chloromethane using an MBfR. By using H_2 as the electron donor, the MBfR treated multiple chlorinated solvents in one step. Evidence of inhibition by chloroform (CF) was observed at a concentration > 1 m*M*. Possible inhibition of TCE reduction may have occurred from accumulation of chloroethane or chloromethane [113].

Sequencing biofilm reactors that cycle between a degradation stage and rejuvenation stage are an appealing treatment technology for aerobic cometabolism of low concentrations of TCE and other chlorinated solvents. Segar et al. [114] studied the effect of phenol-feeding strategies for sustaining high rates of TCE removal efficiency. In a sequencing packed-bed reactor, TCE-contaminated water was cycled between a rejuvenation stage, during which phenol (a more readily biodegradable compound) was supplied in a nutrient mix, and a degradation stage, where TCE was added as a contaminant in the absence of phenol. The mixed culture of phenol degraders grew rapidly, attached well to surfaces, and tolerated a range of environmental conditions [114]. The feeding strategies yielded average TCE removals between 70 and 90 percent at an HRT of 14 minutes. The authors suggest that a full-scale sequencing reactor could operate indefinitely with the proper feeding strategy [114].

Cometabolic degradation rates of TCE and 1,1,1-TCA were determined using mixed culture methanotrophic biofilms grown on glass beads in a mixed laboratory reactor sparged with CH_4 [96]. Maximum degradation rates of approximately 302 and 400 μg/l · h were observed for TCA and TCE, respectively. For TCE concentrations < 760 / μg/L, TCA was degraded faster than TCE. Small quantities of a chlorinated metabolite, presumed to be 2,2,2-trichloroethanol, were formed during TCA degradation. The authors calculated that useful reductions of TCA and TCE could be obtained in similar reactors with retention times of 1 to 12 hours [96].

Perchloroethylene (PCE) can be reductively dechlorinated by anaerobic microorganisms to TCE, DCE isomers, and VC [103, 104]. Fathepure and Vogel [3] employed a two-stage anaerobic–aerobic biofilm reactor to degrade a mixture of hexachlorobenzene (HCB), PCE, and CF. The effluent from the anaerobic biofilm column was routed directly to an aerobic column. The extent of reductive dechlorination was maximized when the anaerobic biofilm column received acetate as a primary carbon source. HCB, PCE, and CF were dechlorinated to tri- and dichlorinated products (99, 80, and 32 percent, respectively) when acetate was supplied in the feed. The less-chlorinated compounds were metabolized by the aerobic biofilm. After transport through both columns, the total amount transformed to nonvolatile intermediates and CO_2 was 94, 96, and 83 percent for HCB, TCE, and CF, respectively [3].

A methanogenic fluidized bed reactor fed with lactate and PCE was operated for 14 months to study the effect of electron donor and PCE loading on chloroethene dechlorination rates [115]. Lactate was fed continuously and the influent PCE feed concentration was increased stepwise from 3.5 to 160 μmol/L. Vinyl chloride and ethene accounted for 80 and 20 percent, respectively, of the PCE dechlorination products. Batch tests with various electron donors showed that H_2, propionate, and lactate supported dechlorination of PCE, TCE, c-DCE, and VC [115].

Ohandja and Stuckey [116] used a continuous-flow flat sheet hybrid membrane aerated biofilm reactor (MABR) to treat synthetic wastewater-containing PCE. The reactor biodegraded 70 mg/ L of PCE in 9 hours without accumulation of any intermediates, resulting in a removal rate of 247 mmol of PCE/h/m^3. Some chlorinated ethenes may have adsorbed to the biofilm, or aerobic intermediates of low-chlorinated compounds such as trichloroethanol, dichloroacetyl, and chloroacetaldehyde were produced. Due to their high PCE removal rates, hybrid MABRs could potentially be used to decompose a number of refractory organics which require combined anaerobic/aerobic biological treatment for degradation.

A major limitation to the application of reductive dechlorination for treatment of TCE and PCE is the slow removal of the intermediate VC (which is also a known human carcinogen). Accumulation of VC may therefore hinder complete dechlorination to ethane. Using a methanogenic biofilm reactor containing *Dehalococcoides* spp. as the putative dechlorinating microorganism, Aulenta et al. [1] investigated the kinetics of VC formation from PCE, and VC dechlorination to ethane. Both VC formation and dechlorination increased in rate with increasing H_2 concentration. The maximum VC formation rate was about 10 times greater than the maximum VC dechlorination rate. The authors concluded that direct addition of H_2 was important in achieving high rates of VC dechlorination, thus preventing VC accumulation.

A methanogenic and sulfate-reducing consortium was obtained from anaerobic digested sludge from a wastewater treatment plant. An ascending fixed-bed reactor was inoculated and fed semi-continuously with various PCE loading rates [105]. Dechlorination of PCE was complete for concentrations ranging from 40 to 215 mM of PCE. A feed rate of 215 mM PCE was degraded rapidly at the rate of 3 mmol/L/h and with 98 percent PCE removal. No chlorinated products accumulated in the effluent. Experiments with ^{13}C-labeled PCE indicated that it was completely converted to biomass carbon and CO_2.

19.4 Chlorinated Aromatics

Common chlorinated aromatics include trichlorophenol, tetrachlorophenol and pentachlorophenol. Structures of selected chlorinated aromatics are shown in Figure 19.5.

19.4.1 Degradation of Chlorinated Aromatics in Biofilm Reactors

Complete degradation of 2,4,6-trichlorophenol, 2,3,4,6-tetrachlorophenol and pentachlorophenol (PCP) was reported using a fluidized bed biofilm reactor; the key microbial populations involved included *Pseudomonas* sp. and *Rhodococcus* sp. [117]. A rotating perforated tubes biofilm reactor was developed to decompose 2,4,dichlorophenol (DCP) in synthetic wastewater [94]. The system contained mixed microbial biomass from activated sludge supplemented with DCP-degrading *Pseudomonas putida*. Nearly 100 percent DCP removal was obtained with a feed DCP concentration of 50 mg/L resulting in a toxicity removal efficiency of 79 percent.

In a study by Guiot et al. [118], anaerobic degradation of pentachlorophenol (PCP) was augmented by inoculation of a pure culture of *Desulfitobacterium frappferi* POP-1, an anaerobic Gram-positive bacterium, to an anaerobic upflow sludge bed system. This resulted in a degradation rate of 4 mg PCP/g volatile suspended solids/day, which was attained within 56 days. Organisms attached rapidly to sludge granules and densely colonized the biofilm. With time, a dense outer layer (approx. 50 μm thick) formed.

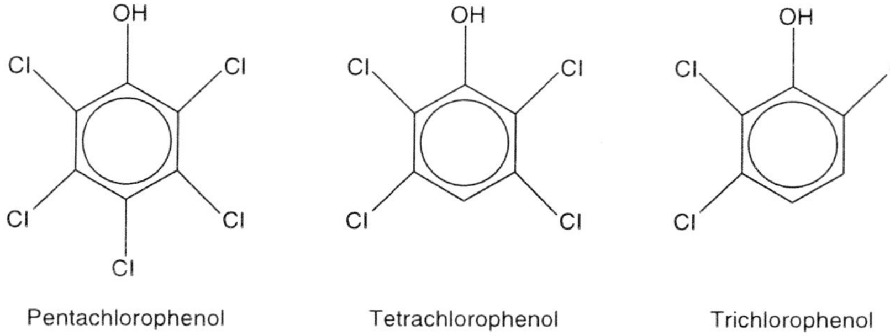

Figure 19.5 Structures of selected chlorinated phenols.

Rapid PCP consumption occurred at the granule periphery, producing a steep PCP gradient within the anaerobic granule, thus leaving the granule core PCP-free. As a result, methanogens and other bacteria could be shielded from PCP toxicity and allowed to thrive in a less toxic environment [118].

The chlorophenol elimination potential of a continuous-flow biofilm reactor (CFBR) and a sequencing batch biofilm reactor (SBBR) was compared by Kaballo et al. [119]. In the CFBR a stratification of biomass occurred, whereas biomass in the SBBR developed uniformly due to the use of a modified fill strategy. Under shock loading, degradation in SBBR was greater than that in CFBR. The elimination rate per biomass unit in the SBBR was fivefold higher than that observed in the CFBR. One reason for the higher efficiency may have been more efficient transport of substrates caused by the smaller internal mass transport resistance within the thinner biofilm.

Using a SBBR Farabegoli et al. [120] measured 99 percent removal efficiencies for both phenol and 2-CP from contaminated groundwater; complete removal was achieved over a react phase of seven hours. During recirculation, the concentration gradient was eliminated and biomass became evenly distributed along the bed height. As a result, the elimination process was accelerated. When the compounds were fed simultaneously, 2-CP removal kinetics improved likely due to cometabolism.

Chang et al. [85] reported the degradation of 2-CP using a hydrogenotrophic biofilm under denitrification, sulfate-reduction, and dechlorination conditions. After four months of acclimation with 2-CP, removal efficiency was 95 percent in denitrification, 94 percent in sulfate-reduction and 95 percent in dechlorination reactors, respectively. H_2 served as an electron donor for dechlorination of 2-CP.

The use of phenol and 4-CP as enrichment substrates for the development of a 4-CP-degrading biofilm population has been shown to be successful. 4-CP was degraded as a co-metabolite in the presence of phenol by *A. eutrophus* [121]. In addition, phenol stimulates the degradation of other aromatic compounds such as *o*-cresol, *p*-cresol and methylphenols [86, 122].

A continuous-stirred, gas-permeable hollow-fiber membrane biofilm reactor was effective for degradation or detoxification of *p*-chloronitrobenzene in water by biotransforming it first to *p*-chloroaniline (nitroreduction) and then to aniline (reductive dechlorination) with H_2 as electron donor [123]. The membrane served as a growth-support medium, providing a large specific surface area for biofilm development, especially when hollow fibers were used [123].

19.4.2 Benefits of Activated Charcoal and Other Organic Matrixes for Biofilm Reactors

Granular activated carbon (GAC) biofilm reactors have proven successful for both biodegradation and sorption of xenobiotic hydrocarbons [124]; the adsorptive capacity and irregular shape of GAC particles provide niches for bacterial colonization protected from fluid forces [40], while the vast range of functional groups on the surface enhance microbial attachment [86, 125]. Treatment of groundwater contaminated with chlorobenzene at concentrations up to 170 mg/L in a GAC fluidized-bed reactor was achieved with more than 99 percent efficiency [86, 126].

A bacterial consortium that can degrade chloro- and nitrophenols was isolated from the rhizosphere of *Phragmitis communis* and placed in a biofilm reactor with GAC as a support medium [88]. This system was used for the decomposition of 4-CP. 4-chlorophenol was not detected in the column effluent, being either adsorbed to the GAC or biodegraded by the consortium. A residual fraction of GAC-bound 4-CP was unavailable to the microbes [88]. Carvalho et al. [86] obtained a bacterial consortium capable of degrading 4-CP from the rhizosphere of *Phragmites australis*. A GAC biofilm reactor was established using the consortium, and the reactor achieved 4-CP removal efficiencies of 69 to 100 percent. Periods of lower performance were attributed to clogging of the column with biomass and the formation of channels. Dynamic population changes were observed throughout the study. One isolate recovered from the biofilm was shown to be capable of degrading 4-CP as a sole carbon and energy source.

Compact matrixes composed of wheat husk and wheat bran were successfully used to immobilize the fungus *Coriolus versicolor* to treat effluents contaminated with dichlorophenol (DCP) and PCP at 50 mg/L. Removal rates of 75 to 80 percent and 100 percent, respectively, were observed within 24 hours [86, 127]. Quintelas et al. [128] used a biofilm of *Arthrobacter viscosus* supported on GAC to remove CP, phenol, and *o*-cresol from aqueous solutions. The order of removal percentage after 15 hours was: phenol > chlorophenol > *o*-cresol [128].

Scrap vehicle tire chips were used as packing material for sequential anaerobic–aerobic biofilm reactors to remove 2,4-DCP and 4-CP [130]. More than 98 percent of DCP was dehalogenated to CP in the anaerobic reactor, 70 to 98 percent of which was subsequently degraded in the aerobic reactor. The amount of biomass that attached to the surface of scrap tires was 3.16 and 3.72 mg volatile suspended solids/cm^2 after 14 and 37 days, respectively [130].

19.5 Polychlorinated Biphenyls (PCBs)

Polychlorinated biphenyls (PCBs) had previously enjoyed many important uses—for example, as components within electrical transformers, capacitors, voltage regulators, switches, bushings, and electromagnets; as oil in motors and hydraulic systems; in fluorescent light ballasts; as thermal insulation material; in oil-based paint; and other uses [131]. PCBs are now known to be highly toxic compounds. PCBs consist of 209 different congeners; the chemical formula is $C_{12}H_{10-n}Cl_n$, where the number of chlorines (n) ranges from 1 to 10 (Figure 19.6). Despite their ban in the United States over three decades ago, PCBs are still detected in soils and sediments worldwide [132]. All congeners are poorly soluble in water, and solubility decreases with the number of chlorine atoms attached to the aromatic ring. Their poor water solubility and reasonably good solubility in organic solvents and in fat are the primary factors controlling the distribution of PCBs in the biosphere [50].

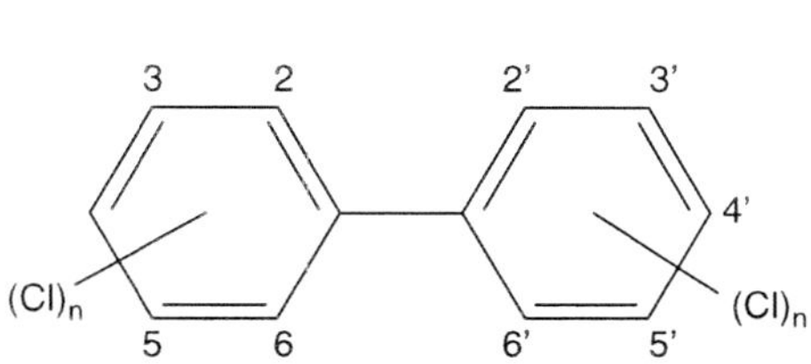

Figure 19.6 Generic structure of a polychlorinated biphenyl molecule.

Few reliable and cost-effective technologies are available for remediation, so dredging, dewatering, and landfilling are most often mandated for sites contaminated with PCBs [133–135]. These technologies are often cost-prohibitive for treating large areas of contamination in rivers, lakes, and coastal sediments. They are also disruptive to environmentally sensitive areas such as marshes and wetlands. In contrast, however, biological remediation has shown promise in a number of cases.

19.5.1 Microbiology of PCB Biodegradation

Bioremediation for complete mineralization of PCBs requires two interrelated processes. The first is anaerobic reductive dechlorination, where chlorine atoms are removed from highly chlorinated congeners using PCBs as electron acceptors [136, 137]. The second process is aerobic degradation, where the biphenyl ring is cleaved and mineralized using the PCB molecule as an electron donor. This process results in the generation of O_2 and H_2O as products [136, 137]. Successful bioremediation consists of identifying PCB dechlorinating and degrading bacterial species and their substrate needs in order to promote decomposition in the contaminated location. Supporting the colonization of specially chosen bacteria in soil or sediment allows for more rapid and environmentally benign cleanup of PCB-contaminated sites [13, 137]. Several microbial types have been identified with the ability to grow on and/or decompose PCBs; these include *Pseudomonas, Achromobacter, Acetobacter, Acinetobacter, Alcalegenes, Klebsiella* [138], *Rhodococcus* [139, 140, 147], and *Janibacter* [92]. A moorland soil contaminated with PCBs revealed species of *Sphingomonas*, the *Acidobacterium* phylum, and members of the genus *Burkholderia* [50]. *Burkholderia* was the main genus in isolates enriched on biphenyl and various chlorobenzoates. The phylum of bacteria termed dechlorinating *Chloroflexi* has been found to reductively dechlorinate highly chlorinated PCB congeners under anaerobic conditions; less-chlorinated structures remain for degradation by aerobes [141].

Payne et al. [142] reported on the impacts of the anaerobic dechlorinating bacterium *Dehalobium chlorocoercia* DF1 in sediment contaminated with weathered PCBs. Total penta- and higher chlorinated PCBs decreased by approximately 56 percent in bioaugmented mesocosms after 120 days compared with no activity observed in unamended controls. Addition of GAC imparted a slight stimulatory effect. The GAC may have also stimulated indigenous dechlorinating microbial communities to dechlorinate other PCBs.

The efficacy of anaerobic halorespiring *Dehalobium chlorocoercia* DF1 and aerobic *Burkholderia xenovorans* LB400 with GAC was determined in mesocosms containing weathered Aroclor-contaminated sediment from Baltimore Harbor, MD [2]. The greatest effect was seen in the mesocosm bioaugmented with both DF1 and LB400 together, which resulted in an 80 percent decrease of PCBs, from 8 mg/kg to less than 2 mg/kg after 120 days. There was no significant increase in lesser-chlorinated congeners, indicating that both anaerobic dechlorination by DF1 and aerobic degradation by LB400 occurred. These results suggest that an in situ treatment employing the simultaneous application of anaerobic and aerobic microorganisms could be an effective, environmentally sustainable strategy to reduce PCBs levels in contaminated sediment [2].

19.5.2 Biofilms and PCB Degradation

A young biofilm developing on PCB droplets revealed a diverse microbial community with species of the genera *Herbaspirillum* and *Bradyrhizobium* as dominant members [142]. The biofilm displayed distinct stages of PCB degradation and biofilm development.

Transmission electron microscopy of ultrathin sectioned biofilms revealed bacteria occurring within clay aggregates [143]. Because all bacteria were present in the center of the aggregates, it was concluded that the bacteria first attach to the substratum and then pick up selectively clay leaflets and iron oxohydroxy complexes from the bulk water, thus successively constructing clay aggregates.

A mature multispecies bacterial biofilm developed in one month on GAC in contact with aquatic sediment but did not hamper PCB adsorption [13]; additionally, PCB adsorption did not influence biofilm formation. It has also been shown that the colonization of AC particles by a bacterial biofilm increases the efficiency of PCB removal from aqueous wastes [144]. Using microcosm experiments Payne et al. [142] demonstrated that GAC did not inhibit microbial dehalogenation of PCBs in sediments when used as a carrier for dispersing PCB halorespiring microorganisms in PCB-impacted sediment. In contrast, however, the development of a bacterial biofilm embedded in exopolymeric substances may affect the affinity of ACs for PCB molecules. According to McDonough et al. [135], the equilibrium adsorption capacities of AC for several PCB congeners decreased with increasing numbers of attached bacteria. AC does not completely remove contaminants within sediment [144]; therefore, it is important to determine the possibility that bacterial biofilms developed on AC may interfere with adsorption of pollutants [13].

Bacterial biomass quantification, bacterial diversity analyses and cryo-SEM observations suggest that biofilm formation on GAC occurs in a three-stage process [145, 146]. Initial adherence of bacteria from sediment onto GAC is rapid; after 2 days of incubation, bacterial numbers may be significant [13]. The second stage of biofilm formation, observed from 2 to 30 days, is characterized by a significant shift in the genetic structure of the bacterial community adsorbed to the GAC. Additionally a non-continuous bacterial accumulation occurs. One hypothesis [146] is that the number of damaged cells increases after 2 weeks of growth, potentially due to an aging process of the biofilm; the damaged cells detach, leading to a decrease of biofilm biomass. In the third stage, by day 30, a maturation stage is characterized by the development of a bacterial biofilm embedded in an exopolymeric matrix [13].

Terminal Restriction Fragment Length Polymorphism (T-RFLP) analysis of Aroclor 1260-saturated GAC showed that the genetic structure of a multispecies bacterial biofilm did not significantly change from 7 to 30 days incubation [13]. These results are in agreement with those of Macedo et al. [145] who showed that microbial diversity of biofilms grown on PCB droplets remained relatively constant after 24 days of growth.

Macedo et al. [145] observed that multispecies biofilm communities developed on GAC after 35 days of incubation in the presence of Aroclor 1242 microdroplets from soil and sediment; these bacterial communities were distinct from the control biofilms developed from the same samples without addition of Aroclor. The advantage of this so-called *biocatalytic* form of GAC is that it both sequesters PCBs bioavailable to microorganisms and actively remediates them by microbial transformation [2].

19.5.3 Degradation of PCBs in Biofilm Reactors

The efficiency of PCB degradation can be improved by facilitating the formation of organohalide-respiring biofilms [137]. Biodegradation of PCBs using a three-phase fluidized bed biofilm reactor was reported by Borja et al. [147]. The biofilm was developed on cement balls and acclimatized to PCBs for two months by feeding the reactor alternately with PCBs and biphenyl. The rate of PCB degradation was influenced by the extended exposure of the biofilm to PCBs and the presence of mixed cultures in the biofilm. Overall, PCB degradation eventually reached 89 to 92 percent.

Since the techniques for delivering GAC to sediments through water columns have been in place for over a decade [148–150], engineered systems involved in PCB bioremediation could greatly benefit from a combination of GAC amendment with biofilm engineering, which has been evaluated on a mesocosm scale for contaminated sediment [5]. Coupling the concept of bioaugmentation with that of GAC sediment amendment, it has been shown that mature biofilms of dehalogenating bacteria can develop on GAC surfaces [13].

Plants can also perform remediation of xenobiotic hydrocarbons via the process termed *phytoremediation*. Certain plants located in contaminated soils can facilitate degradation of PCBs. Plants stimulate microbial activity in soil due to their ability to provide a generous supply of organic carbon and catalytic enzymes. Plants are involved in numerous biological, chemical, and physical processes such as adsorption, accumulation, translocation, and transformation, which markedly affect the adjacent soil and aquatic environment [151, 152]. Rhizoremediation is a term that refers to increased microbial degradation of a contaminant in the root zone (rhizosphere) due to the ability of contaminants to adsorb and bacteria to colonize and form biofilms on plant roots [137, 153].

19.6 Polychlorinated Dibenzodioxins

Polychlorinated dibenzodioxins (PCDDs) and polychlorinated dibenzofurans (PCDFs) are considered among the most toxic pollutants in existence. Both are unintentionally created as byproducts from the manufacture of certain herbicides, the bleaching of wood pulp for paper, and incineration of municipal solid waste [154, 155]. Their chemical structures are shown in Figure 19.7. High-temperature incineration is the preferred method for remediation of PCDD-contaminated media. However, biodegradation has been considered, as recent studies have advanced our understanding of the microorganisms available to metabolize such compounds [30, 156]. Hiraishi et al. [156] studied microbial communities and reported 22 percent degradation of PCDD/Fs within three months in highly polluted soil microcosms [156].

Dehalococcoides mccartyi CBDB1 is capable of transforming a wide variety of halogenated compounds such as biphenyls and PCDDs [137, 157]. A study that employed *Burkholderia* sp. NK8 together with *P. aeruginosa* PA01 showed enhanced ability of dual species biofilms to completely degrade chlorinated benzoates [158]. Biofilms consisting of *Comamonas* sp. Strain KD7 in the rhizosphere of white clover (*Trifolium repens*) showed significant reduction of PCDDs in soil [5, 30].

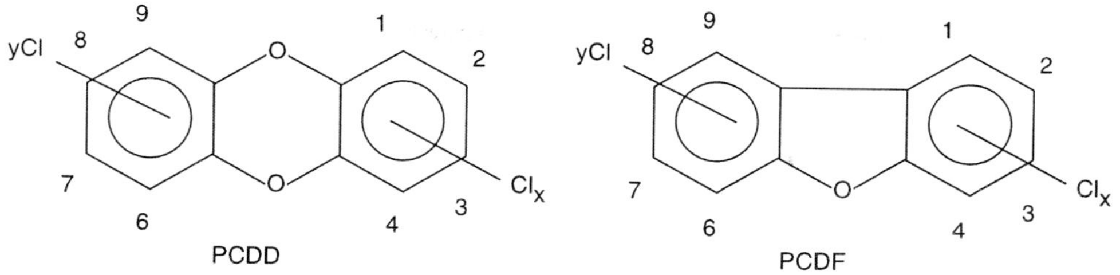

Figure 19.7 Generic structures of a polychlorinated dibenzodioxin and a polychlorinated dibenzofuran.

19.7 Conclusions

Biofilm-based remediation of xenobiotic hydrocarbons continues to draw significant attention for research and development. Based on published literature from recent decades, biofilms have the capability of removing low levels of xenobiotic compounds from industrial wastewaters as well as soil, sediments, and natural waters.

A number of practical benefits may accrue when employing biofilm-based remediation. Biofilm processes can be applied onsite in an industrial facility or in a contaminated environment. Biofilm-based remediation allows for reuse of microbial biomass and is of low cost compared with other bioremediation methods. The utilization of activated charcoal in biofilm reactors has shown promise by providing a surface for bacterial biomass to attach, which increases the efficiency of xenobiotics degradation.

Several challenges, however, must be addressed for biofilm-based remediation to gain further acceptance as a viable remediation technique. For example, microbial transformations of xenobiotics are slow; efficiency must be increased to achieve more rapid results. Microbial populations are significantly affected by variations in their local environment (e.g., pH, salinity, temperature), which negatively impact remediation processes. There remains insufficient knowledge on the identities of all xenobiotics-degrading microorganisms, their abundance in the environment, and how they form and maintain biofilms. Research on these topics is essential in order to enhance the efficiency of biofilm-based remediation.

Many strategies for enhancing biofilm treatment of xenobiotic hydrocarbons have been proposed and/or tested. For example, the bioavailability of hydrophobic compounds is often limited due to strong adsorption to surfaces. Some biofilm-based technologies involve bioaugmentation to supply organisms that can compete for adsorption sites on inert surfaces. Research focusing on gene transfer within biofilms must continue. Bioremediation techniques may be further improved by use of transgenic microbial populations.

References

1 E. Aulenta, C. Di Tomassi, C. Cupo, M. Petrangeli Papini and M. Majone, Influence of hydrogen on the reductive dechlorination of tetrachloroethene (PCE) to ethene in a methanogenic biofilm reactor: Role of mass transport phenomena. *Chem. Technol. Biotechnol.*, 81(9), 1520–1529 (2006).

2 P.B. Payne, S.K. Fagervold, H.D. May and K.R. Sowers, Remediation of polychlorinated biphenyl impacted sediment by concurrent bioaugmentation with anaerobic halorespiring and aerobic degrading bacteria. *Environ. Sci. Technol.*, 47(8), 3807–3815 (2013).
3 B.Z. Fathepure and T.M. Vogel, Complete degradation of polychlorinated hydrocarbons by a two-stage biofilm reactor. *Appl. Environ. Microbiol.*, 57(12), 3418–3422 (1991).
4 J.M. Buth, M. Grandbois, P.J. Vikesland, K. McNeill and W.A. Arnold, Aquatic photochemistry of chlorinated triclosan derivatives: Potential source of polychlorodibenzo-p-dioxins. *Env. Toxicol. Chem.*, 28, 2555–2563 (2009).
5 S.J. Edwards and B.V. Kjellerup, Applications of biofilms in bioremediation and biotransformation of persistent organic pollutants, pharmaceuticals/personal care products, and heavy metals. *Appl. Microbiol. Biotechnol.* 97, 9909-9921 (2013).
6 A.T. Proudfoot, Pentachlorophenol poisoning. *Toxicol. Rev.*, 22, 3–11 (2003).
7 M. Goswami, N. Shivaraman and R.P. Singh, Microbial metabolism of 2-chlorophenol, phenol and *ρ*-cresol by *Rhodococcus erythropolis* M1 in co-culture with *Pseudomonas fluorescens* P1. *Microbiol. Res.*, 160(2), 101–109 (2004).
8 L. Hardell, B. Van Bavel, G. Lindstrom, M. Carlberg, M. Eriksson, A.C. Dreifaldt, H. Wijkstrom, H. Starkhammar, A. Hallquist and T. Kolmert, Concentrations of polychlorinated biphenyls in blood and the risk for testicular cancer. *Int. J. Androl.*, 27, 282–290 (2004).
9 M.J. Moran, J.S. Zogorski and P.J. Squillace, Chlorinated solvents in groundwater in the United States. *Environ. Sci. Technol.*, 41, 74–81.
10 J. Borja, D.M. Taleon, J. Auresenia and S. Gallardo, Polychlorinated biphenyls and their biodegradation. *Process Biochem.*, 40, 1999–2013 (2005).
11 K. Furukawa and H. Fujihara, Microbial degradation of polychlorinated biphenyls: Biochemical and molecular features. *J. Biosci. Bioeng.*, 105, 433–449 (2008).
12 The National Academies Press. *A Risk Management Strategy for PCB-Contaminated Sediments*. National Academy Press, Washington, DC, 2001.
13 A. Mercier, G. Wille, C. Michel, J. Harris-Hellal, L. Amalric, C. Morlay and F. Battaglia-Brunet, Biofilm formation vs. PCB adsorption on granular activated carbon in PCB-contaminated aquatic sediment. *J. Soils Sediments*, 13, 793–800 (2013).
14 B. Huang, C. Lei, C. Wei and G. Zeng, Chlorinated volatile organic compounds (Cl-VOCs) in environment—Sources, potential human health impacts, and current remediation technologies. *Environment International.*, 71, 118–138 (2014).
15 J. Pichtel, *Fundamentals of Site Remediation for Metal- and Hydrocarbon-Contaminated Soils, 2nd edition*. Government Institutes, Inc., Rockville, MD, 2007.
16 US Environmental Protection Agency. Final Report. Contaminated Sediment Remediation Guidance for Hazardous Waste Sites; EPA-540-R-05–012. Washington, DC, 2005.
17 J.R. de Lipthay, N. Tuxen, K. Johnsen, L.H. Hansen, H.J. Albrechtsen, P.L. Bjerg and J. Aamand, In situ exposure to low herbicide concentrations affects microbial population composition and catabolic gene frequency in an aerobic shallow aquifer. *Appl. Environ. Microbiol.*, 69, 461–467 (2003).
18 L. Petrie, N.N. North, S.L. Dollhopf, D.L. Balkwill and J.E. Kostka, Enumeration and characterization of iron (III)-reducing microbial communities from acidic subsurface sediments contaminated with uranium(VI). *Appl. Environ. Microbiol.*, 69, 7467–7479 (2003).

19 H. von Canstein, S. Kelly, Y. Li and I. Wagner-Döbler, Species diversity improves the efficiency of mercury-reducing biofilms under changing environmental conditions. *Appl. Environ. Microbiol.*, 68, 2829–2837 (2002).

20 S.S. Branda, A. Vik, L. Friedman and R. Kolter, Biofilms: The matrix revisited. *Trends Microbiol.*, 13, 20–26 (2005).

21 M.E. Davey and G.A. O'Toole, Microbial biofilms: From ecology to molecular genetics. *Microbiol. Mol. Biol. Rev.*, 64, 847–867 (2000).

22 B. Vu, M. Chen, R.J. Crawford and E.P Ivanova, Bacterial extracellular polysaccharides involved in biofilm formation. *Molecules.*, 14, 2535–2554 (2009).

23 J.W. Costerton, Introduction to biofilm. *Int. J. Antimicrob. Agents*, 11, 217–221 (1999).

24 H. Horn and E. Morgenroth, Transport of oxygen, sodium chloride, and sodium nitrate in biofilms. *Chemical Engineering Science*, 61, 1347–1356 (2006).

25 C.A. Fux, J.W. Costerton, P.S. Stewart and P. Stoodley, Survival strategies of infectious biofilms. *Trends in Microbiology.*, 13(1), 34–40. (2005).

26 W.G. Characklis, Laboratory biofilm reactors. *In Biofilms*, W.G. Characklis and K.C. Marshall (Eds.), John Wiley, New York. (1990).

27 J.V. Headley, J. Gandrass, J. Kuballa, K.M. Peru and Y. Gong, Rates of sorption and partitioning of contaminants in river biofilm. *Environ. Sci. Technol.*, 32, 3968–3973 (1998).

28 R.G. Luthy, D.A. Dzombak, C.A. Peters, S.B. Roy, A. Ramaswami, D. Nakles and B.R. Nott, Remediating tar-contaminated soils at manufactured gas plant sites. *Environ. Sci. Technol.*, 28, 266–274 (1994).

29 Y. Seo, W.-H. Lee, G. Sorial and P.L. Bishop, The application of a mulch biofilm barrier for surfactant enhanced polycyclic aromatic hydrocarbon bioremediation. *Environmental Pollution*, 157, 95–101 (2009).

30 Y. Wang and H. Oyaizu, Enhanced remediation of dioxins-spiked soil by a plant–microbe system using a dibenzofuran-degrading *Comamonas* sp. and *Trifolium repens*. *Chemosphere*, 85, 1109–1114 (2011).

31 R.J.C. McLean and T.J. Beveridge. metal binding capacity of bacterial surfaces and their ability to form mineralized aggregates, in *Microbial Mineral Recovery*, H.C. Ehrlich and C.L. Brierley (Eds.), McGraw-Hill, New York, 1990.

32 H. Gross, V.O. Stockwell, M.D. Henkels, B. Nowack-Thompson, J.E. Loper and W.H. Gerwick, The genomisotopic approach: A systematic method to isolate products of orphan biosynthetic gene clusters. *Chem. Biol.*, 14, 53–63 (2007).

33 S.K. Shukla, N. Mangwani, T.S. Rao and S. Das, Biofilm-mediated bioremediation of polycyclic aromatic hydrocarbons, in *Microbial Biodegradation and Bioremediation*, S. Das (Ed.). Elsevier, New York, 2014.

34 H.J. Heipieper, H. Keweloh and H.J. Rehm, Influence of phenols on growth and membrane permeability of free and immobilized *Escherichia coli*. *Appl. Environ. Microbiol.*, 57, 1213–1217 (1991).

35 T.J. Beveridge, S.A. Makin, J.L. Kadurugamuwa and Z. Li, Interactions between biofilms and the environment. *FEMS Microbiol Rev.*, 20, 291–303 (1997).

36 L. Hall-Stoodley, J.W. Costerton and P. Stoodley, Bacterial biofilms: From the natural environment to infectious diseases. *Nat. Rev. Microbiol.*, 2, 95–108 (2004).

37 J.-M. Monier and S.E. Lindow, Differential survival of solitary and aggregated bacterial cells promotes aggregate formation on leaf surfaces. *Proc. Natl. Acad. Sci. USA.*, 100, 15977–15982 (2003).

38 M.R. Parsek and C. Fuqua, Biofilms 2003: Emerging themes and challenges in studies of surface-associated microbial life. *J. Bacteriol.*, 186, 4427–4440 (2004).

39 D.G. Davies, M.R. Parsek, J.P. Pearson, B.H. Iglewski, J.W. Costerton and E.P. Greenberg, The involvement of cell-to-cell signals in the development of a bacterial biofilm. *Science*, 280, 295–298 (1998).

40 B.E. Christensen and W.G. Characklis, Laboratory biofilm reactors, in *Biofilms*, W.G. Characklis and K.C. Marshall (Eds.), John Wiley, New York, 1990.

41 J.-M. Ghigo, Natural conjugative plasmids induce bacterial biofilm development. *Nature*, 412, 442–445 (2001).

42 J.E.M. Stach and R.G. Burns, Enrichment versus biofilm culture: A functional and phylogenetic comparison of polycyclic aromatic hydrocarbon-degrading microbial communities. *Environmental Microbiology*, 4(3), 169–182 (2002).

43 B.R. Boles, M. Thoendel and P.K. Singh, Self-generated diversity produces 'insurance effects'. *Proc. Natl. Acad. Sci.*, 101, 16630–16635 (2004).

44 R. Singh, D. Paul and R.K. Jain, Biofilms: Implications in bioremediation. *Trends in Microbiology*, 14(9), 389–397 (2006).

45 A. Sivan, M. Szanto and V. Pavlov, Biofilm development of the polyethylene-degrading bacterium *Rhodococcus ruber*. *Appl. Microbiol. Biotechnol.*, 72, 346–352 (2006).

46 Y. Ho, M. Jackson, Y. Yang, J.G. Mueller and P.H. Pritchard, Characterization of fluoranthene- and pyrene-degrading bacteria isolated from PAHs-contaminated soils and sediments and comparison of several Sphingomonas spp. *J. Ind. Microbiol. Biotechnol.*, 2, 100–112 (2000).

47 A.K. Haritash and C.P. Kaushik, Biodegradation aspects of polycyclic aromatic hydrocarbons (PAHs): A review. *J. Hazard. Mater.*, 169, 1–15 (2009).

48 A. Grostern and E.A. Edwards, Growth of *Dehalobacter* and *Dehalococcoides* spp. during degradation of chlorinated ethanes. *Appl. Environ. Microbiol.*, 72(1), 428–436 (2006).

49 R. Galli and P.L. McCary, Biotransformation of 1,1,1-trichloroethane, trichloromethane, and tetrachloromethane by a Clostridium sp. *Appl. Environ. Microbiol.*, 55(4), 837–844, April 1989.

50 W.R. Abraham, H. Lünsdorf, C. Strömpl, B. Nogales, E.R.B. Moore and K.N. Timmis, Microbial communities in composite biofilms participating in the degradation of PCB. *Water Air Soil Pollut. Focus Bioremediation*, 3, 57–64 (2003).

51 C.A. Peters and R.G. Luthy, Coal tar dissolution in water-miscible solvents: Experimental evaluation. *Environ. Sci. Technol.*, 27, 2831–2843 (1993).

52 S. Rodriguez and P. Bishop, Enhancing the biodegradation of polycyclic aromatic hydrocarbons: Effects of nonionic surfactant addition on biofilm function and structure. *J. Environ. Eng.*, 134, 505–512 (2008).

53 European Commission. Directive 2000/60/EC of the European Parliament and of the Council of 23 October, 2000, Establishing a framework for community action in the field of water policy. L 327, 72. 2000.

54 M. Cobas, L. Ferreira, T. Tavares, M.A. Sanromán and M. Pazos, Development of permeable reactive biobarrier for the removal of PAHs by *Trichoderma longibrachiatum*. *Chemosphere*, 91, 711–716 (2013).

55 F. Volkering, A.M. Breure, A. Sterkenburg and J.G. van Andel, Microbial degradation of polycyclic aromatic hydrocarbons: Effect of substrate availability on bacterial growth kinetics. *Appl. Microbiol. Biotechnol.*, 36, 548–552 (1992).

56 H. Harms and T.N.P. Bosma, Mass transfer limitation of microbial growth and pollutant degradation. *J. Ind. Microbiol.*, 18, 97–105 (1997).
57 T.N.P. Bosma, P.J.M. Middeldorp, G. Schraa and A.J.B. Zender, Mass transfer limitation of biotransformation: Quantifying bioavailability. *Environ. Sci. Technol.*, 31, 248–252 (1997).
58 A.R. Johnsen and U. Karlson, Evaluation of bacterial strategies to promote the bioavailability of polycyclic aromatic hydrocarbons. *Appl. Microbiol. Biotechnol.*, 63, 452–459 (2004).
59 T. Krell, J. Lacal, J. Reyes-Darias, C. Jimenez-Sanchez, R. Sungthong and J.J. Ortega-Calvo, Bioavailability of pollutants and chemotaxis. *Curr. Opin. Biotechnol.*, 24, 451–456 (2013).
60 N. Mangwani, S.K. Shukla, S. Kumari, T.S. Rao and S. Das, Characterization of *Stenotrophomonas acidaminiphila* NCW-702 biofilm for implication in the degradation of polycyclic aromatic hydrocarbons. *Journal of Applied Microbiology*, 117, 1012–1024 (2014).
61 A. Tiehm, Degradation of polycyclic aromatic hydrocarbons in the presence of synthetic surfactants. *Appl. Environ. Microbiol.*, 60, 258–263 (1994).
62 A.C. Rodrigues, S. Wuertz, A.G. Brito and L.F. Melo, Three-dimensional distribution of GFP-labeled *Pseudomonas putida* during biofilm formation on solid PAHs assessed by confocal laser scanning microscopy. *Water Science and Technology*, 47(5), 139–142 (2003).
63 J.J. Ortega-Calvo and M. Alexander, Roles of bacterial attachment and spontaneous partitioning in the biodegradation of naphthalene initially present in nonaqueous-phase liquids. *Appl. Environ. Microbiol.*, 61, 2643–2646 (1994).
64 L.K. Wick, T. Colangelo and H. Harms, Kinetics of mass transfer-limited bacterial growth on solid PAHs. *Environ. Sci. Technol.*, 35, 354–361 (2001).
65 L.Y. Wick, A. Ruiz de Munain, D. Springael and H. Harms, Responses of Mycobacterium sp. 501T to the low bioavailability of solid anthracene. *Appl. Microbiol. Biotechnol.*, 58, 378–385 (2002).
66 L.Y. Wick, N. Pasche, S.M. Bernasconi, O. Pelz and H. Harms, Characterization of Multiple-Substrate Utilization by Anthracene-Degrading *Mycobacterium frederiksbergense* LB501T. *Appl. Environ. Microbiol.*, 69(10), 6133–614, Oct. 2003.
67 S. Itoh and T. Suzuki, Effect of rhamnolipids on growth of a *Pseudomonas aeruginosa* mutant deficient in n-paraffin-utilizing ability. *Agric. Biol. Chem.*, 36, 2233–2235 (1972).
68 A. Oberbremer and R. Müller-Hurtig, Aerobic stepwise hydrocarbon degradation and formation of biosurfactants by an original soil population in a stirred reactor. *Appl. Microbiol. Biotechnol.*, 31, 582–586 (1989).
69 M. Kästner, M. Breuer-Jammali and B. Mahro, Enumeration and characterization of the soil microflora from hydrocarbon contaminated soil sites able to mineralize polycyclic hydrocarbons (PAHs). *Appl. Microbiol. Biotechnol.*, 41, 267–273 (1994).
70 J.G. Mueller, R. Devereux, D.L. Santavy, S.E. Lantz, S.G. Willis and P.H. Pritchard, Phylogenetic and physiological comparisons of PAHs-degrading bacteria from geographically diverse soils. *Antonie Van Leeuwenhoek*, 71, 329–343 (1997).
71 L. Bastiaens, D. Springael, P. Wattiau, H. Harms, R. deWachter, H. Verachtert and L. Diels, Isolation of adherent polycyclic aromatic hydrocarbon (PAHs)-degrading bacteria using PAHs sorbing carriers. *Appl. Environ. Microbiol*, 66, 1834–1843 (2000).

72 A.R. Johnsen, A. Winding, U. Karlson and P. Roslev, Linking of micro-organisms to phenanthrene metabolism in soil by analysis of 13C-labelled cell-lipids. *Appl. Environ. Microbiol.*, 68, 6106–6113 (2002).

73 K. Kawahara, U. Seydel, M. Matsuura, H. Danbara, E.T. Rietschel and U. Zaehringer, Chemical structure of glycosphingolipids isolated from *Sphingomonas paucimobilis. FEBS Lett.*, 292, 107–110 (1991).

74 S. Kawasaki, R. Moriguchi, K. Sekiya, T. Nakai, E. Ono, K. Kume and K. Kawahara, The cell-envelope structure of the lipopolysaccharide-lacking Gram-negative bacterium *Sphingomonas paucimobilis. J. Bacteriol.*, 176, 284–290 (1994).

75 E. Déziel, G. Paquette, R. Villemur, F. Lepine and J.G. Bisaillon, Biosurfactant production by a soil Pseudomonas strain growing on polycyclic aromatic hydrocarbons. *Appl. Environ. Microbiol.*, 62, 1908–1912 (1996).

76 P.A. Willumsen and U. Karlson, Screening of bacteria, isolated from PAHs-contaminated soils, for production of biosurfactants and bioemulsifiers. *Biodegradation*, 7, 415–423 (1997).

77 D. Roy, R.R. Kommalapati, S.S. Mandava, K.T. Valsaraj and W.D. Constant, Soil washing potential of a natural surfactant. *Environ. Sci. Technol.*, 31, 670–675 (1997).

78 B. Guieysse, I. Bernhoft, B.E. Andersson, T. Henrysson, S. Olsson and B. Mattiasson, Degradation of acenaphthene, phenanthrene and pyrene in a packed-bed biofilm reactor. *Appl. Microbiol. Biotechnol.*, 54(6), 826–831 (2000).

79 B.G. Plosz, C. Vogelsang, K. Macrae, H.H. Heiaas, A. Lopez, H. Liltved and K.H. Langford, The BIOZO process—A biofilm system combined with ozonation: Occurrence of xenobiotic organic micropollutants in and removal of polycyclic aromatic hydrocarbons and nitrogen from landfill leachate. *Water. Sci. Technol.*, 61, 3188–3197 (2010).

80 D. Backhus, F. Picardal, S. Johnson, T. Knowles, R. Collins, A. Radue and S. Kim, Soil and surfactant reductive dechlorination of carbon tetrachloride in the presence of *Shewanella putrefaciens* 20. *J. Contam. Hydrol.*, 28, 337–361 (1997).

81 F. Volkering, A.M. Breure, J.G. Andel and W.H. Rulkens, Influence of nonionic surfactants on bioavailability and biodegradation of polycyclic aromatic hydrocarbons. *Appl. Environ. Microbiol.*, 61, 1699–1705 (1995).

82 F. Volkering, A.M. Breure and W.H. Rulkens, Microbiological aspects of surfactant use for biological soil remediation. *Biodegradation*, 8, 401–417 (1998).

83 T. Barkay, S. Navon-Venezia, E.Z. Ron and E. Rosenberg, Enhancement of solubilization and biodegradation of polyaromatic hydrocarbons by the bioemulsifier alasan. *Appl. Environ. Microbiol.*, 65, 2697–2702 (1999).

84 I.T. Yeom, M.M. Ghosh and C.D. Cox, Kinetic aspects of surfactant solubilization of soil-bound polycyclic aromatic hydrocarbons. *Environ. Sci. Technol.*, 35, 1589–1595 (1996).

85 C.-C. Chang, S.-K. Tseng, C.-C. Chang and C.-M. Ho, Degradation of 2-chlorophenol via a hydrogenotrophic biofilm under different reductive conditions. *Chemosphere*, 56(10), 989–997 (2004).

86 M.F. Carvalho, A GAC biofilm reactor for the continuous degradation of 4-chlorophenol: Treatment efficiency and microbial analysis. *Appl. Microbiol. Biotechnol.*, 52, 722–729 (2001).

87 D.V. Savant, R. Abdul-Rahman and D.R. Ranade, Anaerobic degradation of adsorbable organic halides (AOX) from pulp and paper industry wastewater. *Bioresource Technology*, 97(9), 1092–1104 (2006).

88 M. Caldeira, S.C. Heald, M.F. Carvalho, I. Vasconcelos, A.T. Bull and P.M. Castro, 4-Chlorophenol degradation by a bacterial consortium: Development of a granular activated carbon biofilm reactor. *Appl. Microbiol. Biotechnol.*, 52(5), 722–729 (1999).

89 D.M. Mackay and J.A. Cherry, Groundwater contamination: Pump-and-treat remediation. *Environ. Sci. Technol.*, 23, 630–636 (1989).

90 A.J. Macedo, K.N. Timmis and W.R. Abraham, Widespread capacity to metabolize polychlorinated biphenyls by diverse microbial communities in soils with no significant exposure to PCB contamination. *Environ. Microbiol.*, 9, 1890–1897 (2007).

91 T.J. Phelps, J.J. Niedzielski, R.M. Schram, S.E. Herbes and D.C. White, Biodegradation of trichloroethylene in continuous-recycle expanded-bed bioreactors. *Appl. Environ. Microbiol.*, 56, 1702–1709 (1990).

92 J.A. Field and R. Sierra-Alvarez, Biodegradability of chlorinated solvents and related chlorinated aliphatic compounds. *EuroChlor*. Dec. 2004. http://www.eurochlor.org/media/14948/sd8-biodegradabilityp1-final.pdf

93 R.J. Jonge, A.M. Breure and J.G. van Andel, Bioregeneration of powdered activated carbon (PAC) loaded with aromatic compounds. *Water Res.*, 30, 875–882 (1996).

94 F. Kargi and S. Eker. Removal of 2,4-dichlorophenol and toxicity from synthetic wastewater in a rotating perforated tube biofilm reactor. *Process Biochemistry*, 40, 210–2111 (2004).

95 E. Arvin, Biodegradation kinetics of chlorinated aliphatic hydrocarbons with methane oxidizing bacteria in an aerobic fixed biofilm reactor. *Water Res.*, 25, 873-882 (1991).

96 S.E. Strand, J.V. Wodich and H.D. Stensel, Biodegradation of chlorinated solvents in a sparged, methanotrophic biofilm reactor. *Research Journal of the Water Pollution Control Federation*, 63(6), 859–867 (1991).

97 T. Vogel, C.S. Criddle and P.L. McCarty, Transformation of halogenated aliphatic compounds. *Environ. Sci. Technol.*, 21, 722–736 (1987).

98 J.C. Hage, R.T. van Houten, J. Tramper and S. Hartmans, Membrane-aerated biofilm reactor for the removal of 1,2-dichloroethane by *Pseudomonas sp.* strain DCA1. *Appl. Microbiol. Biotechnol.*, 64, 718–725 (2004).

99 A. Boudreau, *Degradation and Inhibition of Chlorinated Ethanes and Ethenes in a Two Bed, Sequential Bioremediation System*. MS Thesis. Louisiana State University and Agriculture and Mechanical College, Baton Rouge, LA. 2013.

100 M.D. Mikesell and S.A. Boyd, Dechlorination of chloroform by Methanosarcina strains. *Appl. Environ. Microbiol.*, 56, 1198–1201 (1990).

101 W.W. Mohn and J.M. Tiedje, Microbial reductive dehalogenation. *Microbiol. Rev.*, 56, 482–507 (1992).

102 C.E. Castro, Biodehalogenation: The kinetics and rates of the microbial cleavage of carbon-halogen bonds. *Environ. Toxicol. Chem.*, 12, 1609–1618 (1993).

103 T. Vogel and P.L. McCarty, Biotransformation of tetrachloroethylene to trichloroethylene, dichloroethylene, vinyl chloride and carbon dioxide under methanogenic conditions. *Appl. Environ. Microbiol.*, 49, 1080–1083 (1985).

104 B.Z. Fathepure and S.A. Boyd, Reductive dechlorination of perchloroethylene and the role of methanogens. *FEMS Microbiol. Lett.*, 49, 149–156 (1988).

105 N. Cabirol, F. Jacob, J. Perrier, B. Fouillet and P. Chambon, Complete degradation of high concentrations of tetrachloroethylene by a methanogenic consortium in a fixed-bed reactor. *Jour. Biotechnol.*, 62, 133–141 (1998).

106 L.M. Freitas dos Santos and A.G. Livingston, Extraction and biodegradation of a toxic volatile organic compound (1,2-dichloroethane) from waste-water in a membrane bioreactor. *Appl. Microbiol. Biotechnol.*, 42(2-3), 421–431 (1994).

107 J. Chung, R. Krajmalnik-Brown and B. Rittman, Bioreduction of trichloroethene using a hydrogen-based membrane biofilm reactor. *Environ. Sci. Technol.*, 42, 477–483 (2008).

108 K.C. Lee and B.E. Rittmann, Applying a novel autohydrogenotrophic hollow-fiber membrane biofilm reactor for denitrification of drinking water. *Water Res.*, 36, 2040–2052 (2002).

109 R. Nerenberg and B.E. Rittmann, Hydrogen-based, hollow-fiber membrane biofilm reactor for reduction of perchlorate and other oxidized contaminants. *Water Sci. Technol.*, 49, 223–230 (2004).

110 B.E. Rittmann, R. Nerenberg, K.C. Lee, I. Najm, T.E. Gillogly, G.E. Lehman and S.S. Adham, Hydrogen-based hollow-fiber membrane biofilm reactor (MBfR) for removing oxidized contaminants. *Water Supply*, 4, 127–133 (2004).

111 B.E. Rittmann, The membrane biofilm reactor is a versatile platform for water and wastewater treatment. *Environ. Eng. Res.*, 12(4), 157–175 (2007).

112 J. Chung, R. Nerenberg and B.E. Rittmann, Bio-reduction of soluble chromate using a hydrogen-based membrane biofilm reactor. *Water Res.*, 40, 1634–1642 (2006).

113 J. Chung and B.E. Rittmann, Simultaneous bio-reduction of trichloroethene, trichloroethane, and chloroform using a hydrogen based membrane biofilm reactor. *Wat. Sci. Tech.*, 58, 495–501 (2008).

114 R.L. Segar Jr., S.L. DeWys and G.E. Speitel, Jr., Sustained trichloroethylene cometabolism by phenol-degrading bacteria in sequencing biofilm reactors. *Water Environ. Res.*, 67, 764–774 (1995).

115 B.S. Ballapragada, H.D. Stensel, J.A. Puhakka and J.F. Ferguson, Effect of hydrogen on reductive dechlorination of chlorinated ethenes. *Environ. Sci. Technol.*, 31, 1728–1734 (1997).

116 D.G. Ohandja and D.C. Stuckey, Biodegradation of PCE in a Hybrid Membrane Aerated Biofilm Reactor. *Jour. Environ. Engin.*, 133(1), 20–27 (2007).

117 J.A. Puhakka, E.S. Melin, K.T. Järvinen, P.M. Koro, J.A. Rintala, P. Hartikaine, W.K. Shieh and J.F. Ferguson, Fluidized-bed biofilms for chlorophenol mineralization. *Water Sci. Technol.*, 31(1), 227–235 (1995).

118 S.R. Guiot, B. Tartakovsky, M. Lanthier, M.-J. Levesque, M.F. Manual, R. Beaudet et al., Strategies for augmenting the pentachlorophenol degradation potential of UASB anaerobic granules. *Water Sci. Technol.*, 45(10), 35–41 (2002).

119 H.-P. Kaballo, Y. Zhao and P.A. Wilderer, Elimination of p-chlorophenol in biofilm reactors – A comparative study of continuous flow and sequenced batch operation. *Water Sci. Technol.*, 31(1), 51–60 (1995).

120 G. Farabegoli, A. Chiavola and E. Rolle, Remediation of chlorophenol and phenol-contaminated groundwater by a sequencing batch biofilm reactor. *Water Sci. Tech.* 58, 295–230 (2008).

121 G.A. Hill, B.J. Milne and P.A. Nawrocki, Cometabolic degradation of 4-chlorophenol by *Alcaligenes eutrophus*. *Appl. Microbiol. Biotechnol.*, 46, 163–168 (1996).

122 J.-P. Arcangeli and E. Arvin, Biodegradation rates of aromatic contaminants in biofilm reactors. *Water. Sci. Technol.*, 31, 117–128 (1995).

123 H. Li, Z. Zhang, X. Xu, J. Liang and S. Xia, Bioreduction of parachloronitrobenzene in a hydrogen-based hollow-fiber membrane biofilm reactor: Effects of nitrate and sulfate, *Biodegradation*, 25(2), 205–215 (2014).

124 Ö Aktaş and F. Çeçen, Bioregeneration of activated carbon: A review. *Int. Biodeter. Biodeg.*, 59, 257–272 (2007).

125 W.J. Weber, Jr., M. Pirbazari and G.L. Melson, Biological growth on activated carbon: An investigation by scanning electron microscopy. *Environ. Sci. Technol.*, 12, 817–819 (1979).

126 G.M. Klecka, S.G. McDaniel, P.S. Wilson, C.L. Carpenter, J.E. Clarck, A. Thomas and J.C. Spain, Field-evaluation of a granular activated carbon fluid-bed bioreactor for treatment of chlorobenzene in groundwater. *Environ. Prog.*, 15, 93–107 (1996).

127 M.A. Ullah, H. Kadhim, R.A. Rastall and C.S. Evans, Evaluation of solid substrates for enzyme production by *Coriolus versicolor*, for use in bioremediation of chlorophenols in aqueous effluents. *Appl. Microbiol. Biotechnol.*, 54, 832–837 (2000).

128 C. Quintelas, E. Sousa, F. Silva, S. Neto and T. Tavares, Competitive biosorption of ortho-cresol, phenol, chlorophenol and chromium(VI) from aqueous solution by a bacterial biofilm supported on granular activated carbon. *Process Biochem.*, 41, 2087–2091 (2006).

129 L.R. Radovic and F. Rodrígues-Reinoso, Carbon materials in catalysis, in P.A. Throweir (Ed.), *Chemistry and Physics of Carbon*, vol. 25. Marcel Dekker, New York, 1997.

130 H.S. Shin, K.S. Yoo and J.K. Park, Removal of polychlorinated phenols in a sequential anaerobic–aerobic biofilm reactors packed with tire chips. *Water. Environ., Res.* 71, 363–367 (1999).

131 US Environmental Protection Agency. 2013 *Polychlorinated biphenyls*. http://www3.epa.gov/epawaste/hazard/tsd/pcbs/about.htm

132 O. Roots, A. Roose, A. Kull, I. Holoubek, P. Cupr and J. Klanova, Distribution pattern of PCBs, HCB and PeCB using passive air and soil sampling in Estonia. *Environ. Sci. Pollut. Res. Int.*, 17, 740–749 (2010).

133 US Environmental Protection Agency. Record of decision and responsiveness summary, Hudson River PCBs site, New York, 2016, https://www3.epa.gov/hudson/rod.htm

134 US Environmental Protection Agency. Record of decision operable units 3, 4, and 5 lower Fox River and Green Bay, WI, 2003, http://foxrivercleanup.com/bmos-resources/rod2.pdf.

135 K.M. McDonough, J.L. Fairey and G.V. Lowry, Adsorption of polychlorinated biphenyls to activated carbon: Equilibrium isotherms and a preliminary assessment of the effect of dissolved organic matter and biofilm loadings. *Water. Res.*, 42, 575–584 (2008).

136 B.V. Kjellerup, P. Paul, U. Ghosh, et al., Spatial distribution of PCB dechlorinating bacteria and activities in contaminated soil. *Appl. Environ. Soil. Sci.*, 2012 [online]. http://www.hindawi.com/journals/aess/2012/584970/.

137 M. Horwat, M. Tice and B.V. Kjellerup, Biofilms at work: Bio-, phyto- and rhizoremediation approaches for soils contaminated with polychlorinated biphenyls. *AIMS Bioengineering*, 2(4), 324–334 (2015).

138 J.T. Cookson, *Bioremediation Engineering: Design and Application*. McGraw-Hill, New York, 1995.

139 G. Mukerjee-Dhar, M. Shimura and K. Kimbara, Degradation of polychlorinated biphenyl by cells of *Rhodococcus opacus* strain TSP203 immobilized in alginate and in solution. *Enzyme Microb. Technol.*, 22, 34–41 (1998).

140 I. Sierra, J.L. Valera and M.L. Marina, Study of the biodegradation process of polychlorinated biphenyls in liquid medium and soil by a new isolated aerobic bacterium *(Janibacter* sp.). *Chemosphere*, 53, 609–616 (2003).

141 S.K. Fagervold, J.E. Watts, H.D. May and K.R. Sowers, Sequential reductive dechlorination of meta-chlorinated polychlorinated biphenyl congeners in sediment microcosms by two different *Chloroflexi* phylotypes. *Appl. Environ. Microbiol.*, 71, 8085–8090 (2005).

142 R.B. Payne, H.D. May and K.R. Sowers, Enhanced reductive dechlorination of polychlorinated biphenyl impacted sediment by bioaugmentation with a dehalorespiring bacterium. *Environ. Sci. Technol.*, 45, 8772–8779 (2011).

143 H. Lünsdorf, R.W. Erb, W.R. Abraham and K.N. Timmis, "Clay hutches": A novel interaction between bacteria and clay minerals. *Environ. Microbiol.*, 2, 161–2,168 (2000).

144 U. Ghosh, A. Weber, S. Jensen, J.N. Smith and R. John, Granular activated carbon and biological activated carbon treatment of dissolved and sorbed polychlorinated biphenyls. *Wat. Environ. Res.*, 71, 232–240 (1999).

145 J.W. Costerton, L. Montanaro and C.R. Arciola, Bacterial communications in implant infections: A target for an intelligence war. *Int. J. Artif. Organs.*, 30, 757–763, 2007.

146 A.J. Macedo, U. Kuhlicke, T.R. Neu, K.N. Timmis and W.R. Abraham, Three stages of a biofilm community developing at the liquid–liquid interface between polychlorinated biphenyls and water. *Appl. Environ. Microbiol.*, 71, 7301–7309 (2005).

147 J. Borja, J. Auresenia and S.M. Gallardo, Kinetics of polychlorinated biphenyl biodegradation using biofilm grown on biphenyl. *Asean Jour. Chem. Engin.*, 6(1), 44–52 (2006).

148 P.B. McLeod, M.J. van den Heuvel-Greve, R.M. Allen-King, S.N. Luoma and R.G. Luthy, Effects of particulate carbonaceous matter on the bioavailability of benzo[a] pyrene and 2,2′,5,5′-tetrachlorobiphenyl to the clam, *Macoma balthica. Environ. Sci. Technol.*, 38, 4549–4556 (2004).

149 G. Cornelissen, G.D. Breedveld, K. Naes, A.M. Oen and A. Ruus, Bioaccumulation of native polycyclic aromatic hydrocarbons from sediment by a polychaete and a gastropod: Freely dissolved concentrations and activated carbon amendment. *Environ. Toxicol. Chem.*, 25, 2349–2355 (2006).

150 D. Werner, U. Ghosh and R.G. Luthy, Modeling polychlorinated biphenyl mass transfer after amendment of contaminated sediment with activated carbon. *Environ. Sci. Technol.*, 40, 4211–4218 (2006).

151 R.B. Meagher, Phytoremediation of toxic elemental and organic pollutants. *Curr. Opin. Plant Biol.*, 3, 153–162 (2000).

152 M. Mackova, T. Macek, J. Ocenaskova, et al., Biodegradation of polychlorinated biphenyls by plant cells. *Int Biodeterior Biodegradation*, 39(4), 317–325 (1997).

153 L. Passatore, S. Rossetti, A.A. Juwarkar, et al., Phytoremediation and bioremediation of polychlorinated biphenyls (PCBs): State of knowledge and research perspectives. *J. Hazard. Mat.*, 278, 189–202 (2014).

154 H. Glasser, D.P. Chang and D.C. Hickman, An analysis of biomedical waste incineration. *J. Air Waste Manag. Assoc.*, 41, 1180–1188 (1991).

155 K. Lohman and C. Seigneur, Atmospheric fate and transport of dioxins: Local impacts. *Chemosphere*, 45, 161–171 (2001).

156 A. Hiraishi, H. Miyakoda, B.R. Lim, H.Y. Hu, K. Fujie and J. Suzuki, Toward the bioremediation of dioxin-polluted soil: Structural and functional analyses of in situ microbial populations by quinone profiling and culture-dependent methods. *Appl. Microbiol. Biotechnol.*, 57, 248–256 (2001).

157 M. Cooper, et al., Anaerobic microbial transformation of halogenated aromatics and fate prediction using electron density modelling. *Environ. Sci. Technol.*, 49, 6018–6028 (2015).

158 S. Yoshida, N. Ogawa, T. Fujii and S. Tsushima, Enhanced biofilm formation and 3-chlorobenzoate degrading activity by the bacterial consortium of *Burkholderia* sp. NK8 and *Pseudomonas aeruginosa* PAO1. *J. Appl. Microbiol.*, 106, 790–800 (2009).

20

Plant Pathogenic Bacteria: Role of Quorum Sensing and Biofilm in Disease Development

Deepak Dwivedi, Mayuri Khare, Himani Chaturvedi and Vinod Singh

Department of Microbiology, Barkatullah University, Bhopal, India

20.1 Introduction

Plant pathogenic bacteria cause severe economically damaging diseases, ranging from spots, mosaic patterns or pustules on leaves and fruits, or smelly tuber rots to plant death. Some cause hormone-based distortion of leaves and shoots called fasciation, or crown gall, a proliferation of plant cells producing a swelling at the intersection of stem and soil and on roots [1]. Most of the plant pathogenic bacteria are either Gram-positive, classified within the Phylum *Actinobacteria*, or Gram-negative, in the phylum *Proteobacteria*. In order to be able to colonize the plant they have specific pathogenicity factors. Five main types of bacterial pathogenicity factors are uses of cell wall–degrading enzymes, toxins, effector proteins, phytohormones and exopolysaccharides. Pathogens such as *Erwinia* species use cell wall–degrading enzymes to cause soft rot. *Agrobacterium* species change the level of auxins to cause tumors with phytohormones. Exopolysaccharides are produced by bacteria and block xylem vessels, often leading to the death of the plant [2]. The differentiation is based on chemical or physiological characteristics (e.g., cell wall composition, enzyme production, substrate utilization, etc.). Molecular characterization of 16S ribosomal RNA also may distinguish bacteria from one another. Ribosomes are coded by a highly conserved part of the bacterial chromosome and represent only a small part of the genome.

Plants are colonized by bacteria on their leaves, roots, seeds, and internal vasculature. Each tissue type has unique chemical and physical properties that represent challenges and opportunities for microbial colonists. Many microorganisms in the natural environment exist in multicellular aggregates generally described as biofilms, associated with solid surfaces and in intimate contact with other microbial cells [3, 4]. These cells can be differentiated from their suspended counterparts by generation of an extracellular polymeric substance (EPS) matrix, reduced growth rates, and the up- and down- regulation of specific genes. Attachment is a complex process regulated by diverse characteristics of the growth medium, substratum, and cell surface.

Noncellular materials such as mineral crystals, corrosion particles, clay or silt particles, or blood components, depending on the environment in which the biofilm has developed, may also be found in the biofilm matrix. Biofilm-associated organisms also differ from their planktonic (freely suspended) counterparts with respect to the genes that are transcribed. Biofilms may form on a wide variety of surfaces, including living tissues, indwelling medical devices, industrial or potable water system piping, or natural aquatic systems [5].

The interactions of bacteria are characterized by the interplay of plant immune responses after pathogen perception and the evasion of their responses. The ability to form biofilms in a particular host depends on the deployment of the plant immune response on contact with the bacterial pathogen or the perception of a particular plant signal that triggers the expression of genes that promote biofilm formation. The newly identified potato soft rotter *Pectobacterium carotovorum* subsp. *Brasiliense* was capable of forming biofilm-like structures in the xylem of susceptible potato plants, whereas in disease-tolerant plants, bacterial cells exhibited a planktonic motile behavior, with no evident formation of aggregates inside the vascular tissue [6]. The roots of *Arabidopsis thaliana* form mutant lines that accumulate high concentrations of salicylic acid (SA), a key molecule in the activation of systemic defense responses, showed a reduced bacterial colonization and biofilm formation by *P. aeruginosa*. Many species of bacteria use quorum sensing to coordinate gene expression according to the density of their local population. The nature of these relationships can be amicable, as characterized by symbiotic bacteria, or adversarial, as seen with pathogenic bacteria. There are numerous bacteria that have components of a quorum sensing system, which maintains the phenotype regulation [7].

Biofilms are bacterial communities in which cells are embedded in a matrix of extracellular polymeric compounds attached to a surface [8]. Living in biofilms, which helps protect bacteria from deleterious conditions [9], and the formation of biofilms appears to be an important factor in the disease cycle of bacterial pathogens in both animals and plants.

After the formation of microcolonies, the production of quorum sensing signals occurs, forming mature biofilm [10]. EPSs provide the architectural form of biofilms and stabilize their three-dimensional structure. This biofilms are often permeated by channels that act as a circulatory system, allowing the bacteria to exchange water, nutrients, enzymes, and signals, dispose of potentially toxic metabolites, and display enhanced metabolic cooperativity [11, 12]. The dispersal of biofilms allows bacteria to colonize other surfaces or substrates, thus completing a sequential developmental process of causing disease in plants.

20.2 Mechanism of Biofilm Formation

Biofilms are complex bacterial assemblages with a defined three-dimensional architecture, attached to solid surfaces, and surrounded by self-produced extracellular polymeric substances (EPS) a matrix composed of exopolysaccharides, proteins, lipids, and extracellular DNA. Bacteria live in aggregates and attach to solid surfaces, along with intimate contact to other bacterial cells either in natural or artificial surface, both biotic and abiotic environment. Biofilms have distinct physiological structures, cells within it vary from each other, up- and downregulation of genes also vary from cell to cell that are very responsive to various functions of their surroundings such as they modulate their metabolic functions, respond to nutrient products, waste product gradients, engage cell–cell communication, and contact with adjacent cells. Biofilms have great practical importance in medical, industrial, and agricultural settings, exhibiting both beneficial and detrimental activities.

EPS plays a key role in structure forming and functioning of biofilm communities in diverse environments. The matrix acts as an anion exchanger, which restricts the access of antimicrobial agents into the biofilm. Compounds surrounding the biofilm, which may enter by diffusion are also restricted in this way. The nature of both agents—compounds and biofilm along with EPS matrix—are the main determinants of this restriction. Toxins and different metals ions or cations are also reported to sequester in the presence of EPS and protect bacteria from environmental stresses (UV radiation, pH shift, desiccation, and osmotic shock).

20.2.1 Biofilm Formation *in Vitro* in Plants

Biofilm formation is a nearly universal bacterial trait, and biofilms are found on almost all natural and artificial surfaces [13, 14]. Biofilms have tremendous practical importance in industrial, medical, and agricultural settings, exhibiting both beneficial and detrimental activities. Although most fundamental work on microbial biofilms has focused on abiotic surfaces, it is clear that biofilms can and do form on biotic surfaces during host–microbe interactions. Most plant–bacterial associations rely on the physical interaction between bacteria and plant tissues. Direct observations of bacteria adhered to plant surfaces have revealed multicellular assemblies variably described as microcolonies, aggregates, and cell clusters [15–17].

The terrestrial environment harbors abundant and diverse microbial populations that can compete for and modify resource pools. In this complex and competitive environment, plants offer protective oases of nutrient-rich tissues. Plants are colonized by bacteria on their leaves, roots, seeds, and internal vasculature. Each tissue type has unique chemical and physical properties that represent challenges and opportunities for microbial colonists. Biofilms may form on association or at later stages, with significant potential to direct or modulate the plant–microbe interaction. Water availability and saturation levels in terrestrial environments vary considerably. Plant-associated bacteria experience different levels of hydration depending on the colonization site, prevailing climate conditions and soil composition. Water limitation has dramatic effects on biofilm structure and, therefore, the saturation level of a particular environment and a specific tissue will profoundly affect biofilm growth.

20.2.1.1 Gram-Negative Bacteria

Root-associated pseudomonads have been studied extensively, and many of these promote the growth of host plants or are used as biocontrol agents [18]. Species of *Pseudomonas* form dense biofilms on both abiotic and biotic surfaces, and are a primary model in biofilm research. *Pseudomonas putida* can respond rapidly to the presence of root exudates in soils, converging at root colonization sites and establishing stable biofilms [19].

Species of *Agrobacterium* and genera of symbiotic rhizobia not only cause neoplasia and symbiotic nodules on roots but also are effective root colonizers. Rhizobia preferentially associate with legume root hairs, stimulate root hair curling, infection thread elongation, and nodule formation on the appropriate host plant [20, 21]. Microscopy of rhizobial cells within curled root hairs reveals small biofilm-type aggregates that provide the inocula for root invasion; the rhizobial cells migrate down infection threads as biofilm-like filaments toward the root interior [22]. *Agrobacterium tumefaciens* and rhizobia can form dense, structurally complex biofilms on root surfaces, extensively coating the epidermis and root hairs, and these bacteria also form elaborate biofilms on abiotic surfaces [23, 24].

20.2.1.2 Gram-Positive Bacteria

Gram-positive microbes also effectively colonize the rhizoplane and are well represented in soil populations [25]. Many studies have been conducted to understand the molecular mechanisms of biofilm formation by *Bacillus subtilis*. It has been observed that in order to proliferate in the soil, *Bacillus subtilis* requires a nutrient source such as decaying organic material or plant roots [26]. The rhizosphere is rich in plant secretions that can provide bacteria with nutrients [27–29]. Bacteria in the rhizosphere can benefit the plant, and *Bacillus* spp.—including *B. subtilis*—are sold commercially as biological control agents for agriculture [28, 30, 31]. *Bacillus* spp. can promote growth and protect plants from infections by pathogenic bacteria, fungi and even nematodes. This protection is due to the secretion of antimicrobial compounds by *B. subtilis* coupled with induced systemic resistance in the plant (in response to *B. subtilis*), which enhances the capacity of the plant to resist various pathogens [32–36]. *B. subtilis* is readily isolated from the rhizosphere of plants, and the majority of root-associated strains are capable of forming robust biofilms in laboratory conditions [37, 38].

In addition, several other *Bacillus* spp. form biofilms on plant roots [39–41]. Biofilm formation on plant roots parallels *in vitro* biofilm formation in that the matrix exopolysaccharide (EPS) is required. Figure 20.1a shows the roots of six-day-old *Arabidopsis thaliana* seedlings 24 hours after inoculation with wild-type or EPS synthesis (*eps*)-mutant *B. subtilis* constitutively expressing YFP. Similarly, the master regulator Spo0A and the antirepressor SinI are required for root colonization. In many wild *B. subtilis* isolates, the presence of these genes, and thus the capacity of the organism to form a biofilm on the root, is also required for the strain to exert its maximal biocontrol effect [38]. *B. subtilis* colonization of *A. thaliana* roots also requires the production of surfactin, a lipopeptide antimicrobial that is also important for biofilm formation *in vitro* [39] (see Figure 20.1b). The production of surfactin and other lipopeptides by *Bacillus* spp. cells is one of the main mechanisms for plant biocontrol because these molecules can induce systemic resistance as well as strongly inhibit the growth of common plant pathogens such as *Pseudomonas syringae* [38, 39, 41]. To recruit *B. subtilis*, plants secrete small molecules. For example, when *A. thaliana* is infected with *P. syringae*, the plant secretes malic acid, and this enhances *B. subtilis* biofilm formation on the root (see Figure 20.1b). Furthermore, 54 root exudates from *P. syringae*-infected plants, or purified malic acid induces matrix gene expression in *subtilis* [40].

This phenomenon is not specific to *A. thaliana*; malic acid is also found in tomato root exudates (it is constitutively secreted in the rhizosphere by tomato plants) and, at high concentrations, can stimulate matrix gene expression and biofilm formation *in vitro* [43]. Tomato root exudates stimulate matrix gene expression in a manner that is dependent on the Spo0A kinase KinD. Mutants specifically lacking the extracellular CACHE domain of KinD are less-efficient colonizers of tomato roots.

Bacterial adherence to seeds is a process that strongly influences rhizosphere colonization. Suppliers often deliberately coat their seed stocks with microbial biofilms to inoculate the developing rhizosphere. Additionally, biofilms on seeds and sprouts used

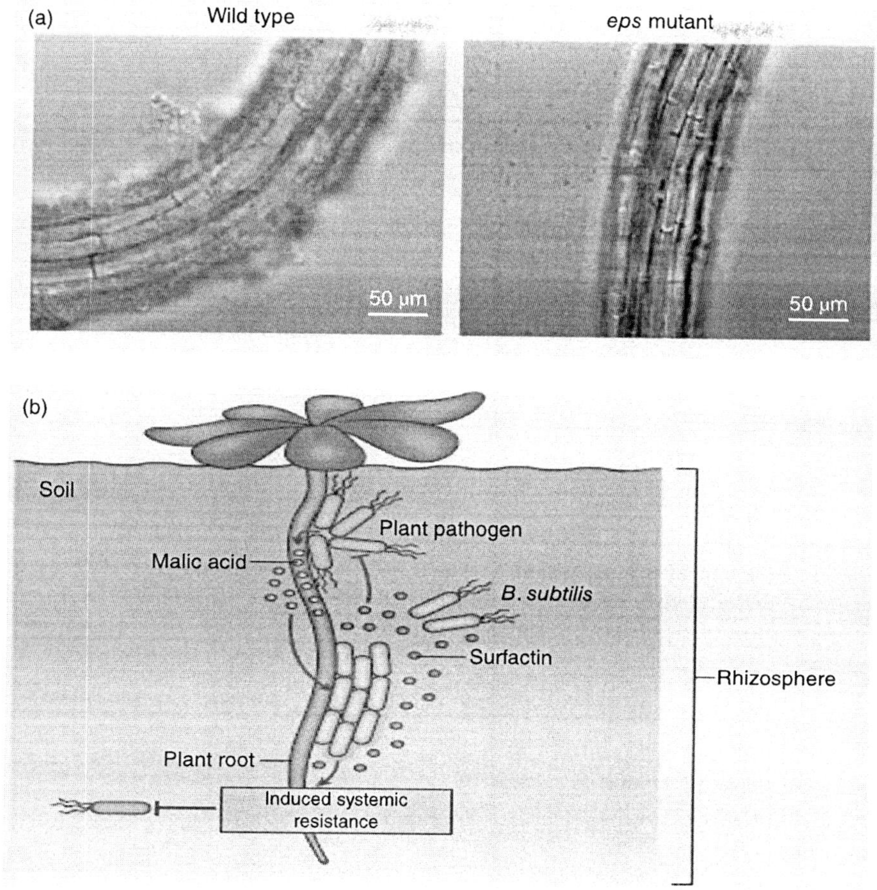

Figure 20.1 (a) Bacterial interaction with the roots of *A. thaliana* plant. (b) Biofilm formation and its interaction with plant root. Image source: nature reviews microbiology 2013 [42]. (*See color plate section for the color representation of this figure.*)

for human consumption are common sources of infection. *P. putida* adheres effectively to seeds and will subsequently colonize the rhizosphere [44]. Several *P. putida* mutants, including one in the lapA homolog of *P. fluorescens*, are deficient in seed adherence and biofilm formation on inert surfaces, emphasizing the overlap between these activities.

20.3 Quorum Sensing Mechanism

The term *quorum sensing* (QS) describes a well-understood mechanism of bacterial cell–cell communication and conveys the concept that certain traits are only expressed when bacteria are crowded together. The language used for this intercellular communication is based on small, self-generated signal molecules called autoinducers. Quorum sensing involves the exchange of low molecular weight, diffusible signal molecules between members of a localized population. If signal production by the population is greater than its loss by diffusion or inactivation, the signal accumulates to a threshold level and activates cognate receptor proteins. A key requirement for quorum sensing is growth of cells in close proximity, as in a biofilm or when confined in an enclosed, diffusion-limited environment. Either condition allows localized signal buildup to occur. The paradigm for quorum sensing is called autoinduction, it is the signal-mediated activation of bioluminescence. Through the use of autoinducers, bacteria can regulate their behavior according to population density.

20.3.1 Quorum Sensing Regulated Virulence Factors

Quorum sensing is thought to afford pathogenic bacteria mechanism to minimize host immune responses by delaying the production of tissue-damaging virulence factors until sufficient bacteria have amassed and are prepared to overwhelm host-defense mechanisms and establish infection. Quorum sensing systems are studied in a large number of Gram-negative bacterial species belonging to α, β, and γ subclasses of proteobacteria. Quorum sensing is considered as a potential novel target for antimicrobial therapy to control multi/all drug-resistant infections.

20.3.1.1 Mechanisms in Gram-Negative Bacteria

Quorum sensing is a mechanism used by pathogenic bacteria not only to modulate virulence factor production but also to adapt to the metabolic demands of living in community [45, 46]. Several Gram-negative bacteria use acylated homoserine lactone (HSL) signals in quorum sensing. Quorum sensing in *Pseudomonas aeruginosa,* an opportunistic human pathogen responsible for persistent and often incurable infections in immunocompromised people and individuals with cystic fibrosis, has been well studied. Expression of a number of extracellular virulence factors produced by *P. aeruginosa* is controlled by quorum sensing. Two QS systems, the *las* and *rhl* systems, have been identified in *P. aeruginosa*. In the *las* quorum sensing system, the *las*I gene product directs the formation of the diffusible extracellular signal, *N*-(3-oxododecanoyl)-l-HSL ($3OC_{12}$-HSL) [47], which interacts with *las*R [45, 48] to activate a number of virulence genes, including *las*B, *las*A, *apr*, *tox*A, and *las*I itself [48, 49]. Synthesis of the siderophore pyoverdine also is activated by the *las* system [50]. *P. aeruginosa* strains lacking a functional *las*R are avirulent in animal models [51]. Although $3OC_{12}$-HSL is diffusible, it appears to partition into cell membranes, and *P. aeruginosa* efflux pumps aid in the movement of this signal to the external environment [52].

The *rhl*I product catalyzes the synthesis of *N*-butyryl-l-HSL (C_4-HSL) [53, 54]. This diffusible signal [55], in conjunction with *rhl*R, activates expression of the *rhlAB* rhamnolipid synthesis genes, *rhl*I, and to some extent *las*B [56, 57]. Other virulence factors and secondary metabolites, including pyocyanin, cyanide, and chitinase, are positively regulated by the *rhl* system [58, 59], although direct transcriptional regulation of the genes involved in synthesis of these compounds has not been shown. A quorum sensing hierarchy exists with the *las* system controlling expression of the transcriptional activator *rhl*R [7, 60, 61]. Therefore, genes controlled by the *rhl* system require a functional *las* system for full activation.

It is becoming apparent that in addition to AHLs, alternative Gram-negative signaling molecules do exist. For example, the plant pathogen *Ralstonia solanacearum* produces 3-hydroxy-palmitic acid methyl ester as a novel signaling molecule, which together with

AHLs is used to regulate virulence [62]. *Xanthomonas campestris,* a cabbage pathogen, produces a diffusible extra cellular factor, which has yet to be chemically characterized, but is not an AHL [63]. In *Pseudomonas aeruginosa,* a third AI, designated PQS (Pseudomonas Quinolone Signal), was identified that is different from the other two AHL AIs produced by this organism, in that it is a 2-heptyl-3-hydroxy-4-quinolone [64, 65]. The structural similarity between PQS and antimicrobial quinolones is quite intriguing, although preliminary studies have showed no antimicrobial activity associated with PQS [66]. Butyrolactones have been isolated from *Pseudomonas aureofaciens* culture supernatants, and recently, a novel family of signaling compounds, identified as diketopiperazines (DKP), were discovered in cell-free supernatants of *P. aeruginosa, P. fluorescens, P. alcaligenes, Enterobacter agglomerans,* and *Citrobacter freundii* [67].

The infection by Enterohemorrhagic *Escherichia coli* (EHEC) O157:H7 can lead to potentially severe gastroenteritis and other extraintestinal manifestations, including fever, meningitis, and septicemia. The EHEC also express Shiga toxin (Stx) in the intestine, which also has its receptors located in the kidneys and the CNS. Stx is a potent inhibitor of protein synthesis and can get absorbed systemically leading to hemolytic uremic syndrome (HUS), seizures, cerebral edema, and coma. More so, in the management of EHEC infections, antimicrobial agents and antimotility agents are contraindicated, as they promote the expression and release of Stx, thereby increasing the occurrence and severity of HUS and CNS involvement. For all these reasons, it becomes more imperative, the need for innovative and cost-effective treatment modalities [68–70].

EHEC senses three signals to activate transcription of virulence genes: a bacterial aromatic AI (AI-3), produced by normal GI flora and two hormones (epinephrine/norepinephrine), produced by the host. Any of these signal molecules can trigger the QseC membrane bound sensor activity, ultimately leading to transcription of virulence genes. The QseC also activates the expression of second sensor QseE, which also helps in fine tuning the signaling cascade. All these transcription events lead to formation of attachment and effacing lesions in the intestine and the production of Stx [71, 72].

Rasko et al. in their study identified a lead structure, LED 209 (N-phenyl-4-(((phenylamino) thioxomenthyl)amino)-benzenesulfonamide), which selectively blocked binding of signals (AI-3/epinephrine and NE) to QseC, consequently inhibiting QseC-mediated activation of virulence gene expression. Significantly, LED 209 did not lead to killing of EHEC cell, as that would have initiated Stx production. LED 209 was also not toxic to the host cell and only inhibited key virulence traits of EHEC [73].

QseC homologs are present in at least 25 important human and plant pathogens, and AI-3/epinephrine/NE QseC receptor signaling system plays a central role in virulence of many important pathogens and qseC mutants of EHEC; *Salmonella typhimurium* and*Francisella tularensis* have been shown to be attenuated in animal models of infection [74, 75].

20.3.1.2 Mechanisms in Gram-Positive Bacteria

A number of Gram-positive organisms are known to employ quorum sensing systems. The nature of the signal molecules used in these systems varies from those of Gram-negative organisms, and thus far no Gram-positive bacteria have been shown to produce AHLs. Gram-positive QS systems typically make use of small post-translationally processed peptide signal molecules. These peptide signals interact with the sensor element of a histidine kinase two-component signal transduction system. Quorum sensing is used to regulate the development of bacterial competence in *Bacillus subtilis* and

Streptococcus pneumoniae, conjugation in *Enterococcus faecalis*, and virulence in *Staphylococcus aureus* [76, 77].

The virulence in *S. aureus* is dependent on temporal expression of a diverse array of virulence factors, including both cell-associated products, such as protein A, collagen, and fibronectin-binding protein and secreted products including lipases, proteases, alpha-toxin, toxin-I, beta-hemolysin, and enterotoxin [77]. During the early stages of *S. aureus* infection, surface proteins involved in attachment (collagen and fibronectin-binding protein) predominate. So is the case with protein involved in defense (protein A). However, once a high cell density is achieved at the site of infection, expression of *S. aureus* surface proteins is decreased and secreted proteins are preferentially expressed. Also, the *S. aureus* is known to exist in two phenotypes, an adhesive colonizer phenotype, which is tolerated by the host, and the severe invasive infective phenotype that can damage the host tissues and is responsible for the disease manifestations [70, 78].

The genetic basis for this temporal gene expression depends on two pleiotropic regulatory loci called *agr* (accessory gene regulator) [79] and *sar* (staphylococcal accessory gene regulator) [80]. The *agr* locus consists of two divergently transcribed operons, RNAII and RNAIII [81, 82]. The RNAII operon contains the *agrBDCA* genes that encode the signal transducer (AgrC) and response regulator (AgrA), and AgrB and AgrD, which are involved in generating the QS signal molecule, the autoinducing cyclic thiolactone peptide (AIP) [82, 83]. The RNAIII operon encodes a hemolysin and is itself a regulatory RNA that plays a key role in agr response.

During *S. aureus* quorum sensing, the AgrC signal transducer is autophosphorylated in response to the octapeptide signal molecule, which, in turn, leads to phosphorylation of the AgrA response regulator [84] Phosphorylated AgrA stimulates transcription of RNAIII and this, in turn, upregulates expression of numerous *S. aureus* exoproteins as well as the *agrBDCA* locus [85]. The latter leads to a rapid increase in the synthesis and export of the octapeptide signal molecule. At the second regulatory locus, the *sar* gene product (SarA) functions as a regulatory DNA-binding protein to induce expression of both the RNAII and RNAIII operons of *agr* locus [86].

The survival of *S. aureus* is also dependent on genes regulated by a second QS system, RAP/TRAP. Here, the proposed AI, RNAIII activating protein (RAP) is believed to be secreted by an as-yet-unknown mechanism. This reenters the cell and activates the target of RAP (TRAP). The activated TRAP upregulates *agr* expression and promotes cellular adherence, which is essential for biofilm formation [87].

This system has the potential to be exploited as another alternative to antibiotics. This can be accomplished by using a heptapeptide, which inhibits the activity of TRAP. The heptapeptide RNAIII inhibiting peptide (RIP) has been shown to inhibit phosphorylation of TRAP and *agr* expression. It was shown that RIP is effective against some strains of MRSA and a nonpeptide analog of RIP, hamamelitannin, was identified that has been shown to prevent device-associated MRSA infections in a concentration dependent manner [87, 88].

20.3.2 Biofilm Formation in *Candida*

The process of biofilm formation by *Candida albicans* involves three main steps: the initial colonization of the substratum by the yeast cells, growth and hypha formation, and the production of an extracellular matrix, primarily composed of β-1,3-glucan. The mature biofilm consists of yeasts, hyphae, and pseudohyphae; however, eventually, the yeast cells leave the biofilm. In *C. albicans* also the QS can modulate all stages of biofilm formation (i.e., attachment, maturation, and dispersal) [89, 90].

The best-characterized QS molecule produced by *C. albicans* is farnesol, [91] which regulates the inter-conversion between its yeast and filamentous form. In *in vitro* experiments, farnesol has been shown to reduce the size of biofilms. The other QS molecule that may also alter biofilm development in *C. albicans* is "tyrosol" [92]. In experiments, farnesol has been shown to repress hyphal growth by inhibiting the Ras1-adenylate cyclase-protein kinase A signaling pathway. However, the role of farnesol in multicellular population can be better understood by discovering mutants with altered farnesol response and farnesol production.

20.4 Plant Pathogenic Bacteria Diversity and Plant Diseases

Microbes associated with plants have a strong influence on plant growth development and yield. Natural microflora present in environment can associate with plants and can have beneficial or deleterious effects on them. Pathogenic bacteria can enter plants through natural openings or wounds and can colonize the host, causing diseases and leading to loss of entire yield.

Bacterial plant diseases are most severe in tropical and subtropical countries, where bacteria receive ideal climatic conditions for their growth resulting in more crop yield losses in these countries. Pathogenic bacteria incite diseases in plants by penetrating into host tissues through natural openings, such as hydathodes, stomata, lenticels, stigma, nectarthodes, or through wounds, and bacteria are directly deposited by insect vectors [93]. Leaf and fruit spots, blights, cankers, vascular wilts, rots, and tumors are characteristic symptoms of bacterial diseases.

The bacterial pathogens of plants are the bacteria of the genera *Pseudomonas, Erwinia, Ralstonia,* and *Xanthomonas*. Certain bacteria *Streptomyces* spp. and *Clavibacter* spp. also cause diseases in plants. The pathogenicity and virulence factors used by phytopathogenic bacteria cause diseases in plants [94].

As the plant bacterial pathogens are extracellular, they deploy a delivery of secreted virulence factors to interfere with host cell processes from outside plant cells. These include production of protein virulence factors (effectors), which are directly injected into host plant cell cytoplasm via a specialized type III secretion path [95, 96], secretion of low-molecular-weight phytotoxins, which are produced into apoplast [97], production of exopolysaccharides [98], and cell wall–degrading enzymes [99, 100].

Table 20.1 summarizes a list of important pathogens and diseases caused by them.

20.5 Blocking Quorum Sensing and Virulence in Combating Phytopathogen

Microbial success critically depends on the ability to perceive and respond rapidly to changes in the local environment. For any individual bacterium, one of the most important of these environmental factors will be the number and growth status of its fellows within its immediate vicinity. Such information may allow the bacterium to anticipate

Table 20.1 List of Bacterial Diseases in Various Crops.

Crop	Disease	Causal Organism
Rice	Bacterial blight	*Xanthomonas oryzae* pv. *oryzae*
	Bacterial leaf streak	*Xanthomonas oryzae* pv. *oryzicola*
	Foot rot	*Erwinia chrysanthemi*
	Grain rot	*Burkholderia glumae*
Wheat	Sheath brown rot	*Pseudomonas fuscovaginae*
	Bacterial mosaic	*Clavibacter michiganensis* subsp. *Tessellarius*
	Bacterial leaf blight	*Pseudomonas syringae* subsp. *Syringae*
	Bacterial sheath rot	*Pseudomonas fuscovaginae*
	Basal glume rot	*Pseudomonas syringae* pv. *Atrofaciens*
	Black chaff	*Xanthomonas campestris* pv. *Translucens*
	Pink seed	*Erwinia rhapontici*
	Spike blight	*Rathayibacter tritici*
Soyabean	Bacterial blight	*Pseudomonas amygdali* pv. *glycinea*
	Bacterial pustules	*Xanthomonas axonopodis* pv. *glycines*
	Bacterial tan spot	*Curtobacterium flaccumfaciens* pv. *flaccumfaciens*
	Bacterial wilt	*Curtobacterium flaccumfaciens* pv. *flaccumfaciens*
	Wildfire	*Pseudomonas syringae* pv. *Tabaci*
Sugarcane	Gumming disease	*Xanthomonas axonopodis* pv. *Vasculorum*
	Leaf scald	*Xanthomonas albilineans*
	Mottled stripe	*Herbaspirillum rubrisubalbicans*
	Ratoon stunting disease	*Leifsonia xyli* subsp. *Xyli*
	Red stripe (top rot)	*Acidovorax avenae* subsp. *avenae*
Cotton	Bacterial blight of cotton	*Xanthomonas citri* subsp. *malvacearum*
	Crown gall	*Agrobacterium tumefaciens*
	Lint degradation	*Erwinia herbicola*
Potato	Bacterial wilt	*Ralstonia solanacearum*
	Dickeya solani	*Dickeya solani*
	Black leg and soft rot	*Pectobacterium carotovorum* subsp. *Atrosepticum*
	Pink eye	*Pseudomonas fluorescens*
	Ring rot	*Clavibacter michiganensis* subsp. *sepedonicus*
	Zebra chip	'*Candidatus* Liberibacter solanacearum'
Tomato	Bacterial canker of tomato	*Clavibacter michiganensis* subsp. *Michiganensis*
	Bacterial speck	*Pseudomonas syringae* pv. *Tomato*
	Bacterial spot	*Xanthomonas campestris* pv. *Vesicatoria*
	Bacterial stem rot and fruit rot	*Erwinia carotovora* subsp. *Carotovora*
	Bacterial wilt	*Ralstonia solanacearum*

Table 20.1 (Continued)

Crop	Disease	Causal Organism
	Pith necrosis	*Pseudomonas corrugata*
	Syringae leaf spot	*Pseudomonas syringae* pv. *Syringae*
Mango	Bacterial black spot	*Xanthomonas campestris* pv. *Mangiferaeindicae*
	Bacterial fruit rot	*Pectobacterium carotovorum* subsp. *carotovorum*
	Crown gall hi	*Agrobacterium tumefaciens*
Citrus	Bacterial spot	*Xanthomonas campestris* pv. *citrumelo*
	Black pit (fruit)	*Pseudomonas syringae*
	Blast	*Pseudomonas syringae*
	Citrus canker	*Xanthomonas axonopodis*
	Citrus variegated chlorosis	*Xylella fastidiosa*
	Huanglongbing	*Candidatus Liberibacter asiaticus*
Onion	Bacterial blight of leek	*Pseudomonas syringae pv. Porri*
	Bacterial leaf streak and bulb rot	*Pseudomonas viridiflava*
	Bacterial soft rot	*Dickeya chrysanthem i (syn. Erwinia chrysanthemi*
	Center rot	*Pantoea ananatis (syn. Erwinia ananatis), P. agglomerans (syn. E. herbicola)*
	Enterobacter bulb decay	*Enterobacter cloacae*
	Slippery skin	*Burkholderia gladioli pv. alliicola*
	Sour skin	*Burkholderia cepacia*
	Leaf blight	*Xanthomonas axonopodis pv. allii*
Chickpea	Bacterial blight	*Xanthomonas campestris* pv. *Cassia*
	Bacterial leaf spot	*Burkholderia andropogonis*
	Bacterial wilt	*Xanthomonas campestris (Pam.) Dowson*
Millets	Bacterial spot	*Pseudomonas syringae*
	Bacterial leaf streak	*Xanthomonas campestris* pv. *Pennamericanum*
	Bacterial leaf stripe	*Acidovorax avenae*
Tea	Bacterial canker	*Xanthomonas campestris* pv. *theicola*
	Bacterial shoot blight	*Pseudomonas avellanae* pv. *Theae*
	Crown gall	*Agrobacterium tumefaciens*
Coffee	Bacterial blight of coffee	*Pseudominas syringae pv garcae*
Tobacco	Angular leaf spot	*Pseudomonas amygdali* pv. *Tabaci*
	Granville wilt	*Ralstonia solanacearum*
	Hairy roots	*Agrobacterium rhizogenes*
	Hollow stalk	*Erwinia carotovora* subsp. *carotovora*
	Leaf gall	*Rhodococcus fascians*

future availability of nutrients or build up of toxins. Quorum sensing is a signaling system that occurs in the pathogenic kingdom to sense its own population density and synchronize the expression of the virulence gene via the secretion of small, diffusible signal molecules, such as *N*-acyl-homoserine lactone (AHL), termed autoinducers [101]. Autoinducers play a critical role in triggering virulence gene expression in QS-dependent pathogens, such as in the production of rotting enzyme. Interfering with the microbial QS system by quorum quenching (QQ) has been suggested as a potential strategy for disease controls [102] because QQ aims to shut down the virulence expression in pathogenic bacteria rather than restrict cell growth and has shown potential to overcome drug toxicities, complicated super-infections, and antibiotic resistance [103].

Plants have evolved strategies to interfere with the bacteria's AHL signaling system to prevent them from initiating a pathogenic attack. Such interference could include the production of signal mimics, signal blockers, or signal-degrading enzymes or the production of compounds that block the activity of the AHL-producing enzymes. Examples of compounds with signal-inhibiting properties are known, and there is evidence for the production of AHL inhibitor molecules by bacteria, algae, and plants. In the simplest case, an AHL produced by one bacterial species may be antagonistic to the activity of an AHL used by a second species. This is seen in the case of *Chromobacterium violaceum*, where the cognate AHL contains an acyl side chain of six carbons. The presence of this molecule or closely related analogues induces the production of the purple pigment violacein. However, related molecules with acyl side chains of 10 or more carbons do not activate violacein production and actually inhibit the normal response to the cognate molecule [104]. More complex blocking molecules are produced by the Australian marine alga, *Delirea pulchra*. This macroalga produces halogenated furanones, which have some structural similarity to AHLs. It appears that *D. pulchra* uses these AHL blockers in vivo to inhibit bacterial sell swarming and attachment responses, thus preventing the buildup of bacterial biofilms on the algal surfaces [105]. Teplitski et al. [106] reported AHL inhibitor activities in exudates from pea seedlings. The compounds responsible have not been identified, but they preferentially partition into polar solvents (unlike the AHL molecules themselves).

Over the last decade, many microbes capable of degrading QS molecules have been documented. The first report of such degradation was the isolation of *Bacillus* sp. 240B [107]. The strain 240B can produce lactonase, cleave the lactone ring from the acyl moiety of AHLs, and render the AHLs inactive in signal transduction. The expression of the *aiiA* gene encoding AHL-lactonase in transformed *Pectobacterium carotovorum* has been shown to significantly reduce the formed QS molecule AHL, thereby quenching potato soft-rot by *Pec. carotovorum*. Soon after, the QQ microbes *Variovorax paradoxus* and *Ralstonia* sp. XJ12B were isolated, which are able to secrete the AHL-degrading enzyme with acylase activity [108, 109]. Subsequent database searches for the homologs of the QQ enzyme in complete bacterial genomes have shown the existence of related enzymes in a wide range of species. Most of the characterized microbes are spread among the QS-mediated pathogens, and a few data related to AHL-degradation are from non-QS bacteria [110]. Many efforts are being invested to search for potential quorum quenchers and their roles in the QS-mediated mechanism in pathogens.

QS can be quenched by degrading AHL signal molecules using QQ enzymes to cause interference with the expression of AHL-regulated traits [110, 111]. The QQ enzyme shows high specificity toward QS signal molecules, but no influence on other molecules

[112]. Some microbes not only produce QQ enzymes as a defense strategy against their competitors but also utilize AHL and its enzymatic degradation products as the sole carbon and nitrogen sources for cell growth [113]. When AHL-acylase from *Streptomyces* sp. was applied to a *P. aeruginosa* culture, a reduction of virulence factor production but not cellular growth was observed [114]. The *comamonas* strain D1 harboring AHL-acylase can enzymatically inactivate the QS signal molecule AHLs [115]. It degrades AHL with acyl-side chains ranging from 4 to 16 carbons with or without 3-oxo or 3-hydroxy substitutions. When co-cultured with other pathogens, some QS-dependent functions, such as violacein production by *C. violaceum* and pathogenicity and antibiotic production in *Pectobacterium*, can be quenched. Recombinant *E. coli* with AiiA lactonase activity was shown to attenuate the pathogenicity of *E. carotovora* when co-cultured together [116]. The expression of *aiiA* in the insecticide *B. thuringiensis* could confer the strain with a strong biocontrol capacity against the AHL-dependent pathogen *E. carotovora* when co-inoculated with the pathogen [107]. The *aac* gene encodes an AHL-acylase from *R. solanacearum* [117]. Its heterologous expression in *C. violaceum* CV026 effectively inhibited violacein and chitinase activity, which were regulated by the QS mechanism, indicating that the acylase Aac could control AHL-dependent pathogenicity. The expression of the AHL-lactonase from *B. thuringiensis* in the phytopathogen *E. carotovora* resulted in substantially reduced levels of AHL via the enzymatic degradation of QS signal molecules, leading to decreased pectolytic enzyme activities and attenuated *E. carotovora* disease symptoms on potatoes and cabbage [107]. To determine the capability of the QQ enzymes to block pathogenicity and toxoflavin production by the QS pathogen *Bur. glumae*, which causes rice grain rot, the AHL-lactonase gene *aiiA* was introduced into this bacteria [118]. The results showed that the AHL level in the transformants was reduced significantly, and that the severity of the soft rot caused by *Pec. carotovorum* sp. *Carotovorum* could be decreased when co-cultured with the recombinant *Bur. glumae*. The rice seedling or rice grain rot could not be shut down in the *aiiA*-transformant, suggesting that the gene *aiiA* encoding enzyme did not affect the virulence or toxoflavin production in *Bur. glumae*.

Other types of QS signal molecules in addition to AHL-like molecules are presumed to occur in *Bur. glumae* that regulate virulence production. The opportunistic pathogen *P. aeruginosa* can cause high mortality rates and typically occurs in immune-compromised patients and cases of hospital-acquired infections [119]. The strain *P. aeruginosa* PAO1 and its closely related pseudomonad are able to degrade and utilize AHL with long-chains (≥8 carbons) but not short-chains as the sole carbon and nitrogen sources for cell growth.

The QQ enzyme expressed in *P. aeruginosa* has been confirmed to reduce the accumulation of the long AHL signal 3-oxo-C12-HSL and prevent the accumulation of the short AHL signal C4-HSL, which results in a decrease in the swarming motility and virulence factor production [120]. The expression of the AHL-acylase *aiiD* in *P. aeruginosa* PAO1 changed the QS-regulated phenotypes, *i.e.*, attenuated its ability to produce elastase and pyocyanin, paralyze nematodes and form a biofilm. When the AHL-acylase gene *pvdQ* was transformed in *P. aeruginosa* PAO1, the overproduced PvdQ was shown to be less virulent than the wild-type strain in a *Caenorhabditis elegans* infection model [121]. More than 75 percent of nematodes exposed to the transformed strain survived and continued to grow in a fast-acting paralysis assay when using this strain as a food source. Hypothetically, AHL-acylase enables *P. aeruginosa* PAO1 to modulate its own

QS-dependent pathogenic potential. A very limited number of QQ enzymes have been characterized, and future work should elucidate the diversity of this class of enzymes and its role in microbial cell–cell communication.

20.6 Conclusion

Bacteria adhere to environmental surfaces in multicellular assemblies described as biofilms. Plant-associated bacteria interact with host tissue surfaces during pathogenesis and symbiosis, and in commensal relationships. Observations of bacteria associated with plants increasingly reveal biofilm-type structures that vary from small clusters of cells to extensive biofilms. The surface properties of the plant tissue, nutrient and water availability, and the proclivities of the colonizing bacteria strongly influence the resulting biofilm structure. Quorum sensing has been found to play a major role in regulating the virulence factors utilized by phytopathogens in forming biofilms. Blocking quorum sensing is an intelligent mechanism to stop this cascade. Many plants have evolved strategies to block the signaling system of these pathogens, which is known as quorum quenching. Many bacterial species have also been documented which are capable of degrading quorum sensing molecules.

With the emergence of antibiotic-resistant strains of bacteria, the available options for treating bacterial infections have become very limited, and the search for a novel general antibacterial therapy has received much greater attention. Quorum quenching can be used to control disease in a quorum sensing system by triggering the pathogenic phenotype. From a biocontrol point of view, a combination of the QQ approach with other treatments, such as antibiotics, to obtain a synergistic effect is a potential strategy that could potentially increase the susceptibility of bacteria to antibiotic treatment.

References

1 A. K. Vidaver, and P. A. Lambrecht, Bacteria as plant pathogens, *The Plant Health Instructor* (2004).
2 Creamer, H. Rebecca, A. Hubble and Lewis, Curtovirus Infection of Chile Pepper in New Mexico, *Plant Diseases* 89 (5), 480–486 (2005).
3 P. Stoodley, K. Sauer, D. G. Davies and J. W. Costerton, Biofilms as complex differentiated communities, *Annu Rev Microbiol.*, 56: 187-209 (2002).
4 J. S. Webb, M. Givskov and S. Kjelleberg, Bacterial biofilms: prokaryotic adventures in multicellularity. *Curr Opin Microbiol.*, 6: 578–585(2003).
5 M. Rodney and Donlan, Biofilms: Microbial Life on Surfaces, Emerg Infect Dis, 8(9): 881–890 (2002).
6 G. C. Kubheka, T. A. Coutinho, N. Moleleki and L. N. Moleleki, Colonization patterns of and Cherry-Tagged Pectobacterium carotovorum subsp. brasiliense strain in potato plants, Phytopathology, 103: 1268–1279 (2013).
7 R. Teresa, Kievit and Barbara H. Iglewski, Bacterial Quorum Sensing in Pathogenic Relationships, Infect. Immun. September, 68: 9:4839–4849 (2000).

8 S. S. Branda, S. Vik, L. Friedman, and R. Kolter, Biofilms: The matrix revisited, *Trends Microbiol.*, 13: 20–26 (2005).
9 M. E. Davey, and G. A. O'Toole, Microbial biofilms: From ecology to molecular genetics. Microbiol., Mol. Biol. Rev. 64: 847–867 (2000).
10 L. V. Rinaudi, and W. Giordano, An integrated view of biofilm formation in rhizobia. *FEMS Microbiol. Lett*, 304: 1–11 (2010).
11 J. W. Costerton, Z. Lewandowski, D. E. Caldwell, D. R. Korber and H. M. Lappin-Scott, Microbial biofilms. *Annu. Rev. Microbiol.*, 49: 711–745 (1995).
12 N. R. Stanley, and B. A. Lazazzera, Environmental signals and regulatory pathways that influence biofilm formation, *Mol. Microbiol.*, 52: 917–924 (2004).
13 L. Hall-Stoodley, J. W. Costerton, P. Stoodley, Bacterial biofilms: from the natural environment to infectious diseases. *Nature Rev. Microbiol.* 2, 95–108 (2004).
14 P. S. Stewart, & M. Franklin, J. Physiological heterogeneity in biofilms, Nature Rev. Microbiol., 6, 199–210 (2008).
15 M.R. Parsek, C. Fuqua, Biofilms: Emerging themes and challenges in studies of surface-associated microbial life, *J Bacteriol.*, 186:4427–4440 (2003).
16 C.E. Morris, J.M. Monier, The ecological significance of biofilm formation by plant-associated bacteria, *Annu Rev Phytopathol*, 41:429–453 (2004).
17 J.M. Monier, S.E. Lindow, Frequency, size, and localization of bacterial aggregates on bean leaf surfaces, *Appl Environ Microbiol.*, 70:346–355 (2004).
18 G.V. Bloemberg, B.J.J. Lugtenberg, *Bacterial biofilms on plants: relevance and phenotypic aspects.* In Microbial Biofilms, Edited by Ghannoum M, O'Toole GA, ASM Press, 2004.
19 B.J.J. Lugtenberg, L. Dekkers, G.V. Bloemberg, Molecular determinants of rhizosphere colonization by *Pseudomonas*, *Annu Rev Phytopathol*, 39:461–490 (2001).
20 M. Espinosa-Urgel, R. Kolter, J.L. Ramos, Root colonization by *Pseudomonas putida*: love at first sight, *Microbiology*, 148:341–343 (2002).
21 G. Smit, S. Swart, B.J.J. Lugtenberg, J.W. Kijne, Molecular mechanisms of attachment of *Rhizobium* bacteria to plant roots, *Mol Microbiol.*, 6:2897–2903 (1992).
22 D.J. Gage, Infection and invasion of roots by symbiotic, nitrogen-fixing rhizobia during nodulation of temperate legumes, *Microb Mol Biol Rev*, 68:280–300 (2004).
23 D.J. Gage, Analysis of infection thread development using Gfp- and DsRed-expressing *Sinorhizobium meliloti*, *J Bacteriol.*, 184:7042–7046 (2002).
24 B.E. Ramey, A.G. Matthysse, C. Fuqua, The FNR-type transcriptional regulator SinR controls maturation of *Agrobacterium tumefaciens* biofilms, *Mol Microbiol.*, 52:1495–1511 (2004).
25 A.G. Matthysse, S. McMahan, The effect of the *Agrobacterium tumefaciens* attR mutation on attachment and root colonization differs between legumes and other dicots, *Appl Environ Microbiol.*, 67:1070–1075 (2001).
26 T. Boureau, M.A. Jaques, R. Berruyer, Y. Dessaux, H. Dominguez, C.E. Morris,. Comparison of the phenotypes and genotypes of biofilms and solitary epiphytic bacterial populations on broad-leaved endive. *Microb Ecol.*, 47:87–95 (2004).
27 A. M. Earl, R. Losick & R. Kolter, Ecology and genomics of *Bacillus subtilis*, *Trends Microbiol.*, 16, 269–275 (2008).
28 T. Danhorn, & C. Fuqua, Biofilm formation by plant associated bacteria, *Annu. Rev. Microbiol.*, 61, 401–422 (2007).
29 B. Lugtenberg, & F. Kamilova, Plant-growth-promoting rhizobacteria, *Annu. Rev. Microbiol.*, 63, 541–556 (2009).

30 B. E. Ramey, M. Koutsoudis, S. B. von Bodman, Fuqua, C, Biofilm formation in plant-microbe associations, *Curr. Opin. Microbiol.*, 7, 602–609 (2004).

31 M. Morikawa , Beneficial biofilm formation by industrial bacteria *Bacillus subtilis* and related species, *J. Biosci. Bioeng*, 101, 1–8 (2006).

32 E. A. Emmert & J Handelsman, Biocontrol of plant disease: a (Gram-) positive perspective, *FEMS Microbiol. Lett*, 171, 1–9 (1999).

33 J. W. Kloepper, C. M. Ryu, & S. Zhang, Induced systemic resistance and promotion of plant growth by *Bacillus* spp, *Phytopathology*, 94, 1259–1266 (2004).

34 C. M. Ryu, J. F. Murphy, K. S. Mysore, & J. W. Kloepper,. Plant growth-promoting rhizobacteria systemically protect *Arabidopsis thaliana* against Cucumber mosaic virus by a salicylic acid and NPR1-independent and jasmonic acid-dependent signaling pathway, *Plant J*, 39, 381–392 (2004).

35 D. K Choudhary & B. N. Johri, Interactions of *Bacillus* spp. and plants – with special reference to induced systemic resistance (ISR), Microbiol. *Res.*, 164, 493–513 (2009).

36 A. Perez-Garcia, D. Romero, & A. de Vicente, Plant protection and growth stimulation by microorganisms: biotechnological applications of Bacilli in agriculture, *Curr. Opin. Biotechnol.*, 22, 187–193 (2011).

37 J. K Vessey, Plant growth promoting rhizobacteria as biofertilizers, *Plant Soil* 255, 571–586 (2003).

38 R. Fall, R. F. Kinsinger, & K. A Wheeler , A simple method to isolate biofilm-forming *Bacillus subtilis* and related species from plant roots. *Syst. Appl. Microbiol.*, 27, 372–379 (2004).

39 Y. Chen, *Biocontrol of tomato wilt disease by Bacillus subtilis isolates from natural environments depends on conserved genes mediating biofilm formation.* Environ. Microbiol. (2012).

40 H. P. Bais, R. Fall, & J. M Vivanco, Biocontrol of *Bacillus subtilis* against infection of *Arabidopsis* roots by *Pseudomonas syringae* is facilitated by biofilm formation and surfactin production. *Plant Physiol.* 134, 307–319 (2004).

41 T. Rudrappa, K. J Czymmek, P. W. Pare & H. P Bais, Root-secreted malic acid recruits beneficial soil bacteria. *Plant Physiol.*, 148, 1547–1556 (2008).

42 M. Ongena, Surfactin and fengycin lipopeptides of *Bacillus subtilis* as elicitors of induced systemic resistance in plants, *Environ. Microbiol.*, 9, 1084–1090 (2007).

43 H. Vlamakis, Y. Chai, P. Beauregard, R. Losick, R. Kolter, *Sticking together: building a biofilm the Bacillus subtilis way, Nature Reviews Microbiology*, 1–13 (2013).

44 M. Espinosa-Urgel, A. Salido, J.L. Ramos, Genetic analysis of functions involved in adhesion of *Pseudomonas putida* to seeds, *J Bacteriol.*, 182:2363–2369 (2000).

45 V.E. Wagner, D. Bushnell, L. Passador, A.I. Brooks, B.H. Iglewski, Microarray analysis of*Pseudomonas aeruginosa Quorum sensing* regulons: Effects of growth phase and environment, *J Bacteriol.*, 185:2080–95 (2003).

46 C.D. Sifri, *Quorum sensing*: Bacteria Talk Sense, *Clin Infect Dis*, 47:1070–6 (2008).

47 M. Schuster, C.P. Lostroh, T. Ogi, E.P, Greenberg, Identification, timing, and signal specificity of *Pseudomonas aeruginosa quorum*-controlled genes: A transcriptome analysis, *J Bacteriol.*, 185:2066–2079 (2003).

48 P.M. Dunman, E. Murphy, S. Haney, D. Palacios, G. Tucker-Kellogg, S Wu, et al., Transcription profiling-based identification of Staphylococcus aureus genes regulated by the *agr* and/ or *sarA* loci, *J Bacteriol.*, 183:7341–7353 (2001).

49 C. Arevalo-Ferro, M. Hentzer, G. Reil, A. Görg, S. Kjelleberg, M. Givskov, et al., Identification of *Quorum sensing* regulated proteins in the opportunistic pathogen Pseudomonas aeruginosa by proteomics, *Environ Microbiol.*, 5: 1350–1369 (2003).

50 H. B. Tang, E. DiMango, R. Bryan, M. Gambello, B.H. Iglewski, J.B. Goldberg, A. Prince, *Infect Immun.*, 64:37–43 (1996).

51 K. Evans, L. Passador, R. Srikumar, E. Tsang, J. Nezezon, K. J. Poole, *Bacteriol.*, 180:5443–5447 (1998).

52 J. P. Pearson, C Van Delden, B. H. Iglewski, *J Bacteriol.*, 181:1203–1210 (1999).

53 M. K. Winson, M. Camara, A. Latifi, M. Foglino, S.R. Chhabra et al., *Proc Natl Acad Sci USA*, 92:9427–9431(1995).

54 J P Pearson, E C Pesci, B H Iglewski, *J Bacteriol.*, 179, 5756–5767 (1997).

55 J P Pearson, L Passador, B H Iglewski, E P Greenberg, *Proc Natl Acad Sci USA*, 92, 1490–1494 (1995).

56 A Latifi, K M Winson, M Foglino, B W Bycroft, G S A B Stewart, A Lazdunski, P Williams, *Mol Microbiol.*, 17, 333–344 (1995).

57 U A Ochsner, J. Reiser, *Proc Natl Acad Sci USA*, 92, 6424–6428 (1995).

58 U A Ochsner, A K Koch, A Fiechter, J Reiser, *J Bacteriol.*, 176, 2044–2054 (1994).

59 E C Pesci, J P Pearson, P C Seed, B H Iglewski, *J Bacteriol.*, 179, 3127–3132 (1997).

60 A Latifi, M Foglino, K Tanaka, P Williams, A Lazdunski, *Mol Microbiol.*, 21, 1137–1146 (1996).

61 AB Flavier, SJ Clough, MA Schell, TP Denny, Identification of 3-hydroxypalmitic acid methyl ester as a novel autoregulator controlling virulence in *Ralstonia solanacearum*, *Mol Microbiol.*, 26, 251–259 (1997).

62 CE Barber, JL Tang, JX Feng, MQ Pan, TJ Wilson, H Slater, et al., A novel regulatory system required for pathogenicity of *Xanthomonas campestris* is mediated by a small diffusible signal molecule, *Mol Microbiol.*, 24, 555–66 (1997).

63 EC Pesci, JB Milbank, JP Pearson, S McKnight, AS Kende, EP Greenberg, et al., Quinolone signaling in the cell-to-cell communication system of *Pseudomonas aeruginosa*, *Proc Natl Acad Sci USA*, 96, 11229–11234 (1999).

64 P Gamard, F Sauriol., N Benhamou, RR Belanger, TC Paulitz, Novel butyrolactones with antifungal activity produced by *Pseudomonas aureofaciens* strain 63-28, *J Antibiot.* , 50, 742–749 (1997).

65 E Déziel, F Lépine, S Milot, J He, MN Mindrinos, RG Tompkins, et al., Analysis of *Pseudomonas aeruginosa* 4-hydroxy-2-alkylquinolines (HAQs) reveals a role for 4-hydroxy-2-heptylquinoline in cell-to-cell communication, *Proc Natl Acad Sci USA*, 101, 1339–1344 (2004).

66 LA Gallagher, SL McKnight, MS Kuznetsova, EC Pesci, C Manoil, *Functions required for extracellular quinolone signaling by Pseudomonas aeruginosa, J Bacteriol.*, 184, 6472–6480 (2002).

67 J Njoroge, V Sperandio, Jamming bacterial communication, New approaches for the treatment of infectious diseases, *EMBO Mol Med.*, 1, 201–210 (2009).

68 PT Kimmitt, CR Harwood, MR Barer, Toxin gene expression by shiga toxin-producing *Escherichia coli*: The role of antibiotics and the bacterial SOS response, *Emerg Infect Dis*, 6, 458–465 (2000).

69 V Sperandio, AG Torres, B Jarvis, JP Nataro, JB Kaper, Bacteria-host communication: The language of hormones, *Proc Natl Acad Sci USA*, 100, 8951–8956 (2003).

70 JB Kaper, JP Nataro, HL Mobley, Pathogenic *Escherichia coli*, *Nat Rev Microbiol.*, 2, 123–140 (2004).

71 M Walters, V Sperandio, Quorum sensing in Escherichia coli and Salmonella, *Int J Med Microbiol.*, 296, 125–131 (2006).

72 DA Rasko, CG Moreira, de R Li, NC Reading, JM Ritchie, MK Waldor, et al., *Targeting QseC signaling and virulence for antibiotic development*, *Science*, 321, 1078–1080 (2008).

73 C Wagner, S Zimmermann, G Brenner-Weiss, F Hug, B Prior, U Obst, et al., The Quorum sensing molecule N-3-oxododecanoyl homoserine lactone (3OC12-HSL) enhances the host defence by activating human polymorphonuclear neutrophils (PMN), *Anal Bioanal Chem.*, 387, 481–487 (2007).

74 M Kleerebezem, LE Quadri, OP Kuipers, WM De Vos, Quorum sensing by peptide pheromones and two component signal transduction systems in Gram positive bacteria, *Mol Microbiol.*, 24, 895–904 (1997).

75 GM Dunny, H Hirt, S Erlandsen, Multiple roles for enterococcal sex pheromone peptides in conjugation, plasmid maintenance and pathogenesis, In: England R, Hobbs G, Bainton N, Roberts DM, editors, *Microbial signalling and communication*, Cambridge United Kingdom: University Press, 117–138, 1999.

76 GM Dunny, AB Leonard, Cell–cell communication in Gram positive bacteria, *Annu Rev Microbiol.*, 51, 527–564 (1997).

77 EA George, TW Muir, Molecular mechanisms of agr Quorum sensing in virulent staphylococci, *Chembiochem.*, 8, 847–855 (2007).

78 E Geisinger, TW Muir, RP Novick, agr receptor mutants reveal distinct modes of inhibition by staphylococcal autoinducing peptides, *Proc Natl Acad Sci U S A*, 106, 1216–1221 (2009).

79 AL Cheung, JM Koomey, CA Butler, SJ Projan, VA Fischetti, Regulation of exoprotein expression in *Staphylococcus aureus* by a locus (*sar*) distinct from agr, *Proc Natl Acad Sci USA*, 89, 6462–6466 (1992).

80 G Ji, RC Beavis, RP Novick, Cell density control of staphylococcal virulence mediated by an octapeptide pheromone, *Proc Natl Acad Sci USA*, 92, 12055–12059 (1995).

81 RP Novick, SJ Projan, J Kornblum, HF Ross, G Ji, B Kreiswirth et al., The agr P2 operon: An autocatalytic sensory transduction system in *Staphylococcus aureus*, *Mol Gen Genet.*, 248, 446–458 (1995).

82 G Lina, S Jarraud, G Ji, T Greenland, A Pedraza, J Etienne, et al., Transmembrane topology and histidine protein kinase activity of AgrC, the agr signal receptor in *Staphylococcus aureus*, *Mol Microbiol.*, 28, 655–662 (1998).

83 E Morfeldt, L Janzon, S Arvidson, S Lofdahl, Cloning of a chromosomal locus (exp) which regulates the expression of several exoprotein genes in *Staphylococcus aureus*, *Mol Gen Genet.*, 211, 435–440 (1988).

84 TM Rechtin, AF Gillaspy, Ma Schumacher, RG Brennan, MS Smeltzer, BK Hurlburt, Characterization of the SarA virulence gene regulator of *Staphylococcus aureus*, *Mol Microbiol.*, 33, 307–316 (1999).

85 G Ji, RC Beavis, RP Novick, Bacterial interference caused by autoinducing peptide variants, *Science*, 276, 2027–2030 (1997).

86 RP Novick, Autoinduction and signal transduction in the regulation of staphylococcal virulence, *Mol Microbiol.*, 48, 1429–1449 (2003).

87 MD Kiran, A Giacometti, O Cirioni, N Balaban, Suppression of biofilm related, device-associated infections by staphylococcal Quorum sensing inhibitors, *Int J Artif Organs*, 31, 761–770 (2008).

88 JW Costerton, PS Stewart, EP Greenberg, Bacterial biofilms: A common cause of persistant infections, *Science*, 284, 1318–1322 (1999).

89 A Deveau, DA Hogan, Linking Quorum sensing regulation and biofilm formation by *Candida albicans*, *Methods Mol Biol.*, 692, 219–233 (2011).

90 JM Hornby, EC Jensen, AD Lisec, JJ Tasto, B Jahnke, R Shoemaker, et al., Quorum sensing in the dimorphic fungus *Candida albicans* is mediated by farnesol, *Appl Environ Microbiol.*, 67, 2982–2992 (2001).

91 H Chen, M Fujita, Q Feng, J Clardy, GR Fink, Tyrosol is a Quorum sensing molecule in *Candida albicans*, *Proc Natl Acad Sci USA*, 101, 5048–5052 (2004).

92 DH Navarathna, JM Hornby, N Krishnan, A Parkhurst, GE Duhamel, KW Nickerson, Effect of farnesol on a mouse model of systemic candidiasis, determined by use of a DPP3 knockout mutant of *Candida albicans*, *Infect Immun.*, 75, 1609–1618 (2007).

93 M Melotto and B N Kunkel, Virulence Strategies of Plant Pathogenic Bacteria, The Prokaryotes– Prokaryotic Physiology and Biochemistry, E Rosenberg, E Stackebrand, E F DeLong, F Thompson, S Lory (eds.), Springer-Verlag, Berlin, 61–82, 2013.

94 K Prasannath, Pathogenicity and Virulence Factors of Phytobacteria, Scholars Academic Journal of Biosciences (SAJB) Sch. *Acad. J. Biosci*, 1(1), 24–33 (2013)

95 A Block and J R Alfano, *Plant targets for Pseudomonas syringae type III effectors: Virulence targets or guarded decoys, Current Opinion in Microbiology*, 14(1), 39–46 (2011).

96 M Lindeberg, S Cunnac, A Collmer, *Pseudomonas syringae* type III effector repertoires: last words in endless arguments, *Trends in Microbiology*, 20(4), 199–208 (2012).

97 C L Bender, F Alarcon-Chaidez and D C Gross, *Pseudomonas syringae phytotoxins: mode of action, regulation and biosynthesis by peptide and polyketide synthetases, Microbiology and Molecular Biology Review*, 63(2), 266–292 (1999).

98 T P Denny, Involvement of Bacterial Polysaccharides in Plant Pathogenesis, *Annual Review of Phytopathology*, 33, 173–197 (1995).

99 J Boch and U Bonas, Gram-negative plant pathogenic bacteria, In: Emerging Bacterial Pathogens, I Muhldorfer and K P Schafer (eds.), *Karger, Basel*, 186–196, 2001.

100 G P C Salmond, Secretion of Extracellular Virulence Factors by Plant Pathogenic Bacteria, *Annual Review of Phytopathology*, 32, 181–200 (1994).

101 T.R. De Kievit, B.H. Iglewski, Bacterial quorum sensing in pathogenic relationships, *Infect. Immun.*, 68, 4839–4849 (2000).

102 Y.H. Dong, J.L. Xu, X.Z. Li, L.H. Zhang, AiiA- an enzyme that inactivates the acylhomoserine lactone quorum-sensing signal and attenuates the virulence of *Erwinia carotovora*, *Proc. Natl. Acad. Sci. USA.*, 97, 3526–3531 (2000).

103 T.B. Rasmussen, M Givskov, Quorum-sensing inhibitors as antipathogenic drugs, *Int. J. Med. Microbiol.*, 296, 149–161 (2006).

104 KH McClean, MK Winzon, L Fish, A Taylor, SR Chhabra, M Camara, et al., Quorum sensing and *Chromobasterium violaceum*: exploitation of violacein production and inhibition for the detection of N-acylhomoserine lactones, *Microbiology-UK*, 143, 3703–3711 (1997).

105 M Givskov, R DeNyz, M Manefield, L Gram, R Maximilien, L Eberl, Eukaryotic interference with homoserine lactone-mediated prokaryotic signaling, *Journal of Bacteriology*, 178, 6618–6622 (1996).

106 M Teplitski, JB Robinson, WD Bauer, Plants secrete substances that mimic bacterial N-acyl homoserine lactone signal activities and affect population density-dependent behaviors in associated bacteria, *Molecular Plant-Microbe Interactions* 13, 637–648 (2000).

107 Y.H. Dong, X.F. Zhang, J.L. Xu, L.H Zhang, Insecticidal *Bacillus thuringiensis* silences *Erwinia carotovora* virulence by a new form of microbial antagonism, signal interference, *Appl. Environ. Microbiol.*, 70, 954–960 (2004).

108 Y.H. Lin, J.L. Xu, J. Hu, L.H. Wang, S.L. Ong, J.R. Leadbetter, L.H Zhang, Acyl-homoserine lactone acylase from *Ralstonia* strain XJ12B represents a novel and potent class of quorum-quenching enzymes, *Mol. Microbiol.*, 47, 849–860 (2003).

109 J.R. Leadbetter, E.P Greenberg, Metabolism of acyl-homoserine lactone quorum-sensing signals by *Variovorax paradoxus*, *J. Bacteriol.*, 182, 6921–6926 (2000).

110 Y. Han, F. Chen, N. Li, B. Zhu, X Li, *Bacillus marcorestinctum* sp. nov., a novel bacterium quenching acylhomoserine lactone quorum-sensing signal from soil, *Int. J. Mol. Sci*,. 11, 507–520 (2010).

111 C.F. Sio, L.G. Otten, R.H. Cool, S.P. Diggle, P.G. Braun, R. Bos, et al., Quorum quenching by an *N*-acyl-homoserine lactone acylase from *Pseudomonas aeruginosa* PAO1, *Infect. Immun.*, 74, 1673–1682 (2006).

112 L.H. Wang, L.X. Weng, Y.H. Dong, L.H Zhang, Specificity and enzyme kinetics of the quorum-quenching *N*-acyl homoserine lactone lactonase (AHL lactonase), *J. Biol. Chem.*, 279, 13645–13651 (2004).

113 N.T. Tinh, N. Asanka, R.A.Y.S. Gunasekara, N. Boon, K. Dierckens, P. Sorgeloos, P Bossier, *N*-Acylhomoserine lactone degrading microbial enrichment cultures isolated from *Penaues vannamei* shrimp gut and their probiotic properties in *Brachionus plicatilis* cultures, *FEMS Microbiol. Ecol.*, 62, 45–53 (2007).

114 S.Y. Park, H.O. Kang, H.S. Jang, J.K. Lee, B.T. Koo, D.Y. Yum, Identification of extracellular *N*-acylhomoserine lactone acylase from a *Streptomyces* sp. and its application to quorum quenching, Appl. *Environ. Microbiol.*, 71, 2632–2641 (2005).

115 S. Uroz, P. Oger, S.R. Chhabra, M. Cámara, P. Williams, Y Dessaux, *N*-Acyl homoserine lactones are degraded via an amidolytic activity in *Comamonas* sp. strain D1, *Arch. Microbiol.*, 187, 249–256 (2007).

116 S.J. Lee, S.Y. Park, J.J. Lee, D.Y. Yum, B.T. Koo, J.K Lee, Genes encoding the *N*-acyl homoserine lactone-degrading enzyme are widespread in many subspecies of *Bacillus thuringiensis*, *Appl. Environ. Microbiol.*, 68, 3919–3924 (2002).

117 B.M. Mole, D.A. Baltrus, J.L. Dangl, S.R Grant, Global virulence regulation networks in phytopathogenic bacteria, *Trends Microbiol.*, 15, 363–371 (2007).

118 J.Y. Park, Y.H. Lee, K.Y. Yang, Y.C. Kim, AiiA-mediated quorum quenching does not affect virulence or toxoflavin expression in *Burkholderia glumae* SL2376, *Lett. Appl. Microbiol.*, 51, 619–624 (2010).

119 Lyczak, J.B.; Cannon, C.L.; Pier, G.B. Establishment of *Pseudomonas aeruginosa* infection: Lessons from a versatile opportunist, *Microbe. Infect.*, 2, 1051–1060 (2000).

120 C. Reimmann, N. Ginet, L. Michel, C. Keel, P. Michaux, V. Krishnapillai, M. Zala, K. Heurlier, K. Triandafillu, H. Harms, et al., Genetically programmed autoinducer destruction reduces virulence gene expression and swarming motility in *Pseudomonas aeruginosa* PAO1, *Microbiology*, 148, 923–932 (2002).

121 E. Papaioannou, M. Wahjudi, P. Nadal-Jimenez, G. Koch, R. Setroikromo, W.J. Quax, Quorum-quenching acylase reduces the virulence of *Pseudomonas aeruginosa* in a *Caenorhabditis elegans* infection model, *Antimicrob. Agents Chemother.*, 53, 4891–4897 (2009).

21

Biofilm Instigation of Plant Pathogenic Bacteria and Its Control Measures

A. Robert Antony, R. Janani and V. Rajesh Kannan

Rhizosphere Biology Laboratory, Department of Microbiology, Bharathidasan University, Tiruchirappalli, India

21.1 Introduction

Agriculture incorporates thousands of years of practices. It is the only source to meet global food demands. Agriculture is a basic way of life for the majority of the world, and it is still the landscape through which disease spreads. This imposes new demands on our understanding of epidemiology if we are to control disease efficiently, whether by means of biological, chemical, cultural, and/or genomically. Structurally, the nature of biofilms can protect the cells against antimicrobial agents and the host's defense. The consequences of biofilm formation have been documented in many different environments, as they constitute a protected mode of growth that allows microorganisms to survival in hostile environments, being their physiology and behavior significantly different from their planktonic counterparts. Biofilms are difficult to eradicate due to their resistant phenotype. So, more efforts need to learn about the impact of microbial biofilms and their recovery responses to damage, as microorganisms can develop resistance and subsequently survive previously effective control procedures. In this review attempt to search for new control strategies, biofilm could be the basis for the strong biological solutions, and specificity seem to be a solid step ahead in overcoming the biofilm impairment issues.

21.2 Plant Pathogens

Traditionally, plant health and welfare have been the concern of pathologists, agronomists, and growers, whose main interest has been crops under cultivation. Besides bacterial pathogens were significant factors that reduce the yield of agriculturally important crops worldwide. Research has generated a vast body of information on the mechanisms that govern plant–microbe interaction, but these advances in the understanding of pathogenicity of the plant pathogenic bacteria were almost entirely restricted to Gram-negative organisms, because they were easier to handle. Most of the plant pathogenic bacteria were Gram-negative; however, some Gram-positive plant pathogens are quite important (Table 21.1), since they cause great losses in the cultivation of crop plants [1–3].

Table 21.1 Common Bacterial Plant Pathogens and Their Gram Grouping.

Sl. No.	Bacterial Pathogens	Gram Grouping
1	*Pseudomonas syringae pathovars*	Gram negative
2	*Ralstonia solanacearum*	Gram negative
3	*Bacillus megaterium*	Gram positive
4	*Bacillus circulans*	Gram positive
5	*Bacillus polymyxa*	Gram positive
6	*Bacillus subtilis*	Gram positive
7	*Agrobacterium tumefaciens*	Gram negative
8	*Xanthomonas oryzae* pv. *Oryzae*	Gram negative
9	*Xanthomonas campestris pathovars*	Gram negative
10	*Xanthomonas axonopodis* pv. *Manihotis*	Gram negative
11	*Clostridium butyricum*	Gram positive
12	*Clostridium botulinum*	Gram positive
13	*Erwinia amylovora*	Gram negative
14	*Clavibacter michiganensis*	Gram positive
15	*Clavibacter rathayi*	Gram positive
16	*Clavibacter tritici*	Gram positive
17	*Clavibacter toxicus*	Gram positive
18	*Clavibacter xyli*	Gram positive
19	*Xylella fastidiosa*	Gram negative
20	*Dickeya (dadantii and solani)*	Gram negative
21	*Pectobacterium carotovorum* (and *P. atrosepticum)*	Gram negative

Plant pathogens can transmit horizontally via vectors such as soil, water, and insects, as well as vertically via mother plants [4]. Some pathogens were specific for certain plant species, while others that were nonspecific have a broad range of hosts. Hence, plant pathogens may play different roles among species within a community [5]. The interactions between plants and pathogens can promote or exclude species coexistence. Plant pathogens can influence community stabilities when their impact on plant growth relies on host relative abundance [6].

21.2.1 Importance and Impact of Plant Pathogenic Bacteria

In terms of species number, bacteria are the major causal agents of animal and human diseases. In addition, plant pathogenic bacteria affect about 100 species [7]. Most important agricultural crops suffer from at least one bacterial disease, and for some crops, a bacterial disease was the main came of yield losses. As an example of a bacterial disease of economic importance, black rot disease of crucifer plants, caused by *Xanthomonas campestris* pathovar (pv.) *campestris*, was considered the most serious disease of cultivated brassica and radishes worldwide. Another bacterial species belonging to *Xanthomonas* genus, *X. citri*, leads to the eradication of millions of citrus trees in Florida, Sao Paulo, and other parts of the world. *Xanthomonas oryzae* pv. *oryzae* causes bacterial leaf blight (BLB) of rice. That was one of the most important diseases of rice in many parts of the world and was very destructive in Japan, India, and other parts of Asia. As more than 50 percent of the world population relies on rice for basic nutrition, damage to rice production caused by BLB poses significant economic and social risks. *Erwinia amylovora* are responsible for the fire blight disease in a wide range of plants from the Rosaceae family. This destructive disease could entirely prevent the cultivation of apples and pears in some parts of the world. Similarly, the European grape, *Vitis vinifera*, would not be cultivated in certain regions of the United States due to the bacterium *Xylella fastidiosa*, the causal agent of Pierce's disease of grapevines [8].

Agrobacterium tumefaciens is the source of bacterial crown gall disease, in which bacterium affects woody and herbaceous plants belonging to more than 100 genera. *Agrobacterium tumefaciens* induces the production of galls (tumors) on the stem and/or roots of the plant. Infected plants often grow poorly and with reduced yields. However, the threat of this disease is not limited to its plant pathological damage. During the infection process, *A. tumefaciens* introduces a fragment of its own DNA into the host plant cell. The transferred DNA fragments contain genes that are further expressed in the plant host, an essential step for gall formation and pathogenesis. Much research has been done to understand the mechanism of DNA transfer by *A. tumefaciens*. Based on the knowledge acquired from that, researchers have developed methods for genetic modification of plants using this bacterium. Indeed, *A. tumefaciens* serves today as the main tool for generation of transgenic plants for biotechnological and agricultural purposes, as well as for basic research investigations [9].

21.2.2 Plant Pathology and Plant Bacteriology: Historical Background

Plant diseases have had profound effects on mankind through the centuries, as evidenced by biblical references to blasting and mildew of plants. The Greek philosopher Theophrastus (370–286 BC) was first to described maladies of tress, cereals, and legumes that we today classify as leaf scorch, rots, scab, and cereal rust. The Romans were also aware of rust disease of their grain crops. They celebrated the holiday of Robigalia, which involved sacrifices of reddish-colored dogs and cattle in an attempt to appease the rust god Robigo [10].

With the invention of the microscope in the seventeenth century, fungi and bacteria associated with plants were investigated. In 1665, Robert Hooke published the first illustration of rust on a rose leaf. Advances in the study of disease life cycles were hampered by the widely held beliefs of *spontaneous generation*. This theory, held by most people in the mid-eighteenth century, considered pathogenic microorganisms as products of disease rather than causal agent of disease. The devasting epidemics of late blight of potato in Ireland in 1845 and 1846 dramatized the effect of plant disease on mankind. Tragically, these epidemics caused famine and death for over a million people and resulted in a loss of nearly a third of Ireland's population between 1845 and 1860. In 1861, a German botanist Anton De Bary proved that a fungus *Phytophthora infestans* was the causal agent of late blight of potato. That was a milestone in the study of plant disease, since it showed that a fungal pathogen was indeed the cause of late blight of potato rather than an organism simply associated with the disease. Two years later in

1863, Luis Pasteur proposed his germ theory of disease that finally disproved the theory of spontaneous generation. This milestone essentially changed the way modern science investigated the disease of all living.

21.2.3 Classification of Plant Pathogenic Bacteria

21.2.3.1 Rhizosphere Pathogen

Plant root exudates were influenced by pathogenic fungi, bacteria, and nematodes in various ways. Root exudates may also contain inhibitory substances, preventing the establishment of pathogens. The balances between the rhizosphere microflora and plant pathogens are important in the host–pathogenic relationship. In that context, the biochemical qualities of root exudates and the presence of antagonistic microorganisms play an important role in the proliferation and survival of root-infecting pathogens in the soil or rhizosphere [11].

21.3 Plant Physiological Alteration by Plant Pathogens

Pathogen-infected plants physiology obviously disrupted on whole genome. Pathogens have evolved a way to interact and/or interfere with almost every aspect of a plant's growth and development. The major plant functions affected by pathogens are detailed in this section.

21.3.1 Photosynthesis

Pathogens can interfere with photosynthesis by physical obstruction or by chemical inhibition. Many sooty and snow molds cover leaf surfaces with a dense mantle colonization, which prevent light from reaching the chloroplasts. Some can passively survive on leaf guttation without actually infecting the leaf cells. Many pathogens physically disrupt photosynthesis simply by killing photosynthetic cells, as is the case with many leaf spots and blights. Other pathogens elicit chemical inhibitors of photosynthesis. The invasion by pathogens can also affect water balance in the plant, causing stomata to close, and, in turn, limiting photosynthesis.

21.3.2 Vascular Function

Colonization of roots and vascular tissues can impede vascular function. Whether the pathogen is directly obstructing the vascular vessels and or indirectly limiting vascular flow by killing or parasitizing root tissues, the end result is wilting and nutrient deprivation in the host plant [12].

21.3.3 Root Function

Bacterial infections inhibit root functions which leads to a multitude of physiological maladies. Pathogens that parasitize and kill root hairs greatly reduce the root systems' ability to pull water and nutrients from soil. Pathogens that degrade cortical and vascular root tissues not only reduce energy stores and nutrient transport, but they also weaken the plant's physical support. Many plants exhibit lodging and other support problems when root rot diseases are present.

21.3.4 Respiration

When plants are infected with a pathogen, respiration usually increases. The increase in respiration produces energy that the plant cells can use for manufacturing and mobilizing their defense compounds. Depending on the pathogen and its host's genetics, the rise in respiration can be mild or dramatic. Prolonged high levels of respiration can deplete cell energy reserves and cause oxygen levels in the tissues to fall. Roots in saturated soils and fruits with low surface area to volume ratios can sustain damage from hypoxia [7].

21.3.5 Transpiration

The amount of moisture lost by a plant can increase substantially when a plant is heavily diseased. Often in diseased tissues the protective cuticle or epidermal layers have been ruptured or stomata response has been inhibited. Leaves covered with rust pustules can lose water at high enough rates as to cause temporary wilting. Heavy transpiration associated with disease may cause the formation of tyloses and gums in vascular vessels as a means of preventing continued high water loss [7, 12].

21.4 Virulence Strategies of Plant Pathogenic Bacteria

Mechanisms of pathogenicity in plants by bacterial pathogens are becoming well known [13,14]. Virulence and pathogenicity genes may be harbored in different replicons (independent replicating units), such as spread throughout the chromosomes or in specialized areas termed genomic or pathogenicity islands [15], in bacterial viruses integrated in the chromosome or in a carrier state, and on one or more extra-chromosomal elements (plasmids). The functions of most genes, including those on extra-chromosomal elements, are not known. It is estimated that each bacterium has about 40 percent of its genome devoted to unique genes.

Population development must normally occur for many bacteria to survive and infect plants. Infectious doses normally are in the millions of cells. In several cases, and perhaps all, the cells communicate chemically with one another (quorum sensing) and with other species. These chemical sensing molecules are under intensive study [16]. In some cases, and perhaps most, microorganisms are organized in dense growths to form biofilms that tightly adhere to surfaces, serving as protectants against the elements and enabling cells to produce a favorable environment for survival and spread.

Some structures used by bacteria to insert chemical compounds into plant cells are well studied, such as so-called type III secretion system (five types are currently known). The type III system operates somewhat like a syringe and plunger to transport pathogen-produced proteins that effect disease or trigger defense (Figure 21.1) [17]. These mechanisms having plant pathogens sometimes shown surprising and unexpected similarity to those found in animal and human pathogens [18], as cross-kingdom plant pathogenic bacteria (Table 21.2) [3]. There are even a few strains of bacteria that cross

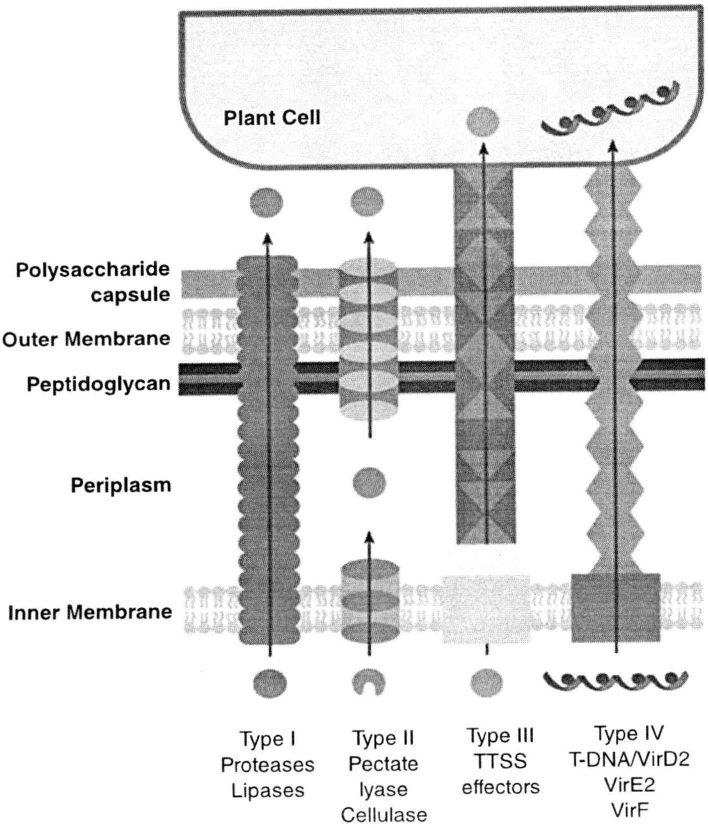

Figure 21.1 Secretory systems used by bacteria to introduce pathogenic compounds into plant cell [115]. (*See color plate section for the color representation of this figure.*)

kingdoms: They can infect both plants and humans. The genetic basis for such novelty is of immense interest and significance regarding the basis of infectious disease.

21.5 Biofilm Formations

Biofilm is an accumulation of microorganisms embedded in a matrix of polysaccharide that is attached to a surface, such as a surface of a sphere, and the associated extracellular substances produced by one or more of the attached microorganisms. These microbial group bacterial biofilms are prevalent on most wet surfaces in nature. Many biofilms are sufficiently thick to be visible to the naked eye. Bacterial biofilms are integrated, multi-species communities of cells that adhere to almost any surface and are fundamental to the ecology and biology of bacteria (Figure 21.2). Biofilms constitute a protected mode of growth that allows survival in a hostile environment. The structures that form in biofilm contain channels in which nutrients can circulate, and cells in different regions of the biofilm exhibit different patterns of gene expression. The complexity of the biofilm structure and metabolism has led to the analogy of biofilm to tissue of higher organism.

Table 21.2 Cross-Kingdom Plant Pathogenic Bacteria [116].

Sl. No.	Pathogen	Plant Host/Niche	Human Disease/Condition
1	*Enterobacter cloacae*	Macadamia, dragon fruit, orchids,	Respiratory/skin/urinary infection, septicaemia
2	*Enterococcus faecalis*	*Arabidopsis thaliana*	Urinary/abdominal/cutaneous infections, septicaemia
3	*Burkholderia ambifaria*	Soil, maize roots	Unknown
4	*Burkholderia cenocepacia*	Soil, maize roots, onion	Septicemia
5	*Burkholderia cepacia*	Soil, rice, maize, wheat, onion	Septicemia
6	*Burkholderia gladioli*	Onion, gladiolus, iris, rice	Septicemia
7	*Burkholderia glathei*	Soil	Unknown
8	*Burkholderia glumae*	Rice	Chronic granulomatous disease
9	*Burkholderia mallei*	Soil	Melioidosis/Glanders
10	*Burkholderia plantarii*	Rice, gladiolus, iris	Melioidosis
11	*Burkholderia pseudomallei*	Tomato	Melioidosis/Glanders
12	*Burkholderia pyrrocinia*	Soil	Melioidosis/Glanders
13	*Pantoea agglomerans*	Crown/root gall	Arthritis/septicaemia
14	*Pantoea ananatis*	Eucalyptus, maize, rice	Septicemia
15	*Pantoea citrea*	Pineapple	Septicemia
16	*Pantoea dispersa*	Seeds	Septicaemia
17	*Pantoea punctata*	Japanese mandarin oranges	Unknown
18	*Pantoea septic*	Unknown	Septicemia
19	*Pantoea stewartii*	Maize	Unknown
20	*Pantoea terrea*	Japanese mandarin oranges	Unknown
21	*Salmonella enterica*	Tomato, *Arabidopsis thaliana*	Gastroenteritis/typhoid fever
22	*Serratia marcescens*	Squash, pumpkin	Septicemia, urinary tract infection

These sessile biofilm communities can give rise to nonsessile individuals, planktonic bacteria that can rapidly multiply and disperse. The biofilm consist of microbial/bacterial colonies on a surface, and within these microbial colonies develop into organized communities with functional heterogeneity. The growth of the biofilm is slow, in one or more localizations, and biofilm formations are often slow to produce overt symptoms.

21.5.1 Mechanism of Biofilm Formation

Pioneers studies of Zobell and Henrici described for the first time that bacteria could attach to and thrive on surfaces [19]. The existence of a surface is perhaps the most

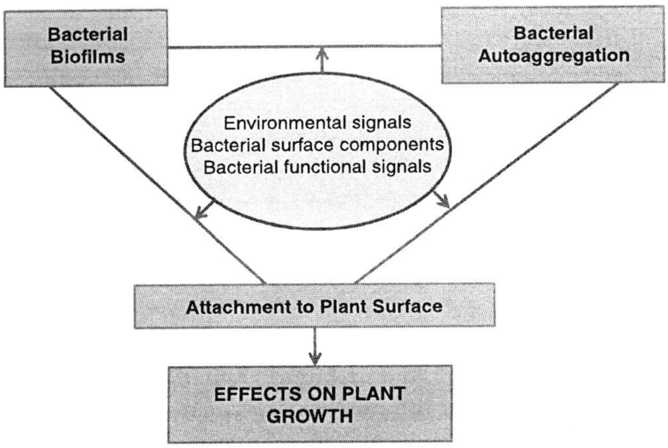

Figure 21.2 Bacterial biofilm developmental conditions [117].

important prerequisite for biofilm formation, as it involves the bacterial detection of a surface. The general rule of thumb is that bacteria will preferentially colonize on surfaces that are hydrophobic, have surface roughness on the nano and micro scale, and are exposed to a conditioning layer in contrast to smooth and hydrophilic surfaces.

21.5.2 Molecular Insights on Biofilm Formation

The biofilm matrix is also commonly referred as glycocalyx; the extracellular substances are typically polymeric substances [20]. Davies et al. [21] note that as bacterial cell density within a biofilm increases, the bacteria may communicate with each other, in cell-to-cell signals. This can lead to the secretion of low-molecular-weight molecules that send a signal when the population has reached a critical threshold. This process, called quorum sensing, is responsible for the expression of virulence factors [22]. For example, *Pseudomonas aeruginosa* produces destructive proteases when the number of these bacteria reaches a high enough density in the biofilm infection that the host defense mechanism cannot ward it off [23].

In case of biofilm development, a series of genes (40 to 60 percent of the prokaryotic genome) are modulated (activated/inhibited) by complex cell-to-cell signaling mechanisms and the biofilm cells become phenotypically distinct from their counterpart-free cells, being more resistant to stress conditions [24].

21.5.3 Structural and Functional Components Involved in Biofilm Formation

Bacterial surface components and extracellular compounds of primarily flagella, lipopolysacharides (LPSs), and exopolysaccharides (EPSs) in combination with environmental and quorum-sensing signals are crucial for auto aggregation and biofilm development [25]. In generally accepted models of biofilm formation, environmental signals trigger the process, and flagella are required for the biofilm community to approach and move across the surface [26]. The initial steps of attachment are mediated

Table 21.3 Examples of Biofilm Forming Plant-Associated Bacteria [119].

	Bacteria	Host Plant	Interaction	Colonization Site
Shoot Colonizers	*Pseudomonas fluorescens*	Diverse	Mutualist	Leaves
	Pseudomonas syringae pv *syringae*	Beans	Pathogen	Leaves
	Methylobacterium spp.	Diverse	Commensal	Leaves (Stomates)
	Erwinia amylovora	Fruit trees	Pathogen	Fruit leaves and flowers
Vascular Colonizers	*Pantoea stewartii*	Maize	Pathogen	Xylem vessels
	Xylella fastidiosa	Citrus and grape	Pathogen	Xylem vessels
	Xanthomonas campestris pv. *campestres*	Crucifers	Pathogen	Xylem vessels
	Ralstonia solanacearum	Diverse	Pathogen	Root to xylem
	Clavibacter michiganensis pv. *sepedonicus*	Potato	Pathogen	Xylem vessels
	Leifsonia xyli subsp. *Xyli*	Sugar cane	Pathogen	Xylem vessels
	Spiroplasma spp.	Diverse	Pathogen	Phloem vessels
Root Colonizers	*Agrobacterium tumefaciens*	Diverse dicots	Pathogen	Root and crown tissue
	Azospirillum brazilense	Cereals	Mutualist	Root hairs
	Bacillus cereus	Diverse	Commensal	Roots
	Burkholderia cenocepacia	Onions	Pathogen	Roots
	Pseudomonas spp.	Diverse	Commensal	Roots
	Rhizobium spp.	Legumes	Mutualist	Root hairs and nodules

by outer membrane proteins (e.g., calcium-binding proteins), pili, or LPSs. After the formation of microbial colonies, the production of quorum sensing signals is required for the formation of a mature biofilm [27]. Exopolysaccharides provide the architectural form of biofilms and stabilize their three-dimensional structure. Biofilms are often permeated by channels that act as a circulatory system, allowing the bacteria to exchange water, nutrients, enzymes, and signals, dispose of potentially toxic metabolites and display enhanced metabolic cooperatively [28].

The major components of biofilm are typically water and the bacterial cells, followed by the EPSs of the matrix [29], which provides: (i) a physical barrier against the diffusion of antibiotics, defense substance, or other important compounds from the host; and (ii) protection against environmental stress factors, such as UV radiation, pH levels, osmotic stress, and desiccation [30, 31]. In *Agrobacterium tumefaciens*, a plant pathogen that persists as surface-associated populations on plants or soil particles and cellulose overproduction results in increased biofilm formation on roots [32]. Minor biofilm components include macromolecules such as

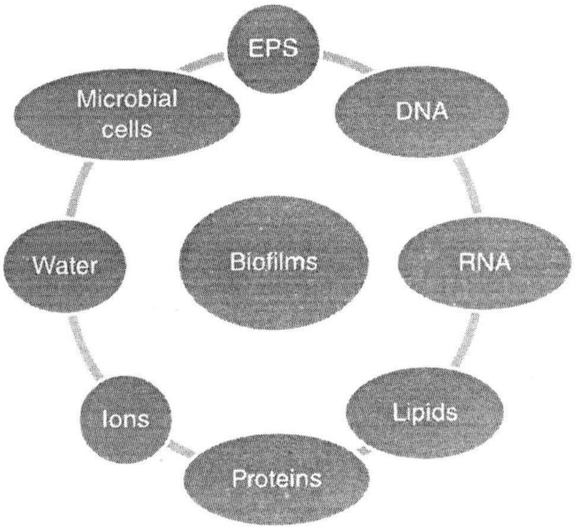

Figure 21.3 Diagrammatic representation of various components of bacterial biofilms [118].

proteins, DNA, and various lysis products, which affects the overall properties of the biofilm (Figure 21.3).

21.5.3.1 Surface Bacterial Factors

Lipopolysaccharide is an important surface structural component of Gram-negative bacteria and covers about 75 percent of the surface area of the outer membrane. It is positioned among the proteins and phospholipids of the outer bacterial membrane and contributes to the structural properties of the membran (e.g., they act as a permeability barrier against various types of molecules). Changes of LPS structure usually affect adhesive forces among bacteria, possibly through alteration of cell surface hydrophobicity. For example, in rhizobacteria such as *Rhizobium leguminosarum* and *R. etli*, LPS modifications typically alter the autoaggregation phenotype [33, 34]. The rhamnose-rich *O*-antigen in the outermost part of the LPS of the xylem-limited plant pathogen of *Xylella fastidiosa* is involved in cell–cell aggregation [35]. The autoaggregative ability of *X. fastidiosa* appears to be an important virulence mechanism because the bacterial clusters block the passage of water and nutrients from the roots to the leaves of the host plant [36].

21.5.3.2 Extracellular Factors Involved in Bacterial Autoaggregation

Extracellular polymeric materials have been shown to act as molecular glue that initiates and maintains contact between cells and causing flocculation. The main extracellular compounds are EPSs, which are linear or branched molecules formed by one repeated sugar (homopolysaccharides) or by a mixture of different sugars (heteropolysaccharides). An example is galactoglucan (EPS II) from the symbiotic rhizobacterium *Ensifer meliloti* [37] and *Streptococcus thermophilus* [38]. Most of the EPSs are polyanionic molecules because of uronic acids presence and sugar having substituents such as pyruvate, sulfate, and or phosphate.

21.5.4 Factors Favoring Biofilm Formation

Besides the bacterial inoculum and exposure time, both host plants and bacterial properties were influenced by the efficacy by which the enteric pathogens attach to plants. Attachment to basil, lettuce, or spinach leaves differs among *Salmonella enterica* serovars, as *S. enftenberg* and *S. typhimurium* showed higher attachment compared with *S. agona* or *S. arizonae*. Microscopic observations of three *Salmonella* serovars attached to tomato fruits show that although all investigated serovars were attached to tomatoes with similar efficiencies, serovars *enftenberg* and *typhimurium* adhered to the fruits in an aggregative pattern, while serovars Thompson adhered in a diffuse pattern [39]. Enteric pathogens such as *E. coli*, *Salmonella*, and *Listeria* adhered more effectively to the peach fruit than plum surfaces attributed to the increased surface area of the peach fruits due to the presence of trichomes [40]. Also, in line with epidemiological data, the affinity of *Salmonella* serovars to lettuce was significantly two to three folds higher than to cabbage [41]. Lettuce is very often associated with foodborne outbreaks, whereas outbreaks associated with cabbage are rare.

The adhesions of pathogen are washing out water in fresh cucumber surfaces, depend on temperature, and are less extensive at lower temperatures. The effect of dewaxing of fruits on adhesion depends on the bacteria. While adhesion of *Listeria* to dewaxed fruits was higher than to waxed fruits, the opposite was reported for *S. typhimurium* and *Staphylococcus aureus* [42]. Several environmental conditions were shown to have an impact on biofilm production. For example, biofilm formation of *Salmonella* has been reported to be maximal under reduced nutrient availability, aerobic conditions, low osmolality, and optimum temperatures [43].

Upon attachment of *Salmonella* cells to the leaf, the bacteria are exposed to environmental conditions (temperature below 30°C, atmospheric oxygen, etc.) that trigger expression of regulatory sRNAs and proteins such as RpoS, CsgD, and SirA. Expression of these proteins and sRNAs are enhanced by stress signals existing on the leaf surface, such as low availability of nutrients and activity of antimicrobial compounds produced by the plant or indigenous microorganisms. The induced regulatory proteins activate the genes involved in production of components of the biofilm matrix such as cellulose, curli, BapA, and capsules (CP), leading to the development of biofilms on the leaf surface. While the biofilm structure stabilizes the colonization on the plant and provides protection from different stresses, its components also contribute to the induction of the local and systemic plant defense response. As part of the plant response, triggered by both single bacteria and biofilms, the plant produces and secretes different signal and antimicrobial compounds, such as reactive oxygen species (ROS) compounds, salicylic acid, jasmonic acid, and sterols. Some of these compounds kill free and biofilm-associated bacteria and/or inhibit the process of biofilm formation [42].

21.6 Biofilm Controlling Strategies in Plant Pathogens

Virulence and pathogenicity of microorganisms is often enhanced when growing as a biofilm; therefore, time to time and based on the pathogens, new controlling strategies are required to be successful in the biofilm formation and development. Many pathogenic microorganisms reside within biofilms, which biofilms cause additional problems when designing new antimicrobial agents [44]. Accordingly, alternative curative and prophylactic approaches for tackling microbial infections within a biofilm are required. Novel strategies are necessary because of the limitations to these current treatments, such as inadequate control supply, potential for disease transfer, and compliance issues. The capability and high resistance of sessile microorganisms to inhibitors, and eradication of biofilm often requires high concentration of disinfectants or antibiotics, cause severe environmental damages, multi-resistance emergence, and nosocomial infections. Public health concerns, as well the economic loss associated with biofilm formation, raise an urgent need for developing biofilm resistant systems [45].

The strategy that combines a broad spectrum of bacterial repellent agent with a surface coating that impairs bacterial growth has been investigated, this primary could be obtained through the modification of the surface by antibacterial compound, which reduces or prevents the biofilm formation by either inhibiting bacterial adhesion and/or killing bacterial cells that have adhered. Various antibacterial substances such antibiotics, antiseptics, and/or metals and enzymes were grafted on various materials, and these surfaces were shown to display antibacterial activities.

21.7 Main Targets and Some Potential Tools to Modify Biofilms

Biofilm formation is a targeted process driven by sophisticated regulatory mechanisms, both at the single-cell level and at the level of cell populations (Table 21.4). Eradication of biofilm can be achieved without intervention of regulatory mechanism of biofilm

Table 21.4 Strategies for Biofilm Control [45].

Target	Technology	Level of Efficacy	Conditions	Disadvantages
Bacteria	Antimicrobial or modified antimicrobial	Good	*In vitro* and *in vivo*	Resistance, efficacy limited
	Antimetabolites	Little	*In vitro*	Low efficacy, high cost
	Antibodies	Little	*In vitro*	Difficult to control
	Phages	Little	*In vitro*	Low efficacy, high cost
	Peptides	Little	*In vitro*	Good efficacy, high cost
	Natural products	Little	*In vitro*	Low efficacy, high cost
Immune response	Antibodies	Low	*In vitro*	Little evidence in *in vivo*
Surfaces	Metal coating	Good	*In vitro* and *in vivo*	Reduced mean life
	Polymers	Good	*In vitro*	Little evidence in *in vivo*

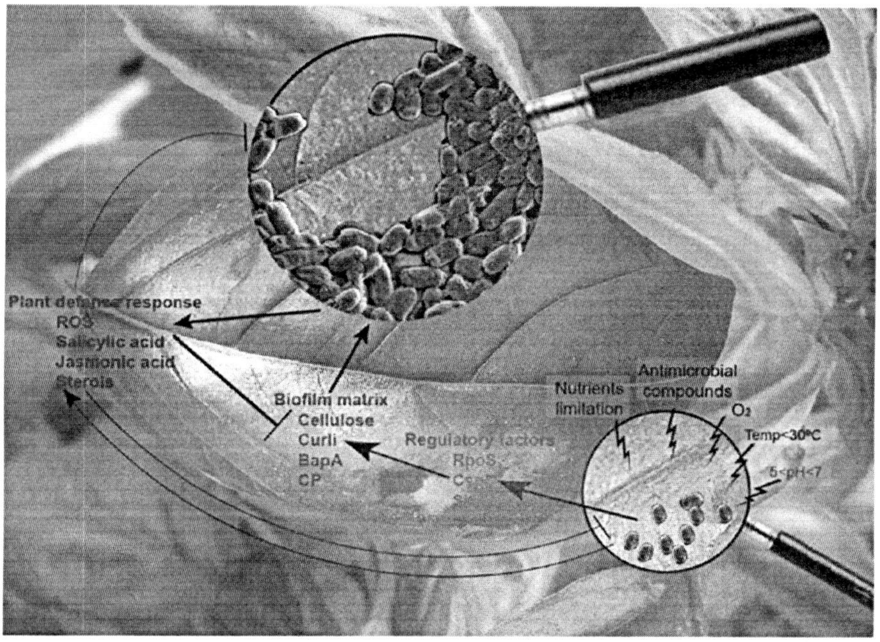

Figure 21.4 Biofilm formation by *Salmonella* cells on a leaf [42]. (*See color plate section for the color representation of this figure.*)

formation. There are several tools and methods available to modify and eradicate biofilms, which can be achieved by making changes in surface properties of biofilm carriers by modifying the roughness, hydrophobicity, charge, and surface conditioning. The mechanical stability of the biofilms can be altered by using liquid flow rate and shear force. Biofilm architecture was altered by making changes in nutrient composition, modification of rhamnolipid concentration, modification of alginate concentration, and application of ultrasound surfactants of nanoparticles. By degrading exopolysaccharides with some hydrolytic enzymes and surface active compounds. Biofilm formation can be reduced by interfered the quorum sensing signals by N-Acyl homoserine lactones (AHLs) and other regulation molecules. Further, cyclic di-GMP of the biofilms can be disrupted by interfering compounds against its synthesis and activity. Finally, the cell viability can be reduced by applying antibiotics and disinfectants [46]. These methods—physical, chemical, biological—for targeting biofilm formation are discussed in more detail in the following sections.

21.8 Physical Tools for Modifying Biofilms

21.8.1 Modification of Biofilm Surfaces

The existence of a surface is perhaps the most important prerequisite for biofilm formation. It involves the bacterial detection of a surface. The physico-chemical factors govern the initial attachment and adhesion of bacteria to surface. The general rule of thumb is that bacteria will preferentially colonize surfaces that are hydrophobic, have surface roughness on the nano and micro scale, and are exposed to a conditioning layer in contrast to smooth, hydrophilic surfaces. The authors affirm that the key challenge in this area is the prevention of the formation of a conditioning layer that passives the exposed surface chemistry and provides a site of attachment for bacteria [47].

All materials are subjected to bacterial contaminations when exposed to air, humidity, or diverse environmental conditions. To overcome these problems, several strategies have been used to create coatings that are either antimicrobial or nonbiofouling. Cheng et al. [48] report a coating that combines both properties, switching from antimicrobials to nonfouling upon hydrolysis, or specifically, the application of polymethacrylate derivate with cationic side chain that becomes zwitterionic upon conversion of a terminal ester to carboxylate. In this technology, 99.9 percent of the bacteria that attached after 1 hour of exposure to the initially prepared coating were dead. Over the course of the next 2 to 8 days, the coating slowly hydrolyzed, releasing 98 percent of the microbial cells. The nonfouling nature of the hydrolyzed coating prevents further attachments of microbial cells and formation of a biofilm.

21.8.2 Hydrophobicity, Surface Roughness, and Surface Charge

Formation of a biofilm begins with the attachment of free-floating cells to a surface. These first colonists adhere to the surface initially through weak and reversible adhesion. If the colonists are not separated immediately from the surface, they can anchor themselves more permanently using cell adhesion structures such as pili. Hydrophobicity also plays an important role in determining the ability of bacteria to form biofilms. Some species are not able to attach to a surface and are sometimes able to establish themselves directly to earlier colonists [49]. On the other hand, some bacteria are unable to successfully form biofilms due to their limited motility. Nonmotile bacteria cannot detect the surface or aggregate together as easily as motile bacteria.

Modification of the surface charge of polymers has also proven to be an effective means of biofilm prevention. Based on the principles of electrostatics, charged particles will repel other particles of like charge. The hydrophobicity and the charge of polymeric chains can be controlled by using several backbone compounds and antimicrobial agents. Positively charged polycationic chains enable the molecule to stretch out and generate bactericidal activity.

Surface roughness can also affect biofilm adhesion. Rough and high-energy surfaces are more conducive to biofilm formation and maturation, while smooth surfaces are less susceptible to biofilm adhesion. The roughness of a surface can affect the hydrophobicity or hydrophilicity of the contacting substance, which, in turn, affects its ability to adhere [50]. It is, thus, desirable to maintain a smooth surface on any products that may come in contact with bacteria [51].

21.8.3 Exopolysaccharides

Extracellular polymeric substances (EPSs) determine the physico-chemical properties of biofilm by establishing the functional and structural integrity of the biofilms [52]. Most of the EPSs are composed of polysaccharides and proteins; they also include some other DNA, lipids, and humic substances. Subsequently, exopolysaccharides generally consist of monosaccharide and some noncarbohydrate substituents. These exopolysaccharides have numerous applications in various food and pharmaceutical industries due to their wide variety of composition [45].

21.8.4 Applications of Hydrolytic Enzymes

A biofilm matrix is composed of various polymeric substances that include polysaccharides and proteins. Hydrolytic enzymes can be easily produced by several bacteria that have deleterious effects on various polysaccharide structures, *Pseudomonas flourescens* produces hydrolytic enzymes in their respective medium, causing destruction in the biofilm [53]. Johansen et al. [54] have reported that a complex mixture of polysaccharide-hydrolyzing enzymes was able to remove bacterial biofilm, and also that combining oxidoreductase with polysaccharide hydrolyzing enzymes resulted in good biofilm removal activity and also exhibited considerable antibacterial activity.

21.8.5 Applications of Surface Active Compounds and Natural Products

Nature continues with its creatures of novel compounds with interesting structures and biological activity. Researchers currently focus on the discovery of many novel natural compounds to inhibit the growth of bacterial biofilms [55]. Many antibiofilm compounds have been identified from diverse natural sources—for example, brominated furanones [56], curcumin [57], garlic [58], ursine triterpenes [59], corosolic acid, and asiatic acid [60], ginseng [61], and 3-indolylacetonitrile [62]. Other than the natural compounds, some of the derivatives are used for the inhibition of biofilms, such as imidazole derivatives and indole derivatives [55]. Similarly, some synthetic organic compounds and lipid compounds are derived from soft coral. *Eunicea* sp. [63] has also been used for biofilm inhibition. Subsequently, plant-derived compounds also play a major role in biofilm inhibition of *P. aeruginosa* biofilms [64]:

- Flavonoids inhibit *E. coli* biofilms.
- Phloretin are active against *E. coli* biofilms [65].
- Ginger extracts reduce the *P. aeruginosa* biofilms [66].
- Hyperforin from *Hypericum perforatum* inhibit growth of *S. aureus* and *Enterococcus faecalis*.
- 7-Epiclusianone from *Rheedia brasiliensi* inhibit the *Streptococcus mutans* biofilm development of isolimonic acid.
- A secondary metabolite from *Citrus* species inhibits *E. coli* biofilm formation [67].

Apart from these compounds, many compounds are studied to inhibit the biofilm formation. However, the majority of the antibiofilm molecules that have been developed are inspired by natural products or are modifications of signaling molecules that regulate biofilm formation [55].

21.8.6 Quorum Quenching

Quorum quenching (QQ) refers to the process in which quorum sensing (QS) is disrupted. Quorum quenching can be achieved in several ways, such as through the enzymatic destruction of QS signal molecules, the development of antibodies to QS signal molecules, or via agents that block QS. In this context, both the QS signal synthase and sensor or response regulator proteins are the primary targets [68]. Quorum quenching may be achieved by inhibiting the biosynthesis of the signal molecules, by inhibiting signal detection by blocking the signal receptor, or by enzyme-catalyzed degradation or modification of the signal molecules. The majority of QQ enzymes characterized have been isolated from bacteria with the ability to degrade AHL signal molecules. The plant pathogen of *Ralstonia solanacearum* regulates the production of virulence-associated extracellular polysaccharides and proteins. Many bacterial strains are able to degrade the diffusible signaling factor (DSF) fatty acid signal molecule, which is produced by *Xanthomonas* spp. and *Xylella fastidiosa* have been isolated, which upon co-inoculation with *X. campestris* into model plants reduced the severity of disease. But, the biochemical mechanism is still not fully understood [69].

21.8.6.1 Compound Interfering Systems of AHLs

The mechanisms of AHL degradation include both enzymatic and production of antagonists. There are three main enzymes that degrade the QS signal molecules: AHL lactonase, AHL acylases, and AHL oxidoreductases. Apart from that, enzymes active toward AI-2 and enzymes active toward 2-alkyl-4(1H) quinolone-type signal molecules are other important molecules that interfere with QS. In the AHL lactonase family there are three proteins, that play a major role in AHL degradation of metallo-β-lactamase like lactonases which includes phosphotri esterase like lactonases (PLLs) and paraoxonases (PONs) [70]. Four different reactions can be responsible for the degradation of AHL [71] among them two reactions that break the HSL ring, and they can be mediated by lactonase or decarboxylase enzymes. The remaining two reactions separate the homo serine lactone (HSL) ring and the acyl chain and can be catalyzed by acylases (amidases) or deaminases (Figure 21.5).

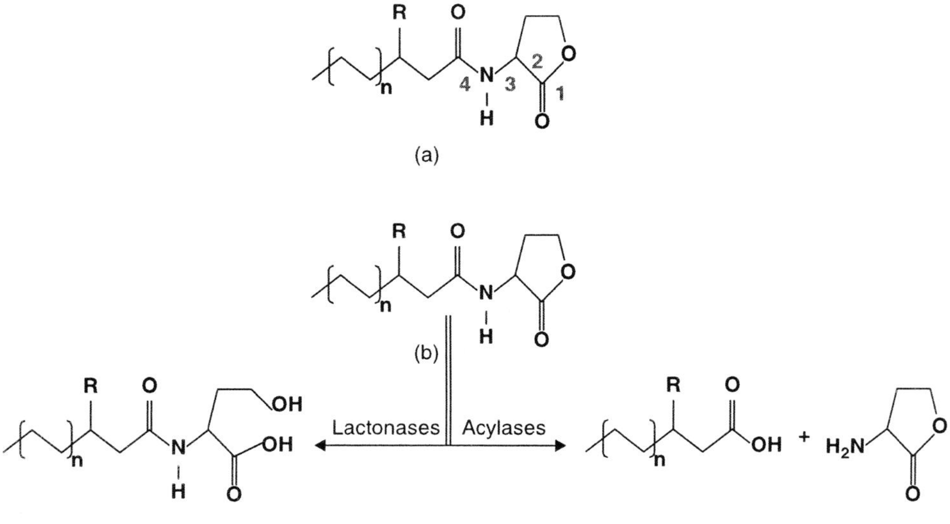

Figure 21.5 AHL degradation mechanism [71].

21.8.6.2 Compound Interfering with Regulation Molecules

The evidence of QS disruption is not always strong and many compounds claimed to be QS inhibitors might prove to be false positives when studying them in more depth. However, several QS inhibitor compounds are present in current scenario to degrade the QS signal. For example, Ajoene inhibit the fluorescence in *P. aeruginosa* [72], brominated thiophenone TF310 inhibit the bioluminescence of *Vibrio harveyi* [73], caffine inhibit *Chromobacterium violaceum* [74], curcumin inhibit bioluminescence in *V. harveyi* MTCC 3438 [75], honaucin A inhibit bioluminescence of *V. harveyi* BB120 [76], hymenialdisin inhibits bioluminescence in *E. coli* pSB1075 iberin inhibits fluorescence of *P. aeruginosa* [72], kojicacid inhibits bioluminescence in *E. coli* pSB1075 [77], limonoids inhibits bioluminescence in *V. harveyi* BB886 [67], microcolins A and B inhibits bioluminescence in *E. coli* pSB1075 [77], naringenin inhibits violacein of *C. violaceum* CV026 [78], protoanemonin inhibits fluorescence in *P. aeruginosa* [79], pyrogallol inhibits bioluminescence in *V. harveyi* [80] thiazolidinediones inhibits bioluminescence in *V. harveyi* BB170 [81].

21.8.6.3 Action of 3-Indolyl Acetyl Nitrile

Small molecules inhibiting biofilm formation reduce *Xanthomonas citri* subsp. *citri* infection and enhance the control of citrus canker disease. D-leucine and 3-indolylacetonitrile (IAN) were found to prevent biofilm formation by *X. citri* subsp. *citri* on different abiotic surfaces and host leaves at a concentration lower than the minimum inhibitory concentration (MIC). Quantitative reverse transcription-polymerase chain reaction (qRT-PCR) analysis indicated that IAN repressed expression of chemotaxis/motility related genes in *X. citri* subsp. *citri* [82].

21.9 Chemical Methods

21.9.1 Inhibitors of Nucleotide Biosynthesis and DNA Replication as Antibiofilm Agents

Cyclic di-GMP, play a pivotal role as signal molecules for biofilm regulation, accumulation of c-di-GMP stimulates production of adhesion factors via a variety of different mechanisms (i.e., allosteric activation of protein activity, protein stabilization, or regulation of gene expression at the transcriptional and translational levels) [83–85]. It is possible that reduction of intracellular c-di-GMP levels by sulfathiazole depends on inhibition of tetrahydrofolate biosynthesis, in turn, affecting thymidine intracellular pools and DNA synthesis. It has recently been reported that fluorouracil, which blocks DNA replication through inhibition of nucleotide biosynthesis, can prevent biofilm formation at concentrations not affecting planktonic cell growth [86, 87]. This demonstrates that nucleotide biosynthesis inhibitors also show antibiofilm activity and suggests that a decrease in cellular nucleotide pools negatively affects biofilm formation. Another possibility might be that an even partial inhibition of nucleotide biosynthesis, such as observed at sulfathiazole or fluorouracil concentrations not affecting bacterial growth, might result in shortage of deoxyribo nucleotides for DNA replication. The bacterial cells might then react by abolishing nonessential DNA synthesis, such as production of extracellular DNA. Indeed, extracellular DNA is an essential component of the biofilm matrix in both Gram-positive and Gram-negative bacteria [88] and treatment with DNase can prevent biofilm formation [89], a fact that suggests exploitable weaknesses in the biofilm matrix.

21.9.2 Effect of Salicylic Acid on Biofilms

It has been shown that a plant-produced phenolic compound salicylic acid (SA) alters biofilm formation and motility in *Pseudomonas aeruginosa*, an organism causing chronic infections in patients with cystic fibrosis. Salicylic acid is widely known as a primary plant immune response signal [90], also attenuates biofilm formation, swimming motility, and AHL production by different plant pathogens: *Erwinia amylovora* (causative agent of fire blight, a disease of Rosaceae family), *Pseudomonas corrugata* (tomato pith necrosis), *Pseudomonas syringae* pv *syringae* (a variety of necrotic diseases of fruits), *Xanthomonas campestris* pv *campestris* (black rot of crucifers), and *Pectobacterium carotovorum* (soft rot of fruits and vegetables).

21.9.3 N-acetyl Cysteine Effects on Biofilm

Despite the importance of biofilm formation for the pathogenicity of plant bacterial pathogens, strategies to control the biofilm-forming bacteria have been very poorly explored in agriculture. In contrast, strategies to control bacteria-producing biofilms in medicine are extensively studied because this type of bacterial growth is the main cause of important human diseases [91]. On approach used in medicine to impair biofilm formation is the use of N-acetyl cysteine (NAC). This compound is a cysteine analogue, mostly known as a mucolytic agent that has antibacterial properties and inhibits biofilm formation in a variety of Gram-negative and Gram-positive bacteria [92–94]. Muranaka et al. [95] have first reported the antibacterial property of NAC against a plant bacterial pathogen of *Xylella fastidiosa*, which mechanism of pathogenicity involves biofilm formation in xylem vessels. The use of NAC may be a more sustainable alternative strategy for controlling *X. citri* subsp. *citri* by breaking down the biofilm. NAC is an analogue of cysteine that disrupts disulphide bonds in bacterial extracellular proteins and, as a consequence, it not only affects biofilm formation but also disrupts the mature biofilm [92, 96, 97].

21.10 Biological Methods

21.10.1 Biosurfactants as Antibiofilm Agents

Biosurfactants are amphiphilic compounds that reduce the free energy of the system by replacing the bulk molecules of higher energy at an interface. These surfactants are classified as cationic, anionic, zwitter ionic, and nonionic, and these are made synthetically from hydrocarbons, lignosulfonates, or triglycerides (Mulligan). These biosurfactants contain a hydrophobic portion with little affinity for bulk medium and a hydrophobic group that is attracted to the bulk medium.

Biosurfactants are a chemically diverse group of molecules that comprise glycolipids, lipopeptides, lipoproteins, fatty acids, phospholipids, neutral lipids, polymerics, and particulate biosurfactants. Though biosurfactants are having versatile applications, their antibiofilm activity has received enormous attention in recent years due to their potential in health care, agriculture, and industrial applications (98, 99,100).

Biosurfactants' surface-modifying properties effectively affect the microbial colonization and subsequent biofilm formation. They selectively reduce the hydrophobicity of bacterial cell walls. Hydrophobicity directly corresponds to the pathogen's biofilm-forming ability; reduction of it will have direct implications on biofilm formation. Apart from efficiently controlling the biofilm formation, biosurfactants' proficiently disrupt the preformed biofilms [101–104]. Certain biosurfactants have pronounced antibiofilm effects (Table 21.5), which are capable of downregulating, the biofilm and virulence genes in biofilm initiators, in addition to the phenotypic suppression [105–107].

Table 21.5 Biosurfactants as Antibiofilm Agents.

Biosurfactants	Pathogen(s)	References
Glycolipid from *Brevibacterium casei*	Mixed biofilm bacteria	[102]
Glycolipid from Coral associated bacteria	*P. aeruginosa*	[121]
Lipopeptide from *Bacillus circulans*	*E. coli*,	[122]
	Mycobacterium flavus,	
	Proteus vulgaris,	
	Serratia marcescens, *Citrobacter freundii*,	
	Klebsiella aerogenes, *Alcaligens faecalis*,	
	S. typhimurium	
Rufisan from *C. lipolytica* UCP 0988	*Lactobacillus casei*,	[123]
	L. reuteri,	
	Streptococcus mutans,	
	S. oralis,	
	S. sanguis,	
	Rothia dentocariosa,	
	S. salivarius,	
	P. aeruginosa,	
	S. aureus,	
	S. epidermidis,	
Lipopeptide from *B. subtilis* AR2	*Candida albicans*	[124]
Lipopeptide from *B. cereus* NK1	*P. aeruginosa*,	[125]
	S. epidermidis	
Biosurfactant from	*Acinetobacter baumannii*,	[126]
Lactobacillus jensenii,	*E. coli*,	
Lactobacillus rhamnosus	MRSA	
Di-rhamnolipid from	*C. albicans*	[104]
Pseudomonas aeruginosa DSVP20		

(Continued)

Table 21.5 (Continued)

Biosurfactants	Pathogen(s)	References
Rhamnolipid *P. aeruginosa* PAO1	*Bordetella bronchiseptica*	[101]
		[127]
Lactobacillus fermentum derived biosurfactant	*S. mutans*	[105]
Lipopeptide from *B. tequilensis* CH	*E. coli*, *S. mutans*	[128]
Lipopeptide from *Paenibacillus polymyxa*	Mixed species biofilm	[129]
Lunasan from *C. sphaerica* UCP 0995	*P. aeruginosa*, *S. agalactiae*, *S. Sanguis*	[130]
Glycolipid from *Lysinibacillus fusiformis* S9	*E. coli*, *S. mutans*	[128]
Glycolipid from *Nocardiopsis* sp. MSA13A	*Vibrio alginolyticus*	[120]
Putisolvin I and II from *P. putida*	*Pseudomonas* sp.	[131]
Rhamnolipid	*Yarrowia lipolytica*	[103]
Biosurfactant from *Streptococcus thermophiles*	*Candida* spp.	[132]
Surfactin	*S. typhimurium*	[133]
Lipopeptide biosurfactant with antibiotics	*E. coli* CFT073	[134]
Glycolipid- type biosurfactant from *Trichosporon montevideense* CLOA72	*C. albicans*	[135]
Lipopeptide from *B. subtilis* and *B. licheniformis*	*E. coli*, *S. aureus*	[136]
Pseudofactin II from *Pseudomonas fluorescens* BD5	*E. coli*, *E. faecalis*, *E. hirae*, *S. epidermidis*, *P. mirabilis*	[137]

21.10.2 Phage Mediated Biocontrol as Antibiofilm Agents

The interaction between phage and biofilms are rather complex process [29]. Theoretically, a biofilm could be rapidly infested because cells are more close to each other, which can enhance phage replication when compared to the less-accessible bacteria of planktonic cultures [108]. On the other hand, the structure and composition of the biofilm, as well as the physiology of its cells, may impose some limitations to biofilm infection. In fact, there is an evidence, that some bacterial strains that were tolerant to phage infection and it increases their biofilm-formation ability. Further, the biofilm phenotype might be an additional strategy of those bacteria to escape from phage infection [109].

Under suitable conditions, the presence of bacterial cells in biofilm does not probe access to and infection by phage particles. Flemming et al. [110] have drawn attention to the fact that the polymeric matrix in which the microbial cells are embedded consists of various polymers, in addition to exopolysaccharides. Thus, the phage must either be able to poleaxe the matrix by diffusion or enzymes associated with the viral particle and must destroy, if they are to release the embedded bacteria and make them vulnerable to the phage themselves. Their effectiveness will depend not only on the extent to which the matrix is composed of polysaccharides but also on the way in which that interacts with all other polymers and ions in the matrix.

The ability of polysaccharides to interact synergistically to enhance biofilm structure has been suggested. In the example of the plant pathogen *Pseudomonas syringae*, Laue et al. [111] suggested that both alginate and the poly-limbo loan contributed to biofilm formation by this bacterium, although a third unidentified polysaccharide might also play a significant role. A role has also been proposed genetic materials released through the action of autolysins in the establishment of biofilms by *Staphylococcus epidermis* [112]. Similar examples may well be found in other bacteria pathogenic for humans and animals. Both physiology and genetically alterations within bacteria can make significant modifications to the biofilm structure and then to the effectiveness or otherwise of phage muck.

In clinical situations, single-species biofilms will present problems, but when multi-species biofilm occur, the interactions between phage and bacterium may be much more complicated and there may also be antagonistic interactions resulting from bacteriocin production [113]. Certainly, Tait et al. [114] have found that attempts to eliminate the biofilms formation from dual enterobacterial species with phage and associated polysaccharides depolymerased were unsuccessful. The phage and bacteria could apparently coexist stably within these particular biofilms.

21.11 Future Prospects of Antibiofilm

Antibiofilm agents could be recognized as a good, promising tool for the development of therapeutics to control the annoying qualities of bacterial populations. However, currently there are few or no precise antibiofilm products available in the market. Nevertheless, continuing research in the area of antibiofilm agents might lead to the identification of new strategies that may be developed as an unambiguous product and may be available in the future.

Particularly, exopolysaccharides and quorum quenching approaches are major antibiofilm agents that can be used several years for biofilm control, which may play a major role in the future. Subsequently, phage preparations are now available in some places for the control of bacterial infections; these bacteriophages may also play a vital role in biofilm control in the future. However, the effectiveness of the biofilm control in application levels may vary, so maintaining the effectiveness of the antibiofilm control agents must be a major concern in biofilm control, and it has been taken as an important factor in antibiofilm agent development.

21.12 Conclusion

Most of the intractable contaminations/infections are caused by the biofilm-forming bacterial communities; these biofilm-forming bacterial communities have shown resistance against the traditional antimicrobial therapies. As explained in this chapter, there are several encouraging prospects to overcoming these biofilm-forming associated/infected bacteria in plant systems. However, these biofilm control agents are not currently capable of remediating the biofilm associated/infected bacterial growth. Gradual progress continues to be made in biofilm control aspects, though, which has led to the development of many biofilm control agents such as chemical compounds, biosurfactants, quorum quenching, and phages. These biofilm control strategies show promising capability to control the biofilms; however, some questions raised in biofilm control strategies remain to be answered.

References

1 K.H. Gartemann, O. Kirchner, J. Engemann, I. Grafen, R. Eichenlaub and A. Burger, *Clavibacter michiganensis* subsp. *michiganensis*: first steps in the understanding of virulence of a Gram-positive phytopathogenic bacterium, *Journal of Biotechnology*, 106(2), 179–191 (2003).

2 J. Mansfield, S. Genin, S. Magori, V. Citovsky, M. Sriariyanum, P. Ronald and I.A.N. Toth, Top 10 plant pathogenic bacteria in molecular plant pathology, *Molecular plant pathology*, 13(6), 614–629 (2012).

3 V. Rajesh Kannan, K. Kurtulas Bastas and R. Sangeetha devi, *Scientific and economic impact of plant pathogenic bacteria. Sustainable approaches to controlling plant pathogenic bacteria, Edited by* Rajesh Kannan V, Kurtulas Bastas K. CRC press, USA, 369–392 (2015).

4 M. Lipsitch, S. Siller and M.A. Nowak, The evolution of virulence in pathogens with vertical and horizontal transmission, *Evolution*, 1729–1741 (1996).

5 J. He, R.L. Baldini, E. Deziel, M. Saucier, Q. Zhang, N.T. Liberati and L.G. Rahme, The broad host range pathogen *Pseudomonas aeruginosa* strain PA14 carries two pathogenicity islands harboring plant and animal virulence genes, *Proceedings of the national academy of sciences of the United States of America*, 101(8), 2530–2535 (2004).

6 E.A. Mordecai, Pathogen impacts on plant communities: unifying theory, concepts, and empirical work. *Ecological Monographs*, 81(3), 429–441 (2011).

7 G.N. Agrios, *Plant Pathology*, 5th Ed. Elsevier Academic Press, San Diego, California, 2005.

8 P.H. Williams, Black rot, *Rimmer RS, Shattuk VI & Buchwaldt L (Eds.)*, 60–62 (2007).

9 L. De La Fuente and S. Burdman, Pathogenic and beneficial plant-associated bacteria, *Encyclopedia of life support systems* (2011).

10 C.L. Campbell and L.V. Madden, Introduction to plant disease epidemiology. *John Wiley and Sons* (1990).

11 D.M. Weller, Biological control of soilborne plant pathogens in the rhizosphere with bacteria, *Annual review of phytopathology*, 26(1), 379–407 (1988).

12 G.N. Agrios, *Plant Pathology*, 4th Ed. Elsevier Academic Press, San Diego, California, 1997.

13 J. Ahlemeyer and R. Eichenlaub, Genetics of phytopathogenic bacteria, *Progress in botany.* 62, 98–113 (2001)
14 A. Burger and R. Eichenlaub, Genetics of phytopathogenic bacteria. *Progress in Botany,* 64, 98–114 (2003).
15 D.L. Arnold, A. Pitman and R.W. Jackson, Pathogenicity and other genomic islands in plant pathogenic bacteria, *Molecular plant pathology,* 4, 407–420 (2003).
16 M.J. Federle and B.L. Bassler, Interspecies communication in bacteria, *Journal of clinical investigation,* 112, 1291–1299 (2003).
17 G. Ponciano, H. Ishihara, S. Tsuyumu and J.E. Leach, Bacterial effectors in plant disease and defense: keys to durable resistance? *Plant Disease,* 87, 1271–1282 (2003).
18 H. Cao, R.L. Baldini and L.G. Rahme, Common mechanisms for pathogens of plants and animals, *Annual Review of phytopathology,* 39, 259–284 (2001).
19 C.E. Zobell, D.Q. Anderson and W.W. Smith, The bacteriostatic and bactericidal action of great salt lake water, *Journal of Bacteriology,* 33, 253–262 (1937).
20 L. Ma, K.D. Jackson, R.M. Landry, M.R. Parsek and D.J. Wozniak, Analysis of *Pseudomonas aeruginosa* conditional Psl variants reveals roles ror the Psl polysaccharide in adhesion and maintaining biofilm structure post attachment, *Journal of Bacteriology,* 188, 8213–8221 (2006).
21 D.G. Davies, M.R. Parsek, J.P. Pearson, B.H. Iglewski, J.W. Costerton and E.P. Greenberg, The involvement of cell-to-cell signals in the development of a bacterial biofilm, *Science,* 280, 295–308 (1998).
22 H.C. Flemming and J. Wingender, The biofilm matrix, *Natural Reviews on Microbiology,* 8, 623–633 (2010).
23 W.C. Fuqua and E.P. Greenberg, Quorum sensing in bacteria: The Luxr-Luxi family of cell density-responsive transcriptional regulators, *Journal bacteriology,* 176, 269–275 (1994).
24 F. Blot, G. Nitenberg, E. Chachaty, B. Raynard, N. Germann, S. Antoun, A. Laplanche, C. Brun-Buisson and C. Tancrede, Diagnosis of catheter-related bacteraemia: a prospective comparison of the time to positivity of hub-blood versus peripheral-blood cultures, *The Lancet,* 354, 1071–1077 (1999).
25 M.A. Schembri, G. Christianesen and P. Klemn, Fim H mediated autoaggregation of *Escheriachia coli, Molecular Microbiology.* 41, 1419–1430 (2001).
26 J.W. Costerton, Z. Lewandowski, D.E. Caldwell, D.R. Korber and H.M. Lappin-Scott, *Annual review of microbiology.* 49, 711–745 (1995).
27 L.V. Rinaudi and W. Giordano, An integrated view of biofilm formation in rhizobia, *FEMS microbiology letters,* 304(1), 1–11 (2010).
28 N.R. Stanley, B.A. Lazazzera, Environmental signals and regulatory pathways that influence biofilm formation, *Molecular Microbiology,* 52, 917–924 (2004).
29 I.W. Sutherland, Biofilm exopolysaccharides: A strong and sticky framework, *Microbiology,* 147, 3–9 (2001).
30 H.C. Flemming, Biofilms and environmental protection, *Water Science and Technology,* 27:1:10 (1993).
31 A.G. Matthysse, M. Marry, L. Krall, M. Kaye, B.E. Ramey, C. Faqua and A.R. White, The effect of cellulose overproduction on binding and biofilm formation on roots by *Agrobacterium tumefaciens, Molecular Plant Microbe Interaction,* 18, 1002–1010 (2005).
32 M. Van de Mortel and L.J. Halverson, Cell envelope components contributing to biofilm growth and survival of *Pseudomonas putida* in low-water-content habitats. *Molecular Microbiology.* 52, 735–750 (2004).

33 U.B. Priefer, Genes involved in lipopolysacharides production and symbiosis are clustered on the chromosome of *Rhizobium leguminosarum* biovar *viciae* VF39, *Journal of Bacteriology*, 171, 6161–6168 (1989).
34 J.R. Cava, P.M. Elias and K.D. Noel, *Rhizobium leguminosarum* CFN42 genetic region encoding lipopolysaccharide structures essential for complete nodule development on bean plants. *Journal of bacteriology.* 171, 8–15 (1989).
35 J.C. Clifford, J.N. Rapicavoli, and M.C. Roper, A rhamnose-rich *O*-antigen mediates adhesion, virulence and host colonization for the xylem-limited phytopathogen, *Xylella fastidiosa, Molecular plant microbe interactions*, 26, 76–685 (2013).
36 D.L. Hopkins, Seasonal concentration of the Pierce's disease bacterium in grapevine stems, petioles, and leaf veins, *Phytopathology*, 71, 415–418 (1981).
37 C. Ryder, M. Byrd and D.J. Wozniak, Role of exopolysaccharides in *Pseudomonas aeruginosa* biofilm development, *Current opinion microbiology*, 10, 644–648 (2007).
38 F. Vaningelgem, M. Zamfir, F. Mozzi, T. Adriany, M. Vancanneyt, J. Swings and L. de Vuyst, Biodiversity of exopolysaccharides produced by *Streptococcus thermophilus* strains is reflected in their production and their molecular and functional characteristics, *Applied Environmental Microbiology*, 70, 900–912 (2004).
39 R.K. Shaw, C.N. Berger, S.V. Sodha, P.M. Griffin and D. Pink, Fresh fruit and vegetables as vehicles for the transmission of human pathogens, *Environmental Microbiology*, 12:2385–2397 (2010).
40 S.L. Collignon, Attachment and colonization by *Escherichia coli* O157:H7, *Listeria monocytogenes, Salmonella enteric* subsp, *enterica* servor *typhimurium*, and *Staphylococcus aureus* on stone fruit surfaces and survival through a simulated commercial export chain, *Journal of food protection.* 120, 1212–1397 (2010).
41 J. Patel and M. Sharma, Differences in attachment of *Salmonella* enterica serovars to cabbage and lettuce leave, *International journal of food microbiology*, 139, 41–47 (2010).
42 L.D. Reina, H.P. Fleming and F. Breidt, Bacterial contamination of cucumber fruit through adhesion, *Journal of food protection*, 65, 1881–1887 (2002).
43 U. Gerstel and U. Romling, Oxygen tension and nutrient starvation are major signals that regulate *agfD* promoter activity and expression of the multicellular morphotype in *Salmonella typhimurium, Environmental microbiology*, 3, 638–648 (2001).
44 R. Sharp, G. Hughes, A. Hart and J.T. Walker, WIPO Patent Application No. PCT/GB2004/000073, 2004.
45 M.E. Cortes, J. Consuegra and R. Sinisterra, Biofilm formation, control and novel strategies for eradication. *Science against microbial pathogens: Communicating current research and technological advances*, 2, 896–905 (2011).
46 J. Masák, A. Čejková, O. Schreiberová and T. Řezanka, Pseudomonas biofilms: possibilities of their control, *FEMS microbiology ecology*, 89(1), 1–14 (2014).
47 G.C.L. Wong and G.A. O'Toole, All Together Now: Integrating biofilm research Across disciplines, *MRS Bulletin*, 36, 339–345 (2011).
48 G. Cheng, H. Xite, Z. Zhang, S. Chen and S. Jiang, A Switchable biocompatible polymer surface with self-sterilizing and nonfouling capabilities, *Angewandte chemie international edition*, 47, 8831–8844 (2008).
49 B. Jansen and W. Kohnen, Prevention of biofilm formation by polymer modification, *Journal Indian Microbiology*, 15, 391–396 (1995).
50 T. Meiron and I. Saguy, Adhesion modeling on rough low linear density polythenes, *Journal of food Science*, 72, 485–791 (2007).

51 J. Jass, S. Surman, T. James, Medical biofilms: detection, prevention and control, *Top academic publications*. 2, 305 (2009).

52 M. Simoes, L.C. Simoes and M.J. Vieira, A review of current and emergent biofilm control strategies, *LWT-Food science and technology*, 43(4), 573–583 (2010).

53 B. Orgaz, J. Kives, A.M. Pedregosa, I.F. Monistrol, F. Laborda and C. San Jose, Bacterial biofilm removal using fungal enzymes, *Enzyme and microbial technology*, 40(1), 51–56 (2006).

54 C. Johansen, P. Falholt and L. Gram, Enzymatic removal and disinfection of bacterial biofilms, *Applied and environmental microbiology*, 63(9), 3724–3728 (1997).

55 N. Rabin, Y. Zheng, C. Opoku-Temeng, Y. Du, E. Bonsu and H.O. Sintim, Agents that inhibit bacterial biofilm formation, *Future medicinal chemistry*, 7(5), 647–671 (2015).

56 M. Hentzer, K. Riedel, T.B. Rasmussen, A. Heydorn, J.B. Andersen, M.R. Parsek and S. Kjelleberg, Inhibition of quorum sensing in *Pseudomonas aeruginosa* biofilm bacteria by a halogenated furanone compound, *Microbiology*, 148(1), 87–102 (2002).

57 I.A. Packiavathy, S. Priya, S.K. Pandian and A.V. Ravi, Inhibition of biofilm development of uropathogens by curcumin an anti-quorum sensing agent from *Curcuma longa. Food chemistry*, 148, 453–460 (2014).

58 T. Bjarnsholt, P.O. Jensen, T. B. Rasmussen, L. Christophersen, H. Calum, M. Hentzer and N. Hoiby, Garlic blocks quorum sensing and promotes rapid clearing of pulmonary *Pseudomonas aeruginosa* infections, *Microbiology*., 151(12), 3873–3880 (2014).

59 J.F. Hu, E. Garo, M.G. Goering, M. Pasmore, H.D. Yoo, T. Esser and Y.S.L. Lee, Bacterial biofilm inhibitors from *Diospyrosdendo, Journal of natural products*, 69(1), 118–120 (2006).

60 E. Garo, G.R. Eldridge, M.G. Goering, E.D. Pulcini, M.A. Hamilton, J.W. Costerton and G.A. James, Asiatic acid and corosolic acid enhance the susceptibility of *Pseudomonas aeruginosa* biofilms to tobramycin, *Antimicrobial agents and chemotherapy*, 51(5), 1813–1817 (2007).

61 H. Wu, B. Lee, L. Yang, H. Wang, M. Givskov, S. Molin and Z. Song, Effects of ginseng on *Pseudomonas aeruginosa* motility and biofilm formation, *FEMS Immunology and medical microbiology*, 62(1), 49–56 (2011).

62 J.H. Lee, M.H. Cho and J. Lee, 3-Indolylacetonitrile decreases *Escherichia coli* O157: H7 Biofilm Formation and *Pseudomonas aeruginosa* virulence, *Environmental microbiology*, 13(1), 62–73 (2011).

63 Y.M. Diaz, V.L. Gina, R.G. Luis, H.M. Wandurragac, Catalina Arevalo-Ferro d, Freddy Ramos Rodriguez b, Carmenza Duque Beltranb, Leonardo Castellanos Hernandez, *Biofilm inhibition activity of compounds isolated from two Eunicea species collected at the Caribbean Sea. Revista Brasileira de Farmacognosia*, 25, 605–611 (2015).

64 X. Ding, B. Yin, L. Qian, Z. Zeng, Z. Yang, H. Li and S. Zhou, Screening for novel quorum-sensing inhibitors to interfere with the formation of *Pseudomonas aeruginosa* biofilm, *Journal of medical microbiology*, 60(12), 1827–1834 (2011).

65 J.H. Lee, S.C. Regmi, J.A. Kim, M.H. Cho, H. Yun, C.S. Lee and J. Lee, Apple flavonoid phloretin inhibits *Escherichia coli* O157: H7 biofilm formation and ameliorates colon inflammation in rats, *Infection and immunity*, 79(12), 4819–4827 (2011).

66 H.S. Kim and H.D. Park, Ginger extract inhibits biofilm formation by *Pseudomonas aeruginosa* PA14, *PloS one*, 8(9), 76106 (2013).

67 A. Vikram, P.R. Jesudhasan, S.D. Pillai and B.S. Patil, Isolimonic acid interferes with *Escherichia coli* O157: H7 biofilm and TTSS in QseBC and QseA dependent fashion, *BMC microbiology*, 12(1), 261 (2012).

68 K.G. Chan, S. Atkinson, K. Mathee, C.K. Sam, S.R. Chhabra, M. Camara and P. Williams, Characterization of N-acylhomoserine lactone-degrading bacteria associated with the *Zingiber officinale* (ginger) rhizosphere: Co-existence of quorum quenching and quorum sensing in *Acinetobacter* and *Burkholderia*, *Biomed Central Microbiology*, 11(1), 51 (2011).

69 K.L. Newman, S. Chatterjee, K.A. Ho, and S.E. Lindow, Virulence of plant pathogenic bacteria attenuated by degradation of fatty acid cell-to-cell signaling factors. *Molecular plant-microbe interactions*. 21(3), 326–334 (2008).

70 S. Fetzner, Quorum quenching enzymes DOI:10.1016/j.jbiotec.2014.09.001#_blank#Opens new window , *Journal of Biotechnology*, 201, 2–14 (2015).

71 M. Romero, L. Acuna and A. Otero, Patents on quorum quenching: interfering with bacterial communication as a strategy to fight infections, *Recent patents on biotechnology*, 6(1), 2–12 (2012).

72 T.H. Jakobsen, M. van Gennip, R.K. Phipps, M.S. Shanmugham, L.D. Christensen, M. Alhede and P.O. Jensen, Ajoene, a sulfur-rich molecule from garlic, inhibits genes controlled by quorum sensing. *Antimicrobial agents and chemotherapy*, 56(5), 2314–2325 (2012).

73 T. Defoirdt, G. Brackman and T. Coenye, Quorum sensing inhibitors: how strong is the evidence? *Trends in Microbiology*, 21(12), 619–624 (2013).

74 S.N.M. Norizan, W.F. Yin and K.G. Chan, Caffeine as a potential quorum sensing inhibitor, *Sensors*, 13(4), 5117–5129 (2013).

75 I.A. Packiavathy, P. Sasikumar, S.K. Pandian and A.V. Ravi, Prevention of quorum-sensing-mediated biofilm development and virulence factors production in *Vibrio spp.* by curcumin, *Applied Microbiology and Biotechnology*, 97(23), 10177–10187 (2013).

76 S. Choi and J. Chae, Optimal biofilm formation and power generation in a micro-sized microbial fuel cell (MFC). *Sensors and Actuators A: Physical*, 195, 206–212 (2013).

77 S. Dobretsov, M. Teplitski, M. Bayer, S. Gunasekera, P. Proksch and V.J. Paul, Inhibition of marine biofouling by bacterial quorum sensing inhibitors. *Biofouling*, 27(8), 893–905 (2011).

78 O.M. Vandeputte, M. Kiendrebeogo, T. Rasamiravaka, C. Stevigny, P. Duez, S. Rajaonson and M. El Jaziri, The flavanone naringenin reduces the production of quorum sensing-controlled virulence factors in *Pseudomonas aeruginosa* PAO1, *Microbiology*, 157(7), 2120–2132 (2011).

79 R.A.B. Fazzini, M.E. Skindersoe, P. Bielecki, J. Puchałka, M. Givskov and V.A. Martins dos Santos, Protoanemonin: a natural quorum sensing inhibitor that selectively activates iron starvation response, *Environmental Microbiology*, 15(1), 111–120 (2013).

80 N. Ni, G. Choudhary, M. Li and B. Wang, Pyrogallol and its analogs can antagonize bacterial quorum sensing in *Vibrio harveyi*, *Bioorganic and Medicinal Chemistry Letters*, 18(5), 1567–1572 (2008).

81 G. Brackman. A.A. Al Quntar, C.D. Enk, I. Karalic, H.J. Nelis, S. Van Calenbergh and T. Coenye, Synthesis and evaluation of thiazolidinedione and dioxazaborocane analogues as inhibitors of AI-2 quorum sensing in *Vibrio Harvey*,. *Bioorganic and Medicinal Chemistry*, 21(3), 660–667 (2013).

82 Jinyun Li and Nian Wang, Foliar application of biofilm formation–inhibiting compounds enhances control of citrus canker Caused by *Xanthomonas citri* subsp. *Citri*, *Phytopathology*, 104, 134–142 (2013).

83 H. Weinhouse, S. Sapir, D. Amikam, Y. Shilo, G. Volman, P. Ohana and M. Benziman, c-di-GMP-binding protein, a new factor regulating cellulose synthesis in *Acetobacter xylinum, FEBS Letters*, 416, 207–211 (1997).

84 R. Simm, M. Morr, A. Kader, M. Nimtz and U. Romling, GGDEF and EAL domains inversely regulate cyclic di-GMP levels and transition from sessility to motility, *Molecular Microbiology*, 53, 1123–1134 (2004).

85 H. Weber, C. Pesavento, A. Possling, G. Tischendorf and R. Hengge, Cyclic-di-GMP-mediated signalling within the sigma network of *Escherichia coli, Molecular Microbioliology*, 62, 1014–1034 (2006).

86 C. Attila, A. Ueda and T.K. Wood, 5-Fluorouracil reduces biofilm formation in *Escherichia coli* K-12 through global regulator AriR as an anti-virulence compound, *Applied Microbiology and Biotechnology*, 82, 525–533 (2009).

87 S. Ueda, K. Isogami, S. Kobayashi, Uncomplicated covering technique for preventing the recurrence in the thoracoscopic surgery for pneumothorax, *The Japanese Journal of Thoracic Surgery*, 62(5), 381–384 (2009).

88 P.S. Guiton, C.S. Hung, K.A. Kline, R. Roth, A.L. Kau, E. Hayes, J. Heuser, K.W. Dodson, M.G. Caparon, S.J. Hultgren, Contribution of autolysin and Sortase A during *Enterococcus faecalis* DNA-dependent biofilm development, *Infect Immunology*, 77, 3626–3638 (2009).

89 C. B. Whitchurch, T. Tolker-Nielsen, P.C. Ragas, J.S. Mattick, Extracellular DNA required for bacterial biofilm formation, *Science*, 295, 1487 (2002).

90 I. Raskin, Role of salicylic acid in plants, *Annual Review Plant Physiology*, 43, 439–463 (1992).

91 T. Zhao and Y. Liu, N-acetylcysteine inhibits biofilms produced by *Pseudomonas aeruginosa, BMC Microbiology*, 10, 140 (2010).

92 C. Perez-Giraldo, A. Rodriguez-Benito, F.J. Moran, Influence of N-acetylcysteine on the formation of biofilm by *Staphylococcus epidermidis, Journal of Antimicrobial Chemother*, 39, 643–646 (1997).

93 B.K. Rubin, Mucolytics, expectorants, and mucokinetic medications, *Respiration Care*, 52, 859–865 (2007).

94 M.M. Hafez, M.M. Aboulwafa, M.A. Yassien, N.A. Hassouna, Activity of some mucolytics against bacterial adherence to mammalian cells, *Applied Biochemical Biotechnology*, 158, 97–112 (2009).

95 L.S. Muranaka, T.E. Giorgiano, M.A. Takita, M.R. Forim, L.F. Silva, H.D. Coletta-Filho and A.A. de Souza, N-Acetylcysteine in Agriculture, a Novel Use for an Old Molecule: Focus on Controlling the Plant–Pathogen Xylella fastidiosa, *PloS one*, 8(8), e72937 (2013).

96 H.Q. Huynh, R.T. Couper, C.D. Tran, L. Moore, R. Kelso, R.N. Butler, N-Acetylcysteine, a novel treatment for *Helicobacter Pylori* infection, *Digestive Disease Science*, 49, 1853– 1861 (2004).

97 S.Y. Quah, S. Wu, J.N. Lui, C.P. Sum, K.S. Tan, N-Acetylcysteine inhibits growth and eradicates biofilm of *Enterococcus faecalis, Journal endodontics*, 38, 81–85 (2012).

98 D.P. Sachdev and S.S. Cameotra, Biosurfactants in agriculture. *Applied Microbiology and Biotechnology*, 97(3), 1005–1016 (2013).

99 I.M. Banat, D.A.M. De Rienzo and A.G. Quinn, Microbial biofilms: biosurfactants as antibiofilm agents, *Applied Microbiology and Biotechnology*, 98(24), 9915–9929 (2014).

100 G.S. Kiran, A.S. Ninawe, A.N. Lipton, V. Pandian and J. Selvin, Rhamnolipid biosurfactants: evolutionary implications, applications and future prospects from untapped marine resource, *Critical Reviews in Biotechnology*, 1–17 (2015).

101 Y. Irie, G.A. O'toole and M.H. Yuk, *Pseudomonas aeruginosa* rhamnolipids disperse *Bordetella bronchiseptica* biofilms, *FEMS Microbiology Letters*, 250(2), 237–243 (2005).

102 G. S. Kiran, B. Sabarathnam and J. Selvin, Biofilm disruption potential of a glycolipid biosurfactant from marine *Brevibacterium casei*, *FEMS Immunology and Medical Microbiology*, 59(3), 432–438 (2010).

103 D.H. Dusane, S. Dam, Y.V. Nancharaiah, A.R. Kumar, V.P. Venugopalan and S.S. Zinjarde, Disruption of *Yarrowia lipolytica* biofilms by rhamnolipid biosurfactant. *Aquatic Biosystems*, 8(1), 1 (2012).

104 N. Singh, S.C. Pemmaraju, P.A. Pruthi, S.S. Cameotra and V. Pruthi, *Candida bio*film disrupting ability of di-rhamnolipid (RL-2) produced from *Pseudomonas aeruginosa* DSVP20, *Applied Biochemistry and Biotechnology*, 169(8), 2374–2391 (2013).

105 A. Tahmourespour, R. Salehi and R.K. Kermanshahi, *Lactobacillus acidophilus*-derived biosurfactant effect on *gtfB* and *gtfC* expression level in *Streptococcus mutans* biofilm cells, *Brazilian Journal of Microbiology*, 42(1), 330–339 (2011).

106 R. Salehi, O. Savabi and M. Kazemi, Effects of *Lactobacillus reuteri* derived biosurfactant on the gene expression profile of essential adhesion genes (*gtfB*, *gtfC* and *ftf*) of *Streptococcus mutans*, *Advanced Biomedical Research*, 3, 169 (2014).

107 O. Savabi, M. Kazemi, S. Kamali, A.R. Salehi, G. Eslami, A. Tahmourespour and R. Salehi, Effects of biosurfactant produced by *Lactobacillus casei* on *gtfB*, *gtfC* and *ftf* gene expression level in *S. mutans* by real-time RT-PCR, *Advanced Biomedical Research*, 3, 231 (2014).

108 B.A. Wiggins, M. Alexander, Minimal bacterial density for bacteriophage replication: implications for significance of bacteriophages in natural ecosystems, *Applied Environmental Microbiology*, 49, 19–23 (1985).

109 A. Lacqua, O. Wanner, T. Colangelo, M.G. Martinotti and P. Landini, Emergence of biofilm-forming subpopulations upon exposure of *Escherichia coli* to environmental bacteriophages, *Applied and Environmental Microbiology*, 72(1), 956–959 (2006).

110 H.C. Flemming, T. R. Neu and D.J. Wozniak, The EPS matrix: the house of biofilm cells, *Journal of Bacteriology*, 189(22), 7945–7947 (2007).

111 H. Laue, A. Schenk, H. Li, L. Lambertsen, T.R. Neu, S. Molin and M.S. Ullrich, Contribution of alginate and levan production to biofilm formation by *Pseudomonas syringae*, *Microbiology*, 152(10), 2909–2918 (2006).

112 Z. Qin, Y. Ou, L. Yang, Y. Zhu, T. Tolker-Nielsen, S. Molin and D. Qu, Role of autolysin-mediated DNA release in biofilm formation of *Staphylococcus epidermidis*. *Microbiology*, 153(7), 2083–2092 (2007).

113 K. Tait, I. Joint, M. Daykin, D.L. Milton, P. Williams and M. Camara, Disruption of quorum sensing in seawater abolishes attraction of zoospores of the green alga *Ulva* to bacterial biofilms, *Environmental Microbiology*, 7(2), 229–240 (2005).

114 K. Tait, L.C. Skillman and I.W. Sutherland, The efficacy of bacteriophage as a method of biofilm eradication. *Biofouling*, 18(4): 305–311. (2002).

115 G. Ponciano, H. Ishihara, S. Tsuyumu, S and J.E. Leach. *Bacterial effectors in plant disease and defense: keys to durable resistance? Plant Dis.* 87:1271–1282. (2003).

116 M.W.B Kirzinger, G. Nadarasah, J. Stavrinides. Insights into cross-kingdom plant pathogenic bacteria. *Genes*, 2: 980–997. (2011).

117 C. Pablo, M. Bogino, O. de lad Mercedes, G. Fernando, G. Sorroche and G. Walter, The role of bacterial biofilms and surface components in plant-bacterial associations, *International Journal of Molecular Sciences*, 14, 15838–15859 (2013).

118 V. Ranganathan, Biofilms: Microbial cities of scientific significance, *Journal of Microbiology and Experimentation*, 1(3), 1137–1196 (2014).

119 J.M. Monier and S.E. Lindow, Frequency, size, and localization of bacterial aggregates on bean leaf surfaces, *Applied Environmental Microbiology*, 70(1), 346–55 (2004).

120 G.S. Kiran, L.A. Nishanth, S. Priyadharshini, K. Anitha, J. Selvin, Effect of Fe nanoparticle on growth and glycolipid biosurfactant production under solid state culture by marine *Nocardiopsis* sp. MSA13A, *BMC Biotechnology*, 14(1), 48 (2014).

121 A.R. Padmavathi, D. Bakkiyaraj and S.K. Pandian, Significance of biosurfactants as antibiofilm agents in eradicating phytopathogens, *In Springer international publishing: Bacterial metabolites in sustainable agroecosystems*, 319–336 (2015).

122 P. Das, S. Mukherjee, R. Sen, Anti-adhesive action of a marine microbial surfactant: Colloids and surfaces, *Biointerfaces*, 71(2), 183–186 (2009).

123 R.D. Rufino, J.M. Luna, L.A. Sarubbo, L.R.M. Rodrigues, J.A.C. Teixeria and G.M. Campos-Takaki, Antimicrobial and anti-adhesive potential of a biosurfactant Rufisan produced by *Candida lipolytica* UCP 0988, *Colloids and surfaces B: Biointerfaces*, 84(1), 1–5 (2011).

124 R. Rautela, A.K. Singh, A. Shukla, and S.S. Cameotra, Lipopeptides from *Bacillus* strain AR2 inhibits biofilm formation by *Candida albicans*, *Antonie Van Leeuwenhoek*, 105(5), 809–821 (2014).

125 M.I. Sriram, K. Kalishwaralal, V. Deepak, R. Gracerosepat, K. Srisakthi and S. Gurunathan, Biofilm inhibition and antimicrobial action of lipopeptide biosurfactant produced by heavy metal tolerant strain *Bacillus cereus* NK1, *Colloids and Surfaces B: Biointerfaces*, 85(2), 174–181 (2011).

126 K. Sambanthamoorthy, X. Feng, R. Patel, S. Patel and C. Paranavitana, Antimicrobial and antibiofilm potential of biosurfactants isolated from *Lactobacilli* against multi-drug-resistant pathogens, *BMC Microbiology*, 14(1), 197 (2014).

127 A. Raya, M. Sodagari, N.M. Pinzon, X. He, B.M.Z. Newby, and L.K. Ju, Effects of rhamnolipids and shear on initial attachment of *Pseudomonas aeruginosa* PAO1 in glass flow chambers, *Environmental Science and Pollution Research*, 17(9), 1529–1538 (2010).

128 A.K. Pradhan, N. Pradhan, G. Mall, H.T. Panda, L.B. Sukla, P.K. Panda and B.K. Mishra, Application of lipopeptide biosurfactant isolated from a halophile: *Bacillus tequilensis* CH for inhibition of biofilm, *Applied Biochemistry and Biotechnology*, 171(6), 1362–1375 (2013).

129 G.A. Quinn, A.P. Maloy, S. McClean, B. Carney and J.W. Slater, Lipopeptide biosurfactants from *Paenibacillus polymyxa* inhibit single and mixed species biofilms, *Biofouling*, 28(10), 1151–1166 (2012).

130 J.M. Luna, R.D. Rufino, L.A. Sarubbo, L.R. Rodrigues, J.A. Teixeira and G.M. de Campos-Takaki, Evaluation antimicrobial and anti adhesive properties of the biosurfactant lunasan produced by *Candida sphaerica* UCP 0995, *Current Microbiology*, 62(5), 1527–1534 (2011).

131 I. Kuiper, E.L. Lagendijk, R. Pickford, J.P. Derrick, G.E. Lamers, J.E. Thomas-Oates and G.V. Bloemberg, Characterization of two *Pseudomonas putida* lipopeptide biosurfactants, putisolvin I and II, which inhibit biofilm formation and break down existing biofilms, *Molecular Microbiology*, 51(1), 97–113 (2004).

132 H.J. Busscher, C.G. Van Hoogmoed, G.I. Geertsema-Doornbusch, M. Van der Kuijl-Booij and H.C. Van der Mei, *Streptococcus thermophilus* and its biosurfactants inhibit adhesion by *Candida* spp. on silicone rubber, *Applied and Environmental Microbiology*, 63(10), 3810–3817 (1997).

133 J.R. Mirele, A. Toguchi and R.M. Harshey, *Salmonella enterica* serovar *Typhimurium* swarming mutants with altered biofilm-forming abilities: surfactin inhibits biofilm formation, *Journal of Bacteriology*, 183(20), 5848–5854 (2001).

134 F. Rivardo, M.G. Martinotti, R.J. Turner and H. Ceri, Synergistic effect of lipopeptide biosurfactant with antibiotics against *Escherichia coli* CFT073 biofilm, *International journal of Antimicrobial Agents*, 37(4), 324–331 (2011).

135 A.S. Monterio, T.T. Miranda, I. Lula, A.M. Denadai, R.D. Sinisterra, M.M. Santoro and V.L. Santos, Inhibition of *Candida albicans* CC biofilms formation in polystyrene plate surfaces by biosurfactant produced by *Trichosporon montevideense* CLOA72, *Colloids and Surfaces B: Biointerfaces*, 84(2), 467–476 (2011).

136 F. Rivardo, R.J. Turner, G. Allegrone, H. Ceri and M.G. Martinotti, Anti-adhesion activity of two biosurfactants produced by *Bacillus* spp. prevents biofilm formation of human bacterial pathogens, *Applied Microbiology and Biotechnology*, 83(3), 541–553 (2009).

137 T. Janek, M. Lukaszewicz and A. Krasowska, Anti adhesive activity of the biosurfactant pseudofactin II secreted by the Arctic bacterium *Pseudomonas fluorescens* BD5, *BMC Microbiology*, 12(1), 24 (2012).

22

Applications of Biofilm and Quorum Sensing Inhibitors in Food Protection and Safety

Ashraf A. Khan[1*], *John B. Sutherland*[1], *Mohammad Shavez Khan*[2], *Abdullah S. Althubiani*[3] *and Iqbal Ahmad*[2]

[1] *Division of Microbiology, National Center for Toxicological Research, U.S. Food and Drug Administration, Jefferson, Arkansas, USA*
[2] *Department of Agricultural Microbiology, Faculty of Agricultural Sciences, Aligarh Muslim University, Aligarh, India*
[3] *Department of Biology, Faculty of Applied Science, Umm Al-Qura University, Makkah, Saudi Arabia*

22.1 Introduction

Growth of microorganisms in foods can result in a number of changes in food, including deterioration and spoilage [1]. To prevent these changes, several strategies are applied, which may include preventing contamination, removing microorganisms from the food, or adjusting conditions of storage to prevent their growth [2]. A variety of pathogenic microorganisms associated with foods may contaminate them at various stages of food processing [3, 4]. Pathogenic microorganisms have become a critical problem for food quality, safety, and public health [4].

In the last few decades, bacterial quorum sensing (QS) and biofilm formation by foodborne microorganisms have been studied, and it is now realized that they have significant effects on the food industry. Excellent review articles have been written on the significance of biofilms for food safety and on new strategies for controlling biofilms [5–7]. Here we have attempted to address the significance of biofilm formation by foodborne pathogens and strategies used to prevent microbial contamination and spoilage in food during and after processing. Natural products and new chemicals that have been identified as antibiofilm and anti-QS agents will be reviewed.

22.2 Biofilm Formation by Foodborne Pathogens

Biofilm formation by bacteria is a multistep process [8]. The steps that occur on food materials (solids and liquids) begin with preconditioning of the adhesion surface for attachment of planktonic cells and lead step by step to colonization, growth, maturation, and dispersal of the biofilm. The physiology and molecular biology of biofilm formation have been well described by various authors [7, 9, 10].

In food and food environments, attachment of microorganisms is most common on surfaces that have rough texture, hydrophobicity, or surface-conditioning films [11].

Several microbial cell structures and products play important roles in biofilm processes. These structures include flagella, pili or fimbriae, prosthecae, stalks, and holdfasts [12]. Flagella, when present, are important for adhesion by providing cells with motility to reach the surface for attachment [12]. Motility and biofilm formation appear to be two contrasting properties, but motility is important not only for initial biofilm formation [13, 14] but also for dispersal [15]. Recent investigations show that the role of bacterial motility is more complex than previously known [15]. Extracellular polymeric substances (EPS) of microbial cells influence the formation and regulation of biofilms by binding cells and other particulate matter together and to the surface [16–18]. The EPS also protects biofilm microorganisms by acting as a physical barrier; tolerance of the biofilm community to physical stress and biocides is mainly due to the barrier functions of EPS [19]. The roles of other structural constituents of EPS, like DNA and protein, have been increasingly recognized in recent years [20, 21].

Bacterial QS is a global gene regulatory mechanism that responds to cell density; several excellent reviews have been written on this subject [22–24]. Both in single and in mixed-species biofilm communities, cell-to-cell communication plays an important role in adhesion and biofilm formation [25–31]. The QS-mediated microbial activities in many Gram-negative bacteria include the production of extracellular enzymes, biofilms, biosurfactants, EPS, and virulence factors [32–34]. For intraspecies signaling, most bacteria utilize QS systems that use one of two types of small molecule autoinducers [35, 36]. Gram-negative bacteria use N-acylated-L-homoserine lactones (AHLs), which are produced by LuxI-type synthase enzymes and bind to cytoplasmic LuxR-type reporters to induce output [36]. Gram-positive bacteria use cyclic peptides as the major autoinducers, which can be recognized by either membrane-associated histidine kinases or cytoplasmic receptors [37]. Mutants with altered QS genes may still form biofilms of varying complexity [29, 38].

22.3 Significance of Biofilms in Food and Food Environments

Foodborne diseases are considered an important global public health problem [39]. From 1996 to 2010, they caused approximately 1,000 reported disease outbreaks [40], 48 million illnesses, 128,000 hospitalizations, and 3,000 deaths annually in the United States [7, 41]. A large number of food-related outbreaks have been associated with biofilms [7, 42–44]. It is now realized that around 80 percent of persistent bacterial infections are related to biofilms [45]. Biofilms provide protection to pathogenic bacteria by enhancing resistance to disinfectants and other cleaning agents [46, 47]. Dairy, poultry, meat, brewing, and seafood processing are all affected by biofilm development [5, 7, 48]. The contamination of food may occur at any stage during harvesting, collection, processing, packaging, or consumption [49]. The attachment of the microorganisms to food products or product contact surfaces leads to serious hygiene problems and economic losses due to food spoilage [50–52]. Bacteria can form biofilms under suitable conditions on almost all types of abiotic food-contact surface materials, including stainless steel, glass, rubber, polyurethane, Teflon, nitrile, butyl rubber, and wood [7, 53, 54]. Recent emphasis has been given to organisms like *Listeria monocytogenes*, *Yersinia enterocolitica*, *Campylobacter jejuni*, and *Escherichia coli* O157:H7 [10, 52, 55, 56].

The main reasons for contamination of milk and related products are improper cleaning and insufficient disinfection of equipment [19]. The formation of biofilms on dairy equipment can result in serious economic losses [57, 58]. In the dairy industry, commonly encountered biofilm-associated bacteria belong to genera that include *Bacillus, Enterobacter, Lactobacillus, Listeria, Micrococcus, Pseudomonas,* and *Streptococcus* [5, 43, 59–63].

In addition to the dairy industry, biofilms also pose serious problem in the produce industry, where a number of outbreaks due to *L. monocytogenes* in whole cantaloupes [64] and intact alfalfa sprouts [65] have been reported.

In the fish-processing industry, both equipment and water are major sources of contamination. Well-known biofilm formers associated with fish products are *Vibrio cholerae* and other species of *Vibrio* [66]. *Listeria monocytogenes, Salmonella* spp., *Bacillus* spp., *Aeromonas* spp., and *Pseudomonas* spp. also form biofilms in fish- and seafood-processing environments [67].

Several reports on the poultry industry show that biofilm-forming microbial contaminants originate from dusty surfaces, feces, poultry feed, and the transportation and processing of live poultry. The major pathogens in poultry processing that receive attention are species of *Salmonella*, such as *S. enterica* ser. Typhimurium, and *Campylobacter* spp. [68–70].

Microbial contamination and biofilm formation have also been reported in the meat industry. Organic residues in meat-processing equipment provide an important niche for bacterial contaminants [6, 71, 72]. *E. coli* O157:H7 and *Acinetobacter calcoaceticus* can form monospecific biofilms as well as mixed-species biofilms on meat samples [73, 74].

In recent years, ready-to-eat products have gained much popularity and may pose a high risk under improper storage conditions [75]. *L. monocytogenes* and *E. coli* O157:H7 strains could be a concern in a wide variety of food products since they form biofilms [76, 77]. Formation of biofilms by foodborne microbial pathogensis is expected to be of higher magnitude in developing countries [78].

22.4 Biofilm Control Strategies in the Food Industry

Several articles have reviewed biofilm control and removal strategies [6, 7, 79]. The fundamental strategies usually involve the removal of biofilms by regular cleaning followed by disinfection to kill the remaining cells and prevent their attachment [6, 7, 80, 81]. Various techniques have been adopted for biofilm control in food processing and food preservation (Table 1). A hygienically planned layout with careful design of equipment, choice of material, and selection of detergent and disinfectants, coupled with physical methods, can be applied for controlling biofilm formation on food contact surfaces [5]. Biofilm detectors may be used to detect and monitor surface colonization by bacteria to allow for control at an early stage [82, 83].

Advances in physical methods of biofilm removal or prevention have exploited the use of magnetic fields, ultrasound, and high-pulsed electric fields in combination with low-pulsed electric fields [5, 7]. Other approaches to prevent attachment and biofilm formation, by incorporation of antimicrobial products into material surfaces, have been used for biomedical applications [84, 85]. Antimicrobial surfaces also might be developed for dairy equipment in valves and dead ends, where biofilms are difficult to control [6]. Recently, copper incorporating specialized stainless steel (304CuSS) has been shown to significantly inhibit biofilm formation by *Staphylococcus aureus* as well as decrease EPS formation and cell adhesion [86]. Selenium nanoparticles, alone or coated on paper towels, have proved effective against foodborne pathogens like *Pseudomonas aeruginosa*, *E. coli*, *Staphylococcus epidermidis*, *S. aureus*, *Salmonella enterica* ser. Typhimurium, and *S. enterica* ser. Enteritidis, suggesting their possible use as effective antibiofilm agents [87, 88].

Food debris and other residues that could promote growth of microbes may be removed through proper cleaning, using mechanical action [89–91]. Some strongly acidic and strongly alkaline cleaning-in-place (CIP) cleaning agents [ECLIN 143, 160, 180] show a significant cleaning effect with successful removal of *E. coli* and *S. aureus* biofilms [92].

After cleaning, disinfection with chemical agents (Table 22.1) must be used to kill the remaining microbes on the surfaces. The selection of disinfectants is based on effectiveness, safety, and economy [6]; their efficiency in preventing biofilm formation may be influenced by organic material, pH, temperature, water hardness, concentration of chemical inhibitors, and contact time [6, 93, 94]. Sodium hypochlorite (NaClO) is

Table 22.1 Strategies Used in Biofilm Control in Foods and the Food Industries.

Strategies	Means	Nature/Mode of Action	Reference
Physical Methods	Super-high magnetic fields	Formation of metastable pores in microbial membranes by magnetic fields	[110, 111]
	Ultrasound treatment	Generates sufficient cavitational bubble activity to remove biofilms from metallic, glass, ceramic and plastic surfaces	[112, 113]
	High pulsed electrical fields	Local instabilities in the membranes of microorganisms are formed by electromechanical compression and electric field-induced tension, which causes pores to form in the membrane	[114–116]
	Biofilm detectors	Able to detect biofilms in the early stages of development	[83]
	Metals and Specialized Materials		
	Selenium nanoparticles and coated paper towel	Metal nanoparticles: Nanoparticles alone coated on paper proved to possess antibiofilm activity against broad spectrum of food pathogenic bacteria	[87, 88]
	304CuSS	Steel incorporated with copper and other specialized treatment shows antibiofilm activity by decreasing microbial cell adhesion to surface	[117]
	Functionalized polycaprolactam	Papain covalently crosslinked on polycaprolactam, papain affects the functional groups, thereby decreasing the biofilm forming ability of *E. coli*.	[105]
	Cold plasma	Atmospheric cold plasma-attenuated quorum sensing regulated mechanism, ultimately inhibiting biofilm formation by foodborne pathogens.	[118]

Table 22.1 (Continued)

Strategies	Means	Nature/Mode of Action	Reference
Chemical Methods	Sodium hypochlorite	Disinfectant (oxidizer): Oxidation of thiol groups in enzymes and proteins, inhibition of DNA synthesis	[98, 119]
	Sodium hydroxide	Alkaline cleaner: Removes extracellular polymeric substances (EPS) produced by biofilm bacteria	[120]
	Hydrogen peroxide	Disinfectant (oxidizer): Hydroxyl radicals attack membrane lipids, DNA, and other essential cell components	[121, 122]
	Ozone	Disinfectant (oxidizer): Free radicals reacting with thiol groups of enzymes and proteins, DNA strand breakage	[123, 124]
	Peracetic acid	Disinfectant (oxidizer): Free radical mediated disruption of proteins, DNA, etc.	[125, 126]
	Chlorine	Disinfectant (oxidizer): Oxidation of thiol groups in enzymes and proteins, inhibition of DNA synthesis	[127]
	Iodine	Iodine rapidly penetrates into microorganisms and attacks key groups of proteins, nucleotides, and fatty acids	[128]
	Glycerol monolaurate	Emulsifier: inhibits the production of toxins and enzymes	[129–131]
	Cetylpyridinium chloride	Antiseptic (quaternary ammonium compound): Membrane damage, leaking of cellular constituents, reduction of the activity of nitrate reductases, which are enzymes that are important for normal biofilm growth.	[132, 133]
	Silver	Antimicrobial: Ag^+ causes the release of K^+ ions from microorganisms; the microbial plasma or cytoplasmic membrane, with which are associated many important enzymes	[134]
	Anhydrides	Act as carriers for biofilm inhibiting compounds	[135]
	Benzoyl chloride	Source of benzoic acid	[136]
	Lauricidin	Monoglyceride: involved in disruption of the cell membrane permeability barrier and inhibition of amino acid uptake	[137, 138]
	Lactic acid	Organic acid: inhibition of *Salmonella* biofilms	[97]
	cis-2-Decenoic acid	Unsaturated fatty acid: dispersion of preformed biofilm	[101]
	Acid-anionic tensioactive cleaning plus peracetic acid	Disinfectant + special cleaning procedure: Removal of biofilm by combined effect of oxidizer and cleaning procedure, imparting alteration in surface tension of given surface	[98]

(Continued)

Table 22.1 (Continued)

Strategies	Means	Nature/Mode of Action	Reference
	Sugar fatty acid esters	Long chain fatty acid esters: Inhibit the initial attachment of *Staphylococcus aureus* cells to the abiotic surface	[92]
	Levulinic acid plus sodium dodecyl sulfate	Sanitizers: biofilm inhibitory effect by inactivating bacterial cells in biofilm, changing the permeability of the cell membrane and aggregation of the cytoplasmic components	[100]
	Malic acid plus ozone	Sanitizers: Combination of oxidizer and organic acid prevents and eradicates biofilm on biotic and abiotic surfaces	[99]
	Octenidine hydrochloride	Sanitizer: Action similar to quaternary ammonium compounds, effectively eliminates planktonic as well as biofilm cells of *Listeria. monocytogenes*	[96]
	ECLIN 143, 160, 180	Strongly alkaline CIP reagent: effective in removing *S. aureus* biofilms	[139]
	Mg^{2+} ions	Cation: $Mg2^{+}$ ions could affect the signal transduction for biofilm formation pathway	[140]
Biological Methods	Nisin	Cationic peptide (bacteriocin): forms membrane pores that cause the membrane to be conductive to ions and other molecular species, ultimately leading to cell lysis	[141, 142]
	Reuterin	3-hydroxypropionaldehyde complex: modifies thiol groups in proteins and small molecules	[143]
	Pediocin	Cationic peptide (bacteriocin): targets cell membrane, thereby mediating leakage through the membrane	[144]
	Sonorensin	Peptide (bacteriocin): Inhibition of biofilms	[102]
	Colanic acid-degrading enzymes	Proteins: degrade colanic acid, a constituent of biofilm matrix (EPS)	[103]
	Proteolytic enzymes	Proteins: cleave various proteins associated with EPS, cell membranes and bacterial adhesion	[145, 146]
	α-Amylases	Proteins: degrade several model exopolysaccharides, including alginate, hyaluronic acid, polygalacturonic acid, and chitosan	[146, 147]
	Oxidoreductases + polysaccharide-hydrolyzing enzymes	Protein: combination of bactericidal effect (oxidoreductases) and EPS-degrading activity (polysaccharide-hydrolyzing enzymes) effectively inhibit and remove biofilms	[148]
	Phages	Lytic phages: Lysis of bacterial cells, leading to biofilm disruption	[7, 149]

effective at low pH against various foodborne microorganisms; hydrogen peroxide (H_2O_2) and ozone have also been used [6]. Potassium sorbate and sodium nitrite show significant biofilm inhibition against *Streptococcus mutans*; however, their effectiveness at killing the cells is negligible [95]. Octenidine hydrochloride, a commonly used clinical disinfectant, is effective against *L. monocytogenes* biofilms as well as planktonic cells over a broad range of temperatures [4°–37°C] in the presence or absence of organic matter [96]. Sucrose monopalmitate and other sugar fatty acid esters with long-chain (C14 to C16) fatty acid residues effectively inhibit initial attachment by *S. aureus* and biofilm formation by *S. mutans* and *L. monocytogenes* [92]. The mildly acidic conditions in many food products may affect the formation of biofilms. O'Leary et al. [97] examined the effect of lactic acid on the survival of biofilms formed by *S. enterica* ser. Typhimurium and found that at pH 5, this bacterium showed differential regulation of various virulence-related genes and a marked reduction in biofilm formation [97]. The use of various sanitizers commonly used in the food industry was assessed by da Silva Fernandes et al.; peracetic acid successfully removed *Enterococcus faecium* and *E. faecalis* biofilms [98].

Different chemicals may also be used in combination for effective prevention of microbial adhesion and biofilm development. A combination of malic acid with ozone-reduced biofilm formation by *S. enterica* ser. Typhimurium on turnips, plastic bags, and PVC pipes; malic acid alone was effective for the complete inhibition of biofilms on carrots and some food contact surfaces [99]. Likewise, the combined bactericidal activity of levulinic acid (LVA) with sodium dodecyl sulfate (SDS) in biofilms for cells of *L. monocytogenes, S. enterica* ser. Typhimurium, and Shiga toxin-producing *E. coli* was evaluated, with the highest concentrations (3% LVA + 2% SDS) providing the greatest log reduction [100]. Sepehr et al. showed that combined treatments with both *cis*-2-decenoic acid and antibiotics or biocides caused almost complete eradication of preestablished biofilms formed by *S. aureus, Bacillus cereus, S. enterica,* and *E. coli* [101].

New strategies exploiting biological agents, such as bacteriocins, enzymes, and phages [7], are being investigated for controlling biofilms in the food industry. The use of bacteriocins (bactericidal polypeptides) to prevent adhesion of bacteria on food-contact surfaces is well established [5]. The FDA has approved Nisin, a polycyclic bacteriocin produced by *Lactococcus lactis*, for control of pathogenic bacteriain processed cheese spreads [5]. Recently, Chopra et al. [102] demonstrated the antibiofilm potential of sonorensin, a bacteriocin produced by *Bacillus sonorensis*, against *S. aureus* and showed that sonorensin-coated films in chicken and meat samples inhibited the growth of spoilage microorganisms. The use of enzymes for removal of EPS from the biofilm matrix is an important strategy of biological control [5, 7, 103, 104]. The plant enzyme papain, covalently crosslinked on polycaprolactam in a wrapper for packaging cottage cheese, has been found effective in preventing *E. coli* biofilms [105]. Bacteriophage application for controlling pathogenic microbes is used with medical biofilms [106], and a similar strategy is expected to be effective in the food industries [7]. The lytic phages, such as T-series coliphages, can effectively control *E. coli* biofilms [107]. Likewise, phages such as ϕIBB-PF7A can be used in controlling *Pseudomonas fluorescens* biofilms [108]. Lu et al. demonstrated the strong antibiofilm potential of engineered bacteriophages that produce polysaccharide depolymerases [109].

22.5 Natural Products as Antibiofilm Agents and Their Potential Applications

Each process used for biofilm control, including disinfection, has its limitations. Increased biofilm resistance to conventional treatments [150] has triggered a search for new agents and strategies to control biofilms. Several natural products that have been shown to be antibiofilm agents and/or anti-QS agents [151] have potential uses for food preservation (Tables 22.2 and 22.3).

Table 22.2 Plant-Derived Products with Promising Antibiofilm Activity and Possible Biofilm Control Agents.

Plant Products	Foodborne Pathogens	Role in Preservation and Food Safety	Reference
Essential Oils (EOs) / Plant Extracts			
Marjoram EO	*Pichia anomala, Bacillus cereus*, and *Pseudomonas putida;* and *P. putida* + *E. coli* (mixed culture)	Essential oils as potential disinfectants and food preservatives	[167]
Lemon EO	*E. coli* and *P. putida* + *E. coli* (mixed culture)		
Cinnamon EO	*P. putida* + *E. coli* (mixed culture)		
Coriander EO	*Staphylococcus aureus* and *E. coli*	A new antibiofilm agent	[170]
Oregano EO	*S. aureus* and *Pseudomonas aeruginosa*	Inhibition of biofilms of various food pathogens and food spoilage organisms	[168]
Clove EO, cinnamon EO	*S. aureus*	Applications in sanitization in the food industry	[166]
Spearmint EO	*Vibrio* spp.	A natural antibiotic and seafood preservative against *Vibrio* contamination	[169]
Rosemary water extract (RWE)	*Bacillus* and *Pseudomonas*	Anti-adhesive and antibiofilm agent against food related *Bacillus* and *Pseudomonas*	[172]
Phenolic extract of cloudberry fruit	*Obesumbacterium proteus*	Inhibits biofilm formation capability of bacteria isolated from brewery filling process	[182]
Phytocompounds			
trans-Cinnamaldehyde, carvacrol, thymol, eugenol	*Listeria monocytogenes*	Controlling biofilms in food-processing environments	[174]

Table 22.2 (Continued)

Plant Products	Foodborne Pathogens	Role in Preservation and Food Safety	Reference
Muscadine polyphenols	*S. aureus*	Natural food preservatives, potential antibiotic replacements, and/or natural sanitizers for processing equipment where foodborne pathogens reside	[175]
Citral, cinnamaldehyde, tea polyphenols	*S. aureus* + *Salmonella enterica* ser. Enteritidis (mixed biofilm)	Application of natural food additives to control biofilm formation of foodborne bacteria	[178]
Resveratrol-hydroxypropyl-γ-cyclodextrin inclusion complexes	*Campylobacter jejuni, Campylobacter coli* and *Arcobacter butzleri*	A new antibiofilm agent and QS inhibitor to enhance the shelf life and safety of foods	[181]
Pinostrobin	*E. coli* and *P. aeruginosa*	Flavonoid with antibiofilm effect against foodborne pathogens	[179]
Quercetin-3-glucoside	*S. aureus, E. coli, P. aeruginosa* and *Bacillus subtilis*	Flavonoid from *Scutellaria oblonga* as potential antibiofilm agent	[180]
Carvacrol, citral, limonene	*S. aureus*	Food additives generally recognized as safe, their anti-biofilm properties may lead to important new applications, such as sanitizers, in the food industry	[176]
Gallic acid	*E. coli, Streptococcus mutans*	Inhibitory effect of gallic acid against foodborne pathogen biofilms over a broad range of physiochemical conditions	[177]
Isomontanolide	*Candida albicans, C. krusei*	Practical applications in food industry as food preservatives to retard the growth of food-spoilage microbes	[171]

Many naturally occurring products, such as essential oils from edible and medicinal plants, herbs, spices, and their isolated constituents, possess antimicrobial and antibiofilm functions and could be used as food additives [152–156]. Essential oils are plant extracts containing secondary metabolites, some of which have been used in medicine, perfumes, and cosmetics or have been added to foods [157]. Several studies have reported inhibitory effects of essential oils and their components on foodborne microbes and food spoilage pathogens [157–160]. They may be used as disinfectant agents [161, 162] and as coating materials on foods [163–165]. Hence, it has been assumed that some essential oils may help prevent the formation of biofilms and could be used in

Table 22.3 Natural Quorum Sensing Inhibitors as Potential Biofilm Control Agents.

	Anti-Quorum Sensing Agents	Biosensor Strain(S) Used	Potential Application in Food Industry	Reference
Essential oils	Tea tree essential oil, rosemary essential oil, resveratrol	*Chromobacterium violaceum*	Potential to be good preservatives	[189]
	Oregano essential oil-pectin edible film	*C. violaceum*	Packaging material with antibacterial efficacy	[188]
	Murraya koenigii essential oil	*C. violaceum, Pseudomonas aeruginosa* PA01	Application to enhance shelf life and food safety	[151]
	Spice oil nanoemulsions	*C. violaceum* and *P. aeruginosa* PA01	Management of foodborne pathogens and biofilm formation in food industries	[190]
Phytocompounds	*Nymphaea tetragona* polyphenols	*C. violaceum*	Promising new components from waterlily as anti-QS agent	[191]
	Punicalagin	*C. violaceum, Salmonella* strains	To be developed as an alternative or supplemental agent for prevention of *Salmonella* infection	[86]
	Quercetin	*C. violaceum, P. aeruginosa* PUFSTb04, *Yersinia enterocolitica* PUFSTb09	Novel QS-based antibacterial and antibiofilm drug to manage foodborne pathogens	[192]
	Resveratrol-hydroxypropyl-γ-cyclodextrin inclusion complexes	*C. violaceum*	New antibiofilm agent and QS inhibitor to enhance shelf life and safety of foods	[181]

sanitization and food preservation [166]. *Origanum majorana* (marjoram) essential oil inhibits *Bacillus cereus, Pichia anomala,* and *Pseudomonas putida* and mixed-culture biofilm formation by *P.putida* and *Escherichia coli*; it also shows a strong QS inhibitor effect on *Chromobacterium violaceum* [167]. Essential oils obtained from *Origanum vulgare* subsp. *hirtum* (oregano) were evaluated for their antibiofilm potential against *S. aureus* and *P. aeruginosa,* showing significant activity against biofilm formation by these food spoilage organisms at sub-MIC levels of 200, 100, and 50 μg/ml [168]. Budri et al. [166] demonstrated the potential application of the essential oils of *Syzygium aromaticum* (clove) and *Cinnamomum zeylanicum* (cinnamon) and their major components, eugenol and cinnamaldehyde, respectively, for *S. aureus* biofilm inhibition on different surfaces. The results showed significant inhibition of biofilm production by clove oil on polystyrene and stainless steel surfaces (69.4 and 63.6 percent, respectively) [166]. Similarly, Snoussi et al. evaluated the effect of *Mentha spicata* (spearmint) essential oil on *Vibrio* spp. biofilm inhibition and eradication by using the XTT assay for cell viability [169]. The terpenoid carvone reduced the *Vibrio* spp. biofilm, suggesting its use as a seafood preservative against *Vibrio* contamination. Inhibition of biofilm growth in a preformed biofilm of *S. aureus* and *Escherichia coli* was effectively achieved by *Mentha* × *piperita* (peppermint), *Coriandrum sativum* (coriander), and *Pimpinella anisum* (anise) essential oils, which have high levels of oxygenated monoterpenoids, particularly linalool [170]. Potential interactions among the constituent compounds might be responsible for the high antibiofilm activity of coriander essential oil, which is even higher than peppermint oil [170]. Extracts of underground parts of three species of the Apiaceae commonly used as functional foods, *Laserpitium latifolium*, *L. zernyi*, and *L. ochridanum*, were investigated, and isomontanolide (a principal component of the extracts of *L. zernyi* and *L. ochridanum*) had significant antibiofilm activity against the yeasts *Candida albicans* and *C. krusei* [171]. The authors advocated their potential use as food additives to check the growth of spoilage organisms. Similarly, Elhariry et al. found that a hot water extract of *Rosmarinus officinalis* (rosemary) leaves rich in benzoic, ellagic, gallic, and rosmarinic acids, as well as the isoflavonoids daidzein and genistein, significantly reduced the biofilm formation and biotic and abiotic surface adherence potential of *Pseudomonas* spp. and *Bacillus* spp. [172].

Phenols are broadly distributed in plants and form part of the plant defense response against pathogens [173]. They not only have antioxidant activity but also possess potent anti-infective activity against a broad range of microbes, including foodborne pathogens. Phenolic compounds, polyphenol enriched fractions, and numerous constituents of essential oils reduce biofilm formation on surfaces related to food processing [174–177]. Citral, cinnamaldehyde, and tea polyphenols are effective against mixed-species biofilms of *S. aureus* and *S. enterica* ser. Enteritidis [178]. Flavonoids like pinostrobin and quercetin-3-glucoside significantly inhibit or eradicate biofilms formed by food and food processing-related microbial contaminants over a broad range of physiochemical conditions [179, 180]. In a 2015 study, Duarte et al. emphasize the use of cyclodextrin as a delivery agent to encapsulate *trans*-resveratrol, an antibacterial phytochemical, for control of species of *Campylobacter* and *Arcobacter* that form biofilms. The resveratrol-hydroxypropyl-γ-cyclodextrin inclusion complexes carry resveratrol deep into the biofilm matrix to inhibit QS and biofilm formation [181].

22.6 Role of QS Inhibitors in Biofilm Control

Interfering with or inhibiting QS is a novel anti-infective strategy [183–186]. Some QS inhibitors may be effective in preventing biofilm formation and increasing shelf life of foods (Table 22.3) by preventing expression of virulence genes [187]. An overview of possible modes of action of QS inhibitors/ biofilm inhibitors is depicted in Figure 22.1.

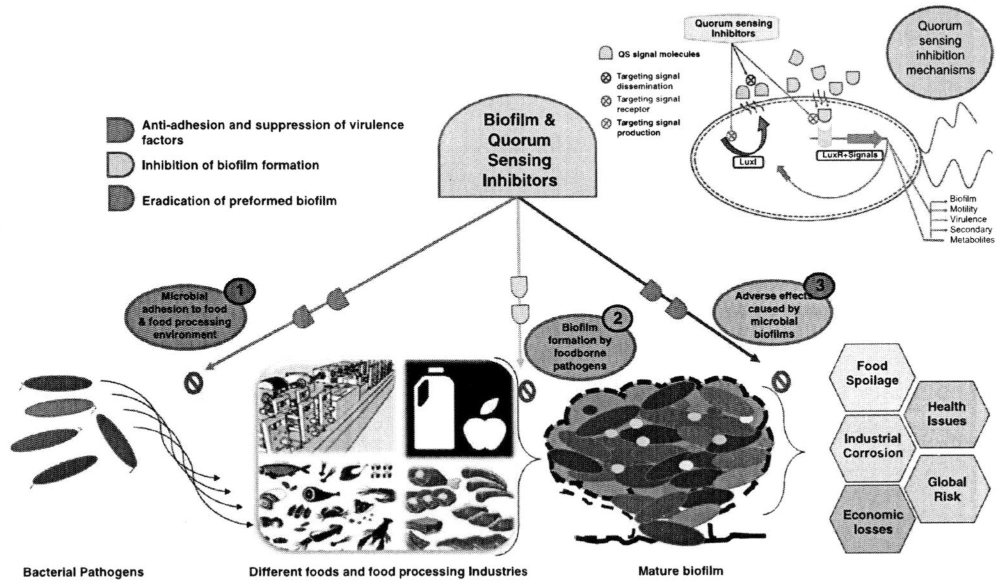

Figure 22.1 Different modes of action of biofilm/quorum-sensing inhibitors. (*See color plate section for the color representation of this figure.*)

Various natural products and chemical compounds of plant or bacterial origin possess QS-inhibiting activity [184, 185, 187]. The role of plant essential oils, including some that are commonly used in food products, as QS-inhibiting agents has been demonstrated [167, 188]. The potential of essential oils for use as food preservatives may be due to their anti-QS properties [151, 189]. Likewise, essential oils are used as primary components in nanoemulsions and as specialized films used in the food industry for prevention of microbial biofilm formation and spoilage, due to their ability of disrupting QS systems in pathogens [188, 190]. Polyphenol-rich fractions of *Nymphaea tetragona*, a waterlily, which show anti-QS potential and synergistic antibacterial effects with tylosin and streptomycin, prove to be effective anti-*Salmonella* agents [191]. The flavonol quercetin and an ellagitannin, punicalagin, demonstrate potent QS-inhibiting activity against *Y. enterocolitica* and *Salmonella* spp., thus indicating their possible use for management of foodborne infections, microbial biofilms, and spoilage in food-related industries [86, 192].

22.7 Conclusions

In the last few decades, extensive research has been conducted on biofilms related to the food industries. Various strategies have been applied to prevent and control biofilms, but there is no universal strategy sufficient to control microbial contamination, adherence, and biofilm formation. Therefore, new approaches to control biofilms must be explored and tested in combination with existing strategies to combat the problem. Improvements in cleaning and disinfection, through the use of enzymes and recently discovered antibiofilm and anti-QS agents, may allow more effective treatment of resistant biofilms. The use of new agents combined with existing strategies must be compatible, cost-effective, and satisfy the needs of food quality and food safety. Natural products and new anti-QS compounds should be investigated further to combat specific biofilm problems in the food industry.

Acknowledgments

We thank Drs. M. S. Nawaz and Y. Sanad for their comments and Dr. Carl E. Cerniglia for his support. The views presented in this article do not necessarily reflect those of the United States Food and Drug Administration.

References

1 J.M. Jay, Intrinsic and extrinsic parameters of foods that affect microbial growth, *in* Modern food microbiology, J.M. Jay, M.J. Loessner and D.A. Golden (Eds.) 38–66, Springer US (1995).
2 W.C. Frazier and D.C. Westhoff, Food Microbiology (4th Ed.) Tata McGraw-Hill Publishing Company Limited, New Delhi (1995).
3 L.R. Beuchat, Pathogenic microorganisms associated with fresh produce, *J. Food Protect.*, 59(2), 204–216 (1996).

4 S.L. Burnett and L.R. Beuchat, Human pathogens associated with raw produce and unpasteurized juices, and difficulties in decontamination, *J. Ind. Microbiol. Biotechnol.*, 25(6), 281–287 (2000).

5 C.G. Kumar and S.K. Anand, Significance of microbial biofilms in food industry: a review, *Int. J. Food Microbiol.*, 42(1), 9–27 (1998).

6 M. Simões, L.C. Simões and M.J. Vieira, A review of current and emergent biofilm control strategies, *LWT-Food Sci. Technol.*, 43(4), 573–583 (2010).

7 S. Srey, I.K. Jahid and S.D. Ha, Biofilm formation in food industries: a food safety concern, *Food Control*, 31(2), 572–585 (2013).

8 J.A. Gaddy and L.A. Actis, Regulation of *Acinetobacter baumannii* biofilm formation, *Future Microbiol.*, 4, 273–278 (2009).

9 G. O'Toole, H.B. Kaplan and R. Kolter, Biofilm formation as microbial development, *Annu. Rev. Microbiol.*, 54(1), 49–79 (2000).

10 X. Shi and X. Zhu, Biofilm formation and food safety in food industries, *Trends Food Sci. Technol.*, 20(9), 407–413 (2009).

11 M. S. Chae, H. Schraft, L. Truelstrup and R. Mackereth, Effects of physicochemical surface characteristics of *Listeria monocytogenes* strains on attachment to glass, *Food Microbiol.*, 23, 250–259 (2006).

12 R.S. Harbron and C.A. Kent, Aspects of cell adhesion, *in* Fouling Science and Technology, L.F. Melo, T.R. Bott, and C. A. Bernardo (Eds.) 125–140, Springer Netherlands (1988).

13 S. Bouttier, C. Linxe, C. Ntsama, G. Morgant, M.N. BellonFontaine and J. Fourniat, Attachment of *Salmonella choleraesuis* to beef muscle and adipose tissues, *J. Food Protect.*, 60, 16–22 (1997).

14 R. Van Houdt and C.W. Michiels, Role of bacterial cell surface structures in *Escherichia coli* biofilm formation, *Res. Microbiol.*, 156(5-6), 626–633 (2005).

15 S.B. Guttenplan and D.B. Kearns, Regulation of flagellar motility during biofilm formation, *FEMS Microbiol. Rev.*, 37 (6) 849–871 (2013).

16 W. G. Characklis and P.A. Wilderer, Glossary, in Structure and function of biofilms, W.G. Characklis and P.A. Wilderer (Eds.), 369–371, Wiley (1989).

17 I.W. Sutherland, The biofilm matrix—an immobilized but dynamic microbial environment, *Trends Microbiol.*, 9, 222–227 (2001).

18 D.G. Allison, The biofilm matrix, *Biofouling*, 19, 139–150 (2003).

19 M. Simões, M.O. Pereira, and M.J. Vieira, Effect of mechanical stress on biofilms challenged by different chemicals, *Water Res.*, 39(20), 5142–5152 (2005).

20 E.E. Mann and D.J. Wozniak, *Pseudomonas* biofilm matrix composition and niche biology, *FEMS Microbiol. Rev.*, 36(4), 893–916 (2012).

21 S.U. Gerbersdorf and S. Wieprecht, Biostabilization of cohesive sediments: revisiting the role of abiotic conditions, physiology and diversity of microbes, polymeric secretion, and biofilm architecture, *Geobiology*, 13(1), 68–97 (2015).

22 M.B. Miller and B.L. Bassler, Quorum sensing in bacteria, *Annu. Rev. Microbiol.*, 55, 165–199 (2001).

23 L.C.M. Antunes, R.B.R. Ferreira, M.M.C. Buckner and B.B. Finlay, Quorum sensing in bacterial virulence, *Microbiology*, 156(8) 2271–2282 (2010).

24 M. Schuster, D.J. Sexton, S.P. Diggle and E.P. Greenberg, Acyl-homoserine lactone quorum sensing: from evolution to application, *Annu. Rev. Microbiol.*, 67, 43–63 (2013).

25 C. Fuqua and E.P. Greenberg EP, Listening in on bacteria: acyl-homoserine lactone signalling, *Nat. Rev. Mol. Cell. Biol.* 3:685–695 (2002).

26 M.R. Parsek and E.P. Greenberg, Sociomicrobiology: the connections between quorum sensing and biofilms, *Trends Microbiol.*, 13(1), 27–33 (2005).

27 C.D. Sifri, Quorum sensing: bacteria talk sense, *Clin. Infect. Dis.*, 47(8), 1070-1076 (2008).

28 T. Abee, A.T. Kovács, O.P. Kuipers and S. Van der Veen, Biofilm formation and dispersal in Gram-positive bacteria, *Curr. Opin. Biotechnol.*, 22(2), 172–179 (2011).

29 H.S. Joo and M. Otto, Molecular basis of *in vivo* biofilm formation by bacterial pathogens, *Chem. Biol.*, 19(12), 1503–1513 (2012).

30 C. Solano, M. Echeverz and I. Lasa, Biofilm dispersion and quorum sensing, *Curr. Opin. Microbiol.*, 18, 96–104 (2014).

31 C. Nunes-Alves, Microbiome: Taking advantage of quorum sensing, *Nat. Rev. Microbiol.*, 13(3), 252 (2015).

32 S.T. Rutherford and B.L. Bassler, Bacterial quorum sensing: its role in virulence and possibilities for its control, *Cold Spring Harbor Perspect. Med.*, 2(11), a012427 (2012).

33 Z.R. Suárez-Moreno, J.F. González, G. Devescovi and V. Venturi, Quorum sensing in Gram-negative bacteria: signals, role and networking, *in* Bacterial Regulatory Networks, A.A.M. Filloux (Ed.), 87, Caister Academic Press, UK (2012).

34 C. Neyen and B. Lemaitre, Sensing Gram-negative bacteria: a phylogenetic perspective, *Curr. Opin. Immunol.*, 38, 8–17 (2016).

35 K.K. Sivakumar, P.R. Jesudhasan, and S.D. Pillai. Detection of autoinducer (AI-2)-like activity in food samples, *Methods Mol. Biol*, 692, 71–82 (2010).

36 W.R.J.D. Galloway, J.T. Hodgkinson, S.D. Bowden, M. Welch, and D.R. Spring, Quorum sensing in Gram-negative bacteria: small-molecule modulation of AHL and AI-2 quorum sensing pathways. *Chem. Rev.* 111, 28–67 (2011).

37 M. Thoendel, J.S. Kavanaugh, C.E. Flack, and A.R. Horswill, Peptide signaling in the Staphylococci. *Chem. Rev.*, 111, 117–151 (2011).

38 S. Kjelleberg, and S. Molin, Is there a role for quorum sensing signals in bacterial biofilms? *Curr. Opin. Microbiol.*, 5(3), 254–258 (2002).

39 WHO. Food safety and foodborne illness, World Health Organization, Retrieved 26.03.12, from. http://www.who.int/mediacentre/factsheets/ fs237/en/ (2007).

40 CDC. Vital signs: Incidence and trends of infection with pathogens transmitted commonly through food, Foodborne diseases active surveillance network, 10 U.S. Sites, 1996e2010. Morbidity and Mortality Weekly Report (MMWR), Retrieved 28.03.12, from. http://www.cdc.gov/mmwr/preview/mmwrhtml/mm6022a5. htm?s_cid¼mm6022a5_w. (2011).

41 E. Scallan, Foodborne illness acquired in the United States-unspecified agents, *Emerg. Infect. Dis.*, 17(1), 16–22 (2011).

42 G.A. Dykes, B. Sampathkumar and D.R. Korber, Planktonic or biofilm growth affects survival, hydrophobicity and protein expression patterns of a pathogenic *Campylobacter jejuni* strain, *Int. J. Food Microbiol.*, 89(1), 1–10 (2003).

43 A. Lapidot, U. Romling and S. Yaron, Biofilm formation and the survival of *Salmonella typhimurium* on parsley, *Int. J. Food Microbiol.*, 109(3), 229–233 (2006).

44 K. Aarnisalo, J. Lundén, H. Korkeala and G. Wirtanen, Susceptibility of *Listeria monocytogenes* strains to disinfectants and chlorinated alkaline cleaners at cold temperatures, *LWT-Food Sci. Technol.*, 40(6), 1041–1048 (2007).

45 J.C.A. Janssens, H. Steenackers, S. Robijns, E. Gellens, J. Levin, H. Zhao, K. Hermans, D. De Coster, T.L. Verhoeven, K. Marchal, J. Vanderleyden, D.E. De Vos and S.C. De

Keersmaecker, Brominated furanones inhibit biofilm formation by *Salmonella enterica* serovar Typhimurium, *Appl. Environ. Microbiol.*, 74(21), 6639–6648 (2008).

46 C. K. Bower, J. McGuire, and M.A. Daeschel, The adhesion and detachment of bacteria and spores on food-contact surfaces, *Trends Food Sci. Technol.*, 7(5), 152–157 (1996).

47 M.S. Sidhu, S. Langsrud and A. Holck, Disinfectant and antibiotic resistance of lactic acid bacteria isolated from the food industry, *Microb. Drug Resist.*, 7(1), 73–83 (2001).

48 P. Teixeira and D. Rodrigues, Antibiofilm strategies in the food industry, *in Antibiofilm Agents*, K.P. Rumbaugh and I. Ahmad (Eds.) 359-381, Springer, Berlin Heidelberg (2014).

49 H.L.M. Lelieveld, J. Holah and D. Napper (Eds.), Hygiene in food processing: principles and practice, 2nd ed., Elsevier (2014).

50 L.K. Strawn, E.D. Fortes, E.A. Bihn, K.K. Nightingale, Y.T. Gröhn, R.W. Worobo, M. Wiedmann and P.W. Bergholz, Landscape and meteorological factors affecting prevalence of three food-borne pathogens in fruit and vegetable farms, *Appl. Environ. Microbiol.*, 79(2), 588–600 (2013).

51 M.H. Larsen, M. Dalmasso, H. Ingmer, S. Langsrud, M. Malakauskas, A. Mader, T. Møretrø, S.S. Možina, K. Rychli, M. Wagner, R.J. Wallace, J. Zentek and K. Jordan, Persistence of foodborne pathogens and their control in primary and secondary food production chains, *Food Control*, 44, 92–109 (2014).

52 A. Bridier, P. Sanchez-Vizuete, M. Guilbaud, J. C. Piard, M. Naïtali and R. Briandet, Biofilm-associated persistence of food-borne pathogens, *Food Microbiol.*, 45, 167–178 (2015).

53 S. Stepanović, I. Ćirković, V. Mijač and M. Švabić-Vlahović, Influence of the incubation temperature, atmosphere and dynamic conditions on biofilm formation by *Salmonella* spp., *Food Microbiol.*, 20, 339–343 (2003).

54 C. Mariani, N. Oulahal, J.-F. Chamba, F. Dubois-Brissonnet, E. Notz and R. Briandet, Inhibition of *Listeria monocytogenes* by resident biofilms present on wooden shelves used for cheese ripening, *Food Control*, 22(8), 1357–1362 (2011).

55 S.P. Oliver, B.M. Jayarao and R.A. Almeida, Food borne pathogens in milk and the dairy farm environment: food safety and public health implications, *Foodborne Pathog. Dis.*, 2(2), 115–129 (2005).

56 J.A. Khan, R.S. Rathore, S. Khan, and I. Ahmad, *In vitro* detection of pathogenic *Listeria monocytogenes* from food sources by conventional, molecular and cell culture methods, *Braz. J. Microbiol.*, 44(3), 751–758 (2013).

57 P.J. Bremer, S. Fillery and A.J. McQuillan, Laboratory scale clean-in-place (CIP) studies on the effectiveness of different caustic and acid wash steps on the removal of dairy biofilms, *Int. J. Food Microbiol.*, 106, 254–262 (2006).

58 L. Gram, D. Bagge-Ravn, Y.Y. Ng, P. Gymoese and B.F. Vogel, Influence of food soiling matrix on cleaning and disinfection efficiency on surface attached *Listeria monocytogenes*, *Food Control*, 18, 1165–1171 (2007).

59 M. Sharma and S.K. Anand, Characterization of constitutive microflora of biofilms in dairy processing lines, *Food Microbiol.*, 19, 627–636 (2002).

60 S. Marchand, J. De Block, V. De Jonghe, A. Coorevits, M. Heyndrickx and L. Herman, Biofilm formation in milk production and processing environments; influence on milk quality and safety, *Comp. Rev. Food Sci. Food Saf.*, 11(2), 133–147 (2012).

61 K.V. Mogha, N.P. Shah, J.B. Prajapati and A.R. Chaudhari, Biofilm a threat to dairy industry, *Indian J. Dairy Sci.*, 67(6) (2014).

62 B. Seale, P. Bremer, S. Flint, J. Brooks and J. Palmer, Overview of the problems resulting from biofilm contamination in the dairy industry, *in* Biofilms in the Dairy Industry, K.H. Teh, S. Flint, J. Brooks and G. Knight (Eds.), John Wiley & Sons, UK (2015).

63 S. Flint, N.M. Jamaludin, B. Somerton, J. Palmer and J. Brooks, The effect of milk composition on the development of biofilms, *in* Biofilms in the Dairy Industry, K.H. Teh, S. Flint, J. Brooks and G. Knight (Eds.), John Wiley & Sons, Ltd., Chichester, UK (2015).

64 Caron, C. CDC: Cantaloupe *Listeria* outbreak deadliest in a decade, ABC News. Retrieved 21.11.12, from http://abcnews.go.com/Health/cdc-cantaloupelisteria-outbreak-deadliest-decade/story?id¼14622507#.UKx7K-SE3pw. (2011).

65 M. Gandhi, S. Golding, S. Yaron and K.R. Matthews, Use of green fluorescent protein expressing *Salmonella* Stanley to investigate survival, spatial location, and control on alfalfa sprouts, *J. Food Protect.*, 64(12), 1891–1898 (2001).

66 S.M. Faruque, K. Biswas, S.M.N. Udden, Q.S. Ahmad, D.A. Sack, G.B. Nair and J.J. Mekalanos, Transmissibility of cholera: *in vivo*-formed biofilms and their relationship to infectivity and persistence in the environment, *Proc. Natl. Acad. Sci. U.S.A.*, 103(16), 6350–6355 (2006).

67 K.T. Rajkowski, Biofilms in fish processing, *in* Biofilms in the food and beverage industries, P.M. Fratamico, B.A. Annous and N.W. Gunther (Eds.), 499-516, Woodhead Publishing (2009).

68 I. Hanning, R. Jarquin and M. Slavik, *Campylobacter jejuni* as a secondary colonizer of poultry biofilms, *J. Appl.Microbiol.*, 105(4), 1199–1208 (2008).

69 C. Marin, A. Hernandiz and M. Lainez, Biofilm development capacity of *Salmonella* strains isolated in poultry risk factors and their resistance against disinfectants, *Poult. Sci.*, 88(2), 424–431 (2009).

70 S.H. Park, R. Jarquin, I. Hanning, G. Almeida and S.C. Ricke, Detection of *Salmonella* spp. survival and virulence in poultry feed by targeting the *hilA* gene, *J. Appl. Microbiol.*, 111(2), 426–432 (2011).

71 L. McLandsborough, A. Rodriguez, D. Pérez-Conesa and J. Weiss, Biofilms: at the interface between biophysics and microbiology, *Food Biophysics*, 1(2), 94–114 (2006).

72 J. D. Brooks and S.H. Flint, Biofilms in the food industry: problems and potential solutions, *Int. J. Food Sci. Technol.*, 43(12), 2163–2176 (2008).

73 P.S. Stewart and M.J. Franklin, Physiological heterogeneity in biofilms, *Nat. Rev. Microbiol.*, 6(3), 199–210 (2008).

74 O. Habimana, E. Heir, S. Langsrud, A.W. Asli and T. Møretrø, Enhanced surface colonization by *Escherichia coli* O157:H7 in biofilms formed by an *Acinetobacter calcoaceticus* isolate from meat-processing environments, *Appl. Environ. Microbiol.*, 76(13), 4557–4559 (2010).

75 J.N. Sofos and I. Geornaras, Overview of current meat hygiene and safety risks and summary of recent studies on biofilms, and control of *Escherichia coli* O157:H7 in nonintact, and *Listeria monocytogenes* in ready-to-eat, meat products, *Meat Sci.*, 86(1), 2–14 (2010).

76 K. Silagyi, S.H. Kim, Y.M. Lo, and C.I. Wei, Production of biofilm and quorum sensing by *Escherichia coli* O157:H7 and its transfer from contact surfaces to meat, poultry, ready-to-eat deli, and produce products, *Food Microbiol.*, 26(5), 514–519 (2009).

77 A. Di Pinto, L. Novello, F. Montemurro, E. Bonerba and G. Tantillo, Occurrence of *Listeria monocytogenes* in ready-to-eat foods from supermarkets in Southern Italy, *New Microbiol.*, 33(3), 249–252 (2010).

78 P. Kaur, A. Chakraborti, and A. Asea. Enteroaggregative *Escherichia coli*: an emerging enteric food borne pathogen, *Interdiscipl. Persp. Infect. Dis.*, 254159, 10 (2010).

79 D. Kregiel, Advances in biofilm control for food and beverage industry using organo-silane technology: A review, *Food Control*, 40, 32–40 (2014).

80 G. Midelet and B. Carpentier, Impact of cleaning and disinfection agents on biofilm structure and on microbial transfer to a solid model food, *J. Appl.Microbiol.*, 97, 262–270 (2004).

81 M. Simões, L.C. Simões, I. Machado, M.O. Pereira and M.J. Vieira, Control of flow-generated biofilms using surfactants—evidence of resistance and recovery, *Food Bioprod. Process.*, 84, 338–345 (2006).

82 R. Philip-Chandy, P.J. Scully, P. Eldridge, H.J. Kadim, M.G. Grapin, M.G. Jonca, M.G. D'Ambrosio and F. Colin, An optical fiber sensor for biofilm measurement using intensity modulation and image analysis, *IEEE J. Sel. Top. Quantum Electron.*, 6, 764–772 (2000).

83 A. Pereira, J. Mendes and L.F. Melo, Using nanovibrations to monitor biofouling, *Biotechnol. Bioeng.*, 99(6), 1407–1415 (2008).

84 F. Paladini, M. Pollini, A. Sannino, and L. Ambrosio. Metal-based antibacterial substrates for biomedical applications, *Biomacromolecules*, 16, 1873–1885 (2015).

85 F. Rafii and M. Hart. Antimicrobial resistance in clinically important biofilms. *World J. Pharmacol.*, 4, 31–46 (2015).

86 G. Li, C. Yan, Y. Xu, Y. Feng, Q. Wu, X. Lv, B. Yang, X. Wang and X. Xia, Punicalagin inhibits *Salmonella* virulence factors and has anti-quorum-sensing potential, *Appl. Environ. Microbiol.*, 80 (19) 6204–6211 (2014).

87 G.M. Khiralla and B.A. El-Deeb, Antimicrobial and antibiofilm effects of selenium nanoparticles on some foodborne pathogens, *LWT-Food Sci. Technol.*, 63(2), 1001–1007(2015).

88 Q. Wang, P. Larese-Casanova and T.J. Webster, Inhibition of various gram-positive and gram-negative bacteria growth on selenium nanoparticle coated paper towels, *Int. J. Nanomed.*, 10, 2885–2894 (2015).

89 R. Srinivasan, P.S. Stewart, T. Griebe, C.I. Chen, and X. Xu, Biofilm parameters influencing biocide efficacy, *Biotechnol. Bioeng.*, 46 (6), 553–560 (1995).

90 S.J. Forsythe and P.R. Hayes, Food hygiene, microbiology and HACCP (3rd Eds.) Aspen Publishers (1998).

91 J. Maukonen, J. Mättö, G. Wirtanen, L. Raaska, T. Mattila-Sandholm and M. Saarela, Methodologies for the characterization of microbes in industrial environments: a review, *J. Ind. Microbiol. Biotechnol.*, 30, 327–356 (2003).

92 S. Furukawa, Y. Akiyoshi, G.A. O'Toole, H. Ogihara and Y. Morinaga, Sugar fatty acid esters inhibit biofilm formation by food-borne pathogenic bacteria, *Int. J. Food Microbiol.*, 138(1), 176–180 (2010a).

93 P. J. Bremer, I. Monk and R. Butler, Inactivation of *Listeria monocytogenes/Flavobacterium* spp. biofilms using chlorine: impact of substrate, pH, time and concentration, *Lett. Appl. Microbiol.*, 35, 321–325 (2002).

94 T. Kuda, T. Yano, and M.T. Kuda, Resistances to benzalkonium chloride of bacteria dried with food elements on stainless steel surface, *LWT-Food Sci. Technol.*, 41, 988–993 (2008).

95 A. Al-Ahmad, M. Wiedmann-Al-Ahmad, T. M. Auschill, M. Follo, G. Braun, E. Hellwig and N. B. Arweiler, Effects of commonly used food preservatives on biofilm formation of *Streptococcus mutans in vitro*, *Arch. Oral Biol.*, 53(8), 765–772 (2008).

96 M.A.R. Amalaradjou, C.E. Norris and K. Venkitanarayanan, Effect of octenidine hydrochloride on planktonic cells and biofilms of *Listeria monocytogenes*, *Appl. Environ. Microbiol.*, 75(12), 4089–4092 (2009).

97 D. O'Leary, E.M. McCabe, M.P. McCusker, M. Martins, S. Fanning, and G. Duffy, Acid environments affect biofilm formation and gene expression in isolates of *Salmonella enterica* Typhimurium DT104, *Int. J. Food Microbiol.*, 206, 7–16 (2015).

98 F.M. da Silva Fernandes, D.Y. Kabuki, and A.Y. Kuaye, Biofilms of *Enterococcus faecalis* and *Enterococcus faecium* isolated from the processing of ricotta and the control of these pathogens through cleaning and sanitization procedures, *Int. J. Food Microbiol.*, 200, 97–103 (2015).

99 R. Singla, H. Goel and A. Ganguli, Novel synergistic approach to exploit the bactericidal efficacy of commercial disinfectants on the biofilms of *Salmonella enterica* serovar Typhimurium, *J. Biosci.Bioeng.*, 118(1), 34–40 (2014).

100 D. Chen, T. Zhao and M.P. Doyle, Control of pathogens in biofilms on the surface of stainless steel by levulinic acid plus sodium dodecyl sulfate, *Int. J. Food Sci. Technol.*, 207, 1–7 (2015).

101 S. Sepehr, A. Rahmani-Badi, H. Babaie-Naiej and M.R. Soudi, Unsaturated fatty acid, *cis*-2-decenoic acid, in combination with disinfectants or antibiotics removes pre-established biofilms formed by food-related bacteria, *PLoS One*, 9(7) e101677 (2014).

102 L. Chopra, G. Singh, K.K. Jena and D.K. Sahoo, Sonorensin: A new bacteriocin with potential of an anti-biofilm agent and a food bio preservative, *Sci. Rep.*, 5 (2015).

103 M.M.P. van Speybroeck, G. Bruggeman, J. van Poele, K.L.I. van Peel and E.J. van Damme, Exopolysaccharide-degrading enzyme and use of the same, PCT Patent Appl., WO 9631610 (1996).

104 Y. Lequette, G. Boels, M. Clarisse and C. Faille, Using enzymes to remove biofilms of bacterial isolates sampled in the food-industry, *Biofouling*, *26*(4), 421–431 (2010).

105 V. Prabhawathi, T. Boobalan, P.M. Sivakumar and M. Doble, Functionalized polycaprolactam as an active food package for antibiofilm activity and extended shelf life, *Colloids Surf. B Biointerfaces*, 123, 461–468 (2014).

106 J.J. Curtin and R.M. Donlan, Using bacteriophages to reduce formation of catheter-associated biofilms by *Staphylococcus epidermidis*, *Antimicrob. Agents Chemother.*, 50(4), 1268-1275 (2006).

107 B.D. Corbin, R.J.C. McLean and G.M. Aron, Bacteriophage T4 multiplication in a glucose-limited *Escherichia coli* biofilm, *Can. J. Microbiol.*, 47(7), 680–684 (2001).

108 S. Sillankorva, P. Neubauer, and J. Azeredo, Phage control of dual species biofilms *of Pseudomonas fluorescens* and *Staphylococcus lentus*, *Biofouling*, 26(5), 567–575 (2010).

109 T.K. Lu and J.J. Collins, Dispersing biofilms with engineered enzymatic bacteriophage, *Proc. Natl. Acad. Sci. U.S.A.*, 104 (27), 11197–11202 (2007).

110 T.E. Vaughan and J.C. Weaver, Molecular change due to biomagnetic stimulation and transient magnetic fields: mechanical interference constraints on possible effects by cell membrane pore creation via magnetic particles, *Bioelectrochem.Bioener.*, 46(1), 121–128 (1998).

111 M.F. San Martín, F.M. Harte, H. Lelieveld, G.V. Barbosa-Cánovas and B.G. Swanson, Inactivation effect of an 18-T pulsed magnetic field combined with other technologies on *Escherichia coli*, *Innov. Food Sci. Emerg. Technol.*, 2(4), 273–277 (2001).

112 N. Oulahal-Lagsir, A. Martial-Gros, E. Boistier, L.J. Blum and M. Bonneau, The development of an ultrasonic apparatus for the noninvasive and repeatable removal of fouling in food processing equipment, *Lett. Appl. Microbiol.*, 30(1), 47–52 (2000).

113 F. Chemat, Zill-e-Huma and M.K. Khan, Applications of ultrasound in food technology: processing, preservation and extraction, *Ultrason. Sonochem.*, 18(4), 813–835 (2011).

114 A.J. Castro, G.V. Barbosa-Canovas and B.G. Swanson, Microbial inactivation of foods by pulsed electrical fields, *J. Food Process. Preserv.*, 17, 47–73 (1993).

115 U.R. Pothakamury, H. Vega, Q. Zhang, G.V. Barbosa-Canovas and B.G. Swanson, Effect of growth stage and processing temperature on the inactivation of *E. coli* by pulsed electric fields, *J. Food Prot.*, 59, 1167–1171 (1996).

116 V.G. Barbosa-Canovas, U.R. Pothakamury, M.M. Gongora-Nieto and B.G. Swanson, Preservation of foods with pulsed electric fields, *in Emerging Technologies for Food Processing*, D.W. Sun (Ed.) Academic Press (1999).

117 L. Nan, K. Yang and G. Ren, Anti-biofilm formation of a novel stainless steel against *Staphylococcus aureus*, *Mater. Sci. Eng. C* 51, 356–361 (2015).

118 D. Ziuzina, D. Boehm, S. Patil, P.J. Cullen and P. Bourke, Cold plasma inactivation of bacterial biofilms and reduction of quorum sensing regulated virulence factors, *PLoS One*, 10(9), e0138209 (2015).

119 T. Møretrø, E. Heir, L.L. Nesse, L.K. Vestby and S. Langsrud, Control of *Salmonella* in food related environments by chemical disinfection, *Food Res. Int.*, 45(2), 532–544 (2012).

120 A. Antoniou and J.F. Frank, Removal of *Pseudomonas putida* and associated extracellular polymeric substances from stainless steel by alkali cleaning, *J. Food Prot.*, 68, 277–281 (2005).

121 T. Christina, V. Penna, P.G. Mazzola, A. Maria and S. Martins, The efficacy of chemical agents in cleaning and disinfection programs, *BMC Infect. Dis.*, 1(16) (2001).

122 K. Toté, T. Horemans, D. VandenBerghe, L. Maes and P. Cos, Inhibitory effect of biocides on the viable masses and matrices of *Staphylococcus aureus* and *Pseudomonas aeruginosa* biofilms, *Appl. Environ. Microbiol.*, 76(10), 3135–3142 (2010).

123 A. K. Greene, B.K. Few and J. C. Serafini, A comparison of ozonation and chlorination for the disinfection of stainless steel surfaces, *J. Dairy Sci.*, 76, 3617–3620 (1993).

124 M. Tachikawa, K. Yamanaka and K. Nakamuro, Studies on the disinfection and removal of biofilms by ozone water using an artificial microbial biofilm system, *Ozone-Sci. Eng.*, 31(1), 3–9 (2009).

125 J.T. Holah, C. Higgs, S. Robinson, D. Worthington and H. Spenceley, A conductance-based surface disinfection test for food hygiene, *Lett. Appl. Microbiol.*, 11, 255–259 (1990).

126 T.K. Cabeça, A.C. Pizzolitto and E.L. Pizzolitto, Assessment of action of disinfectants against *Listeria monocytogenes* biofilms, *J. Food Nutr.*, 17(2), 121–125 (2008).

127 A. Lomander, P. Schreuders, E. Russek-Cohen and L. Ali, Evaluation of chlorine's impact on biofilms on scratched stainless steel surfaces, *Bioresour. Technol.*, 94(3), 275–283 (2004).

128 K.L. Cargill, B.H. Pyle, R.L. Sauer and G.A. McFeters, Effects of culture conditions and biofilm formation on the iodine susceptibility of *Legionella pneumophila*, *Can. J. Microbiol.*, 38, 423–429 (1992).

129 D.H. Oh and D.L. Marshall, Effect of pH on the minimum inhibitory concentration of monolaurin against *Listeria monocytogenes*, *J. Food Prot.*, 55, 449–450 (1992).

130 D.H. Oh and D.L. Marshall, Influence of temperature, pH and glycerol monolaurate on growth and survival of *Listeria monocytogenes*, *J. Food Prot.*, 56, 744–749 (1993a).

131 D.H. Oh and D.L. Marshall, Antimicrobial activity of ethanol, glycerol monolaurate or lactic acid against *Listeria monocytogenes*, *Int. J. Food Microbiol.*, 20, 239–246 (1993b).

132 P.J. Breen, H. Salari and C.M. Compadre, Elimination of *Salmonella* contamination from poultry tissues by cetylpyridinium chloride solutions, *J. Food Prot.*, 60, 1019–1021 (1997).

133 J.J. Harrison, R.J. Turner, D.A. Joo, M.A. Stan, C.S. Chan, N.D. Allan, H.A. Vrionis, M.E. Olson and H. Ceri, Copper and quaternary ammonium cations exert synergistic bactericidal and antibiofilm activity against *Pseudomonas aeruginosa*, *Antimicrob. Agents Chemother.*, 52(8), 2870–2881 (2008).

134 J. Rogers, A.B. Dowsett and C.W. Keevil, A paint incorporating silver to control mixed biofilms containing *Legionella pneumophila*, *J. Ind.Microbiol.*, 15, 377–382 (1995).

135 Y.M. Weng and J.H. Hotchkiss, Anhydrides as antimycotic agents to polyethylene films for food packaging, *Pack. Technol. Sci.*, 6, 123–128 (1993).

136 Y.M. Weng, M.J. Chen and W. Chen, Benzoyl chloride modified ionomer films as antimicrobial food packaging materials, *Int. J. Food Sci. Technol.*, 32, 229–234 (1997).

137 M. Dufour, R.S. Simmonds and P.J. Bremer, Development of a laboratory scale clean-in-place system to test the effectiveness of "natural" antimicrobials against dairy biofilms, *J. Food Prot.*, 67, 1438–1443 (2004).

138 N. Kato and I. Shibasaki, Combined effect of citric and polyphosphoric acid on the antibacterial activity of monoglycerides, *J. Antibact. Antifung. Agents*, 4, 254–261 (1976).

139 S. Furukawa, Y. Akiyoshi, M. Komoriya, H. Ogihara and Y. Morinaga, Removing *Staphylococcus aureus* and *Escherichia coli* biofilms on stainless steel by cleaning-in-place (CIP) cleaning agents, *Food Control*, 21(5), 669–672 (2010b).

140 H. Oknin, D. Steinberg and M. Shemesh, Magnesium ions mitigate biofilm formation of *Bacillus* species via downregulation of matrix genes expression, *Front. Microbiol.*, 6, 907 (2015).

141 B. Ray, Nisin of *Lactococcuslactis* ssp. *lactis* as a food biopreservative, *in* Food biopreservatives of microbial origin, B. Ray and M. Daeschel (Eds.), 207–264, CRC Press, Boca Raton, Florida, USA (1992).

142 J.N. Hansen and W.E. Sandine, Nisin as a model food preservative, *Crit. Rev. Food Sci. Nutr.*, 34(1), 69–93 (1994).

143 L. Schaefer, T.A. Auchtung, K.E. Hermans, D. Whitehead, B. Borhan and R.A. Britton, The antimicrobial compound reuterin (3-hydroxypropionaldehyde) induces oxidative stress via interaction with thiol groups, *Microbiology*, 156(6), 1589–1599 (2010).

144 S. Quintavalla and L. Vicini, Antimicrobial food packaging in meat industry, *Meat Sci.*, 62(3), 373–380 (2002).

145 S.G. Parkar, S.H. Flint and J.D. Brooks, Evaluation of the effect of cleaning regimes on biofilms of thermophilic bacilli on stainless steel, *J. Appl. Microbiol.*, 96, 110–116 (2004).

146 K. Ivanova, M.M. Fernandes, A. Francesko, E. Mendoza, J. Guezguez, M. Burnet and T. Tzanov, Quorum quenching and matrix degrading enzymes in multilayer coatings

synergistically prevent bacterial biofilm formation on urinary catheters, *ACS Appl. Mater. Interfaces*, 7(49), 27066–27077 (2015).
147 B. Craigen, A. Dashiff and D.E. Kadouri, The use of commercially available alpha-amylase compounds to inhibit and remove *Staphylococcus aureus* biofilms, *Open Microbiol. J.*, 5, 21–31 (2011).
148 C. Johansen, P. Falholt and L. Gram, Enzymatic removal and disinfection of bacterial biofilms, *Appl. Environ. Microbiol.*, 63(9), 3724–3728 (1997).
149 S.T. Abedon, Ecology of anti-biofilm agents II: bacteriophage exploitation and biocontrol of biofilm bacteria, *Pharmaceuticals*, 8(3), 559–589 (2015).
150 N. Høiby, T. Bjarnsholt, M. Givskov, S. Molin, and O. Ciofu, Antibiotic resistance of bacterial biofilms, *Int. J. Antimicrob. Agents*, 35, 322–332 (2010).
151 A.J. Bai and R.R. Vittal, Quorum sensing inhibitory and anti-biofilm activity of essential oils and their *in vivo* efficacy in food systems, *Food Biotechnol.*, 28(3), 269–292 (2014).
152 D. Stojković, M. Soković, J. Glamočlija, A. Džamić, A. Ćirić, M. Ristić and D. Grubišić, Chemical composition and antimicrobial activity of *Vitexagnus-castus* L. fruits and leaves essential oils, *Food Chem.*, 128, 1017–1022 (2011).
153 D. Stojković, J. Petrović, M. Soković, J. Glamočlija, J. Kukić-Marković and S. Petrović, *In situ* antioxidant and antimicrobial activities of naturally occurring caffeic acid, *p*-coumaric acid and rutin, using food systems, *J. Sci. Food Agric.*, 93(13), 3205–3208 (2013a).
154 D. Stojković, F. Reis, I.C.F.R. Ferreira, L. Barros, J. Glamočlija, A. Ćirić, M. Nikolić, T. Stević, A. Giweli and M. Soković, *Tirmania pinoyi*: chemical composition, *in vitro* antioxidant and antibacterial activities and in situ control of *Staphylococcus aureus* in chicken soup, *Food Res. Int.*, 53, 56–62 (2013b).
155 D. Bukvički, D. Stojković, M. Soković, L. Vannini, C. Montanari, B. Pejin, A. Savić, M. Veljić, S. Grujić and P.D. Marin, *Saturejahorvatii* essential oil: *In vitro* antimicrobial and antiradical properties and *in situ* control of *Listeria monocytogenes* in pork meat, *Meat Sci.*, 96, 1355–1360 (2014).
156 C.M. Saviuc, V. Drumea, L. Olariu, M.C. Chifiriuc, E. Bezirtzoglou and V. Lazar, Essential oils with microbicidal and antibiofilm activity, *Curr. Pharm. Biotechnol.*, 16(2), 137–151 (2015).
157 M. Nikolić, D. Stojković, J. Glamočlija, A. Ćirić, T. Marković, M. Smiljković and M. Soković, Could essential oils of green and black pepper be used as food preservatives? *J. Food Sci. Technol.*, 52(10), 6565–6573 (2015).
158 C. Niu and E.S. Gilbert, Colorimetric method for identifying plant essential oil components that affect biofilm formation and structure, *Appl. Environ. Microbiol.*, 70(12), 6951–6956 (2004).
159 A. Rahman and S.C. Kang, *In vitro* control of food-borne and food spoilage bacteria by essential oil and ethanol extracts of *Lonicera japonica*, *Food Chem.*, 116, 670–675 (2009).
160 P.O. Angienda, D.M. Onyango and D.J. Hill, Potential application of plant essential oils at sub-lethal concentrations under extrinsic conditions that enhance their antimicrobial effectiveness against pathogenic bacteria, *J. Microbiol.*, 4, 1678–1684 (2010).
161 G.T. Gündüz, Ş.A. Gönül and M. Karapınar, Efficacy of oregano oil in the inactivation of *Salmonella typhimurium* on lettuce, *Food Control*, *21*, 513–517 (2010).

162 N. Karagözlü, B. Ergönüland D. Özcan, Determination of antimicrobial effect of mint and basil essential oils on survival of *E. coli* O157:H7 and *S. typhimurium* in fresh-cut lettuce and purslane, *Food Control*, 22, 1851–1855 (2011).

163 B.J. Min and J.H. Oh, Antimicrobial activity of catfish gelatin coating containing origanum (*Thymus capitatus*) oil against gram-negative pathogenic bacteria, *J. Food Sci.*, 74, 143–148 (2009).

164 J. Gómez-Estaca, A. López de Lacey, M.E. López-Caballero, M.C. Gómez-Guillén and P. Montero, Biodegradable gelatine-chitosan films incorporated with essential oils as antimicrobial agents for fish preservation, *Food Microbiol.*, 27, 889–896 (2010).

165 K.D. Vu, R.G. Hollingsworth, E. Leroux, S. Salmieri and M. Lacroix, Development of edible bioactive coating based on modified chitosan for increasing the shelf life of strawberries, *Food Res. Int.*, 44, 198–203 (2011).

166 P.E. Budri, N.C.C. Silva, E.C.R. Bonsaglia, A. FernandesJúnior, J.P. AraújoJúnior, J.T. Doyama, J.L. Gonçalves, M.V. Santos, D. Fitzgerald-Hughes and V.L.M. Rall, Effect of essential oils of *Syzygium aromaticum* and *Cinnamomum zeylanicum* and their major components on biofilm production in *Staphylococcus aureus* strains isolated from milk of cows with mastitis, *J. Dairy Sci.*, 98(9), 5899–5904 (2015).

167 E. B. Kerekes, É.T.M. Deák, R. Tserennadmid, T. Petkovits, C. Vágvölgyi and J. Krisch, Anti-biofilm-forming and anti-quorum sensing activity of selected essential oils and their main components on food-related micro-organisms, *J. Appl.Microbiol.*, 115(4), 933–942 (2013).

168 D. Schillaci, E.M. Napoli, M.G. Cusimano, M. Vitale and A. Ruberto, *Origanum vulgare* subsp. *hirtum* essential oil prevented biofilm formation and showed antibacterial activity against planktonic and sessile bacterial cells, *J. Food Prot.*, 76(10), 1747–1752 (2013).

169 M. Snoussi, E. Noumi, N. Trabelsi, G. Flamini, A. Papetti and V. De Feo, *Menthaspicata* essential oil: chemical composition, antioxidant and antibacterial activities against planktonic and biofilm cultures of *Vibrio* spp. strains, *Molecules*, 20(8), 14402–14424 (2015).

170 M.M. Bazargani and J. Rohloff, Antibiofilm activity of essential oils and plant extracts against *Staphylococcus aureus* and *Escherichia coli* biofilms, *Food Control*, 61, 156–164 (2016).

171 V. Popović, D. Stojković, M. Nikolić, A. Heyerick, S. Petrović, M. Soković and M. Niketić, Extracts of three *Laserpitium* L. species and their principal components laserpitine and sesquiterpene lactones inhibit microbial growth and biofilm formation by oral *Candida* isolates, *Food Funct.*, 6(4), 1205–1211 (2015).

172 H. Elhariry, A.A. Abuzaid, G. M. Khiralla and Y. Gherbawy, Antibiofilm formation and anti-adhesive (to HEp-2 cells) effects of rosemary water extract against some food-related pathogens, *Int. J. Food Sci. Technol.*, 49, 1132–1141 (2014).

173 V. Cheynier. Phenolic compounds: from plants to foods. *Phytochem. Rev.*, 11, 153–177 (2012).

174 A. Upadhyay, I. Upadhyaya, A. Kollanoor-Johny, and K. Venkitanarayanan, Antibiofilm effect of plant derived antimicrobials on *Listeria monocytogenes*, *Food Microbiol.*, 36(1), 79–89 (2013).

175 C. Xu, Y. Yagiz, W.Y. Hsu, A. Simonne, J. Lu and M.R. Marshall, Antioxidant, antibacterial, and antibiofilm properties of polyphenols from muscadine grape

(*Vitis rotundifolia* Michx.) pomace against selected foodborne pathogens, *J. Agric. Food Chem.*, 62(28), 6640–6649 (2014).

176 L. Espina, R. Pagán, D. López and D. García-Gonzalo, Individual constituents from essential oils inhibit biofilm mass production by multi-drug resistant *Staphylococcus aureus*, *Molecules*, 20(6), 11357–11372 (2015).

177 D. Shao, J. Li, J. Li, R. Tang, L. Liu, J. Shi, Q. Huang and H. Yang, Inhibition of gallic acid on the growth and biofilm formation of *Escherichia coli* and *Streptococcus mutans*, *J. Food Sci.*, 80, M1299–1305 (2015).

178 H. Zhang, W. Zhou, W. Zhang, A. Yang, Y. Liu, Y. Jiang, S. Huang and J. Su, Inhibitory effects of citral, cinnamaldehyde, and tea polyphenols on mixed biofilm formation by foodborne *Staphylococcus aureus* and *Salmonella* Enteritidis, *J. Food Prot.*, 77(6), 927–933 (2014).

179 L.R. Christena, S. Subramaniam, M. Vidhyalakshmi, V. Mahadevan, A. Sivasubramanian and S. Nagarajan, Dual role of pinostrobin-a flavonoid nutraceutical as an efflux pump inhibitor and antibiofilm agent to mitigate food borne pathogens, *RSC Adv.*, 5(76), 61881–61887 (2015).

180 N. Rajendran, S. Subramaniam, L.R. Christena, M.S. Muthuraman, N.S. Subramanian, B. Pemiah and A. Sivasubramanian, Antimicrobial flavonoids isolated from Indian medicinal plant *Scutellaria oblonga* inhibit biofilms formed by common food pathogens, *Nat. Prod. Res.*, 30(17), 2002–2006 (2016). DOI: 10.1080/14786419.2015.1104673.

181 A. Duarte, A.C. Alves, S. Ferreira, F. Silva and F.C. Domingues, Resveratrol inclusion complexes: Antibacterial and anti-biofilm activity against *Campylobacter* spp. and *Arcobacter butzleri*, *Food Res. Int.*, 77, 244–250 (2015).

182 O. Priha, V. Virkajärvi, R. Juvonen, R. Puupponen-Pimiä, L. Nohynek, S. Alakurtti, M. Pirttimaa and E. Storgårds, Quorum sensing signalling and biofilm formation of brewery-derived bacteria, and inhibition of signalling by natural compounds, *Curr. Microbiol.*, 69(5), 617–627 (2014).

183 M. Hentzer and M. Givskov, Pharmacological inhibition of quorum sensing for the treatment of chronic bacterial infections, *J. Clin. Invest.*, 112(9), 1300–1307 (2003).

184 T. B. Rasmussen and M. Givskov, Quorum-sensing inhibitors as anti-pathogenic drugs, *Int. J. Med. Microbiol.*, 296(2), 149–161 (2006).

185 V.C. Kalia, Quorum sensing inhibitors: an overview, *Biotechnol. Adv.*, 31(2), 224–245 (2013).

186 F.M. Husain and I. Ahmad, Marine organisms as source of quorum sensing inhibitors, *in* Quorum sensing vs quorum quenching: a battle with no end in sight, V.C. Kalia (Ed.) 259–268, Springer India (2015).

187 P. Truchado, M. Larrosa, I. Castro-Ibáñez, and A. Allende, Plant food extracts and phytochemicals: Their role as quorum sensing inhibitors, *Trends Food Sci. Technol.*, 43, 189–204 (2015).

188 M. V. Alvarez, L. A. Ortega-Ramirez, M. M. Gutierrez-Pacheco, A. T. Bernal-Mercado, I. Rodriguez-Garcia, G. A. Gonzalez-Aguilar and J. F. Ayala-Zavala, Oregano essential oil-pectin edible films as anti-quorum sensing and food antimicrobial agents, *Front. Microbiol.*, 5, 699 (2014).

189 M. V. Alvarez, M. R. Moreira and A. Ponce, Anti quorum sensing and antimicrobial activity of natural agents with potential use in food, *J. Food Safety*, 32(3), 379–387 (2012).

190 G. Venkadesaperumal, R. Sambrey, S. Kothandapani and P.H. Shetty, Anti-quorum sensing activity of spice oil nanoemulsions against food borne pathogens, *LWT-Food Sci. Technol.*, 66, 225–231 (2016).

191 M.A. Hossain, J.Y. Park, J.Y. Kim, J.W. Suh and S.C. Park, Synergistic effect and antiquorum sensing activity of *Nymphaea tetragona* (water lily) extract, *Biomed. Res. Int.*, 562173 (2014).

192 V. Gopu, C.K. Meena and P.H. Shetty, Quercetin influences quorum sensing in food borne bacteria: *in-vitro* and *in-silico* evidence, *PLoS One*, 10(8), e0134684 (2015).

23

Biofilm Inhibition by Natural Products of Marine Origin and Their Environmental Applications

Alwar Ramanujam Padmavathi[1,2], *Dhamodharan Bakkiyaraj*[1,3] *and Shunmugiah Karutha Pandian*[1]

[1] *Department of Biotechnology, Alagappa University, Karaikudi, India*
[2] *Nanotec-PSU Center of Excellence on Drug Delivery System and Department of Pharmaceutical Technology, Faculty of Pharmaceutical Sciences, Prince of Songkla University, Hat Yai, Songkhla, Thailand*
[3] *Department of Microbiology, Faculty of Science, Prince of Songkla University, Hat Yai, Songkhla, Thailand*

23.1 Introduction

Microorganisms are omnipresent, and their existence has been well documented in all possible parts of biosphere [1–4]. It is believed that microbes are the first life evolved in this universe, and they outnumber the total human population on the earth [5]. Microbes play dual role in promoting and deteriorating the health of human and other organisms. Pathogenic microbes garnered more attention due to their ill effects and subsequent economic impact. These microbial infections have been effectively controlled since the discovery of penicillin and other antibiotics [6]. Different classes of antibiotics have been employed to incapacitate these infectious agents [7–9]. However, being the first organism evolved on the Earth, they continue to adopt several evolutionary mechanisms to evade antibiotics [10] (Figure 23.1).

Besides these resistant strategies, biofilm formation is considered as the foremost resistant mechanism that shields microbes from destructive agents and other hostile circumstances. Biofilm, a complex, three-dimensional microbial aggregate, is formed by unicellular microbes to act as a multicellular entity, to protect them from unfavorable environments [11]. They inherently possess and quickly acquire immunity against antimicrobials and host defensive mechanisms. Formation and maintenance of intact biofilms are very crucial for the pathogenesis and persistence of infections. Bacterial and fungal cells encased in biofilm are many fold (up to 1,000) resistant to antibiotics and hence unaffected by them [12]. Microbes form biofilms in every possible substratum, and their associated infections have been observed everywhere, right from plants, animals to human beings [10, 13–15]. In addition to these infections, they play destructive roles in corrosion of industrially important pipelines and other monuments [16–18]. This prompted extensive research into this rapidly expanding field to device appropriate control measures and to overcome the losses incurred. Various sources have been explored in search of novel antibiofilm agents to impede biofilm formation and to exterminate persistent biofilms.

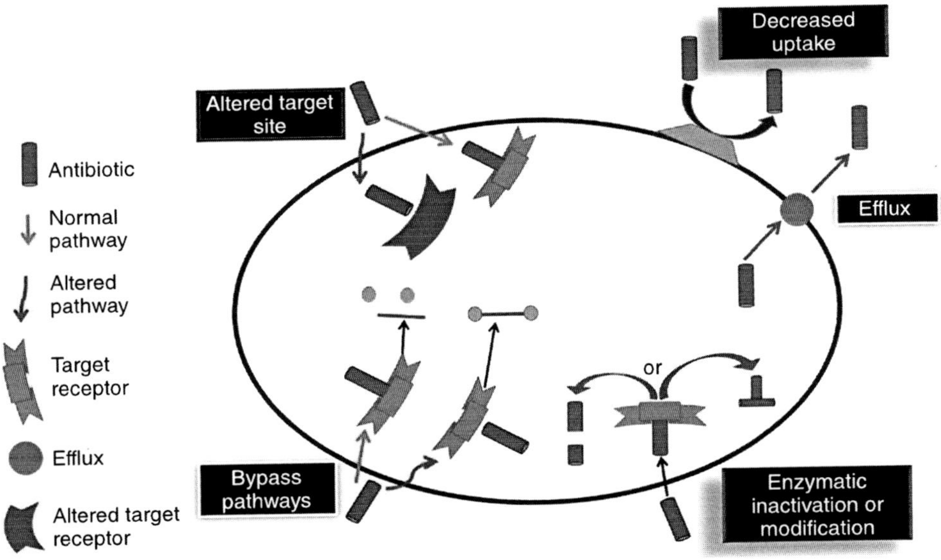

Figure 23.1 Bacterial resistant strategies to evade antibiotics.

This chapter discusses the role of marine natural products in combating biofilms and their environmental applications.

23.2 Unity Is Strength: Benefits of Biofilm Formers

Biofilm mode of growth provides numerous advantages to indwelling microbes. They rapidly develop resistance towards any antimicrobial agent and antibiotics are incapable to penetrate the dense biofilm and failed to eradicate the biofilm formers (Figure 23.2). Biofilm confers broad-spectrum resistance to indwelling microbes and facilitate exchange of genetic information [19]. It provides stable growth environment, which are less likely to be affected by external cues. It acts as storehouse of excessive energy and provides protection from harsh physical and chemical factors [20]. Biofilm formers easily hoodwink the host immune system [21]. Due to these numerous advantages,

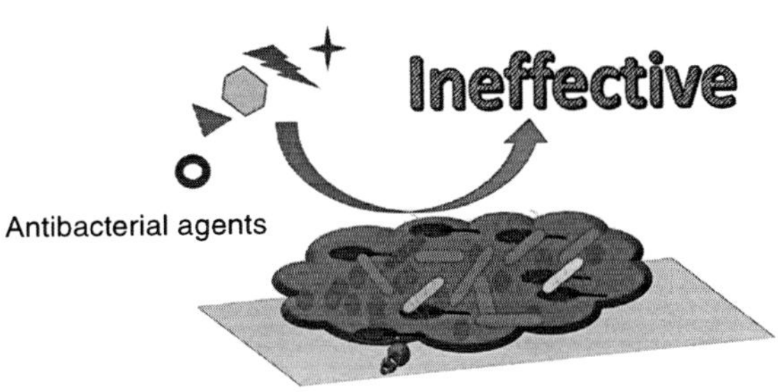

Figure 23.2 Biofilm resistance to antimicrobial agents.

bacteria choose to grow as biofilms over planktonic life style, especially when the conditions are unfavorable.

23.3 Transition of Slimy Film to Persistent Biofilm

Though biofilm and free living are two different states of bacteria, biofilms are instigated from the attachment of free living/planktonic cells. The motile cells, due to various environmental cues, settle to a substratum that initiate the formation of biofilms. There are five developmental stages delineated from the cycle of biofilms:

1) Initial adhesion
2) Irreversible attachment
3) Microcolony formation
4) Maturation of biofilm
5) Dispersion to distinct sites to facilitate newer biofilms [20]

These stages convert the sporadic biofilm to confluent one. Biofilms are strengthened by the extracellular polymeric substances (EPS) produced by microbes. EPS act as stabilizing agent of biofilms and are composed of lipids, polysaccharides, proteins, and nucleic acids [19]. Several studies exemplify the failure of EPS mutants to form biofilm or reveal the crucial role of EPS in biofilm formation and its integrity [22].

23.4 Biofilm-Related Infections in Plants

Microbes form biofilms on almost all solid substratum. Plant diseases caused by biofilms result in decreased productivity (30 to 50 percent annual loss) that leads to famine and serious consequences. The role between biofilm and plants is different based on the place and organism forming biofilm. Biofilms play both beneficial and detrimental roles in plant systems. *Bacillus subtilis* forms biofilms that prevent the colonization and adherence of other pathogens, thus preventing the plant from infections [23, 24]. Nitrogen fixers colonize and form biofilms on plant roots, aiding plant health [25]. These are few examples of beneficial biofilms found in plants. Apart from these beneficial biofilms, numerous pathogens form biofilms on various parts of plants affecting the productivity and leading to economic loss (Table 23.1).

23.5 Need for Antibiofilm Agents

Due to the failure of antibiotics and emergence of multidrug resistance in pathogens, an alternate treatment strategy is needed to control the bacterial infections in plants. Antibiofilm agents gained more attention due to their selective role of disentangling the biofilm cells rather than killing them. This eliminates the risk of resistance development in bacterial and fungal pathogens. Antibiofilm agents exert two different activities on biofilm cells, either they inhibit the biofilm formation or disrupt the preformed mature biofilms, and sometimes they can act both ways, depending on the concentrations used.

Table 23.1 Biofilm Forming Plant Pathogens and Their Detrimental Effects on Plants.

Sl. No	Pathogen	Disease	Host Plant(s)	Reference(s)
1	*Agrobacterium rhizogenes*	Hairy root	Dicotyledonous plants	[26]
2	*A. rhizogenes*	Root mat	Tomato	[27]
3	*A. tumefaciens*	Crown gall	Dicotyledonous plants	[28]
4	*Burkholderia cenocepacia*	Banana finger-tip Rot	Banana	[29]
5	*B. cepacia*	Rot	Onion	[30]
6	*Clavibacter michiganensis* subsp. *sepedonicus*	Brown rot and ring rot	Potato	[31]
7	*Leifsonia xyli* subsp. *xyli*	Ratoon stunting disease	Sugarcane	[32]
8	*Enterococcus faecalis*	Rotting	*Arabidopsis thaliana*	[33]
9	*Erwinia amylovora*	Fire blight	Rosaceae family	[34]
11	*E. carotovora*	Rot	Potato	[35]
12	*Dickeya dadantii*	Soft rot and wilt disease	Tubers	[36]
16	*Pantoea agglomerans*	Leaf blight	Rice	[37]
17	*P. ananatis*	White spot disease	Maize	[38]
18	*P. stewartii*	Wilt and leaf blight	Maize	[39]
19	*P. stewartii* subsp. *stewartii*	Stewart's wilt	Sweet corn	[40]
20	*Pseudomonas syzygii*	Sumatra disease	Clove	[41]
21	*P. aeruginosa*	Root rot	Sweet basil, *A. thaliana*	[42]
22	*P. syringae* pv *phaseolicola*	Halo blight	Bean	[43]
23	*P. syringae* pv *syringae*	Holcus spot	Corn	[44]
24	*P. syringae* pv *syringae*	Brown spot and Halo blight	*Phaseolus vulgaris*	[43]
25	*P. syringae* pv. *tomato*	Bacterial speck	Tomato	[45]
26	*Ralstonia solanacearum*	Lethal wilt	Tomato	[46]
27	*Spiroplasma kunkelii*	Corn stunt	Zea mays	[47]
30	*Xanthomonas campestris* pv. *vesicatoria*	Bacterial spot	Pepper and tomato	[48]
31	*X. axonopodis* pv. *citri*	Canker	Citrus	[49]
31	*X. campestris* pv *phaseoli*	Common bacterial blight	*Phaseolus vulgaris*	[50]
32	*X. campestris* pv. *campestris*	Black rot	Cruciferous plants	[51]

Table 23.1 (Continued)

Sl. No	Pathogen	Disease	Host Plant(s)	Reference(s)
33	*X. oryzae* pv *oryzae*	Bacterial leaf blight	Rice	
34	*Xylella fastidiosa*	Variegated chlorosis	Citrus	[53–56]
35	*X. fastidiosa*	Pierce's disease	Grapevine	[57]

Antibiofilm activity can be achieved by enzymes, small molecules, phage particles, EPS degrading agents, and combinatorial therapies [58].

23.6 Natural Products of Marine Origin as Antibiofilm Agents

Recently, bacterial and plant secondary metabolites gained attention as novel antibiofilm agents. Even though several natural and synthetic chemicals have shown to inhibit biofilm formation in various organisms, antibiofilm agents of marine origin are of special interest, as many of them are expected to be novel. Marine environment is an untapped reserve for novel bioactivities and more than 99 percent of marine microbes are yet to be discovered hence providing insights on their hidden wealth. Until now, exploration of marine environment delivered potential antibiofilm agents against various bacterial and fungal pathogens, which are tabulated in Table 23.2.

23.7 Semi-synthetic Antibiofilm Agents Inspired by Marine Natural Products

Apart from numerous antibiofilm agents produced by bacteria, fungi, and other organisms of marine origin, chemically modified marine natural products are also available, which have more pronounced antibiofilm activity owing to their modified side chains. Repertoire of synthetic and semi-synthetic compounds have been evaluated for their selective antibiofilm activities (Table 23.3). The identified compounds could serve as a foundation to develop novel antiadhesive polymers to reduce the fouling in marine environments and can be used as therapeutics, which can aid in eradicating the bacterial biofilms in both medical and industrial settings [92].

23.8 Environmental Applications of Antibiofilm Agents

Antibiofilm agents from marine sources are expected to be stable and active in harsh environments like high salinity, temperature, and so on, and hence can be used to treat plants and crops being cultivated in any part of the world. Unlike pesticides, these antibiofilm agents are specific to the target organism and will not have any ill effects on the plant products or humans who consume those products. Apart from their direct

Table 23.2 Antibiofilm Agents of Marine Origin.

Sl. No.	Organism	Pathogen(s)	Antibiofilm Agent	Reference
1	*Marinomonas* sp.	*Vibrio cholerae* O1	Indole-3-carboxaldehyde	[59]
2	*Bacillus amyloliquefaciens*	*Listeria monocytogenes*	Cyclo(L-Leucyl- L-Prolyl)	[60]
3	Brown Alga *Halidrys siliquosa*	Human pathogens	Methanolic extract	[61]
4	*B. amyloliquefaciens*	MRSA	Cyclo(L-Leucyl- L-Prolyl)	[62]
5	*Marinobacter litoralis*	*Pseudomonas aeruginosa* PA01	Lipopolysaccharide	[63]
6	*Pseudoalteromonas* sp. 3J6	*Vibrio tapetis*	Exoproduct	[64]
7	*Brevibacterium casei* MSI04	*V. vulnificus*, *V. fischeri*, *V. parahaemolyticus*, *V. alginolyticus*, *V. harveyi.*	Poly-hydroxy butyrate biopolymer	[65]
8	*Aspergillus flavipes* AIL8	*Staphylococcus aureus*, *Bacillus subtilis*	Phenyl derivatives	[66]
9	*Neosartorya paulistensis*, *N. laciniosa*, *N. tsunodae*	*S. aureus*, *B. subtilis*, *Escherichia coli* and *P. aeruginosa* and MDR *S. aureus*	Tryptoquivalines and meroditerpenes	[67]
10	*Oceanobacillus iheyensis*	*S. aureus*	Extracellular polymeric substances	[68]
11	*B. amyloliquefaciens*	*Streptococcus mutans*	Cyclo(L-Leucyl- L-Prolyl)	[69]
12	Marine fungi	*S. epidermidis*	Mevalonolactone	[70]
13	*B. licheniformis*	*Candida albicans* BH, *P. aeruginosa* PAO1, *B. pumilus* TiO1	Extracellular protein	[71]
14	*B. subtilis* S8-18	*V. cholerae*, MRSA	α-amylase	[72]
15	*B. firmus* and *V. parahaemolyticus*	MRSA	Ethyl acetate extract	[73]
16	*Delftia tsuruhatensis* and *Stenotrophomonas maltophilia*	*S. marcescens*	Ethyl acetate extract	[74]
17	*Pseudoalteromonas* sp. D41	Mixed biofilm	Culture supernatant	[75]
18	*Vibrio* sp. QY101	*P. aeruginosa* and *S. aureus*	Exopolysaccharide	[76]
19	*Bacillus* sp. SS4	*P. aeruginosa* PAO1	Ethyl acetate extract	[77]
20	*S. phocae* PI80	Gram positive and Gram negative bacterial pathogens	Exopolysaccharide	[78]
21	*B. pumilus* S8-07, *B. indicus* S6-01	*Vibrio* spp.	Culture supernatants	[79]

Table 23.2 (Continued)

Sl. No.	Organism	Pathogen(s)	Antibiofilm Agent	Reference
22	*B. pumilus, B. indicus, B. arsenicus, Halobacillus trueperi, Ferrimonas balearica* and *Marinobacter hydrocarbonoclasticus*	*P. aeruginosa* PAO1	Ethyl acetate extract	[80]
23	*B. casei* MSA19	Mixed pathogenic biofilm	Glycolipid	[81]
24	*Pseudoalteromonas* sp. strain 3J6	*P. aeruginosa, S. enterica* and *E. coli*	Culture supernatant	[82]
25	*Streptomyces akiyoshiensis*	*S. aureus*	Methanolic extract	[83]
26	Coral-associated actinomycetes	*S. pyogenes*	Culture supernatants	[84]
27	Coral-associated bacteria	*S. pyogenes*	Ethyl acetate extract	[85]
28	*Paracentrotus lividus*	*S. epidermidis* DSM 3269 and *S. aureus* ATCC 29213	Beta-thymosin	[86]
29	Sponge-associated marine bacteria	*V. halioticoli* and by *L. hongkongensis*	Culture extract	[87]
30	Marine natural product	*S. mutans*	2-amino-imidazole/ triazole conjugate	[88]
31	*Pseudoalteromonas* sp. 129-1	*P. aeruginosa* PAO1	Alkaline protease	[89]
32	*V. alginolyticus* strain G16	*Serratia marcescens*	Phenol, 2,4-bis (1,1-dimethylethyl)	[90]
33	*Dysidea avara*	*P. aeruginosa* PAO1	Sesquiterpene hydroquinone avarol	[91]

Table 23.3 Semi-synthetic Antibiofilm Agents of Marine Origin

Sl. No.	Antibiofilm Agent	Pathogen(s)	Reference
1	AuNPs and AgNPs of seaweed *Turbinaria conoides* extract	*E. coli, Salmonella* sp., *S. liquefaciens* and *Aeromonas hydrophila*	[93]
2	AgNPs of deep sea *P. aeruginosa* JQ989348	*P. aeruginosa* and *S. aureus*	[94]
3	AgNPs *of Yarrowia lipolytica* NCYC 789	*S. paratyphi* MTCC 735	[95]
4	Analogs of makaluvamine	*S. mutans*	[96]
5	Marine pyrrole alkaloid derivatives	MRSA, MSSA and *S. epidermidis*	[97]
6	Marine alkaloid oroidin library	*P. aeruginosa* and *A. baumannii*	[98]
7	4-thiazolidinones derivatives of marine bromopyrrole alkaloids	*S. aureus* and *S. epidermidis*	[99]

application, these antibiofilm agents could be used to treat the surfaces of fruit or seed storage containers, storehouses, processing environments, and transporting devices to minimize or control the contamination or infection of the products by the disease-causing pathogens.

In addition to their applications in agroindustry, biofilms can combat biofouling, a serious problem in marine environment regarding manmade structures. Antibiofilm agents can prevent the initial attachments of microbial biofilms. Antibiofilm agents of marine origin hold several advantages over synthetic and other sources. They are biodegradable and, as they belong to that niche, will not alter the nature of marine environment, and hence are nontoxic to marine animals. Marine antibiofilm agents can be incorporated into paint and other polymeric material to coat the surface of ships and other manmade structures in the sea.

23.9 Conclusion

Marine environment—a seldom-explored reserve for bioactive agents, is now the choice of interest for researchers. The marine environment is a highly complex environment where there is a constant rivalry between organisms for their food and survival, so it is important for the marine organisms to produce antagonistic agents of diverse activities. Exploration and exploitation of these marine bioactive agents is the need of hour to solve persisting problems. Antibiofilm agents, in particular, are viable alternatives for existing antimicrobial strategies in plant pest control. Widespread application of these natural antibiofilm agents for control of plant-associated diseases will be made economically feasible in near feature.

References

1 G. Horneck, D.M Klaus and R.L. Mancinelli, Space microbiology, *Microbiol. Mol. Biol. Rev.*, 74, 121–156 (2010).

2 P. Pednekar and R. Jain, G. Mahajan, Anti-infective Potential of Hot-spring Bacteria, *J. Glob. Infect. Dis.*, 3, 241–245 (2011).

3 B.C. Christner, J.C. Priscu, A.M. Achberger, C. Barbante, S.P. Carter, K. Christianson, A.B. Michaud, J.A. Mikucki, A.C. Mitchell, M.L. Skidmore and T.J. Vick-Majors, A microbial ecosystem beneath the West Antarctic ice sheet, *Nature*, 512, 310–313 (2014).

4 Fox D, Lakes under the ice: Antarctica's secret garden, *Nature*, 512, 244–246 (2014).

5 Editorial, Microbiology by numbers, *Nat. Rev. Microbiol.*, 9, 628 (2011).

6 A. Fleming, On the antibacterial action of cultures of a *Penicillium*, with special reference to their use in the isolation of *B. influenza*, *British J. Expt. Pathol.* 10, 226–236 (1929).

7 L. Katz and G.W. Ashley, Translation and protein synthesis: macrolides. *Chem. Rev.*, 105, 499–528 (2005).

8 K. Drlica, M. Malik, R.J. Kerns and X. Zhao, Quinolone-mediated bacterial death. *Antimicrob. Agents Chemother.*, 52, 385–392 (2008).

9 H. Cho, T. Uehara and T.G. Bernhardt, Beta-lactam antibiotics induce a lethal malfunctioning of the bacterial cell wall synthesis machinery, *Cell*, 159, 1300–1311(2014).

10 A.R. Padmavathi, D. Bakkiyaraj and S.K. Pandian, Significance of Biosurfactants as Antibiofilm Agents in Eradicating Phytopathogens. *Bacterial Metabolites in Sustainable Agroecosystem*, 12, 319–336 (2015a).
11 A.R. Padmavathi, M. Periyasamy and S.K. Pandian, Assessment of 2,4-Di-tert-butylphenol induced modifications in extracellular polymeric substances of *Serratia marcescens. Bioresour. Technol.*, 188,185–189 (2015b).
12 T.F Mah, Biofilm-specific antibiotic resistance, *Future Microbiol.*, 7, 1061–1072 (2012).
13 D. Lebeaux, J.M. Ghigo and C. Beloin, Biofilm-related infections: bridging the gap between clinical management and fundamental aspects of recalcitrance toward antibiotics, *Microbiol. Mol. Biol. Rev.*, 78, 510–543 (2014).
14 L. Chalupowicz, E.M. Zellermann, M. Fluegel, O. Dror, R. Eichenlaub, K.H. Gartemann, A. Savidor, G. Sessa, N. Iraki, I. Barash and S. Manulis-Sasson, Colonization and movement of GFP-labeled *Clavibacter michiganensis* subsp. *michiganensis* during tomato infection, *Phytopathol.* 102, 23–31 (2011).
15 A.L. Clutterbuck, E.J. Woods, D.C. Knottenbelt, P.D. Clegg, C.A. Cochrane and S.L. Percival, Biofilms and their relevance to veterinary medicine, *Vet. Microbiol.*, 121, 1–17 (2007).
16 T.C. Dakal and S.S Cameotra, Microbially induced deterioration of architectural heritages: routes and mechanisms involved, *Environ Sci Europe*, 24, 36 (2012).
17 K.S. Ibrahim, D. Bakkiyaraj, R. James, T.G. Babu and S.K. Pandian, Isolation and sequence analysis of a small cryptic plasmid pRK10 from a corrosion inhibitor degrading strain *Serratia marcescens* ACE2, *Plasmid*, 62, 183–190 (2009).
18 I.B. Beech, Corrosion of technical materials in the presence of biofilms—current understanding and state-of-the art methods of study, *Int. Biodeter. Biodegr.*, 53,177–183 (2004).
19 H.C. Flemming and J. Wingender, The biofilm matrix, *Nat. Rev. Microbiol.*, 8, 623–633 (2010).
20 L. Hall-Stoodley, J.W. Costerton and P. Stoodley, Bacterial biofilms: from the natural environment to infectious diseases, *Nat. Rev. Microbiol.*, 2, 95–108 (2004).
21 J.G. Leid, M.E. Shirtliff, J.W. Costerton and P. Stoodley, Human leukocytes adhere to, penetrate, and respond to *Staphylococcus aureus* biofilms, *Infect. Immun.*, 70, 6339–6345 (2002).
22 P.N. Danese, L.A. Pratt and R. Kolter, Exopolysaccharide production is required for development of *Escherichia coli* K-12 biofilm architecture, *J. Bacteriol.*, 182, 3593–3596 (2000).
23 P.B. Beauregard, Y. Chai, H. Vlamakis, R. Losick and R. Kolter, *Bacillus subtilis* biofilm induction by plant polysaccharides, *Proc. Natl. Acad. Sci. USA.*, 110, E1621–1630 (2013).
24 S. Gao, H. Wu, W. Wang, Y. Yang, S. Xie, Y. Xie and X. Gao, Efficient colonization and harpins mediated enhancement in growth and biocontrol of wilt disease in tomato by *Bacillus subtilis, Lett. Appl. Microbiol.*, 57, 526–533 (2013).
25 N. Zhang, D. Yang, D. Wang, Y. Miao, J. Shao, X. Zhou, Z. Xu, Q. Li, H. Feng, S. Li, Q. Shen and R. Zhang, Whole transcriptomic analysis of the plant-beneficial rhizobacterium *Bacillus amyloliquefaciens* SQR9 during enhanced biofilm formation regulated by maize root exudates, *BMC Genomics*, 16, 685 (2015a).
26 M. Cardarelli, D. Mariotti, M. Pomponi, L. Spano, I. Capone and P. Costantino, *Agrobacterium rhizogenes* T-DNA genes capable of inducing hairy root phenotype, *Mol. Gen. Genet.* 209, 475–480 (1987).

27 S.A. Weller, D.E. Stead, T.M. O'Neill and P.S. Morley, Root mat of tomato caused by rhizogenic strains of *Agrobacterium* biovar 1 in the UK, *Plant Pathol.*, 49, 799 (2000).
28 B.C. Heckel, A.D. Tomlinson, E.R. Morton, J.H. Choi and C. Fuqua, *Agrobacterium tumefaciens* exoR controls acid response genes and impacts exopolysaccharide synthesis, horizontal gene transfer, and virulence gene expression. *J. Bacteriol.*, 196, 3221–3233 (2014).
29 Y.A. Lee and C.W. Chan, Molecular Typing and Presence of Genetic Markers Among Strains of Banana Finger-Tip Rot Pathogen, *Burkholderia cenocepacia*, in Taiwan, *Phytopathol.*, 97, 195–201 (2007).
30 J.L. Jacobs, A.C. Fasi, A. Ramette, J.J. Smith, R. Hammerschmidt and G.W. Sundin, Identification and onion pathogenicity of *Burkholderia cepacia* complex isolates from the onion rhizosphere and onion field soil. *Appl. Environ. Microbiol.*, 74, 3121–3129 (2008).
31 E. Biondi, S. Blasioli, A. Galeone, F. Spinelli, A. Cellini, C. Lucchese and I. Braschi, Detection of potato brown rot and ring rot by electronic nose: from laboratory to real scale. *Talanta*, 129, 422–430 (2014).
32 S.M. Brumbley, L.A. Petrasovits, R.G. Birch and P.W. Taylor, Transformation and transposon mutagenesis of *Leifsonia xyli* subsp. *xyli*, causal organism of ratoon stunting disease of sugarcane. *Mol. Plant Microbe Interact.*, 15, 262–268 (2002).
33 A.K. Jha, H.P. Bais and J.M. Vivanco, *Enterococcus faecalis* mammalian virulence-related factors exhibit potent pathogenicity in the *Arabidopsis thaliana* plant model, *Infect. Immun.*, 73, 464–475 (2005).
34 F. Rezzonico and B. Duffy, The role of luxS in the fire blight pathogen *Erwinia amylovora* is limited to metabolism and does not involve quorum sensing, *Mol. Plant Microbe Interact.*, 20, 1284–1297 (2007).
35 P.E. Axelrood, M. Rella and M.N. Schroth, Role of antibiosis in competition of *Erwinia* strains in potato infection courts, *Appl. Environ. Microbiol.*, 54, 1222–1229 (1988).
36 X. Wu, Q. Zeng, B.J. Koestler, C.M. Waters, G.W. Sundin, W. Hutchins and C.H. Yang, Deciphering the components that coordinately regulate virulence factors of the soft rot pathogen *Dickeya dadantii*, *Mol. Plant Microbe Interact.*, 27, 1119–1131 (2014).
37 H.B. Lee, J.P. Hong and S.B. Kim, First Report of Leaf Blight Caused by *Pantoea agglomerans* on Rice in Korea, *Plant Dis.*, 94, 1372–1372 (2010).
38 A.M. Miller, J.E. Figueiredo, G.A. Linde, N.B. Colauto and L.D. Paccola-Meirelles, Characterization of the inaA gene and expression of ice nucleation phenotype in *Pantoea ananatis* isolates from Maize White Spot disease, *Genet. Mol. Res.*, 15 (2016).
39 J.A. Asselin, J. Lin, A.L. Perez-Quintero, I. Gentzel, D. Majerczak, S.O. Opiyo, W. Zhao, S.M. Paek, M.G. Kim, D.L. Coplin, J.J. Blakeslee and D. Mackey, Perturbation of maize phenylpropanoid metabolism by an AvrE family type III effector from *Pantoea stewartii*. *Plant Physiol.*, 167, 1117–1135 (2015).
40 A. Wensing, S. Zimmermann and K. Geider, Identification of the corn pathogen *Pantoea stewartii* by mass spectrometry of whole-cell extracts and its detection with novel PCR primers. *Appl. Environ. Microbiol.*, 76, 6248–6256 (2010).
41 S.J. Roberts, S.J. Eden-Green, P. Jones and D.J. Ambler, *Pseudomonas syzygii*, sp. nov., the Cause of Sumatra Disease of Cloves, *Syst. Appl. Microbiol.*, 13, 34–43(1990).
42 T.S. Walker, H.P. Bais, E. Deziel, H.P. Schweizer, L.G. Rahme, R. Fall and J.M. Vivanco. *Pseudomonas aeruginosa*-plant root interactions. Pathogenicity, biofilm formation, and root exudation. *Plant Physiol.* 134, 320–331 (2004).

43 A.J. Gonzalez, E. Landeras and M.C. Mendoza, Pathovars of *Pseudomonas syringae* causing bacterial brown spot and halo blight in *Phaseolus vulgaris* L. are distinguishable by ribotyping, *Appl. Environ. Microbiol.* 66, 850–854 (2000).

44 G. Surico and J.E. DeVay, Effect of syringomycin and syringotoxin produced by *Pseudomonas syringae* pv. *syringae* on structure and function of mitochondria isolated from holcus spot resistant and susceptible maize lines, *Physiol Plant Pathol.*, 21, 49–53 (1982).

45 S.P. Thapa and G. Coaker, Genome Sequences of two *Pseudomonas syringae* pv. *tomato* Race 1 Strains, Isolated from Tomato Fields in California. *Genome Announc.*, 4, e01671-15 (2016).

46 J. Tans-Kersten, H. Huang and C. Allen, *Ralstonia solanacearum* needs motility for invasive virulence on tomato, *J. Bacteriol.*, 183, 3597–3605 (2001).

47 C.G. Summers and J.J. Stapleton, Management of corn leafhopper (Homoptera: Cicadellidae) and corn stunt disease in sweet corn using reflective mulch. *J. Econ. Entomol.*, 95, 325–330 (2002).

48 A. Obradovic, A. Mavridis, K. Rudolph, J.D. Janse, M. Arsenijevic, J.B. Jones, G.V. Minsavage and J. Wang, Characterization and PCR-based Typing of *Xanthomonas campestris* pv. *vesicatoria* from Peppers and Tomatoes in Serbia, *European J. Plant Pathol.* 110, 285–292 (2004).

49 Y. Zhang, N. Jalan, X. Zhou, E. Goss, J.B. Jones, J.C. Setubal, X. Deng and N. Wang, Positive selection is the main driving force for evolution of citrus canker-causing *Xanthomonas, ISME J.*, 9, 2128–2138 (2015b).

50 R. Lopez, C. Asensio and R.L. Gilbertson, Phenotypic and Genetic Diversity in Strains of Common Blight Bacteria (*Xanthomonas campestris* pv. *phaseoli* and *X. campestris* pv. *phaseoli* var. *fuscans*) in a Secondary Center of Diversity of the Common Bean Host Suggests Multiple Introduction Events, *Phytopathol*, 96, 1204–1213 (2006).

51 J.G. Vicente and E.B. Holub, *Xanthomonas campestris* pv. *campestris* (cause of black rot of crucifers) in the genomic era is still a worldwide threat to brassica crops. *Mol. Plant Pathol.* 14, 2–18 (2013).

52 Y. Liu, J. Long, D. Shen and C. Song, *Xanthomonas oryzae* pv. *oryzae* requires H-NS-family protein XrvC to regulate virulence during rice infection. *FEMS Microbiol. Lett.*, fnw067 (2016).

53 S.M. Fry, J.S. Huang and R.D. Milholland, Isolation and preliminary characterization of extracellular proteases produced by strains of *Xylella fastidiosa* from grapevines, *Phytopathol*, 84, 357–363 (1994).

54 T. Koide, J.F. da Silva Neto, S.L. Gomes and M.V. Marques, Insertional transposon mutagenesis in the *Xylella fastidiosa Citrus* Variegated Chlorosis strain with transposome. *Curr. Microbiol.*, 48, 247–250 (2004a).

55 T. Koide, P.A. Zaini, L.M. Moreira, R.Z. Vencio, A.Y. Matsukuma, A.M. Durham, D.C. Teixeira, H. El-Dorry, P.B. Monteiro, A.C. da Silva, S. Verjovski-Almeida, A.M. da Silva and S.L. Gomes, DNA microarray-based genome comparison of a pathogenic and a nonpathogenic strain of *Xylella fastidiosa* delineates genes important for bacterial virulence, *J. Bacteriol.*, 186, 5442–5449 (2004b).

56 M.R. Guilhabert and B.C. Kirkpatrick, Identification of *Xylella fastidiosa* antivirulence genes: hemagglutinin adhesins contribute a biofilm maturation to *X. fastidios* and colonization and attenuate virulence, *Mol. Plant Microbe Interact.*, 18, 856–868 (2005).

57 R. Nascimento, H. Gouran, S. Chakraborty, H.W. Gillespie, H.O. Almeida-Souza, A. Tu, B.J. Rao, P.A. Feldstein, G. Bruening, L.R. Goulart and A.M. Dandekar, The Type II Secreted Lipase/Esterase LesA is a Key Virulence Factor Required for *Xylella fastidiosa* Pathogenesis in Grapevines, *Sci. Rep.*, 6, 18598 (2016).

58 M.S.A. Khan, I. Ahmad, M. Sajid and S.S. Cameotra, Current and Emergent Control Strategies for Medical Biofilms. *In Antibiofilm Agents: From Diagnosis to Treatment and Prevention*, 8, 117–159 (2014).

59 M. Rajalaxmi, R.B. Shafreen, P.M. Iyer, R.S. Vino, K. Balamurugan and S.K. Pandian. *In silico*, *in vitro* and *in vivo* investigations of indole-3-carboxaldehyde identified from the seawater bacterium *Marinomonas* sp. as an antibiofilm agent against *Vibrio cholera* O1. *Biofouling*, 32, 1–12 (2016).

60 S. Gowrishankar, M. Sivaranjani, A. Kamaladevi, A.V. Ravi, K. Balamurugan and S.K. Pandian, Cyclic dipeptide-cyclo (L-leucyl-L-prolyl) from marine *Bacillus amyloliquefaciens* mitigates biofilm formation and virulence in *Listeria monocytogenes*. *Pathog. Dis.*, pii: ftw017. doi: 10.1093/femspd/ftw017 (2016).

61 A. Busetti, T.P. Thompson, D. Tegazzini, J. Megaw, C.A. Maggs and B.F. Gilmore, Antibiofilm Activity of the Brown Alga *Halidrys siliquosa* against Clinically Relevant Human Pathogens, *Mar. Drugs*, 13, 3581–3605 (2015).

62 S. Gowrishankar, A. Kamaladevi, K.S. Ayyanar, K. Balamurugan and S.K. Pandian, *Bacillus amyloliquefaciens*-secreted cyclic dipeptide—cyclo(L-leucyl-L-prolyl) inhibits biofilm and virulence production in methicillin-resistant *Staphylococcus aureus*, *RSC. Adv.* 5, 95788–95804 (2015).

63 R.K. Sardar, K. Kavita and B. Jha, Lipopolysaccharide of *Marinobacter litoralis* inhibits swarming motility and biofilm formation in *Pseudomonas aeruginosa* PA01, *Carbohydr. Polym.*, 123, 468–475 (2015).

64 S. Rodrigues, C. Paillard, A. Dufour and A. Bazire, Antibiofilm Activity of the Marine Bacterium *Pseudoalteromonas* sp. 3J6 Against *Vibrio tapetis*, the Causative Agent of Brown Ring Disease, *Probiotics Antimicrob. Proteins.*, 7, 45–51 (2015).

65 G.S. Kiran, A.N. Lipton, S. Priyadharshini, K. Anitha, L.E. Suarez, M.V. Arasu, K.C. Choi, J. Selvin and N.A. Al-Dhabi, Antiadhesive activity of poly-hydroxy butyrate biopolymer from a marine *Brevibacterium casei* MSI04 against shrimp pathogenic vibrios. *Microb. Cell Fact.*, 13, 114 (2014).

66 Z.Q. Bai, X. Lin, Y. Wang, J. Wang, X. Zhou, B. Yang, J. Liu, X. Yang and Y. Liu, New phenyl derivatives from endophytic fungus *Aspergillus flavipes* AIL8 derived of mangrove plant *Acanthus ilicifolius*. *Fitoterapia*, 95, 194–202 (2014).

67 N.M. Gomes, L.J. Bessa, S. Buttachon, P.M. Costa, J. Buaruang, T. Dethoup, A.M. Silva and A. Kijjoa, Antibacterial and antibiofilm activities of tryptoquivalines and meroditerpenes isolated from the marine-derived fungi *Neosartorya paulistensis*, *N. laciniosa*, *N. tsunodae*, and the soil fungi *N. fischeri* and *N. siamensis*, *Mar. Drugs*, 12, 822–839 (2014).

68 K. Kavita, V.K. Singh, A. Mishra and B. Jha, Characterisation and anti-biofilm activity of extracellular polymeric substances from *Oceanobacillus iheyensis*. *Carbohydr. Polym.*, 101, 29–35 (2014).

69 S. Gowrishankar, B. Poornima and S.K. Pandian (2014) Inhibitory efficacy of cyclo(L-leucyl-Lprolyl) from mangrove rhizosphere bacterium-*Bacillus amyloliquefaciens* (MMS-50) toward cariogenic properties of *Streptococcus mutans*. *Res. Microbiol.*, 165, 278–289.

70 M. Scopel, W.R. Abraham, A.L. Antunes, A.T. Henriques and A.J. Macedo, Mevalonolactone: an inhibitor of *Staphylococcus epidermidis* adherence and biofilm formation, *Med. Chem.*, 10, 246–251 (2014).

71 D.H. Dusane, S.R. Damare, Y.V. Nancharaiah, N. Ramaiah, V.P. Venugopalan, A.R. Kumar and S.S. Zinjarde, Disruption of microbial biofilms by an extracellular protein isolated from epibiotic tropical marine strain of *Bacillus licheniformis*, *PLoS One*, 8, e64501 (2013).

72 B.J. Kalpana, S. Aarthy and S.K. Pandian, Antibiofilm activity of alpha-amylase from *Bacillus subtilis* S8-18 against biofilm forming human bacterial pathogens. *Appl. Biochem. Biotechnol.* 167, 1778–1794 (2012).

73 S. Gowrishankar, N.D. Mosioma and S.K. Pandian, Coral-associated bacteria as a promising antibiofilm agent against methicillin-resistant and -susceptible *Staphylococcus aureus* biofilms. *Evid. Based Complement. Alternat. Med.*, 2012, 862374 (2012).

74 D. Bakkiyaraj, C. Sivasankar and S.K. Pandian, Inhibition of quorum sensing regulated biofilm formation in *Serratia marcescens* causing nosocomial infections. *Bioorg. Med. Chem. Lett.*, 22, 3089–3094 (2012)..

75 G.L. Klein, E. Soum-Soutera, Z. Guede, A. Bazire, C. Compere and A. Dufour, The anti-biofilm activity secreted by a marine *Pseudoalteromonas* strain, *Biofouling*, 27, 931–940 (2011).

76 P. Jiang, J. Li, F. Han, G. Duan, X. Lu, Y. Gu and W. Yu, Antibiofilm activity of an exopolysaccharide from marine bacterium *Vibrio* sp. QY101, *PLoS One*, 6, e18514 (2011).

77 K.S. Musthafa, V. Saroja, S.K. Pandian, A.V. Ravi, Antipathogenic potential of marine *Bacillus* sp. SS4 on N-acyl-homoserine-lactone-mediated virulence factors production in *Pseudomonas aeruginosa* (PAO1), *J. Biosci.*, 36, 55–67 (2011).

78 P. Kanmani, R. Satish Kumar, N. Yuvaraj, K.A. Paari, V. Pattukumar and V.Arul, Production and purification of a novel exopolysaccharide from lactic acid bacterium *Streptococcus phocae* PI80 and its functional characteristics activity *in vitro*, *Bioresour. Technol.*, 102, 4827–4833 (2011).

79 C. Nithya and S.K. Pandian, The *in vitro* antibiofilm activity of selected marine bacterial culture supernatants against *Vibrio* spp. *Arch. Microbiol.*, 192, 843–854 (2010b).

80 C. Nithya, M.F. Begum and S.K. Pandian, Marine bacterial isolates inhibit biofilm formation and disrupt mature biofilms of *Pseudomonas aeruginosa* PAO1. *Appl. Microbiol. Biotechnol.*, 88, 341–358 (2010a).

81 G.S. Kiran, B. Sabarathnam and J. Selvin, Biofilm disruption potential of a glycolipid biosurfactant from marine *Brevibacterium casei*, *FEMS Immunol. Med. Microbiol.*, 59, 432–438 (2010).

82 A. Dheilly, E. Soum-Soutera, G.L. Klein, A. Bazire, C. Compere, D. Haras and A. Dufour, Antibiofilm activity of the marine bacterium *Pseudoalteromonas* sp. strain 3J6, *Appl. Environ. Microbiol.*, 76, 3452–3461 (2010).

83 D. Bakkiyaraj and S.K. Pandian, *In vitro* and *in vivo* antibiofilm activity of coral associated actinomycete against drug resistant *Staphylococcus aureus* biofilms, *Biofouling*, 26, 711–717 (2010).

84 P. Nithyanand, R. Thenmozhi, J. Rathna, S.K. Pandian, Inhibition of *Streptococcus pyogenes* biofilm formation by coral-associated actinomycetes. *Curr. Microbiol.*, 60, 454–460 (2009).

85 R. Thenmozhi, P. Nithyanand, J. Rathna and S.K. Pandian, Antibiofilm activity of coral-associated bacteria against different clinical M serotypes of *Streptococcus pyogenes, FEMS Immunol. Med. Microbiol.*, 57, 284–294 (2009).
86 D. Schillaci, V. Arizza, N. Parrinello, V. Di Stefano, S. Fanara, V. Muccilli, V. Cunsolo, J.J. Haagensen and S. Molin, Antimicrobial and antistaphylococcal biofilm activity from the sea urchin *Paracentrotus lividus, J. Appl. Microbiol.*, 108, 17–24 (2009).
87 S. Dash, C. Jin, O.O. Lee, Y. Xu and P.Y. Qian, Antibacterial and antilarval-settlement potential and metabolite profiles of novel sponge-associated marine bacteria, *J. Ind. Microbiol. Biotechnol.*, 36, 1047–1056 (2009).
88 W. Pan, M. Fan, H. Wu, C. Melander and C. Liu, A new small molecule inhibits *Streptococcus mutans* biofilms *in vitro* and *in vivo, J. Appl. Microbiol.*, 119, 1403–1411(2015).
89 S. Wu, G. Liu, D. Zhang, C. Li and C. Sun, Purification and biochemical characterization of an alkaline protease from marine bacteria *Pseudoalteromonas* sp., *J. Basic Microbiol.*, 55, 1427–1434 (2015).
90 A.R. Padmavathi, B. Abinaya and S.K. Pandian, Phenol, 2,4-bis(1,1-dimethylethyl) of marine bacterial origin inhibits quorum sensing mediated biofilm formation in the uropathogen *Serratia marcescens, Biofouling*, 30, 1111–1122 (2014).
91 B. Pejin, C. Iodice, G. Tommonaro, B. Stanimirovic, A. Ciric, J. Glamoclija, M. Nikolic, S. De Rosa and M. Sokovic, Further *in vitro* evaluation of antimicrobial activity of the marine sesquiterpene hydroquinone avarol., *Curr. Pharm. Biotechnol.*, 15, 583–588 (2014).
92 K. Glinel, P. Thebault, V. Humblot, C.M. Pradier and T. Jouenne, Antibacterial surfaces developed from bio-inspired approaches. *Acta Biomater.*, 8, 1670–1684 (2012).
93 S.R. Vijayan, P. Santhiyagu, M. Singamuthu, N. Kumari Ahila, R. Jayaraman and K. Ethiraj, Synthesis and characterization of silver and gold nanoparticles using aqueous extract of seaweed, *Turbinaria conoides*, and their antimicrofouling activity, *Scient. World J.*, 2014, 938272 (2014).
94 V. Ramalingam, R. Rajaram, C. PremKumar, P. Santhanam, P. Dhinesh, S. Vinothkumar and K. Kaleshkumar, Biosynthesis of silver nanoparticles from deep sea bacterium *Pseudomonas aeruginosa* JQ989348 for antimicrobial, antibiofilm, and cytotoxic activity, *J. Basic Microbiol.*, 54, 928–936 (2014).
95 M. Apte, D. Sambre, S. Gaikawad, S. Joshi, A. Bankar, A.R. Kumar and S. Zinjarde, Psychrotrophic yeast *Yarrowia lipolytica* NCYC 789 mediates the synthesis of antimicrobial silver nanoparticles via cell-associated melanin, *AMB Express*, 3, 32 (2013).
96 B. Nijampatnam, D.H. Nadkarni, H. Wu and S.E. Velu, Antibacterial and antibiofilm activities of makaluvamine analogs, *Microorganisms*, 2,128–139 (2014).
97 R.A. Rane, N.U. Sahu, C.P. Shah and N.K. Shah, Design, synthesis and antistaphylococcal activity of marine pyrrole alkaloid derivatives, *J. Enzyme Inhib. Med. Chem.*, 29, 401–407 (2014).
98 T.E. Ballard, J.J. Richards, A.L. Wolfe and C. Melander. Synthesis and antibiofilm activity of a second-generation reverse-amide oroidin library: a structure-activity relationship study, *Chemistry*, 14, 10745–10761(2008).
99 R.A. Rane, N.U. Sahu and C. P. Shah, Synthesis and antibiofilm activity of marine natural product-based 4-thiazolidinones derivatives, *Bioorg. Med. Chem. Lett.*, 22, 7131–7134 (2012).

24

Plant-Associated Biofilms Formed by Enteric Bacterial Pathogens and Their Significance

Meenu Maheshwari[1], Mohammad Shavez Khan[1], Iqbal Ahmad[1], Ashraf A. Khan[2], John B. Sutherland[2] and Abdullah S. Althubiani[3]

[1] *Department of Agricultural Microbiology, Faculty of Agricultural Sciences, Aligarh Muslim University, Aligarh, India*
[2] *Division of Microbiology, National Center for Toxicological Research, U.S. Food and Drug Administration, Jefferson, Arkansas, USA*
[3] *Department of Biology, Faculty of Applied Science, Umm Al-Qura University, Makkah, Saudi Arabia*

24.1 Introduction

Foodborne disease outbreaks arising from the consumption of fresh and fresh-cut produce often are due to the persistence of human pathogenic bacteria [1, 2]. A variety of bacterial pathogens have been isolated from fresh fruits and vegetables, including *Salmonella enterica*, *Listeria monocytogenes*, *Escherichia coli*, *Shigella* spp., *Campylobacter* spp., and *Yersinia* spp. [3]. Plant–microbe interactions that are currently being studied include the fitness of human bacterial pathogens on crop plants, their survival on plant surfaces, and their growth and resource utilization in the phyllosphere and rhizosphere [4–6]. The relationship of human pathogenic bacteria on plants to foodborne illness has encouraged new research on the ecology of these pathogens on plant surfaces [7].

Formerly, plants were typically considered not to support the persistence of enteric pathogens, especially *E. coli* O157:H7 and *S. enterica*, as it was thought that they would survive poorly in the harsh environmental conditions, more suited to spoilage organisms, on plant surfaces [8, 9]. The emergence of produce-borne outbreaks caused by enteric bacteria has challenged conventional ideas and evoked intensive studies on attachment and internalization of enteric pathogens on fresh produce [1, 10]. These studies indicate that contamination of fresh produce with enteric pathogens may occur frequently. Since fruits, vegetables, and leafy greens are often consumed raw, the chances of disease transmission from contaminated food sources are relatively high [11].

Foodborne illnesses and outbreaks due to enteric pathogenic bacteria have been linked to a wide variety of fresh produce, including lettuce [12–14], cilantro [15], cantaloupes [16], alfalfa sprouts [17], unpasteurized apple juice [18], unpasteurized orange juice [19], tomatoes [20], potatoes [21], melons, mangoes, and celery [8], and parsley [22]. Colonization and persistence of *S. enterica* serovar Typhimurium and *L. monocytogenes*, *L. ivanovii*, and *L. innocua* on barley roots [23], of *S. enterica* serovars on lettuce and cabbage leaves [24, 25], and of *Escherichia coli* O157:H7 on lettuce [26] have also been reported.

The ecological success of most enteric bacteria in different environments is enhanced both by aggregation and by formation of biofilms [27]. Biofilms are communities of microorganisms in which cells are attached to a surface and to each other, and are embedded in a self-composed matrix of microbial exopolysaccharide (EPS) [28]. Most of the biofilm studies of enteric bacteria have emphasized biofilms associated with infectious diseases. However, reports of foodborne illnesses caused by enteropathogenic bacteria that have colonized and persisted on vegetables, fruits, and other plants have drawn attention to biofilm formation in these situations. In this chapter, we describe the prevalence of enteric pathogens in the plant environment and the mechanisms involved in biofilm formation by enteropathogenic bacteria, primarily *E. coli* and *Salmonella* strains. We also address the relevance of enterobacterial biofilms to food safety and human health.

24.2 Enteric Pathogens in the Plant Environment

Outbreaks of foodborne illnesses due to enteric pathogens (Table 24.1) have often been associated with the consumption of contaminated fruits and vegetables [10], although occasionally the origin of the contamination remains unknown. Produce can become contaminated with bacterial pathogens in the field from soil, human excreta, contaminated irrigation water, biosolids, manure and compost, pesticides, fertilizers, dust, insects, and animals [29, 30]. It may also become contaminated by harvesting equipment, transport containers, and food handling in food-service establishments [11, 29, 31]. A schematic diagram of the various routes of contamination that may introduce enteric pathogens into fresh produce is presented in Figure 24.1.

Enteric bacteria can survive for long periods of time in manure over a broad range of temperatures [32]. Manure can serve as a source of enteric pathogens on plants when it is applied in the field to fertilize soils [33]. Poor hygiene practices by field workers and a lack of on-site sanitation facilities provided by the grower may lead to the spread of foodborne illnesses associated with enteric bacteria [34]. Wastewater used for crop irrigation and runoff from animal pastures are primary sources of contamination by enteric bacteria in the field [35]. Frequency of contamination is particularly high in arid areas where fresh water is scarce and irrigation water is provided by effluents from wastewater treatment plants, as *E. coli* and *S. enterica* can easily survive in wastewater sediments [36, 37].

Insect vectors, such as the Mediterranean fruit fly (*Ceratitis capitata*) and the common fruit fly or vinegar fly (*Drosophila melanogaster*), may transmit pathogenic *E. coli* strains to apples [38, 39]. Plant pathogenic bacteria have been shown to be disseminated by honeybees among apple and pear flowers in fruit orchards [40], and epiphytic bacteria are transmitted by insects on wet leaf surfaces [41]. The ubiquitous presence of insects on manure piles and in fruit and vegetable fields suggests that this type of transmission must be considered as a factor in the preharvest contamination of fresh produce with enteric bacteria [39].

Plants include both suitable and hostile habitats for enteric bacteria. Nutrients are abundant in the rhizosphere, which contains the remnants of lysed root cells and root exudates, including mucilage, which helps to retain water for rhizosphere microorganisms [9]. In contrast, nutrients are limited in the phyllosphere and their distribution is highly heterogeneous. However, modifications to the plant surfaces by indigenous

Table 24.1 Occurrence of Various Enteric Pathogens in Crops and Crop Products.

Plant-Associated Enteric Bacteria	Name of Crop or Product	References
E. coli O26	Clover sprouts	[47]
E. coli O157:H7	Romaine lettuce	[48]
E. coli O104	Fenugreek seeds	[49]
E. coli O145	Lettuce	[50]
E. coli O157:H7	Lettuce	[51]
E. coli O157:H7	Spinach	[52]
E. coli O157:H7	Spinach	[53]
E. coli O145	Romaine lettuce	[54]
S. enterica sv. SaintPaul	Alfalfa sprouts	[55]
S. enterica sv. Newport	Alfalfa sprouts	[56]
S. enterica sv. Bovismorbificas	Alfalfa seeds	[57]
S. enterica sv. Stanley	Alfalfa sprouts	[58]
S. enterica sv. Weltevreden	Alfalfa sprouts	[59]
S. enterica sv. Baildon	Tomato	[60]
S. enterica sv. Anatum	Basil	[61]
S. enterica sv. Senftenberg	Basil	[62]
Salmonella sv. Thompson	Cilantro	[15]
Salmonella sv. Saintpaul	Cantaloupe	[63]
Salmonella sv. Saintpaul	Jalapeño and serrano peppers	[64]
Salmonella sv. Thompson	Rocket leaves	[65]
Salmonella sv. Enteritidis, *Salmonella* sv. Typhimurium	Filipino ulam	[66]
S. Typhimurium sv. Saintpaul	Orange juice	[67]
Salmonella spp.	Fresh leafy	[68]
Salmonella spp., *Shigella* spp.	Water leaf	[69]
Shigella sonnei, entero-toxigenic *Escherichia*	Parsley	[22]
Listeria monocytogenes	Alfalfa sprouts	[70]
Helicobacter pylori	Salads	[71]

bacteria may facilitate the establishment and survival of enteric pathogens [4]. The survival of *Salmonella* strains in lettuce and cilantro leaves has been related to temperature and leaf age [5, 42]. Younger leaves are at greater risk of contamination with human pathogens; enteric pathogens can multiply on lettuce leaves and *E. coli* O157:H7 and *Salmonella* populations are consistently larger on young lettuce leaves than on the middle or older leaves from mature lettuce heads [42].

Members of the Brassicaceae, including radish, turnip, and broccoli, have a higher prevalence of *Salmonella* contamination than carrot, lettuce or tomato [43]. *L. monocytogenes* may also be found in raw salad vegetables [44]. N-acyl homoserine lactone

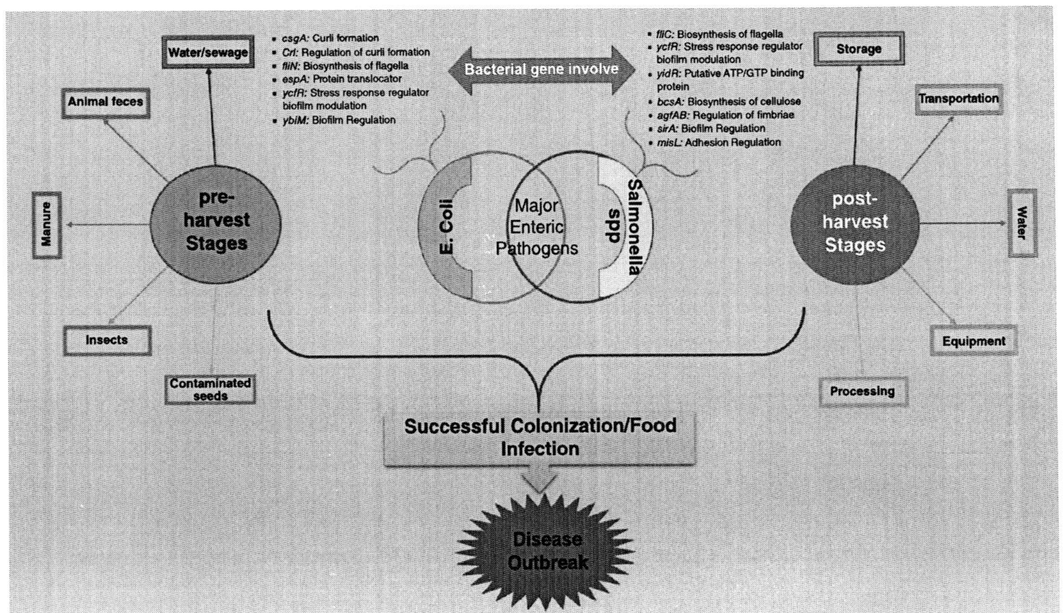

Figure 24.1 Post and pre-harvest routes of contamination of enteric bacteria in crop and crop products.

(AHL)-producing strains of the human pathogen *Enterobacter asburiae* have been isolated from lettuce leaves [45]; production of AHL by bacteria is considered a contributing factor in biofilm formation [46].

24.3 Colonization and Biofilm Formation by Enteric Bacteria on Plant Surfaces

Development of bacterial biofilms on surfaces typically involves several stages that are also likely to occur on the surfaces of plants. Attachment is the first step in the colonization and establishment of bacteria on the plant surface [35]. Initially, attachment is a weak, reversible, and nonspecific binding that depends on hydrophobic and electrostatic interactions. Later on, this binding becomes strong and irreversible [72]. Several specific virulence genes may be involved in the adhesion of enteropathogenic bacteria to plant tissues [6, 73, 74]. An investigation of the roles of flagella and type I pili in biofilm formation, using *E. coli* as a model system, showed that mutations in the genes *fliC* and *flhD* that regulate normal flagellar function impaired the formation of biofilms [75]. Flagella and pili can serve as adhesion factors that promote attachment of a cell to the surface and movement along it, although chemotaxis is not involved in biofilm formation [75, 76]. The major phase-variable outer membrane protein of *E. coli*, known as Antigen 43, is required for biofilm formation in glucose-minimal medium and may play a direct role in the interaction of the bacterial cell with a surface [77].

The long, aggregative amyloid fibers known as curli [78] play roles in the binding of diarrheagenic strains of *E. coli* to plant surfaces [79]. The diarrheagenic strains are able to bind not only to alfalfa sprouts but also to the open seed coats of tomato and *Arabidopsis thaliana* seedlings. Curli, which are required for virulence in *S. enterica*, mediate binding to, and invasion of, epithelial cells [78]. Laboratory K-12 strains of *E. coli* failed to show significant binding to plant surfaces, but they were able to bind to alfalfa sprouts after acquiring either a plasmid containing the curli biosynthesis regulator gene *mlrA* or two plasmids containing the curli synthesis genes *csgA-G* [79].

Type I pili, which are also essential for the initial attachment of *E. coli*, contain a mannose-specific adhesin, FimH, which promotes adhesion to some nonliving surfaces [75]. In many cases, attachment also involves bacterial cellulose fibers. Mutants deficient in cellulose production perturb the ability of *E. coli* O157:H7 to attach to alfalfa sprouts [80] and the ability of *S.* Typhimurium to attach to tomato fruits [81]. The gene *ycfR* encodes a membrane protein that functions in attachment of *S.* Typhimurium and *S.* Saintpaul to plants; deletion of *ycfR* in both serovars under chlorine stress led to a decrease in attachment of the cells to spinach leaves and grape tomatoes [50, 82].

Virulence genes not only cause human disease but also have a role in attachment to plant surfaces. They may be produced as common strategies used by enteropathogenic bacteria for adhesion to animal and human epithelial cells as well as to plant tissue. The HecA adhesin of the plant pathogen *Erwinia chrysanthemi* (= *Dickeya dadantii*), which contributes to its virulence, aggregation, and attachment to tobacco leaves, shares homology with the hemagglutinins of enteric animal pathogens [83].

An *E. coli* O104:H4 strain that caused an outbreak in Germany, with many cases of hemolytic uremic syndrome including 54 fatalities, in 2011 [84, 85] was found to form a more stable biofilm than *E. coli* O157:H7 by overexpressing the *aggR* and *pgaA* genes.

These genes are required for aggregation [86] and β-1, 6-*N*-acetyl-D-glucosamine exopolysaccharide (EPS) production [87], respectively [85]. Moreover, this strain possesses multiple virulence factors, and formation of biofilms *in vivo* contributes to enhanced virulence gene expression [85]. The *E. coli* strains isolated from plants form significantly more biofilms, extracellular matrix, cellulose, and curli than *E. coli* strains isolated from humans and other animals [88].

Producing an extracellular matrix helps with maintaining the structure of a biofilm by providing mechanical stability [89]. The matrix, required for the interactions of cells with the surface and with each other to develop the complex architecture of the biofilm, is composed of polysaccharides, proteins, nucleic acids, and other components [89, 90]. Synthesis of the exopolysaccharide colanic acid in *E. coli* [77] is induced upon attachment of the bacteria to a surface. Colanic acid is required not for initial attachment to an abiotic surface but rather for establishing the three-dimensional structure of biofilms. *E. coli* K-12 strains defective in colanic acid production form biofilms of reduced thickness [77]. The extracellular matrix of mature biofilms acts as a shield against environmental stresses, antimicrobial agents and plant defenses [28]. When mucoid strains of *E. coli, Acinetobacter calcoaceticus,* and the plant pathogen *Erwinia stewartii* (= *Pantoea stewartii*) were compared to nonmucoid variants in their resistance to desiccation, there was better survival of the EPS-producing mucoid strains under conditions of dehydration [91]. Microcolonies that form in the rhizospheres of plants form biofilms with increased cell–cell contact and conjugation [90]. They develop into mature biofilms with complex three-dimensional structures [92].

Enteropathogenic bacteria may adhere and form biofilms on plant surfaces as a mechanism for survival [9, 93]. On parsley plants, resistance to chlorination treatments was greater in biofilm-producing *S. typhimurium* than in a biofilm-deficient isogenic mutant, when washing was conducted a week after inoculation. However, no differences were observed when the treatment was carried out a few hours after inoculation [93].

The final phase of biofilm development is development of a complex structure. Detachment and dispersal of bacterial populations in search of new niches take place during this step [9].

24.4 Biofilm Regulation in Enteric Bacteria

Biofilm-grown cells show greater cell densities but are exposed to increased osmolarity and decreased levels of oxygen [94]. Environmental signals, surface properties of the bacteria and plant tissue, and quorum sensing regulate different stages of biofilm development [94–97].

A regulatory pathway affecting biofilm formation was found during the analysis of *E. coli* K-12 mutants [98]. Expression of the *csgA* gene, encoding curli synthesis, was regulated by EnvZ–OmpR, a two-component regulatory system important for transcriptional regulation of the *ompC* and *ompF* porin genes in response to differences in external osmolarity [99]. The *csgA* gene was overexpressed in the presence of the *ompR* gene, resulting in increased production of curli, which are responsible for adherence [78]. Strains of *E. coli* that form biofilms lose their capability for adherence with knockout mutations affecting either *ompR* or *csgA* [98].

In *E. coli*, the autoinducer AI-2 plays an important role in regulating the formation of biofilms by enhancing activity of the flagella [96]. Quorum-sensing regulatory genes that play important roles in cell-to-cell signaling, like *pepT*, are highly expressed in the biofilm mode of growth as compared to the planktonic mode of growth and are highly induced in the late exponential growth phase [94]. The putative stress regulatory genes *ycfR* (a surface membrane protein), *sirA* (a secretion protein) and *yigG* (probably an inner membrane protein) are expressed in *S.* Typhimurium under chlorine stress [50, 100].

Environmental fluctuations affect the regulation of gene expression during biofilm formation. A substantial fraction of genes in *E. coli* (~38%) are differentially expressed in biofilms [94]. Analysis of gene expression in *E. coli* following pilus-mediated adhesion to a surface revealed the transcriptional activation of a sensor-regulator protein in the iron-starvation response. Pilus attachment mediates expression of a protein (BarA, also referred to as AirS) that is essential for the acquisition of iron [101].

24.5 Influence of Plant Defense on Survival and Biofilm Formation by Enteropathogens

In response to bacteria, plants react in various ways by activating mechanisms for protection against pathogens. The plant releases phytochemicals and defense enzymes, which may be local or systemic. Some have broad-range antimicrobial activity and others appear only in the presence of a pathogen [9]. It is evident that plants respond to enteric bacteria in ways similar to how they respond to some plant pathogenic bacteria [9]. For instance, plants can detect pathogen-associated molecular patterns (PAMPs), such as bacterial lipopolysaccharides, flagellin, or glycoproteins, on the surfaces of microbial cells. This interaction elicits signaling cascades that lead to the transcriptional activation of immune response genes [9]. Two possible responses to PAMPs by the plant are releasing reactive oxygen species and strengthening the plant cell walls with callose [9]. Many cell surface structures of *E. coli* and *Salmonella* strains, such as flagella, curli, pili, fimbriae, and lipopolysaccharides, are recognized like the PAMPs of bacterial plant pathogens [9]. Mutants of enteric bacteria that are deficient in flagella, curli, or other surface structures, may show even better survival or colonization on plants than wild-type strains [102]. *Salmonella* mutants missing either flagella or part of the type III secretion system, which secretes effector proteins into eukaryotic cells, may colonize alfalfa plants in greater numbers than the wild type [102, 103]. The lipopolysaccharide of *Salmonella* is also reported to act as an elicitor in tobacco plants [104]. These bacterial surface molecules are also components of the biofilm matrix.

In some cases, biofilm formation by enteropathogens may mask the underlying bacterial surface, providing protection against plant defense responses. An *E. coli* O157:H7 mutant that produces a great amount of exopolysaccharides and a thick capsule exhibits a better survival pattern on *Arabidopsis* than the wild type [102]. Bacteria utilize adaptive strategies to improve their fitness in the plant habitat. Pathogen effector proteins may be produced to counteract PAMP-triggered immunity [105]. Most Gram-negative human pathogens have a type III secretion system that enables the bacteria to release the effector proteins into the plant cells, suppressing the PAMP defense responses [104]. *Arabidopsis thaliana* reacts to *Salmonella* by inducing the activation of mitogen-activated protein kinase (MAPK) cascades and enhancing the expression of pathogenesis-related genes [106]. However, *Salmonella* overcomes plant defense mechanisms and enters and proliferates inside various *Arabidopsis* tissues, causing wilting, chlorosis and eventually death of the infected organs [106]. In contrast, other bacterial endophytes colonize the interior of plants without causing disease or forming symbiotic structures. *Klebsiella pneumoniae* 342 enhances plant growth and nutrition, fixes N_2 and increases maize yield in the field [107, 108].

Plants are also protected by other defense responses, such as systemic acquired resistance [109] and induced systemic resistance, which is induced during the colonization of roots by rhizosphere bacteria and activated by salicylic acid, jasmonic acid, and ethylene [110]. Some of these compounds have a role in preventing not only the persistence of enteric pathogens in plants but also biofilm formation. For instance, ethylene and salicylic acid produced by alfalfa decrease the endophytic colonization of *Salmonella* strains [103]. In tomatoes, jasmonic acid and its precursors, produced as plant responses to bacteria, strongly reduce the expression of a gene required for bacterial capsule synthesis, *yihT*, and thus inhibit biofilm formation [111].

Many plant-produced compounds have broad-spectrum antimicrobial activity and inhibit biofilm formation at low concentrations. The essential oil from *Mentha piperita* (peppermint) and its menthol component inhibit quorum sensing and biofilm formation by Gram-negative bacteria at sub-MIC concentrations [112]. Similarly, lemon essential oil inhibits biofilm formation by *E. coli* and mixed cultures [113]. The formation of *E. coli* O157:H7 biofilms is inhibited by β-sitosterol glucoside from citrus peel, apparently by repressing regulatory genes for the activity of flagella [9, 114]. *Carex* (sedge) extracts inhibit *E. coli* O157:H7 and *Pseudomonas aeruginosa* biofilm formation; one of the active anti-biofilm compounds in these extracts has been identified as ε-viniferin [115].

24.6 Plant-Associated Enteric Bacteria in Food Safety and Human Health

Pathogenic enteric bacteria acquired from food are a serious threat to public health. In 2012, the incidence of these infections in the USA per 100,000 population for each enteropathogen was *Salmonella* (16.42), *Campylobacter* (14.30), *Shigella* (4.50), Shiga toxin-producing *E. coli* non-O157 (1.16), Shiga toxin-producing *E. coli* O157 (1.12), *Yersinia* sp. (0.33), and *Listeria* sp. (0.25) [116].

The attachment and internalization of pathogenic enteric bacteria on plants [24, 50, 73] have caused many disease outbreaks due to consumption of fresh produce. For example, in 2006, an outbreak of *E. coli* O157:H7 across several US states was reported as arising from contaminated fresh spinach [117]. An international outbreak of Shiga toxin-producing *Escherichia coli* O157 infection in 2007 was due to contaminated lettuce that had been shredded and packed in a Dutch food processing plant [51]. In 2005–2006, more than 450 cases of salmonellosis associated with contaminated tomatoes were observed in Canada and the United States in four separate outbreaks [118]. Similarly, in 2007, a multinational outbreak of *S. enterica* serovar Paratyphi associated with spinach and vegetable salad was reported in northern Europe [119]. The

2011 outbreak in Germany of *E. coli* O104:H4, involving cases of hemolytic uremic syndrome [84], was traced to seeds of fenugreek imported from Egypt [120].

Invasion of human enteric pathogens into roots could lead to systemic infection of the plant and contamination of seeds and fruits [121], which would allow transmission of the pathogenic bacteria to animals and humans. The formation of biofilms on plant surfaces increases the risk of product contamination with *E. coli* O157:H7, *L. monocytogenes* and *Salmonella* spp. [9]. Enteric bacteria may persist for a long time in soil and on plant surfaces [35, 122]. For example, when *Salmonella* strains were inoculated into stems of tomato plants by injection at flowering and early stages of fruit development, their survival could be observed through fruit ripening [123]. The bacterial contamination of fresh produce is a major issue in food-processing environments and increases the risk of foodborne illness [124, 125], so various means of control are still being developed [126–128].

The attachment and persistence of pathogenic microorganisms on food-contact surfaces is a matter of concern for both human health and food safety. Improper handling of vegetables before and during processing may lead to transmission of enteropathogenic microorganisms, so it is necessary to develop proper handling, cleaning, and disinfection methods and control systems in food-processing environments. Whereas washing vegetables is widely reported to reduce risks of microbial contamination and to be important for food safety [129], some studies have found that washing of fresh produce with antimicrobial solutions, such as chlorine, fails to significantly reduce the load of attached pathogens [127, 130, 131]. Enteric pathogens are less susceptible to chlorine treatment than the indigenous microorganisms, suggesting that after produce is sanitized, these pathogens may be able to survive and reestablish on the fresh produce with less competition [128].

24.7 Conclusions

Because biofilms play a crucial role in the persistence and long-term survival of pathogenic enteric bacteria on fruits and vegetables and provide them with greater resistance to disinfection processes, a better understanding of the mechanisms involved in colonization and biofilm formation by enteric bacteria on plants should facilitate the development of new strategies for inhibiting biofilm formation, thus reducing disease outbreaks and economic losses.

References

1 U.M. Abdul-Raouf, L.R. Beuchat and M.S. Ammar, Survival and growth of *Escherichia coli* O157:H7 on salad vegetables, *Appl. Environ. Microbiol.*, 59, 1999–2006 (1993).
2 J. Zheng, S. Allard, S. Reynolds, P. Millner, A. Gabriella, R.J. Blodgett and E.W. Brown, Colonization and internalization of *Salmonella enterica* in tomato plants, *Appl. Environ. Microbiol.*, 79, 2494–2502 (2013).
3 L.R. Beuchat, Pathogenic microorganisms associated with fresh produce, *J. Food. Protect.*, 59, 204–216 (1996).

4 J.H.J., Leveau and S.E. Lindow, Appetite of an epiphyte: quantitative monitoring of bacterial sugar consumption in the phyllosphere, *Proc. Natl. Acad. Sci. USA*, 98, 3446–3453 (2001).

5 M.T. Brandl and R.E. Mandrell, Fitness of *Salmonella enterica* serovar Thompson in the cilantro phyllosphere, *Appl. Environ. Microbiol.*, 68, 3614–3621 (2002).

6 J.D. Barak, L. Gorski, P. Naraghi-Arani and A.O. Charkowski, *Salmonella enterica* virulence genes are required for bacterial attachment to plant tissue, *Appl. Environ. Microbiol.* 71, 5685–5691 (2005).

7 M.M. Klerks, E. Franz, M. van Gent-Pelzer, C. Zijlstra, A.H. van Bruggen, Differential interaction of *Salmonella enterica* serovars with lettuce cultivars and plant–microbe factors influencing the colonization efficiency, *ISME J.*, 1, 620–631 (2007).

8 S.L. Burnett and L.R. Beuchat, Human pathogens associated with raw produce and unpasteurized juices, and difficulties in decontamination, *J. Ind. Microbiol. Biotechnol.*, 25, 281–287 (2000).

9 S. Yaron and U. Römling, Biofilm formation by enteric pathogens and its role in plant colonization and persistence, *Microb. Biotechnol.*, 7, 496–516 (2014).

10 L.J. Harris, J.N. Farber, L.R. Beuchat, M.E. Parish, T.V. Suslow, E.H. Garrett and F.F. Busta, Outbreaks associated with fresh produce: incidence, growth, and survival of pathogens in fresh and fresh-cut produce, *Comp. Rev. Food Sci. Food Saf.*, 2 (Suppl.), 78–141 (2003).

11 M.T. Brandl, Fitness of human enteric pathogens on plants and implications for food safety, *Annu. Rev. Phytopathol.*, 44, 367–392 (2006).

12 M.L. Ackers, B.E. Mahon, E. Leahy, B. Goode, T. Damrow, P.S. Hayes, W.F. Bibb, D.H. Rice, T.J. Barrett, L. Hutwagner and P.M. Griffin, An outbreak of *Escherichia coli* O157:H7 infections associated with leaf lettuce consumption, *J. Infect. Dis.*, 177, 1588–1593 (1998).

13 E.D. Hilborn, J.H. Mermin, P.A. Mshar, J.L. Hadler, A. Voetsch, C. Wojtkunski, M. Swartz, R. Mshar, M.A. Lambert-Fair, J.A. Farrar and M.K. Glynn, A multistate outbreak of Escherichia coli O157: H7 infections associated with consumption of mesclun lettuce, *Arch. Intern. Med.*, 159, 1758–1764 (1999).

14 B.F. Buss, M.V. Joshi, A.L. O'Keefe, C.D. Allensworth, A. Garvey, K. Obbink, S. Mandernach and T.J. Safranek, Regional investigation of a cyclosporiasis outbreak linked to imported Romaine lettuce – Nebraska and Iowa, June–August 2013, *Epidemiol. Infect.*, 22, 1–11 (2015).

15 J.V. Campbell, J. Mohle-Boetani, R. Reporter, S. Abbott, J. Farrar, M. Brandl, R. Mandrell and S.B. Werner, An outbreak of Salmonella serotype Thompson associated with fresh cilantro, *J. Infect. Dis.*, 183(6), 984–987 (2001).

16 J.C. Mohle-Boetani, R. Reporter, S.B. Werner, S. Abbott, J. Farrar, S.H. Waterman and D.J. Vugia, An outbreak of *Salmonella* serogroup Saphra due to cantaloupes from Mexico, *J. Infect. Dis.*, 180, 1361–1364 (1999).

17 J.C. Mohle-Boetani, J.A. Farrar, S.B. Werner, D. Minassian, R. Bryant, S. Abbott, L. Slutsker and D.J Vugia, *Escherichia coli* O157 and *Salmonella* infections associated with sprouts in California, 1996-1998, *Ann. Intern. Med.*, 135, 239–247 (2001).

18 S.H. Cody, M.K. Glynn, J.A. Farrar, K.L. Cairns, P.M. Griffin, J. Kobayashi, M. Fyfe, R. Hoffman, A.S. King, J.H. Lewis and B. Swaminathan, An outbreak of *Escherichia coli* O157:H7 infection from unpasteurized commercial apple juice, *Ann. Intern. Med.*, 130, 202–209 (1999).

19 K.A. Cook, T.E. Dobbs, W.G. Hlady, J.G. Wells, T.J. Barrett, N.D. Puhr, G.A. Lancette, D.W. Bodager, B.L. Toth, C.A. Genese, A.K. Highsmith, K.E. Pilot, L. Finelli and D.L. Swerdlow, Outbreak of *Salmonella* serotype Hartford infections associated with unpasteurized orange juice, *Jama*, 280, 1504–1509 (1998).

20 S.D. Bennett, K.W. Littrell, T.A. Hill, M. Mahovic and C.B. Behravesh, Multistate foodborne disease outbreaks associated with raw tomatoes, United States, 1990–2010: a recurring public health problem, *Epidemiol. Infect.*, 143, 1352–1359 (2015).

21 N. Launders, M.E. Locking, M. Hanson, G. Willshaw, A. Charlett, R. Salmon, J. Cowden and G.K. Adak, A large Great Britain-wide outbreak of STEC O157 phage type 8 linked to handling of raw leeks and potatoes, *Epidemiol. Infect.*, 144, 171–181 (2016).

22 T.S. Naimi, J.H. Wicklund, S.J. Olsen, G. Krause, J.G. Wells, J.M. Bartkus, D.J. Boxrud, M. Sullivan, H. Kassenborg, J.M. Besser, E.D. Mintz, M.T. Osterholm and C.W. Hedberg, Concurrent outbreaks of *Shigella sonnei* and enterotoxigenic *Escherichia coli* infections associated with parsley: implications for surveillance and control of foodborne illness, *J. Food Prot.*, 66, 535–541 (2003).

23 S. Kutter, A. Hartmann, and M. Schmid, Colonization of barley *(Hordeum vulgare)* with *Salmonella enterica* and *Listeria* spp, *FEMS Microbiol. Ecol.*, 56, 262–271 (2006).

24 J. Patel and M. Sharma, Differences in attachment of *Salmonella enterica* serovars to cabbage and lettuce leaves, *Int. J. Food Microbiol.*, 139, 41–47 (2010).

25 J. Ponniah, T. Robin, S. Radu, Y.K. Cheah, F.M.Ghazali, M. Nishibuchi, Y. Nakaguchi and P. Malakar, Characterisation of *L. monocytogenes* strains isolated from salad vegetables, *Int J Food Safety, Nutrition and Public Health*, 5 (2) (2015).

26 Z. Hou, R.C. Fink, M. Sugawara, F. Diez-Gonzalez and M.J. Sadowsky, Transcriptional and functional responses of *Escherichia coli* O157:H7 growing in the lettuce rhizoplane, *Food Microbiol.*, 35, 136–142 (2013).

27 L. Hall–Stoodley, J.W. Costerton and P. Stoodley, Bacterial biofilms: from the natural environment to infectious diseases, *Nat. Rev. Microbiol.*, 2, 95–108 (2004).

28 J.W. Costerton, Z. Lewandowski, D.E. Caldwell, D.R. Korber and H.M. Lappin-Scott, Microbial biofilms. In: *Annual Review of Microbiology*, L.N. Ornston, A. Ballows and E.P. Greenberg (Eds.), Palo Alto, CA: *Annu. Rev. Inc.*, 711–745 (1995).

29 R. Carr, Excreta-related infections and the role of sanitation in the control of transmission, *in Water Quality: Guidelines, Standards and Health*, L. Fewtrell and J. Bartram (eds.), World Health Organization (WHO), IWA Publishing, London, UK (2001).

30 Y. Jung, H. Jang and K.R. Matthews, Effect of the food production chain from farm practices to vegetable processing on outbreak incidence, *Microb. Biotechnol.*, 7, 517–527 (2014).

31 R.C. Beier, S.D. Pillai, T.D. Philips and R.L. Ziprin (eds.), *Preharvest and Postharvest Food Safety: Contemporary Issues and Future Directions*, Ames, IA: Blackwell (2004).

32 I.T. Kudva, K. Blanch and C.J. Hovde, Analysis of *Escherichia coli* O157:H7 survival in ovine or bovine manure and manure slurry, *Appl. Environ. Microbiol.*, 64, 3166–3174 (1998).

33 S. Himathongkham, S. Bahari, H. Riemann and D. Cliver, Survival of *Escherichia coli* O157:H7 and *Salmonella* Typhimurium in cow manure and cow manure slurry, *FEMS Microbiol. Lett.*, 178, 251–257 (1999).

34 M.L. Hutchison, L.D. Walters, A. Moore, K.M. Crookes and S.M. Avery, Effect of length of time before incorporation on survival of pathogenic bacteria present in livestock wastes applied to agricultural soil, *Appl. Environ. Microbiol.*, 70, 5111–5118 (2004).

35 B.M. Martínez-Vaz, R.C. Fink, F. Diez-Gonzalez, M.J. Sadowsky, Enteric pathogen-plant interactions: molecular connections leading to colonization and growth and implications for food safety. *Microbes Environ.*, 29, 123–135 (2014).
36 C.W. Hendricks, Increased recovery rate of salmonellae from stream bottom sediments versus surface waters, *Appl. Microbiol.*, 21, 379–380 (1971).
37 C.P. Gerba and J.S. McLeod, Effect of sediments on the survival of *Escherichia coli* in marine waters, *Appl. Environ. Microbiol.*, 32, 114–120 (1976).
38 W.J. Janisiewicz, W.S. Conway, M.W. Brown, G.M. Sapers, P. Fratamico, R.L. Buchanan, Fate of *Escherichia coli* O157:H7 on fresh-cut apple tissue and its potential for transmission by fruit flies, *Appl. Environ. Microbiol.*, 65, 1–5 (1999).
39 T. Sasaki, M. Kobayashi and N. Agui, Epidemiological potential of excretion and regurgitation by *Musca domestica* (Diptera: Muscidae) in the dissemination of *Escherichia coli* O157:H7 to food, *J. Med. Entomol.*, 37, 945–949 (2000).
40 K.B. Johnson, V.O. Stockwell, T.L. Sawyer and D. Sugar, Assessment of environmental factors influencing growth and spread of *Pantoea agglomerans* on and among blossoms of pear and apple, *Phytopathology*, 90, 1285–1294 (2000).
41 S.S. Hirano and C.D. Upper, Bacteria in the leaf ecosystem with emphasis on *Pseudomonas syringae*—a pathogen, ice nucleus, and epiphyte, *Microbiol. Mol. Biol. Rev.*, 64, 624–653 (2000).
42 M.T. Brandl and R. Amundson, Leaf age as a risk factor in contamination of lettuce with *Escherichia coli* O157:H7 and *Salmonella enterica*, *Appl. Environ. Microbiol.*, 74, 2298–2306 (2008).
43 J.D. Barak, A. Liang and K.E. Narm, Differential attachment to and subsequent contamination of agricultural crops by *Salmonella enterica*, *Appl. Environ. Microbiol.*, 74, 5568–5570 (2008).
44 J. Ponniah, T. Robin, M.S. Paie, S. Radu, F.M. Ghazali, C.Y. Kqueen, M. Nishibuchi, Y. Nakaguchi and P.K. Malakar, *Listeria monocytogenes* in raw salad vegetables sold at retail level in Malaysia, *Food Control*, 21, 774–778 (2010).
45 Y.Y. Lau, J. Sulaiman, J.W. Chen, W.F. Yin and K.G. Chan, Quorum sensing activity of *Enterobacter asburiae* isolated from lettuce leaves, *Sensors*, 13, 14189–14199 (2013).
46 Y.V. Zaitseva, A.A. Popova and I.A. Khmel, Quorum sensing regulation in bacteria of the family Enterobacteriaceae, *Russ. J. Genet.*, 50, 323–340 (2014).
47 CDC, Multistate outbreak of shiga toxin-producing *Escherichia coli* 026 infections linked to raw clover sprouts at Jimmy John's restaurants (Final Update), URL http://www.cdc.gov/ecoli/2012/o26-02-12/index.html, (2012).
48 R.B. Slayton, G. Turabelidze, S.D. Bennett, C.A. Schwensohn, A.Q. Yaffee, F. Khan, C. Butler, E. Trees, T.L. Ayers, M.L. Davis and A.S. Laufer, Outbreak of Shiga toxin-producing *Escherichia coli* (STEC) O157: H7 associated with romaine lettuce consumption, 2011, *PLoS One*, 8(2), 5300 (2013).
49 G. F Gault, X. Weill, P. Mariani-Kurkdjian, N. Jourdan-da Silva, L. King, B. Aldabe, M. Charron et al, Outbreak of haemolytic uraemic syndrome and bloody diarrhoea due to *Escherichia coli* O104: H4, south-west France, June 2011, *Euro. Surveill.*, 16, 26 (2011).
50 J.K. Salazar, K. Deng, M.L. Tortorello, M.T. Brandl, H. Wang and W. Zhang, Genes *ycfR*, *sirA* and *yigG* contribute to the surface attachment of *Salmonella enterica* Typhimurium and Saintpaul to fresh produce, *PLoS ONE* 8, 57272 (2013).
51 I. Friesema, G. Sigmundsdottir, K. van der Zwaluw, A. Heuvelink, B. Schimmer, C. de Jager, B. Rump, H. Briem, H. Hardardottir, A. Atladottir, E. Gudmundsdottir and W. van

Pelt, An international outbreak of Shiga toxin-producing *Escherichia coli* O157 infection due to lettuce, September – October 2007, *Euro Surveill.*, 13 (50), 3029–3035 (2008).
52 CDC, Update on multi-state outbreak of *E. coli* O157: H7 infections from fresh spinach, October 6, 2006, *Morb. Mortal. Wkly. Rep.*, 55, 1045–1046 (2006).
53 A.M. Wendel, D.H. Johnson, U. Sharapov, J. Grant, J.R. Archer, T. Monson, C. Koschmann and J.P. Davis, Multistate outbreak of *Escherichia coli* O157: H7 infection associated with consumption of packaged spinach, August–September 2006: the Wisconsin investigation, *Clin. Infect. Dis.*, 48(8), 1079–1086 (2009).
54 E.V. Taylor, T.A. Nguyen, K.D. Machesky, E. Koch, M.J. Sotir, S.R. Bohm, J.P. Folster, R. Bokanyi, A. Kupper, S.A. Bidol and A. Emanuel, Multistate outbreak of *Escherichia coli* O145 infections associated with romaine lettuce consumption, 2010, *J. Food Protect.*, 76(6), 939–944 (2013).
55 CDC, Outbreak of *Salmonella* serotype saintpaul infections associated with eating alfalfa sprouts—United States, 2009, *Morb. Mortal. Wkly. Rep.*, 58, 500–503 (2009).
56 CDC, Multistate Outbreak of Human *Salmonella* Newport Infections Linked to Raw Alfalfa Sprouts (Final Update) URL http://www.cdc.gov/salmonella/2010/newport-alfalfa-sprout-6-29-10.html, (2010).
57 R. Rimhanen-Finne, T. Niskanen, T. Lienemann, T. Johansson, M. Sjöman, T. Korhonen, S. Guedes, H. Kuronen, M.J. Virtanen, J. Mäkinen and J. Jokinen, A nationwide outbreak of *Salmonella Bovismorbificans* associated with sprouted alfalfa seeds in Finland, 2009, *Zoonoses Public Health*, 58(8), 589–596 (2011).
58 S. Werner, K. Boman, I. Einemo, M. Erntell, B. De Jong, A. Lindqvist, M. Löfdahl, S. Löfdahl, A. Meeuwisse, G. Ohlen, and M. Olsson, Outbreak of *Salmonella Stanley* in Sweden associated with alfalfa sprouts, July–August 2007, *Euro. Surveill.*, 12(10), 3291(2007).
59 K.E. Emberland, S. Ethelberg, M. Kuusi, L. Vold, L. Jensvoll, B.A. Lindstedt, K. Nygård, C. Kjelsø, M. Torpdahl, G. Sørensen and T. Jensen, Outbreak of *Salmonella weltevreden* infections in Norway, *Euro. Surveill.*, 12(11), 4 (2007).
60 K. Cummings, E. Barrett, J.C. Mohle-Boetani, J.T. Brooks, J. Farrar, T. Hunt, A. Fiore, Komatsu, K., S.B. Werner and L. Slutsker, A multistate outbreak of *Salmonella enterica* serotype Baildon associated with domestic raw tomatoes, *Emerg. Infect. Dis.*, 7(6), 1046 (2001).
61 J. Pakalniskiene, G. Falkenhorst, M. Lisby, S.B. Madsen, K.E. Olsen, E.M. Nielsen, A. Mygh, J. Boel and K. Mølbak, A foodborne outbreak of enterotoxigenic *E. coli* and *Salmonella anatum* infection after a high-school dinner in Denmark, November 2006, *Epidemiol. Infect.*, 137, 396–401 (2009).
62 L. Pezzoli, R. Elson, C.L. Little, H. Yip, I. Fisher, R. Yishai, et al., Packed with *Salmonella* – investigation of an international outbreak of *Salmonella senftenberg* infection linked to contamination of prepacked basil in 2007, *Foodborne Pathog. Dis.*, 5, 661–668 (2008).
63 S.A. Munnoch, K. Ward, S. Sheridan, G.J. Fitzsimmons, C.T. Shadbolt, J.P. Piispanen, Q. Wang, T.J. Ward, T.L. Worgan, C. Oxenford, J.A. Musto, J. McAnulty, D.N. Durrheim, A multi-state outbreak of *Salmonella Saintpaul* in Australia associated with cantaloupe consumption, *Epidemiol. Infect.*, 137, 367–374 (2009).
64 CDC, Multistate outbreak of *Salmonella saintpaul* infections linked to raw produce (final update), URL, http://www.cdc.gov/salmonella/2008/raw-produce-8-28-2008.html, (2008).

65 K. Nygard, J. Lassen, L. Vold, Y. Andersson, I. Fisher, S. Lofdahl, J. Threlfall, I. Luzzi, T. Peters, M. Hampton, M. Torpdahl, G. Kapperud and P. Aavitsland, Outbreak of *Salmonella thompson* infections linked to imported rucola lettuce, *Foodborne Pathog. Dis.*, 5, 165–173 (2008).

66 M.S. Najwa, Y. Rukayadi, A. Ubong, Y.Y. Loo, W.S. Chang, Y.L. Lye, T.Y. Thung, S.A. Aimi, T.T.H. Malcolm, S.G. Goh and C.H. Kuan, Quantification and antibiotic susceptibility of Salmonella spp., *Salmonella enteritidis* and *Salmonella typhimurium* in raw vegetables (ulam), *Int. Food Res. J.*, 22(5) (2015).

67 S. Jain, S.A. Bidol, J.L. Austin, E. Berl, F. Elson, M.L. Williams, M. Deasy, M.E. Moll, V. Rea, J.D. Vojdani and A.Y. Patricia, Multistate outbreak of *Salmonella typhimurium* and Saintpaul infections associated with unpasteurized orange juice—United States, 2005, *Clin. Infect. Dis.*, 48(8), 1065–1071 (2009).

68 M.N. Losio, E. Pavoni, S. Bilei, B. Bertasi, D. Bove, F. Capuano, S. Farneti, G. Blasi, D. Comin, C. Cardamone and L. Decastelli, Microbiological survey of raw and ready-to-eat leafy green vegetables marketed in Italy, *Int. J. Food Microbiol.*, 210, 88–91 (2015).

69 J.A. Lennox, M. Eja, U. Edeghor, C. Okoro and E.C. Okpako, Incidence of *Salmonella* and *Shigella* species on some selected fruits and vegetables obtained from open area markets in Calabar Metropolis, *Int. J. Curr. Microbiol. App. Sci.*, 4(5), 262–268 (2015).

70 L. Gorski, J.D. Palumbo and K.D. Nguyen, Strain-specific differences in the attachment of *Listeria monocytogenes* to alfalfa sprouts, *J. Food Protect.*, 67(11), 2488–2495 (2004).

71 S. Atapoor, F.S. Dehkordi and E. Rahimi, Detection of *Helicobacter pylori* in various types of vegetables and salads, *Jundishapur J. Microb.*, 7(5) (2014).

72 W.M. Dunne, Bacterial adhesion: seen any good biofilms lately? *Clin. Microbiol. Rev.*, 15, 155–166 (2002).

73 K. Takeuchi, C.M. Matute, A.N. Hassan and J.F. Frank, Comparison of the attachment of *Escherichia coli* O157:H7, *Listeria monocytogenes*, *Salmonella Typhimurium*, and *Pseudomonas fluorescens* to lettuce leaves, *J. Food Prot.*, 63, 1433–1437 (2000).

74 J.D. Barak, L.C. Whitehand and A.O. Charkowski, Differences in attachment of *Salmonella enterica* serovars and *Escherichia coli* O157:H7 to alfalfa sprouts, *Appl. Environ. Microbiol.*, 68, 4758–4763 (2002).

75 L.A. Pratt and R. Kolter, Genetic analysis of *Escherichia coli* biofilm formation: roles of flagella, motility, chemotaxis and type I pili, *Mol. Microbiol.*, 30, 285–293 (1998).

76 P.C. Bogino, M.D.L.M. Oliva, F.G. Sorroche and W. Giordano, The role of bacterial biofilms and surface components in plant-bacterial associations, *Int. J. Mol. Sci.*, 14, 15838–15859 (2013).

77 P.N. Danese, L.A. Pratt, S.L. Dove and R. Kolter, The outer membrane protein, Antigen 43, mediates cell-to-cell interactions within *Escherichia coli* biofilms, *Mol. Microbiol.*, 37, 424–432 (2000).

78 M.M. Barnhart and M.R. Chapman, Curli biogenesis and function, *Annu. Rev. Microbiol.*, 60, 131–147 (2006).

79 C. Jeter and A.G. Matthysse, Characterization of the binding of diarrheagenic strains of *E. coli* to plant surfaces and the role of curli in the interaction of the bacteria with alfalfa sprouts, *Mol. Plant Microbe. In.*, 18, 1235–1242 (2005).

80 A.G. Matthysse, R. Deora, M. Mishra and A.G. Torres, Polysaccharides cellulose, poly-β-1, 6-*N*-acetyl-D-glucosamine, and colanic acid are required for optimal binding of *Escherichia coli* O157:H7 strains to alfalfa sprouts and K–12 strains to plastic but not for binding to epithelial cells, *Appl. Environ. Microbiol.*, 74, 2384–2390 (2008).

81 R.K. Shaw, I. Lasa, B.M. Garcia, M.J. Pallen, J.C. Hinton, C.N. Berger and G. Frankel, Cellulose mediates attachment of *Salmonella enterica* serovar Typhimurium to tomatoes, *Environ. Microbiol. Rep.*, 3, 569–573 (2011).

82 K. Deng, S. Wang, X. Rui, W. Zhang and M.L. Tortorello, Functional analysis of ycfR and ycfQ in *Escherichia coli* O157: H7 linked to outbreaks of illness associated with fresh produce, *Appl. Environ. Microbiol.*, 77, 3952–3959 (2011).

83 C.M. Rojas, J.H. Ham, W.L. Deng, J.J. Doyle and A. Collmer, HecA, a member of a class of adhesins produced by diverse pathogenic bacteria, contributes to the attachment, aggregation, epidermal cell killing, and virulence phenotypes of *Erwinia chrysanthemi* EC16 on *Nicotiana clevelandii* seedlings, *Proc. Natl. Acad. Sci.*, 99, 13142–13147 (2002).

84 C. Frank, D. Werber, J.P. Cramer, M. Askar, M., Faber, M. an der Heiden, H. Bernard, A. Fruth, R. Prager, A. Spode and M. Wadl, Epidemic profile of Shiga-toxin-producing *Escherichia coli* O104:H4 outbreak in Germany, *N. Engl. J. Med.*, 365, 1771–1780 (2011).

85 R. Al Safadi, G.S. Abu-Ali, R.E. Sloup, J.T. Rudrik, C.M. Waters, K.A. Eaton and S.D Manning, Correlation between *in vivo* biofilm formation and virulence gene expression in *Escherichia coli* O104:H4, *PLoS ONE* 7, 41628 (2012).

86 J.P. Nataro, D. Yikang, D. Yingkang and K. Walker, AggR, a transcriptional activator of aggregative adherence fimbria I expression in enteroaggregative *Escherichia coli, J. Bacteriol.*, 176, 4691–4699 (1994).

87 X. Wang, J.F. Preston and T. Romeo, The *pgaABCD* locus of *Escherichia coli* promotes the synthesis of a polysaccharide adhesin required for biofilm formation, *J. Bacteriol.* 186, 2724–2734 (2004).

88 G. Méric, E.K. Kemsley, D. Falush, E.J. Saggers and S. Lucchini, Phylogenetic distribution of traits associated with plant colonization in *Escherichia coli, Environ. Microbiol.*, 15, 487–501 (2013).

89 S.S. Branda, A. Vik, L. Friedman and R. Kolter, Biofilm: the matrix revisited, *Trends Microbiol.*, 13, 20 (2005).

90 M.E. Davey and G.A. O'Toole, Microbial biofilms: from ecology to molecular genetics, *Microbiol. Mol. Biol. Rev.*, 64, 847–867 (2000).

91 T. Ophir and D.L. Gutnick, A role for exopolysaccharides in the protection of microorganisms from desiccation, *Appl. Environ. Microbiol.* 60, 740–745 (1994).

92 R.I. Aminov, Horizontal gene exchange in environmental microbiota, *Front. Microbiol.*, 2, 158 (2011).

93 A. Lapidot, U. Römling and S. Yaron, Biofilm formation and the survival of *Salmonella typhimurium* on parsley, *Int. J. Food. Microbiol.*, 109, 229–233 (2006).

94 C. Prigent-Combaret, O. Vidal, C. Dorel and P. Lejeune, Abiotic surface sensing and biofilm-dependent regulation of gene expression in *Escherichia coli, J. Bacteriol.*, 181, 5993–6002 (1999).

95 B.E. Ramey, M. Koutsoudis, S.B. Bodman, and C. Fuqua, Biofilm formation in plant–microbe associations, *Curr. Opin. Microbiol.*, 7, 602–609 (2004).

96 A.F. González Barrios, R. Zuo, Y. Hashimoto, L. Yang, W.E. Bentley and T.K. Wood, Autoinducer 2 controls biofilm formation in *Escherichia coli* through a novel motility quorum-sensing regulator (MqsR, B3022), *J. Bacteriol.*, 188, 305–316 (2006).

97 I. Ahmad, E. Wigren, S. Le Guyon, S. Vekkeli, A. Blanka, Y, Mouali, N. Anwar, M.L. Chuah, H. Lünsdorf, R. Frank and M. Rhen, The EAL-like protein STM1697 regulates virulence phenotypes, motility and biofilm formation in *Salmonella* Typhimurium, *Mol. Microbiol.*, 90, 1216–1232 (2013).

98 O. Vidal, R. Longin, C. Prigent-Combaret, C. Dorel, M. Hooreman and P. Lejeune, Isolation of an *Escherichia coli* K-12 mutant strain able to form biofilms on inert surfaces: involvement of a new *ompR* allele that increases curli expression, *J. Bacteriol.*, 180, 2442–2449 (1998).

99 S.J. Cai and M. Inouye, EnvZ–OmpR interaction and osmoregulation in *Escherichia coli*, *J. Biol. Chem.*, 277, 24155–24161 (2002).

100 M. McClelland, K.E. Sanderson, J. Spieth, S.W. Clifton, P. Latreille, L. Courtney, S. Porwollik, J. Ali, M. Dante, F. Du and S. Hou, Complete genome sequence of *Salmonella enterica* serovar Typhimurium LT2, *Nature*, 413, 852–856 (2001).

101 J.P. Zhang and S. Normark, Induction of gene expression in *Escherichia coli* after pilus-mediated adherence, *Science*, 273, 1234–1236 (1996).

102 S. Seo and K.R. Matthews, Influence of the plant defense response to Escherichia coli O157:H7 cell surface structures on survival of that enteric pathogen on plant surfaces, *Appl. Environ. Microbiol.*, 78, 5882–5889 (2012).

103 A.L. Iniguez, Y. Dong, H.D. Carter, B.M. Ahmer, J.M. Stone and E.W. Triplett, Regulation of enteric endophytic bacterial colonization by plant defenses, *Mol. Plant–Microbe Interact.*, 18, 169–178 (2005).

104 N. Shirron and S. Yaron, Active suppression of early immune response in tobacco by the human pathogen *Salmonella* Typhimurium, *PLoS ONE*, 6, 18855 (2011).

105 S.T. Chisholm, G. Coaker, B. Day and B.J. Staskawicz, Host–microbe interactions: shaping the evolution of the plant immune response, *Cell*, 124, 803–814 (2006).

106 A. Schikora, A. Carreri, E. Charpentier and H. Hirt, The dark side of the salad: *Salmonella* Typhimurium overcomes the innate immune response of *Arabidopsis thaliana* and shows an endopathogenic lifestyle, *PLoS ONE*, 3, 2279 (2008).

107 M.K. Chelius and E.W. Triplett, Immunolocalization of dinitrogenase reductase produced by *Klebsiella pneumoniae* in association with *Zea mays*, *Appl. Environ. Microbiol.*, 66, 783–787(2000).

108 P.J., Riggs, M.K. Chelius, A.L. Iniguez, S.M. Kaeppler and E.W. Triplett, Enhanced maize productivity by inoculation with diazotrophic bacteria, *Aust. J. Plant Physiol*, 28, 829–836 (2001).

109 J. Ryals, S. Uknes and E. Ward, Systemic acquired resistance, *Plant Physiol.*, 104, 1109–1112 (1994).

110 C.M. Pieterse, S.C. Van Wees, J.A. Van Pelt, M. Knoester, R. Laan, H. Gerrits, P.J. Weisbeek and L.C. Van Loon, A novel signaling pathway controlling induced systemic resistance in *Arabidopsis*, *Plant Cell*, 10, 1571–1580 (1998).

111 M. Marvasi, C.E. Cox, Y. Xu, J.T. Noel, J.J. Giovannoni and M. Teplitski, Differential regulation of *Salmonella* Typhimurium genes involved in O-antigen capsule production and their role in persistence within tomato fruit, *Mol. Plant–Microbe Interact.*, 26, 793–800 (2013).

112 F.M., Husain, I. Ahmad, M.S. Khan, E. Ahmad, Q. Tahseen, Khan, M.S. and N.A. Alshabib, Sub-MICs of *Mentha piperita* essential oil and menthol inhibits AHL mediated quorum sensing and biofilm of Gram-negative bacteria, *Front. Microbiol.*, 6 (2015).

113 E.B. Kerekes, É. Deák, M. Takó, R. Tserennadmid, T. Petkovits, C. Vágvölgyi and J. Krisch, Anti-biofilm forming and anti-quorum sensing activity of selected essential oils and their main components on food-related micro-organisms, *J. Appl. Microbiol.*, 115, 933–942 (2013).

114 A. Vikram, G.K. Jayaprakasha, R.M. Uckoo and B.S. Patil, Inhibition of *Escherichia coli* O157:H7 motility and biofilm by beta-sitosterol glucoside, *Biochim. Biophys. Acta.*, 1830, 5219–5228(2013).

115 H.S. Cho, J.H. Lee, S.Y. Ryu, S.W. Joo, M.H. Cho and J. Lee, Inhibition of *Pseudomonas aeruginosa* and *Escherichia coli* O157:H7 biofilm formation by plant metabolite epsilon viniferin, *J. Agric. Food Chem.*, 61, 7120–7126 (2013).

116 CDC, Centers for Disease Control and Prevention, Incidence and trends of infection with pathogens transmitted commonly through food – Foodborne Diseases Active Surveillance Network, 10 U.S. Sites, 1996–2012. *MMWR Morb Mortal Wkly Rep* 62, 283–287 (2013).

117 Anon, Ongoing multistate outbreak of *Escherichia coli* serotype O157:H7 infections associated with consumption of fresh spinach – United States, *J. Am. Med. Assoc.*, 296, 2195–2196 (2006).

118 CDC, "Multistate outbreaks of *Salmonella* infections associated with raw tomatoes eaten in restaurants—United States, 2005-2006." *MMWR. Morbidity and mortality weekly report* 56, 909 (2007).

119 J. Denny, J. Threlfall, J. Takkinen, S. Löfdahl, T. Westrell, C. Varela, B. Adak, N. Boxall, S. Ethelberg, M. Torpdahl, M. Straetemans and W. van Pelt, Multinational *Salmonella paratyphi* B variant Java (*Salmonella* Java) outbreak, August–December 2007, *Euro surveill.*, 12, 3332 (2007).

120 P. Mariani-Kurkdjian and E. Bingen, *Escherichia coli* O104:H4: a hybrid pathogen, *Arch. Pediatr.*, 19 (Suppl. 3), S97–S100 (2012).

121 X. Guo, M.W. van Iersel, J. Chen, R.E. Brackett and L.R. Beuchat, Evidence of association of salmonellae with tomato plants grown hydroponically in inoculated nutrient solution, *Appl. Environ. Microbiol.*, 68, 3639–3643 (2002).

122 S.B. Baloda, L. Christensen and S. Trajcevska, Persistence of a *Salmonella enterica* serovar Typhimurium DT12 clone in a piggery and in agricultural soil amended with *Salmonella* contaminated slurry, *Appl. Environ. Microbiol.*, 67, 2859–2862 (2001).

123 X. Guo, J. Chen, R.E. Brackett and L.R. Beuchat, Survival of salmonellae on and in tomato plants from the time of inoculation at flowering and early stages of fruit development through fruit ripening, *Appl. Environ. Microbiol.*, 67, 4760–4764 (2001).

124 K. Takeuchi and J.F. Frank, Expression of red-shifted green fluorescent protein by *Escherichia coli* O157:H7 as a marker for the detection of cells on fresh produce, *J. Food Prot*, , 64, 298–304 (2001).

125 X. Shi and X. Zhu, Biofilm formation and food safety in food industries, *Trends Food Sci. Technol.*, 20, 407–413 (2009).

126 W.N. Wade, A.J. Scouten, K.H. McWatters, R.L. Wick, A. Demirci, W.F. Fett and L.R. Beuchat, Efficacy of ozone in killing *Listeria monocytogenes* on alfalfa seeds and sprouts and effects on sensory quality of sprouts, *J. Food. Prot.*, 66, 44–51(2003).

127 N. Kondo, M. Murata and K. Isshiki, Efficiency of sodium hypochlorite, fumaric acid, and mild heat in killing native microflora and *Escherichia coli* O157:H7, *Salmonella* Typhimurium DT104, and *Staphylococcus aureus* attached to fresh-cut lettuce, *J. Food Prot.*, 69, 323–329 (2006).

128 N. Shirron, G. Kisluk, Y. Zelikovich, I. Eivin, E. Shimoni and S. Yaron, A comparative study assaying commonly used sanitizers for antimicrobial activity against indicator bacteria and a *Salmonella* Typhimurium strain on fresh produce, *J. Food Prot.*, 72, 2413–2417(2009).

129 E.E. Natvig, S.C. Ingham, B.H. Ingham, L.R. Cooperband and T.R. Roper, *Salmonella enterica* serovar Typhimurium and *Escherichia coli* contamination of root and leaf vegetables grown in soils with incorporated bovine manure, *Appl. Environ. Microbiol.*, 68, 2737–2744(2002).

130 L.R. Beuchat, Comparison of chemical treatments to kill *Salmonella* on alfalfa seeds destined for sprout production, *Int. J. Food Microbiol.*, 34, 329–333 (1997).

131 M. Gandhi, S. Golding, S. Yaron and K.R. Matthews, Use of green fluorescent protein expressing *Salmonella* Stanley to investigate survival, spatial location, and control on alfalfa sprouts, *J. Food Prot.*, 64, 1891–1898 (2001).

25

Anti-QS/Anti-Biofilm Agents in Controlling Bacterial Disease: An *in silico* Approach

K. Ahmad[1], M.H. Baig[2], Fohad Mabood Husain[3], Iqbal Ahmad[4], M.E. Khan[5], M. Oves[6], Inho Choi[2] and Nasser Abdulatif Al-Shabib[3]

[1] *Department of Biosciences, Integral University, Lucknow, India*
[2] *Department of Medical Biotechnology, Yeungnam University, Republic of Korea*
[3] *Department of Food Science and Nutrition, College of Food and Agricultural Sciences, King Saud University, Saudi Arabia*
[4] *Department of Agricultural Microbiology, Faculty of Agricultural Sciences, Aligarh Muslim University, Aligarh, India*
[5] *School of Chemical Engineering, Yeungnam University, Republic of Korea*
[6] *Centre of Excellence in Environmental Studies, King Abdul Aziz University, Jeddah, Saudi Arabia*

25.1 Introduction

Plant diseases caused by bacterial pathogens constitute one of the major constraints on crop production and result in significant annual losses on a global scale. Therefore, identification and effective disease management is key for sustainability of crop production needed for the increasing global human population.

Various classical and integrated approaches are applied in different parts of the globe for bacterial disease management. However, management strategies for plant bacterial diseases are highly knowledge based, since success of an integrated approach depends on the availability of suitable host plant cultivar, efficacious chemicals, and the animals and plants involved. Antibiotics such as streptomycin, tetracyclin, kasugamycin, and gentamycin are used in controlling disease caused by *Erwinia amylovora* (fire blight), *Xanthomonas* spp., *Pseudomonas* spp. [1]. However, development of resistance to antibiotics among phytopathogenic bacteria due to long-term use has been documented. Therefore, many countries discourage the use of clinical antibiotics for plant disease management.

Other effective chemical agents include copper bactericides and antibacterial peptides. Strategies like biocontrol agents and use of inducer of systemic resistance (SAR) have been proposed as an effective means to control bacterial diseases [2, 3].

New developments in the understanding of bacterial–host interaction and the role of quorum sensing in global gene regulation of virulence and biofilm as surviving strategy by bacteria in the last decade has led to assumptions regarding the potential for controlling pathogens through antagonizing QS and biofilm. QS and biofilms are now considered as anti-infective target to control bacterial infection [4]. Interfering quorum sensing in pathogenic bacteria may effectively decrease its ability to establish infection and could not develop resistance to QS inhibition. Some successful experiments were conducted on transgenic plants by application AHL-degrading enzymes in transgenic plant [5, 6].

Lindow et al. [7] demonstrated another strategy, termed *pathogen confusion*, by stimulation of QS controlled traits at inapproachable timing in transgenic plants.

Another strategy currently being studied is the use of metal nanoparticles in bacterial disease management [8–10]. However, bacterial pathogens are known to form biofilm after gene expression in the biofilm state, since biofilm provides bacteria protective advantages against environmental stress, host defense, and antimicrobials, and is an important virulence factor of many pathogenic bacteria. Therefore, development of biofilm inhibitors is an attractive strategy to control bacterial adherence and disease (food and medical settings).

Attempts have been made to use biofilm inhibitors against plant pathogenic disease. For example, D-aminoacid and indole derivatives are biofilm inhibitors that are used to form early biofilm formation, resulting in reduced disease symptoms caused by animal and plant pathogens and increased bacterial sensivity to antibiotics and copper [11–13]. Similarly, Chua et al. [14], demonstrated that a decrease in c-di GMP level can trigger biofilm dispersal, which can be used as promising strategy in biofilm control. Bis [3, 5] cyclic dimeric GMP is a bacterial secondary messenger that participates in regulating virulence of many bacterial species.

Other compounds include 2-aminoimidazole (2AI), which inhibits biofilm of *Xanthomonas euvesicatoria* in *in vitro* and shows synergistic interaction with copper hydroxide in suppression of bacterial infection on tomato in field conditions [12]. Similarly, 3-indoylacenitrile and D-leucine inhibit biofilm formation by *Xanthomonas citrii*. N-acetylcystein also reduces adhesion [15].

In this chapter, a brief description of the role of QS and biofilms in disease control is presented. Basic *in silico* approaches are described for target identification, modeling, and data extraction. Further, virtual screening for identification of potent anti-QS agents is also elaborated.

25.2 Biofilm and Its Significance

In 1978, the concept of biofilm was coined by Bill Costerton. According to him, "A biofilm is an adhesion of bacterial cells to form a structured community that is enclosed in its own polymeric matrix and is adherent to an inert or living surface." The process of biofilm formation and biofouling originated in oligotrophic systems about 3.25 billions years ago [16]. Formation of this biofilm by bacteria is a primary weapon in the development of infections, thereby causing a severe health saddle. Many factors are important for inducing microbes to form biofilms, such as attachment sites on a surface, nutritional factors, detergents, and sub-inhibitory concentration of antibiotics [17]. According to the Centers for Disease Control and Prevention, 65 percent of all infections in the developed countries are caused by biofilms [18]. Microbial biofilms occurs on any surface by immigration and gathering of micro and macrofoulers on immersed structures.

Biofilm formation and QS-associated features direct to biofouling. The key step in biofouling process is biofilm formation. O'Toole and his colleagues [19] proposed microbial development concept for biofilm formation. According to Martin Dworkin, *development* refers to a sequence of changes occurring in a cell in response to different environmental stimuli that help the cells to acclimatize and live in their dynamic environment [20]. Initially, microbiologists mistreated the sociobiology concept, but the studies on cooperative behavior in Myxobacteria and quorum sensing/biofilm formation in *Pseudomonas aeruginosa* provoke biofilm research [21]. Halan et al. said that microbes are found on almost every surface and interfaces exposed to oil, water, or air [22]. They form biofilm on both biotic and abiotic surfaces [23, 24] and on human hosts [25].

The biofilm provides a more protective mode of bacterial growth in nature [25]. Bacteria in biofilm mode easily escape environmental stresses such as pH, temperature, osmolarity, UV damage [26], desiccation [27], predation [18], and antibiotics [29], but biofilm in planktonic form do not.

P. aeruginosa is the bacteria in which the *in vivo* biofilm formation has most been studied. Research shows that microcolonies form because of initial attachment of *P. aeruginosa* to a surface, which in turn take the shape of a larger structure. Observations based on transcriptomics have revealed that gene expression of biofilm cell differs from most stages in growth of planktonic bacteria [30]. Although the conditions in which experiments were performed would differ, a core set of genes must be expressed for biofilm formation program, regardless of the experimental conditions. To date, a large number of reports as well as recent transcriptomic studies of *P. aeruginosa* biofilms suggest the existence of multiple pathways by which a biofilm is formed. The role of quorum sensing in biofilm formation is depicted in Figure 25.1.

Biofilm formation begins with a transition of bacteria from the planktonic to its genetically distinct attached state [23, 31, 32]. The physiological and genetic transition occurs across the life cycle of the biofilm, which indicated (i) quorum sensing, (ii) EPS and

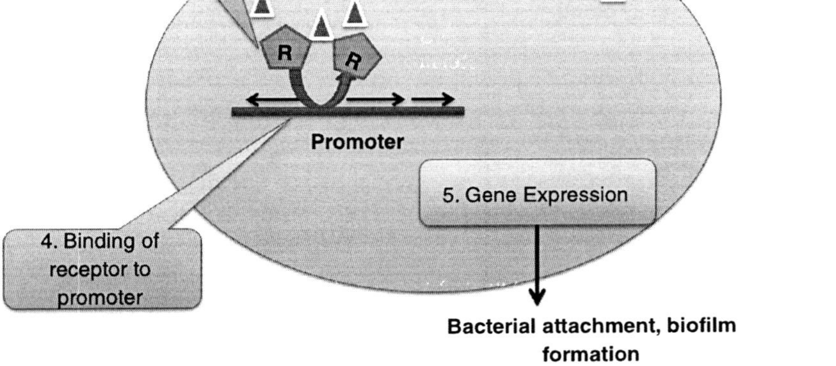

Figure 25.1 Role of QS in biofilm formation.

microcolony formation, (iii) maturation of biofilms, and (iv) migration and reattachment of planktonic cells.

25.3 Bioinformatics Approaches in Drug Target Identification and Drug Discovery

Identifying drug targets is the initial step, playing an essential role toward successful drug design in combating diseases. In the past, the time and cost of developing new drugs have soared significantly. In general, it takes about 15 years and up to $800 million to convert a promising new compound into a drug in the market [33]. In the procedure of drug discovery, the identification of drug targets is the first and one of the most important steps. With the therapeutic targets, the optimal compounds with expected effects can be designed and new indications of old drugs may be discovered [34, 35].

The biological drug targets could be a cellular/membrane receptors or any other molecules that are critically involved in a disease process. The major aim in drug discovery is to either inhibit or activate a drug target by drug molecules, ranging from small organic molecule to bigger antibodies therapeutic protein. The physical attachment of a drug molecule to its respective drug target triggers a cascade of biological reactions, followed by a cellular reaction [36].

Drug discovery is a costly and time-consuming process, as it takes several years along with a huge amount of money to produce new drug potentials and enlarge the scope of diseases applications [37]. High throughput screening and virtual screening are the two different methods that are widely used for drug discovery in the pharmaceutical industry.

Drug discovery is considered to be one of the most crucial components of the research and development (R&D) process of the pharmaceutical industry and is the first essential step in the development of any robust, inventive drug pipeline [38].

In high throughput screening (HTS), the synthesized series of compounds is screened against cell-based or protein-based assays. This is the most commonly used process of drug discovery in all major pharmaceutical industries. However, the large expense on synthesis of each compounds and their *in vitro* testing imposes huge problems for pharmaceutical industries. Efforts are being made within the industry to cut down the timeline and costs. The process to identify compounds featuring a desired phenotype using HTS fails several times, as a large number of drugs do not pass in the clinical development because of insufficient specificity of the compounds, which may either be the intolerable side effects or poor pharmacokinetic characteristics [39]. A very highly efficient screening method is needed for researchers to test millions of compounds within a short time period for finding a novel hit. *In silico* screening procedures came into existence to solve this problem, where computational approaches are used for the screening of these big data sets of compounds [40].

25.4 Target Identification Using *in silico* Technologies

Although many targets derived from sequenced genomes have been identified in this modern era of next-generation sequencing and whole genome sequencing, there is still a need to understand and be updated regarding which *in silico* methodologies are available for identification of other new, potential drug targets. These methodologies include gene selection, gene and protein annotation, and prioritization.

Gene selection methodologies comprise three main areas, including computational tools and approach, database searching, and data mining. The computational approach consists of similarity searching, gene finding, EST identification, and splice variant construction [41]. Database searching typically depends on the quality and amount of data available in organized manner—that is, database of available resources (commercial, public, or internal databases). Data mining is an important process to fetch relevant data from various databases. While identifying a potential target through the data-mining approach, expert scientists include factors such as detail of disease, differential gene and protein expression data, literature survey, and studies of SNPs [42].

The next step after selecting a potential gene target is the annotation the encoded protein. This step may include the structure as well as function prediction of the target, their homology searches, various public annotations, and pattern searches. Assessment and prioritization of the selected target is the last step of target identification process. For the examination of various issues associated with the selected target such as their disease association and druggability, a decision-making structure is used that is a combination of computational as well as experimental results in addition to the biological experience. Soon after the selection of potential target molecules, on the basis of particular criteria, this decision-making system is assigned to begin its search for known compounds and inhibitors that may bind and have potential correlation with the disease to create a final prioritized target list.

25.5 Data Resources for Drug Target Identification

The process of drug target identification involves the understanding of a specific disease state at the molecular level. This may include the analysis of molecular pathway, gene association, and knowledge of gene sequence information, the structure of the encoded protein structures, various protein interactions, and metabolic pathways [43]. The ultimate goal of the drug identification process is to discover a suitable target that could be used as a target for drug discovery and whose biological activity can be directly linked to a pathological process. In the modern genomics era, one must incorporate and integrate a wide range of data ranging from gene sequence, expression, and polymorphism data. There is a large list of publically available biological databases providing a plethora of functional information for drug discovery. It will be very useful to integrate the already-existing data retrieved from public databases to portray the systematic analysis architecture for surmising the basic genes interaction and for gaining an insights of the structural pathway with which drug targets interact.

25.6 Homology Modeling

The next step after target identification is to search for the structure of the selected target. In computer-aided drug design, the structure of the target protein is required for drug designing [44]. Since the experimental determination of the structure of protein through X-ray crystallography is a very costly and time-consuming process, homology modeling approach came into existence to provide a fast and stress-free means to generate protein structure for further studies [45].

There are several protein/enzymes whose structures has not been experimentally solved by X-ray crystallography or NMR. In this case, we have to use the amino acid sequence information of those protein/enzymes to model its 3D structure, using the structural information of one or more sufficiently similar proteins. This computational process is known as *homology modeling* and is a very useful approach to carry out computer-aided drug designing (CADD). This method is also termed as *comparative modeling*. In this method, an attempt for constructing a three-dimensional model of the target protein on the basis of its sequential information (amino acid sequence) using the available 3D structures of related homologous proteins as templates [46]. The process of homology modeling is based on the concept that proteins showing similarity at their sequence level must have similar 3D structures.

Homology modeling is a four step process:

1) *Template identification:* In this step we have to identify one or more related proteins with available experimental structures that can serve as a template. The most-used techniques to identify templates are BLAST, PSI-BLAST [47], and FFAS [48].
2) *Sequence alignment:* In this step, we have to align the sequence of our target protein with the sequence of the template. We have to use multiple sequence alignment tools like ClustalW and ClustalX [49].
3) *Model generation:* In this step, the structure of target protein is modeled, taking the 3D structure of template as a reference. Commonly used programs for this step are Modeller and Swiss model [50, 51].
4) *Model evaluation/validation:* In this step, the quality of modeled protein is evaluated on the basis of various parameters. The quality of the modeled structure of target protein is generally correlated with the structure of template. Analysis of the Ramachandran plot is one of the validation steps for assessing the quality of a modeled structure. The distribution of backbone bond angles is shown. The quality of a homology model can also be observed by scrutinizing the distribution of hydrophilic and lipophilic residues inside as well as outside the protein. Commonly used programs in this step are SAVS and Procheck.

25.7 Docking

Molecular docking is the process for determining the binding affinity between a protein structure and a ligand using computational methods. In these methods, there is a proficient sampling of all the possible poses of ligands within the binding pocket of target protein/enzyme to ease the optimal binding conformation, as measured by the scoring function [52,53]. There are three different ways to perform the molecular docking of small molecules:

1) *Rigid docking:* In this method, the target molecule and ligand are kept rigid.
2) *Semi-flexible or flexible ligand docking:* In this method, the target molecule is kept rigid while a certain degree of freedom (flexibility) is allowed to the ligand.
3) *Flexible docking:* In this method, both the target as well as ligand are provided flexibility [54].

In molecular docking, protocols can also be defined as a blend of a search algorithm and a scoring function [34, 55]. Many scoring functions and algorithms are currently available. The search algorithm is supposed to provide support and freedom to the protein–ligand coordination to enable accurate and sufficient sampling, including the binding modes. Logically, the search algorithm is supposed to have good speed and effectiveness, while the scoring function must be able to analyze physicochemical properties of molecules and thermodynamics of interaction. The complexity of docking increases in the order of rigid docking, flexible ligand docking, and flexible docking [54].

25.8 Virtual Screening

Virtual screening (VS) is an *in silico* approach for the discovery of new drugs that has successfully complemented HTS for hit detection [64]. The objective of virtual screening approach is to use a computational approach for a speedy, accurate, and cost-effective assessment of large virtual compound databases to screen out few novel potential leads that can be subjected for wet lab synthesis and further be experimentally examined for their biological activity [65]. Unlike HTS, VS does not rely on brute force search and is instead based on starting information of the receptor under inspection or its active ligands [66]. VS methods can be divided into two different categories, structure-based and ligand-based. Table 25.1 shows some widely used programs for virtual screening.

25.9 Application of Bioinformatics in Development of Anti-QS/anti-biofilm Agents

Quorum sensing interruption is a novel and efficient approach to manage different bacteria-associated infections that are pathogenic to animals and plants. Quorum sensing comprises a novel target for the directed drug design to identify new chemical entities. Therefore, identification of compounds having inhibitory property for QS is a promising way to control biofilm formation where the antibiotics are not effective. Many researchers have reported QS inhibitory property of different compounds derived naturally and/or synthetically, but many of these studies were applied for inhibiting the QS process in biosensor strains [67].

Nowadays, treatment of several chronic infections related to bacteria is based on synthetic or natural drug compounds that intend to kill or inhibit the bacterial growth [68]. Development of anti-QS agents has been a promising strategy to battle pathogenic infections caused by bacteria [69]. From human and veterinary medicine, to agriculture, and aquaculture, the need of quorum sensing–inhibitory compounds is increasing day by day in many fields, including and coupled with commercial interests. Many biotechnology companies are developing anti–quorum-sensing and anti-biofilm drugs (QS signal molecules) using state-of-the-art techniques. Gram-negative bacteria typically use N-acyl homoserine lactones (AHLs) as a cognate signal molecule. Several strategies for breaking circuits of bacterial quorum sensing are possible, including inhibition of the AHL signal generation, dissemination, and reception [68, 70].

Table 25.1 Some Widely Tools for Virtual Screening.

Software	Developer	Strategy used	Availability	Link	References
Autodock	Scripps Research Institute†	GA/MC	Free	http://autodock.scripps.edu	[56]
Dock	University of California	IC	Free	http://dock.compbio.ucsf.edu	[57]
Gold	Cambridge Crystallographic Data Centre	GA	Commercial	http://www.ccdc.cam.ac.uk/products/life_sciences/gold	[58]
Glide	Schrodinger, Inc.	Hybrid	Commercial	http://www.schrodinger.com/Glide	[59]
FlexX	BioSolveIT GmbH	IC	Commercial	https://www.biosolveit.de	[60]
ICM	Molsoft LLC.	MC	Commercial	http://www.molsoft.com/docking.html	[61]
Surflex	Tripos Inc.	IC	Commercial	http://www.tripos.com/index.php	[62]
Ligandfit	Accelrys Inc.	MC	Commercial	http://accelrys.com/products/discovery-studio	[63]

The computational drug designing has more possibilities of success because of having multiple targets in a microorganism. If a compound has the potential to inhibit multiple proteins, it is likely that it inhibits at least one. A number of bioinformatics tools and approaches are available for development and discovery of potent novel anti-QS agents from plant-based natural compounds and synthetic chemical compounds, including structure-based drug design (SBDD), ligand-based drug design (LBDD), molecular docking, pharmacophore modeling, and virtual screening (structure-based virtual screening and ligand-based virtual screening).

25.10 Virtual Screening for Identification of QS Inhibitors

The QS system present in *Staphylocoocus spp.* consists of RAP (AI RNAIII activating protein) and TRAP (target molecule for RAP). The inhibition of RAP using *in silico* approaches by RNAIII-inhibiting peptide (RIP) results in reduction of virulence. A non-peptide analogue was identified through virtual screening of a database having 300,000 commercially available small-molecule compounds as QSI [71].

Zeng et al. [72] reported that they used the automated docking program DOCK 5.3.0 for screening out QS inhibitory compounds of *P. aeruginosa* from a database having 51 active compounds of traditional Chinese medicines having antibacterial property and finally suggested that baicalein can be used as a potent lead for further wet lab experiments in drug development processes for *P. aeruginosa.*

In 2013, Sean Yang-Yi Tan et al. reported their study, which includes virtual screening approach (structure-based) for the discovery of novel QSI candidates [73]. Molecular docking simulation study was done using the QS receptor LasR as a target against a library of 3,040 natural compounds. Based on docking scores, the top 22 compounds were selected for further analysis. Finally, five compounds were found having potential to inhibit QS-regulated gene expression in *P. aeruginosa* in a dose-dependent manner. G1 was found to be the most potent compound, significantly affecting the abundance of 46 proteins in *P. aeruginosa* PAO1. This compound was also found capable of reducing the release of extracellular DNA and further inhibiting the LasR-regulated secretion of the virulence factor, elastase. These findings reveal the utility of a structure-based VS approach for the discovery of target-specific QSIs.

In 2006, Riedel et al. reported the computer-aided design of NCEs (new chemical entities) having capabilities of inhibiting the quorum sensing system of *B. cenocepacia* [74]. A compound database was virtually screened and the initial screened out leads, which were further tested in several bioassays. Focused combinatorial libraries were generated on the basis of these results. Finally, three potent compounds were identified carrying inhibitory potential for quorum sensing for *B. cenocepacia,* snooping with many QS-regulated functions without affecting bacterial growth.

A variety of approved drugs have been reported to carry anti-QS properties along with the biological effectiveness for which they are sold. For instance, some macrolide antibiotics have the ability to inhibit AHL-mediated QS, possibly at the ribosomal level. Nonmacrolide antibiotics are also shown to having inhibitory property against QS systems in Gram-negative bacteria. Skindersoe et al. reported that the antibiotics ciprofloxacin and ceftazidime are capable of lowering the expression of a range of QS-regulated virulence factors in *P. aeruginosa* at sub-inhibitory concentrations [75]. Yang et al.

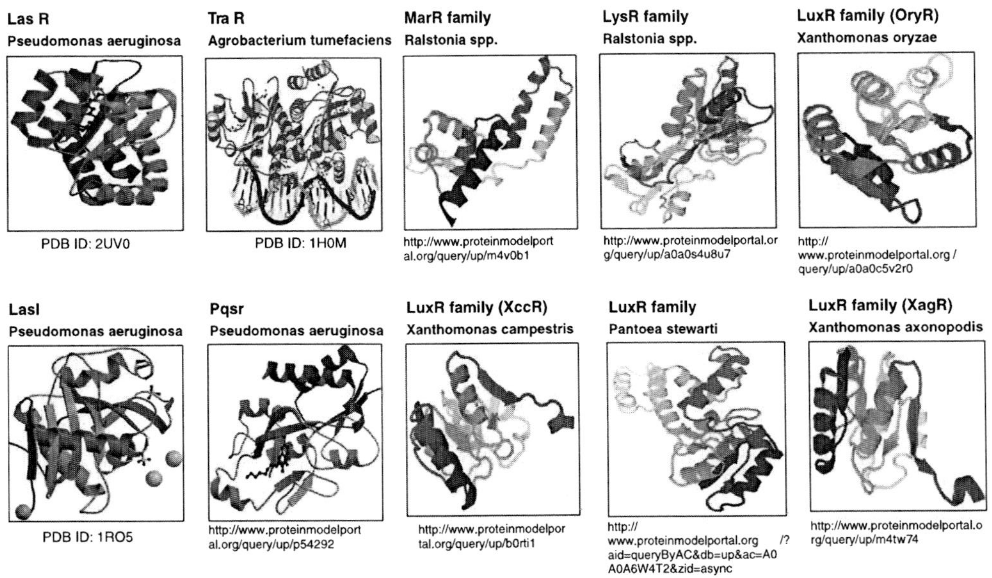

Figure 25.2 Potent plant pathogenic bacterial protein targets for QS inhibitors. (*See color plate section for the color representation of this figure.*)

identified three potent compounds (chlorzoxazone, nifuroxazide, and salicylic acid) through virtual screening of drugs against *P. aeruginosa* QS [76]. These compounds were reported to show their effectivity via their interaction with the LasR protein. In another study, Husain et al. demonstrated broad spectrum inhibition of AHL-regulated virulence and biofilm against three bacterial pathogens by sub-inhibitory concentrations of ceftazidime [70]. It is considered that that competitive binding of these inhibitors to LasR takes place with the protein in a more open interspecies communication among Gram-negative and Gram-positive bacteria. Certain possible LasR and other QS-related protein targets of various different plant pathogenic bacteria are presented in Figure 25.2.

25.11 Conclusion

Considering the progress made in understanding biofilm role in virulence and pathogenicity of bacterial pathogens and the role of biofilm inhibitors in controlling bacterial disease alone or in combination with antibiotic and other agents, it is expected that high throughput and *in silico* approaches might be useful in identifying the most effective anti-biofilm agents. Furthermore, compounds identified for medical application through this approach should also be tested for controlling bacterial diseases to assess their efficacy.

References

1 L.F. Castiblanco and G.W. Sundin, New insights on molecular regulation of biofilm formation in plant-associated bacteria, *J. Integr. Plant Biol,*. 58, 362–372 (2015).

2 A. Kerr, Biological control of crown gall. *Australas. Plant Pathol.* 45, 15–18 (2016).

3 G.W. Sundin, L.F. Castiblanco, X. Yuan, Q. Zeng, and C.H. Yang, Bacterial disease management: Challenges, experience, innovation, and future prospects. *Mol. plant pathol.* (2016).

4 M. Hentzer and M. Givskov, Pharmacological inhibition of quorum sensing for the treatment of chronic bacterial infections. *J. Clin. Invest.*, 112(9), 1300–1307 (2003).

5 Y.H. Dong, L.H. Wang, J.L. Xu, H.B. Zhang, X.F. Zhang, and L.H. Zhang, Quenching quorum-sensing-dependent bacterial infection by an N-acyl homoserine lactonase. *Nature*, 411, 813–817 (2001).

6 L. Molina, F. Constantinescu, L. Michel, C. Reinmann, B. Duffy, and G. Defago, Degradation of pathogen quorum-sensing molecules by soil bacteria: a preventive and curative biological control mechanism, *FEMS Microbiol. Ecol.*, 45, 71–81 (2003).

7 S. Lindow, K. Newman, S. Chatterjee, C. Baccarai, A.T. Iavarone, and M. Ionescu, Production of Xylella fastidiosa diffusible signal factor in transgenic grape causes pathogen confusion and reduction in severity of Pierce's disease. *Mol. Plant–Microbe Interact.*, 27, 244–254 (2014).

8 C.N. Lok, C.M. Ho, R. Chen, Q.Y. He, W.Y. Yu, H. Sun, P.K. Tam, J.F. Chiu, and C.M. Che, Silver nanoparticles: partial oxidation and antibacterial activities. *J. Biol. Inorg. Chem.*, 12, 527–534 (2007).

9 I. Ocsoy, M.L. Paret, M.A. Ocsoy, S. Kunwar, T. Chen, M.X. You, and W.H. Tan, Nanotechnology in plant disease management: DNA-directed silver nanoparticles on graphene oxide as an antibacterial against *Xanthomonas perforans*. *ACS Nano*, 7, 8972–8980. (2013).

10 M.L. Paret, G.E. Vallad, D.R. Averett, J.B. Jones, and S.M. Olson, Photocatalysis: effect of light-activated nanoscale formulations of TiO(2) on *Xanthomonas perforans* and control of bacterial spot of tomato. *Phytopathology*, 103, 228–236 (2013).

11 G. Brackman, P.Cos, L. Maes, H.J. Nelis, and T. Coenye, Quorum sensing inhibitors increase the susceptibility of bacterial biofilms to antibiotics *in vitro* and *in vivo*. *Antimicrob. Agents Chemother*, 55, 2655–2661. (2011).

12 R.J. Worthington, S.A. Rogers, R.W. Huigens, C. Melander, and D.F. Ritchie, Foliar-applied small molecule that suppresses biofilm formation and enhances control of copper-resistant *Xanthomonas euvesicatoria* on pepper. *Plant Dis.*, 96, 1638–1644. (2012).

13 J. Liand, N. Wang, Foliar application of biofilm formation-inhibiting compounds enhances control of citrus canker caused by *Xanthomonas citri* subsp. *citri*. *Phytopathology*, 104, 134–142. (2014).

14 S.L. Chua, L.D. Hultqvist, M. Yuan, M. Rybtke, T.E Nielsen, M. Givskov, T. Tolker-Nielsen, and L. Yang, *In vitro* and *in vivo* generation and characterization of *Pseudomonas aeruginosa* biofilm-dispersed cells via c-di-GMP manipulation. *Nat. Protoc*, 10, 1165–1180 (2015).

15 C. Yang, C.A. Powell, Y. Duan, R. Shatters, and M. Zhang, Antimicrobial nanoemulsion formulation with improved penetration of foliar spray through citrus leaf cuticles to control citrus huanglongbing, *PLoS One*, 10 (2015).

16 J. Wimpenny, W. Manz, and U. Szewzyk, Heterogeneity in biofilms. *FEMS Microbiol. Rev.*, 24, 661–671 (2000).

17 L.R. Hoffman, D.A. D'Argenio, M.J. MacCoss, Z. Zhang, R.A. Jones, and S.I. Miller, Aminoglycoside antibiotics induce bacterial biofilm formation. *Nature*, 436, 1171–1175 (2005).

18 K. Lewis, Persister cells, dormancy and infectious disease. *Nat. Rev. Microbiol.*, 5, 48–56 (2007).

19 G. O'Toole, H.B. Kaplan, and R. Kolter, Biofilm formation as microbial development. *Ann. Rev. in Microbiol.*, 54, 49–79 (2000).

20 M. Dworkin, and D. Kaiser, Cell interactions in myxobacterial growth and development. *Science*, 230, 18–24 (1985).

21 V. C. Kalia, and H. J. Purohit, Quenching the quorum sensing system: potential antibacterial drug targets, *Critical Reviews in Microbiology*, 37, 121–140 (2011).

22 B. Halan, K. Buehler, and A. Schmid, Biofilms as living catalysts in continuous chemical syntheses. *Trends in Biotechnology*, 30, 453–465 (2012).

23 L. Hall-Stoodley, J. W. Costerton, and P. Stoodley, Bacterial biofilms: from the natural environment to infectious diseases. *Nat. Rev. Microbiol.*, 2, 95–108 (2004).

24 G. Lear, and G. D. Lewis, Microbial biofilms: current research and applications. *Horizon Scientific Press*. (2012).

25 K.K. Jefferson, What drives bacteria to produce a biofilm? *FEMS microbial. Let.*, 236, 163–173 (2004).

26 M.O. Elasri, and R.V. Miller, Study of the response of a biofilm bacterial community to UV radiation. *App. and Env. Microbiol.*, 65, 2025–2031 (1999).

27 W.S. Chang, M. van de Mortel, L. Nielsen, G.N. Li. X. de Guzman, and L.J. Halverson, Alginate production by *Pseudomonas putida* creates a hydrated microenvironment and contributes to biofilm architecture and stress tolerance under water-limiting conditions. *Journal of Bacteriology*, 189, 8290-8299 (2007).

28 C. Matz, D. McDougald, A.M. Moreno, P.Y. Yung, F.H. Yildiz, and S. Kjelleberg, Biofilm formation and phenotypic variation enhance predation-driven persistence of Vibrio cholerae. *PNAS*, 102, 16819–16824 (2005).

29 P. S. Stewart, and J. W. Costerton, Antibiotic resistance of bacteria in biofilms. *The Lancet*, 358, 135–138 (2001).

30 Á. Hentzer, Á. Eberl, and Á. Givskov, Transcriptome analysis of *Pseudomonas aeruginosa* biofilm development: anaerobic respiration and iron limitation. *Biofilms*, 2, 37–61 (2005).

31 R. Singh, D. Paul, and R.K. Jain, Biofilms: implications in bioremediation. *Trends in Microbiology*, 14, 389–397 (2006).

32 M.R. Parsek, and T. Tolker-Nielsen, Pattern formation in *Pseudomonas aeruginosa* biofilms. *Current Opinion in Microbiology*, 11, 560–566 (2008).

33 F.T. Rothaermel, and D. L. Deeds, Exploration and exploitation alliances in biotechnology: A system of new product development. *Strategic Management Journal*, 25, 201–221 (2004).

34 M.H. Baig, K. Ahmad, S. Roy, J.M. Ashraf, M. Adil, M.H. Siddiqui, and I. Choi, Computer Aided Drug Design: Success and Limitations. *Current Pharmaceutical Design*, 22, 572–581 (2015).

35 K.A. O'Connor, and B.L. Roth, Finding new tricks for old drugs: an efficient route for public-sector drug discovery. *Nat. Rev. Drug Discov.* 4, 1005-1014 (2005).

36 G. Saito, J.A. Swanson, and K.D. Lee, Drug delivery strategy utilizing conjugation via reversible disulfide linkages: role and site of cellular reducing activities. *Advanced Drug Delivery Rev.*, 55, 199–215 (2003).

37 V.S. Rao, and K. Srinivas, Modern drug discovery process: an in silico approach. *Journal of Bioinformatics and Sequence Analysis*, 2, 89–94 (2011).

38 D. Taylor, The Pharmaceutical Industry and the Future of Drug Development. (2015).

39 H. Kubinyi, R. Mannhold, and H. Timmerman, Virtual screening for bioactive molecules (Vol. 10). H. J. Böhm, & G. Schneider (Eds.). John Wiley & Sons (2008).

40 H.R. Noori, and R. Spanagel, In silico pharmacology: drug design and discovery's gate to the future. *In silico Pharmacology*, 1 (2013).

41 A.I. Nesvizhskii, Proteogenomics: concepts, applications and computational strategies. *Nature Methods*, 11, 1114–1125 (2014).

42 M.G. Kann, Protein interactions and disease: computational approaches to uncover the etiology of diseases. *Briefings in Bioinformatics*, 8, 333–346 (2007).

43 Z. Jiang, and Y. Zhou, Using gene networks to drug target identification. *Journal of Integrative Bioinformatics*, 2 (2005).

44 J. Greer, J.W. Erickson, J.J. Baldwin, and M.D. Varney, Application of the three-dimensional structures of protein target molecules in structure-based drug design. *Journal of Medicinal Chemistry*, 37, 1035–1054 (1994).

45 Z. Xiang, Advances in homology protein structure modeling. *Current Protein & Peptide Science*, 7 (2006).

46 N.B. Centeno, J. Planas-Iglesias, and B. Oliva, Comparative modelling of protein structure and its impact on microbial cell factories. *Microbial Cell Factories*, 4 (2005).

47 S.F. Altschul, T.L. Madden, A.A. Schäffer, J. Zhang, Z. Zhang, W. Miller, and D.J. Lipman, Gapped BLAST and PSI-BLAST: a new generation of protein database search programs. *Nucleic Acids Research*, 25, 3389–3402 (1997).

48 L. Jaroszewski, Z. Li, X.H. Cai, C. Weber, and A. Godzik, FFAS server: novel features and applications. *Nucleic Acids Research*, 39, W38–W44 (2011).

49 J.D. Thompson, T. Gibson, and D.G. Higgins, Multiple sequence alignment using ClustalW and ClustalX. *Current Protocols in Bioinformatics*, 2–3 (2002).

50 R. Sánchez, and A. Šali, Comparative protein structure modeling: introduction and practical examples with modeller. Protein structure prediction: *Methods and protocols*, 97–129 (2000).

51 M. Biasini, S. Bienert, A. Waterhouse, K. Arnold, G. Studer, T. Schmidt, and T. Schwede, SWISS-MODEL: modelling protein tertiary and quaternary structure using evolutionary information. *Nucleic acids research*, gku340 (2014).

52 M.K. Gilson, and H.X. Zhou, Calculation of protein-ligand binding affinities*. *Ann. Rev. of Biophysics and Biomolecular Structure*, 36 (2007).

53 L.L. Looger, M.A. Dwyer, J.J. Smith, and H.W. Hellinga, Computational design of receptor and sensor proteins with novel functions. *Nature*, 423, 185–190 (2003).

54 G. Jones, P. Willett, R.C. Glen, A.R. Leach, and R. Taylor, Development and validation of a genetic algorithm for flexible docking. *Journal of Molecular Biology*, 267, 727–748 (1997).

55 A.N. Jain, Effects of protein conformation in docking: improved pose prediction through protein pocket adaptation. *Journal of Computer-Aided Molecular Design*, 23, 355–374. (2009).

56 D.S. Goodsell, G.M. Morris, and A.J. Olson, Automated docking of flexible ligands: applications of Auto Dock. *Journal of Mol. Recog.*, 9, 1–5 (1996).

57 D.T. Moustakas, P.T. Lang, S. Pegg, E. Pettersen, I.D. Kuntz, N. Brooijmans, and R.C. Rizzo, Development and validation of a modular, extensible docking program: DOCK 5. *Journal of Computer-Aided Molecular Design*, 20, 601–619 (2006).

58 M.L. Verdonk, J.C. Cole, M.J. Hartshorn, C.W. Murray, and R.D. Taylor, Improved protein–ligand docking using GOLD. *Proteins: Structure, Function, and Bioinformatics*, 52, 609–623 (2003).

59 R.A. Friesner, J.L. Banks, R.B. Murphy, T.A. Halgren, J.J. Klicic, D.T. Mainz, D.E. Shaw, Glide: a new approach for rapid, accurate docking and scoring. 1. Method and assessment of docking accuracy. *Journal of medicinal chemistry*, 47, 1739–1749 (2004).

60 I. Schellhammer, and M. Rarey, FlexX-Scan: Fast, structure-based virtual screening. *PROTEINS: Structure, Function, and Bioinformatics*, 57, 504–517 (2004).

61 R. Abagyan, M. Totrov, and D. Kuznetsov, ICM—a new method for protein modeling and design: applications to docking and structure prediction from the distorted native conformation. *Journal of computational chemistry*, 15, 488–506 (1994).

62 A.N. Jain, Surflex: fully automatic flexible molecular docking using a molecular similarity-based search engine. *Journal of medicinal chemistry*, 46, 499–511 (2003).

63 C.M. Venkatachalam, X. Jiang, T. Oldfield, and M. Waldman, Ligand Fit: a novel method for the shape-directed rapid docking of ligands to protein active sites. *Journal of Molecular Graphics and Modelling*, 21, 289–307 (2003).

64 G. Klebe, Virtual ligand screening: strategies, perspectives and limitations. *Drug discovery today*, 11, 580–594 (2006).

65 W.P. Walters, M.T. Stahl, and M.A. Murcko, Virtual screening—an overview. *Drug Discovery Today*, 3, 160–178 (1998).

66 P. Tosco, and T. Balle, Brute-force pharmacophore assessment and scoring with Open3DQSAR. In 6th *German Conference on Chemoinformatics* (2011).

67 T. Defoirdt, G. Brackman, and T. Coenye, Quorum sensing inhibitors: how strong is the evidence? *Trends in microbiology*, 21, 619–624 (2013).

68 M. Hentzer, H. Wu, J.B. Andersen, K. Riedel, T.B. Rasmussen, N. Bagge, and M. Manefield, Attenuation of *Pseudomonas aeruginosa* virulence by quorum sensing inhibitors. *The EMBO journal*, 22, 3803–3815 (2003).

69 C.L. Koh, C.K. Sam, W.F. Yin, L.Y. Tan, T. Krishnan, Y.M. Chong, and K.G. Chan, Plant-derived natural products as sources of anti-quorum sensing compounds. *Sensors*, 13, 6217–6228 (2013).

70 F.M. Husain, I. Ahmad, M.H. Baig, M.S. Khan, M.S. Khan, I. Hassan, and N.A. Al-Shabib, Broad-spectrum inhibition of AHL-regulated virulence factors and biofilms by sub-inhibitory concentrations of ceftazidime. *RSC Advances*, 6, 27952–27962 (2016).

71 M.D. Kiran, N.V. Adikesavan, O. Cirioni, A. Giacometti, C. Silvestri, G. Scalise, and N. Balaban, Discovery of a quorum-sensing inhibitor of drug-resistant staphylococcal infections by structure-based virtual screening. *Molecular Pharmacology*, 73, 1578–1586 (2008).

72 Z. Zeng, L. Qian, L. Cao, H. Tan, Y. Huang, X. Xue, S. Zhou, Virtual screening for novel quorum sensing inhibitors to eradicate biofilm formation of *Pseudomonas aeruginosa*. *Applied microbiology and biotechnology*, 79, 119–126 (2008).

73 S.Y.Y. Tan, S.L. Chua, Y. Chen, S.A. Rice, S. Kjelleberg, T.E. Nielsen, and M. Givskov, Identification of five structurally unrelated quorum-sensing inhibitors of *Pseudomonas aeruginosa* from a natural-derivative database. *Antimicrobial Agents and Chemotherapy*, 57, 5629–5641 (2013).

74 K. Riedel, M. Köthe, B. Kramer, W. Saeb, A. Gotschlich, A. Ammendola, and L. Eberl, Computer-aided design of agents that inhibit the cep quorum-sensing system of *Burkholderia cenocepacia*. *Antimicrobial Agents and Chemotherapy*, 50, 318–323 (2006).

75 M.E. Skindersoe, M. Alhede, R. Phipps, L. Yang, P.O. Jensen, T.B. Rasmussen and M. Givskov, Effects of antibiotics on quorum sensing in *Pseudomonas aeruginosa*. *Antimicrobial Agents and Chemotherapy*, 52, 3648–3663 (2008).

76 L. Yang, M.T. Rybtke, T.H. Jakobsen, M. Hentzer, T. Bjarnsholt, M. Givskov, and T. Tolker-Nielsen, Computer-aided identification of recognized drugs as *Pseudomonas aeruginosa* quorum-sensing inhibitors, *Antimicrobial Agents and Chemotherapy*, 53, 2432–2443 (2009).

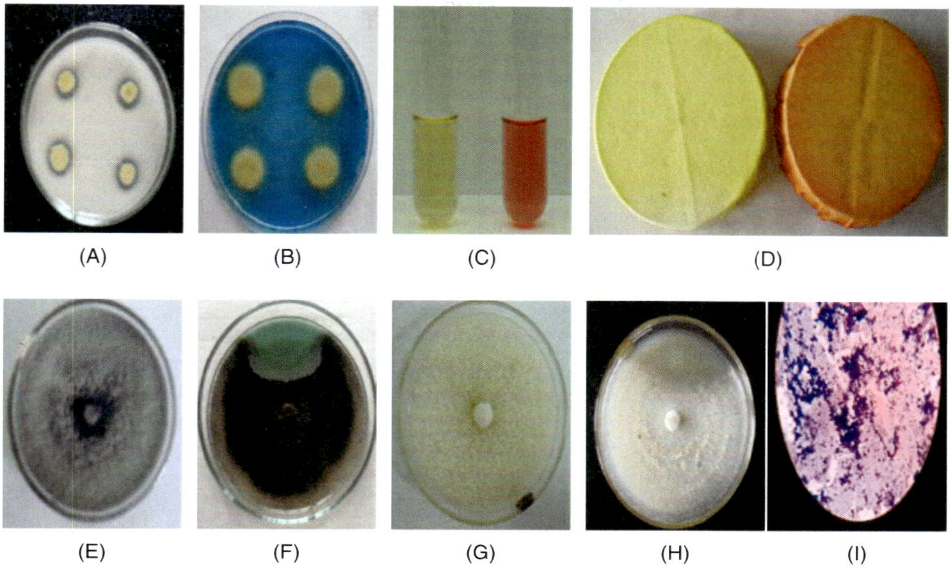

Figure 2.1 Schematic diagram showing various plant growth–promoting Attributes.

Figure 4.1 Schematic diagram showing the life cycle of *Bacillus* biofilm formation on the root surface f wheat (*Triticum aestivum*) plant. (a–b) Attachment of microorganism to sterile surface and suppression of motility. (c) Production of extracellular matrix, irreversible attachment and aggregation of planktonic cells. (d) Maturation and development of biofilm architecture. (e) Dispersion of planktonic cells from biofilm.

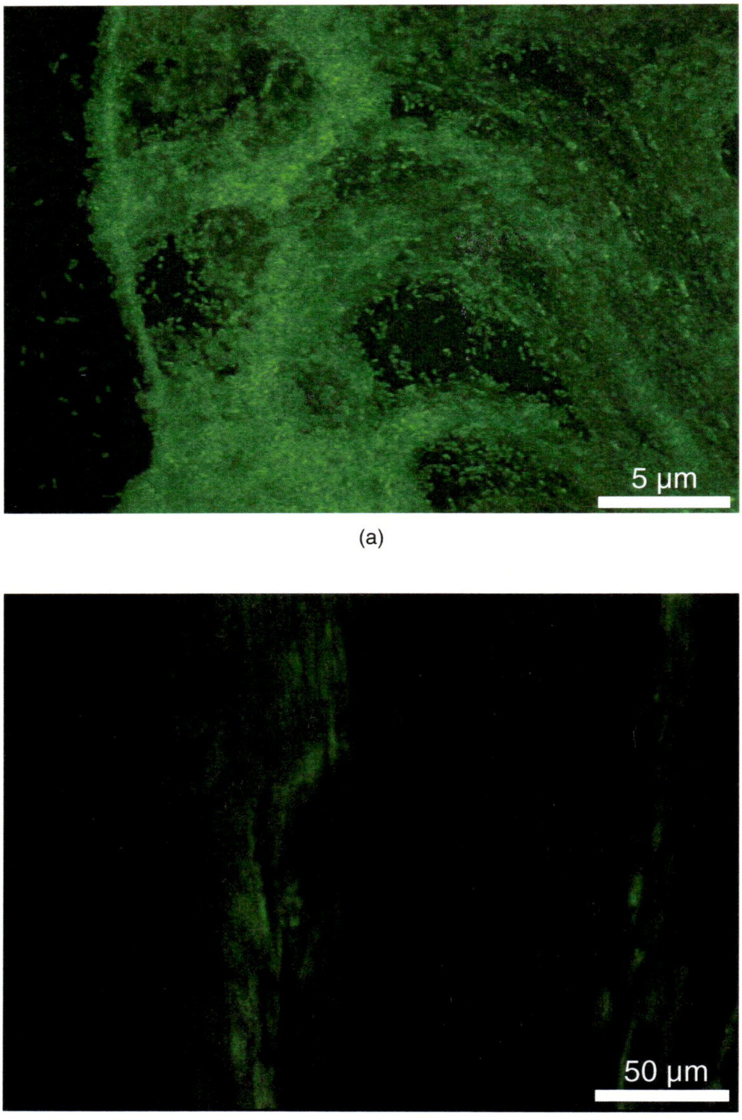

Figure 4.3 Confocal laser scanning microscopy images of *Bacillus subtilis* biofilm formed on glass coverslip (a), and chickpea root (b). Biofilms were stained with acridine orange.

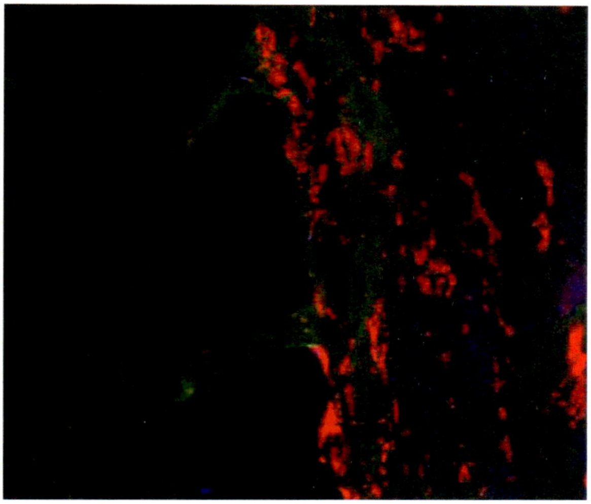

Figure 6.1 Confocal laser scanning micrograph of bacterium *Serratia liquefaciens* MG44 labeled with red probe Eub-338-I-Cy3, forming biofilm on the roots of *Arabidopsis thaliana*.

Figure 6.3 Tobacco transgenic plant leaf transformed with the AHL-synthase LuxI from *Burkholderia graminis*. Production of violacein (purple color) in the agar plate is due to the detection of the QSM by the strain CV026 of *Chromobacterium violaceum*.

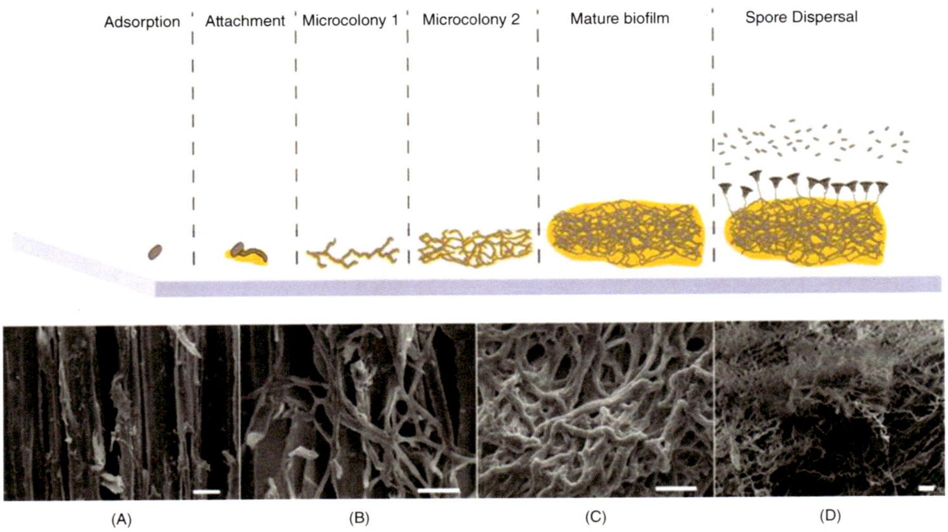

Figure 8.1 Stages of filamentous fungal biofilm formation. The upper panel shows a diagrammatic representation of six stages of fungal biofilm development. From left to right, the **Adsorption stage** involves propagules making contact with a suitable substrate or host. The deposition is reversible until the **Attachment stage** where secretion of adhesive substances from spores and germlings makes the attachment irreversible. This is followed by the **Microcolony I stage**, where initial formation of a microcolony occurs via apical elongation and branching of hyphae. Extracellular matrix is produced. **The Microcolony II** stage includes maturation of the colony where compacted hyphal networks and pervasive hyphal–hyphal adhesion resulting in hyphal bundles and cords. In submerged biofilms, fluid channels appear during this stage, via hydrophobic repulsion between hyphae. The **Mature biofilm stage** often has aerial growth and three-dimensional colony layering with near-complete encasement within the extracellular matrix. The final stage involves **Spore dispersal** via formation of fruiting bodies or sporogenous cells often formed apically on aerial hyphae. These structures produce spores or propagules for survival or dispersal and would include spores, sporangia, infective hyphal fragments or other structures. These detached cells can be translocated to new substrates or hosts and reinitiate the cycle. Scanning electron micrographs in the lower panel show *Didymella bryoniae* biofilms representing Microcolony I (A), Microcolony II (B), Mature biofilm (C) and Spore Dispersal (D). Images A-C were collected from *D. bryoniae* biofilms produced on wood surfaces in a static multiwall plate biofilm reactor (BEST™ Assay). The aerial hyphae and sporegenous cells shown in (D) were observed as part of a *D. bryoniae* biofilm on symptomatic cucumber fruit. Scale bars = 20 μm.

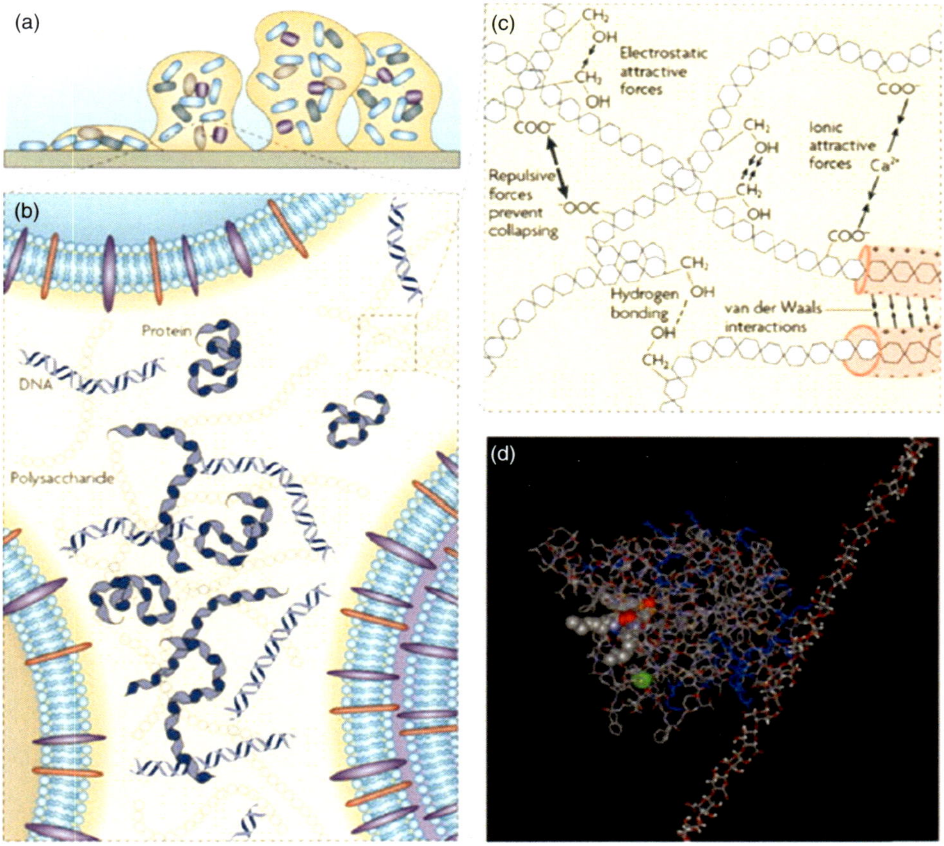

Figure 9.1 The extracellular polymeric substances matrix at different levels of resolution. (a) A model of a bacterial biofilm attached to a solid surface. Biofilm formation starts with the attachment of a cell to a surface. A microcolony forms through division of the bacterium, and production of the biofilm matrix is initiated. Other bacteria can then be recruited as the biofilm expands owing to cell division and the further production of matrix components. (b) The major matrix components—polysaccharides, proteins, and DNA—are distributed between the cells in a nonhomogeneous pattern, setting up differences between regions of the matrix. (c) The classes of weak physicochemical interactions and the entanglement of biopolymers that dominate the stability of the EPS matrix. (d) Amolecular modeling simulation of the interaction between the exo-polysaccharide alginate (right) and the extracellular enzyme lipase (left) of *Pseudomonas aeruginosa* in aqueous solution. The starting structure for the simulation of the lipase protein was obtained from the Protein Data Bank. The coloured spheres represent 1,2-dioctylcarbamoylglycero-3-O-octylphosphonate in the lipase active site (which was present as a part of the crystal structure), except for the green sphere, which represents a $Ca2^{+}$ ion. The aggregate is stabilized by the interaction of the positively charged amino acids arginine and histidine (indicated in blue) with the polyanionic alginate. Water molecules are not shown. Image courtesy of H. C. Flemming, CAM-D Technologies, Essen, Germany. (With copyright permission, from [Colloids and Surfaces B: Biointerfaces, Volume 86, Issue 2, 2011, 251–259]).

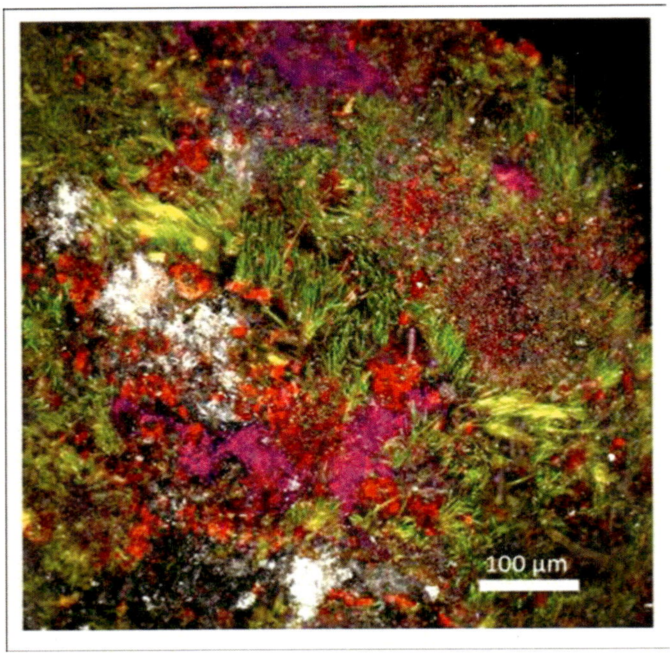

Figure 12.2 Maximum intensity projection of a multichannel dataset showing a complex river biofilm landscape. The sample was stained for nucleic acids, lectin-specific glycoconjugates, and β-D-glucans (with a 1→3 or 1→4 linkage). For multichannel imaging of reflection, autofluorescence and fluorescence staining, the data were recorded in five separate detector channels in combination with colocalization. Color allocation: green, nucleic acids; red, glycoconjugates; blue, algae; pink, cyanobacteria; yellow, β-D-glucans; white, reflection. Reproduced from Neu and Lawrence, *Trends in Microbiol.*, 23, 233–242 (2015) withpermission from Elsevier.

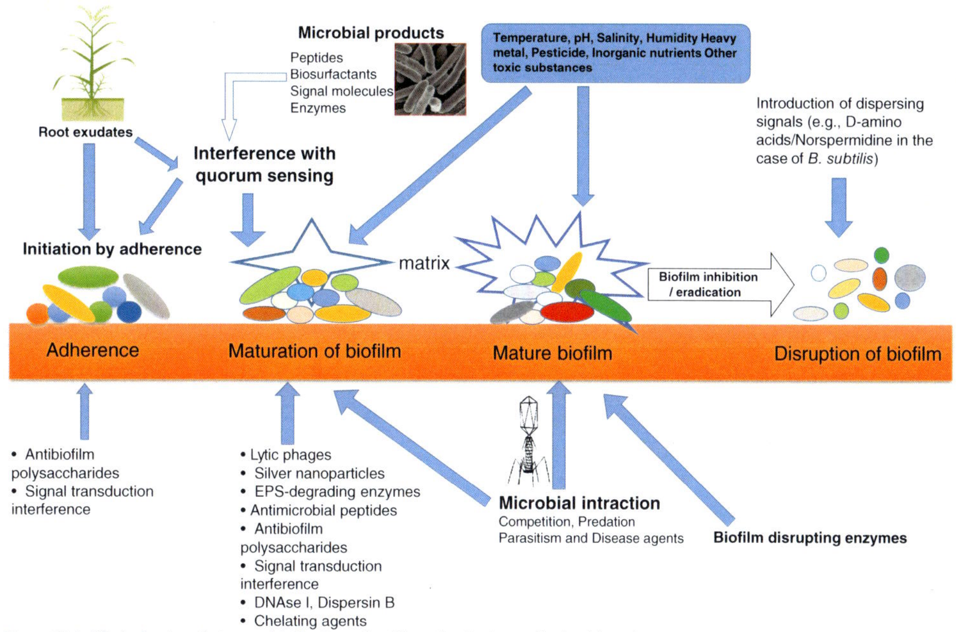

Figure 15.1 Effect of various factors on biofilm formation/disruption *in vitro* and in the rhizosphere.

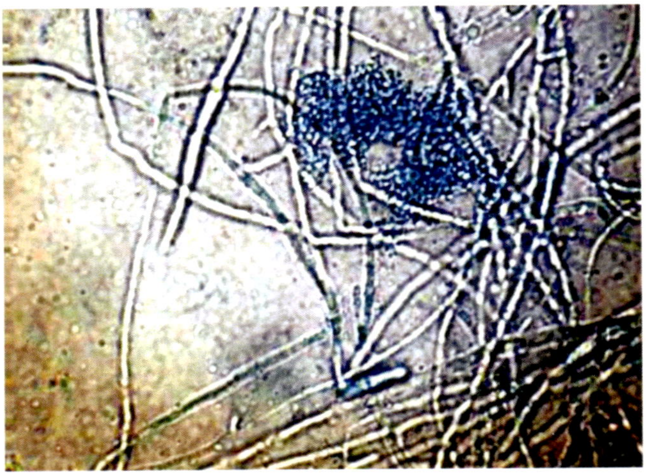

Figure 17.1 *Colletotrichum* sp. mycelium colonized by *Azotobacter* sp. forming fungal–bacterial biofilms (FBBs), when developed under *in vitro* conditions and stained with lactophenol cotton blue. Magnification, x 400.

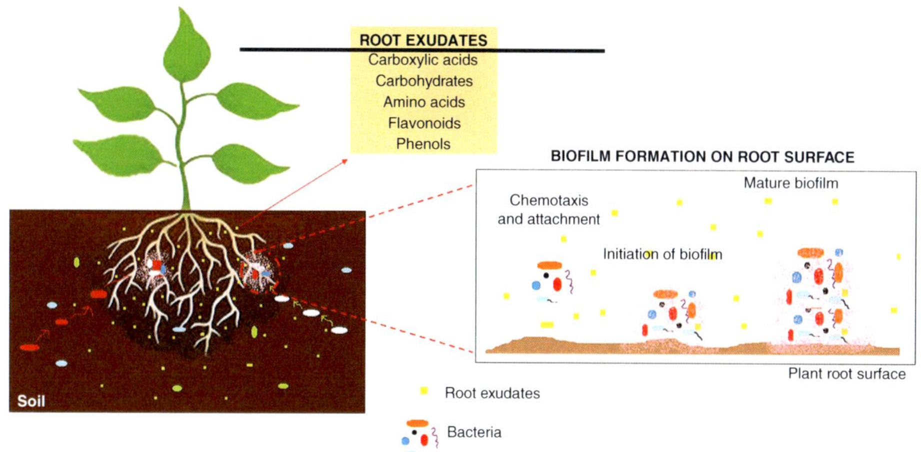

Figure 18.2 Intricate correlation between root exudates, biofilm formation, and chemotaxis.

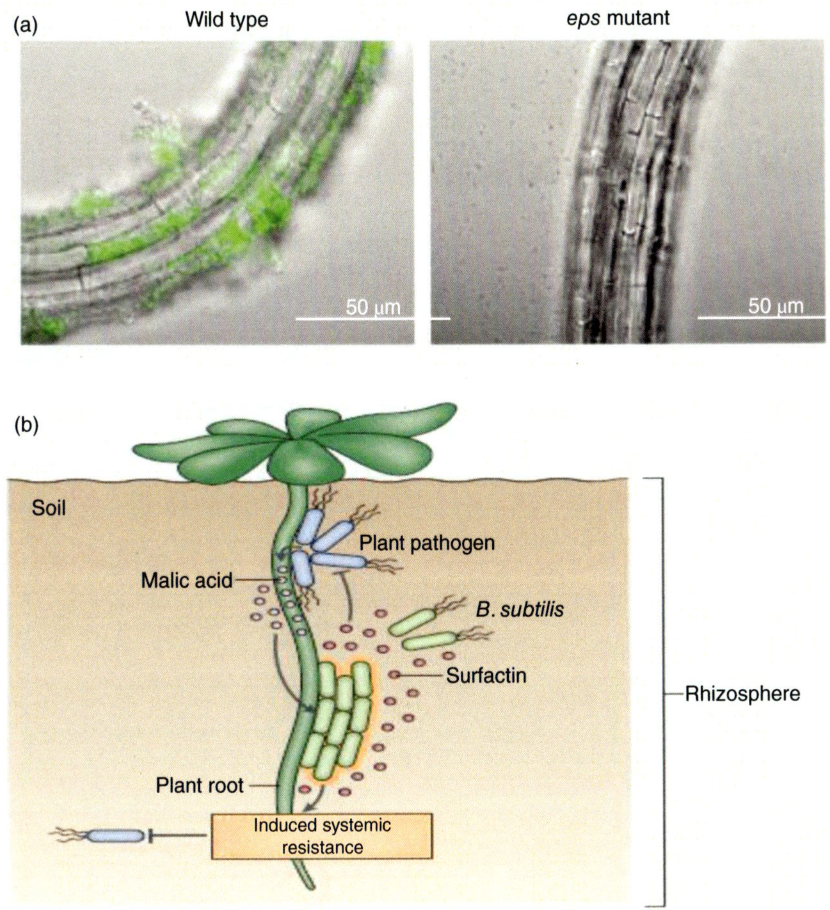

Figure 20.1 Image source: nature reviews microbiology 2013 [42].

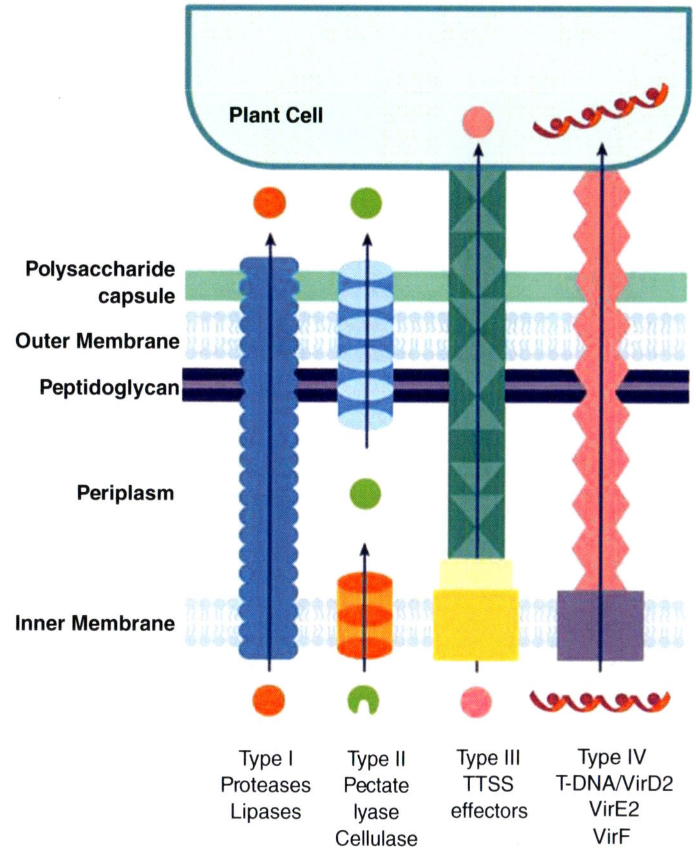

Figure 21.1 Secretory systems used by bacteria to introduce pathogenic compounds into plant cell [115].

Figure 21.4 Biofilm formation by *Salmonella* cells on a leaf [42].

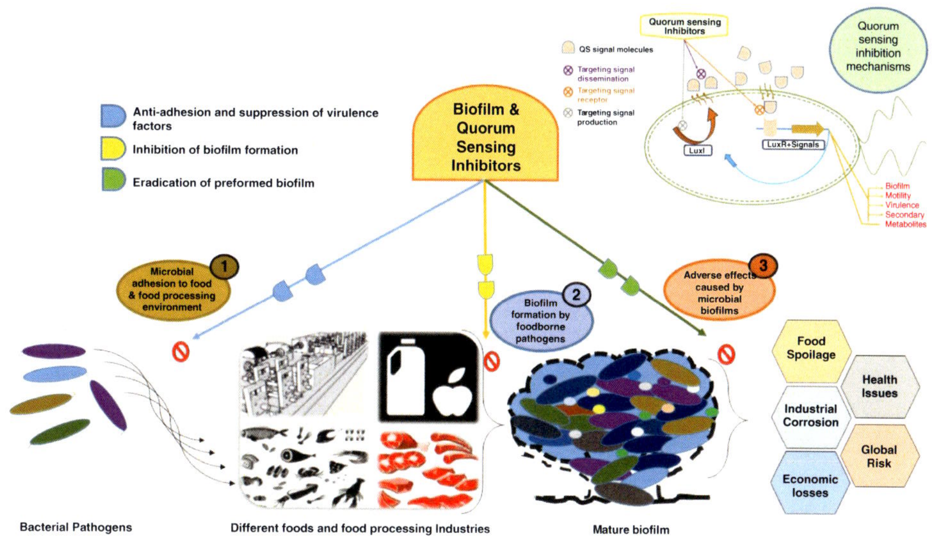

Figure 22.1 Different modes of action of biofilm/quorum-sensing inhibitors.

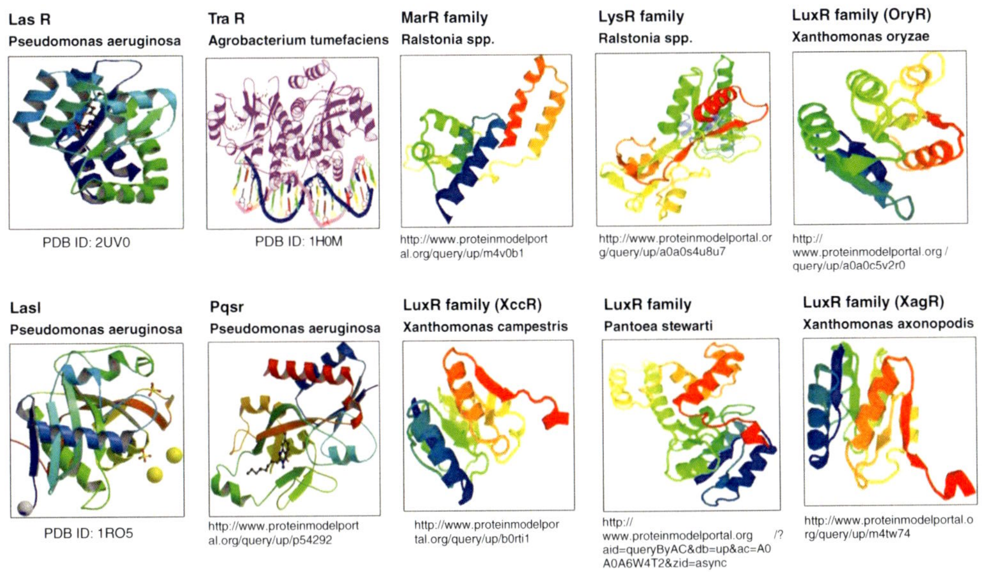

Figure 25.2 Potent plant pathogenic bacterial protein targets for QS inhibitors.

Index

Note: figures are indicated by italic page numbers, table by bold numbers.

n